T0213204

# Chemical Processing of Ceramics

**Second Edition**

# MATERIALS ENGINEERING

# Chemical Processing of Ceramics

## Second Edition

edited by

## Burtrand Lee
## Sridhar Komarneni

CRC Press
Taylor & Francis Group
Boca Raton London New York

CRC Press is an imprint of the
Taylor & Francis Group, an **informa** business

CRC Press
Taylor & Francis Group
6000 Broken Sound Parkway NW, Suite 300
Boca Raton, FL 33487-2742

First issued in paperback 2019

© 2005 by Taylor & Francis Group, LLC
CRC Press is an imprint of Taylor & Francis Group, an Informa business

No claim to original U.S. Government works

ISBN-13: 978-1-57444-648-7 (hbk)
ISBN-13: 978-0-367-39249-9 (pbk)

Library of Congress Card Number 2004065504

---

### Library of Congress Cataloging-in-Publication Data

---

Chemical processing of ceramics.–2nd ed. / edited by Burtrand Lee and Sridhar Komarneni.
    p. cm.– (Materials engineering ; 28)
Includes bibliographical references and index.
ISBN 1-57444-648-7 (alk. paper)
    1. Ceramics–Analysis. 2. Ceramic materials.  I. Lee, Burtrand Insung. II. Komarneni, Sridhar. III. Title. IV. Series: Materials engineering (Marcel Dekker, Inc.) ; 28.

TP810.5.C48 2005

---

**Visit the Taylor & Francis Web site at**
**http://www.taylorandfrancis.com**

**and the CRC Press Web site at**
**http://www.crcpress.com**

# Foreword

The progress of civilization is often marked by naming ages, for example, the Stone Age, the Bronze Age, the Iron Age, the Steel Age. We are at the threshold of change from the Silicon or Information Age to the Age of Biology. A vital question is "What is the role of ceramics in an age of biology?" It is a challenging question and the future growth of the ceramics industry may well depend on how cleverly we approach the needs of the biological revolution and an aging population. Large segments of the population will need replacement parts before death. Can ceramics provide the additional survivability of these prostheses? Can bioactive ceramics with controlled release of ions and growth factors be used to turn on the body's own regenerative potential? We are also at the threshold of a change from an energy-rich society to an energy-declining society. How will ceramics help industry respond to this need? Can we create recyclable ceramic products that are affordable? Can we make products with substantially lower power requirements? Again, these questions are difficult, but the creativity of our responses may determine the quality of life for the new age.

This new edition of *Chemical Processing of Ceramics* offers a scientific and technological framework for achieving creative solutions to the questions posed above. It has been 20 years since the first Ultrastructure and Chemical Processing Conference proceedings were published. Enormous progress has been made in understanding the process mechanisms for chemical-based processing of new materials. The theoretical foundations are now well established and are being applied to an expanding range of materials. New process methods are being discovered. These new developments are all discussed in this new edition. The editors have made thoughtful selections from the leading researchers in the field. Their success makes this book a must for every serious investigator in the field of ceramic processing.

**Larry L. Hench**
*Professor of Ceramic Materials*
*Imperial College, London*

# Preface

Despite many recent advances in materials science and engineering, the performance of ceramic components in severe conditions is still far below the ideal limits predicted by theory. The emphasis on the relation between processing, structure, and behavior has been fruitful for ceramic scientists for several decades. It has been recently realized, however, that major advances in ceramics during the next several decades will require an emphasis on molecular-level or nanoscale control. Organic chemistry, once abhorred by ceramic engineers trained to define ceramics as "inorganic-nonmetallic materials," has become a valuable asset in designing and synthesizing new ceramics. It has recently been established that as the structural scale in ceramics is reduced from macro- to micro- to nanocrystalline regimes, the basic properties are drastically altered. Some brittle ceramic materials have been shown to be partially ductile. Quantum dot semiconducting ceramic particles emit different colors, depending on their size, and this property can be useful in various applications.

The impetus and the ultimate goal in chemical processing of ceramic materials is to control physical and chemical variability by the assemblage of uniquely homogeneous structures, nanosized second phases, controlled surface compositional gradients, and unique combinations of dissimilar materials to achieve desired properties. Significant improvements in environmental stability and performance should result from such nanoscale or molecular design of materials.

A number of books are available that deal with the chemical processing aspect of ceramic materials, but most of them are conference proceedings. This revised edition of *Chemical Processing of Ceramics* is written to update, enhance, and expand the topics in the first edition published in 1994. Many authors who are actively involved in the field of chemical processing of ceramic materials from all over the world contributed to the first edition. The authors in this edition are also from the international community—Australia, Japan, Germany, Korea, France, Russia, Switzerland, and the U.S.—practicing chemical principles in the fabrication of superior ceramic materials.

Thus this book presents current developments and concepts in the chemical techniques for production and characterization of state-of-the-art ceramic materials in a truly interdisciplinary fashion. The 27 chapters are divided into five parts reflecting topical groups. The first part discusses the starting materials—how to prepare and modify them in the nanoscale range. Powders are the most heavily used form of starting ceramic materials. The synthesis, characterization, and behavior of ceramic powders are presented in parts I and II. In the third part, processing of ceramic films via the sol-gel technique is discussed. Fabrication of nonoxide ceramics is covered in part IV. In the last part, several specific examples of classes of ceramic materials fabricated by chemical processing, including thin films, membranes, ferroelectrics, bioceramics, dielectrics, batteries, and super-conductors, are presented. These classes of examples are chosen on the basis of the current demand and active research. The topics on basic principles of the sol-gel technique, sintering, and postsintering processes are not included in this volume because there are other excellent books dealing solely with these topics.

Although this book is edited, it is organized to reflect the sequence of ceramics processing and the coherent theme of chemical processing for advanced ceramic materials. Hence this book is suitable as a supplementary textbook for advanced undergraduate and graduate courses in ceramic science and materials chemistry, as well as an excellent reference book for practicing ceramists, chemists, materials scientists, and engineers. As shown by the data presented in this book, some of the interesting results from chemical processing have not yet made their way into real applications of ceramic materials. We are optimistic that, through further research, the full potential of chemical processing for high-performance ceramic materials can be realized. It is hoped that this book, through the authors' and editors' contributions, will bring researchers and engineers in the ceramics and chemical fields closer together to produce superior ceramic materials.

**Burtrand I. Lee**
**Sridhar Komarneni**

# Editors

**Burtrand Insung Lee, Ph.D.** is a professor in the School of Materials Science & Engineering at Clemson University, Clemson, South Carolina. He obtained his B.S. degree in chemistry from Southern Adventist University, Tennessee, and Ph.D. degree in materials science and engineering from the University of Florida, Gainesville, Florida in 1986. His industrial working experience includes Biospherics Inc., Gel Tech Inc., Kemet Electronics Materials Corp., Hitachi Ltd., and Samsung Electronics.

Dr. Lee was a Fulbright Professor at Norwegian Institute of Technology in Norway in 1989. In 1993 he spent a sabbatical year at Hitachi Research Laboratory as a senior visiting researcher. He has published over 170 technical papers and other books on ceramic and polymer processing as well as several U.S. patents. He has co-organized many national/international technical symposia on materials and nano-processing. Dr. Lee received the MRS Award in 1986, Fulbright Scholar Award in 1989, Clemson Board of Trustee Faculty Excellence Awards in 2001 and 2004, and he was selected as a Lady Davis Fellow in 2004.

Dr. Lee teaches colloidal and surface science as well as general materials processing at Clemson University. He is also director of Nanofabritech®. His current research activities are focused on chemical processing of ceramic and polymeric materials, paying particular attention to surface and interfacial chemistry.

**Sridhar Komarneni, Ph.D.** is a professor of clay mineralogy at The Pennsylvania State University, University Park, Pennsylvania. He conducts research on synthesis and processing of nanophases and nanocomposites by sol-gel and hydrothermal processing and on both basic and applied aspects of clay minerals. He has published over 415 refereed papers and edited or written 13 books during his career and received numerous awards. Dr. Komarneni was elected to The World Academy of Ceramics, The European Academy of Sciences and Fellows of The American Association for the Advancement of Science, The  Royal Society of Chemistry, The American Society of Agronomy, The Soil Science Society of America, and The American Ceramic Society. He serves as the editor-in-chief of *Journal of Porous Materials.*

# Contributors

**André Ayral**
Universite Montpellier II
Montpellier, France

**Prerak Badheka**
Clemson University
Clemson, South Carolina

**John Ballato**
Clemson University
Clemson, South Carolina

**Dunbar P. Birnie, III**
Rutgers University
Piscataway, New Jersey

**Guozhong Cao**
University of Washington
Seattle, Washington

**Jin-Ho Choy**
Ewha Womans University
Seoul, Korea

**Philippe Colomban**
CNRS and Université P.
& M. Curie (Paris 6)
Thiais, France

**Amit Daga**
University of Florida
Gainesville, Florida

**Moni K. Datta**
Carnegie Mellon University
Pittsburgh, Pennsylvania

**Deepa Dey**
Tokyo Institute of Technology
Yokohama, Japan

**Jeffrey DiMaio**
Clemson University
Clemson, South Carolina

**Harold Dobberstein**
University of Cambridge
Cambridge, England

**Aaron C. Dodd**
University of Western Australia
Crawley, Australia

**Toshiya Doi**
Kagoshima University
Kagoshima, Japan

**Eric Forsythe**
Army Research Lab
Adelphi, Maryland

**Christian Guizard**
Universite Montpellier II
Montpellier, France

**Masashi Inoue**
Kyoto University
Kyoto, Japan

**Anne Julbe**
Universite Montpellier II
Montpellier, France

Masato Kakihana
IMRAM, Tohoku University
Sendai, Japan

Hendrik K. Kammler
Swiss Federal Institute of Technology
Zurich, Switzerland

Takeo Katoh
Hanyang University
Seoul, South Korea

Masanori Kikuchi
National Institute for Materials
  Science
Tsukuba, Japan

Eung Soo Kim
Kyonggi University
Suwon, South Korea

Il-Seok Kim
Carnegie Mellon University
Pittsburgh, Pennsylvania

Sang Kyun Kim
Hanyang University
Seoul, South Korea

Sridhar Komarneni
Pennsylvania State University
University Park, Pennsylvania

Prashant N. Kumta
Carnegie Mellon University
Pittsburgh, Pennsylvania

Burtrand I. Lee
Clemson University
Clemson, South Carolina

Lutz Mädler
Swiss Federal Institute of Technology
Zürich, Switzerland

Jeffrey P. Maranchi
Carnegie Mellon University
Pittsburgh, Pennsylvania

David Morton
Army Research Lab
Adelphi, Maryland

Masaki Narisawa
Osaka Prefecture University
Osaka, Japan

Tomoji Ohishi
Shibaura Institute of Technology
Tokyo, Japan

Nickolay N. Oleynikov
Moscow State University
Moscow, Russia

Masataka Ozaki
Yokohama City University
Yokohama, Japan

Ungyu Paik
Hanyang University
Seoul, South Korea

Jea Gun Park
Hanyang University
Seoul, South Korea

Man Park
Kyungpook National University
Teagu, South Korea

Valery Petrykin
IMRAM, Tohoku University
Sendai, Japan

Georgios Pyrgiotakis
University of Florida
Gainesville, Florida

**Lai Qi**
Clemson University
Clemson, South Carolina

**Rustum Roy**
Pennsylvania State University
University Park, Pennsylvania

**Theodor Schneller**
Institut fur Werkstoffe der
 Elektrotechnik II
Aachen, Germany

**Robert Schwartz**
University of Missouri
Rolla, Missouri

**Oleg A. Shlyakhtin**
Moscow State University
Moscow, Russia

**Wolfgang Sigmund**
University of Florida
Gainesville, Florida

**Shigeyuki Somiya**
Tokyo Institute of Technology and
 Teikyo University of Science and
 Technology
Tokyo, Japan

**Koji Tomita**
IMRAM, Tohoku University
Sendai, Japan

**Yuri D. Tretyakov**
Moscow State University
Moscow, Russia

**Oleg I. Velikokhatnyi**
Carnegie Mellon University
Pittsburgh, Pennsylvania

**Li-Qiong Wang**
Pacific Northwest National
 Laboratory
Richland, Washington

**Xinyu Wang**
University of Minnesota
Minneapolis, Minnesota

**Ying Wang**
University of Washington
Seattle, Washington

**Rainer Waser**
RWTH Aachen University of
 Technology
Aachen, Germany

**Markus Weinmann**
Max-Planck-Institut fur
 Metallforschung
Stuttgart, Germany

**D.H. Yoon**
Clemson University
Clemson, South Carolina

**Ki Hyun Yoon**
Yonsei University
Seoul, Korea

# Table of Contents

## SECTION II  *Powder Processing at Nanoscale*

## SECTION III  *Sol-Gel Processing*

# SECTION IV    Ceramics Via Polymers

# SECTION V    Processing of Specialty Ceramics

# Section I

## Powder Synthesis and Characterization

# 1 Hydrothermal Synthesis of Ceramic Oxide Powders*

*Shigeyuki Somiya, Rustum Roy, and Sridhar Komarneni*

## CONTENTS

---

\* Reproduced with permission from the Indian Academy of Sciences. This article was originally published in the *Bull. Mat. Sci.*, 23, 453–460 and the current version is slightly modified.

# I. INTRODUCTION

Inorganic powders play a key role in many fields—ceramics, catalysts, medicines, food, etc.—and many papers and books discuss powder preparation.[1-5] Powder preparation is a very important step in the processing of ceramics. Table 1.1 presents the methods used for preparing fine ceramic powders.[6]

---

**TABLE 1.1**
**Methods for Producing Fine Ceramic Powders**

1. Mechanical
   a. Ball milling (powder mixing)
   b. Attrition milling
   c. Vibration milling
2. Thermal decomposition
   a. Heating (evaporation)
   b. Spray drying
   c. Flame spraying
   d. Plasma spraying
   e. Vapor phase (CVD)
   f. Freeze-drying (cryochemical)
   g. Hot kerosene drying
   h. Hot petroleum drying
   i. Combustion
   j. Laser beam
   k. Electron beam
   l. Sputtering
3. Precipitation or hydrolysis
   a. Neutralization and precipitation
   b. Homogeneous precipitation
   c. Coprecipitation
   d. Salts solution
   e. Alkoxides
   f. Sol-gel
4. Hydrothermal
   a. Precipitation (coprecipitation)
   b. Crystallization
   c. Decomposition
   d. Oxidation
   e. Synthesis
   f. Electrochemical
   g. Mechanochemical
   h. Reactive electrode submerged arc (RESA)
   i. Microwave
   j. Ultrasonic
5. Melting and rapid quenching

---

## II. HYDROTHERMAL SYNTHESIS

The term hydrothermal comes from the earth sciences, where it implies a regime of high temperatures and water pressures. Table 1.2 shows the types of hydrothermal synthesis.[7–18] The major differences between hydrothermal processing and other technologies are shown in Table 1.3.[4,6,17,19] For typical hydrothermal research one needs a high-temperature, high-pressure apparatus called an autoclave or bomb. A great deal of early experimental work was done using the

### TABLE 1.2
### Hydrothermal Synthesis

Hydrothermal crystal growth
Hydrothermal treatment
Hydrothermal alteration
Hydrothermal dehydration
Hydrothermal extraction
Hydrothermal reaction sintering
Hydrothermal sintering
Corrosion reaction
Hydrothermal oxidation
Hydrothermal precipitation—hydrothermal crystallization
Hydrothermal decomposition
Hydrothermal hydrolysis—hydrothermal precipitation
Hydrothermal electrochemical reaction
Hydrothermal mechanochemical reaction
Hydrothermal + ultrasonic
Hydrothermal + microwave

### TABLE 1.3
### Major Differences of Powders and Processing Between Hydrothermal and Other Technologies for Powder Preparation

1.  Powders are formed directly from solution.
2.  Powders are anhydrous, crystalline, or amorphous depending on the hydrothermal temperature.
3.  Particle size controlled by hydrothermal temperature.
4.  Particle shape controlled by starting materials.
5.  Ability to control chemical composition, stoichiometry, etc.
6.  Powders are highly reactive in sintering.
7.  In many cases, powders do not need calcination.
8.  In many cases, powders do not need a milling process.

Based on studies mainly by W.J. Dawson, D. Segal, D.W. Johnson, and S. Somiya.

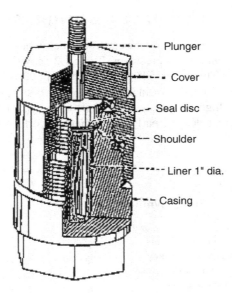

**FIGURE 1.1** Autoclave with flat plate closure. (Morey, G.W., Hydrothermal synthesis, *J. Am. Ceram. Soc.*, 36, 279, 1953.)

Morey bomb[7] and Tuttle-Roy test tube bomb (made by Tem-Press), which are shown in Figure 1.1 and Figure 1.2. Hydrothermal synthesis involves water both as a catalyst and occasionally as a component of solid phases in synthesis at elevated temperatures (greater than 100°C) and pressures (more than a few atmospheres). At present, one can get many kinds of autoclaves to cover different pressure–temperature ranges and volumes. In the U.S., there are three companies:

- Tem-Press—They are the best source for research vessels of all kinds, including test tube bombs and gas intensifiers for specialized gases such as argon, hydrogen, oxygen, ammonia, etc.
- Autoclave Engineers—They make a complete line of laboratory-scale valves, tubing, collars, fittings for connections, etc., and they also make very large autoclaves (1–3 m) for quartz and other chemical processes.
- Parr Instrument Co.—They make simple, low-pressure (1000 bars), low-temperature (300°C) laboratory-scale autoclaves (50 ml to 1 L) for low-temperature reactions, including vessels lined with Teflon.

For hydrothermal experiments the requirements for starting materials are an accurately known composition that is as homogeneous, pure, and fine as possible.

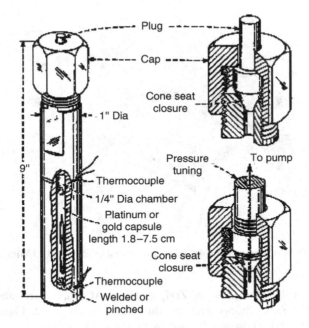

**FIGURE 1.2** Reaction vessel with a cold-cone seat closure (Tem-Press).

## A. Some Results in Different Categories (Hydrothermal Decomposition)

### 1. Ilmenite

Ilmenite ($FeTiO_3$) is a very stable mineral. Extraction of titanium dioxide ($TiO_2$) from such ores has potential. Using 10 M KOH or 10 M NaOH mixed with ilmenite in a ratio of 5:3 (ilmenite:water) under reaction conditions of 500°C and 300 kg/cm$^2$, ilmenite was decomposed completely after 63 h.[20] If the ratio of ilmenite to water is 5:4, under the same conditions, 39 h is needed to decompose the ilmenite. Reactions were as follows, in the case of KOH solution:

$$3FeTiO_3 + H_2O \rightarrow Fe_3O_4 + 3TiO_2 + H_2$$

$$nTiO_2 + 2KOH \rightarrow K_2O(TiO_2)_n + H_2O, \quad n = 4 \text{ or } 6.$$

## B. Hydrothermal Metal Oxidation

### 1. Zirconium Metal

Zirconium metal powder (10–50 g) was reacted with water to form $ZrO_2$:[21]

$$Zr + 2H_2O \text{ (at 300°C)} \rightarrow ZrO_2 + ZrH_x \text{ (400°C)} \rightarrow ZrO_2 + 2H_2.$$

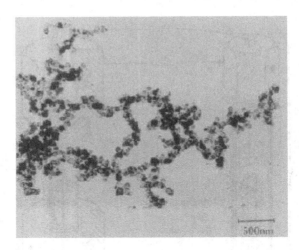

**FIGURE 1.3** TEM of zirconia powder by hydrothermal oxidation (100 MPa at 500°C for 3 h).

At 300°C under 98 MPa, $ZrO_2$ and $ZrH_x$ appeared. At temperatures above 400°C under 98 MPa, $ZrH_x$ disappeared and only $ZrO_2$ was formed. Figure 1.3 and Figure 1.4 show $ZrO_2$ powders formed by hydrothermal oxidation.

## 2. Aluminum Metal

Aluminum metal was reacted with water under 100 MPa at temperatures between 200 and 700°C for up to 6 h.[22] AlOOH appeared at 100°C and $\alpha$-$Al_2O_3$ appeared at between 500 and 700°C.

## 3. Titanium Metal

Titanium metal powder was reacted with water in a ratio of 1:2 in a gold capsule under hydrothermal conditions of 100 MPa for 3 h at temperatures of up to 700°C.[23] Figure 1.5 shows the results.

## C. HYDROTHERMAL REACTIONS

There are many papers on hydrothermal reactions.[24–28] Hydrated zirconia was formed when $ZrCl_4$ solution was reacted with $NH_4OH$. Then it was washed with

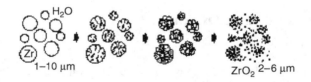

**FIGURE 1.4** Schematic illustration of hydrothermal oxidation of zirconia powder.

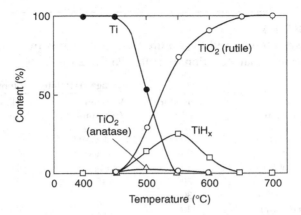

**FIGURE 1.5** Variation in product amount with temperature in hydrothermal oxidation of titanium in a closed system under 100 MPa for 3 h.

distilled water and dried for 48 h at 120°C. This starting material was placed into a platinum or gold tube with various solutions under 100 MPa at 300°C for 24 h (Figure 1.6). The results[24,25] are shown in Table 1.4.

## D. Hydrothermal Precipitation or Hydrothermal Hydrolysis

### 1. Alumina

One of the industrial applications of hydrothermal precipitation is ordinary alumina production. The Bayer process is shown in Figure 1.7.[29] T

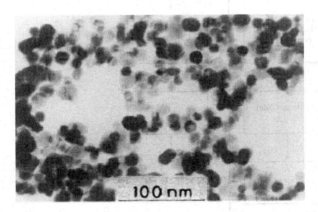

**FIGURE 1.6** TEM of monoclinic zirconia powder using hydrothermal reaction (100 MPa at 400°C for 24 h) using 8 wt% KF solution.

**TABLE 1.4**
**Phases Present and Crystallite Size of Products by Hydrothermal Reaction at 100 MPa for 24 h**

| Mineralizer | Temperature (°C) | Average crystallite size (nm) Tetragonal $ZrO_2$ (nm) | Monoclinic $ZrO_2$ (nm) |
|---|---|---|---|
| KF (8 wt%) | 200 | Not detected | 16 |
| KF (8 wt%) | 300 | Not detected | 20 |
| NaOH (30 wt%) | 300 | Not detected | 40 |
| $H_2O$ | 300 | 15 | 17 |
| LiCI (15 wt%) | 300 | 15 | 19 |
| KBi (10 wt%) | 300 | 13 | 15 |

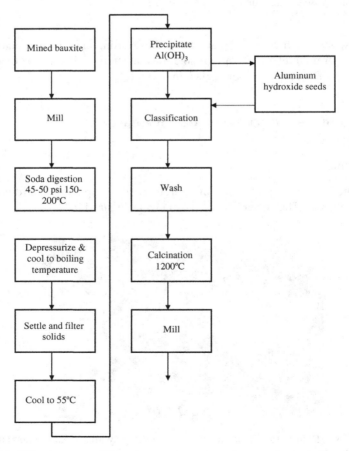

**FIGURE 1.7** Bayer process (Riman).

**TABLE 1.5**
**Typical Characteristics of ZrO$_2$ Powders by Hydrothermal Homogeneous Precipitation**

| Powder | ZY30 | ZY80 | ZP20 |
|---|---|---|---|
| Chemical composition (wt%) | | | |
| ZrO$_2$ | 94.7 | 86.0 | >99.9 |
| Y$_2$O$_3$ | 5.2 | 13.9 | — |
| Al$_2$O$_3$ | 0.01 | 0.01 | 0.005 |
| SiO$_2$ | 0.01 | 0.01 | 0.005 |
| Fe$_2$O$_3$ | 0.005 | 0.005 | 0.005 |
| Na$_2$O | 0.001 | 0.001 | 0.001 |
| Cl$^-$ | <0.01 | <0.01 | <0.01 |
| Ignition loss | 1.5 | 1.5 | 8.0 |
| Crystallite size (nm) | 22 | 22 | 20 |
| Average particle size (μm)[a] | 0.5 | 0.5 | 1.5 |
| Specific surface area (m$^2$/g)[b] | 20 | 25 | 95 |
| Sintered specimens | 1400°C × 2 h | 1500°C × 2 h | |
| Bulk density (g/cm$^3$) | 6.05 | 5.85 | |
| Bending strength (Mpa)[c] | 1000 | 300 | |
| Fracture toughness (Mpam$^{1/2}$)[d] | 6.0 | 2.5 | |
| Vicker's hardness (GPa) | 12.5 | 11 | |
| Thermal expansion 20°C–1000°C (10$^{-6}$/°C) | 11.0 | 10.6 | |

[a] Photo sedimentation method.
[b] BET method (N$_2$).
[c] Three-point bending method.
[d] MI method.

## 2. Zirconia

Hydrothermal homogeneous precipitation is one of the best ways to produce zirconia powders. The process, properties of the powders, and microstructure of the sintered body are shown in Table 1.5 and Figure 1.8, Figure 1.9, and Figure 1.10.[30,31]

### E. HYDROTHERMAL ELECTROCHEMICAL METHOD

Figure 1.11 shows an apparatus used in the hydrothermal electrochemical method. For preparing BaTiO$_3$, titanium and platinum plates are used as anodes and cathodes, respectively. A solution of barium nitrate 0.1 N or 0.5 N and temperatures up to 250°C were used for the experiment. The current density was 100 mA/cm$^2$. Under these conditions we were able to produce BaTiO$_3$ powder. The BaTiO$_3$ powder produced by this process is shown in Figure 1.12. ZrO$_2$ was also produced by this method. In the case of ZrO$_2$, Zr plates were used.[32,33]

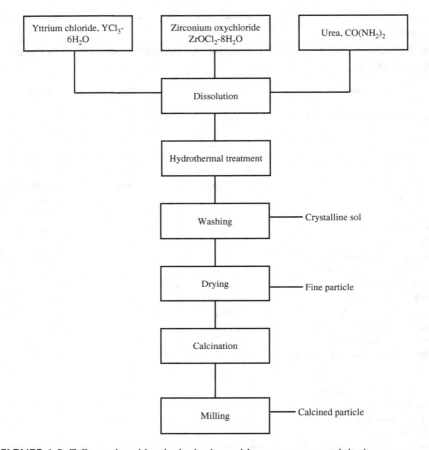

**FIGURE 1.8** $ZrO_2$ produced by the hydrothermal homogeneous precipitation process.

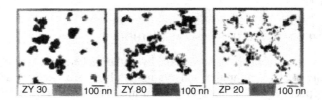

**FIGURE 1.9** TEM of different grades of zirconia powder using hydrothermal homogeneous precipitation (Chichibu Onoda Cement Corp.) (see Table 1.5.)

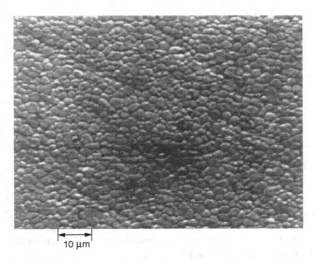

**FIGURE 1.10** TEM of zirconia sintered at 1400°C for 2 h.

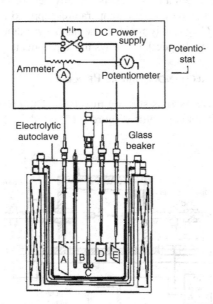

**FIGURE 1.11** Schematic of the electrochemical cell and circuit arrangements for anodic oxidation of a titanium metal plate under hydrothermal conditions. (A) Counter electrode (platinum plate), cathode; (B) thermocouple; (C) stirrer; (D) reference electrode (platinum plate); (E) working electrode (titanium plate), anode.

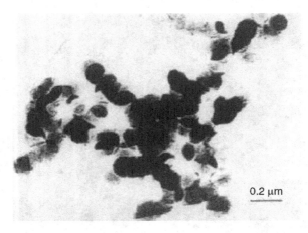

**FIGURE 1.12** TEM of BaTiO₃ powders prepared by the hydrothermal electrochemical method (250°C, 0.5 N Ba(NO₃)₂, titanium plate).

## F. REACTIVE ELECTRODE SUBMERGED ARC

Reactive electrode submerged arc (RESA) is a totally new process for making powders.[34,35] RESA produces extremely high temperatures (approximately 10,000 K) with a pressure of 1 atm $H_2O$ (possibly more in the nanoenvironment). It allows one to change liquids very easily. Figure 1.13 shows the apparatus to produce powders.

## G. HYDROTHERMAL MECHANOCHEMICAL PROCESS

Ba(OH)₂ and FeCl₃ were used as starting materials. The precipitate was crystallized hydrothermally in an apparatus (Figure 1.14) combined with an attritor and

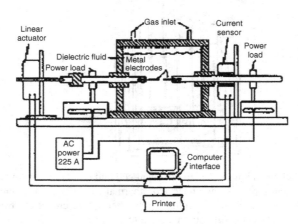

**FIGURE 1.13** Schematic of microprocessor-controlled RESA apparatus for fine powder preparation (A. Kumar and R. Roy).

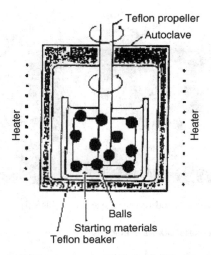

**FIGURE 1.14** Experimental apparatus for hydrothermal mechanochemical reactions.

ambient water pressure. The starting solutions with the precipitate and stainless steel balls (5 mm diameter) were placed in Teflon beakers. A Teflon propeller was rotated in the beaker under 200°C and 2 MPa. The speed of the propeller was from 0 to 107 rpm. The number of stainless steel balls was 200, 500, and 700. X-ray diffraction profiles are shown in Figure 1.15.[36]

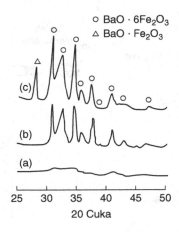

**FIGURE 1.15** X-ray diffraction profiles of (a) starting materials, (b) material fabricated at 200°C under 2 MPa for 4 h without rotation, and (c) material fabricated at 200°C for 4 h using 200 balls at 37 rpm.

**FIGURE 1.16** Microwave-assisted reaction system (MARS 5).

## H. MICROWAVE HYDROTHERMAL PROCESS

Microwave-assisted hydrothermal synthesis is a novel powder processing technology for the production of a variety of ceramic oxides and metal powders under closed-system conditions. Komarneni et al. developed this hydrothermal process into which microwaves are introduced.[37–48] This closed-system technology not

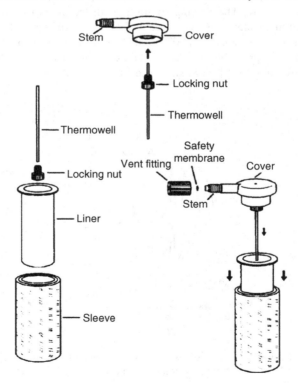

**FIGURE 1.17** Components of reaction vessel used in the MARS-5 unit.

only prevents pollution during the synthesis of lead-based materials, but also saves energy, and thus could substantially reduce the cost of producing many ceramic powders. Hydrothermal microwave treatment of 0.5 M $TiCl_4$ was done in 1 M HCl to form rutile. The system (Figure 1.16) operated at a 2.45 GHz. The vessel is lined with Teflon (Figure 1.17) and the system is able to operate up to 250°C. The parameters used are temperature, pressure, time, concentration of the metal solution, pH, etc. The key result is crystallization reactions, which lead to faster kinetics by one or two orders of magnitude compared to conventional hydrothermal processing. The use of microwaves in both solid and liquid states is gaining in popularity for many reasons, but especially because of the potential energy savings. The use of microwaves under hydrothermal conditions can accelerate the synthesis of anhydrous ceramic oxides such as titania, hematite, barium titanate, lead zirconate titanate, lead titanate, potassium niobate, and metal powders such as nickel, cobalt, platinum, palladium, gold, silver, etc., and this is expected to lead to energy savings. The term "microwave-hydrothermal" processing was first coined by us for reactions taking place in solutions that are heated to temperatures greater than 100°C in the presence of microwaves. The value of this technique has been demonstrated in rapid heating to the temperature of treatment, which can save energy; increasing the reaction kinetics by one to two orders of magnitude; forming novel phases; and eliminating metastable phases. Figure 1.18 shows a nanophase powder of hematite.

**FIGURE 1.18** Hematite synthesized from 0.02 M ferric nitrate at 100°C under microwave-hydrothermal conditions.

## I.   HYDROTHERMAL SONOCHEMICAL METHOD

Ultrasonic waves are often used in analytical chemistry for dissolving powder into solution.[49] The hydrothermal sonochemical method is a new method for synthesizing materials.[50]

# III. IDEAL POWDERS AND REAL POWDERS

The characteristics of ideal powders and real powders produced by hydrothermal processing are shown in Table 1.6 and Table 1.7. Hydrothermal powders are close to ideal powders.

---

## TABLE 1.6
### Characteristics of an Ideal Powder

Fine powder less than 1 μm
Soft or no agglomeration
Narrow particle size distribution
Morphology: sphere
Chemical composition controllable
Microstructure controllable
Uniformity
Free flowing
Fewer defects, dense particle
Less stress
Reactivity, sinterability
Crystallinity
Reproducibility
Process controllable

---

## TABLE 1.7
### Characteristics of Hydrothermal Powders

Fine powder less than 1 μm
No or weak agglomeration
Single crystal in general; depends on preparation temperature
Flow ability: forming is good
Good homogeneity
Good sinterability
No pores in grain
Narrow particle size distribution
Ability to synthesize low-temperature form and/or metastable form
Ability to make composites such as organic and inorganic mixtures
Ability to make a material that has a very high vapor pressure

## REFERENCES

1. Veale, C.R., *Fine Powders: Preparation, Properties and Uses*, Applied Science Publishers, London, 147, 1972.
2. Kato, A., and Yamaguchi, T., *New Ceramic Powder Handbook*, Tokyo Science Forum, Tokyo, 558, 1983.
3. Vincenzini, P., Ed., *Ceramic Powders*, Elsevier Scientific, Amsterdam, 1025, 1983.
4. Segal, D., *Chemical Synthesis of Advanced Ceramic Materials*, Cambridge University Press, Cambridge, 182, 1989.
5. Ganguli, D., and Chatterjee, M., *Ceramic Powder Preparation: A Handbook*, Kluwer Academic, Dordrecht, The Netherlands, 1997, 221.
6. Somiya, S., and Akiba, T., *Trans. MRS-J*, 24, 531, 1999.
7. Morey, G.W., Hydrothermal synthesis, *J. Am. Ceram. Soc.*, 36, 279, 1953.
8. Walker, A.C., *J. Am. Ceram. Soc.*, 36, 250, 1953.
9. Eitel, W., *Silicate Science*, vol. IV, Academic Press, New York, 149, 1966.
10. Laudise, R.A., *Hydrothermal Growth: The Growth of Single Crystals*, Prentice Hall, Englewood Cliffs, NJ, 275, 1970.
11. Lobachev, A.N., Ed., *Hydrothermal Synthesis of Crystals*, Consultant Bureau, New York, 1971, 152.
12. Somiya, S., Ed., *Proceedings of the First International Symposium on Hydrothermal Reactions*, Gakujutsu Bunken Fukyu Kai, Tokyo, 965, 1983.
13. Somiya, S., Ed., *Hydrothermal Reactions for Material Science and Engineering: An Overview of Research in Japan*, Elsevier Applied Science, London, 505, 1989.
14. Somiya, S., *Advanced Materials: Frontiers in Materials Science and Engineering*, vol. 19B, Elsevier Science, Amsterdam, 1105, 1993.
15. Rabenau, A.A., *Chem. Int. Ed. Engl.*, 24, 1026, 1985.
16. Brice, L.C., *Hydrothermal Growth, Crystal Growth Processes*, Blackie Halsted Press, Glasgow, 194, 1986.
17. Dawson, W.J., Hydrothermal synthesis of advanced ceramic powders, *Am. Ceram. Soc. Bull.*, 67, 1673, 1988.
18. Byrappa, K., Ed., *Hydrothermal Growth of Crystals, Progress in Crystal Growth and Characterization of Materials*, Pergamon Press, Oxford, 1991.
19. Johnson, D.W., Jr., *Advances in Ceramics*, vol. 21, *Innovations in Ceramic Powder Preparation*, G.L. Messing et al., Eds., American Ceramic Society, Westerville, OH, 3, 1987.
20. Ismail, M.G.M.U., and Somiya, S., *Proceedings of the International Symposium on Hydrothermal Reactions*, Gakujutsu Bunken Fukyu Kai, Tokyo, 669, 1983.
21. Yoshimura, M., and Somiya, S., *Rep. Res. Lab. Eng. Mat. Tokyo Inst. Technol.*, 9, 53, 1984.
22. Toraya, H. et al., *Advances in Ceramics*, vol. 12, *Science and Technology of Zirconia II*, American Ceramic Society, Westerville, OH, 806, 1984.
23. Yoshimura, M. et al., *Rep. Res. Lab. Eng. Mat. Tokyo Inst. Technol.*, 12, 59, 1987.
24. Tani, E., Yoshimura, M., and Somiya, S., Hydrothermal preparation of ultrafine monoclinic $ZrO_2$ powder, *J. Am. Ceram. Soc.*, 64, C181, 1981.
25. Nishizawa, H. et al., *J. Am. Ceram. Soc.*, 65, 343, 1982.
26. Komarneni, S. et al., *Advanced Ceramic Materials*, 1, 87, 1986.
27. Haberko, K. et al., *J. Am. Ceram. Soc.*, 74, 2622, 1991.
28. Haberko, K. et al., *J. Am. Ceram. Soc.*, 78, 3397, 1995.

29. Riman, R., *The Textbook of Ceramic Powder Technologies,* American Ceramic Society, Westerville, OH, 1999.
30. Hishinuma, K. et al., *Advances in Ceramics,* vol. 24, *Science and Technology of Zirconia III,* Somiya, S., Yamamoto, N., and Hanagida, H., Eds., American Ceramic Society, Westerville, OH, 201, 1988.
31. Somiya, S. et al., *Hydrothermal Growth of Crystals,* vol. 21, *Progress in Crystal Growth and Characterization of Materials,* Byrappa, K., Ed., Pergamon Press, Oxford, 195, 1991.
32. Yoo, S.E., Yoshimura, M., and Somiya, S., Preparation of $BaTiO_3$ and $LiNbO_3$ powders by hydrothermal anodic oxidation, in Sintering '87, vol. l, 4th International Symposium on the Science and Technology of Sintering, Nov. 4–6, Somiya S. et al., Eds., Elsevier, New York, 108, 1988.
33. Yoshimura, M. et al., *Rep. Res. Lab. Eng. Mat. Tokyo Inst. Technol.,* 14, 21, 1989.
34. Kumar, A., and Roy, R., *J. Mater. Res.,* 3, 1373, 1988.
35. Kumar, A., and Roy, R., *J. Am. Ceram. Soc.,* 72, 354, 1989.
36. Yoshimura, M. et al., *J. Ceram. Soc. Jap. Int. Ed.,* 97, 14, 1989.
37. Komarneni, S., Roy, R., and Li Q.H., *Mater. Res. Bull.,* 27, 1393, 1992.
38. Komarneni, S., Li, Q.H., Stefasson, K.M. and Roy, R., *J. Mater. Res.,* 8, 3176, 1993.
39. Komarneni, S., and Li, Q.H., *J. Mater. Chem.,* 4, 1903, 1994.
40. Komarneni, S., Pidugu, R., Li, and Roy, R., *J. Mater. Res.,* 10, 1687, 1995.
41. Komarneni, S., Hussen, M.Z., Liu, C., Breval, E., and Malla, P.B., *Eur. J. Solid State Inorg. Chem.,* 32, 837, 1995.
42. Komarneni, S., Novel microwave–hydrothermal processing for synthesis of ceramic and metal powders, in *Novel Techniques in Synthesis and Processing of Advanced Materials,* Singh, J., and Copley, S.M., Eds., Minerals, Metals, and Materials Society, Warrendale, PA, 103, 1995.
43. Komarneni, S., and Menon, V.C., *Mater. Letts.,* 27, 313, 1996.
44. Katsuki, H., Furuta, S., and Komarneni, S., *J. Am. Ceram. Soc.,* 82, 2257, 1999.
45. Komarneni, S., Rajha, P.K., and Katsuki, H., *Mater. Chem. Phys.,* 61, 50, 1999.
46. Katsuki, H., Furuta, S., and Komarneni, S., *J. Porous Mater.,* 8, 5, 2001.
47. Katsuki, H., and Komarneni, S., *J. Am. Ceram. Soc.,* 84, 2313, 2001.
48. Komarneni, S., Li, D., Newalkar, B. Katsuki, H., and Bhalla, A.S., Microwave-polyol process for Pt and Ag nanoparticles, *Langmuir,* 18, 5959, 2002.
49. Milia, A.M., *Sonochemistry and Cavitation,* Gordon and Breach Publishers, Luxembourg, 543, 1995.
50. Meskin, P.E., Barantchikov, A.Y., Ivanov, V.K., Kisterev, E.V., Burukhin, A.A., Churagulov, B.R., Oleynikov, N.N., Komarneni, S., Tretyakov, Yu. D., *Doclady Chem.,* 389, 207, 2003.

# 2 Solvothermal Synthesis

## Masashi Inoue

## CONTENTS

# I. INTRODUCTION

Metal oxides are usually prepared by calcinations of suitable precursors such as hydroxides, nitrates, carbonates, carboxylates, etc. This process usually gives oxides with pseudomorphs of the starting materials. When large amounts of thermal energy are applied for the decomposition of the precursors, it facilities sintering of the product particles and therefore aggregated particles are obtained. When mixed oxides such as spinel, perovskite, and pyrochlore are the desired products, heat treatment at higher temperatures is required.

For the preparation of inorganic materials with well-defined morphologies, liquid phase syntheses are preferred. These synthetic reactions proceed at relatively lower temperatures and therefore require lower energies. The sol-gel (alkoxide) method is one of these methods;[1,2] however, this method usually gives amorphous products, and calcination of the products is required to obtain crystallized products. In this chapter, solvothermal methods are dealt with, which are convenient for the synthesis of a variety of inorganic materials. General considerations for solvothermal reactions are discussed first and then the solvothermal synthesis of metal oxides is reviewed.

Recently, use of organic media for inorganic synthesis has garnered much attention. Since 1984, we have been exploring the synthesis of inorganic materials in organic media at elevated temperatures (200 to 300°C) under autogenous pressure of the organics.[3] This technique is now generally called the "solvothermal" method.[4] The term "solvothermal" means reactions in liquid or supercritical media at temperatures higher than the boiling point of the medium. Hydrothermal reactions are one type of solvothermal reaction. To carry out reactions at temperatures higher than the boiling point of the reaction medium, pressure vessels (autoclaves) are usually required. Some researchers favor the use of sealed ampoules of glass or silica, but these experiments should be carried out with great care because the ampoules are easily broken by the internal pressure of the reaction medium. To avoid explosion of the ampoules, they may be placed in an autoclave together with a suitable medium to create a vapor pressure to balance the internal pressure of the ampoule.

It must be noted that the liquid structure of the solvent is essentially unchanged at above or below the boiling point because the compressibility of the liquid is quite small. (Note that near the critical point, the structure of the solvent is drastically altered by changes in the solvent density.) Higher pressure may increase or decrease the reaction rate; it depends on the relative volume of the activated complex at the transition state to the volume of the starting molecule(s). However, it is known that to measure the effect of reaction pressure, GPa-scale pressure is required. This means that the autogenous pressure created by the vapor pressure of the solvent has only a minor effect on the reaction rate. Therefore there is no need to differentiate the reactions at the temperatures above and below the boiling point. Consequently "solvothermal" reaction should be defined more loosely as the reaction in a liquid (or supercritical) medium at high temperatures. Reactions in a closed system using autoclaves or sealed ampoules and in an open system using a flask equipped with a reflux condenser sometimes give completely different results, especially when a low boiling point byproduct such as water is formed.

Various compounds have been prepared by solvothermal reactions: metals,[5,6] metal oxides,[7,8] chalcogenides,[9,10] nitrides,[4,9,11] phosphides,[12] open-framework structures,[13,14] oxometalate clusters,[15,16] organic-inorganic hybrid materials,[14,17,18] and even carbon nanotubes.[19,20] Most of the solvothermal products are nano- or microparticles with well-defined morphologies. The distribution of the particle size of the product is usually quite narrow, and formation of monodispersed particles is frequently reported.[21,22] When the solvent molecules or additives are preferentially adsorbed on (or have a specific interaction with) a certain surface of the products, growth of the surface is prohibited and therefore products with unique morphologies may be formed by the solvothermal reaction.[9,23,24] Thus nanorods,[24] wires,[25] tubes,[26] and sheets[27] of various types of products have been obtained solvothermally.

## II. CHOICE OF THE REACTION MEDIUM

### A. INORGANIC MEDIUM

### 1. Water

Water (boiling point [bp], 100°C; critical temperature [Tc], 374°C; critical pressure [Pc], 218 atm) is the most widely examined reaction medium for solvothermal reactions. Geochemists first applied this technique to explore the formation mechanism of minerals and thus quite long reaction periods were applied to examine the equilibrium conversion of minerals. Today researchers seek more rapid conversion to synthesize materials, and therefore adequate synthesis of the precursors by, for example, the coprecipitation method and the sol-gel method become more important. Addition of salt, acid, or base may facilitate the reaction or alter the morphology of the products. These materials are called mineralizers. Fluoride ions sometimes have a drastic effect on the hydrothermal synthesis of materials. Besides the excellent article by Somiya, Roy, and Komarneni in this book, many review articles have appeared on hydrothermal synthesis;[28–32] therefore this technique will not be discussed further in this chapter.

## 2.  Ammonia

Besides water for hydrothermal reactions, liquid ammonia (bp, 78°C; Tc, 132°C; Pc, 113 atm) is also used for the solvothermal synthesis of nitrides. Metastable or otherwise unobtainable nitride materials were reported to be formed by this method.[33–35] Ammonium and amide ($NH_2$) ions are the strongest acid and base, respectively, for the liquid ammonia system, and therefore ammonium salt acts as the acid mineralizer,[36] while amide ion can be prepared by addition of alkali metals to the solvent. Since ammonia has a low boiling point, the reaction pressure is usually quite high.

## 3.  Other Inorganics

Sulfur dioxide (bp, −10°C; Tc, 157.5°C; Pc, 78 atm) is another interesting inorganic solvent. This compound has a high dielectric constant and low basicity (actually, it acts as an acid). To the best of my knowledge, there have been no articles that apply this solvent for the solvothermal synthesis of inorganic materials. However, the highly corrosive nature of this solvent may limit its use in autoclaves.

Hydrofluoric acid (bp, 19.5°C; Tc, 188°C; Pc, 64 atm), nitrogen dioxide (bp, 21°C; Tc, 158.2°C; Pc, 100 atm; in equilibrium with $N_2O_4$), sulfuric acid (decomposition at 280°C), and polyphosphoric acid are candidates for solvents in solvothermal reactions, and the reactions of these solvents will produce a variety of products that cannot be prepared by any other methods. For example, Bialowons et al.[37] reported that solvothermal treatment of $(O_2)_2Ti_7F_{30}$ in anhydrous HF at 300°C yielded single crystals of $TiF_4$. Solvothermal reactions in these solvents may produce fruitful results and a new field seems to be awaiting many researchers.

## B.  ORGANIC MEDIUM

## 1.  General Considerations

Various organic solvents have been applied for the synthesis of inorganic materials. Because most of inorganic synthesis researchers are not familiar with organic solvents, some important features are summarized here.

Since organic compounds easily react with oxygen in highly exothermic reactions (i.e., combustion of organics), the gas phase in the reaction vessel must be completely purged with an inert gas such as nitrogen. From a thermodynamic point of view, all organic compounds have an inherent tendency to decompose into carbon, hydrogen (or water, if the organic compound has oxygen atoms), nitrogen, and so on, at high temperatures. Methane is the most stable aliphatic hydrocarbon, however, it can decompose into carbon and hydrogen at temperatures above 360°C in the presence of a suitable catalyst such as nickel or iron.[38] Therefore most of the organic compounds, even though they may be stable at room temperature, act as reducing agents at high temperatures. To avoid reduction during the course of the reaction, the reaction can be carried out in the presence of air in the gas phase in the autoclave. For example, microcrystalline $\alpha$-LiFe$_5$O$_8$ powders were directly synthesized by a reaction of FeCl$_2 \cdot$4H$_2$O and lithium metal in ethylenediamine at 120°C.[39] In this reaction, one-fifth of the iron ions must be oxidized by air. In this reaction system, however, the gas phase in the autoclave can be an explosive mixture. Since an electric spark is required to ignite the explosive mixture, the reaction can be carried out without any problems. However, even if one run of the experiment is conducted without any troubles, one should not think that another run of the same experiment can be carried out safely, because static electricity may supply sufficient energy to explode the gas mixture. Nor should one scale up the reaction, as an increase in the gas phase volume drastically increases the risk of explosion.

Carbon nanotubes can be prepared using solvothermal reactions,[19,20] however, inherent decomposition of organic compounds into carbon and hydrogen is not yet utilized in the solvothermal synthesis of nanotubes, but hexachlorobenzene was reduced by alkali metal according to the following reaction:

$$C_6Cl_6 + K \rightarrow KCl + 6/nC_n \text{ (carbon nanotubes).}$$

Absolutely dried organic solvents are highly hygroscopic, even though the solubility of water in organic solvent is quite low. The solubility of water increases with an increase in the reaction temperature.

## 2. Paraffins

Paraffins have low dielectric constants and therefore are essentially inert in inorganic synthesis. Because of the relatively low solubility of water, the activity of water in these solvents easily approaches unity, and when inorganic particles are present in the medium, water is easily adsorbed on the particles, where water may facilitate hydrothermal conversion of the inorganic particles (see Section III.A.1 in this chapter).

*n*-Hexane (this is the common name; the International Union of Pure and Applied Chemistry [IUPAC] name is hexane) is one of the most common aliphatic solvents; however, this compound has a specific toxicity that is not shown by branched C6 hydrocarbons nor by other straight-chain paraffins such as pentane and heptane. The origin of this toxicity is well established. Through the metabolic system of human beings, this compound is oxidized by oxygenase to 2,5-hexanedione, which attacks the peripheral nervous system.[40] Therefore the author recommends that readers use heptane in place of hexane.

$n$-Paraffins with various carbon numbers are commercially available. Since the boiling point of the organic solvent usually increases with an increase in molecular weight, one can carry out the solvothermal reaction at relatively low pressure by using higher paraffins such as tetradecane (bp = 253°C) or hexadecane (bp = 287°C). However, higher straight-chain paraffins have quite low autoignition points (201°C for hexadecane), thus contact of heated paraffins with air or leaking of these solvents from the autoclave will cause spontaneous ignition.

When higher hydrocarbons are desired for use as reaction solvents, mineral oil (bp = 260 to 330°C) is recommended because the autoignition temperature of mineral oil (260 to 370°C) is much higher than those of the $n$-paraffins. Since it is produced in large quantity, it is quite cheap, and is a mixture of various branched aliphatic hydrocarbons.

## 3. Aromatic Hydrocarbons

Aromatic hydrocarbons are relatively inert and have slightly higher base and dielectric constants as compared with the paraffins. Benzene is the simplest aromatic hydrocarbon and is thermally stable at high temperatures. Some researchers favor the use of benzene for the medium of solvothermal reactions. For example, the reaction of $GaCl_3$ with $LiN_3$ in benzene at 300°C yields gallium nitride (GaN).[41] Similarly the reaction of $TiCl_4$ with $NaN_3$ yields TiN.[42] The authors of these reports called this reaction procedure a "benzene-thermal" reaction. However, benzene is highly toxic and it causes fatal damage to hemopoietic organs. The author strongly recommends that readers use toluene in place of benzene. Although toluene also attacks the nervous system, the methyl group in toluene is easily oxidized by oxygenase and the thus-formed benzoic acid reacts with glutathione and is digested.

Solvothermal reaction (thermal decomposition) of metal alkoxides in toluene usually yields the corresponding metal oxides (see Section III.B.1). Aromatic hydrocarbons are favored for this reaction over aliphatic hydrocarbons because of the higher solubility of the precursors in the former than those in the latter. Xylenes (dimethylbenzenes) are also suitable solvents for the solvothermal synthesis.

Naphthalene is solid at room temperature, but is known to be a good solvent for various organic reactions. When this solvent is used for the synthesis of inorganic materials, it may cause problems in washing out the solvent by other low boiling point solvents such as acetone and methanol. However, naphthalene easily sublimes, and therefore this solvent can be eliminated by sublimation. In this regard, this method may avoid the coagulation of particles during the drying stage, which is caused by the surface tension of the liquid remaining between the product particles.

## 4. Alcohols

Solvothermal reactions in alcohols are sometimes called "alcohothermal" reactions; this word is derived from alcoholysis based on the similarity between "hydrothermal" and "hydrolysis." Alcohols are the most common solvents for sol-gel synthesis. Primary alcohols are fairly stable at higher temperatures (up to 360°C) and therefore are widely used for the solvothermal reactions.[43] For example, amorphous gel derived by hydrolysis of metal alkoxides can be crystallized by solvothermal treatment in alcohols. Since lower alcohols (methanol, ethanol, and 1-propanol) are completely miscible with water, water molecules present in the precursor gel may be replaced with the solvent alcohols. Therefore the precursor gel is easily dispersed in the solvent, where crystallization takes place. Detailed mechanisms for the formation of crystals are not yet fully elucidated. Crystallization of metal oxides is usually reported to take place by dissolution-recrystallization mechanisms, but the mechanism seems to depend on the gel structure. Moreover, water molecules dissolved from the gel in the reaction medium may facilitate crystallization of the product. More discussion is given in Section III.D of this chapter.

Note that the primary alcohols corrode aluminum,[44] and therefore aluminum cannot be used as the sealing material for the autoclave when solvothermal reactions in primary alcohols are performed.

Since noble metal ions have a high tendency to be reduced to the metallic state, noble metal particles are easily formed by heating noble metal compound in alcohols at relatively low temperatures.[45] In these reactions, alcohols are oxidized to the corresponding aldehydes and then to the carboxylic acids.

Secondary alcohols such as 2-propanol (the common name is isopropyl alcohol; some researchers use "isopropanol," but this name is created by confusion of the common and IUPAC names) and 2-butanol are easily dehydrated at temperatures of 200 to 300°C, liberating olefins and water. Metal alkoxide can be hydrolyzed by the thus-formed water. This method was first reported by Fanelli and Burlew[46] (see Section III.B.8). Since water is formed homogeneously in the reaction system, this method can be regarded as the sol-gel version of the homogeneous precipitation method.[47] The usual precipitation method uses the precipitation reagent (acid or alkali), and addition of these reagents to the reaction system causes heterogeneity in pH in the system because of a large difference in pH between the system and the reagent. On the other hand, in the homogeneous precipitation method, pH of the system is homogeneously changed by, for example, the hydrolysis of urea intentionally added to the reaction system.

Dehydration of alcohols proceeds by heterolytic cleavage of the C–O bond, yielding carbocation and hydroxide anion, and the dehydration rate is determined by the stability of the thus-formed carbocation. Therefore tertiary alcohols such as *tert*-butyl alcohol (2-methyl-2-propanol) are more easily dehydrated. When these solvents are used for the solvothermal reaction, the essential nature of the reaction may be identical to that of the hydrothermal reaction.

## 5. Glycols

The solvothermal reaction in glycols is called a "glycothermal" reaction. Ethylene glycol (1,2-ethanediol) is the simplest glycol. This compound is stable at high temperatures. The hydroxyl group is an electron-withdrawing group (electrons are withdrawn through the C–O covalent bond) and the carbocation formed by the cleavage of a C–O bond is destabilized by the intramolecular hydroxyl group. Therefore dehydration of ethylene glycol barely proceeds.

This molecule has two hydroxyl groups that can donate their lone-pair electrons to those that are electron deficient, such as a metal cation. Thus it forms stable chelates with many metal ions and easily dissolves various metal salts. A number of glycolato complexes have been prepared and their crystal structures determined.[48–50]

One of the most significant results in solvothermal chemistry reported thus far is the synthesis of silica-sodalite, reported by Bibby and Dale,[13] which has never been prepared in aqueous systems. Ethylene glycol was used as the solvent in this synthesis, and it was shown that ethylene glycol is trapped in the cage of the sodalite.[51]

Ethylene glycol can be used to reduce metal ions to the metallic state. All the noble metals can be easily reduced by heating the metal salt in glycol at temperatures of 120 to 200°C,[5,6,52,53] and therefore reactions can be carried out in an open system. Nickel is also reduced, but iron is not reduced to the metallic state. This method was first reported by Figlarz et al.,[5] and they called this method the polyol process.

The author synthesized various mixed oxides using 1,4-butanediol. A more detailed discussion of this solvent is given in Sections III.B.9 and III.C.

## 6. Cyclic Ethers

Tetrahydrofuran (THF) and 1,4-dioxane are the most common cyclic ethers. Since the basicity of cyclic ethers is approximately 1 $pK_a$ unit higher than acyclic ethers, these compounds are widely used as solvents in many organic reactions. However, it should be noted that peroxide is formed by the reaction with molecular oxygen. The peroxides formed from these solvents are relatively stable and therefore can accumulate during storage. Once the reagent bottle is opened, the remaining solvent should be discarded. Old reagent should not be used because accumulated peroxide easily explodes on heating. It should be noted that tetrahydrofuran can polymerize with acid compounds in an exothermic reaction, and therefore acid compounds should be avoided in the reaction system. Some mixed metal oxides such as silica-alumina and titania-silica have a highly acidic surface and the reaction systems which may produce these oxides in tetrahydrofuran should be avoided.

## 7. Carboxylic Acids and Esters

Carboxylic acids can be used as the media for the solvothermal synthesis of inorganic materials. Synthesis of magnetite by thermal decomposition of an iron extract from the aqueous solution with a carboxylic acid has been reported (see Section III.B.4 of this chapter).[54] In organic chemistry, it is well known that carboxylic salts of divalent metal cations decompose into ketone yielding carbon dioxide.

Essential oils and vegetable oils are carboxylic acid esters of glycerin. In an inert atmosphere, these compounds are fairly stable at high temperatures. When they are heated in an open system, however, they are gradually oxidized by air and begin to decompose before they evaporate. When aluminum hydrogels are dropped into heated vegetable oils, alumina can be formed with evaporation of water. In this method, hydrothermal conditions are partially formed; thus boehmite particles are formed which combine with primary particles of amorphous alumina, and after calcinations, $\gamma$-alumina granules with sufficient mechanical strength are formed.

## 8. Amines

Ammonia as the solvothermal medium was described in Section II.A.2. Alkyl amines are similar to alcohols: lone-pair electrons on the nitrogen atom can stabilize metal cations, while protons on the nitrogen atom can stabilize anions by the formation of hydrogen bonding. The C–N bonds are more stable than the C–O bonds, and alkyl amines have higher thermal stabilities than the corresponding alcohols. Various amines are commercially available and a variety of amines have been used as templates (structure-directing agents) in the hydrothermal crystallization of zeolites and open-framework metal oxides and phosphates.[55,56]

### a. Hydrazine hydrate ($H_2N–NH_2 \cdot H_2O$)

It was reported that tetragonal SnO powders with a disc-like morphology were solvothermally prepared from stannous oxalate ($SnC_2O_4$) in hydrazine hydrate at 50°C–200°C.[57] However, hydrazine hydrate decomposes on heating or exposure to ultraviolet light to form ammonia, hydrogen, and nitrogen, therefore great care must be taken to prevent overheating of the reaction system. Moreover, contact with cadmium, gold, brass, molybdenum, and stainless steel containing greater than 0.5% molybdenum may cause rapid decomposition of hydrazine.[58] A Teflon-lined autoclave should be used and this solvent should not be used at high temperatures. The autoignition temperature is reported to be 280°C. Similarly, all the compounds that contain N–N, N–O, N–Cl, O–O, and Cl–O bonds can explosively decompose, and therefore great care should be taken when using these compounds.

### b. Ethanolamine (2-aminoethanol), diethanolamine (2,2'-iminobisethanol), and triethanolamine (2,2',2''-nitrilotrisethanol)

All these compounds are thermally stable and therefore are good solvents for solvothermal reactions. Since the amino group also donates its lone-pair electrons to Lewis acid (metal cation), metal cations are stabilized and triethanolamine is a tetradentate ligand. Laine et al.[59] reported that some metal oxides and hydroxides such as silica and aluminum hydroxide are completely dissolved in ethylene glycol in the presence of triethanolamine by refluxing the solvent with continuous distilling of the water produced by formation of triethanolamine complex of the metal ions. They used dissolved complex as the precursor for mixed oxides via spray pyrolysis.[59–62]

### c. Ethylenediamine, diethylenetriamine, triethylenetetramine

These solvents dissolve the chalcogen elements sulfur, selenium, and tellurium, and therefore can be used for the solvothermal synthesis of metal chalcogenides by the reaction of metal oxalate with chalcogens at 120°C–200°C.[63–65] It has been reported that the reaction also proceeds in tetrahydrofuran or pyridine, but the reaction in polyamines having more than two chelating atoms proceeds more completely.[65]

## 9. Other Nitrogen-Containing Compounds

### a. Acetonitrile

Acetonitrile has medium basicity and medium polarity, and therefore this compound is used for the synthesis of inorganic compounds such as polyoxometalates.[15,66–68]

### b. Nitromethane

This solvent is unique because it is poorly basic but has a high polarity. However, the nitro group is thermally unstable and therefore this solvent should not be heated. Nitrobenzene has a similar specificity; moreover, this compound is more toxic than nitromethane. Another compound having similar properties is sulforane. This compound is synthesized by the reaction of 1,3-butadiene and sulfur dioxide followed by hydrogenation of the C=C double bond and is now widely used for extraction of aromatic compounds from the residue of steam cracking for the synthesis of lower olefins. The thermal stability of sulforane (decomposes slightly at 285°C) is much higher than that of nitromethane.[69]

## 10. Dipolar Aprotic Solvents

Dimethyl sulfoxide (DMSO), $N,N$-dimethylformamide (DMF), and hexamethylphosphoramide ($C_6H_{18}N_3OP$; HMPA) are highly basic and have high dielectric constants. Because of the high basicity of these solvents, cations are highly solvated but anions are left unsolvated.[70] Therefore anions in these solvents have high reactivities. Only a few papers have dealt with the effect of these solvents in inorganic synthesis.

Since these solvents have high affinities to protein, they have a high toxicity to liver. Because of the relative low vapor pressures of these solvents at room temperature, there is relatively low risk when these solvents are used at lower temperatures. However, solvothermal reactions use these solvents at high temperatures, and therefore small leaks from the autoclave can cause severe damage to the researcher. The thermal stability of DMSO is slightly lower than that of the other two solvents, and its decomposition point is reported to be 189°C.[71]

Finally, the author strongly recommends that the reader refer to the material safety data sheet before performing solvothermal reactions in any type of solvent. The material safety data sheet can be obtained from the maker of the chemicals or can be found on the Internet.

## III. SOLVOTHERMAL SYNTHESIS OF METAL OXIDES

As discussed in Section I of this chapter, the prefix "solvo-" means any kinds of solvent; in this section, however, "solvothermal" is used to refer to reactions in organic solvents. When alcohols and glycols are used as the reaction media, the reactions are called "alcohothermal" and "glycothermal" reactions, respectively.

Metal oxides can be synthesized by various solvothermal techniques: solvothermal dehydration of metal hydroxides, solvothermal decomposition of metal alkoxides, solvothermal synthesis of mixed oxides, solvothermal crystallization of amorphous oxides, solvothermal ion exchange or solvothermal intercalation, and solvothermal oxidation of metals. Solvothermal techniques for synthesizing crystalline particles of metal oxide are summarized in Figure 2.1. Although some of the reaction products discussed in this section are not oxides, but hydroxides, oxyhydroxides, acetates, or organic-inorganic hybrid materials, oxides can be prepared by calcination of these products. As discussed later, most of the solvothermal products have fine particle sizes and therefore the oxides derived from these products have physical properties significantly different from those of metal oxides obtained by conventional methods. In this section, various solvothermal routes for the synthesis of metal oxides are reviewed, addressing the differences and similarities between solvothermal and hydrothermal reactions.

### A. SOLVOTHERMAL DEHYDRATION

Hydrothermal synthesis of metal oxides from the corresponding hydroxides usually requires high temperatures because of the limitations of equilibrium:

$$M(OH)_n \rightleftharpoons MO_{n/2} + n/2 \; H_2O. \tag{2.1}$$

Solvothermal dehydration can avoid this limitation, and dehydration may proceed at a temperature much lower than that required by the hydrothermal reaction. However, thermal dehydration may compete with the solvothermal reaction. When dehydration of hydroxide starts, water formed by dehydration of the starting material may facilitate hydrothermal transformation of the starting material. Therefore complicated reactions may occur simultaneously. As an example, the reaction of gibbsite (a polymorph of aluminum hydroxide, $Al(OH)_3$) in alcohols at 250°C is explained.[72]

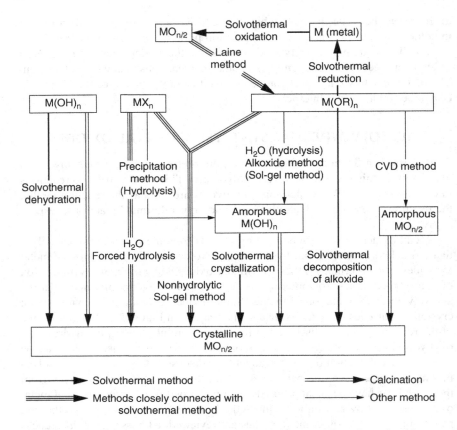

**FIGURE 2.1** Solvothermal routes to synthesize crystalline particles of metal oxide.

## 1. Solvothermal Dehydration of Aluminum Hydroxide in Alcohols

Solvothermal reaction of gibbsite in ethanol gives boehmite (AlOOH; a polymorph of aluminum oxyhydroxide), which is also obtained by hydrothermal reaction of the same starting material. The product is comprised of randomly oriented, thin, small crystals of boehmite. This result suggests that a dissolution-recrystallization mechanism takes place during the conversion.

The reaction of gibbsite in methanol occurs by a different route. The product is AlO(OCH$_3$) and has a boehmite-like structure. Since the product has an essentially identical morphology to that of the starting material, Kubo and Uchida[73] concluded that the reaction takes place by means of a solid state reaction in which methanol diffuses into the gibbsite structure with the aid of the Hedvall effect.[74] Since the molecular sizes of ethanol and higher alcohols are larger than that of methanol, the former solvents cannot diffuse into the gibbsite structure.

With an increase in the carbon number of the alcohol, the yield of boehmite decreases and the yield of $\chi$-alumina increases.[72] The latter compound is the thermal dehydration product of gibbsite,[75] and the highest yield of $\chi$-alumina is observed in pentanol. The morphology of the product obtained in pentanol is identical to that of the starting material. This feature (pseudomorphism) is typical for thermal dehydration of metal hydroxides and suggests that a solid state transformation mechanism takes place.[76] Since higher alcohols have progressively lower dielectric constants, the dissolution rate of the starting material in higher alcohols decreases and thermal dehydration of gibbsite becomes significant in these solvents. Therefore higher alcohols have essentially no effect on product formation. However, the solvent decreases the fugacity of water and prevents the hydrothermal conversion of gibbsite, as discussed later.

The reaction in higher alcohols is greatly affected by the particle size of the starting materials (Table 2.1).[72] When coarse gibbsite is used, the yield of boehmite increases. If boehmite were to be formed by the dissolution-recrystallization mechanism, the yield of boehmite would decrease with an increase in the particle size of the starting material; this is not the case. Formation of boehmite can be explained by the intraparticle hydrothermal reaction mechanism, originally proposed by de Boer et al.[77,78] for the formation of boehmite by thermal dehydration of gibbsite. It is well known that when coarse gibbsite is thermally dehydrated, boehmite is formed, and that formation of boehmite stops when formation of $\chi$-alumina starts. According to the mechanism, thermal dehydration starts at the "hot spot" of the particle and water molecules formed by dehydration cannot diffuse out from the particles. Therefore a hydrothermal condition is achieved inside the particle and boehmite is formed. When thermal dehydration yielding $\chi$-alumina starts, pore systems are formed inside the originating particle and water that facilitates the hydrothermal conversion into boehmite diffuses out from the particles; therefore formation of boehmite ceases. Formation of boehmite from coarse gibbsite under the solvothermal conditions in higher alcohols can be explained similarly.

Further increases in the carbon number of alcohol result in an increase in the yield of boehmite.[72] Since the reaction was carried out in a closed system, the water formed by the dehydration of gibbsite is adsorbed on the surface of the starting material, where water facilitates the hydrothermal conversion of gibbsite into boehmite. Actually, when the same starting material is heated in a closed vessel in the absence of any solvent, complete conversion into boehmite is observed. This result shows that the alcoholic solvent somehow retards the hydrothermal conversion of gibbsite. Solvent molecules (higher alcohols) lower the fugacity (activity) of water by forming strong hydrogen bonds between alcohol and water molecules. This effect becomes less significant with an increase in the carbon number of alcohol, and therefore the yield of boehmite increases with an increase in the carbon number of alcohol. The reaction mechanisms for the alcohothermal dehydration of gibbsite are summarized in Table 2.1.[79]

The reaction in an open system gives a completely different result. Since the boiling point of water is much lower than that of higher alcohols, water formed

**TABLE 2.1**
**Solvothermal Reaction of Gibbsite in Alcohols and Mineral Oil at 250°C for 2 h**

| Solvent | Product | Yield (%) of the Product from Gibbsite with a Particle Size of | | Mechanism for the Formation of the Product |
|---|---|---|---|---|
| | | 80 μm | 8 μm | |
| Water | Boehmite | 100 | 100 | DP(sol)[a] |
| Ethanol | Boehmite | 100 | 100 | DP(sol) |
| Propanol | Boehmite | 99 | 70 | DP(sol), (IPH)[b] |
| | χ-alumina | 0 | 24 | Solid state reaction |
| Butanol | Boehmite | 98 | 56 | DP(sol), IPH |
| | χ-Alumina | 0 | 38 | Solid state reaction |
| Pentanol | Boehmite | 51 | 21 | IPH |
| | χ-Alumina | 16 | 41 | Solid state reaction |
| Hexanol | Boehmite | 51 | 27 | IPH, HYD[c] |
| | χ-Alumina | 0 | 30 | Solid state reaction |
| Octanol | Boehmite | 86 | 41 | HYD, IPH |
| | χ-Alumina | 3 | 53 | Solid state reaction |
| Mineral oil | Boehmite | 100 | 100 | HYD |

[a] Dissolution-precipitation mechanisms: dissolution of aluminum species in the solvent used for the solvothermal reaction.
[b] Intraparticle hydrothermal reaction originally proposed by de Boer et al.[77,78] for the formation of boehmite during thermal dehydration of coarse gibbsite.
[c] Hydrothermal reaction takes place because of the water adsorbed on the particles surface: Therefore crystallization of the product occurs by means of the dissolution-precipitation mechanism.

by partial thermal dehydration of gibbsite will escape from the reaction system, and therefore simple thermal dehydration will occur.[72]

## 2. Alcohothermal Dehydration of Hydroxides of Metals Other Than Aluminum

Although the solvothermal dehydration of metal hydroxides depends on the solubility of the precursor in the reaction medium and the ease of thermal dehydration of the precursors, the essential features are similar to those for aluminum hydroxide discussed in the previous section. Yin et al.[80] examined the phase transformation of $H_2TiO_9 \cdot nH_2O$ (prepared from fibrous $K_2Ti_4O_9$) by solvothermal reaction in methanol and ethanol at 200 to 325°C and compared them with those by calcination in air or hydrothermal reaction. Under the solvothermal conditions, it transformed to monoclinic $TiO_2$ and then anatase, while under hydrothermal conditions or calcination in air it transforms into $H_2Ti_8O_{17}$ before transformation into monoclinic $TiO_2$. All the products retained a fibrous morphology similar to that of $K_2Ti_4O_9$, which was used as the starting material.

However, the microstructure of the fibers changed significantly; therefore they concluded that the phase transformation proceeded by a dissolution-reprecipitation mechanism. They also reported that anatase obtained by alcohothermal reactions had much higher photocatalytic activity than Degussa P25 titania.[81,82]

## 3. Solvothermal Dehydration of Aluminum Hydroxide in Glycols and Related Solvents

When a solvent having an electron-donating group, such as glycol, amino alcohol, or alkoxyethanol, is used as the solvent for the solvothermal dehydration of gibbsite, a completely different reaction occurs. The product is the alkyl (glycol) derivative of boehmite $(AlO(OH)_x(O(CH_2)_nX)_y$; $x + y = 1$; $X = OH$, $NR_2$, or $OR)$, in which solvent molecules are incorporated between the boehmite layers through the covalent bonds.[17,83] The fact that the product has Al–O–C bonds[17,83] indicates that equilibrium occurs between aluminum hydroxide and alkoxide[84] and that alkoxide can be formed from hydroxide at high temperatures:

$$M(OH)n + nROH \rightleftharpoons M(OR)n + nH_2O \quad \Delta H > 0, \Delta S > 0. \quad (2.2)$$

Formation of alkoxide from hydroxide is a reverse reaction of hydrolysis of alkoxide, which proceeds easily at room temperature and is a highly exothermic reaction (therefore Equation 2.2 has a positive reaction enthalpy). However, metal hydroxide is usually solid and has a polymeric M–(OH)–M network, while metal alkoxide usually has oligomeric structure. Therefore the former compound has lesser freedom (lower entropy). Consequently the unfavorable enthalpy term is overcome by the entropy term at high temperatures and equilibrium is attained.

The difference between the reactions in simple alcohols and in glycols can be attributed to the stabilization effects of the intramolecular electron-donating group on the intermediate and to the decrease in the activity (fugacity) of water due to solvation by the highly hydrophilic solvent molecules.

The particle size of the starting material has a significant effect on this reaction.[85] When gibbsite with a particle size less than 0.2 μm is used, complete conversion into boehmite derivatives is attained, while the use of coarser gibbsite results in an increase in the yield of well-crystallized boehmite formed by the intraparticle hydrothermal reaction. An increase in the particle size of gibbsite also increases recovery of the starting material. Randomly oriented thin plates of boehmite derivatives are formed on the surface of the pseudomorphs of the originating gibbsite particles. An important point here is that the product is not formed in the bulk of the solvent. Two explanations are possible: the gibbsite surface acts as the nucleation site of the product, or the gradient of the concentration of water, which declines from the surface of the particle to the bulk solvent, causes precipitation of product particles near the surface of the originating particles. The product particles formed on the originating gibbsite crystals prohibit dissolution of the starting material and a certain amount of the starting material remains unconverted even with prolonged reaction times.

## 4. Glycothermal Synthesis of α-Alumina

When microcrystalline gibbsite (less than 0.2 μm) is allowed to react in higher glycols such as 1,4-butanediol, 1,5-pentanediol, or 1,6-hexanediol at higher temperatures (approximately 300°C), α-alumina is formed.[7] This was the first example of thermodynamically stable metal oxide crystallized in organic solvent at a temperature lower than that required by the hydrothermal reaction. The α-alumina is formed from the glycol derivative of boehmite as an intermediate and product particles are hexagonal plates with a relatively large surface area. This indicates that the dissolution-recrystallization mechanism occurs.

Hydrothermal synthesis of α-alumina has been well studied. Since the hydrothermal reaction of aluminum compound yields boehmite at relatively low temperatures (approximately 200°C),[86] transformation of boehmite was examined and it was reported that more than 10 hours is required for complete conversion into α-alumina, even with a reaction at 445°C in a 0.1 N NaOH solution and in the presence of seed crystals.[87] On the other hand, under glycothermal conditions, α-alumina is formed at 285°C for 4 h. The equilibrium point between diaspore (another polymorph of AlOOH) and α-alumina under the saturated vapor pressure of water was determined to be 360°C.[88] However, near the equilibrium point, the transformation rate is very sluggish, and only a small conversion of diaspore is observed. Therefore complete conversion of diaspore into α-alumina requires a much higher temperature. Since boehmite is slightly less stable than diaspore, the hypothetical equilibrium point between boehmite and α-alumina would be lower than that for diaspore-alumina. However, α-alumina would not be formed by a hydrothermal reaction at such a low temperature as has been achieved in the glycothermal reaction.

The difference between two reactions may be attributed to the activity of water present in the reaction system, since the overall reaction is the dehydration reaction (Equation 2.1). However, intentional addition of a small amount of water caused enhancement of α-alumina formation rather than the retardation expected from the equilibrium point of view.[89] Another important factor is the difference in the thermodynamic stabilities of the intermediates between glycothermal and hydrothermal reactions; that is, the glycol derivative of boehmite vs. well-crystallized boehmite. The latter compound is fairly stable and therefore conversion of this compound into α-alumina has only a small driving force. On the other hand, the glycol derivative of boehmite has Al–O–C bonds and therefore is more unstable with respect to α-alumina. Thus conversion of this compound into α-alumina has a much larger driving force. The smaller crystallite size of the glycol derivative of boehmite also contributes to the instability of the intermediate.

The reaction is strongly affected by the particle size of the starting material.[89] When gibbsite with a particle size less than 0.2 μm is used, complete conversion into α-alumina is attained. When coarse gibbsite is used for the reaction, the product is $\chi$-alumina together with well-crystallized boehmite. $\alpha$-Alumina was not formed at all.[89] As mentioned in the previous section, formation of the intermediate, the glycol derivative of boehmite, ceases at a certain point in the reaction because the originating gibbsite particles are covered with the crystals of the glycol derivative of boehmite. Higher reaction temperatures cause thermal dehydration of gibbsite into $\chi$-alumina, which seems to facilitate epitaxial decomposition of the glycol derivative of boehmite into a transition alumina. Therefore, when coarser gibbsite is used as the starting material, $\alpha$-alumina cannot be obtained, even at higher temperatures.

There are two other important points in this reaction.[89] Water formed as a byproduct in the formation of the glycol derivative of boehmite facilitates crystallization of $\alpha$-alumina by increasing the dissolution rate. Intentional addition of a small amount of water actually lowers the $\alpha$-alumina crystallization temperature; however, excess water causes the formation of boehmite, which is fairly stable under the reaction conditions and contaminates the product. Another point is the presence of $\alpha$-alumina nuclei in the starting gibbsite sample. The starting gibbsite is commercially produced by milling coarser gibbsite. It is known that prolonged grinding or milling of gibbsite particles causes mechanochemical transformation into $\alpha$-alumina.[90] Therefore $\alpha$-alumina-like domains are formed at this stage and act as the nuclei of $\alpha$-alumina in the glycothermal reaction. Bell et al.[91] also addressed the importance of seed particles on the size of hexagonal $\alpha$-alumina platelets.

Although the glycol derivative of boehmite is also prepared from aluminum alkoxides (see Section III.B.9), this product cannot transform into $\alpha$-alumina because of the absence of $\alpha$-alumina nuclei and water. However, the reaction of a mixture of gibbsite (microcrystalline) and aluminum alkoxide in the presence of a small amount of water gives $\alpha$-alumina with larger crystallite sizes than those obtained from gibbsite alone.

Cho et al.[92] reported that $\alpha$-alumina is formed from aluminum hydroxide prepared by precipitation with potassium hydroxide. However, when alkaline hydroxide is used as the precipitation agent, alkali cations are incorporated into the product, and commercial gibbsite samples are always contaminated with a small amount of sodium ions. Therefore their starting material seems to be contaminated with potassium, and the presence of potassium ions in their precursor seems to play an important role in the nucleation of $\alpha$-alumina. They also reported that hydroxyl ions, acetic acid, and pyridine added to the glycothermal reaction system affect the morphology of the $\alpha$-alumina particles because of their preferential adsorption to a specific surface.[93]

## B. Solvothermal Decomposition of Metal Alkoxides

### 1. Metal Alkoxide in Inert Organic Solvents

Thermal decomposition of metal alkoxides gives the corresponding metal oxides. This reaction is usually applied for the synthesis of oxide films or

**TABLE 2.2**

**Phases Formed by Solvothermal Decomposition of Alkoxides and Acetylacetonates**

| Starting Material | Reaction Temperature | Product | Reference |
|---|---|---|---|
| Aluminum isopropoxide | 300 | $\chi$-Alumina | 95 |
| Aluminum n-butoxide | 300 | No reaction | 95 |
| Aluminum tert-butoxide | 300 | Amorphous | 95 |
| Zirconium isopropoxide | 300 | Tetragonal zirconia | 96 |
| Zirconium n-propoxide | 300 | No reaction | 96 |
| Zirconium acetylacetonate | 300 | Tetragonal zirconia | 96 |
| Zirconium tert-butoxide | 200 | Amorphous | 96 |
| Titanium isopropoxide | 300 | No reaction | 97 |
| Titanium oxyacetylacetonate | 300 | Anatase | 97 |
| Titanium tert-butoxide | 300 | Anatase | 98 |
| Niobium n-butoxide | 300 | Amorphous | 99, 100 |
| Tantalum n-butoxide | 300 | Amorphous | 101 |
| Iron acetylacetonate | 300 | Magnetite | 102 |
| Iron (n-butoxide) | 300 | Hematite + magnetite | 103 |
| Lanthanum isopropoxide | 300 | Lanthanum hydroxide | 103 |

aerosols by the chemical vapor deposition (CVD) method.[94] Whereas CVD reactions of metal alkoxides under reduced pressure usually produce amorphous products, solvothermal reactions of secondary alkoxides in inert organic solvents such as toluene, in some cases, give crystalline products (Table 2.2). For example, thermal decomposition of aluminum isopropoxide and zirconium isopropoxide in toluene at 300°C yields $\chi$-alumina[95] and tetragonal zirconia,[96] respectively. This means that solvent molecules present in the reaction system facilitate crystallization of the product.

The primary alkoxides of these metals do not decompose at 300°C, as decomposition of these compounds requires much higher temperatures, whereas tertiary alkoxides decompose at much lower temperatures, yielding amorphous products. These results indicate that heterolytic cleavage of the C–O bond yielding carbocation and metaloxo anion (>M–O; Equation 2.3) is the key step, and stability of the carbocation determines the reactivity of the metal alkoxides:[95]

$$M(OR)_n \rightarrow >M-O^- + R^+ \tag{2.3}$$

However, this reaction is strongly affected by the metal cation of the alkoxide; thus $Nb(OBu)_5$ decomposes in toluene at 573 K yielding amorphous $Nb_2O_5$ powders,[100] whereas $Ti(O-^iPr)_4$ is not decomposed under the same reaction conditions.[97] The factors of metal cations controlling decomposition of alkoxides are not yet fully elucidated, but electronegativity as well as the oligomeric structure of the alkoxide seem to determine the thermal decomposition reactivity of the alkoxides.

Solvothermal decomposition of titanium *tert*-butoxide[98] and titanium oxyacetylacetonate $(TiO(acac)_2)$[97] in toluene at 300°C yields nanocrystalline anatase. It should be noted that the lowest temperature required for the formation of crystalline titania by CVD synthesis was reported to be 400 to 450°C.[104–106]

One of the limitations of this method (thermal decomposition of alkoxide in inert organic solvent) is the prerequisite of purification of the starting alkoxides (see for comparison, the glycothermal reaction described in Section III.B.9), since the alkoxide is easily hydrolyzed in moist air. Note, however, that fairly good reproducibility may be obtained by using a fresh reagent from a brand-new reagent bottle and discarding the remaining reagent.

This method is closely related to the nonhydrolytic sol-gel method.[107] For example, titania is prepared by etherolysis/condensation of $TiCl_4$ by diisopropyl ether (Equation 2.4) or by direct condensation between $TiCl_4$ and $Ti(O-^iPr)_4$ (Equation 2.5). Detailed chemistry of the reaction was examined by means of nuclear magnetic resonance (NMR), and it has been reported that the true precursors are titanium chloroisopropoxides in equilibrium through fast ligand exchange reactions.[108] A variety of metal oxides,[109,110] nonmetal oxides,[111] multicomponent oxides[112,113] were studied, and the nonhydrolytic sol-gel method was surveyed by Vioux and Leclercq.[107]

$$TiX_4 + 2ROR \rightarrow TiO_2 + 4RX \qquad (2.4)$$

$$TiX_4 + Ti(OR)_4 \rightarrow 2TiO_2 + 4RX \qquad (2.5)$$

Trentler et al.[114] reported synthesis of "hydroxyl-free" anatase nanocrystals by the nonhydrolytic sol-gel method (Equation 2.4). Titanium halide was mixed with distilled trioctylphosphine oxide (TOPO) in heptadodecane and heated to 300°C under dry nitrogen and a titanium alkoxide was then rapidly injected into the hot solution. Anatase with a crystallite size less than 10 nm in diameter was obtained, although there was considerable size distribution. They reported that the reaction rate dramatically increased with greater branching of R, and suggested that an $S_N1$ mechanism takes place, consistent with similar low-temperature reactions.[115] This result is also consistent with our conclusion that heterolytic cleavage of the C–O bond (Equation 2.3) is the key step for the solvothermal decomposition of metal alkoxide. However, from a physical organic chemistry viewpoint, it is widely accepted that high temperatures favor an elimination reaction (Equation 2.1) over a substitution reaction $(S_N1)$, suggesting formation of olefins rather than alkyl halide. Although titanium isopropoxide does not decompose in inert organic solvents at the reaction temperature (300°C),[95] the presence of a small amount of hydrogen halide formed by decomposition of $TiX_4$ or RX possibly catalyzes the decomposition of metal alkoxide. Thus this reaction procedure does not ensure the formation of "hydroxyl-free" product because of the following reactions:

$$Ti(OR)_4 \rightarrow Ti(OH)_4 + olefins \tag{2.6}$$

$$Ti(OH)_4 \rightarrow TiO_2 + 2H_2O. \tag{2.7}$$

Rather, the observed lack of surface hydroxyl groups may be due to the open reaction system, where surface hydroxyl groups are dehydrated as water is eliminated from the reaction system.

Trentler et al.[114] also proposed that nanocrystalline products are obtained at elevated temperatures because $TiX_4$ serves as a crystallization agent as well as a reactant. They pointed out the importance of a chemical reversibility, that is, Ti–O bond breaking and forming, that would erase defects incorporated into growing titania crystals. This statement seems to be made because they do not know that the solvothermal reaction of titanium *tert*-butoxide and titanium oxyacetylacetonate in toluene yields nanocrystalline anatase.[97,98] However, this point is closely connected with one of the important features of solvothermal products, and therefore will be discussed here.

In hydrothermal reactions, dissolution-deposition equilibrium takes place. Dissolution of adatoms on the surface occurs preferentially, while preferential adsorption of ions at the vacancies of the surface proceeds. Therefore a nearly perfect surface is formed. Since a nearly perfect growing surface is created, crystals formed by the hydrothermal reaction usually contain fewer defects than the crystals formed by other methods. On the other hand, under solvothermal conditions, dissolution of oxide materials into the organic solvent barely takes place, and therefore the product usually contains various types of crystal defects.

## 2. Metal Alkoxides in Inert Organic Solvent: Synthesis of Mixed Oxides

Thermal decomposition of two starting materials in inert organic solvent may provide a convenient route for the synthesis of mixed oxides or precursor of mixed oxides. For example, when a mixture of aluminum isopropoxide and tetraethoxysilane (tetraethylorthosilicate) in a 3:1 ratio is decomposed in toluene, an amorphous product is obtained.[116] Note that thermal decomposition of the former compound alone yields $\chi$-alumina,[95] while the latter compound alone does not decompose at the reaction temperature.[116] Mullite is crystallized by calcination of the product at 900°C.[116] It is known that the crystallization behavior of mullite from the precursor gel depends on the homogeneity of mixing of aluminum and silicon atoms in the precursor: when the precursor gel has atomic scale homogeneity, mullite is crystallized at 900°C, and the gel with homogeneity in a nano scale causes crystallization of silicon-aluminum spinel at around 900°C, while heterogeneous gel requires 1300°C for crystallization of mullite.[117,118] Therefore atomic scale homogeneity is attained in the solvothermal product, even though the reaction procedure is quite simple.

The solvothermal decomposition of a mixture of La(O-$^i$Pr)$_3$ and Fe(OBu)$_3$ in toluene yields an amorphous product, whereas the reaction of individual starting materials yielded crystalline La(OH)$_3$ and a mixture of $\alpha$-Fe$_2$O$_3$ and Fe$_3$O$_4$, respectively.[103] Calcination of the amorphous product at 550°C yields crystalline LaFeO$_3$ (perovskite). Low crystallization temperature also suggests high homogeneity of the solvothermal product.

## 3. Metal Acetylacetonate in Inert Organic Solvent

Besides alkoxides, acetylacetonates are also used as the starting materials for the synthesis of oxides. Titania (anatase) is obtained by decomposition of titanium oxyacetylacetonate (TiO(acac)$_2$) in toluene at 300°C.[97] Similarly solvothermal treatment of Fe(III) acetylacetonate in toluene yields microcrystalline magnetite.[102] One of the drawbacks of the use of acetylacetonate may be formation of various high boiling point organic by-products via aldol-type condensation of the acetylacetone. Actually more than 50 compounds are detected by gas chromatography-mass spectrometry (GC-MS) analysis of the supernatant of the reaction, some of which are phenolic compounds and are hardly removed from the oxide particles by washing with acetone.[97]

## 4. Metal Carboxylates

Konishi et al.[54] reported thermal decomposition of iron carboxylate: Fe(III) was extracted from an aqueous solution using Versatic 10 (tertiary monocarboxylic acids) and the organic layer was diluted with Exxsol D80 (aliphatic hydrocarbons: bp 208 to 243°C). The organic solution was then filtered through glass filter paper and passed through phase separating paper to remove physically entrained water. Then the organic solution was solvothermally treated. Magnetite particles about 100 nm in size were formed when the solution was heated at 245°C, but in the presence of intentionally added water, hematite ($\alpha$-Fe$_2$O$_3$) contaminated in the product, while pure hematite was formed at a lower temperature in the presence of a larger amount of water. The carboxylic acid serves as a reducing agent and is partially decomposed into carbon dioxide:

$$RCO_2^- \rightarrow R\cdot + CO_2 + e^-. \tag{2.8}$$

The strategy of their research is solvent extraction (hydrometallurgy) from mineral resources followed by thermal decomposition of the extracts directly. Therefore they used a rather special carboxylic acid, Versatic 10.

Solvothermal decomposition of stannous oxalate ($SnC_2O_4$) yielding tetragonal SnO powders was also reported.[57]

## 5.  Cupferron Complexes

Rockenberger et al.[119] described a general route for the synthesis of dispersible nanocrystals of transition metal oxides. Their route involves the decomposition of cupferron (N-nitroso-N-phenylhydroxylamine) complexes of metal ions such as $Fe^{3+}$, $Cu^{2+}$, and $Mn^{3+}$ in the high-temperature solvent trioctylamine at 250 to 300°C, obtaining the oxides γ-$Fe_2O_3$, $Cu_2O$, and $Mn_3O_4$, respectively, of 4 to 10 nm in diameter. The products are crystalline and are dispersible in organic solvent. By addition of a threefold volume excess of methanol, the nanocrystals can be reprecipitated. Particles with the smallest average size (4 nm) were synthesized by lowering the reaction temperature and/or lowering the injected precursor concentration, and average particle diameter was controlled by subsequent injection of the precursor. The decompositions takes place at sufficiently low temperatures, and a capping agent such as a long-chain amine can be employed to prevent sintering and to prepare well-dispersed nanoparticles.

Thimmaiah et al.[120] and Guatam et al.[121] extended the method of Rockenberger et al. by replacement of relatively expensive (and toxic) amines with toluene using a closed system, and showed that mixed transition metal oxides such as the spinel $CoFe_2O_4$ can also be prepared.

## 6.  Solvothermal Decomposition of Alkoxide Followed by Removal of Organic Media in a Supercritical or Subcritical State

Removal of the organic solvent at the solvothermal temperature is an interesting modification of this type of reaction. Since inert organic solvent usually has a relatively low critical point, the reaction temperature may be in the supercritical or subcritical region. The removal of the organic phase directly from the reaction vessel at the reaction temperature gives well-divided powders of the product.[122,123]

When a product is washed with water and then dried, coagulation of the product particles occurs in the drying stage. The surface tension of the water remaining between the product particles pulls the particles closer as the drying proceeds,[124] producing tightly coagulated products. When water is replaced with an organic solvent, this coagulation may be loose because the surface tension of organic solvent is less than that of water. However, when product is dried with supercritical fluid, coagulation of the particles can be avoided.[125] This process is called supercritical drying and the product is called an aerogel.[126] Thus the procedure provides a convenient method for the synthesis of oxide powders using only one reaction vessel, combining solvothermal synthesis and the supercritical drying process.

## 7. Metal Alkoxide in Alcohols

When metal alkoxides are allowed to react in alcohols, an alkoxyl exchange reaction takes place at lower temperatures:[84]

$$M(OR)_n + nR'OH \rightleftharpoons M(OR')_n + nROH. \qquad (2.9)$$

Therefore the composition of alkyl groups in the coordination sites of the metal is determined by the relative number of the two alkyl groups in the reaction system and by the relative volatility of the two alcohols. When the alcohol derived from the alkoxide has a low boiling point, the alcohol will evaporate and escape from the reaction system to the gas phase in the reaction vessel, and therefore the corresponding alkyl groups are completely expelled from the alkoxide.

The reaction of aluminum isopropoxide in primary alcohols yields the alkyl derivative of boehmite, in which the alkyl moieties derived from the solvent remain while expelling all the alkyl groups derived from the alkoxide.[127] Straight-chain primary alcohols with a carbon number up to 12 were examined, and a linear relationship between the basal spacing and carbon number of the alcohols was observed. Note that when the reaction is carried out in 2-ethyl-1-hexanol, the product is $\chi$-alumina, indicating that this solvent behaves as an inert solvent.[127]

As described in Section III.B.1, decomposition of primary alkoxides in inert organic solvents requires temperatures much higher than 300°C, but in alcohols they may decompose at relatively low temperatures. The carbocation formed by the heterolytic cleavage of the C–O bond is only poorly solvated in the inert organic solvent; therefore the reaction barely proceeds. On the other hand, in alcohols, carbocation is solvated, which lowers the activation energy for the decomposition of alkoxide. For example, aluminum ethoxide does not decompose in toluene at 300°C, while it does decompose in ethanol, yielding the alkyl derivative of boehmite.

Barj et al.[43] and Pommier et al.[128] examined thermal decomposition of $Mg[Al(O\text{-}sec\text{-}Bu)_4]_2$ in ethanol: decomposition starts at 283°C, yielding an essentially amorphous product, while the reaction at 360°C yields partially crystallized $MgAl_2O_4$ spinel. The product consisted of secondary particles with a size of 3 µm, which is easily disaggregated by ultrasonic treatment into 0.02 µm primary particles. They also reported that thermal dehydration of ethanol becomes significant at temperatures higher than 360°C. Therefore crystallization of this product seems to occur with the aid of water formed by dehydration of the solvent.

Wang et al.[129] reported the reaction of $TiCl_4$ in various alcohols at 100°C and 160°C. Methanol, ethanol, 1-propanol, and 2-propanol gave anatase, and butanol and octanol yielded rutile, while ethylene glycol yielded a mixture of rutile and anatase. It is known that $TiCl_2(OR)_2$ can be readily formed when

$TiCl_4$ is introduced into alcohols with liberation of HCl.[130] The products consist of spherical particles, but agglomerated tenuous fibers were obtained in octanol (rutile) and 2-propanol (anatase). The quite low crystallization temperature of titania as well as the low decomposition temperature of the alkoxides derived from $TiCl_4$ are rather surprising. However, water, formed by dehydration and etherification of alcohols (the authors detect ethers by GC-MS) with the aid of dissolved HCl as a catalyst, possibly hydrolyzes the alkoxide and facilitates hydrothermal crystallization of the product. Moreover, HCl may catalyze the cleavage of Ti–O–R bonds (Equation 2.10) and also mediates the crystallization of titania through the bond breaking and forming equilibrium shown in Equation 2.11. They also noted that the phase formed by the reaction is affected by the concentration of HCl, with a lower concentration of HCl favoring the formation of anatase.

$$\equiv TiOR + H^+ \rightarrow \equiv Ti\text{-}O(H^+)\text{-}R \rightarrow \equiv TiOH + R^+ \qquad (2.10)$$

$$\equiv Ti\text{-}O\text{-}Ti + HX \rightleftharpoons \equiv TiOH + XTi \qquad (2.11)$$

## 8. Reaction of Alkoxide in Secondary Alcohols

Reaction of alkoxide in secondary alcohols gives a completely different route than that occurring in inert organic solvents. Secondary alcohols are easily dehydrated at higher temperatures, yielding water and olefins, and water can hydrolyze the alkoxide.[46] In the CVD reaction, Takahashi et al.[131] observed that decomposition of titanium isopropoxide in the presence of isopropyl alcohol occurs at lower temperatures because of the formation of water from alcohol. This hydrolysis route may compete with the thermal decomposition route; however, even when *tert*-alkoxide is used as the starting material, the alkoxyl exchange reaction proceeds at lower temperatures and therefore the hydrolysis route seems to be the predominant one. Water formed by dehydration of solvent alcohols also facilitates hydrothermal crystallization of the hydrolysis product of alkoxide (see Section III.E). When the reaction in primary alcohol is carried out at high temperatures (greater than 360°C) and/or in the presence of acid catalyst, a similar route can be expected.

Fanelli and Burlew[46] first applied this method for the formation of alumina by the reaction of aluminum *sec*-butoxide in 2-butanol at 250°C and reported the formation of alumina with a quite large surface area. They pointed out that this method could be regarded as the sol-gel version of the homogeneous precipitation method (see Section II.B.4).

Courtecuisse et al.[132] reported a flow reactor for the formation of titania by decomposition of titanium isopropoxide in 2-propanol at 260 to 300°C. They reported that the rate-determining step is the thermal dehydration of titanium hydroxide formed by the hydrolysis of the precursor alkoxide.[133] They also found that an increase in the supercritical fluid density decreased the overall reaction rate, but adequate explanation was not given by the authors.

## 9. Reaction of Alkoxide in Glycols

The reaction of aluminum alkoxides in glycol yields the glycol derivative of boehmite.[17,134,135] The crystallite size of the product increased in the following order: $HO(CH_2)_2OH < HO(CH_2)_3OH < HO(CH_2)_6OH < HO(CH_2)_4OH$. The physical properties of the products and the aluminas derived by calcination thereof varied according to this order.[134,136] This result suggests that development of the product structure is controlled by the heterolytic cleavage of the O–C bond of the glycoxide intermediate, $>Al-O-(CH_2)_nOH$, formed by alkoxyl exchange between aluminum alkoxide and the glycol used as the medium. The presence of an electron-withdrawing group, that is, a hydroxyl group, near the O–C bond retards the formation of carbocation, thus only poorly crystallized product is obtained in ethylene glycol. On the other hand, the largest crystallite size, obtained by the use of 1,4-butanediol (1,4-BG), is interpreted by the ease of the cleavage of the O–C bond due to participation of the intramolecular hydroxyl group,[137] which yields an aluminate ion ($>Al-O$) and protonated tetrahydrofuran (Equation 2.12). A similar medium effect was also observed for the glycothermal treatment of other alkoxides:

$$>M-O-CH_2-CH_2-CH_2-CH_2-OH \rightarrow M-O^- + \underset{\underset{OH^+}{\diagup}}{\overset{\overset{\displaystyle CH_2-CH_2}{\diagup \quad \diagdown}}{CH_2 \qquad CH_2}} \quad (2.12)$$

It must be noted that both the reactions of aluminum hydroxide and alkoxide in glycol yielded the glycol derivative of boehmite with identical morphology.[17,83,134] As discussed in Section III.A.3, equilibrium between hydroxide-alcohol and alkoxide-water (Equation 2.2) is attained at high temperatures.[84] Because of this equilibrium, glycoxide (a kind of alkoxide) is generated from the hydroxide under glycothermal conditions, and therefore partially hydrolyzed alkoxide can be used for the reaction. (Complete hydrolysis of aluminum alkoxides yields microcrystalline boehmite [pseudoboehmite], the structure of which is fairly stable, and therefore fully hydrolyzed alkoxide does not give the desired product by the glycothermal reaction.) Actually the glycothermal reaction does not require any precautions for handling alkoxides, and reproducible results are obtained even without purification of the starting alkoxides.[138] (As for zirconium alkoxide, the polycondensation reaction of the Zr-OH group proceeds rapidly as compared with the alkoxyl exchange reaction; as a result, partial hydrolysis of zirconium alkoxide severely affects the physical properties of the glycothermal products.)[96,138–141]

Various oxides such as $ZrO_2$,[96,141] $TiO_2$,[85] $ZnO$,[142] and $Nb_2O_5$[99] have been prepared by the glycothermal reaction of the corresponding alkoxides. Kang et al.[143] reported that anatase (122 $m^2$/g) obtained by glycothermal (1,4-butanediol) treatment of titanium isopropoxide at 300°C has considerably higher photocatalytic activity than the catalyst prepared by the sol-gel method.

## C. GLYCOTHERMAL SYNTHESIS OF MIXED METAL OXIDES

### 1. Rare Earth Aluminum Garnets

Since metaloxo anion (>M-O) is expected to be formed by decomposition of the glycoxide intermediate derived from alkoxide and 1,4-butanediol, the presence of metal cation that gives basic oxides would give M–O–M bonds. According to this working hypothesis, we examined the reaction of aluminum isopropoxide (AIP) with yttrium acetates in 1,4-BG at 300°C and found the formation of crystalline yttrium aluminum garnet (YAG).[8] The hydrothermal reaction of pseudoboehmite (hydrolyzed product of AIP) with yttrium acetate at 300°C yielded boehmite together with a small amount of YAG. Single-phase YAG was not obtained, even with prolonged reaction time.[8,144] The difference between glycothermal and hydrothermal reactions can be attributed to the different stabilities of the intermediate phases, that is, the glycol derivative of boehmite vs. well-crystallized boehmite, which is easily formed by hydrothermal reaction of aluminum compounds (see Section III.A.4).

Similarly the reaction of the stoichiometric mixture of AIP and rare earth (RE) acetate (Gd–Lu) gives the corresponding rare earth aluminum garnet (REAG) in single phase.[144] Synthesis of single-phase REAG by the reaction of mixed alumina and RE oxide powders normally requires a temperature higher than 1600°C with a prolonged heating period.[145] Homogeneous mixing of aluminum and RE atoms in the starting materials can lower the crystallization temperature of REAG,[146–149] but these materials still require calcination temperatures higher than 800°C to crystallize the REAG phase.

The glycothermal reaction of rare earth acetate alone yields $RE(OH)_2(OAc)$, $REO(OAc)$ (two polymorphs), and $RE(OH)(OAc)_2$, depending on the ionic size of the RE ion.[150,151] The acetate ions are not completely expelled from the coordination sites of the RE ion. However, in the presence of aluminum alkoxide as the starting material, acetate ions are fully eliminated from the product. Therefore anionic species (that is, >Al-O) facilitate cleavage of the bond between acetate and RE ions.

The reaction of samarium or europium acetate with AIP produced SmAG or EuAG, although the product was contaminated with RE acetate oxide $(RE(CH_3COO)O)$.[144] The reaction of AIP with neodymium acetate gave only $Nd(CH_3COO)O$ as the sole crystalline product. The thermodynamic stabilities of the garnet phases depend on the ionic size of the RE element, and REAGs were reported to be thermodynamically stable for all the RE elements from terbium to lutetium.[152] Therefore all the thermodynamically stable REAGs were prepared by the glycothermal method. GdAG (metastable phase) has been prepared by many researchers, but synthesis of SmAG and RuAG has never been reported by any other methods.

Hydrothermal synthesis of single-phase REAG requires higher temperatures (350 to 600°C) and pressures (70–175 MPa),[153,154] although Mill' reported that the lower temperature limit for the formation of YAG was 280°C.[153] He also reported that with an increase in the ionic size of the RE element, the lower temperature limit increased. The REAG with the largest RE ion size that has been thus far prepared hydrothermally is TbAG,[153] and it was reported that 420 to 450°C was required for the formation of this garnet. Therefore there seems to be no possibility that SmAG and EuAG can be prepared by the hydrothermal method, because the ionic size of these elements is much larger than that of Tb.

## 2. Rare-Earth (Nd-Lu) Gallium Garnets

The reaction of RE (Nd-Lu) acetates with $Ga(acac)_3$ in 1,4-BG at 300°C yielded the corresponding RE gallium garnets (REGGs).[21,156] The garnet phases were reported to be thermodynamically stable for all the RE elements from samarium to lutetium,[157] and all of the stable REGGs were prepared by the glycothermal method. The reaction at 270°C also gave phase pure REGGs for Sm-Lu, but the reactions at 250°C gave amorphous products except for the reaction of yttrium, which gave yttrium gallium garnet (YGG). Hydrothermal reaction of $Ga(acac)_3$ with RE acetate under conditions identical to the glycothermal reaction (except reaction medium) gave -$Ga_2O_3$ together with a small amount of the garnet phase.[21]

The reaction of praseodymium and cerium acetates with $Ga(acac)_3$ gave $RECO_3(OH)$ as the sole crystalline product. When the reaction was carried out in the presence of gadolinium gallium garnet (GGG) seed crystals, garnets were crystallized in spite of the fact that unit cell parameters of these garnets are much larger than those of GGG.[21] These results suggest that under glycothermal conditions, nucleation is the most difficult process, and that once nucleation takes place, crystal growth proceeds easily.

The particles of gallium garnets with large RE ions are spherical (0.5–2 μm) and the surface of the particles is smooth. On the other hand, the surface of the particles (0.1–0.3 μm) of the gallium garnets of terbium and RE elements having ionic sizes smaller than terbium is rough, with apparent polycrystalline outlines. However, high-resolution images of the latter type of particles show that a whole particle is covered with a single lattice fringe, indicating that each particle is a single crystal grown from only one nucleus. The authors concluded that the latter type of morphology is formed by quite rapid crystal growth.[21]

Monodispersed particles are formed for garnet with smaller RE ions.[21] Monodispersed particles can be prepared if a burst of nuclei is formed at the early stage of the reaction and if nucleation does not take place during the crystal growth stage.[158] Once nucleation occurs in the glycothermal synthesis of gallium garnets, the concentration level decreases, which is determined by the balance between the dissolution rate and rate of consumption of the reactants in the solution by crystal growth. For the small RE ions, crystals grow rapidly, and therefore the concentration of the reactants becomes low, prohibiting nucleation during the crystal growth stage.

## 3. Metastable Hexagonal REFeO$_3$

The reactions of RE acetates with Fe(acac)$_3$ yielded three different types of the product, depending on the ionic size of the RE element. For Nd-Gd, the product was RE(CH$_3$COO)O, and no binary oxide was formed. The product obtained from Er-Lu acetate was a novel phase of REFeO$_3$ having a hexagonal crystal system ($ao$ = 6.06, $co$ = 11.74).[159] The hexagonal phase was also formed in the reaction at 220°C, but the product had quite a low crystallinity. The product is isomorphous to the hexagonal YMnO$_3$ ($a_o$ = 6.125, $c_o$ = 11.41, P6$_3$cm).[160] In the structure, iron atoms are in trigonal bipyramidal coordination surrounded by five oxygen atoms. Hydrothermal reaction of a mixture of Fe(acac)$_3$ with RE acetate yields Fe$_2$O$_3$ (hematite) together with an amorphous RE species.[159,161] In the glycothermal reaction, metal alkoxide (glycoxide) or acetylacetonate is a starting material or may be formed as an intermediate, and the reaction proceeds by thermal decomposition of these alkoxides (glycoxides) instead of the hydrolysis of the alkoxide process, which usually yields amorphous products. In the alkoxide process, part of the free energy of the starting materials is consumed in the hydrolysis stage, whereas in the glycothermal reaction, instability of the intermediate phase gives a large driving force to product formation. Therefore crystal growth proceeds rapidly and metastable phases can be formed by the glycothermal reaction.

## 4. Other Mixed Oxides

Glycothermal reaction of two starting materials gives various crystalline mixed oxides, and some of the mixed oxides synthesized by the glycothermal method are summarized in Table 2.3.[8,99,142,144,159,162,163] According to our working hypothesis, the reaction between metaoxo anion and metal cation is expected. In other words, the glycothermal synthesis of mixed oxides can be regarded as an acid-base reaction. Metal ions of highly electronegative elements form acidic oxides, and the hydroxyl groups in the coordination sites of these metal ions can be easily deprotonated, yielding metaoxo ions. On the other hand, metal cations with elements that are less electronegative form basic oxide, and the hydroxyl groups in the coordination sites of these metal ions can be liberated, yielding hydroxide ion and metal cation. As shown in Table 2.3, combining these two types of elements gives crystalline oxides. Although combining titanium and RE ions resulted in the formation of amorphous products, the glycothermal reaction of each starting material yields a crystalline product. Therefore, in the

## TABLE 2.3
## Mixed Oxide Formed by Glycothermal Reactions

| Acidic Oxide | Metal Cation that Forms Basic Oxide | | | | |
|---|---|---|---|---|---|
| | $RE^{3+}$ | $Zn^{2+}$ | $Ca^{2+}$ | $Li^+$ | $Ba^{2+}$ |
| $P_2O_5$ | $REPO_4$ | $Zn_3(PO_4)_2$ | $Ca_5(OH)(PO_4)_3$ | | |
| $Nb_2O_5$ | $RE_3NbO_7$ | $ZnNb_2O_6$ | $CaNb_2O_6$ | $LiNbO_3$ | |
| $Al_2O_3$ | $RE_3Al_5O_{12}$ | $ZnAl_2O_4$ | | | |
| $Ga_2O_3$ | $RE_3Ga_5O_{12}$ | $ZnGa_2O_4$ | | | |
| $Fe_2O_3$ | $REFeO_3$ | $ZnFe_2O_4$ | | | |
| $TiO_2$ | Amorphous | $Zn_2TiO_4$ | | $LiTi_2O_4$ | $BaTiO_3$ |
| $Ta_2O_5$ | Unknown | Amorphous | $Ca_3Ta_2O_7$ | | |

amorphous product, titanium and RE ions somehow strongly influenced each other.[138]

For another possible reaction mechanism for the formation of mixed oxides, binary alkoxide (glycoxide) may be formed prior to the formation of the mixed oxide. In fact, various binary glycoxides have been prepared and their crystal structures have been elucidated.[164–166] Formation of $BaTiO_3$ by the glycothermal reaction may be explained by this mechanism,[167] because Ba-Ti binary alkoxide is well established.[166,168] However, formation of garnet phases (Sections III.C.1 through III.C.3) cannot be explained by this mechanism because addition of seed crystals, in some cases, gives a product with a chemical composition completely different from that of the product obtained without the addition of the seed crystals.

Besides acid-base chemistry, combining two elements that have similar properties sometimes gives mixed oxides. An example is formation of -$Al_2O_3$-$Ga_2O_3$ by the reaction of aluminum alkoxide and gallium acetylacetonate.[169] A linear relationship between the lattice parameter of the solid solution and the composition of the two starting materials is observed.[169,170] Since the products have particle sizes in nano scale with quite large surface areas, reaction mechanisms are not yet fully elucidated, but statistical decomposition of the intermediates seems to be most plausible.

Some other important properties for the formation of mixed oxides are summarized below:

- In most cases, the particles of the glycothermally prepared mixed oxide are spherical and each particle is a single crystal grown from a nucleus.
- The nucleation step is the most difficult process; addition of seed crystals has a significant effect on product formation. When nucleation takes place, crystal growth proceeds quite rapidly.[161]

- Only one mixed oxide is crystallized for the system where many oxide phases are present in the phase diagram, and the mixed oxide can be crystallized from a starting composition far from the stoichiometric one.
- Addition of a small amount of water (up to 5% by volume) usually facilitates the reaction, but larger amounts of water may disturb the formation of single-phase product by the formation of stable intermediates.
- The use of ethylene glycol usually yields amorphous products (see Section III.C.5).[171,172]
- The particle size of the mixed oxide is usually larger than simple oxide prepared by the glycothermal method.
- When product with a large crystal size is formed, the product particles contain a significant number of crystal defects.
- The surface of the product particles is covered with organic moieties attached through covalent bonding. The organic moieties can be eliminated by heat treatment at 250 to 300°C in an air flow.

## 5. Reaction in Ethylene Glycol

The reaction of metal alkoxides in ethylene glycol at 250 to 300°C usually yields amorphous products.[171,172] This is because cleavage of the C–O bond in the glycoxide is difficult because of the inductive effect of the intramolecular hydroxyl group. Metal cations are fixed in the networks of the gel structure of ethylene glycol molecules. From the amorphous product, obtained by the glycothermal reaction of two suitable starting materials, mixed oxide is crystallized at low temperature.[171,172] For example, the reaction of aluminum isopropoxide with yttrium acetate (Y/Al = 2) yields an amorphous product. By calcination of the product, single-phase monoclinic yttrium aluminum oxide ($Y_4Al_2O_9$) was crystallized at relatively low temperature.[172] From a mixture of the same starting materials with stoichiometric composition for garnet (Y/Al = 3/5), YAG was crystallized at 920°C through the intermediate formation of hexagonal $YAlO_3$.[171,172] This crystallization behavior is essentially identical with that of the gel derived by hydrolysis of yttrium aluminum double alkoxide.[173]

Although solvothermal reaction in ethylene glycol usually yields amorphous products, addition of a small amount of water may give crystalline product. Thus Kominami et al.[174] reported that solvothermal treatment of $TiO(acac)_2$ in ethylene glycol in the presence of sodium laurate and a small amount of water at 300°C yielded microcrystalline brookite having an average size of 14 nm × 67 nm without contamination of other $TiO_2$ phases.

## D. Crystallization of Amorphous Starting Materials

The starting materials are prepared by the sol-gel method[175–181] or the (co)precipitation method,[182–186] and the precursors are solvothermally treated to crystallize the products. Since hydrous gel is highly hydrophilic, solvents miscible with water are usually favored, and lower alcohols and glycols are usually used. Ethylenediamine was also used.[181] Inert organic solvents together with a surfactant to disperse the precursor gel may also work,[183] and may control the morphology of the agglomerated particles. Solvothermal crystallization of sonochemically prepared hydrous yttrium-stabilized zirconia (YSZ) colloid was also reported.[187] A semicontinuous process for $BaTiO_3$ synthesis was reported by Bocquet et al.[176] The first step is the hydrolysis of the alkoxide $BaTi(O-^iPr)_6$ in isopropanol at 100 to 200°C. The second step is a thermal treatment of the formed solids under the supercritical state of the solvent.

An important point is that the precursor gels thus prepared contain significant amounts of water, even if the gels are dried by some suitable method. One of the weak points of solvothermal crystallization may be difficulty in controlling the water content in the precursor gel. Water facilitates hydrothermal crystallization of the precursor gel, and therefore the essential chemistry here may be hydrothermal. Actually, solvothermal crystallization usually requires higher temperatures and more prolonged reaction times than hydrothermal crystallization.[175] Moreover, in the solvothermal crystallization of stabilized zirconia, the presence of an adequate amount of water was reported to be so critical in dissolving the oxide powder that no crystallization occurred in absolute alcohol.[184]

Solvothermal crystallization usually produces products with smaller sizes than those obtained by the hydrothermal method. The formation of polymorphs that cannot be obtained by the hydrothermal method is frequently reported: $PbTiO_3$ with pyrochlore structure/perovskite $PbTiO_3$,[175] cubic $BaTiO_3$/tetragonal-phase $BaTiO_3$,[176,178] cubic $ZrO_2$/tetragonal $ZrO_2$.[177] Formation of these phases may be due to the smaller crystallite size of the solvothermal products.

For the hydrothermal crystallization of amorphous gel, the dissolution-recrystallization mechanism usually, although not always, takes place in which dissolution of amorphous particles occurs followed by nucleation of the product in the solution (homogeneous nucleation) and crystal growth. Ostwald ripening occurs when the product has enough solubility. For solvothermal crystallization, similar dissolution-recrystallization mechanisms are frequently reported to occur.[175,178,183,184] However, since a limited amount of water is present in the reaction system, microscopic chemistry may differ from that of hydrothermal chemistry. Water is adsorbed on the surface of the amorphous particles and dissolves a part of the surface having higher energies. Water molecules transfer the solute species to other parts of the particle surface that have lower energies, which can act as the nucleation site of the product (heterogeneous crystallization). Crystal growth takes place by diffusion of the component with the aid of adsorbed water, finally converting whole amorphous particles into crystals. This mechanism is proposed because solvothermal crystallization of the amorphous precursors usually leads to nanocrystals, in spite of the fact that nucleation of oxide in organic solvent or aqueous organic solvent is difficult as compared with aqueous systems. If homogeneous nucleation were to occur, one should find some reason for the higher nucleation frequency in organic solvent than in the aqueous system. In fact, our experience is that in the glycothermal synthesis of metal oxides, nucleation of oxides is quite difficult or even does not occur in the solvent, whereas in the presence of seed crystals, rapid crystal growth takes place.

Because dissolution of the whole particle is not expected and because one amorphous particle is expected to transform into a crystalline particle, formation of homogeneous precursor gel is essential. On the other hand, if the real dissolution-recrystallization mechanism takes place, synthesis of homogenous gel does not have any meaning, since preferential dissolution of one component may be expected to occur in the solvothermal reaction.

To increase the crystallization rate and to alter the product phase, an alkaline mineralizer is sometimes added to the solvothermal reaction. Some researchers believe that, compared with the hydrothermal process, solvothermal synthesis allows the product to be free from foreign ions because the organic solution, having a low relative permittivity, is free from ionic species. When precursor gels are prepared from alkoxide, one can prepare products free of foreign ions. However, when the precursor gel is prepared by precipitation from salt solutions, or when alkali/acid mineralizer or ionic surfactant is added to the solvothermal crystallization system, the above statement is a myth. In fact, ions are easily adsorbed or occluded in the product particles because of the low dielectric constant of the organic solvent.

Zhao et al.[177] prepared a translucent sol of cubic zirconia particles with a size of 5 nm by addition of ethanol-water to a solution of $Zr(OPr)_4$ in a mixture of ethanol and diethylene glycol followed by solvothermal treatment. Feldmann and Jungk[179] used essentially the same technique and reported the formation of a variety of colloidally stable oxide particles. Their method is as follows: a solution of acetylacetonate or alkoxide in diethylene glycol was heated to 140°C. Subsequently deionized water was added and the mixture was heated to 180°C for 2 h.

Kominami et al.[103] reported that the amorphous product obtained by solvothermal decomposition of $La(OiPr)_3$ and $Fe(OBu)_3$ in toluene crystallized into perovskite-type $LaFeO_3$ using glycothermal treatment, while direct glycothermal reaction of the two starting materials yielded a mixture of $La(OH)_3$ and magnetite.

Yin et al.[188,189] and Inoue et al.[190] reported that fine crystals of anatase prepared with solvothermal treatment of an amorphous gel in methanol possessed much better sinterability and photocatalytic activity than those fabricated by hydrothermal reactions or calcinations in air.

Lu et al.[191] prepared dense ceramic fibers of $Pb_{1-x}La_xTiO_3$ (PLT, x = 0–0.2) by a sol-gel method from lead acetate trihydrate, lanthanum acetate, and titanium isopropoxide in triethanolamine and acetic acid, followed by solvothermal treatment in a mixture of xylene and triethylamine at 200°C for 12 h. They reported that removal of the organic residue prior to the gel-to-ceramic fiber transformation (heat treatment) is a key step in achieving dense ceramic fibers with a reduced grain size, although solvothermal treatment does not give the crystalline material. This result shows that removal of organic residue is an interesting application of the solvothermal method in relation with sol-gel chemistry.

## E. HYDROTHERMAL CRYSTALLIZATION IN ORGANIC MEDIA

Since primary alkoxide of zirconium does not decompose in toluene, and therefore hydrolysis of the alkoxide in inert organic solvent followed by hydrothermal crystallization of the hydrolyzed product is examined.[192] In this method, the alkoxide solution in an inert organic solvent is placed in a test tube, which is then placed in an autoclave. The desired amount of water is placed between the test tube and the autoclave wall. When the autoclave is heated, the water evaporates and is dissolved into the toluene solution from the gas phase, where hydrolysis of the alkoxide takes place, followed by hydrothermal crystallization of the hydrolyzed products.

Using this method, monoclinic zirconia with a large surface area was prepared.[192] Kominami et al.[193-195] further extended the application of this method and found that titania prepared by this method has a high photocatalytic activity because the product has a large surface area as well as a low number of crystal defects, which can act as the recombination sites for holes and electrons generated by photoactivation. They called this method the HyCOM method.[194]

Using this method, amorphous $Nb_2O_5$ powder with a large surface area (greater than 200 $m^2/g$) was obtained from $Nb(OBu)_5$ at 300°C.[196] An increase in the amount of water induced crystallization of $Nb_2O_5$ into the TT phase and reduced the surface area of the product.[196] Amorphous $Ta_2O_5$ with a large surface area (greater than 200 $m^2/g$) was also synthesized from $Ta(OBu)_5$ at 200 to 250°C, while $\beta$ phase of $Ta_2O_5$ was obtained at 300°C.[197] Microcrystalline $\alpha$-$Fe_2O_3$ (hematite) powders were obtained from Fe(III) acetylacetonate, which then was reduced to magnetite after a prolonged reaction time.[102] Note that direct decomposition of the same precursor yields magnetite.[102]

This method may provide a convenient route for the synthesis of nanocrystalline oxide materials from alkoxide using only one reaction vessel; however, because this method uses water dissolved from the gas phase, configuration of the reaction apparatus, that is, the ratio of the surface area of the liquid to the bulk volume of the liquid, may affect the physical properties of the product.

Lee et al.[198] reported that a colloidal solution containing nano-sized $TiO_2$ (anatase) particles was obtained by solvothermal treatment of titanium isopropoxide in 1-butanol at 200°C in the presence of a small amount of aqueous HCl. They found that the particles coated on $\gamma$-alumina showed excellent performances for photocatalytic decomposition of $CHCl_3$.

## F. Solvothermal Ion Exchange and Intercalation

Ion exchange and intercalation are synthetic routes for a variety of oxide materials from starting materials with tunneled and layered structures.[199] Aqueous media are usually used for this purpose, however, organic solvents may produce novel products because ions are less solvated in organic media. Although there are many host lattices that can undergo ion exchange reactions, solvothermal methods have been sparingly investigated.

Li et al.[200] reported that the reaction of $MnO_2$ with $LiOH \cdot H_2O$ in the presence of NaOH in ethanol at 200°C yielded microcrystalline $Li_{1-x}Mn_2O_{4-\sigma}$. (180 nm). Wang et al.[201] pointed out the importance of the tunnel structure of the starting $MnO_2$ in their solvothermal reaction with LiCl in 1-pentanol at 170°C: $MnO_2$ with the $1 \times 1$ tunnel structure yields the spinel lithium manganese oxide, whereas a novel structural $Li_xMnO_2$ is formed from $MnO_2$ with the $2 \times 1$ tunnel structure. In both reactions, alcoholic media acted as the reducing agent.

A new modification of $LiFeO_2$ with a corrugated layer structure is synthesized by the ion exchange reaction between $\gamma$-FeOOH and $LiOH \cdot H_2O$ under hydrothermal conditions.[202] On the other hand, solvothermal ion exchange reaction of these two starting materials in various alcohols at the reflux temperature yields another modification of $LiFeO_2$ with a hollandite-type structure.[203] It was reported that lithium cells consisting of cathodes of this compound and lithium anodes showed good charge and discharge reversibility.[203]

Tabuchi et al.[204] reported that the solvothermal reaction of a mixture of $\alpha$-$NaFeO_2$ (obtained by hydrothermal reaction of $\alpha$-FeOOH with NaOH) and LiCl in ethanol at 220°C for 96 h yielded an $Fe^{2+}$-containing compound with the formula of $Li_{1-x}Fe_{5+x}O_8$ with an inverse spinel structure.

Although a variety of solvents can be used for solvothermal reactions, only alcoholic solvents have been investigated for solvothermal ion exchange. This may be due to the fact that both cation and counteranion are required to be solvated. As for another possible solvent system, a binary system composed of a solvent with a low basicity and a high dielectric constant together with a crown ether that can selectively solvate a cation with a specific size may be designed which poorly solvates the exchanging ions and can capture exchangeable cations from the host lattice into the solvent.

Intercalation of organic molecules into layered host lattice produces a variety of organic-inorganic hybrid materials. The solvothermal method provides a reaction system that allows application of high temperatures and therefore is a powerful technique for preparation of intercalation compounds. Exfoliation of layers may occur because of applied high temperatures. For example, exfoliated polyethylene/montmorillonite nanocomposites were reported to be prepared by solvothermal reaction of organophilic montmorillonite with polyethylene in toluene at 170°C for 2 h.[205]

Pillared inorganics are prepared by intercalation of suitable metalorganics into host layers followed by heat treatment to decompose the guest molecules, yielding inorganic pillars. Liu et al.[206] reported a unique solvothermal method for the synthesis of a pillared compound: octyl amine was intercalated into birnessite(H)-type manganese oxide at room temperature and the product was then treated in tetraethyl orthosilicate (TEOS; bp 165.8°C) liquid at 140°C for 48 h, yielding silica-pillared layered manganese oxide with a surface area of 260 m$^2$/g. The authors suggest that solvothermal treatment of the precursor with TEOS liquid accelerates the ripening of silicate polymer to compact hydrous silica particles.

## G. SOLVOTHERMAL OXIDATION OF METALS

Some metal ions have highly negative reduction potentials and therefore these metals can be easily oxidized in water and alcohols, which is accompanied by formation of hydrogen molecules. Although hydrothermal oxidation of metals was reported to be an efficient way to synthesize metal oxides with quite unique properties, solvothermal oxidation of metals has scarcely been examined.[207,208] Aluminum metal is oxidized by straight-chain primary alcohols with a carbon number as high as 12, yielding the alkyl derivatives of boehmite, and a linear relationship is observed between the basal spacing of the product and the carbon number of the alcohol used as the reaction medium (and as the oxidant as well).[44,127] Similarly magnesium and zinc oxides are obtained by solvothermal oxidation in alcohols.[44]

The most interesting result is the formation of a transparent colloidal solution of ceria with 2 nm particles.[209] Cerium metal tips with the superficial layers of oxide are allowed to react in 2-methoxyethanol at 250 to 300°C, and removal of coarse ceria particles originating from the superficial layers yields the colloidal solution. Addition of water to the solution does not cause any change except dilution of the color of the solution, but addition of a drop of a solution of any kind of salt immediately causes precipitation of ceria particles.[209,210] The reaction mechanism is as follows: The solvent slowly dissolves the superficial layers, and when the solvent reaches the metal, rapid reaction occurs, yielding an alkoxide solution. The concentration of the ceria precursor becomes so high that a burst of nucleation occurs, yielding the colloidal solution. The reaction of cerium acetylacetonate in the same solvent yields ceria particles but does not give a colloidal solution.

Recently Chen et al.[211] reported that hydroxyl-free zinc oxide was prepared by solvothermal oxidation of zinc powders with two equivalents of trimethyl amine $N$-oxide or 4-picoline $N$-oxide as the oxidant in organic solvent toluene, ethylenediamine, $N,N,N,N$-tetramethylethylenediamine at 180°C. The morphology of the product is affected by the oxidant, and trimethylamine $N$-oxide yields rod-like particles, while 4-picoline $N$-oxide produces spherical particles. Solvent affects the particle size of the product and the smallest particles (24 nm, 4-picoline $N$-oxide) are obtained in less-polar toluene. Chen et al. showed that a small amount of water in organic solvent catalyzes the reaction.

## H. SOLVOTHERMAL REDUCTION

As discussed in Section 2, organic solvents have inherent reducing ability. Figlarz et al.[5] first reported the formation of noble metal particles as well as nickel and cobalt particles by the reaction in ethylene glycol. They called this method the "polyol process," and by combining this method with microwaves, Komarneni et al.[212–214] synthesized a variety of metal nanoparticles very rapidly. The reducing abilities of organic solvents are also utilized for the synthesis of metal oxides. A typical example is the synthesis of $Fe_3O_4$ from Fe(III) precursors. Synthesis of $\gamma$-$Mn_2O_3$ by the reaction of $MnO_2$ in ethanol[215,216] or $KMnO_4$ in $CH_3OH$ or $CH_3CH_2OH$[217] has also been reported.

## REFERENCES

1. Jones, R.W., *Fundamental Principles of Sol-gel Technology*, Institute of Metals, Brookfield, VT, 1989.
2. Jeffrey, C.B., *Sol-Gel Science: The Physics and Chemistry of Sol-Gel Processing*, Academic Press, Boston, 1990.
3. Inoue, M., Kondo, Y., and Inui, T., *Chem. Lett.*, 1421, 1986.
4. Demazeau, G., *High Press. Res.*, 18, 203, 2000.
5. Figlarz, M., Fievet, F., and Lagier, J.P., presented at the International Meeting on Advanced Materials, Tokyo, Japan, May 30–June 3, 1988, Nikkan Kogyo Shinbun, Ltd., Tokyo, Japan, 1988.
6. Silvert, P.-Y., Herrera-Urbina, R., and Tekaia-Elhesisen, K., *J. Mater. Chem.*, 7, 293, 1997.
7. Inoue, M., Tanino, H., Kondo, Y., and Inui, T., *J. Am. Ceram. Soc.*, 72, 352, 1989.
8. Inoue, M., Otsu, H., Kominami, H., and Inui, T., *J. Am. Ceram. Soc.*, 74, 1452, 1991.
9. Yu, S.-H., *J. Ceram. Soc. Jpn.*, 109, S65, 2001.
10. Rajamathi, M., and Seshadri, R., *Curr. Opin. Solid State Mater. Sci.*, 6, 337, 2002.
11. Demazeau, G., *J. Mater. Chem.*, 9, 15, 1999.
12. Qian, Y., *Adv. Mater.*, 11, 1101, 1999.
13. Bibby, D.M., and Dale, M.P., *Nature (London)*, 317, 157, 1985.
14. Bowes, C.L., and Ozin, G.A., *Adv. Mater.*, 6, 13, 1996.
15. Day, V.W., Klemperer, W.G., and Yaghi, O.M., *J. Am. Chem. Soc.*, 111, 5959, 1989.
16. Spandl, J., Brüdgam, I., and Hartl, H., *Angew. Chem. Int. Ed.*, 40, 4018, 2001.
17. Inoue, M., Kondo, Y., and Inui, T., *Inorg. Chem.*, 27, 215, 1988.
18. Yu, S.-H., and Yoshimura, M., *Adv. Mater.*, 14, 296, 2002.
19. Jiang, Y., Wu, Y., Zhang, Xu, C., Yu, W., Xie, Y., and Qian, Y., *J. Am. Chem. Soc.*, 122, 12283, 2000.
20. Hu, G., Cheng, M., Ma, D., and Bao, X. *Chem. Mater.*, 15, 1470, 2003.
21. Inoue, M., Nishikawa, T., Otsu, H., Kominami, H., and Inui, T., *J. Am. Ceram. Soc.*, 81, 1173, 1998.
22. Chen, S.-J., Chen, X.-T., Xue, Z., Li, L.-H., and You, X.-Z., *J. Cryst. Growth*, 246, 169, 2002.

23. Stupp, S.I., and Braun, P.V., *Science*, 277, 1242, 1997.
24. Patzke, G.R., Krumeich, F., and Nesper, R., *Angew. Chem. Int. Ed.*, 41, 2446, 2002.
25. Yu, S.-H., Shu, L., Yang, J., Han, Z.-H., Qian, Y.-T., and Zhang, Y.-H., *Chem. Phys. Lett.*, 361, 362, 2002.
26. Wei, C., Deng, Y., Lin, Y.-H., and Nan, C.-W., *Chem. Phys. Lett.*, 372, 590, 2003.
27. Yu, S.-H., Yang, J., Qian, Y.-T., and Yoshimura, M., *Chem. Phys. Lett.*, 361, 362, 2002.
28. Yoshimura, M., and Somiya, S., *Am. Ceram. Soc. Bull.*, 59, 246, 1980.
29. Hirano, S., *Am. Ceram. Soc. Bull.*, 67, 1342, 1987.
30. Dawson, W.J., *Am. Ceram. Soc. Bull.*, 67, 1673, 1988.
31. Walton, R.I., *Chem. Soc. Rev.*, 31, 230, 2002.
32. Yamasaki, N., *J. Ceram. Soc. Jpn.*, 111, 709, 2003.
33. Jacobs, H., and Stuve, C., *J. Less Common Met.*, 96, 323, 1984.
34. Su, H.L., Xie, Y., Li, B., Liu, X.M., and Qian, Y.T., *Solid State Ionics*, 122, 157, 1999.
35. Demazeau, G., Goglio, G., Denis, A., and Largeteau, A., *J. Phys. Condens. Matter*, 14, 11085, 2002.
36. Purdy, A.P., *Chem. Mater.*, 11, 1648, 1999.
37. Bialowons, H., Müller, M., and Müller, B.G., *Anorg. Allg. Chem.*, 621, 1227, 1995.
38. Koerts, T., Deelen, M.J.A.G., and van Santen, R.A., *J. Catal.*, 138, 101, 1992.
39. Li, B., Xie, Y., Su, H., Qian, Y., and Liu, X., *Solid State Ionics*, 120, 251, 1999.
40. Spencer, P.S., Kim, M.S., and Sabri, M.I., *Int. J. Hyg. Environ. Health*, 205, 131, 2002.
41. Xie, Y., Qian, Y.T., Wang, W.Z., Zhang, S.Y., and Zhang, Y.H., *Science*, 272, 1926, 1996.
42. Hu, J.Q., Lu, Q.Y., Tang, Yu, S.H., Qi, Y.T., Zhou, G., and Liu, X.M., *J. Am. Ceram. Soc.*, 83, 430, 2000.
43. Barj, M., Bocquet, J.F., Chhor, K., and Pommier, C., *J. Mater. Sci.*, 27, 2187, 1992.
44. Inoue, M., Kimura, M., and Inui, T., *Adv. Sci. Technol.*, 16, 593, 1999.
45. Gao, S., Zhang, J., Zhu, Y.-F., and Che, C.-M., *New J. Chem.*, 24, 739, 2000.
46. Fanelli, A.J., and Burlew, J.V., *J. Am. Ceram. Soc.*, 69, C-174, 1986.
47. Willard, H.H., and Tang, N.K., *J. Am. Chem. Soc.*, 59, 1190, 1937.
48. Larcher, D., Gérand, B., and Tarascon, J.-M., *J. Solid State Elecrochem.*, 2, 137, 1998.
49. Wang, D., Yu, R., Kumada, N., and Kinomura, N., *Chem. Mater.*, 11, 2008, 1999.
50. Larcher, D., Sudant, G., Patrice, R., and Tarascon, J.-M., *Chem. Mater.*, 15, 3543, 2003.
51. Richardson, J.W., Jr., Pluth, J.J., Smith, J.V., Dytrych, W.J., and Bibby, D.M., *J. Phys. Chem.*, 92, 243, 1988.
52. Liao, W., Wang, J., and Li, D., *Mater. Lett.*, 57, 1309, 2003.
53. Wei, G., Deng, Y., and Nan, C.-W., *Chem. Phys. Lett.*, 367, 512, 2003.
54. Konishi, Y., Kawamura, T., and Asai, S., *Ind. Eng. Chem. Res.*, 32, 2888, 1993.
55. Derouane, E.G., Ed., *Zeolite Microporous Solids: Synthesis, Structure, and Reactivity*, Kluwer, Dordecht, 1992.
56. Suib, S.L., *Annu. Rev. Mater. Sci.*, 26, 131, 1996.
57. Han, Z., Guo, N., Li, F., Zhang, W., Zhao, H., and Qian, Y., *Mater. Lett.*, 48, 99, 2001.

58. Physical and Theoretical Chemistry Laboratory, Safety (MSDS) data for hydrazine hydrate, Oxford University, http://physchem.ox.ac.uk/MSDS/HY/hydrazine_hydrate.html.

59. Laine, R.M., Treadwell, D.R., Mueller, B.L., Bickmore, C.R., Waldner, K.F., and Hinklin, T.R., *J. Mater. Chem.*, 6, 1441, 1996.

60. Waldner, K.F., Laine, R.M., Dhumrongvaraporn, C.R., Tayaniphan, S., and Narayanan, R., *Chem. Mater.*, 8, 2850, 1996.

61. Kansal, P., Laine, R.M., and Babonneau, F., *J. Am. Ceram. Soc.*, 80, 2597, 1997.

62. Sutorik, A.C., Neo, S.S., Treadwell, D.R., and Laine, R.M., *J. Am. Ceram. Soc.*, 81, 1477, 1998.

63. Yu, S.-H., Wu, Y.-S., Yang, Han, Z.-H., Xie, Y., Qian, Y.-T., and Liu, X.M., *Chem. Mater.*, 10, 2309, 1998.

64. Yu, S.-H., Yang, J., Wu, Han, Z.-H., Lu, J., Xie, Y., and Qian, Y.-T., *J. Mater. Chem.*, 8, 1949, 1998.

65. Yu, S.-H., Han, Z.-H., Yang, J., Yang, R.-Y., Xie, Y., and Qian, Y.-T., *Chem. Lett.*, 1111, 1988.

66. Salta, J., Chang, Y.-D., and Zubieta, J., *J. Chem. Soc., Chem. Commun.*, 1039, 1994.

67. Finn, R.C., and Zubieta, J., *J. Cluster Sci.*, 11, 461, 2000.

68. Khan, M.I., Tabussum, S., and Doedens, R.J., *Chem. Commun.*, 532, 2003.

69. Riddick, J.A., Bunger, W.B., and Sakano, T.K., *Organic Solvent: Physical Properties and Methods of Purification*, 4th ed., John Wiley & Sons, New York, 1986.

70. Amis, E.S., and Hinton, J.F., *Solvent Effects on Chemical Phenomena*, Academic Press, New York, 1973.

71. Matheson Tri-Gas, Inc., Material safety data sheet, http://www.matheson-trigas.com/msds/MAT07770.pdf.

72. Inoue, M., Kitamura, K., Tamino, H., Nakayama, H., and Inui, T., *Clays Clay Miner.*, 37, 71, 1989.

73. Kubo, T., and Uchida, K., *Kogyo Kagaku Zasshi (J. Ind. Chem. Soc. Jpn.)*, 73, 70, 1970.

74. Hedvall, J.A., *Adv. Catal.*, 8, 1, 1956.

75. Gitzen, W.H., *Alumina as a Ceramic Material*, American Ceramic Society, Columbus, OH, 1970.

76. Goodman, J.F., *Proc. R. Soc.*, A247, 346, 1958.

77. de Boer, J.H., Fortuin, J.M.H., and Steggerda, J.J., *Konikl Ned. Acad. Wetenschap. Proc.*, 57B, 170, 1954.

78. de Boer, J.H., Fortuin, J.M.H., and Steggerda, J.J., *Konikl Ned. Acad. Wetenschap. Proc.*, 57B, 435, 1954.

79. Inoue, M., Kitamura, K., and Inui, T., *J. Chem. Tech. Biotechnol.*, 46, 233, 1989.

80. Yin, S., Uchida, S., Fujishiro, Y., Aki, M., and Sato, T., *J. Mater. Chem.*, 9, 1191, 1999.

81. Yin, S., Wu, J., Aki, M., and Sato, T., *Int. J. Inorg. Mater.*, 2, 325, 2000.

82. Yin, S., Uchida, S., Aki, M., and Sato, T., *High Press. Res.*, 20, 121, 2001.

83. Inoue, M., Tanino, H., Kondo, Y., and Inui, T., *Clays Clay Miner.*, 39, 151, 1991.

84. Bradley, D.C., *Prog. Inorg. Chem.*, 2, 303, 1960.

85. Inoue, M., Kondo, Y., Kominami, H., Tanino, H., and Inui, T., *Nippon Kagaku Kaishi (J. Chem. Soc. Jpn.)*, 1339, 1991.

86. Ginsberg, H., and Koster, M., *Z. Anorg. Allgem. Chem.*, 293, 204, 1957.

87. Yamaguchi, G., Yanagida, H., and Fujimaru, T., *Bull. Chem. Soc. Jpn.*, 38, 54, 1965.
88. Fyfe, W.S., and Hollander, M.A., *Am. J. Sci.*, 262, 709, 1964.
89. Inoue, M., Kimura, M., Kominami, H., Tanino, H., and Inui, T., 72nd National Meeting of the Chemical Society of Japan, Tokyo, March 1997, abstract no. 1BY09.
90. Arai, Y., Yasue, T., and Yamaguchi, I., *Hippon Kagaku Kaishi (J. Chem. Soc. Jpn.)*, 1395, 1972.
91. Bell, N.S., Cho, S.-B., and Adair, J.H., *J. Am. Ceram. Soc.*, 81, 1411, 1998.
92. Cho, S.-B., Venigalla, S., and Adair, J.H., *J. Am. Ceram. Soc.*, 79, 88, 1996.
93. Bell, N.S., and Adair, J.H., *J. Cryst. Growth*, 203, 213, 1999.
94. Hitchman, M.L., and Jensen, K.F., *Chemical Vapor Deposition: Principles and Applications*, Academic Press, London, 1993.
95. Inoue, M., Kominami, H., and Inui, T., *J. Am. Ceram. Soc.*, 75, 2597, 1992.
96. Inoue, M., Kominami, H., and Inui, T., *Appl. Catal. A: General*, 97, L25, 1993.
97. Inoue, M., Kominami, H., Otsu, H., and Inui, T., *Nippon Kagaku Kaishi (J. Chem. Soc. Jpn.)*, 1364, 1991.
98. Kominami, H., Kato, J., Takada, Y., Doushi, Y., Ohtani, B., Nishimoto, S., Inoue, M., Inui, T., and Kera, Y., *Catal. Lett.*, 46, 235, 1997.
99. Kominami, H., Inoue, M., and Inui, T., *Catal. Today*, 16, 309, 1993.
100. Kominami, H., Oki, K., Kohno, M., Onoue, S., Kera, Y., and Ohtani, B., *J. Mater. Chem.*, 11, 604, 2001.
101. Kominami, H., Miyakawa, M., Murakami, Yasuda, T., Kohno, M., Onoue, S., Kera, Y., and Ohtani, B., *Phys. Chem. Chem. Phys.*, 3, 2697, 2001.
102. Kominami, H., Onoue, S., Matsuo, K., and Kera, Y., *J. Am. Ceram. Soc.*, 82, 1937, 1999.
103. Kominami, H., Inoue, H., Konishi, S., and Kera, Y., *J. Am. Ceram. Soc.*, 85, 2148, 2002.
104. Yokozawa, M., Iwasa, H., and Teramoto, I., *Jpn. J. Appl. Phys.*, 7, 96, 1968.
105. Okuyama, K., Kousaka, Y., Tohge, N., Yamamoto, S., Wu, J.J., Flagan, R.C., and Seinfeld, J.H., *AIChE J.*, 32, 2010, 1986.
106. Komiyama, H., Kanai, T., and Inoue, H., *Chem. Lett.*, 1283, 1984.
107. Vioux, A., and Leclercq, D., *Heterogen. Chem. Rev.*, 3, 65, 1996.
108. Arnal, P., Corriu, R.J.P., Leclercq, D., Mutin, P.H., and Vioux, A., *Chem Mater.*, 9, 694, 1997.
109. Acosta, S., Arnal, P., Corriu, R.J.P., Leclercq, D., Mutin, P.H., and Vioux, A., *Mater. Res. Soc. Symp. Proc.*, 346, 339, 1994.
110. Acosta, S., Corriu, R.J.P., Leclercq, D., Lefevre, P., Mutin, P.H., and Vioux, A., *J. Non-Cryst. Solids*, 170, 234, 1994.
111. Corriu, R.J.P., Leclercq, D., Lefevre, P., Mutin, P.H., and Vioux, A., *J. Non-Cryst. Solids*, 146, 301, 1992.
112. Acosta, S., Corriu, R.J.P., Leclercq, D., Mutin, P.H., and Vioux, A., *J. Sol-Gel Sci. Technol.*, 2, 25, 1994.
113. Andrianainarivelo, M., Corriu, R.J.P., Leclercq, D., Mutin, P.H., and Vioux, A., *J. Mater. Chem.*, 6, 1665, 1996.
114. Trentler, T.J., Denler, T.E., Bertone, J.F., Agrawal, A., and Colvin, V.L., *J. Am. Chem. Soc.*, 121, 1613, 1999.
115. Vioux, A., *Chem. Mater.*, 9, 2292, 1997.

116. Inoue, M., Kominami, H., and Inui, T., *J. Am. Ceram. Soc.*, 79, 793, 1996.
117. Okada, K., Otsuka, N., and Somiya, S., *Am. Ceram. Soc. Bull.*, 70, 1633, 1991.
118. Hoffman, D., Roy, R., and Komarneni, S., *J. Am. Ceram. Soc.*, 67, 468, 1984.
119. Rockenberger, J., Scher, E.C., and Alivisatos, A.P., *J. Am. Chem. Soc.*, 121, 11595, 1999.
120. Thimmaiah, S., Rajamathi, M., Singh, Bera, P., Meldrum, F., Chandrasekhar, N., and Seshadri, R., *J. Mater. Chem.*, 11, 3215, 2001.
121. Gautam, U.K., Ghosh, M., Rajamathi, M., and Seshadri, R., *Pure Appl. Chem.*, 74, 1643, 2002.
122. Praserthdam, P., Inoue, M., Mekasuwandumrong, O., Thanakulrangsan, W., and Phatanasuri, S., *Inorg. Chem. Commun.*, 3, 671, 2000.
123. Mekasuwandumrong, O., Kominsmi, H., Praserthdam, P., and Inoue, M., *J. Am. Ceram. Soc.*, in press.
124. Kistler, S.S., *J. Phys. Chem.*, 36, 52, 1932.
125. Armor, J.N., and Carlson, E.J., *J. Mater. Sci.*, 22, 2549, 1989.
126. Pajonk, G.M., *Appl. Catal.*, 72, 217, 1991.
127. Inoue, M., Kimura, M., and Inui, T., *Chem. Mater.*, 12, 55, 2000.
128. Pommier, C., Chhor, K., Bocquet, J.F., and Barj, M., *Mater. Res. Bull.*, 25, 213, 1990.
129. Wang, C., Deng, Z.-X., Zhang, G., Fan, S., and Li, Y., *Powder Technol.*, 125, 39, 2002.
130. Arnal, P., Coriu, R.J.P., Leclercq, D., Llefevre, P., Mutin, P.H., and Vioux, A., *J. Mater. Chem.*, 6, 1925, 1996.
131. Takahashi, Y., Suzuki, H., and Nasu, N., *J. Chem. Soc. Faraday Trans.*, 81, 3117, 1985.
132. Courtecuisse, V.G., Bocquet, J.F., Chhor, K., and Pommier, C., *J. Supercrit. Fluids*, 9, 222, 1996.
133. Courtecuisse, V.G., Chhor, K., Bocquet, J.-F., and Pommier, C., *Ind. Eng. Chem. Res.*, 35, 2539, 1996.
134. Inoue, M., Kominami, H., and Inui, T., *J. Chem. Soc. Dalton Trans.*, 3331, 1991.
135. Inoue, M., Kominami, H., and Inui, T., *J. Am. Ceram. Soc.*, 73, 1100, 1990.
136. Inoue, M., Kominami, H., and Inui, T., *J. Mater. Sci.*, 29, 2459, 1994.
137. Winstein, S., Allred, E., Heck, R., and Glick, R., *Tetrahedron*, 3, 1, 1958.
138. Inoue, M., *Adv. Sci. Technol.*, 29, 855, 2000.
139. Inoue, M., Sato, K., Nakamura, T., and Inui, T., *Catal. Lett.*, 65, 79, 2000.
140. Kongwudthiti, S., Praserthdam, P., Tanakulrungsank, W., and Inoue, M., *J. Mater. Proc. Technol.*, 136, 186, 2003.
141. Inoue, M., Kominami, H., and Inui, T., *Res. Chem. Intermed.*, 24, 571, 1998.
142. Inoue, M., Otsu, H., Kominami, H., and Inui, T., *Nippon Kagaku Kaishi (J. Chem. Soc. Jpn.)*, 1036, 1991.
143. Kang, M., Lee, S.-Y., Chung, C.-H., Cho, S. M., Han, G. Y., Kim, B.-W., and Yoon, K. J., *J. Photochem. Photobiol, A*, 144, 185, 2001.
144. Inoue, M., Otsu, H., Kominami, H., and Inui, T., *J. Alloys Comp.*, 226, 146, 1995.
145. Abell, J.S., Harris, I.R., Cockayne, B., and Lent, B., *J. Mater. Sci.*, 9, 527, 1974.
146. Messier, D.R., and Gazza, G.E., *Ceram. Bull.*, 51, 692, 1972.
147. Yamaguchi, O., Takeoka, K., Hirota, K., Takano, H., and Hayashida, A., *J. Mater. Sci.*, 27, 1264, 1992.
148. Cinibulk, M.K., *J. Am. Ceram. Soc.*, 83, 1276, 2000.

149. Kakade, M.B., Ramanthan, S., and Ravindran, P.V., *J. Alloys Comp.*, 350, 123, 2003.

150. Inoue, M., Kominami, H., Otsu, H., and Inui, T., *Nippon Kagaku Kaishi (J. Chem. Soc. Jpn.)*, 1254, 1991.

151. Kominami, H., Inoue, M., and Inui, T., *Nippon Kagaku Kaishi (J. Chem. Soc. Jpn.)*, 605, 1993.

152. Cockayne, B., *J. Less-Common Met.*, 114, 199, 1985.

153. Mill', B.V., *Soviet Phys.-Crystal.*, 12, 137, 1967.

154. Kolb, E.D., and Laudise, R.A., *J. Cryst. Growth*, 29, 29, 1975.

155. Takamori, T., and David, L.D., *Am. Ceram. Soc. Bull.*, 65, 1282, 1986.

156. Inoue, M., Otsu, H., Kominam, H., and Inui, T., *J. Mater. Sci. Lett.*, 14, 1303, 1995.

157. Mizuno, M., and Yamada, T., *J. Ceram. Soc. Jpn.*, 97, 1334, 1989.

158. LaMer, V.K., and Dinegar, R.H., *J. Am. Chem. Soc.*, 72, 4847, 1950.

159. Inoue, M., Nishikawa, T., Nakamura, T., and Inui, T., *J. Am. Ceram. Soc.*, 80, 2157, 1997.

160. Bertaut, F., and Mareschal, J., *Compt. Rend.*, 257, 867, 1963.

161. Inoue, M., Nishikawa, T., and Inui, T., *J. Mater. Res.*, 13, 856, 1998.

162. Inoue, M., *Adv. Sci. Tech.*, 14, 41, 1999.

163. Inoue, M., Nakamura, T., Otsu, H., Kominami, H., and Inui, T., *Nippon Kagaku Kaishi (J. Chem. Soc. Jpn.)*, 612, 1993.

164. Cruickshank, M.C., and Dent Glasser, L.S., *Acta Cryst.*, C41, 1041, 1985.

165. Laine, R.M., Blohowiak, K.Y., Robinson, T.R., Hoppe, M.L., Nardi, P., Kampf, J., and Uhm, J., *Nature (London)*, 353, 642, 1991.

166. Day, V.W., Eberspacher, T.A., Frey, M.H., Klemperer, W.G., Liang, S., and Payne, D.A., *Chem. Mater.*, 8, 330, 1996.

167. Kimura, M., Inoue, M., and Inui, T., 72nd National Meeting of the Chemical Society of Japan, Tokyo, March 1997, abstract no. 1BY26.

168. Suyama, Y., and Nagasawa, M., *J. Am. Ceram. Soc.*, 77, 603, 1994.

169. Inoue, M., Inoue, N., Yasuda, T., Takeguchi, T., and Iwamoto, S., *Adv. Sci. Technol.*, 29, 1421, 2000.

170. Takahashi, M., Iwamoto, S., and Inoue, M., *Adv. Tech. Mater. Mater. Proc. J*, 4, 76, 2002.

171. Inoue, M., Otsu, H., Kominami, H., and Inui, T., *Nippon Kagaku Kaishi (J. Chem. Soc. Jpn.)*, 1358, 1991.

172. Inoue, M., Nishikawa, T., and Inui, T., *J. Mater. Sci.* 33, 5835, 1998.

173. Yamaguchi, O., Takeoka, K., and Hayashida, A., *J. Mater. Sci. Lett.*, 10, 101, 1990.

174. Kominami, H., Kohno, M., and Kera, Y., *J. Mater. Chem.*, 10, 1151, 2000.

175. Chen, D., and Xu, R., *J. Mater. Chem.*, 8, 965, 1998.

176. Bocquet, J.F., Chhor, K., and Pommier, C., *Mater. Chem. Phys.*, 57, 273, 1999.

177. Zhao, J., Fan, W., Wu, D., and Sun, Y., *J. Mater. Res.*, 15, 402, 2000.

178. Chen, D., and Jiao, X., *J. Am. Ceram. Soc.*, 83, 2637, 2000.

179. Feldmann, C., and Jungk, H.-O., *Angew. Chem. Int. Ed.*, 40, 359, 2001.

180. Mekelkina, O., Hüsing, N., Pongratz, P., and Schubert, U., *J. Non-Cryst. Solids*, 285, 64, 2001.

181. Yu, D., Yu, W., Wang, D., and Qian, Y., *Thin Sold Films*, 419, 166, 2002.

182. Bae, D.-S., Han, K.-S., Cho, S.-B., and Choi, S.H., *Mater. Lett.*, 37, 255, 1998.

183. Chen, D., Jiao, X., and Chen, D., *Mater. Res. Bull.*, 36, 1057, 2001.

184. Zhang, Y., Xu, G., Yan, Z., Yang, Y., Liao, C., and Yan, C., *J. Mater. Chem.*, 12, 970, 2002.
185. Bae, D.-S., Lim, B., Kim, B.-I., and Han, K.-S., *Mater. Lett.*, 56, 610, 2002.
186. Bae, D.-S., Kim, S.-W., Lee, H.-W., and Han, K.-S., *Mater. Lett.*, 57, 1997, 2003.
187. Pang, G., Sominska, E., Cölfen, H., Mastai, Y., Avivi, S., Koltypin, Y., and Gedanken, A., *Langmuir*, 17, 3223, 2001.
188. Yin, S., Inoue, Y., Uchida, S., Fujishiro, Y., and Sato, T., *Rev. High Press. Sci. Technol.*, 7, 1438, 1998.
189. Yin, S., Inoue, Y., Uchida, S., Fujishiro, Y., and Sato, T., *J. Mater. Res.*, 13, 844, 1998.
190. Inoue, Y., Yin, S., Uchda, S., Ishitsuka, M., Min, E., and Sato, T., *Br. Ceram. Trans.*, 97, 222, 1998.
191. Lu, Q., Chen, D., and Jiao, X., *J. Mater. Chem.*, 12, 687, 2002.
192. Inoue, M., Kominami, H., and Inui, T., *Appl. Catal. A*, 121, L1, 1995.
193. Kominami, H., Takada, Y., Yamagiw, H., Kera, Y., Inoue, M., and Inui, T., *J. Mater. Sci.*, 15, 197, 1996.
194. Kominami, H., Murakami, S., Kera, Y., and Ohtani, B., *Catal. Lett.*, 56, 125, 1998.
195. Kominami, H., Kato, J., Takada, Y., Murakami, S., Inoue, M., Inui, T., Ohtani, B., and Kera, Y., *Ind. Eng. Chem. Res.*, 38, 3925, 1999.
196. Kominami, H., Oki, K., Kohno, M., Onoue, S., and Kera, Y., *J. Mater. Chem.*, 11, 604, 2001.
197. Kominami, H., Miyakawa, M., Murakam, S., Yoshida, T., Kohno, M., Onoue, S., Kera, Y., and Ohtani, B., *Phys. Chem. Chem. Phys.*, 3, 2697, 2001.
198. Lee, S.-H., Kang, M., Cho, S.M., Han, Y., Kim, B.-W., Yoon, K.J., and Chung, C.-H., *J. Photochem. Photobiol., A*, 146, 121, 2001.
199. Whittingham, M.S., and Jacobson, A.J., Eds., *Intercalation Chemistry*, Academic Press, New York, 1982.
200. Li, W.-J., Shi, E.-W., Chen, Z.-Z., Zhen, Y.-Q., and Yin, Z.-W., *J. Solid State Chem.*, 163, 132, 2002.
201. Wang, W., Wu, M., and Liu, X., *J. Solid State Chem.*, 164, 5, 2002.
202. Kanno, R., Shirane, T., Kawamoto, Y., Takeda, Y., Takano, M., Ohashi, M., and Yamaguchi, Y., *J. Electrochem. Soc.*, 143, 2435, 1996.
203. Matsumura, T., Kanno, R., Inaba, Y., Kawamoto, Y., and Takano, M., *J. Electrochem. Soc.*, 149, A1509, 2002.
204. Tabuchi, M., Ado, K., Kobayashi, H., Matsubara, I., Kageyama, H., Wakita, M., Tsutsui, S. Nasu, S., Takeda, Y., Masquelier, C., Hirana, A., and Kanno, R., *J. Solid. State Chem.*, 141, 554, 1989.
205. Song, L., Hu, Y., Wang, S., Chen, Z., and Fan, W., *J. Mater. Chem.*, 12, 3152, 2002.
206. Liu, Z.-H., Ooi, K., Kanoh, H., Tang, W., Yang, X., and Tomida, T., *Chem. Mater.*, 13, 473, 2001.
207. Torkar, H., Worel, H., and Krischner, H., *Monatsh. Chem.*, 91, 653, 1960.
208. Toraya, H., Yoshimura, M., and Somiya, S., *J. Am. Ceram. Soc.*, 65, C-72, 1982.
209. Inoue, M., Kimura, M., and Inui, T., *Chem. Commun.*, 957, 1999.
210. Inoue, M., Kimura, M., and Inui, T., *Ceram. Trans.*, 108, 53, 2000.
211. Chen, S.-J., Li, L.-H., Chen, X.-T., Xue, Z., Hong, J.-M., and You, X.-Z., *J. Cryst. Growth*, 252, 184, 2003.
212. Komarneni, S., Roy, R., and Li, Q.H., *Mater. Res. Bull.*, 27, 1393, 1992.
213. Komarneni, S., Pidugu, R., Li, Q.H., and Roy, R., *J. Mater. Res.*, 10, 1687, 1995.

214. Komarneni, S., Li, D., Newalkar, B., Katsuki, H., and Bhalla, A.S., *Langmuir*, 18, 5959, 2002.
215. Li, W.-J., Shi, E.-W., Chen, Z.-Z., Zhen, Y.-Q., and Yin, Z.-W., *J. Solid State Chem.*, 163, 132, 2002.
216. He, W., Zhang, Y., Zhang, X., Wang, H., and Yan, H., *J. Cryst. Growth*, 252, 285, 2003.
217. Zhang, W., Wang, C., Zhang, X., Xie, Y., and Qian, Y., *Solid State Ionics*, 117, 331, 1999.

# 3 Mechanochemical Synthesis of Ceramics

*Aaron C. Dodd*

## CONTENTS

## I. INTRODUCTION

The kinetics of solid state chemical reactions are ordinarily limited by the rate at which reactant species are able to diffuse across phase boundaries and through intervening product layers. As a result, conventional solid state techniques for manufacturing ceramic materials invariably require the use of high processing temperatures to ensure that diffusion rates are maintained at a high level, thus allowing chemical reaction to proceed without undue kinetic constraint.[1]

Conducting synthesis reactions at high temperatures inevitably leads to the formation of coarse-grained reaction products due to the occurrence of sintering and grain growth during processing. Such coarse-grained materials are generally undesirable for manufacturing advanced engineering ceramics due to their poor sinterability. Furthermore, the high temperatures required for rapid solid state chemical reaction can prevent the successful synthesis of materials that are thermodynamically metastable. Consequently there is considerable interest in alternative synthesis techniques that either reduce the required processing temperatures or eliminate the need for applied heating altogether.[2]

The apparent necessity for high processing temperatures in solid state synthesis reactions can be avoided through the use of mechanochemical processing, which simply entails high-energy milling of a reactant powder charge.[3] This has the effect of inducing chemical changes directly or activating chemical reaction during subsequent low-temperature heat treatment. This chapter presents an overview of mechanochemical processing and its application within the synthesis and processing of ceramic materials.

## A. PROCESS DESCRIPTION

Mechanochemical processing refers to a range of techniques, which can be conveniently classified as mechanical grinding, mechanical alloying, and reaction milling. Although all of these techniques are based on high-energy mechanical processing, they are distinguished from each other by the nature of the reactant powder charge and also by the structural and chemical changes that occur during processing.[4]

Mechanical grinding specifically refers to milling processes where there is no change in the chemical composition of the reactant powder charge. Mechanical alloying refers to the formation of alloys by milling of precursor materials. Finally, the process termed reaction milling uses high-energy mechanical processing to induce chemical reactions.

## B. MILLING SYSTEMS FOR MECHANOCHEMICAL PROCESSING

The most commonly used mill in experimental studies of mechanochemical processing is the vibratory Spex 8000 mixer/mill. In this mill, the reactant powder charge and grinding are contained within a cylindrical vial that undergoes rapid vibratory motion in a "figure-eight" trajectory. The Spex mill is highly energetic, which allows the use of short milling times.

Another type of mill commonly used in studies of mechanochemical processing is the planetary mill. As implied by the name, the milling container is rotated about two separate parallel axes in a manner analogous to the rotation of a planet around the sun. The milling action of a planetary mill is similar to that of a conventional horizontal tumbling mill. However, the velocity of the grinding media is not limited by centrifugal forces.

Attritor mills consist of a stationary container filled with grinding balls that are stirred by impellers attached to a drive shaft. The velocity of the grinding media in attritor mills is significantly lower than that in planetary or Spex-type mills and consequently the energy available for mechanochemical processing is lower. However, unlike planetary and Spex-type mills, attritors are readily amenable to scale-up, which allows mass production of powders through mechanochemical processing.[5]

## C. POWDER CONTAMINATION

A major issue of concern with regard to mechanochemical processing is contamination of the powder charge, since the milling action inevitably results in abrasion of the grinding media and container. The degree and type of contamination experienced during mechanochemical processing has been found to depend on a variety of factors, including the relative hardness of the powder charge and grinding media, the duration of milling, and the chemical nature of the powder charge. In general, the extent of such contamination can be limited by minimizing the milling duration and ensuring that the hardness of the grinding media and container is greater than that of the powder being milled.

One approach that has been taken for avoiding contamination of the powder charge from the grinding media and container is to use the same material for the media and container as at least one of the components of the powder charge. However, this approach is of limited applicability given the restricted range of materials that are suitable for the construction of grinding media. Furthermore, even though contamination by foreign materials is avoided by this method, the stoichiometry of the final powder will be different from that of the starting powder charge.[6]

## II. MECHANICAL GRINDING

Mechanical grinding finds extensive use in mineral processing and powder metallurgy for the purposes of particle size reduction and powder blending, However, mechanical grinding has also found use as a technologically simple means of inducing structural transformations[7] and also for synthesizing nanocrystalline and amorphous materials.[8] In addition, experimental studies have shown that mechanical grinding can also be used to significantly increase the chemical reactivity of materials during subsequent processing, thus allowing the use of lower processing temperatures.

High-energy mechanical milling results in severe microstructural refinement and the accumulation of lattice defects, which can substantially increase the chemical reactivity of the powder charge. This phenomenon, which is known as mechanical activation, can be used to enhance the kinetics of solid state synthesis reactions. For example, Ren et al.[9,10] have developed a process, called integrated mechanical and thermal activation (IMTA), for the synthesis of nanostructured carbide and nitride ceramic powders. In this process, high-energy milling of the reactant mixtures allows the use of comparatively low temperatures and short reaction times during subsequent carbothermic reduction.

## III. MECHANICAL ALLOYING

Mechanical alloying was originally developed in the late 1960s at the International Nickel Company as a means of manufacturing oxide dispersion strengthened alloys for aerospace applications.[6] Since then, the process of mechanical alloying

**TABLE 3.1**

**Examples of Ceramic Powders Synthesized by Mechanical Alloying**

| Product | Reactants | Milling Time (h) | Crystallite Size (nm) | Ref. |
|---|---|---|---|---|
| $Pb(Mg_{1/3}Nb_{2/3})O_3$ | $PbO + MgO + Nb_2O_5$ | 20 | 20–30 | 11 |
| $BaFe_{12}O_{19}$ | $BaCO_3 + 6Fe_2O_3$ | 24 | 100 | 12 |
| $Ca_{10}(PO_4)_6(OH)_2$ | $Ca_3(PO_4)_2 + 3Ca(OH)_2$ | 60 | 22–39 | 13 |
| $NiFe_2O_4$ | $NiO + Fe_2O_3$ | 35 | 6 | 14 |
| $MgO\text{-}ZrO_2$ | $MgO + ZrO_2$ | 24 | 15 | 15 |
| $Al_2O_3\text{-}Y_2O_3\text{-}ZrO_2$ | $Al_2O_3 + ZrO_2 + Y_2O_3$ | 48 | 15 | 16 |

has been applied to the synthesis of a wide range of metastable and nanocrystalline materials. The majority of studies of mechanical alloying have been concerned with the synthesis of metallic materials. However, in recent years there has been increasing interest in the application of mechanical alloying to the synthesis and processing of ceramic materials, as shown in Table 3.1. Two areas of particular interest have been the synthesis of ferroelectric and magnetic ceramics.

## A. FERROELECTRIC PEROVSKITES

The ferroelectric $Pb(Mg_{1/3}Nb_{2/3})O_3$ (PMN) ceramic has been the subject of extensive investigations due to its high dielectric coefficient and high electrostrictive coefficient, which renders it suitable for use in capacitors and electrostrictive actuators. However, the successful exploitation of this material is limited by the difficulty of producing a single-phase material with the perovskite structure. Conventional solid state synthesis techniques invariably result in the formation of one or more pyrochlore phases, which exhibit poor dielectric properties.

Recent work by Wang et al.[11] has shown that formation of undesirable pyrochlore phases can successfully be avoided through the use of mechanical alloying. In their study, it was found that mechanical alloying of the constituent oxides for 20 hours resulted in the formation of a single phase PMN perovskite powder with an average particle size of 20 to 30 nm.

## B. SYNTHESIS OF HIGH-COERCIVITY FERRITE MAGNETS

Ferrite ceramics have found widespread applications as materials for permanent magnets and recording media due to their low cost and attractive magnetic properties. The conventional method for manufacturing ferrite ceramics involves solid state reaction of oxide or carbonate precursors at high temperature. Although technically simple, this method does not readily allow control of the product's microstructure and purity, which is necessary for the attainment of optimal magnetic properties. As a result, there has been considerable research into the development of alternative synthesis techniques.

Numerous studies have examined mechanical alloying as a method of manufacturing high-coercivity magnetic materials. In this process, precursors are milled together and then heat treated to form the final product phase. For example, Ding et al.[12] reported the synthesis of nanocrystalline $BaFe_{12}O_{19}$ by mechanical alloying of $BaCO_3$ with $Fe_2O_3$ followed by heat treatment at the relatively low temperature of 850°C. The resulting material was found to exhibit significantly higher coercivity compared to the same material that was prepared by conventional powder metallurgy techniques.

## IV. REACTION MILLING

In addition to providing a means of inducing solid state transformations and alloying, mechanochemical processing can be used more generally to promote chemical reactions. The underlying mechanism of reaction milling is repeated deformation, fracture, and welding of the powder charge during collisions of the grinding media. Fracture of particles exposes fresh reacting surfaces and welding generates interfaces between reactant phases across which short-range diffusion can occur, thus allowing chemical reactions to occur without kinetic constraint. Diffusion rates are also enhanced by the high concentration of lattice defects, which provide "short circuit" diffusion paths.[3]

### A. REACTION KINETICS

Chemical reaction during milling either occurs gradually or suddenly by mechanically activated combustion. Whether the reaction mechanism is gradual or combustive has been shown to depend primarily on the magnitude of the enthalpy change and the milling conditions. If the enthalpy change associated with the formation of the product phases is low, then the heat generated will be insufficient to significantly influence the reaction kinetics and reaction will proceed in a gradual manner.

Gradual mechanochemical reaction systems have generally been found to exhibit sigmoidal reaction kinetics. The reaction rate initially increases with milling due to increasing activation and microstructural refinement of the reactants. The reaction rate then reaches a maximum at an intermediate milling time before decreasing as the reaction approaches completion due to dilution of the reactants by the product phases.[3]

Combustion during milling is detected experimentally by the temperature spike associated with the sudden release of heat. This is illustrated in Figure 3.1, which shows a representative example of the variation in temperature of a milling vial during processing of a highly exothermic reactant mixture. In the initial stages of milling, the temperature gradually increases up to a steady-state value as a result of heat generated by collisions of the grinding media. Following the critical milling time, the vial temperature suddenly increases due to the heat generated by combustion of the reactants. The temperature then slowly decays to the previous steady-state value.

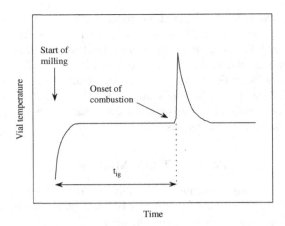

**FIGURE 3.1** Vial temperature as a function of time during milling of a combustive reaction system.

Schaffer and McCormick[17] proposed the concept of an ignition temperature as an explanation for the duration of milling required to initiate a combustive reaction. In the initial stages of milling, the temperature required to ignite a self-propagating reaction is higher than the highest local temperature reached during collisions of the grinding media. However, the ignition temperature decreases with continued milling as a consequence of mechanical activation and micro-structural refinement. Combustive reaction finally occurs when the ignition temperature is reduced to the highest temperature reached during collisions. This model is illustrated schematically in Figure 3.2.

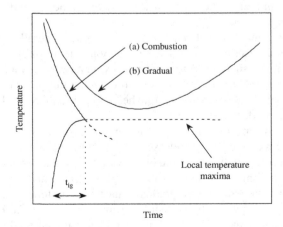

**FIGURE 3.2** Schematic illustration showing the variation in the critical ignition temperature for (a) combustion and (b) gradual reaction.

Combustion during mechanochemical processing of highly exothermic reactant mixtures can be delayed or even completely suppressed through the addition of solid or liquid diluent. The addition of inert diluent to an exothermic reactant mixture inhibits combustive reaction primarily by acting as an additional sink for the heat generated by chemical reaction during collision events. This has the effect of lowering the local reaction temperature. Without a sufficient increase in temperature, the reaction is unable to undergo thermal propagation.

## B. CARBIDE AND NITRIDE SYNTHESIS

Mechanochemical processing has been used to manufacture nanocrystalline powders of nitride and carbide ceramics. The majority of systems involve milling of the metal precursor with a source of carbon or nitrogen. The source of carbon or nitrogen has typically taken the form of the element itself. However, a variety of other reagents have also been used. For example, Zhang et al.[18] reported the synthesis of titanium nitride by milling titanium metal with pyrazine in a benzene solution.

Mechanochemical reaction systems based on the reduction of oxide precursors have been investigated as a means of manufacturing nanocrystalline carbide powders at low temperature. For example, Welham and Llewellyn[19] synthesized nanocrystalline TiC powders by mechanochemical reduction of rutile and ilmenite powders with magnesium metal in the presence of graphite. The by-product phases formed during milling were readily removed by a simple acid leaching process, which allowed recovery of the TiC product. In a similar process, El-Eskandarany et al.[20] prepared nanocrystalline tungsten carbide powders by mechanochemical reaction of tungsten oxide with magnesium and carbon.

## C. MECHANOCHEMICAL SYNTHESIS OF ULTRAFINE POWDERS

Recent experimental studies have shown that suitably designed mechanochemical exchange and reduction reactions can be used to synthesize ceramic powders that are characterized by nanocrystalline particle sizes and minimal hard agglomeration. In this process, chemical precursors undergo reaction, either during milling or during subsequent low-temperature heat treatment, to form a composite powder consisting of nanocrystalline ceramic grains dispersed within a salt matrix. The ultrafine ceramic powder is then recovered by removing the salt matrix through washing with an appropriate solvent.[21-29]

Although numerous processes exist for the synthesis of nanoparticulate materials, the mechanochemical synthesis technique is of particular interest since it is applicable to a wide variety of materials and is capable of producing powders with minimal hard agglomeration. In contrast to other synthesis techniques where particles are able to encounter each other during processing, mechanochemical processing effectively encapsulates each individual powder particle in by-product salt, thus separating the particles and thereby minimizing hard agglomerate formation.[21] As an example of a nonagglomerated powder

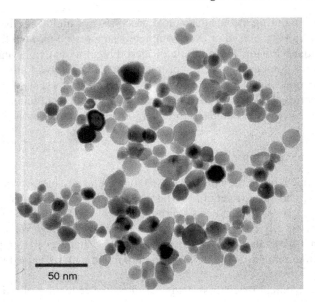

**FIGURE 3.3** Bright-field TEM image of ZnO powder that was synthesized by mechanochemical processing.

synthesized by mechanochemical processing, Figure 3.3 shows a bright-field transmission electron microscopy (TEM) micrograph of a ZnO powder that was prepared by mechanochemical reaction of $ZnCl_2$ with $Na_2CO_3$.

Ultrafine powders consisting of dispersed single-crystal particles are important for a variety of technological applications. For example, the sinterability of ceramic powders is determined by the average agglomerate size. Attaining full density with minimal grain growth therefore requires the use of weakly agglomerated precursor powders. A further example is spectrally selective coatings containing ZnO particles that transmit visible light while absorbing ultraviolet (UV) light. As shown in Figure 3.4, aqueous suspensions of ZnO particles synthesized by mechanochemical processing exhibit high transmittance at visible wavelengths. This is only possible for nonagglomerated powders with a sufficiently small particle size.

The mechanochemical synthesis technique has been shown to offer considerable scope for controlling the average particle and crystallite size of the resultant powder. Any parameter that affects the crystallite size developed during milling will have a corresponding effect on the average particle size of the final washed powder.[21] In addition, postmilling heat treatment of sufficient temperature and duration can be used to coarsen the average particle size through diffusional mass flow and interparticle sintering.[28]

As shown in Table 3.2, the majority of mechanochemical reaction systems that have been developed for the synthesis of compounds are based on reaction of either an oxide or chloride precursor with an exchange reagent. However,

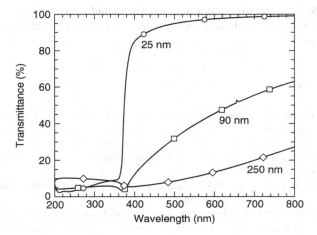

**FIGURE 3.4** Ultraviolet and visible spectra of aqueous suspensions of ZnO particles that were synthesized by mechanochemical processing.

## TABLE 3.2
## Summary of Some Representative Examples of Mechanochemical Reaction Systems Developed for the Synthesis of Ultrafine Ceramic Powders

| Powder | Reactants | Dilution | Heat Treatment (°C) | Particle Size (nm) | Ref. |
|--------|-----------|----------|---------------------|--------------------|------|
| $Al_2O_3$ | $2AlCl_3 + 3CaO$ | None | 350 | 10–20 | 22 |
| $ZrO_2$ | $ZrCl_4 + 2MgO$ | None | 400 | 5 | 23 |
| $Fe_2O_3$ | $2FeCl_3 + 3CaO$ | $5CaCl_2$ | 200 | 10–30 | 24 |
| $Cr_2O_3$ | $Na_2Cr_2O_7 + S$ | $Na_2SO_4$ | 800 | 10–80 | 25 |
| ZnO | $ZnCl_2 + Na_2CO_3$ | $8.6NaCl$ | 400 | 27 | 26 |
| $SnO_2$ | $SnCl_2 + Na_2CO_3$ | NaCl | 700 | 5–30 | 27 |
| $CeO_2$ | $CeCl_3 + 3NaOH$ | $12NaCl$ | 500 | 10 | 28 |
| GaN | $Ga_2O_3 + Mg_3N_2$ | None | 800 | 4–20 | 29 |

Tsuzuki and McCormick[28] have shown that suitably designed reduction reactions can also be used for the synthesis of ultrafine oxide powders. In their study, ultrafine powder particles of $Cr_2O_3$ were prepared by the reduction of $Na_2Cr_2O_7$ with sulfur.

## V. CONCLUSION

The mechanochemical processing technique provides the means to produce chemical and structural changes in a reactant powder charge at low temperatures. This provides a technologically simple method for manufacturing a wide variety of advanced engineering materials with controlled microstructures and chemical compositions.

## REFERENCES

1.  Schmalzried, H., *Chemical Kinetics of Solids*, Wiley-VCH, Weinheim, 1995.
2.  Stein, A., Keller, S.W., and Mallouk, T.E., Turning down the heat: design and mechanism in solid-state synthesis, *Science*, 259, 1558, 1993.
3.  McCormick, P.G., Application of mechanical alloying to chemical refining, *Mater. Trans. JIM*, 36, 161, 1995.
4.  McCormick, P.G., and Froes, F.H., The fundamentals of mechanochemical processing, *J. Metals*, 50, 61, 1998.
5.  McCormick, P.G., Mechanical alloying and mechanically induced chemical reactions, in *Handbook on the Physics and Chemistry of Rare Earths*, vol. 24, Elsevier Science, New York, 1997.
6.  Suryanarayana, C., Mechanical alloying and milling, *Prog. Mater. Sci.*, 46, 1, 2001.
7.  Lin, I.J., and Nadiv, S., Review of the phase transformation and synthesis of inorganic solids obtained by mechanical treatment (mechanochemical reactions), *Mater. Sci. Eng.*, 39, 193, 1979.
8.  Koch, C.C., Synthesis of nanostructured materials by mechanical milling: problems and opportunities, *Nanostruct. Mater.*, 9, 13, 1997.
9.  Ren, R.M., Yang, Z.G., and Shaw, L.L., Synthesis of nanostructured TiC via carbothermic reduction enhanced by mechanical activation, *Scripta Mater.*, 38, 735, 1998.
10. Ren, R.M., Yang, Z.G., and Shaw, L.L., Nanostructured TiN powder prepared via an integrated mechanical and thermal activation, *Mater. Sci. Eng. A*, 286, 65, 2000.
11. Wang, J., Junmin, X., Dongmei, W., and Weibeng, N., Mechanochemical fabrication of single phase PMN of perovskite structure, *Solid State Ionics*, 124, 271, 1999.
12. Ding, J., Miao, W.F., McCormick, P.G., and Street, R., High-coercivity ferrite magnets prepared by mechanical alloying, *J. Alloys Compd.*, 281, 32, 1998.
13. Silva, C.C., Pinheiro, A.G., Miranda, M.A.R., Góes, J.C., and Sombra, A.S.B., Structural properties of hydroxyapatite obtained by mechanosynthesis, *Solid State Sci.*, 5, 553, 2003.
14. Jovalekic, C., Zdujic, M., Radakovic, A., and Mitric, M., Mechanochemical synthesis of $NiFe_2O_4$ ferrite, *Mater. Lett.*, 24, 345, 1995.
15. Michel, D., Faudot, F., Gaffet, E., and Mazerolles, L., Stabilized zirconias prepared by mechanical alloying, *J. Am. Ceram. Soc.*, 76, 2884, 1993.
16. Kwon, N.-H., Kim, G.-H., Song, H.S., and Lee, H.-L., Synthesis and properties of cubic zirconia-alumina composite by mechanical alloying, *Mater. Sci. Eng. A*, 299, 185, 2001.
17. Schaffer, G.B., and McCormick, P.G., The direct synthesis of metals and alloys by mechanical alloying, *Mater. Sci. Forum*, 88–90, 779, 1992.
18. Zhang, F., Kaczmarek, W.A., Lu, L., and Lai, M.O., Formation of titanium nitrides via wet reaction ball milling, *J. Alloys Compd.*, 307, 249, 2000.

19. Welham, N.J., and Llewellyn, D.J., Formation of nanometric hard materials by cold milling, *J. Eur. Ceram. Soc.*, 19, 2833, 1999.
20. El-Eskandarany, M.S., Omori, M., Ishikuro, M., Konno, T.J., Takada, K., Sumiyama, K., Hirai, T., and Suzuki, K., Synthesis of full-density nanocrystalline tungsten carbide by reduction of tungstic oxide at room temperature, *Metal. Mater. Trans. A*, 27, 4210, 1996.
21. McCormick, P.G., Tsuzuki, T., Robinson, J.S., and Ding, J., Nanopowders synthesized by mechanochemical processing, *Adv. Mater.*, 13, 1008, 2001.
22. Ding, J., Tsuzuki, T., and McCormick, P.G., Ultrafine alumina particles prepared by mechanochemical/thermal processing, *J. Am. Ceram. Soc.*, 79, 2958, 1996.
23. Dodd, A.C., Raviprasad, K., and McCormick, P.G., Synthesis of ultrafine zirconia powders by mechanochemical processing, *Scripta Mater.*, 44, 689, 2001.
24. Ding, J., Tsuzuki, T., and McCormick, P.G., Hematite powders synthesized by mechanochemical processing, *Nanostruct. Mater.*, 8, 739, 1997.
25. Tsuzuki, T., and McCormick, P.G., Synthesis of $Cr_2O_3$ nanoparticles by mechanochemical processing, *Acta Mater.*, 48, 2795, 2000.
26. Tsuzuki, T., and McCormick, P.G., ZnO nanoparticles synthesised by mechanochemical processing, *Scripta Mater.*, 44, 1731, 2001.
27. Cukrov, L.M., Tsuzuki, T., and McCormick, P.G., $SnO_2$ nanoparticles prepared by mechanochemical processing, *Scripta Mater.*, 44, 1787, 2001.
28. Tsuzuki, T., and McCormick, P.G., Synthesis of ultrafine ceria powders by mechanochemical processing, *J. Am. Ceram. Soc.*, 84, 1453, 2001.
29. Cai, S., Tszuzuki, T., Fisher, T., and McCormick, P.G., Synthesis of nanocrystalline gallium nitride by mechanochemical reaction, in *Conference on Optoelectronic and Microelectronic Materials and Devices*, Institute of Electrical and Electronic Engineers, New York, 1999.

# 4 Cryochemical Synthesis of Materials

*Oleg A. Shlyakhtin, Nickolay N. Oleynikov, and Yuri D. Tretyakov*

## CONTENTS

## I. INTRODUCTION

### A. MAIN PROCESSES OF CRYOCHEMICAL SYNTHESIS

Various kinds of low-temperature processing have been used in materials science and technology. Among these, the cryochemical method is believed to be the first from the work of Landsberg and Campbell[1] who dealt with the synthesis of nanocrystalline powders of W and W-Re alloys. This new synthesis technique attracted the attention of several groups who used this new method for the production of various electroceramic materials.[2–5] Most of the work during this period was related to various kinds of single and complex oxide powders and materials. The first experiments in cryochemical synthesis, performed by the group of Tretyakov at the Moscow State University (MSU) in the late 1960s, were aimed at the cryochemical synthesis of ferrites.[6–8] Further investigations in this field in Western laboratories and at MSU developed in different directions. European and U.S. groups were oriented toward the search for new applications of cryochemical methods, while the MSU group investigated mostly the fundamental background of the cryochemical synthesis processes.

At present the bibliography of cryochemical synthesis includes more than 700 references related to various types and applications of cryochemical techniques. In spite of the relatively long history of this method, the mean rate of appearance of new papers has remained almost constant over the last 15 to 20 years. All these publications are a testament to the importance of the method, which has great potential for further development. Most synthetic procedures have changed little since the early work of the cryochemical pioneers (Figure 4.1). The first stage of the process is preparation of the single- or multicomponent solution containing the components of the target material in a stoichiometric ratio. In most cases, aqueous solutions are used, though it is possible to use other solvents with reasonable equilibrium pressures over the solid state at low temperatures, for example, tert-butanol, acetic acid, benzene, toluene, etc. However, due to low solubility of many inorganic salts in these solvents, their applications in cryochemical processing are rather limited. Application of many different salts representing various precursor components has some limitations because of their behavior in the cryocrystallization process. Initially the cryochemical synthesis method was based on the application of true solutions only. In this case, various salt components of solutions should not form insoluble precipitates. Similar limitations also applied to the pH of starting solutions.

The prepared solution is then subjected to freezing. Unlike biomedical studies, where slow cooling is used, freezing of aqueous solutions in cryochemical synthesis is performed by fast cooling. This process is usually realized by spraying the solution into liquid nitrogen via a pneumatic nozzle or ultrasonic nebulizer under intense mechanical stirring. The product of freezing consists of soft agglomerated spherical granules 0.01 mm to 0.5 mm in size, and it is separated from the cooling agent by decantation followed by evaporation of the rest of the refrigerant. The easy evaporation of liquid nitrogen reduces the cooling rate of freezing droplets and therefore several authors have proposed the use of cold hexane instead of nitrogen[6,9] or freezing the solution by spraying it on a cold metal surface.[10] Fast heat exchange ensures the maximum possible cooling rate in the last case, though the productivity of this technique is significantly lower. Along with true aqueous solutions, the freezing of colloidal solutions, gels, suspensions, and residues has been used more often in the last several years. The poor chemical homogeneity of these precursors is compensated for by the specific and flexible micromorphology of these synthesis products, which are suitable for a larger number of applications.

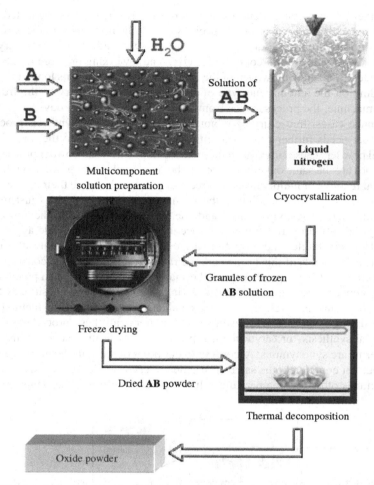

**FIGURE 4.1** Scheme of the freeze-drying synthesis process.

The next crucially important stage of synthesis is the freeze-drying process. Low temperatures and reduced pressures are selected such that the drying time is minimized, while preventing the formation of liquid phase in the macroscale by melting of ice-salt eutectics and by devitrification of glassy frozen states. A considerable advantage of this process over other processes is the availability of a wide range of industrially produced freeze-drying machines, hence this process is especially suitable for small laboratories. The freeze-drying process can easily be scaled up. The fact that this process is actively applied in the food and pharmaceutical industries makes it much easier to apply for other industrial applications. Freeze-drying of gels and residues can be done by using the simplest freeze-drying modules, for example, freeze-drying in flasks attached to a vacuum system without external cooling. Freeze-drying of true solutions, which often have rather low eutectic or devitrification temperatures, is more complicated and usually demands external cooling, especially during the first stages of the process. True solutions are freeze-dried in trays using more advanced freeze-drying machines with temperature-controlled shelves and a working pressure of $5 \times 10^2$ mbar or less. The main drawback of the freeze-drying process is its long duration. Depending on the amount and type of product and the geometry of the freeze-drying machine, the process takes from several hours to several days.

In most cases, freeze-drying is not the final stage of the synthesis process. Its product needs further processing in order to be converted to the final oxide or metal powder. These subsequent thermal decomposition and powder processing techniques are the same or similar to other chemical methods of powder synthesis. Essential features of multicomponent freeze-drying products are their high chemical homogeneity, often leading to the formation of complex oxides just in the course of thermal decomposition, and the low bulk density of the powders. Cryocrystallization of true solutions is accompanied by considerable agglomeration of salt crystallites and formation of the dense three-dimensional salt framework. The fragments of this framework are rather stable and lead to considerable agglomeration of oxide powders after thermal decomposition of such precursors.

The above process involving freeze-drying and thermal decomposition, which is used by many researchers, is usually called the "freeze-drying method" or "freeze-drying synthesis." However, several authors call these procedures "cryochemical synthesis" or "cryochemical processing," with the understanding that these terms are synonymous. According to our point of view, these terms are very close but not completely the same. The scope of cryochemical synthesis is wider and includes other low-temperature solution processing techniques (Figure 4.2).

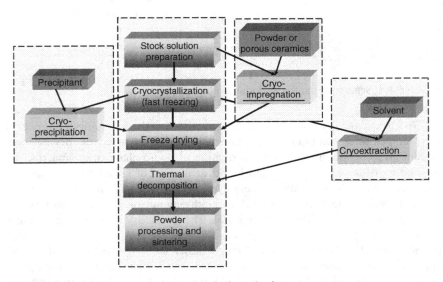

**FIGURE 4.2** Main processes of cryochemical synthesis.

## B. ALTERNATIVE CRYOPROCESSING TECHNIQUES

As was mentioned before, the main disadvantage of freeze-drying is its rather long processing time. Several efforts have been made to develop alternative low-temperature dehydration techniques for frozen aqueous solutions. Interesting results have been obtained using cryoextraction, a process by which ice can be extracted from cryogranules (droplets of frozen solution) while stirring in polar solvents cooled to 240–260 K. The main criteria that must be satisfied by these solvents are (1) a large or infinite solubility of water in them at low temperatures, (2) poor solubility of inorganic salts in them, and (3) low freezing temperature. Along with the above factors, low viscosity of solvents at low temperatures is also essential. The process kinetics are usually diffusion controlled and the rate of cryoextraction is determined mostly by the mass transfer of water from the cryogranules, which in turn is closely related to the dynamic properties of the solvent. Only a few inexpensive organic solvents have these required properties. They are methanol, ethanol, and acetone (Figure 4.3), and hence most cryoextraction experiments were conducted with these. The normal duration of the cryoextraction process is several hours. The process is often accompanied by considerable recrystallization of the frozen salt framework and coarsening of salt crystallites due to the small, but definite solubility of the most popular precursor salts in these solvents, especially in their mixtures with water forming in the course of cryoextraction. Normally the product of cryoextraction consists of higher hydrates of precursor salts. Thermal dehydration of such products leads to further recrystallization and limits the application areas of this method.

Another interesting low-temperature processing technique is the cryoprecipitation process. The main disadvantage of widely used coprecipitation method, mentioned by many authors but only recently confirmed by direct experimental studies,[11] is microinhomogeneity of the coprecipitation products. Simultaneous precipitation of components with substantially different solubility products or different precipitation rates results in enrichment of the first portions of the residue by the less soluble or easily precipitated components, while the other components are concentrated in separate particles or as the shells of precipitated crystallites. Low-temperature coprecipitation from frozen solutions allows minimization of these inhomogeneities both due to reduced differences in solubility products at low temperatures and to spatial localization of the precipitation process in the much smaller volumes. During cryoprecipitation, the frozen droplets obtained by fast freezing (cryocrystallization) of the multicomponent working solution are placed in the solution of precipitant at 230 to 270 K under intense stirring. In this case, the precipitation occurs step by step in the course of dissolution of cryogranules, so that the scale of possible microinhomogeneities is limited by the thickness of the surface layer of frozen granules, which is considerably less than the droplet size in the coprecipitation method. In contrast to freeze-drying synthesis and cryoextraction, cryoprecipitation is accompanied by complete destruction of the cryogranules and hence the dense framework of cryocrystalized salts. The obtained finely grained

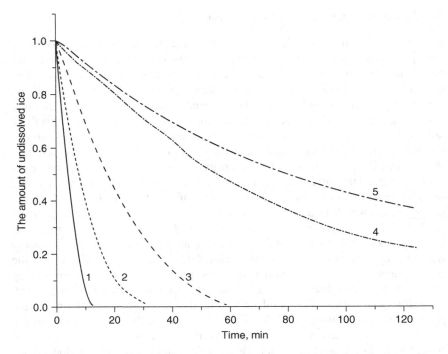

**FIGURE 4.3** The rate of ice cryoextraction from frozen solutions of ammonium trilox-alatoferriate by various solvents at a temperature of 248 K: 1, methanol; 2, acetone; 3, ethanol; 4, isopropanol; 5, propanol.

residue is separated from the supernatant solution by cold filtering followed by freeze or thermal drying.

The number of precipitants that can be used in cryoprecipitation is rather limited due to their relatively high eutectic temperatures. The most popular cryoprecipitant is aqueous ammonia, though its application is hindered by the formation of readily soluble ammoniacates of many transition elements. Other possible hydroxide cryoprecipitants are KOH ($T_{eut}$ = 195 K) and NaOH ($T_{eut}$ = 245 K). Many successful cryoprecipitation experiments have been performed using a solution of oxalic acid or ammonium oxalate in water and ethyl or methyl alcohol. Along with reducing the freezing temperature, application of these alcohols, which have a high affinity for water, and hence a high extraction efficiency, reduces the duration of the process.

One of the most practical modifications of freeze-drying synthesis is the cryoimpregnation process. This process is often considered as an independent synthesis technique. In this case, the object of freeze-drying is a porous matrix of fibers or powders soaked/wetted with a solution of microcomponents and frozen by fast cooling. The purpose of the cryoprocessing method in this case is prevention of macro- and microscopic redistribution of the microcomponents within the substrate, which is usually observed during air or thermal drying.

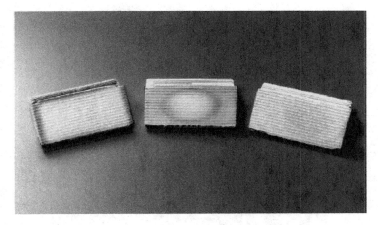

**FIGURE 4.4** Impregnation profiles of nickel catalyst on the cordierite substrate obtained using (left to right) air static drying, freeze-drying for 1 hour followed by static drying, and freeze-drying.[12]

Recent direct comparative analysis of various drying methods during wet impregnation of cordierite substrate confirmed the high efficiency of cryoimpregnation in the immobilization of active components (Figure 4.4). Apart from ordinary freeze-drying synthesis, when all components of the final material are distributed in a precursor powder with maximum possible homogeneity, cryoimpregnation is aimed at the directed placement of the micro- or second component at the surface of grains and grain boundaries of the main component or inert matrix material. Thus this method has proved to be rather efficient for the preparation of supported catalysts.[13–15]

Along with low-temperature solution processing techniques, "cryochemical synthesis" is also used for several preparation methods of metal nanopowders and metal/polymer nanocomposites, based on the low-temperature matrix isolation technique. Simultaneously evaporated metal and chemically reactive monomer, which can be polymerized at low temperatures, are quenched together at the solid surface and cooled by refrigerant (usually liquid nitrogen). Further processes in the quenched system during its warming up can lead, depending on the chemical nature of the metal and monomer, to the formation of the complex compounds of metal with monomer, metal/polymer nanocomposites, and nano-crystalline metal powders encapsulated in the polymer matrix.[16] In most cases, acrylic acid and its derivatives are used as monomers. The possibility of obtaining metal nanopowders depends on the ability of metal to initiate low-temperature polymerization of a particular monomer and on the stability of intermediate metal-monomer complex compounds.

Among low-temperature procedures widely used in the powder production of ductile materials are various kinds of cryodispersion or cryomilling methods based on mechanical comminution of materials at cryogenic temperatures. A decrease of the comminution temperature is accompanied by a corresponding

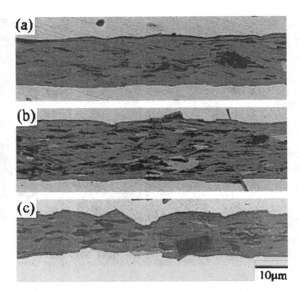

**FIGURE 4.5** SEM micrographs of HTSC tapes in silver sheath obtained using (a) cryogenic pressing and (b and c) usual cold pressing with a deformation rate 11.5% and 20%, respectively.[19]

increase in the interaction energy between defects and dislocations, causing reduced mobility of dislocations followed by microfissure formation.[17] Along with grinding the polymers and pharmaceutical solids, cryodispersion methods have demonstrated their efficiency in preparing powders of plastic metals like zinc and lead. Recent studies have demonstrated the possibility of mechanical alloying at cryogenic temperatures. A nanocrystalline Fe-Al powder obtained by this method showed enhanced morphological stability. It can be used as a chemically and morphologically homogeneous precursor in powder metallurgy applications.[18] In most cases, inexpensive and easily available liquid nitrogen is used as a refrigerant and milling medium, while the kinds of milling devices can be rather different. The plasticity reduction of silver at cryogenic temperatures can be used also for optimization of mechanical processing of high-temperature superconducting (HTSC) composite tapes, which are usually difficult to process because of severe differences in the mechanical properties of the ceramic core and silver sheath (Figure 4.5).[19]

Application of cryogenic liquids to the processing medium has resulted in the development of unusual materials processing techniques in the last several years. Use of liquid nitrogen as a medium for electrophoresis produces an increase (up to 11 kV) in the working voltage compared to the usually applied organic solvents. This has led to a decrease in the duration of electrophoretic deposition of HTSC powders.[20] Processing by the pulsed plasma channel technique in liquid oxygen led to direct single-stage production of $YBa_2Cu_3O_x$ nanopowders, which show a superconducting transition without further thermal treatment. Electric discharge between two electrodes immersed in liquid oxygen is accompanied by intensive electric erosion of electrodes made from $YBa_2Cu_3O_x$ ceramics and led to the formation of superconducting nanopowders.

## II. GENERAL FEATURES OF THE SYNTHESIS PROCESS

Since the discovery of the freeze-drying method in 1965, its application has been based on two main concepts. First, fast freezing ensures preservation of the maximum chemical homogeneity achieved in the starting solution of the multicomponent system in the products of cryocrystallization. The freeze-drying and thermal decomposition processes have little or no influence on the components' distribution, that is, the transformation of the homogeneous starting solution proceeds through homogeneous intermediate states to the final product of synthesis, a homogeneous solid complex oxide. Second, segregation of ice during cryocrystallization is accompanied by complete separation of the system into individual components. The cryocrystallization product consists of a highly dispersed mechanical mixture of the individual compounds, so that chemically homogeneous complex oxide is formed only during thermal processing at the final stage of the synthesis.

Systematic studies of the processes occurring during freeze-drying synthesis demonstrated that the actual situation in a majority of multicomponent systems is better described by the second concept, although the actual component distribution depends on a number of processing variables and the chemical features of the components. In order to forecast the behavior of a particular multicomponent system in the course of synthesis, several general physicochemical aspects of synthesis processes should be taken into account.

### A. CRYOCRYSTALLIZATION

The main process during cryocrystallization that has a decisive influence on the evolution of chemical homogeneity and micromorphology of the system is the crystallization of ice. A number of experimental studies have demonstrated that solidification of pure water at atmospheric pressure can lead to one of three ice polymorphs: hexagonal ice, cubic ice, and amorphous ice; other polymorphs can be observed only at high pressures. In the structure of hexagonal ice, having D6h symmetry, the hexagonal rings formed by water molecules constitute the layers perpendicular to the $c$-axis. Many unusual physicochemical properties of hexagonal ice are produced by the channels in water's framework parallel and perpendicular to the $c$-axis. Cubic ice cannot be obtained by crystallization of water at normal pressure; it is usually produced by condensation of water vapor on substrates cooled down to 133 to 153 K. The metastability of this polymorph is rather limited; therefore warming of cubic ice is accompanied by its conversion to hexagonal ice at 140 to 210 K. A further decrease in the cold trap temperature can lead to the condensation of amorphous ice at temperatures of less than 106 K;

devitrification of the last one at 131 to 138 K results in crystallization of cubic ice. However, the only product of pure liquid water solidification at normal pressure, irrespective of temperature, is crystalline hexagonal ice.

The actual temperature of ice crystallization in real systems is controlled mostly by kinetic factors. The maximum possible degree of water supercooling, usually calculated as $\Delta T/T_f$, where $T_f$ is melting point, is rather small compared to other solvents and doesn't exceed 0.15. During slow cooling of small (10 μm) water droplets, crystallization of ice was observed at 233 K.[21] This temperature can be achieved only under certain conditions which avoid heterogeneous nucleation. With heterogeneous nucleation, crystallization of ice will be observed at 243 to 260 K. Such behavior of water is attributed by various authors to the existence of soft-bonded dynamic clusters of ice-like structure in the water, whose concentration gradually increases at reduced temperatures, especially near the freezing point.

The behavior of aqueous solutions during freezing is much more complicated. At low cooling rates (1 to 5 K/min), crystallization in the "water-salt" system is thermodynamically controlled and can be satisfactorily described by the rules of eutectic crystallization, as most of these systems have eutectic-type phase diagrams (Figure 4.6). If the solution concentration is below the eutectic point (which is the usual case for most of the solutions used), crystallization of ice will be observed at $T \leq T_a$, followed by a systematic increase in concentration of the remaining solution until $T_e$, and simultaneous crystallization of ice and salt at $T \leq T_e$ (Figure 4.6, line 1). For a solution with a eutectic concentration, only the last process will be observed (line 2).

At higher cooling rates (5 to 20 K/min), the simultaneous processes of ice and salt crystallization during solidification of a eutectic mixture will be much more clearly separated into two almost independent crystallization events, clearly distinguishable by low-temperature differential scanning calorimetry (DSC) or differential thermal analysis (DTA) curves. Only part of the ice will be crystallized during the first process, which occurs at almost the same temperature, in spite of the increased cooling rate, due to the poor ability of water to supercool, while the temperature of crystallization of the remaining part of the solution will be progressively decreased. At a definite cooling rate of approximately 20 K/min, the second effect disappears in DTA curves and the curves show that vitrification of the liquids remained after partial ice crystallization occurs. At very high cooling rates (greater than $10^4$ K/min; varies greatly for different salts), the entire solution solidifies without crystallization, with no significant effect on the DTA curves.

Thus the processes occurring at elevated cooling rates cannot be adequately described in terms of the equilibrium crystallization model. The evolution of the system in this case is substantially controlled by the short-range hydration features of the dissolved components. The crystallization behavior of a particular multicomponent water-salt system at high cooling rates, usually applied in freeze-drying synthesis, depends considerably on the solution concentration, and can be substantially modified by cryoprotectants, that is, organic additives that prevent or retard ice crystallization from aqueous solutions. Depending on these features,

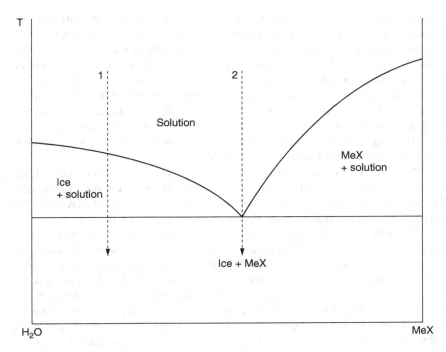

**FIGURE 4.6** Schematic phase diagram of the water-salt (MeX) system.

the cryocrystallization product of the salt solution may consist of (1) a fine mechanical mixture of crystalline hexagonal ice and crystalline salt or its higher hydrate; (2) a mixture of crystalline ice and solid amorphous solution; or (3) an amorphous, chemically homogeneous solid.

The main feature of the nonequilibrium behavior of solutions during cryocrystallization is the appearance of amorphous solids. Generally vitrification of the liquid system depends on the rate of structural relaxation processes, which are substantially determined by the viscosity of the solution. At higher cooling rates and reduced temperatures, the cluster structure of the solution cannot follow the changes, predetermined by the equilibrium behavior of the system, so that even after solidification, the structure of the amorphous solid is very similar to the structure of the solution at low temperatures. According to modern concepts, the amorphous state can be considered as a kind of supercooled liquid with an extremely high viscosity coefficient.

Analysis of solidification processes of various salt solutions demonstrated that the concentrations of the solutions that transform congruently to a solid amorphous state after partial crystallization of ice are almost the same for the particular salt, irrespective of the initial concentration of the solution and, with some limitations, of the cooling rate. If the solution of this concentration, called the critical concentration of solution ($C_c$) is used, the system transforms to the vitreous state congruently, even at moderate cooling rates. This concentration was first considered close to the eutectic one, but further investigations have demonstrated its closeness to the limit of complete hydration (LCH).[22] The term LCH was introduced in solution chemistry to denote the state of the solution when the number of water molecules per one formula unit of salt is equal to the sum of the coordination numbers of cations and anions. An aqueous solution at $C_{LCH}$ is structurally uniform and consists only of ions surrounded by tightly bound water molecules. Such a structure is rather stable and differs considerably from the dynamic structure of pure water. It contains neither ice-like fragments, promoting fast crystallization of hexagonal ice even at low supercooling, nor crystalloid-like clusters, fostering formation of crystalline hydrates.

The ability of the water-salt system to vitrify during cooling is closely related to the coordination chemistry of both the cations and anions constituting the salt. The stability and dynamic features of aquacomplexes are very different for the various ions used in materials synthesis. According to these features, these systems demonstrate rather different crystallization behavior. At a cooling rate of approximately $10^2$ K/sec, often realized during cryocrystallization, the solutions of potassium and ammonium chlorides, sulfates, nitrates, and rare earth sulfates cannot be obtained in an amorphous state at any concentration. On the contrary, the solutions of many salts of lithium, rare earth chlorides, and transition metal nitrates cannot be crystallized at this cooling rate. For several salts, the behavior of solutions during freezing depends on the concentration; more concentrated solutions have higher viscosity and hence a more pronounced tendency to vitrify. Typical features of the crystallization behavior of several salts often used in materials synthesis are summarized in Table 4.1.

As can be seen from this table, in spite of the tendency of many salts to vitrify during cooling, their behavior during heating can be rather different. In addition, the heating process is also rather important. According to the freeze-drying synthesis scheme, the cryocrystallization process is followed by freeze-drying, which can be realized with practically acceptable rates only at temperatures greater than 240 K. This means that analysis of cryocrystallization should include processes both during freezing of the solution droplets in liquid nitrogen and during defrosting of frozen granules before freeze-drying starts.

The evolution of amorphous frozen solutions during defrosting is realized by one of two scenarios. According to the first one, the mobility of ions in the solid product systematically increases and at $T \geq T_g$ the amorphous solid transforms back to liquid solution. This behavior is typical only for salts with a very strong tendency to vitrify (lithium and rare earth chlorides). More typical is when, at some temperature far below the eutectic point (170 to 200 K), the amorphous frozen solution transforms into metastable supercooled liquid. This process is clearly detectable by conductivity measurements and by disappearance of the Mossbauer signal. Further cooling of this very viscous liquid, poorly distinguished from solids by many other parameters, results in low-temperature crystallization of the mixture of crystalline ice and salt or its hydrate, similar to that observed during cooling at low cooling rates. Hence even vitrification during freezing often doesn't prevent the water-salt system from changing further during

**TABLE 4.1**

**Usual Behavior of Some Aqueous Solutions During Cryocrystallization and Following Heating**

| Ion | $NO_3^-$ | | $Cl^-$ | | $SO_4^{2-}$ | |
|-----|---------|---------|---------|---------|---------|---------|
|     | Cooling | Heating | Cooling | Heating | Cooling | Heating |
| $Li^+$ | G | C, M | G | — | G | — |
| $Na^+$ | G | $C_m$, $M_m$ | G | C, M | C | M |
| $K^+$ | C | M | C | M | C | M |
| $NH_4^+$ | C | M | C | M | C | M |
| $Mg^{2+}$ |  | C, M | G | C, M | G | C, M |
| $Sr^{2+}$ | $C_m$ | $M_m$ | C | M | / | / |
| $Ba^{2+}$ | C | M | C | M | / | / |
| $Mn^{2+}$ | G | C, M | G | — | G | C, M |
| $Co^{2+}$ | G | C, M | G | — | G | C, M |
| $Ni^{2+}$ | G | C, M | G | — | G | C, M |
| $Cu^{2+}$ | G | C, M | G | — | G | C, M |
| $Zn^{2+}$ | G | C, M | G | — | G | C, M |
| $Al^{3+}$ | G | C, M | G | C, M | G | C, M |
| $Fe^{3+}$ | G | — | G | — | G | — |

*Note:* G = glassy state (vitrification); C = crystallization; $C_m$ = crystallization of metastable crystallohydrates; M = melting; $M_m$ = melting of the eutectics.

low-temperature processing. Further heating of such a mixture, as well as of the crystalline salt-ice freezing products, leads to their melting at eutectic temperature.

These transformations of water-salt systems during cryocrystallization can have a considerable influence on the properties of the freeze-drying synthesis products, even for binary water-salt systems, used in the synthesis of individual oxide powders. However, their role is especially great in multicomponent systems, where they directly affect the chemical homogeneity of synthesis intermediates. Multicomponent oxide materials often contain cations different in their chemical nature and hence demonstrating substantially different crystallization behavior. Because of a number of technical problems, the amount of direct experimental information on low-temperature processes in multicomponent systems has been rather limited until now. Several studies in this field performed at Moscow State University[23–25] have demonstrated that salt components with different characteristics of low-temperature evolution, being present in comparable amounts, display almost independent crystallization behavior, and cryocrystallization processes are accompanied by significant phase separation of the system moderated by formation of metastable solid solutions. Instead, the systems containing components of a similar chemical nature, or when they are present in highly different amounts, usually exhibit synergistic behavior with little or no tendency for phase separation, even when it is predicted by the equilibrium phase

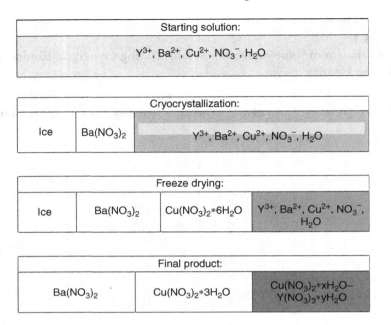

**FIGURE 4.7** Evolution of the phase composition of frozen solution of yttrium, barium, and copper nitrates during cryocrystallization and freeze-drying (dark background = amorphous compounds).

diagrams. Real multicomponent systems used in the synthesis of contemporary oxide materials often demonstrate mixed behavior. Rather typical from this point of view is the evolution of the $Y(NO_3)_3$-$Ba(NO_3)_2$-$Cu(NO_3)_2$-$H_2O$ system, used in the synthesis of HTSCs (Figure 4.7). Fast cooling of the starting solution is accompanied by crystallization of ice and $Ba(NO_3)_2$, which doesn't form crystallohydrates and crystallizes very easily, while copper and yttrium nitrates solidify without crystallization. Further evolution of this system during freeze-drying is accompanied by formation of crystalline $Cu(NO_3)_2 \cdot 3H_2O$ and semicrystalline solid solution of hydrated yttrium and copper nitrates.[25]

A very important factor that can modify considerably the crystallization behavior of aqueous solutions is their pH value. Detailed investigations of the relationship between pH and solidification processes are still in progress, although recent studies have demonstrated that in several water-salt systems the pH can be considerably (for one to four units) changed during both equilibrium and nonequilibrium crystallization.[26] More important from a practical point of view is reverse correlation. True solutions of chlorides and nitrates of many multicharged cations due to severe hydrolysis are stable only at very low pH values, where there is the presence of significant amounts of free nitric or hydrochloric acid. Taking into account that eutectic temperatures for the mixtures of these acids with ice are rather low (207 K for $HNO_3$ and 187 K for $HCl$), freeze-drying of such solutions at 230 to 240 K will be accompanied by the appearance of the liquid phase and collapse of the product microstructure.[24] Numerous observations have demonstrated that at pH < 1, the tendency of solutions to vitrify during freezing is considerably increased.

One of the ways to solve the problem of processing multicharged cations is application of colloidal solutions. Taking into account that low-temperature processing of many multicomponent true solutions results in phase segregation, the chemical homogeneity of frozen colloidal solutions can be the same or even higher than that of true ones. Colloidal solutions, containing multicharged cations, can be effectively obtained from starting acid true solutions by ion-exchange processing.[27–29] During contact of these solutions with anion-exchange resin in the OH⁻ form, ion exchange takes place releasing OH⁻. The OH⁻ lead to hydrolytic polymerization, which in turn lead to the formation of very stable multicomponent colloidal solutions. During this process, reduction of the ionic strength of solution occurs and pH values increase up to 2.5 to 4.0 due to partial neutralization of hydrated protons by released hydroxides.

Low-temperature evolution of such solutions proceeds similarly to the evolution of true solutions with easily crystallized components, although their melting points are different. If the colloidal solution doesn't contain other soluble components, its insoluble colloidal component doesn't form a eutectic mixture with ice. Such a system is equivalent to the mechanical mixture of ice and colloidal particles, so that the liquid phase appears in these products at temperatures around 0°C; this feature is very useful for the intensification of freeze-drying processes. Along with cryoprocessing of true and colloidal solutions, freeze-drying of slurries, pastes, and especially coprecipitated gels and residues has been used more often during the last several years. The chemical homogeneity of the last two types of precursors is also higher than the homogeneity of multicomponent frozen solutions with roughly different components. During freezing of residues, most of the soft-bound water crystallizes into hexagonal ice with little or no effect on the components distribution within the nanoparticles of precipitates. These freeze-dried products are characterized also by specific micromorphologies that are rather useful for a number of applications and are considerably different from those obtained by true frozen solutions.

The micromorphology of the cryocrystallization products demonstrates significant dependence on processing conditions (cooling rate, temperature gradients) and parameters of the starting solution (concentration, pH). The crystallization of ice during cooling followed by squeezing out the solution of critical concentration, typical for the majority of starting solutions, leads to the formation of cellular structures (Figure 4.8A,B). A three-dimensional network of cell walls is formed by solidified critical solution, pushed out by growing ice, while cells are filled with ice crystals. These crystals are often isolated, although in many cases the ice particles also form two- or three-dimensional structures (Figure 4.8A,B) or even a second continuous network penetrating the main one. The size of the ice particles is 1 to 5 µm, while the thickness of the wall elements varies from 50 to 300 nm, depending on the solution concentration and cooling rate. The size of the wall elements is very important for understanding the crucial feature

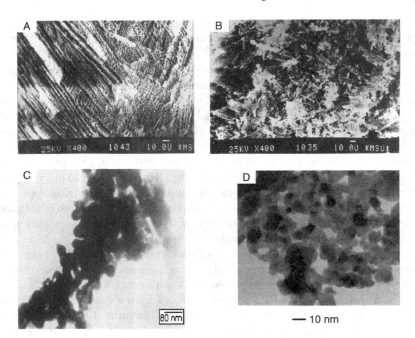

**FIGURE 4.8** Micromorphology of the freeze-dried frozen solutions (A and B) and gels (C and D): (A) Pb(NO₃)₂; (B) NaCl; (C) Al polyphosphate;[173] (D) Fe₃ₓTiₓO₄.[11]

of freeze-drying products. In spite of separate crystallization processes that occur during freezing of many multicomponent systems, the crystallite size of newly formed individual phases cannot exceed the wall thickness or, more probably, is much lower than this size. It means that separation processes occur in very limited volumes and the mean chemical homogeneity of the system remains rather high.

General features of the morphological evolution in gels and slurries during freezing are very similar to the behavior of true solutions: preferential crystallization of ice particles squeezes out and densifies gel particles, although the character of the structures formed in gels is much more diverse (Figure 4.8C,D). Among other reasons, it is mainly caused by a variety of cooling rates during gel freezing, unlike from true solutions, where fast freezing in liquid nitrogen dominates completely. At moderate cooling rates, the micromorphologies of the surface layer and the core of the sample are rather different. In the vicinity of the cold surface the structure is rather uniform, with very fine ice crystals and pores a few nanometers in size. In the core of the sample, ice can form large (up to several millimeters) dendritic crystals,[30] while the hydrated oxides form flat or polygonal fibers or lamellar structures.[31]

Efforts have been made during the last several years to control product micromorphology by directed crystallization processes. Freezing of gel or slurry in the temperature gradient results in preferential orientation of ice crystals, forming large colonies inclined to the cooling surface.[32] Nanoparticles of gels, similar to true solutions, form a cellular framework, although in the case of directed crystallization, the cells are rather anisotropic and orient in the direction of ice growth. The cell walls usually have their own internal structure that leads to the appearance of bimodal porosity in such samples after freeze-drying.[33] The structures formed during the freezing of gels are rather useful as precursors for the production of porous ceramics. One of the most serious problems with these precursors is that they lead to poor mechanical strength and continuity of the porous framework because of expansion of ice during crystallization, which often causes cracks and fissures that cannot be eliminated during further thermal processing. The problem can be solved by modification of the solidification process using cryoprotectants. Introduction of glycerol to the initial slurry results in reducing the size of the growing ice crystals, lower rejection of solid particles, and reduced volumetric expansion of ice during freezing (5% vs. 9% at 20% glycerol).[34,35] Another way to avoid cracking is based on thawing the frozen gel and soaking the sample in an excess of tert-butanol. Tert-butanol, which sublimes easily and has a much lower volume expansion coefficient, substitutes for most of the unbound water. This process is based on the so-called freeze gelation effect—gelification of sols during freezing and formation of a rigid framework that remains stable not only during freeze-drying but also during conventional thermal drying processes. This effect is used in freeze-casting processes for producing net-shaped ceramics with little or no organic additives when ice is used as a temporary and easily removable binder.[36,37] In order to ensure better morphological homogeneity and mechanical strength of the cast, nanocrystalline sols in the precursor mixtures for freeze-casting are often mixed with coarser-grained ceramic powders working as inert filler.

## B. Freeze-Drying

The main technique of solvent elimination from the frozen products is freeze-drying, based on the direct transformation of solid ice into vapor, avoiding liquid phase formation. Formation of liquid usually degrades high chemical and morphological homogeneity of the cryocrystallization products achieved during fast freezing.

Processing conditions of freeze-drying can be evaluated using pressure-temperature (PT) phase diagrams. According to Gibbs' rule, coexistence of three states (ice, liquid, and vapor) in the single-component system can be achieved at a single point (0) (Figure 4.9). Coordinates of this point for water are T = 273.15 K and P = 610.5 Pa. Coexistence of two states in this system are possible along the lines P = f(T). So the coexistence of ice and water vapor necessary for the ice sublimation process is possible along the a0 curve that can be expressed by the following linear equation:

$$\ln P \text{ (Pa)} = 29.06 - 6182/T \qquad (4.1)$$

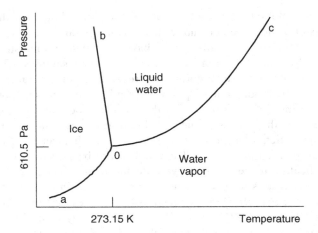

**FIGURE 4.9** Schematic phase diagram of water.

Equation 4.1 sets considerable limitations on the freeze-drying conditions. For instance, the ice can be evaporated (sublimed) without melting under an external heat supply if the pressure in the system doesn't exceed 610.5 Pa (4.58 Torr). If the product of cryocrystallization contains a eutectic salt-ice mixture with a melting point of 263 K, then, according to equation 1, it can be subjected to freeze-drying at P ≤ 261.8 Pa (1.96 Torr). The usual temperature of the product during freeze-drying is 253 K to 273 K, which correspond to equilibrium water vapor pressure values of 100 Pa to 610 Pa.

The water vapor formed during ice sublimation leaves a layer of frozen product under the action of the pressure gradient:

$$\Delta P_{H2O} = (P_s)_{H2O} - (P_c)_{H2O}, \tag{4.2}$$

where $(P_s)_{H2O}$ and $(P_c)_{H2O}$ are water vapor pressures in the zone of sublimation (near the surface of the frozen product) and near the surface of the desublimator (condenser), respectively. The value of $(P_c)_{H2O}$ is also connected with the temperature of the condenser surface via equation 1. Taking into account that the usual temperature of the condenser surface varies from 233 K (simplest dryers) to approximately 200 K (advanced models), $(P_c)_{H2O}$ values range from 0.1 Torr to 0.002 Torr.

Hence the water vapor pressure, $P_{H2O}$, within the freeze-drying chamber can vary by three orders of magnitude, since $P_{H2O} = 610.5$ Pa (maximum pressure in the sublimation zone) to $P_{H2O} \approx 1$ Pa at the condenser area. At the same time, the usual working total pressure, P, of most of the industrially produced freeze-dryers is 7 Pa to 12 Pa, which is close to the maximum possible $P_{H2O}$ over the surface of the condenser. It can be concluded that in the sublimation zone $P_{H2O} \gg P$, while in the desublimation (condensation) zone $P_{H2O} \leq P$, which establishes the direction of vapor transfer in the freeze-drying chamber.

After setting the boundary conditions of the freeze-drying, let us consider several features of this process dealing with localization of sublimation and desublimation zones. $P_{H2O}$ outside the product layer cannot exceed P, while $P_{H2O}$ at the solid-vapor boundary is much higher. It was found that during drying pure ice $P_{H2O}$ usually attains P in the bulk of the product. According to numerous studies, ice has proven to be an ideally porous body where all pores and capillaries are connected with each other and form a continuous system filled with water vapor during freeze-drying. Such a system sustains a quasi-stationary state when the vapor gain from the drying surface and the vapor drain outside the layer compensate each other so that the bulk remains in local equilibrium.

As the sublimation is a considerably endothermic process ($\Delta H_s \sim 52$ kJ/mole), the temperature of the outer ice surface is decreased approaching the quasi-equilibrium temperature ($T_q$), corresponding to the working pressure in the drying chamber if there is no heat supply. The intensity of the drying process without external heat is negligible. Freeze-drying at a reasonable rate needs a permanent external heat supply. The energy necessary for drying can be provided by various means. The most common types of heat supply are (1) conductive (contact), (2) convective, and (3) radiative (radiant).

The conductive energy supply is realized during direct mechanical contact of the drying object with a heat source (usually it is a heatable plate in the freeze-dryer). Convective heat is supplied via gas molecules colliding with the drying body. The contribution of convective heat is decreased at lower total pressures (external convection), although it can be substantially increased during heat exchange within capillary bodies (internal heat exchange). The functioning of commercial freeze-dryers is based on the above two kinds of heat supply.

Radiative heat is supplied by specially designed devices. The energy is usually transferred by electromagnetic radiation with various wavelengths [infrared (IR), radiofrequency (RF), microwave (MW)]. It should be noted that during radiation heating, apart from conductive and convective heat transfer, the energy is consumed not by the surface but by the bulk of the drying body, so that the sublimation zone is delocalized and ice evaporates from the surface of all internal pores and capillaries, irrespective of their position with reference to the surface of the product layer. This feature causes the appearance of two—external and internal—solid-gas boundaries, associated with freeze-drying zones, unlike conductive and convective heat supplies, providing formation of external freeze-drying zones only.

Concerning gas transport through a porous solid, the ice-salt system is more complex than pure ice, as it is not an ideal porous system. The product of cryocrystallization of aqueous solutions has a honeycomb micromorphology where walls are formed by crystalline salt while cells are filled with ice. Each ice granule within such a cell can be considered a potential localized sublimation zone placed into the individual microreactor. An external heat supply to the surface of a drying body with such a micromorphology results in predominantly radial heat distribution in the bulk of the solid through a three-dimensional salt framework. This process is further amplified by emerging breaks in the salt framework with ice microgranules due to intense sublimation at the ice-salt boundaries. The water vapors accumulated in the micropores enhance the convective heat exchange within the microreactor. A further increase in vapor pressure results in partial breaking of the walls of the salt framework and formation of a continuous capillary system. Realization of this scenario leads to the appearance of little or no liquid phase during drying.

According to another scenario, the pressure of water vapor within microreactors is not high enough to break the walls. With an increase in pressure, the internal vapor pressure exceeds the critical limit and formation of the liquid phase occurs. The softer walls of the salt framework are found in the frozen diluted solutions, while harder walls of the salt framework could be observed at higher salt concentrations.

Actual development of the mentioned processes during freeze-drying of real systems can be tracked using the so-called frozen droplet technique, when the mean temperature of the individual frozen droplet is controlled by an immersed thermocouple. Freeze-drying of the ideally porous body (pure ice) results in thermal equilibrium, indicated as a plateau in a diagram of the temperature dependence on drying time, $T = f(\tau)$. Deviation from the isotherm shows the nonideality of the capillary-porous behavior of the system; for most frozen solutions, $T = f(\tau)$ is a monotonically increasing function. This deviation from the ideal behavior indicates the $P_{H2O}$ increases in the enclosed system inside the frozen droplet (Equation 4.1). The last feature causes considerable rearrangement in the microstructure of the salt framework. This rearrangement, in turn, causes shrinkage and densification of the layer of the drying product that promotes a further increase of local $P_{H2O}$ within the drying areas (microreactors) and further realignment of microstructure, etc.

In general, it can be concluded that the main reason for the evolution of the salt framework during freeze-drying is the complicated diffusion of water vapor within the product particles caused by the specific micromorphology of frozen solutions. Such complications take place during ice sublimation from all cryocrystallization products and they can be considered as one of the main features of their freeze-drying processes.

Numerous experiments, performed using the frozen droplet technique, have allowed to us to identify the following general features:

1. For a limited number of salt-water systems, the rate of $P_{H2O}$ increase exceeds the rate of vapor release from channels and this promotes melting of the product even with a moderate heat supply. This is rather typical for amorphous systems without a distinct crystalline salt framework, as in the case of frozen solutions of rare earth and ferrous nitrates.
2. Several easily crystallized cryoproducts, such as frozen solutions of ammonium salts, do not demonstrate a complicated intragranular diffusion.
3. Most of the frozen systems demonstrate an intermediate behavior, that is, moderately complicated intradiffusion. A decrease in solution

concentration for such systems often promotes saving of the initial salt framework during freeze-drying without significant rearrangement.

The above mentioned features promote a better understanding of the experiments on freeze-drying of $K_2SO_4$ solutions of eutectic concentration ($C_{eut} = 6.7$ mass %, $T_{eut} = 1.51°C$). Drying of frozen neutral solutions (pH ≈ 7) results in a salt product with $S_{BET} = 10$ m$^2$/g, while adding either KOH to pH 12 or $H_2SO_4$ to pH 1 leads to a decrease of specific surface area to 3 m$^2$/g. These model experiments clearly illustrate the role of the liquid phase during drying. Indeed, eutectic temperatures in $KOH-H_2O$ and $H_2SO_4-H_2O$ systems are much lower than the value for the $K_2SO_4-H_2O$ system, so that with the usual heat supply both $K_2SO_4-KOH-H_2O$ and $K_2SO_4-H_2SO_4-H_2O$ cryoproducts will contain microlayers of liquid phase promoting structural rearrangement. Such a phenomenon of intense shrinkage of the product during drying accompanied by significant morphological evolution is often considered to be a "collapse" of the drying product.

It is clear that intense processes of structural rearrangement induced by vapor streams do not involve only micromorphology of the drying product. Freeze-drying, often mistakenly considered to be a purely physical process, also causes principal changes of the crystallographic structure of hydrates. The character of these changes has been proved to be dependent on the $P_{H2O}$ in the local drying zone, which in turn is strongly influenced by the mentioned features of intragranular diffusion of water vapor.

As was shown in Section II.A, the processes of cryocrystallization and following warming up of the frozen solution up to freeze-drying temperatures often result in crystallization of salts and their solid solutions. As crystallization occurs at low temperatures, the product of crystallization contains the maximum possible amount of chemically bound water molecules. For the salts forming crystallohydrates, this process corresponds to the formation of equilibrium higher hydrates; in some cases ($NaNO_3$, $Sr(NO_3)_2$) the formation of metastable hydrates, stable only at cryogenic temperatures and containing an unusually high number of water molecules, is observed. Due to the low $P_{H2O}$ in the course of freeze-drying, the process of ice sublimation is always accompanied by considerable dehydration, that is, removal of chemically bound water from crystallohydrates.

The process of thermal dehydration of crystallohydrates at normal pressure proceeds stepwise, that is, it is accompanied by the formation of intermediate crystallohydrates containing fewer water molecules. In this case, the processes of water removal from the hydrate and the corresponding lattice rearrangement occur almost simultaneously and are often indistinguishable. Numerous studies of dehydration of crystallohydrates at reduced pressures have demonstrated that structural rearrangement of salts at low $P_{H2O}$ is rather complicated, so that the general feature of these processes is considerable time delay between the mass transport and structural transformation stages of the dehydration process. The observations of various groups have demonstrated that dehydration under these specific conditions can lead to formation of two kinds of unusual crystallohydrates. According to Quinn et al.[38] and others,[39,40] mechanical stresses caused by complicated structural rearrangement result in break-up of the salt framework of higher hydrates into tiny mutually disordered fragments, and hence amorphization of the product. Others[39,40] have demonstrated that for many other salts, such a break-up doesn't occur, and the salt product of dehydration remains crystalline, keeping the distorted structure of the initial higher hydrate in spite of the smaller number of water molecules in the lattice. This specific kind of dehydration product is called the "lacunary phase."[40]

Analysis of the possible changes in hydrate structures of cryocrystallization products in the course of freeze-drying performed at Moscow State University for a group of metal sulfates, $MeSO_4$ (Me = Mg, Mn, Fe, Co, Ni, Cu, Zn),[41] demonstrated that both kinds of unusual vacuum dehydration products can be observed, even for the same individual salt, depending on processing conditions. Generally the products of freeze-drying can be classified into three main groups: equilibrium lower hydrates, lacunary phases, and amorphous hydrates.

For a limited number of hydrates, characterized by high flexibility of structural fragments, freeze-drying results in formation of usual (equilibrium) lower hydrates, when the actual amount of chemically bound water corresponds to the calculated number of water molecules in the reference structure. As can be anticipated, formation of the equilibrium dehydration products is promoted by an increase in $P_{H2O}$ in the local reaction zone by various means, including enhanced heat supply, a thick product layer, etc. For most of the hydrates and usual freeze-drying conditions, the normal drying products are lacunary phases containing, according to chemical or thermal analysis, the number of water molecules corresponding to the lower hydrate and having the more or less distorted structure of the initial higher hydrate. Further reducing $P_{H2O}$ during dehydration results in formation of amorphous dehydration products, even from rather crystalline higher hydrates formed during cryocrystallization. These products are usually observed at the lowest drying rate, that is, at a minimum heat supply and at the lowest pressures in the drying chamber. Quinn et al.[38] considered amorphous hydrates as an ultrafine form of equilibrium lower hydrates ($MeSO_4*H_2O$ for the $MeSO_4*nH_2O$ series), but later studies[42] demonstrated that, along with significantly higher water content ($n$ = 2.3 to 2.4), all studied amorphous species demonstrated specific dehydration behavior during thermal analysis. Mossbauer spectra of amorphous $FeSO_4*nH_2O$ also cannot be attributed to the equilibrium or lacunary $FeSO_4$ hydrates, so that it is more reasonable to consider amorphous lower hydrates as individual vacuum dehydration products. It should be noted that the tendency for amorphization and the formation of lacunary phases during dehydration is rather different for various salts; for example, for $Na_2HPO_4$ with a rather rigid salt framework, the formation of amorphous dehydration products is observed even under poor vacuum ($P_{H2O}$ = 100 mbar).[43]

The thermodynamic instability of all freeze-drying products at normal pressure and normal humidity (40 to 80%) results in their interaction with atmospheric moisture after freeze-drying, although the results of this interaction can be rather different. A large number of equilibrium lower hydrates demonstrate significant kinetic stability to rehydration, so that a short contact time of the product with

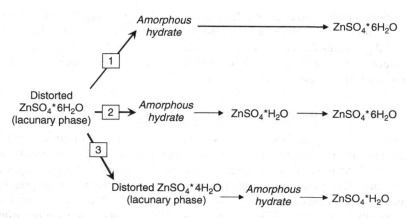

**FIGURE 4.10** Evolution of the phase composition of freeze-dried $ZnSO_4$ solution during aging at various humidities: (1) $P_{H2O} \geq 17.5$ mbar; (2) $P_{H2O} < 12$ mbar; (3) $P_{H2O} \leq 3.5$ mbar.[44]

atmospheric moisture followed by storage at low $P_{H2O}$ has no effect on the phase composition of the product, while continuous storage in air is accompanied by the formation of higher hydrates. Instead, for lacunary hydrates, the processes of direct rehydration are not usual. Due to internal stresses, the interaction of these products with water vapor is usually accompanied by their destruction; the character of the following transformations depends significantly on the $P_{H2O}$ during storage (Figure 4.10). The most typical decomposition product is the amorphous phase, similar to those formed during freeze-drying. At low $P_{H2O}$, the product can remain amorphous for a long time, while storage in air is accompanied by the formation of crystalline lower hydrate ($MeSO_4*H_2O$), even at relatively high $P_{H2O}$. Similar slow crystallization and step-by-step rehydration processes are observed during storage of amorphous hydrates obtained directly by freeze-drying of crystalline products of cryocrystallization.

At the same time, the products obtained by freeze-drying of the glassy frozen solutions are very unstable and hygroscopic at almost any $P_{H2O}$. Their interaction with atmospheric moisture is accompanied in most cases by fast and complete collapse of dried powder that makes their practical application rather problematic, in spite of the tentatively high chemical homogeneity of these products. These rehydration processes cannot be described in the same way as in the previous scheme. At the same time, similar processes are often observed during freeze-drying of organics. It has been shown that the softening temperature ($T_s$) of various glassy freeze-drying products depends on the residual water content and remains far below room temperature.[45] Even small rehydration of such a product leads to a further decrease in $T_s$, so that formation of the sticky mass complicates further processing of this product, even at a rather low $P_{H2O}$.

Apart from glassy hydrates, the evolution of crystalline freeze-drying products during their aging at low to moderate $P_{H2O}$ has little or no effect on the powder micromorphology. Significant recrystallization followed by phase separation in multicomponent products is observed only at high $P_{H2O}$ (80 to 90% relative humidity) due to formation of the liquid phase at the grain surface.[41] The critical $P_{H2O}$ values for safe processing of the particular salt composition can be calculated using the tables of water vapor pressures over saturated aqueous solutions of the corresponding salt components.

## C. THERMAL DECOMPOSITION

Thermal processing of freeze-dried salt precursors will be discussed briefly because most of the features of related processes are not specific for freeze-drying synthesis. Formation of the complex oxides usually occurs at the last stages in the thermal decomposition of salt precursors, although the range of possible formation temperatures is rather wide—200 to 900°C. The temperatures of decomposition of the individual components of salt mixtures are usually shifted to lower temperatures compared to pure individual salts due to formation of solid solutions between components during previous processing and due to reaction between components and intermediates (e.g., between CuO and $Ba(NO_3)_2$ during synthesis of $YBa_2Cu_3O_x$ from nitrate precursors[25]). Completion of synthesis at high temperatures is usually caused by thermodynamic (the existence of a low-temperature stability boundary of the target compound) or kinetic reasons (easy formation of stable intermediate product; in most cases, carbonates and oxycarbonates). It is well known that one of the most efficient instruments for reducing synthesis duration is proper selection of the heating rate. The recommendations on the desirable heating mode can be completely opposite, depending on the reasons for synthesis complications.

For substances with a low-temperature stability boundary close to the decomposition temperature ($Li_2CuO_2$, $BaCeO_3$, $YBa_2Cu_3O_x$, $Bi_2Sr_2CaCu_2O_8$), slow heating results in formation of intermediate compounds or lower homologues when target complex oxides form homological series. Crystallographic ordering and defects annealing in these compounds during further heating cause a significantly lower formation rate of the target complex oxide after reaching its stability region, so that the synthesis of these phases should be performed at the maximum possible heating rate. Instead, the amount of stable intermediate barium, strontium, and lithium carbonates and lanthanum oxycarbonate increases greatly at higher $P_{CO2}$ in the reaction zone, which in turn considerably increases at high heating rates, while heating at 1 K/min to 2 K/min often minimizes or even avoids formation of these undesirable intermediates.

Recent studies have demonstrated that the chemical homogeneity of the freeze-dried precursor in many cases has no decisive influence on its thermolysis mechanism. It is well known that thermal decomposition of solid solutions of salts can lead to the formation of individual oxides instead of complex oxide compounds.[46] Similar phenomena were observed during thermal decomposition of amorphous freeze-dried precursors in spite of their high chemical homogeneity.[47-49] Meanwhile, the small grain size of thermolysis products and their high reactivity resulted in the fast synthesis of complex oxides soon after thermal decomposition during further heating. These facts allow one to conclude that the difference in reactivity between a fine mechanical mixture of oxides and a single phase precursor compound becomes negligible. This conclusion is confirmed by the experiments of Kirchnerova and Klvana,[50] where formation of complex perovskites occurred at the same temperatures[51,52] from freeze-dried suspension that contained one of the components in the solid state and the other ones in solution, and/or from freeze-dried solution precursor.

It should be noted, however, that many parameters of the synthesis product have no direct correlation with synthesis temperature, so that powders of the same chemical composition, obtained at the same temperature, can have rather different physical properties. This feature is related to the influence of chemical and thermal prehistory on the properties of materials, often called the topochemical memory effect.[17] The mechanism of this effect can be rather different for various materials. In most cases the variation of properties of materials with different chemical prehistories deals with different concentrations and type of defects. The formation of defects, in turn, is closely connected with the thermal decomposition mechanism. For the products of freeze-drying synthesis, the effect of topochemical memory usually appears as the influence of the anion composition of the starting solution on the properties of the target materials. Thus the difference in thermolysis mechanisms of $LiCoO_2$ precursors, containing different anions, causes a different rate of crystallographic ordering of $LiCoO_2$ powders during thermal postprocessing. This leads to the different electrochemical performances of these powders in spite of identical thermal treatment.[47] Similarly, the $CoFe_2O_4$ powder obtained by thermal decomposition of oxalate precursor demonstrates excellent catalytic activity with moderate sinterability, while the same powder obtained from sulfate precursor is fairly sinterable, although its catalytic activity is rather poor.[53] In other cases, the topochemical memory effect is realized by the various micromorphologies of powders obtained at the same temperature from different salt precursors. In the case of $LiV_3O_8$, such a difference in micromorphology also has a significant influence on the electrochemical performance of cathode materials obtained from these powders (Figure 4.11).[54]

Generally the evolution of the micromorphology of freeze-dried precursors during thermal decomposition follows the general rules of topochemical transformation of salts in these processes. Dramatic differences in specific molar volumes of salt precursor and final oxide leads to stress generation in the crystallites of intermediate products followed by their decomposition into nanocrystalline fragments. High excess energy and relatively high mobility of the nanocrystalline fragments results in rearrangement of these particles similar to the collapse observed during freeze-drying of amorphous frozen solutions due to their small size and the poor strength of their interparticle bridges.

The rate of such a collapse is very different for various single and complex oxides, while the character of the morphological evolution of a particular system during this process is determined mostly by the aggregate structure of the salt precursor. The aggregate structure, apart from crystallites themselves, changes rather little during thermal decomposition.[55,56] As the aggregate structure of the

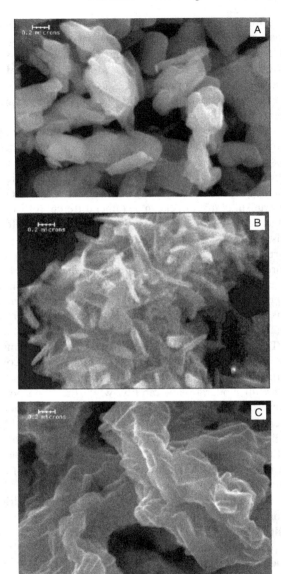

**FIGURE 4.11** SEM micrographs of $LiV_3O_8$ powders obtained by thermal decomposition of freeze-dried solutions of (A) $LiNO_3$ + $NH_4VO_3$; (B) $CH_3COOLi$ + $NH_4VO_3$; (C) $HCOOLi$ + $NH_4VO_3$.

salt precursor is significantly controlled by its synthesis methods, the micromorphology of the oxide powder obtained by its decomposition also keeps the fingerprints of the precursor synthesis technique, and this can be considered as another kind of topochemical memory effect. The characteristic features of the cryocrystallization products are close connection of salt crystallites within primary agglomerates and isolation of separate agglomerates from each other due to formation of the coarse ice particles between them (see Figure 4.8A,B). The main structure is preserved in the course of thermal decomposition, and hence the grain growth during thermal processing of oxide powder occurs readily within the primary agglomerates, while further grain size increase is rather complicated.[57] Thus the typical feature of oxide powders obtained from freeze-dried solutions is a relatively low rate of grain growth at elevated temperatures.

Apart from the internal architecture of the primary agglomerates, which are almost completely controlled by the synthesis method and cannot be changed during further powder processing, the distance between agglomerates can be modified by mechanical or ultrasonic processing at the intermediate stages of synthesis. Mechanical or ultrasonic modification of powder combined with thermal treatment produces powders with different grain sizes starting from the same salt precursor (Figure 4.12). The various methods applied for controlled synthesis of powders with variable grain size are known as "powder engineering techniques."[58–60]

## III. CONTEMPORARY APPLICATIONS OF CRYOCHEMICAL PROCESSING

Since the discovery of freeze-drying synthesis in 1965, the powders obtained by this method have found various applications. Some applications can be considered traditional and some recent applications are related to new breakthroughs in materials and technologies. The existence of traditional applications demonstrates that specific properties of the powders obtained by cryochemical processing (chemical homogeneity, specific micromorphology) fit well with the usual demands imposed by particular kinds of materials (ferrites, superionics, catalysts, etc.). The applications to new technologies (e.g., nanostructured materials and nanopowders) show that there is great potential for further development of cryochemical processing.

Due to the numerous applications of freeze-drying in the biomedical industry and the rapid development of physical organic chemistry, many interesting papers have appeared in the last decade on the behavior of pharmaceuticals during low-temperature crystallization and following freeze-drying; however, the scope of this chapter is limited to cryochemical processing of inorganic materials. Another interesting and important area is freeze-drying of macromolecular substances and materials. A large number of references (more than 500) on the applications of freeze-drying synthesis published before 1997 were systematized and briefly discussed in the monograph, *Cryochemical Technology of Advanced Materials*.[17] Readers may refer to this monograph for additional information. This chapter largely focuses on cryochemical applications that have appeared since 1997.

**FIGURE 4.12** LiCoO$_2$ powders obtained from the single freeze-dried salt precursor using various methods of the powder engineering technique:[58] (A) particle coarsening; (B, C) reducing particle size using mechanical processing of precursors.

## A. Luminescent Materials

Efficient application of most of the inorganic powdery phosphor materials implies that the powder is chemically homogeneous, well crystallized, contains trace amounts of impurities, and has a submicrometer particle size with narrow distribution. Chemical homogeneity is especially important for rare earth doped oxides, as the uniform distribution of dopant is rather critical for high efficiency luminescence. In the last several years, due to the growing demand for high-resolution electroluminescent devices, there has been considerable interest in the development of highly efficient nanocrystalline materials. Preparation of phosphor powders by sol-gel methods is often accompanied by residual contamination of the samples with poorly controlled grain growth, which reduces the luminescence efficiency. Successful cryochemical synthesis of nanocrystalline $Gd_2O_3$:$Eu^{3+}$ red emission phosphor is described by Louis et al.[61] Careful washing and ultrasonic processing of the residue obtained by coprecipitation of europium and gadolinium hydroxides from nitrate solution resulted in the formation of a stable sol. Freeze-drying of the sol followed by thermal processing resulted in the formation of nanocrystalline powder with a very limited tendency for grain growth at temperatures of 1000°C or less. At the same time, the rate of crystallographic ordering at 1000°C is high enough so that annealing of the precursor powder at this temperature results in formation of highly crystalline $Gd_2O_3$:$Eu^{3+}$ powder with a grain size of 40 nm to 50 nm with a high efficiency of red luminescence. The luminescence efficiency of the cryochemically prepared sample is considerably higher than that of the reference sample obtained by the solid state synthesis method. The authors attribute the difference to the high homogeneity of dopant distribution and the nanocrystalline particle size of the freeze-dried phosphor powder. Faster grain growth during annealing of the samples obtained by usual sol-gel methods compared to freeze-dried sols is attributed to intense agglomeration of particles in the thermally dried sol-gel precursors.

## B. Magnetic Materials

Hard magnetic materials with a magnetoplumbite structure, such as barium hexaferrite ($BaFe_{12}O_{19}$), have a complicated phase formation behavior and their magnetic properties can be easily varied. Cryoprocessing methods are highly attractive for these types of materials. The high chemical homogeneity of cryochemical precursors obtained by freeze-drying of acetate solutions allowed Wartewig et al.[62] to ensure homogeneous distribution of cations and to investigate the fundamental magnetic properties of $BaFe_{12-2x}Zn_xTi_xO_{19}$ solid solutions. Similar to spinel ferrites, initially the substitution of $Fe^{3+}$ with diamagnetic cations resulted in enhancement of low-temperature net magnetization, with a maximum at $x = 0.4$. However, further substitution up to $x = 2.0$ was accompanied by a progressive decrease in magnetization. Such a dependence correlates quite well with the results of Rietveld refinement of neutron diffraction data: up to $x = 0.8$, zinc ions are located at the tetrahedral $4f_1$ sites, while titanium is equally distributed at the octahedral $4f_2$ and 12k sites. At $x > 1.2$, zinc and titanium ions appear in the 2b and 2a sites, respectively, which results in $Fe^{3+}$ spin deviation from the collinear structure and hence the net magnetization decreases.

One of the factors that control the magnetic properties of a material is its chemical prehistory. The influence of this factor on the magnetic properties of strontium hexaferrite ($SrFe_{12}O_{19}$) is clearly demonstrated by Calleja et al.[63] Thermal decomposition of precursors obtained by freeze-drying a solution of strontium and iron nitrates or by coprecipitation from chloride solution by $Na_2CO_3$ followed by thermal drying leads to the formation of hexaferrite at temperatures of 800°C or more. At the same time, the functional properties of semicrystalline ferrites, obtained by identical thermal processing at 1000°C, are rather different: the sample obtained from the nitrate precursor demonstrates saturation magnetization up to 50 emu/g and coercivity up to 5700 Oe, while maximum magnetization of the coprecipitated sample is no more than 30 emu/g and coercivity is approximately 1300 Oe. Different magnetization values can be attributed to the different crystallographic ordering rates during annealing. As the coercivity value usually correlates with the number of magnetic pinning sites, a higher coercivity value for a cryochemically processed sample shows that it has a higher concentration of multidimensional defects, which usually lead to a higher pinning force. Hence, in this case, the influence of chemical prehistory is realized by means of various defect structures in the hexaferrite samples, that is, by different kinds and concentrations of defects in the lattice. It should be noted also that the combination of magnetization and coercivity values observed for cryochemically processed samples are rather desirable for various applications of magnetically hard $SrFe_{12}O_{19}$.

Because of the growing interest in the properties of nanostructured and nanocrystalline materials, several papers have looked at the synthesis and investigation of nanocrystalline ferrite powders. Thermal decomposition of freeze-dried nitrate precursor leads to the formation of $BaFe_{12}O_{19}$ powder with a wide particle size distribution (20 nm to 120 nm).[64] The magnetic properties of this powder demonstrate several features, including superparamagnetic behavior, ultimately associated with nanocrystalline magnetic species. At the same time, because of the wide particle size distribution and substantial anisotropy of powder crystallites, the character of the magnetic field-temperature (HT) diagram of the material is rather complex. Observed superparamagnetic behavior of high-density magnetic recording materials like $BaFe_{12}O_{19}$ can be considered as a potentially negative factor from a practical point of view, as it causes the loss of magnetic stability.

A simple and efficient method to obtain considerably smaller particles with a narrow size distribution from usual precursors was proposed by Ding et al.[65] and Shi et al.[66] Precursors obtained by freeze-drying of coprecipitated hydroxides (or by other chemical synthesis methods) are processed in a planetary mill with excess NaCl. Such a treatment reduces the size of the soft agglomerates formed during precipitation and insulates these aggregates from each other. The formation of complex oxide particles during thermal decomposition of the mixture occurs within these isolated aggregates, so that the coalescence of newly formed particles is minimized. After completion of thermal processing, NaCl is removed by repeated washing. Crystallization of $BaFe_{12}O_{19}$ obtained by this method occurs at 800°C, but the powder consists of nonagglomerated plate-like particles with a thickness of 10 nm to 40 nm and a diameter of 50 nm to 100 nm. Its magnetic properties at room temperature (saturation magnetization 67 emu/g; coercivity 5200 Oe) are rather useful for practical applications in high-density magnetic recording media. The authors have found no traces of the superparamagnetic behavior for $BaFe_{12}O_{19}$ (although the low-temperature HT diagram was not studied) or $CoFe_2O_4$ particles obtained by a similar technique (particle size 10 nm to 40 nm; saturation magnetization 65 emu/g; coercivity 1800 Oe). The particles of nickel ferrite processed under the same conditions are considerably smaller. Their mean size (approximately 10 nm) is close to the limit of superparamagnetism in $NiFe_2O_4$, and corresponding features of their magnetic behavior have been demonstrated. Further optimization of processing conditions for $CoFe_2O_4$ reduced the grain size to 10 nm, and similar features were observed, apparently caused by the nanocrystallinity of $CoFe_2O_4$ powders.[66]

The rate of crystallographic ordering of various complex oxides, their ability for grain growth, and hence the possibility of obtaining crystalline nanopowders by thermal decomposition of salt precursors are very different and depend considerably on their composition. The ability of titanium oxide to retard grain growth in $Fe_3O_4$-based solid solutions without significant complication of the crystallization process was used[67,68] for the synthesis of $Fe_{2.5}Ti_{2.5}O_4$ nanoparticles (15 to 20 nm) with a narrow size distribution by thermal processing of coprecipitated hydroxides at 380 to 460°C. Comparison of the average composition of precipitates, studied by the energy dispersive X-ray (EDX) and inductively coupled plasma-atomic emission spectrometry (ICP-AES) methods, and the surface composition of particles, determined from X-ray photoelectron spectroscopy (XPS) results, clearly demonstrate substantial enrichment of the surface by titanium due to different precipitation rates of iron and titanium hydroxides. Thermal processing of the precipitates results in elimination of this composition gradient.[11] High and easy magnetization and relatively low coercivity also characterize these finely dispersed $Fe_{2.5}Ti_{2.5}O_4$ powders, while no traces of superparamagnetic behavior (observed in similar powders obtained by the mechanochemical synthesis method[67]) were observed. This behavior was probably the result of exceeding the size limit of superparamagnetism for these compounds. Hydrogen reduction of nanocrystalline hematite obtained by thermal decomposition of freeze-dried nitrate precursor at 200°C causes formation of nanocrystalline [6 to 18 nm by transmission electron microscopy (TEM)] iron particles.[69] Due to the low threshold of superparamagnetism for $\alpha$-Fe (approximately 4 to 8 nm), this powder demonstrates ferromagnetic behavior with a saturation magnetization of 178 emu/g and a coercivity of 650 Oe at room temperature. An ultrathin oxide layer at the particle's surface ensures its excellent stability in air.

Another recent class of magnetic materials attracting a lot of attention is molecular magnets. Nanocrystalline powders of new molecular magnets,

$Cu_{1.5}[Fe(CN)_6] \cdot 6H_2O$ [Curie temperature $(T_c)$ = 19 K] and $K_{0.8}Ni_{1.1}[Fe(CN)_6] \cdot 4.5H_2O$ ($T_c$ = 28 K), were obtained by freeze-drying their precipitates, formed during mixing of $K_3[Fe(CN)_6]$ and $CuSO_4$ or $Ni(NO_3)_2$ solutions, respectively. Heat treatment of complex cyanides in argon led to rearrangement of the bonding structures and a systematic decrease in lattice parameters until disappearance of ferromagnetism during dehydration at 140 to 150°C.[70,71]

In addition to the above ferrites, which have been synthesized by the freeze-drying method since the late 1960s, new kinds of ferromagnetic materials can also be prepared by cryochemical processing. The discovery of colossal magnetoresistance (CMR) in the substituted rare earth manganites prompted a great deal of interest in the fundamental properties of these materials and their possible practical applications. Several members of the solid solution family $La_{1-x}$ $Me_xMnO_3$ (Me = Sr, Ba, Ag, Na, K) have Curie temperatures greater than 300 K, so that CMR effect can be observed in these materials at room temperature. Application of homogeneous freeze-dried acetate precursors allowed researchers to obtain and study for the first time the structure and magnetic properties of the perovskite solid solution $La_{0.85}(Na_{1-x}K_x)_{0.15}MnO_3$ ($0 \leq x \leq 1$).[72] It was found that substitution of $Na^+$ with $K^+$ leads to cell expansion and to a decrease in structural distortion from the ideal cubic structure. At the same time, the Curie temperatures of these compounds unexpectedly demonstrate no dependence on $x$ ($T_c$ = 333 K along the entire series).

Monotonical dependence of $T_c$ on the composition in $La_{1-x}Me_xMnO_3$ was used for the discovery of a new material with biomedical applications—smart mediators for self-controlled inductive heating.[73] Absorption of radiofrequency radiation by such a material, placed into liquid, causes intense heating of the surrounding medium (slightly below $T_c$). Transition of the absorber into a paramagnetic state causes cessation of absorption followed by its reactivation after cooling of the system, so that the temperature of the liquid is self-controlled without thermal sensors or other external devices (Figure 4.13).

Most of the CMR applications of $La_{1-x}Me_xMnO_3$ powders deal with production of ceramic materials. One of the first papers related to the sintering behavior of fine manganite powders obtained by chemical synthesis methods describes the preparation of dense $La_{0.7}Ca_{0.3}MnO_3$ ceramics from freeze-dried acetate and nitrate precursor.[74] Careful selection of thermal processing and, especially, deagglomeration conditions produced ceramics with a density of more than 95% at 1200°C for a few hours. Corresponding density values for similar powders obtained by the solid state synthesis method can be achieved only by continuous sintering at 1400°C. An important feature of the obtained densification curves is the considerable dedensification, which is pronounced, especially for the most active powders.

Colossal magnetoresistance studies of ceramics have demonstrated no significant influence of ceramic density on the magnetoresistance at H = 1.5T. However, ceramics obtained by the solid state synthesis method and containing inclusions of lanthanum-doped manganese oxide displayed substantially higher magnetoresistance values.[75] This effect was later used for directed enhancement of low field

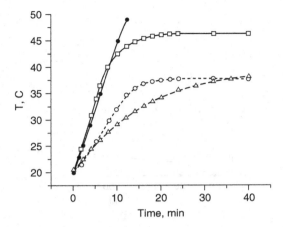

**FIGURE 4.13** Feedback heating rate for various mediators: ● = Dextran-coated $Fe_3O_4$ (commercially available mediator; no feedback); □ = $La_{0.75}Sr_{0.25}MnO_3$; ○ = $La_{0.8}Sr_{0.2}MnO_3$; Δ = $ZnFe_2O_4$.[73]

magnetoresistance of $La_{0.7}Sr_{0.3}MnO_3$ ceramics by introducing inert dopants to ceramics and thus producing CMR composites. Such a dopant should be chemically resistant to the interaction with manganite at elevated temperatures during sintering and generate defects at the interface with the manganite matrix that will act as new flux pinning centers. Comparative analysis of tentative dopant materials chosen from various complex oxides of strontium allowed selection of $SrZrO_3$, which demonstrated the best compatibility with $La_{0.7}Sr_{0.3}MnO_3$. Producing CMR composites from a mechanical mixture of $La_{0.7}Sr_{0.3}MnO_3$ and $SrZrO_3$ powders obtained by freeze-drying synthesis resulted in the maximally observed low field CMR response for manganite ceramics at room temperature ($\Delta R/R(0) = 2.5\%$ at H = 100 Oe) at a manganite:dopant ratio close to the percolation threshold.[76,77] A systematic shift of the enhancement boundary with sintering temperature showed that even $SrZrO_3$ demonstrates only limited chemical stability during sintering (Figure 4.14). Further analysis of the reaction products by EDX nanoprobe detected significant outdiffusion of lanthanum and manganese into $SrZrO_3$ particles, leading to the composition change of the manganite matrix and corresponding modification of its electrical and magnetic properties.[78]

## C. FERROELECTRICS

Further efforts to reduce the sintering temperature of $BaTiO_3$ ceramics, widely used in modern electronics, were based on the application of nanocrystalline (30 to 90 nm; $S_{BET}$ = 60 m²/g) powders. These powders are obtained by reaction of the fresh residue of hydrated titania with barium acetate solution at the boiling point (102 to 105°C) and a pH greater than 13 followed by freeze-drying of carefully washed powder. The formation process of $BaTiO_3$ particles was controlled and modified by various kinds of surfactants. It was found that the minimum agglomeration of $BaTiO_3$ particles during synthesis is realized when anionic

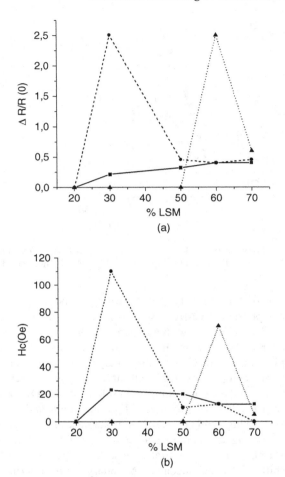

**FIGURE 4.14** Values of magnetoresistance (H = 100 Oe, T = 293 K) and coercivity of $La_{0.7}Sr_{0.3}MnO_3/SrZrO_3$ composites sintered at 1000°C (solid line), 1100°C (dashed line), and 1200°C (dotted line).[76]

surfactant is added before precipitation of hydrated titania particles, so that the most critical stage of the process is $TiO(OH)_2 \cdot yH_2O$ precipitation. The powders can be readily sintered at 1100 to 1200°C. Use of surfactant reduces the sintering temperature by approximately 100°C compared to precipitated powders without surfactant.[79]

The problem of reducing sintering temperature is also crucial for $Ba_{0.5}Sr_{0.5}TiO_3$, which possesses ferroelectric properties but is used more often as a material with a positive temperature coefficient of resistance (PTCR). The citrate synthesis method is often considered as an alternative to freeze-drying synthesis, but in the following case the solution of barium, strontium, and titanium citrates was dried by various methods (i.e., oven drying and freeze-drying).[80] Careful analysis of the thermal decomposition mechanism allowed the authors to observe complete separation of the system and formation of three independent decomposition products: $SrCO_3$, $BaCO_3$, and $TiO_2$, which react to form $Ba_{0.5}Sr_{0.5}TiO_3$ upon heat treatment. In spite of a large difference in specific molar volumes between these citrates and $Ba_{0.5}Sr_{0.5}TiO_3$, causing drastic morphological evolution during thermal decomposition, the freeze-dried $Ba_{0.5}Sr_{0.5}TiO_3$ powder demonstrated considerably less agglomeration compared to oven-dried $Ba_{0.5}Sr_{0.5}TiO_3$ powder. The maximum sinterability was also demonstrated for freeze-dried powder. Finally, in spite of the considerable agglomeration inherent to all citrate-derived powders, dense $Ba_{0.5}Sr_{0.5}TiO_3$ ceramics were obtained at 1250 to 1300°C, and this temperature is rather close to the sintering temperature of nanocrystalline $BaTiO_3$ powder.

One of the main problems during synthesis of $Pb(Mg_{1/3}Nb_{2/3})O_3$ (PMN) for application in relaxor ferroelectric ceramics is coexistence of the target perovskite phase with an undesirable pyrochlore intermediate. Formation of the $Pb_2Nb_2O_7$-based pyrochlore is kinetically preferable, while the formation of perovskite needs complicated $Mg^{2+}$ diffusion into the matrix of lead niobium oxide. For this diffusion process, the chemical homogeneity of the precursor powder is very important. Comparative analysis of the precursors obtained by freeze-drying of an oxalate-nitrate starting solution and by the consecutive precipitation technique demonstrated a substantial advantage for the former. Unlike freeze-dried precursor, thermal decomposition of precipitated hydroxides at 700°C gives almost no perovskite phase.[81] Further annealing leads to an increase in the perovskite:pyrochlore ratio in both cases, but the sintered PMN pellets with a cryochemical prehistory consist of a single phase perovskite, while ceramics from hydroxides still contain some pyrochlore. Such a difference in the phase composition is accompanied by a corresponding difference in the dielectric properties: PMN ceramics obtained from freeze-dried precursors demonstrate a considerably higher relative dielectric permittivity (14542 vs. 8019), higher Curie temperature, and smaller dissipation factor. All these differences are attributed to better chemical homogeneity of the freeze-dried precursor.[81]

The problem of undesirable pyrochlore formation is rather important also for the new generation of lead zirconate titanate (PZT)-based ferroelectric ceramics, usually containing several additional cations. Investigation of phase formation processes in the freeze-dried mixture of coprecipitated lead, lanthanum, zirconium, tin, and titanium hydroxides during thermal processing demonstrated that amorphous decomposition product crystallizes at a temperature of 550°C or less into nanocrystalline pyrochlore (particle size 5 to 8 nm).[82] Conversion of this product into single phase perovskite is completed at 750°C; the process is accompanied by considerable grain coarsening (50 to 200 nm). More detailed analysis of the reaction products demonstrated that in this system the composition of the pyrochlore and perovskite phases are almost identical except for the oxygen content. Nevertheless, the temperature of single phase perovskite formation in this composition, used for the production of actuator $(Pb,La)(Zr,Sn,Ti)O_3$ (PLZST) ceramics, is substantially lower than for many other chemical synthesis methods. Further studies have demonstrated that the temperature of perovskite formation can be further reduced. Because of the 8.5% larger molar volume of the pyrochlore compared to perovskite, preliminary pressing of nanocrystalline pyrochlore powder obtained by thermal decomposition of precursor at 550°C resulted in a considerable increase in the transformation rate. More efficient illustration of this approach was obtained during direct hot pressing experiments when approximately 60% of the perovskite was observed even at 550°C and the transformation was completed at 650°C.[83]

## D. Solid State Ionics

Applications of cryochemical synthesis in the production of materials with high oxygen ionic conductivity are based in most cases on the high chemical homogeneity of precursors, leading to phase formation at lower temperatures, and on the small grain size of oxide powders, which often reduce the sintering temperature. Both features of the cryochemical powders have been used to produce and study the compositional dependence of ionic conductivity in $La_{1-x}Sr_xCoO_3$ and $La_{1-x}Ca_xCoO_3$ ($0.1 \leq x \leq 0.5$) ceramics.[84] Optimization of synthesis conditions allowed the authors to obtain dense (greater than 95%) gas-tight ceramics at 1130°C, compared to the 1200°C to 1300°C usually applied to sintering of cobaltates.

The discovery of high-temperature superconductivity in cuprates attracted additional interest to the properties of these compounds, while some of them demonstrate significant oxygen ionic conductivity. One such compound is $LaSrCuO_4$, belonging to the solid solution series of $La_{2-x}Sr_xCuO_4$. Several members of this series are able to pass into a superconducting state, while all of them possess considerable oxygen nonstoichiometry and oxygen mobility that is especially pronounced at $x = 1$. The synthesis of the final phase in these systems is rather complicated due to low free energy of formation, and is usually completed at temperatures greater than 1000°C. Mazo et al.[85] demonstrated that a substantial part of the synthesis problem deals with formation of very stable intermediate $La_2O_2CO_3$, which forms easily not only during solid state synthesis but also during thermal decomposition of carbon-containing precursors. At the same time, application of carbon-free precursors and thermal decomposition in a $CO_2$-free environment accelerates substantially the formation of $LaSrCuO_4$. Even in this case, careful optimization of thermal processing conditions is needed in order to ensure the synthesis of a single phase nanostructured powder with a domain size of approximately 100 nm at 600°C. However, the sinterability of this powder with a minimum grain size was rather moderate, so that additional optimization efforts were needed in order to obtain dense $LaSrCuO_4$ ceramics at 1100 to 1150°C.[86]

Oxygen transport through the solid state can be caused by the difference in both electrical and chemical potentials. The last effect can be used for the development of oxygen-selective ceramic membranes. The demands of materials for such membranes are similar to the demands for solid electrolytes: high density in order to prevent physical penetration of the gas and a high coefficient of oxygen self-diffusion, usually associated with high oxygen ionic conductivity. In order to study the oxygen permeation kinetics and to determine the limiting stage of the permeation process, powders with a total composition of $SrFeCo_{0.5}O_{3.25}$ have been obtained by thermal decomposition of freeze-dried nitrate precursor followed by annealing at 1090°C. Sintering of these powders after deagglomeration produced ceramic tubular membranes with a density of more than 97%, although further studies demonstrated that sintering results in the formation of a multiphase composition. Investigation of oxygen permeability through these membranes identified bulk diffusion as the dominating oxygen transport mechanism.[87] Similar experiments with $Sm_{0.5}Sr_{0.5}CoO_3$ tubular membrane obtained from freeze-dried nitrate precursor determined the ambipolar oxygen diffusion coefficient ($D_a$) and surface exchange rate coefficient ($k_{i0}$) for this material ($8.6 \times 10^7/cm^2/sec$ and $6 \times 10^5/cm/sec$, respectively, at a temperature of 915°C).[88]

Ceria-based oxygen-conducting complex oxides are considered to be the best prospective materials for solid oxide fuel cells because of their high ionic conductivity at intermediate temperatures (800°C to 900°C). However, the field of their application as electrolytes is considerably limited by the appearance of substantial electronic conductivity at low $P_{O2}$, though this feature makes possible their application as electrode materials. A series of studies has been performed in order to obtain precise and reliable information on the electrochemical performance of various rare earth-substituted ceria materials at different $P_{O2}$. Complex oxide powders for these studies have been obtained by freeze-drying of nitrate solutions, thermal decomposition of powders at 600°C followed by additional annealing at 1200°C for powder crystallization ($Ce_{0.8}Gd_{0.2}O_{1.9}$ and $Ce_{0.8}Sm_{0.2}O_{1.9}$)[89,90] or by thermal decomposition at 900°C ($Ce_{0.8}Gd_{0.2x}Pr_xO_{1.9}$).[91] Sintering of these powders at 1400 to 1600°C resulted in dense (93 to 95%) ceramics used for electrochemical performance studies. It was found that the field of electrolytic application of $Ce_{0.8}Sm_{0.2}O_{1.9}$ can be extended, as its electronic conductivity under reducing conditions is lower than that of undoped ceria.[89] Comparison of conductivity mechanisms of $Ce_{0.8}Gd_{0.2}O_{1.9}$ and $Ce_{0.8}Sm_{0.2}O_{1.9}$ ceramics, obtained by sintering at 1400°C, demonstrated that their total conductivities are almost indistinguishable, and although the bulk conductivity of samaria-doped ceria is higher, its lower grain boundary conductivity has a decisive influence on total conductivity, at least at moderate temperatures.[90] It was found also that introducing a CoO sintering aid by impregnation of $Ce_{0.8}Gd_{0.2}O_{1.9}$ and $Ce_{0.8}Sm_{0.2}O_{1.9}$ before sintering leads to a dramatic decrease of the sintering temperature and produces doped ceria ceramics with a density greater than 93% at 1000°C with almost the same electrochemical parameters (Figure 4.15).[90] Enhanced chemical homogeneity of freeze-dried precursors refined the data on the influence of small (1% to 3%) praseodymium doping on the conductivity of $Ce_{0.8}Gd_{0.2}O_{1.9}$. As has been previously observed by other authors, a decrease in the electronic conductivity in the n-type range and an increase in conductivity in the p-range was found.[91]

In spite of the extensive applications of yttrium-stabilized zirconia (YSZ)-based solid electrolytes, many details of their conductivity mechanism and its

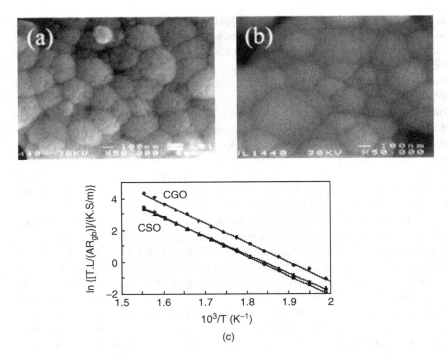

FIGURE 4.15 (a, b) SEM micrographs of $Ce_{0.8}Gd_{0.2}O_{1.9}$ ceramics sintered at 1000°C with various amounts of sintering aids; (c) grain boundary resistance curves for undoped $Ce_{0.8}Gd_{0.2}O_{1.9}$ and $Ce_{0.8}Sm_{0.2}O_{1.9}$ ceramics sintered at 1400°C, and cobalt doped $Ce_{0.8}Gd_{0.2}O_{1.9}$ ceramics sintered at 1000°C.[91]

relationship with the microstructure of these ceramics remain unclear. It is known, however, that at low and middle temperatures (300°C to 950°C), the dominating charge transfer mechanism is grain boundary conductivity. It is also known that grain boundary conductivity is controlled by the structure of the grain boundaries, which in turn is greatly influenced by powder processing and sintering features. In order to make clear the mechanism of correlation between ionic conductivity of YSZ ceramics, their microstructure, and powder synthesis method, a comparative study of the electrochemical behavior of YSZ (9% $Y_2O_3$) ceramics was performed. The YSZ ceramics were obtained from modified commercial YSZ powder and from the product of thermal decomposition of freeze-dried yttrium and zirconium sulfates. The bulk conductivities of the YSZ ceramic samples with close density values and similar grain size, but obtained from different precursor powders, were found to be very close to each other and to the bulk conductivities of YSZ single crystals. At the same time, the grain boundary conductivity of the sample with a cryochemical prehistory is 30 times higher, with a strong tendency to further increase with increasing sintering time and temperature for both ceramics. TEM analysis revealed the presence of thin glassy films at the grain boundaries of the sample with a prehistory of solid state synthesis. The amount of the glassy phase in freeze-dried samples was much smaller, and this phase was concentrated mostly in lens-shaped pockets at triple points having almost no influence on the intergrain contacts.[92] Such a glassy phase is metastable and disappears during continuous sintering at 1600°C. An increase in grain boundary conductivity with sintering temperature correlates also with increasing grain size of the ceramics.

Cryochemical applications in solid state ionics are related to oxygen ion conductors, although many oxide phases demonstrate both oxygen and proton conductivity. One of these newly discovered phases is complex perovskite ($Ba_3Ca_{1.18}Nb_{1.82}O_{8.73}$; BCN18). Its preparation by cryochemical synthesis is complicated by the poor compatibility of the components in solution. A suspension synthesis procedure was applied when precipitate of niobium hydroxide was dispersed in a solution of barium and calcium acetates.[93] Thermal decomposition of freeze-dried precursor by slow heating to 850°C followed by sintering at 1400°C for 2 h resulted in single phase $Ba_3Ca_{1.18}Nb_{1.82}O_{8.73}$ ceramics with a 96% relative density. These samples demonstrated considerable proton conductivity during humidification, although conductivity values did not exceed the level obtained using other synthesis methods.

## E. ELECTRODE MATERIALS

One of the most amenable materials for freeze-drying synthesis applications are cathode materials for secondary lithium batteries. These materials are rather attractive because they can be used in a powder form, which reduces the number of processing steps. In spite of more than a decade of industrial production of $LiCoO_2$-based cathode materials, many fundamental aspects of their synthesis behavior and electrochemical performance became clear only recently. The mechanism of the influence of chemical prehistory on the cyclability and fade rate of $LiCoO_2$ powders obtained by freeze-drying using various combinations of salt precursors was studied.[47] The key point of this influence is the different rate of crystallographic ordering during thermal processing of $LiCoO_2$ powders, which in turn demonstrated good correlation with various residual structural inhomogeneities caused by different thermal decomposition mechanisms for different precursors.

A very efficient alternative synthesis scheme allowed Huang et al.[94] to obtain a crystallographically ordered high-temperature polymorph of $LiCoO_2$ directly by thermal decomposition of freeze-dried hydroxide precursors at 200 to 300°C. This is quite different from the usually applied continuous processing at 750 to 800°C, which is necessary for conversion from low-temperature modification. An unusual feature of well-ordered cathode material deals with its relatively low electrochemical capacity and considerable capacity decrease during battery cycling, which is more typical for low-temperature $LiCoO_2$ polymorphs. This fact was attributed to the presence of impurities, although, according to recent studies, it may be due to a particle size effect.

Due to the growing interest in the properties of nanocrystalline powders, several efforts have been made to obtain nanocrystalline $LiCoO_2$ cathode materials. The task is complicated by the poor cyclability of the low-temperature polymorph, which is easily obtainable in the nanocrystalline form and by significant grain growth during high-temperature processing, which is necessary to obtain high-temperature modification. One of the first successful efforts to overcome this barrier, using the method of Ding et al.[65] and Shi et al.,[66] is described in Shlyakhtin et al.[58,95] Salt precursors obtained by freeze-drying lithium and cobalt acetates have been mixed in the planetary mill with excess $K_2SO_4$ instead of NaCl. Efficient insulation of small agglomerates of $LiCoO_2$ precursor reduced particle coalescence during processing at 800°C and produced 50 nm to 60 nm crystallites of HT modification of $LiCoO_2$ with reasonable cyclability (Figure 4.16).

Along with investigation of currently applied $LiCoO_2$ cathode material, numerous studies are aimed at the search of a more efficient alternative to this material, which has several important drawbacks. Studies of the formation mechanism and electrochemical properties of $Li_2CuO_2$ were performed on various freeze-dried precursors. This material is considered to be an alternative cathode material that shows a complicated phase formation behavior at reasonable synthesis temperatures. However, $Li_2CuO_2$ degrades easily with atmospheric moisture and $CO_2$ during storage, accompanied by fast fading of its electrochemical capacity.[96]

One of the most competitive series of new cathode materials is based on solid solutions of lithium-nickel and lithium-manganese oxides. $LiNi_{0.5}Mn_{0.5}O_2$, being less expensive than $LiCoO_2$, demonstrates reversible electrochemical capacity values up to 140 mAh/g to 150 mAh/g, which is close to the maximum values for $LiCoO_2$. Synthesis of this cathode material from freeze-dried acetate precursor and analysis of the oxidation states of the constituting elements are described in Shaju et al.[97] Further studies of solid solutions in this system demonstrated the possibility of further enhancement of the reversible discharge capacity values up to 180 mAh/g to 190 mAh/g.

One of the main advantages of crystalline cathode materials with a hexagonal structure, similar to $LiCoO_2$, along with their reasonable capacity and good capacity retention during charge-discharge processes, is their ability to discharge at a more or less fixed voltage value. There is a large group of alternative cathode materials that cannot be effectively discharged at fixed voltage and this limits the scope of their possible applications. However, these materials based on amorphous oxides of transition metals, often demonstrate substantially higher total capacities that sometimes are accompanied by high-capacity retention comparable to crystalline cathode materials. Due to the amorphous nature of these products, a detailed understanding of the processes occurring during charge-discharge cycles and systematic optimization of their synthesis processes are rather complicated. The most typical representative of this family is amorphous $MnO_2$, whose electrochemical performance depends greatly on the processing conditions. One of the most efficient synthesis methods is the soft reduction of sodium

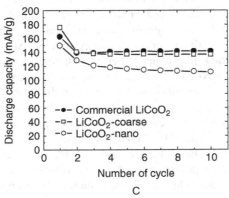

**FIGURE 4.16** SEM micrographs: (A) coarse $LiCoO_2$, and (B) nano $LiCoO_2$. (C) Life cycle performance curves of coarse-grained and nanocrystalline $LiCoO_2$ cathodes.[95]

or potassium permanganates by fumaric acid or its derivatives followed by freeze-drying. This preparation technique produces cryogels with a specific surface area of up to 300 m²/g and capacities up to 300 mAh/g at small discharge rates (C/100). An increase in the discharge current results in a rapid decrease in reversible capacity, although, even at fast discharge rates (2C), capacity remains high (175 mAh/g).[98,99] Direct correlation between specific surface area and electrochemical capacity of amorphous $MnO_2$ shows the kinetic rather than the thermodynamic character of capacity limitation in this kind of material.

Similar electrochemical performance parameters are demonstrated by amorphous vanadium oxide. It was shown recently that capacity of amorphous $V_2O_5$ can be further increased by copper or silver doping. In order to obtain copper-doped $V_2O_5$, the hydrogel of vanadium oxide, obtained by cationic treatment of $NH_4VO_3$ followed by aging, was reacted with copper powder. Further dehydration of this amorphous gel with no traces of metallic copper in a freeze-drier and vacuum drying hood at 100°C resulted in $Cu_{0.1}V_2O_5 \cdot H_2O$ powder capable of accepting up to 2.2 equivalents of lithium during electrochemical intercalation. This material, similar to amorphous $MnO_2$, demonstrated reversible electrochemical capacity up to 300 mAh/g at slow discharge rates (C/22), reduced to 150 mAh/g at high discharge rates (3C), and rather good capacity retention during cycling.[100]

Due to the relatively high charge voltage, most lithium secondary batteries employ nonaqueous liquid electrolytes. The functioning of $MnO_2$-based lithium cathode materials in the battery with aqueous electrolyte (9M KOH) is possible, but the range of working voltages is considerably lower (−1 to +0.3 V vs. Hg/HgO reference electrode). Comparison of the electrochemical performance of bismuth-doped $LiMn_2O_4$, obtained by thermal decomposition of freeze-dried acetates, and λ-$MnO_2$, obtained by acid (diluted HCl) leaching of Bi-$LiMn_2O_4$, demonstrated an obvious advantage (75 mAh/g and 150 mAh/g, respectively), although in both cases, after 20 to 25 cycles a rapid capacity decrease was observed.[101]

Along with lithium secondary batteries, considerable progress has been made recently in the field of primary portable energy sources. Application of $K_2FeO_4$ cathode instead of $MnO_2$ in primary batteries with zinc anode and KOH electrolyte results in an approximate 50% energy capacity increase. This new kind of battery attracted a lot of attention and were given the name "super-iron batteries."[102] During further development, several other ferrates have been obtained ($ZnFeO_4$, $MgFeO_4$) whose capacities exceed that of $K_2FeO_4$. However, $BaFeO_4$ demonstrated a lower capacity but better C-rate performance (capacity retention at high discharge rates). The $MeFeO_4$ compounds for these studies were obtained by freeze-drying of precipitates, formed during mixing of aqueous solutions of $K_2FeO_4$ and corresponding Me acetates. Application of the low-temperature dehydration technique (freeze-drying) prevented partial decomposition of hydrated ferrates.

A small number of papers have appeared recently on sensor applications of freeze-dried materials, although freeze-drying is one of the best possible methods for obtaining nanocrystalline $SnO_2$, a popular semiconductor sensor for toxic and flammable gas detection. One such sensor, obtained by sintering of freeze-dried nanocrystalline $SnO_2$ powders, demonstrated high efficiency in CO detection.[103] An interesting feature observed in this work is dependence of sensitivity of the ceramic sensor on the powder compaction pressure before sintering. This dependence has been attributed to the magnification of Shottky's barriers and the increasing number of sites available for the adsorption of ionic gas species.

## F. High-Temperature Superconductors

After 10 years of intensive studies, the attention given this group of materials has been considerably reduced recently, although freeze-drying remains the popular preparation method for HTSC powders with enhanced reactivity and chemical homogeneity.[59,60,104–115,117–119] Most of these studies dealt with bismuth-based superconductors because of their growing number of applications and their complicated phase formation processes. Direct experimental evaluation of the reactivity of $(Bi,Pb)_2Sr_2Ca_2Cu_3O_x$ precursor powders, obtained from the same nitrate solution by different processing methods (freeze-drying, spray drying, thermal evaporation drying) demonstrated that the rate of target phase formation from pelletized precursors under fixed processing conditions is significantly higher for freeze-dried precursor. The authors attribute this difference to the higher chemical homogeneity and smaller particle size of cryochemically processed precursors.[104] A more quantitative approach to the evaluation of particle size effect on the reactivity of (Bi,Pb)-2223 precursors was demonstrated by Shlyakhtin et al.[59] and Ischenko et al.,[60] where precursor powders with different grain sizes were directly obtained from the same salt precursor using the powder engineering technique.

Another interesting method to accelerate formation of $(Bi,Pb)_2Sr_2Ca_2Cu_3O_x$ was proposed by Badica et al.,[111] where partial substitution of strontium with barium resulted in 80% formation of the target phase during annealing of the precursor powder, although intense formation of this phase occurs only in pressurized species (pellets, thermomechanically processed composite tapes). At the same time, efforts to apply this technique during $(Bi,Pb)_2Sr_2Ca_2Cu_3O_x$ ceramics production demonstrated its very limited efficiency in this case, even in combination with advanced processing methods [reactive field assisted sintering technique (RFAST)].[108,112,114]

A promising direction for further development of bulk HTSC materials is based on the enhancement of intragrain critical current densities by means of artificial creation of magnetic pinning centers using particles of alien phases, creating a number of defects at the interface with the HTSC matrix. Maximum concentration of the pinning centers is created in the HTSC composites with uniformly distributed fine particles of chemically inert materials by using freeze-drying synthesis methods. The efficiency of this approach was demonstrated during preparation of melt-processed composites $Bi_2Sr_2CaCu_2O_x$-$SrZrO_3$. Due to chemical resistance of $SrZrO_3$ to Bi-2212, thermal decomposition of the freeze-dried $Bi_2Sr_2CaCu_2O_x$ salt precursor containing an excess of zirconium and strontium resulted in formation of the fine mixture of target phases. Further melt processing of this powder led to an HTSC composite with uniformly distributed 0.1 μm to 0.2 μm $SrZrO_3$ particles. The comparison of HTSC properties of this composite with undoped Bi-2212 ceramics obtained using the same processing technique demonstrated a threefold enhancement of critical current densities for composite ceramics.[115]

The high chemical stability of $SrZrO_3$ to HTSC cuprates was used for the production of multifilamentary HTSC tapes.[116] One of the most serious problems in making these materials is efficient insulation of HTSC filaments from each other. Along with the chemical stability to silver and HTSC cuprates, the insulation material should not interrupt the formation of superconducting chains within filaments during thermomechanical processing of tapes. Ultrafine (less than 0.1 μm) powder of $SrZrO_3$ obtained by freeze-drying synthesis and mixed with silver-coated HTSC filaments before thermomechanical processing ensured the formation of a continuous and uniform insulating barrier between HTSC filaments. Formation of this barrier substantially reduced alternating current (AC) coupling losses in the multifilamentary HTSC tapes and increased critical current density compared to similar tapes obtained using other insulation materials.

Freeze-drying synthesis of yttrium and rare earth barium cuprates was aimed mostly at the fundamental studies of HTSC properties. $YBa_2Cu_3O_x$ ceramics with a high degree of chemical and morphological homogeneity were used for comparative analysis of intergrain, intragrain, and intrinsic magnetic losses.[117] The effect of substitution of barium with strontium in $RBa_2Cu_4O_8$ (R = Gd, Ho) was investigated using freeze-dried precursors.[118] Fine (120 nm to 200 nm) powders of $YBa_2Cu_3O_x$ with a cryochemical prehistory were found to consist of approximately 40 nm domains with a mutually orthogonal orientation.[119]

## G. CATALYSTS

Various materials for catalytic applications are also traditional objects of freeze-drying synthesis because of the ability of this method to produce finely dispersed particles with an enhanced specific surface area and to promote formation of complex oxides at reduced temperatures. In order to ensure a minimum grain size, the coprecipitation methods are used in most cases, although low-temperature decomposition of the salts of lowest carbonic acids can also be accompanied by the formation of nanoparticles. So thermal decomposition of freeze-dried acetates of cerium and calcium at 300°C produces ceria and calcium-doped ceria powders with a grain size of 7 to 8 nm. These powders are poorly agglomerated and demonstrate good catalytic activity in the reaction of oxidative ethane dehydrogenation. An interesting feature of undoped ceria obtained by this method is very good morphological stability in the reductive reaction environment at 700 to 750°C, while processing of the powder in air at the same temperature is accompanied by significant grain growth.[120]

Formation of complex lanthanum and rare earth manganites during thermal decomposition of acetate precursors occurs at higher temperatures of 600 to 700°C, although in this case the grain size of these powders is rather small (20 to 40 nm).[51,52,121] Interest in the unusual physical properties of these materials, already mentioned in Section III.B, appeared only recently, while their high catalytic activity in oxidation reactions has been studied for many years. The influence of strontium and potassium substitution in lanthanum and neodymium manganites on their catalytic performance in the reaction of $H_2O_2$ decomposition at room temperature was studied.[52] It was shown that high catalytic activity has no direct correlation with experimentally observed surface enrichment of nanopowders with potassium, but correlates quite well with the number of oxygen vacancies. In the case of ethane combustion, potassium substitution causes a decrease in catalytic activity; with a greater potassium content, the selectivity of catalyst is increased.[51,121] The effect of surface enrichment by potassium and manganese was observed for manganite powders obtained at 600°C, although further annealing at 800°C causes homogenization of the material accompanied by significant grain growth.

Another well-known group of oxidation reaction catalysts is the complex cobaltates of lanthanum. However, a large number of their fundamental properties are still under investigation. In order to study oxygen mobility by the isotope exchange method and its role in the mechanism of catalytic oxidation on $La_{1-x}Sr_xCoO_3$ and $LaSrCoO_4$, single phase homogeneous samples of these phases have been obtained by thermal decomposition of nitrate precursors. However, application of carbon-containing precursors was restricted by the demands of the isotope exchange experimental conditions. The obtained materials showed high catalytic activity in methane and CO oxidation reactions.[122] It was also established that catalytic methane oxidation occurred not by adsorbed but by intrinsic surface oxygen atoms of cobaltates and the catalytic activity of these materials was determined by the structural oxygen mobility.[122]

Studies of the catalytic properties of $LaRuO_3$ were complicated by the relatively low stability of $Ru^{3+}$ and the volatility of ruthenium oxides at the phase formation temperature. This problem was solved successfully using freeze-drying synthesis. The phase formation temperature was reduced not so significantly (850 to 900°C), but formation of stable ruthenium-containing intermediates during thermal decomposition prevented $RuO_x$ losses and retained cation stoichiometry during synthesis. A study of the catalytic activity of these materials demonstrated their high efficiency in the low-temperature (100 to 200°C) oxidation of hydrocarbons, CO, CO + NO mixtures, etc.[123]

The traditional scheme of freeze-drying synthesis is based on the freeze-drying of true aqueous solutions, although recently freeze-drying of coprecipitated residues has been applied more frequently. Freeze-drying of suspensions with one of the components as solid particles and the other as a solution was considered undesirable due to poor chemical homogeneity. As was mentioned in Section II.C, the temperatures of phase formation (600 to 700°C) of lanthanum-strontium transition metal-based perovskites (manganites, chromites, aluminates) obtained from suspension- and true solution-derived precursors are almost the same. The catalytic properties of the products obtained from suspensions are also very reasonable.[50] It was found that such a synthesis approach has its own advantages, as the morphology of the product is substantially controlled by the morphology of the lanthanum-containing residue, so that these products are less agglomerated than those obtained from freeze-dried true solutions. This method is especially suitable for the preparation of nickel-cobalt-based perovskites, ensuring not only low-temperature synthesis but also the enhanced thermal stability of the powder catalyst.

Another way to increase the specific surface area of catalytically active materials and to raise their thermal stability at elevated temperatures is the impregnation technique. In the case of cryoimpregnation, the porous substrate before freeze-drying is wetted by solution containing catalyst components, so that coalescence of the catalyst particles during synthesis and following thermal processing is substantially limited. Application of this technique[13] enhanced the catalytic activity of the multicomponent $La_{0.65}Sr_{0.35}Ni_{0.29}Co_{0.69}Fe_{0.02}O_3$ combustion catalyst compared to finely dispersed unsupported perovskite powder. In addition to the considerably enhanced uniform distribution of catalytic components in the bulk of the porous substrate, application of cryoimpregnation instead of thermal drying demonstrated another advantage. Application of freeze-drying reduces the contact time of the solution with the substrate and hence minimizes dissolution, which may lead to undesirable component redistribution.[15] In the case of $MoO_3$ catalyst on alumina substrate, the formation of coarser $MoO_3$ crystallites was observed and this led to a reduction in catalytic activity.

The cryoimpregnation technique can also be efficiently used for the distribution of metal and metal alloy catalysts on the surface of various thermostable substrates. During the preparation of these materials, thermal decomposition of precursor coating is accompanied by hydrogen reduction. This method leads to efficient electrocatalysts for prototypes of room-temperature fuel cells.[14,124] The different stabilities of platinum chlorides and $WO_3$ to reduction at elevated temperatures leads to a composite platinum/$WO_3$ coating on the carbon substrate by using this method, and this composite exhibited higher catalytic activity due to the oxygen spillover effect.[14] In this case, direct thermal processing of freeze-dried wetted substrate in hydrogen is more efficient than reduction after preliminary thermal decomposition, leading to both an enhanced active surface area of catalyst and to a larger number of Pt-$WO_3$ particle contacts necessary for the spillover process. In order to further reduce the grain size of the nanocrystalline coating, nonaqueous solutions of organometallic noble metal compounds can be used.[125] The application of nonaqueous solvents results in the need to perform freeze-drying at low temperatures (220 to 240 K), thus increasing the process duration, although this technique has demonstrated good efficiency in the preparation of nanocrystalline multimetal (alloy) coatings for catalytic applications. It was found that platinum and palladium reduce the decomposition temperatures of other, more stable noble metal components of the catalyst. Similar processes occur during redox processing of unsupported freeze-drying products at reduced temperatures, and hence this method can be used for the preparation of very fine noble metal alloy nanopowders with high chemical homogeneity. Similar results (2 nm platinum powders; platinum/ruthenium and platinum/ruthenium/nickel alloy/composite powders with 2 to 5 nm grain sizes and high catalytic activity) have been demonstrated by using liquid phase reduction of Me chloride solutions by $LiBH_4$ or $NaBH_4$ followed by freeze-drying of aqueous suspension of washed Me nanoparticles.[126,127]

Along with particle size, an important morphological parameter of the powder catalyst is its porous structure. In order to ensure effective functioning of the catalyst, the powder should have extended mesoporosity, usually correlated with high specific surface area, and a considerable number of micropores for fast mass transport of gaseous or liquid reagents to/from the catalytically active surface within particles of catalyst. Several studies on the influence of drying conditions of precipitated hydroxides on the porous structure of oxide powders, performed independently by various authors,[128-132] demonstrated that thermal drying, apart from freeze- and supercritical drying, always results in a considerable decrease in the volume of micropores. Dependence of the mesoporosity on the drying method is more complicated and varies not only with precipitated cations, but also with precipitation method.[129] Sometimes thermal drying leads to powders with larger surface areas compared to freeze-dried powders.[128,129,131] The main reason for this effect can probably be explained by thermal processing of powders, which is necessary for conversion of hydroxides to target oxides. Formation of interparticle bridges during thermal drying stabilizes soft agglomerates of primary crystallites in some cases and retains intragranular mesoporosity while non-bridged, freeze-dried, soft agglomerates are collapsed. Nevertheless, this tendency is not universal, and for a large number of systems, freeze-drying of precipitated residues results in formation of cryogels that are comparable with aerogels not only with respect to specific surface area but also by their stability at elevated temperatures.[132] The formation of a multimodal pore size distribution during pillaring of natural clays by $H_2TiCl_6$ followed by freeze-drying was observed.[133] The $V_2O_5$ catalyst produced using freeze-drying demonstrated a significant increase in catalytic activity (catalytic removal of NO) and enhancement of thermal stability of the catalyst compared to commercial $V_2O_5$ catalyst on $TiO_2$ support.[133]

## H. MATERIALS WITH HIGH POROSITY

One of the fastest developing directions in cryochemical synthesis is the production of porous ceramics. Recent studies have been directed toward design of the porous structure of ceramics for specific applications by means of directed influence on the precursors and intermediates during synthesis (Figure 4.17). A useful instrument for the development of the oriented porous structure in ceramics is directed ice crystallization, caused by unidirectional freezing (see Section II.A).[31-33,134] The mold's micromorphology is kept during freeze-drying and following thermal processing without collapse, leading to the formation of macropores along the direction of ice growth.[33,134] Careful selection of processing conditions allows one to vary independently the porous structure of ceramics and particle morphology.[31] One of the instruments of directed influence on the micromorphology of the molds is freeze gelation, leading to gelification of sols after the freezing and thawing cycle. This process in combination with freeze-drying can be applied for producing biocers, new kind of composites for immobilization and conservation of living microorganisms for biocatalytic purposes.[34] A process of freeze-casting based on the freeze gelation method can be successfully used

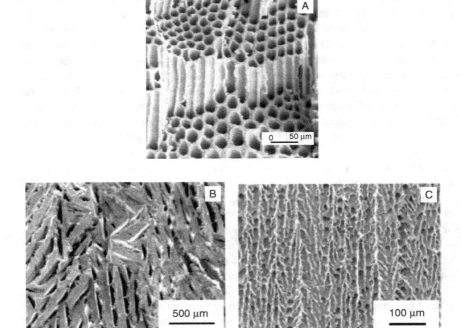

**FIGURE 4.17** Samples of ceramics with oriented porosity obtained using cryochemical processing: (A) TiO$_2$;[135] cross-sections of Al$_2$O$_3$ ceramics perpendicular (B) and parallel (C) to the macroscopic ice growth direction.[33]

for increasing powder loading during production of complex-shaped silica, alumina, and composite ceramics.[30]

One of the main problems during production of porous ceramics is cracking of samples in the course of processing.[31,32,135,136] As was mentioned earlier, substitution of water with tert-butanol reduces mechanical stresses in the dried mold, although the efficiency of the application of this method is rather different for different solvents. Porous SiO$_2$ ceramics obtained using directional freezing followed by soaking with tert-butanol was morphologically uniform.[31] Tert-butanol was used in the formation of porous, low-dielectric constant SiO$_2$ films for ultra-large-scale (ULSI) technologies, although excess of tert-butanol in the starting mixture caused cracking.[137] At the same time, efforts to apply freeze-drying for the processing of Al$_2$O$_3$, TiO$_2$, and hydroxyapatite porous ceramics directionally solidified by the alginate gelation method resulted in cracking during sintering in spite of soaking the gels in tert-butanol.[135] However, crack-free porous hydroxoapatite ceramics for ion exchange treatment of waste water was successfully prepared using freeze-drying without additional tert-butanol processing, while air drying of the same slurry resulted in extended cracking.[138] Similarly, freeze-drying of the porous silicon film obtained at the surface of the silicon substrate by hydrofluoric acid etching caused no cracking without additional precautions. Thermal drying of this film was accompanied by morphological degradation and peeling of the porous layer.[139] This comparison shows that the problem of cracking cannot be solved by a single universal approach; instead, careful selection and optimization of a number of gel preparation and processing variables are needed.

Methods of directed formation of the porous structure can also be used for processing of nonoxide materials. Application of the freeze gelation effect resulted in a molded macroporous silica matrix that was infiltrated by phenolic resin.[140] Carbonization by thermal processing of the obtained composite in argon at 850°C followed by sintering at 1500°C resulted in formation of net-shaped macroporous SiC ceramics. Similarly, porous $Si_3N_4$ ceramics were produced from a slurry containing $Si_3N_4$ powder, sintering aids, and dispersants after casting into a cylindrical mold, freezing unidirectionally, freeze-drying, and sintering in a nitrogen atmosphere at 1700 to 1850°C.[141] These ceramics contained vertically aligned microscopic pores along with a large number of fibrous grains protruding from the internal walls of the $Si_3N_4$ matrix.

Taking into account the high-temperature applications of many kinds of porous ceramics, the thermal stability of the porous structure is rather important for their effective functioning. Analysis of the morphological evolution of the porous $SiO_2$ ceramics obtained from freeze-dried tetraethoxysilane (TEOS) hydrolysis products demonstrated that mobility of $SiO_2$ crystallites in this structure occurs far below the crystallization point (approximately 1100°C): the collapse of the mesopores is observed at temperatures of 800°C or more, while more coarse and stable interagglomerate micropores disappeared at 1000°C.[142] Along with the initial micromorphology, the stability of the porous structure depends also on the diffusion mobility of the ceramic constituents at a given temperature, usually correlating with the melting point of the solids. Porous MgO ceramics ($T_m$ = 2825°C) obtained from freeze-dried $MgSO_4$ solution retained a rather high specific surface area ($S_{BET}$ = 20 m²/g) after 20 hours of sintering at 1300°C.[143]

As mentioned before, applications of freeze-drying in polymer science are numerous and varied, and are beyond the scope of this chapter. Because carbon has a semi-inorganic character, freeze-drying synthesis of mesoporous carbon is briefly presented in this chapter. Freeze-drying is a less expensive alternative to preparation of carbon aerogels. Freeze-drying of resorcinol-formaldehyde polycondensation products followed by simple pyrolysis in $N_2$ at 1000°C led to carbon powders with an $S_{BET}$ greater than 800 m²/g, approaching that of aerogels (1000 to 1200 m²/g).[144-148] An additional advantage of the carbon cryogels is that they have considerable microporosity, which is usually generated during thermal processing of precursors. Similar to inorganic cryogels, rinsing of polycondensation products with tert-butanol promotes better stability of the porous structure of the product. Another similarity with inorganic gels is the high sensitivity of the product's micromorphology to the gel drying conditions.[146]

## I.  REFRACTORY CERAMICS AND COMPOSITES

A very limited number of papers have been published recently on the freeze-drying of oxide refractories. Modern cryoprocessing technique was applied to alumina processing.[35] The freeze-casting method was first used on stabilized $Al_2O_3$ suspension followed by freeze-drying for the preparation of highly dense $Al_2O_3$ ceramics with a uniform microstructure. Unusually high loading of alumina (57.5%) and absence of cracking were achieved by addition of a large amount of glycerol (up to 20%) to the starting slurry. In some cases, freeze-drying was used as part of a solid state synthesis method in order to prevent component redistribution in the slurry after homogenization of the starting mixture in a ball mill. Such an approach also led to dense (93%) $B_4C$ ceramics using TiC as a sintering aid[78] and dense particulate-reinforced ceramic matrix $SiC-TiB_2$ composites.[149]

Highly functional ZnO-based varistor ceramics can be prepared with preferential localization of dopants at the ZnO grain boundaries. In order to obtain such a material by surface doping of ZnO grains, CuO/ZnO composite ceramics were obtained by freeze-drying of ZnO slurry in copper acetate solution. A ZnO-based solid solution was formed with a CuO content of less than 1%, while with a higher copper content, CuO inclusions formed in ZnO grain boundaries, resulting in good nonlinear electrical properties of this ceramic. An increased density of ceramics with increasing amounts of copper at temperatures of 850°C or more showed that CuO acts as a sintering aid.[150–152] Several other electroceramic composites ($Bi_2Sr_2CaCu_2O_x/SrZrO_3$, $La_{0.7}Sr_{0.3}MnO_3/SrZrO_3$, $La_{0.7}Sr_{0.3}MnO_3/SrTiO_3$) were discussed earlier.

## J.  FINE AND NANOCRYSTALLINE POWDERS

A growing number of studies deal with the cryochemical synthesis of various oxide fine powders, regardless of their applications, with a great deal of attention to their particle size, micromorphology, and mechanism of formation. The mechanism of nanocrystalline alumina formation was investigated using freeze-dried precursors obtained from $Al(NO_3)_3$ solution by $NH_4OH$ precipitation or by OH-anionic exchange treatment to a pH of 7. These sol-derived powders showed considerably smaller crystallographic and morphological evolution, which was attributed to their very narrow particle size distribution.[28]

Significant attention was recently focused on the morphological evolution of powders during thermal processing of precursors. One of the model oxides for such studies is ferric oxide. Directed influence on the aggregate structure of its precursors followed by thermal processing led to powders and ceramics with variable grain sizes.[55,57] An important feature of oxide powders obtained by thermal decomposition of freeze-dried salt precursors is their fractal properties. The evolution of the fractal properties during thermal processing of powders and correlation of fractal dimensions of the powder surface with chemical prehistory and the decomposition temperature of the precursor were discussed.[153–156]

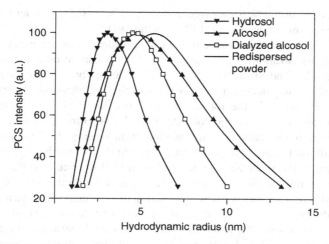

**FIGURE 4.18** The hydrodynamic particle radii distribution of initial $SnO_2$ sols and redispersed powder.[159]

The synthesis of nanocrystalline oxide powders using freeze-drying is based mainly on the processing of various kinds of hydroxide precursors. Intercalation of tetraalkylammonium ions into hydrated manganese oxide followed by washing causes delamination of oxide particles. The product of freeze-drying of this suspension is nanostructured flake-like manganese oxide particles.[157] Thermoreduction of freeze-dried coprecipitated V-Fe hydroxides was accompanied by significant grain growth, although even in this case the size of the $Fe_{3x}V_xO_4$ powders didn't exceed 100 nm.[158]

Applications of nanocrystalline powders demand their ability to be redispersed and to form stable colloidal solutions after drying. A synthesis of fully redispersable $SnO_2$ nanopowders from products of $SnCl_4$ thermohydrolysis in the ethanolic solution have been described (Figure 4.18).[159] In another paper it was shown that resdispersability of cationically modified $SiO_2$ nanopowders correlates with their agglomerate structure.[160] Introduction of substances that are traditionally used as cryoprotectants (glycerol, trehalose) promoted deagglomeration of precipitated particles and resulted in considerable improvement in their suspendibility with a minimum negative influence on their DNA-binding capacity. Modification of the anatase powder surface by $V_2O_5$ using cryoimpregnation also promotes better stability of $TiO_2$ suspensions.[161]

A method to enhance poor flowability of nanocrystalline powders has been proposed.[162] Pneumatic spraying of $SiO_2$ and $TiO_2$ powder suspensions stabilized by organic surfactants into liquid nitrogen followed by freeze-drying yielded 0.2 to 0.7 mm flowable granules with very low strength. The strength of the freeze-dried granules differs greatly from widely used thermal spray drying granulation. The freeze granulation method was applied also for the production of free-flowing ammonium nitrate powder.[163] Boron redistribution during the thermal drying process was prevented by using freeze-drying of $SiO_2$-$B_2O_3$ powder precursors for polarization-maintained optical fiber application.[164]

Along with artificial rearrangement of the powder system by means of granulation, self-organization of nanoparticles in liquids can be used for the design of new particles and structures. It has been shown that self-organization of nanocrystalline $TiO_2$ nanoparticles during freezing of ultrasonically processed suspension can result in the formation of capped microtubes or flat sheets.[165] Similar self-organization of the colloidal nanosheets formed during exfoliation of a layered protonic titanate by intercalation of quaternary alkylammonium ions results in different kinds of lamellar structures that depend on the drying method. Slow heating of the suspension in air leads to a well-ordered lamellar structure, while fast freezing and freeze-drying of the suspension promote formation of flat thin flakes that remain almost unchanged during transformation of hydrated titania to anatase during heating.[166] The crystallization of anatase and the transformation of anatase to rutile were found to be rather sensitive to the structural organization of titania particles. Thus better preservation of the voluminous structure of hydrated titania sol during freeze-drying compared to oven drying not only increases the specific surface area of the powder, but results in anatase crystallization and anatase-to-rutile conversion at higher temperatures.[167,168] It was also mentioned that the presence of residual chlorides in cryogels promotes anatase crystallization at lower temperatures.[167]

Grain growth of zinc oxide occurs at much lower temperatures than in titania, so that production of fairly crystalline ZnO nanopowders needs special precautions. The thermal decomposition of freeze-dried $Zn(NO_3)_2$ at 260 to 270°C results in the formation of poorly crystalline, but coarse-grained ZnO powders.[107] At the same time, insulation of ZnO particles in the inert NaCl matrix by the method described before[65,66] led to no change in the size of the ZnO particles. The ZnO particles were obtained from freeze-dried precipitate of zinc hydroxide of 15 to 20 nm, and this size was maintained at least up to 600°C.[169]

Along with single oxides, several studies have been performed in order to obtain fine powders of complex oxides. The formation of fine $BaTiO_3$ powders from carbon-containing precursors is often complicated by its poor crystallization and easy formation of $BaCO_3$ intermediate due to the presence of large amounts of $CO_2$ and $H_2O$ gaseous products during thermal decomposition. This usually causes an increase in the thermal processing temperature to 800 to 900°C,[48,170] although optimization of thermolysis conditions occurs with a decrease in temperature to 700 to 750°C.[171] The usual size of the $BaTiO_3$ crystallites (50 to 200 nm), consisting of paraelectric cubic polymorph, is determined both by the chemical prehistory of the powder and by the final temperature of thermal processing. Formation of fine, bulky $BaZrO_3$ powders with a skeletal structure from freeze-dried ethylenediaminetetraacetate acid (EDTA) precursors readily occurs at 700°C.[172] An example of the successful combination of various processing techniques is given by Fang et al.,[49] where transformation of the starting solution of lead and titanyl nitrates into a microemulsion reduced the size of $PbTiO_3$ crystallites to about 60 nm, although even in this case the product was significantly agglomerated. In this case, the formation of individual lead and titanium oxides during thermal decomposition of homogeneous precursor was observed, although synthesis of tetragonal $PbTiO_3$ was achieved at 650°C. Apart from $PbTiO_3$, formation of semiamorphous $NiFe_2O_4$ from freeze-dried acetates occurs in a single step at considerably lower temperatures (300°C), so that the nanocrystalline thermolysis product has a specific surface area of about 500 m$^2$/g. Crystallization of ferrite at 400 to 450°C is accompanied by significant grain growth ($S_{BET}$ = 50 m$^2$/g).[56]

During the synthesis of fine powders of salts and nonmetallic compounds, freeze-drying is usually applied as a simple, safe, and efficient extraction method of solid particles from liquid synthesis medium. This approach was used during the synthesis of nanopowders of aluminum polyphosphate,[173] dimers of capped CdS nanoparticles,[174] and fine mixture of $Al_2O_3$ and SiC powders, the latter used as a feedstock material for plasma-sprayed composite coatings.[175] Taking into account the high temperatures of SiC formation, freeze-drying synthesis from $SiO_2$ sol and sugar solution has proved to be a very useful method for the preparation of SiC nanopowders (D = 80 to 100 nm, calculated from $S_{BET}$).[176] One of the traditional application areas of the freeze-drying process is in the safe production of explosives. Thus freeze-drying of the polymer gel obtained by wet mixing of resorcinol, formaldehyde, and hydrazinium perchlorate avoids the potentially harmful procedure of dry mixing the components.[177]

## K.  HOMOGENEOUS PRECURSORS FOR FUNDAMENTAL STUDIES

The relatively high chemical homogeneity of cryochemical precursors and the simplicity of synthesis procedures prompted the appearance of many papers where freeze-dried precursors were used to obtain reliable and reproducible fundamental information about solid state substances and processes. Cryochemical processing was applied to the synthesis of recently discovered substances for detailed study of their structure and magnetic properties. These examples include $Sr_3CoSb_2O_9$, $Sr_2CoSbO_6$,[178] and $La_{0.85}Ca_{0.15}MnO_3$[179] obtained by thermal decomposition of freeze-dried acetate or nitrate-acetate precursors. Freeze-drying led to a large group of ferrates ($SrFeO_4$, $MgFeO_4$, $CaFeO_4$, $ZnFeO_4$, $Ag_2FeO_4$,[102] $Li_2FeO_4$, $Na_2FeO_4$, alkylammonium ferrates[180]) that could be difficult to obtain by other methods due to their potentially low thermal stability. Anionic treatment of aluminum and chromium nitrate solution followed by freeze-drying produces the best-known precursor for ruby powder synthesis ($T_{synt}$ = 950°C).[29] Ammonolysis of freeze-dried precursors was shown to be a useful preparation method of binary (V-Cr, Cr-Mo, Ni-Mo, V-Mo) oxynitrides.[181–183]

The cryochemical method is also rather useful for the preparation of various metastable compounds. Thus thermal decomposition of freeze-dried oxalate complexes resulted in the formation of metastable monoclinic $CuNb_2O_6$ and $CuTa_2O_6$ with trirutile structure.[184] An unusual polymorph of FeOOH—akageneite—was found during analysis of freeze-dried hydrated ferric oxides obtained from different precursors under various precipitation conditions.[185,186] Soft oxidation of the nanocrystalline $V_xFe_{3x}O_4$ ($0 \leq x \leq 2$) powders produces metastable oxides with a structure of cation-deficient spinel.[158] Thermolysis of newly obtained complex cyanide $Cu_{1.5}[Fe(CN)_6]*6H_2O$ in argon leads to the formation of nanocrystalline powders of Cu-Fe alloy, while further heat treatment results in their decomposition to individual body-centered cubic (BCC) iron and face-centered cubic (FCC) copper particles.[70]

As for thermodynamically stable compounds, high chemical homogeneity of freeze-dried precursors reduces kinetic problems during synthesis. It is often used for determination of solid solution boundaries in phase diagrams of solid state systems, especially containing several oxides. In the last several years freeze-drying synthesis has been used for refinement of solid solution ranges of $Ba_2YNbO_6$,[187] $Ba_3Cu_{1+x}Nb_{2-x}O_{9-1.5x}$,[184] $La_{1-x}Cr_xMnO_3$, and $La_{1-x}Co_xMnO_3$.[188] It was found that $SrLaGaO_4$ and $SrLaAlO_4$ form continuous solid solutions with a tetragonal structure. The closeness of the lattice parameters to $YBa_2Cu_3O_x$ makes them a potential substrate material for HTSC films.[189] However, in spite of the homogeneity of salt precursors, no significant solubility of Co in $SrCeO_3$ was found.[190]

Several studies were aimed at the study of phase formation processes during the thermal decomposition of freeze-dried precursors. A precise and detailed analysis of manganese ferrite and La-Sr manganite synthesis from freeze-dried carboxylates can be found in Langbein et al.[191] and Boerger et al.[192] The formation of magnesium aluminum spinel from nitrate precursors is discussed in detail by McHale et al.[193] The excess enthalpy of $MgAl_2O_4$ powder obtained at 800°C compared to coarse-grained product annealed at 1500°C was experimentally evaluated as 40 kJ/mol. Fine $LiFeO_2$ and $Li_{0.5}Fe_{2.5}O_4$ powders obtained from freeze-dried precursors were used for the realization of the newly discovered solid state exchange reaction between lithium ferrites and $ZnSO_4$, which results in the formation of $ZnFe_2O_4$.[194]

## REFERENCES

1. Landsberg, A., and Campbell, T.T., *J. Metals,* 856, 1965.
2. Johnson, D.W., and Schnettler, F.J., *J. Am. Ceram. Soc.*, 53, 440, 1970.
3. Gallagher, P.K., and Schrey, F., *Thermochim. Acta*, 1, 465, 1970.
4. Kim, Y.S., and Monforte, F.R., *Ceram. Bull.*, 50, 532, 1971.
5. Roehrig, F.K., and Wright, T.R., *J. Vac. Sci. Technol.*, 9, 1368, 1972.
6. Anastasyuk, N.V., Oleinikov, N.N., and Tretyakov, Y.D., in *Synthesis of Inorganic Materials in Non-Aqueous Media*, Nauka, Moscow, 1971.
7. Anastasyuk, N.V., Studies of the effectiveness of chemical methods of ferrite preparation, Ph.D. dissertation, Moscow State University, Moscow, 1972.
8. Tretyakov, Y.D., Oleinikov, N.N., Anastasyuk, N.V., and Pershin, V.I., Patent USSR 1937660/29-33, 1973.
9. Schnettler, F.J., Monforte, F.R., and Rhodes, W.W., *Sci. Ceram.*, 4, 79, 1968.
10. Krishnaraj, P., Lelovic, M., Eror, N.G., and Balachandran, U., *Physica C*, 215, 305, 1993.

11. Perriat, P., Fries, E., Millot, N., and Domenichini, B., *Solid State Ionics*, 117, 175, 1999.
12. Vergunst, T., Kapteijn, F., and Moulijn, J.A., *Appl. Catal. A*, 213, 179, 2001.
13. D. Klvana, J. Delval, J. Kirchnerova, J. Chaouki, *Appl. Catal. A*, 165, 171, 1997.
14. Chen, K.Y., Sun, Z., and Tseung, A.C.C., *Electrochem. Solid State Lett.*, 3, 10, 2000.
15. Carrier, X., Lambert, J.-F., Kuba, S., Knozinger, H., Che, M., *J. Mol. Struct.*, 656, 231, 2003.
16. Sergeev, G.B., Nemukhin, A.V., Sergeev, B.M., Shabatina, T.I., and Zagorskii, V.V., *Nanostruct. Mater.*, 12, 1113, 1999.
17. Tretyakov, Y.D., Oleinikov, N.N., and Shlyakhtin, O.A., *Cryochemical Technology of Advanced Materials*, Chapman & Hall, London, 1997.
18. Perez, R.J., Jiang, H.G., Dogan, C.P., and Lavernia, E.J., *Metal. Mater. Trans. A*, 29A, 2469, 1998.
19. Guo, Y.C., Liu, H.K., Liao, X.Z., and Dou, S.X., *Physica C*, 301, 199, 1998.
20. Eggenhoffner, R., Tuccio, A., Masini, R., Diaspro, A., Leporatti, S., and Roland, R., *Supercond. Sci. Technol.*, 10, 142, 1997.
21. Franks, F., *Water and Aqueous Solutions at the Temperatures Below 0°C*, Naukova Dumka, Kiev, 1986.
22. Naumov, S.V., Physico-chemical processes during cryocrystallization of water-salt systems, Ph.D. dissertation, Moscow State University, Moscow, 1982.
23. Naumov, S.V., Mozhaev, A.P., and Tretyakov, Y.D., *Russ. J. Phys. Chem.*, 56, 440, 1982.
24. Golovchansky, A.V., Physico-chemical processes during freezing of water-salt systems with high cooling rates, Ph.D. dissertation, Moscow State University, Moscow, 1987.
25. Shlyakhtin, O.A., Kulakov, A.B., Badun, Y.V., and Tretyakov, Y.D., *Trans. Mater. Res. Soc. Japan*, 14A, 7, 1994.
26. Gomez, G., Pikal, M.J., Rodriguez-Hornedo, N., Effect of initial buffer composition on pH changes during far-from-equilibrium freezing of sodium phosphate buffer solutions, *Pharm. Res.*, 18, 90, 2001.
27. Sharikov, F.Y., Cryochemical synthesis of finely dispersed oxide powders using ion exchange, Ph.D. dissertation, Moscow State University, Moscow, 1991.
28. Mamchik, A.I., Kalinin, S.V., and Vertegel, A.A., *Chem. Mater.*, 10, 3548, 1998.
29. Eliseev, A.A. Lukashin, A.V., and Vertegel, A.A., *Chem. Mater.*, 11, 241, 1999.
30. Koch, D., Andresen, L., Schmedders, T., and Gratwohl, G., *J. Sol-Gel Sci. Technol.*, 26, 149, 2003.
31. Mukai, S.R., Nishihara, H., and Tamon, H., *Microporous Mesoporous Mater.*, 63, 43, 2003.
32. Kisa, P., Fischer, P., Olszewski, A., Nettleship, I., and Eror, N.G., Preprint, Materials Research Society Fall Meeting, 2003.
33. Fukasawa, T., Deng, Z.-Y., Ando, M., Ohji, T., and Goto, Y., *J. Mater. Sci.*, 36, 2523, 2001.
34. Soltmann, U., Bottcher, H., Koch, D., and Gratwohl, G., *Mater. Lett.*, 57, 2861, 2003.
35. Sophie, S.W., and Dogan, F., *J. Am. Ceram. Soc.*, 84, 1459, 2001.
36. Laurie, J., Bagnall, C.M., Harris, B., Jones, R.W., Cooke, R.G., Russell-Floyd, R.S., Wang, T.H., and Hammett, F.W. *J. Non-Cryst. Solids*, 147–148, 320, 1992.

37. Harris, B., Cooke, R.G., Hammett, F.W., and Russell-Floyd, R.S., *Ind. Ceram.*, 18, 33, 1998.
38. Quinn, H.W., Missen, R.W., and Frost, G.B., *Can. J. Chem.*, 33, 286, 1955.
39. Mutin, J.C., Watelle, G., and Dusausoy, Y.I., *J. Solid State Chem.*, 27, 407, 1979.
40. Oswald, H.-R., Gunter, J.R., and Dubler, E., *J. Solid State Chem.*, 13, 330, 1975.
41. Shlyakhtin, O.A., Topochemical processes in the freeze drying synthesis of nickel-zinc ferrites, Ph.D. Thesis, Moscow State University, Moscow, 1985.
42. Shlyakhtin, O.A., Oleinikov, N.N., Pokholok, K.V., and Tretyakov, Y.D., *Inorg. Mater.*, 24, 368, 1988.
43. Pyne, A., Chatterjee, K., and Suryanarayanan, R., *Pharm. Res.*, 20, 802, 2003.
44. Shlyakhtin, O.A., Oleinikov, N.N., and Tretyakov, Y.D., *Inorg. Mater.*, 21, 1690, 1985.
45. Roos, Y., and Karel, M., *Biotechnol. Prog.*, 7, 49, 1991.
46. Brown, M.E., Dollimore, D., and Galwey, A.K., *Reactions in the Solid State*, Elsevier, Amsterdam, 1980.
47. Brylev, O.A., Shlyakhtin, O.A., Kulova, T.L., Skundin, A.M., and Tretyakov, Y.D., *Solid State Ionics*, 156, 291, 2003.
48. Maison, W., Kleeberg, R., Heimann, R.B., Phanichphant, S., *J. Eur. Ceram. Soc.*, 23, 127, 2003.
49. Fang, J., Wang, J., Ng, S.C., Chew, C.H., and Gan, L.M., *J. Mater. Sci.*, 34, 1943, 1999.
50. Kirchnerova, J., and Klvana, D., *Solid State Ionics*, 123, 307, 1999.
51. Lee, Y.N., El-Fadli, Z., Sapina, F., Martinez-Tamayo, E., and Cortes Corberan, V., *Catal. Today*, 52, 45, 1999.
52. Lee, Y.N., Lago, R.M., Fierro, J.L.G., and Gonzalez, J., *Appl. Catal. A*, 215, 245, 2001.
53. Voinov, I.D., Influence of thermal and chemical prehistory on the substructure and catalytic activity of cobalt ferrite, PhD dissertation, Moscow State University, Moscow, 1976.
54. Brylev, O.A., Synthesis and properties of Li-conducting oxide and polymer materials for secondary lithium batteries, Ph.D. dissertation, Moscow State University, Moscow, 2002.
55. Ischenko, V.V., Shlyakhtin, O.A., and Oleinikov, N.N., *Inorg. Mater.*, 33, 931, 1997.
56. Heegn, H., Trinkler, M., and Langbein, H., *Cryst. Res. Technol.*, 35, 255, 2000.
57. Ischenko, V.V., Shlyakhtin, O.A., Oleinikov, N.N., Sokolov, I.S., Altenburg, H., and Tretyakov, Y.D., *Dokl. Chem.*, 356, 217, 1997.
58. Shlyakhtin, O.A., Yoon Y.S., Oh, Y.-J., *J. Eur. Ceram. Soc.*, 23, 1893, 2003.
59. Shlyakhtin, O.A., Vinokurov, A.L., Ischenko, V.V., Altenburg, H., and Oleynikov, N.N., in Balachandran, U., and McGinn, P.J., Eds., *High Temperature Superconductors II. Synthesis, Processing and Applications*, The Minerals, Metals, & Materials Society, Warrendale, PA, 73, 1997.
60. Ischenko, V.V., Shlyakhtin, O.A., Vinokurov, A.L., Altenburg, H., and Oleinikov, N.N., *Physica C*, 282–287, 855, 1997.
61. Louis, C., Bazzi, R., Flores, M.A., Zheng, W., Lebbou, K., Tillement, O., Mercier, B., Dujardin, C., and Perriat, P., *J. Solid State Chem.*, 173, 335, 2003.
62. Wartewig, P., Krause, M.K., Esquinazi, P., Rosler, S., and Sonntag, R., *J. Magn. Magn. Mater.*, 192, 83, 1999.

63. Calleja, A., Tijero, E., Martinez, B., Pinol, S., Sandiumenge, F., and Obradors, X., *J. Magn. Magn. Mater.*, 196–197, 293, 1999.
64. Ol'khovik, L.P., Sizova, Z.I., Golubenko, Z.V., and Kuz'micheva, T.G., *J. Magn. Magn. Mater.*, 183, 181, 1998.
65. Ding, J., Liu, X.Y., Wang, J., and Shi, Y., *Mater. Lett.*, 44, 19, 2000.
66. Shi, Y., Ding, J., and Yin, H., *J. Alloys Compd.*, 308, 290, 2000.
67. Guigue-Millot, N., Begin-Colin, S., Champion, Y., Hytch, M.J., LeGaer, G., and Perriat, P., *J. Solid State Chem.*, 170, 30, 2003.
68. Millot, N., Begin-Colin, S., Perriat, P., and Le Gaer, G., *J. Solid State Chem.*, 139, 66, 1998.
69. Bermejo, E., Becue, T., Lacour, C., and Quarton, M., *Powder Technol.*, 94, 29, 1997.
70. Ng, C.W., Ding, J., Shi, Y., and Gan, L.M., *J. Phys. Chem. Solids*, 62, 767, 2001.
71. Ng, C.W., Ding, J., Wang, L., Gan, L.M., and Quek, C.H., *J. Phys. Chem. A*, 104, 8814, 2000.
72. El-Fadli, Z., Coret, E., Sapina, F., Martinez, E., Beltran, A., Beltran, D., and Lloret F., *J. Mater. Chem.*, 9, 1793, 1999.
73. Kuznetsov, A.A., Shlyakhtin, O.A., Brusentsov, N.A., and Kuznetsov, O.A., *Eur. Cells Mater.*, 3, 75, 2002.
74. Shlyakhtin, O.A., Oh, Y.-J., and Tretyakov, Y.D., *J. Eur. Ceram. Soc.*, 20, 2047, 2000.
75. Shlyakhtin, O.A., Oh, Y.-J., and Tretyakov, Y.D., *Solid State Commun.*, 111, 711, 1999.
76. Shlyakhtin, O.A., Oh, Y.-J., and Tretyakov, Y.D., *Solid State Commun.*, 117, 261, 2001.
77. Shlyakhtin, O.A., Shin, K.H., and Oh, Y.-J., *J. Appl. Phys.*, 91, 7403, 2002.
78. Sigl, L.S., *J. Eur. Ceram. Soc.*, 18, 1521, 1998.
79. Hung, K.-M., Yang, W.-D., and Huang, C.-C., *J. Eur. Ceram. Soc.*, 23, 1901, 2003.
80. Kao, C.-F., and Yang, W.-D., *Appl. Organometall. Chem.*, 13, 383, 1999.
81. Ng, W.B., Wang, J., Ng, S.C., and Gan, L.M., *J. Mater. Chem.*, 8, 2239, 1998.
82. Lee, J.-H., and Chiang, Y.-M., *J. Mater. Chem.*, 9, 3107, 1999.
83. Lee, J.-H., and Chiang, Y.-M., *J. Electroceram.*, 6, 7, 2001.
84. Zipprich, W., Waschilewski, S., Rocholl, F., and Wiemhofer, H.-D., *Solid State Ionics*, 101–103, 1015, 1997.
85. Mazo, G.N., Shlyakhtin, O.A., and Savvin, S.N., *Int. J. Inorg. Mater.*, 3, 31, 2001.
86. Mazo, G.N., Shlyakhtin, O.A., and Savvin, S.N., *High Temperature Superconductors—Crystal Chemistry, Processing and Properties*, Materials Research Society Symposium Proceedings, 659, II9.13.1, 2001.
87. Kim, S., Yang, Y.L., Christoffersen, R., and Jacobson, A.J., *Solid State Ionics*, 109, 187, 1998.
88. Kim, S., Yang, Y.L., Jacobson, A.J., and Abeles, B., *Solid State Ionics*, 106, 189, 1998.
89. Abrantes, J.C.C., Perez-Coll, D., Nunez, P., and Frade, J.R., *Electrochim. Acta*, 48, 2761, 2003.
90. Lubke, S., and Wiemhofer, H.-D., *Solid State Ionics*, 117, 229, 1999.
91. Perez-Coll, D., Nunez, P., Frade, J.R., and Abrantes, J.C.C., *Electrochim. Acta*, 48, 1551, 2003.
92. Petot, C., Filal, M., Rizea, A.D., Westmacott, K.H., Laval, J.Y., Lacour, C., and Ollitrault R., *J. Eur. Ceram. Soc.*, 18, 1419, 1998.

93. Zimmer, E., Scharf, K., Mono, T., Friedrich, J., and Schober, T., *Solid State Ionics*, 97, 505, 1997.

94. Huang, B., Jang, Y.-I., Chiang, Y.-M., and Sadoway, D.R., *J. Appl. Electrochem.*, 28, 1365, 1998.

95. Shlyakhtin, O.A., Jeon, Y.-A., Yoon, Y.S., and Oh, Y.-J., *Electrochim. Acta*, 50(2-3), 509, 2004.

96. Egorov, A.V., Brylev, O.A., Shlyakhtin, O.A., Kulova, T.L., Skundin, A.M., and Tretyakov, Y.D., in *New Trends in Intercalation Compounds for Energy Storage*, NATO Science Series II, vol. 61, Kluwer Academic, Dordrecht, 633, 2002.

97. Shaju, K.M., Subba Rao, G.V., and Chowdari, B.V.R., *Electrochim. Acta*, 48, 1505, 2003.

98. Yang, J., and Xu, J.J., *J. Power Sources*, 122, 181, 2003.

99. Xu, J.J., and Yang, J., *Electrochem. Commun.*, 5, 230, 2003.

100. Coustier, F., Jarero, G., Passerini, S., and Smyrl, W., *J. Power Sources*, 83, 9, 1999.

101. Schlorb, H., Bungs, M., and Plieth, W., *Electrochim. Acta*, 42, 2619, 1997.

102. Licht, S., Wang, B., and Ghosh, S., *Science*, 285, 1039, 1999.

103. Dos Santos, O., Weiller, M.L., Junior, D.Q., and Medina, A.N., *Sensors Actuators*, B75, 83, 2001.

104. Yavuz, M., Maeda, H., Vance, L., Liu, H.K., and Dou, S.X., *Supercond. Sci. Technol.*, 11, 1166, 1998.

105. Yavuz, M., Maeda, H., Vance, L., Liu, H.K., and Dou, S.X., *J. Alloys Compd.*, 281, 280, 1998.

106. Mani, L., Marinkovi, Z., and Milosevi, O., *Mater. Chem. Phys.*, 67, 288, 2001.

107. Nikolic, N., Mancic, L., Marinkovic, Z., Milosevic, O., and Ristic, M.M., *Ann. Chim. Sci. Mater.*, 26, 35, 2001.

108. Badica, P., Aldica, G., Groza, J.R., Bunescu, M.-C., and Mandache, S., *Supercond. Sci. Technol.*, 15, 32, 2002.

109. Badica, P., Aldica, G., Bunescu, M.-C., and Nemyrovsky, A.V., *J. Mater. Sci. Lett.*, 19, 561, 2000.

110. Shlyakhtin, O.A., Vinokurov, A.L., Baranov, A.N., and Tretyakov, Y.D., *J. Supercond.*, 11, 507, 1998.

111. Badica, P., Aldica, G., and Mandache, S., *Supercond. Sci. Technol.*, 12, 162, 1999.

112. Badica, P., and Aldica, G., *J. Supercond.*, 12, 609, 1999.

113. Bunescu, M.-C., Aldica, G., Badica, P., Vasiliu, F., Nita, P., and Mandache, S., *Physica C*, 281, 191, 1997.

114. Fradina, L.A., Alekseev, A.F., Gridasova, T.J., Morozov, V.V., and Jurchenko, D.O., *Physica C*, 311, 81, 1999.

115. Pupysheva, O.V., Shlyakhtin, O.A., Lennikov, V.V., Putlyaev, V.I., and Tretyakov, Y.D., *Inorg. Mater.*, 33, 951, 1997.

116. Huang, Y.B., Witz, G., Giannini, E., Erb, A., Shlyakhtin, O.A., and Flukiger, R., *Supercond. Sci. Technol.*, 12, 411, 1999.

117. Silva, C.C., and McHenry, M.E., *J. Magn. Magn. Mater.*, 226–230, 311, 2001.

118. Matsuda, M., Aihara, Y., Yamashita, K., and Umegaki, T., *J. Solid State Chem.*, 128, 310, 1997.

119. Weyl, C., Lacour, C., Laher-Lacour, F., Thuery, P., and Zhao, J., *Nucl. Instrum. Meth. B*, 122, 606, 1997.

120. Valenzuela, R.X., Bueno, G., Solbes, A., Sapina, F., Martinez, E., and Cortez Corberan, V., *Top. Catal.*, 15, 181, 2001.

121. Lee, Y.N., Lago, R.M., Fierro, J.L.G., Cortes, V., Sapina, F., and Martinez, E., *Appl. Catal. A*, 207, 17, 2001.
122. Borovskikh, L., Mazo, G., and Kemnitz, E., *Solid State Sci.*, 5, 409, 2003.
123. Labhsetwar, N.K., Watanabe, A., and Mitsuhashi, T., *Appl. Catal. B*, 40, 21, 2003.
124. Chen, K.Y., and Tseung, A.C.C., *J. Electroanal. Chem.*, 451, 1, 1998.
125. Even, W.R., Jr., Method for low temperature preparation of a noble metal alloy, U.S. Patent 6,348,431, Feb. 19, 2002.
126. Lee, S.A., Park, K.W., Kwon, B.K., and Sung, Y.E., Method of preparing platinum alloy electrode catalyst for direct methanol fuel cell using anhydrous metal chloride, U.S. Patent application 20,010,027,160, Oct. 4, 2001.
127. Choi, J.-H., Park, K.-W., Kwon, B.-K., and Sung, Y.-E., *J. Electrochem. Soc.*, 150, A973, 2003.
128. Parvulescu, V.I., Parvulescu, V., Endruschat, U., Lehmann, Ch.W., Grange, P., Poncelet, G., Bonnemann, H., *Microporous Mesoporous Mater.*, 44–45, 221, 2001.
129. Parida, K., Quaschning, V., Lieske, E., and Kemnitz, E., *J. Mater. Chem.*, 11, 1903, 2001.
130. Quaschning, V., Auroux, A., Deutsch, J., Lieske, H., and Kemnitz, E., *J. Catal.*, 203, 426, 2001.
131. Kalinin, S.V., Kheifets, L.I., Mamchik, A.I., Knot'ko, A.V., and Vertegel, A.A., *J. Sol-Gel Sci. Technol.*, 15, 31, 1999.
132. Kirchnerova, J., Klvana, D., and Chaouki, J., *Appl. Catal. A*, 196, 191, 2000.
133. Chae, H.J., Nam, I.-S., Yang, H.S., Song, S.L., and Hur, I.D., *J. Chem. Eng. Jpn.*, 34, 148, 2001.
134. Fukasawa, T., Ando, M., Ohji, T., and Kanzaki, S., *J. Am. Ceram. Soc.*, 84, 230, 2001.
135. Dittrich, R., Tomandl, G., and Mangler, M., *Adv. Eng. Mater.*, 4, 487, 2002.
136. Moritz, T., Werner, G., and Tomandl, G., *J. Porous Mater.*, 6, 111, 1999.
137. Hyun, S.H., Kim, T.Y., Kim, G.S., and Park, H.H., *J. Mater. Sci. Lett.*, 19, 1863, 2000.
138. Suzuki, S., Itoh, K., Ohgaki, M., Ohtani, M., and Ozawa, M., *Ceram. Int.*, 25, 287, 1999.
139. Amato, G., Brunetto, N., and Parisini, A., *Thin Solid Films*, 297, 73, 1997.
140. Vix-Guterl, C., McEnaney, B., and Ehrburger, P., *J. Eur. Ceram. Soc.*, 19, 427, 1999.
141. Fukasawa, T., Deng, Z.-Y., and Ando, M., *J. Am. Ceram. Soc.*, 85, 2151, 2002.
142. Choi, S.J., Park, H.C., and Stevens, R., *J. Mater. Sci.*, 39, 1037, 2004.
143. Yokota, T., Takahata, Y., Katsuyama, T., and Matsuda, Y., *Catal. Today*, 69, 11, 2001.
144. Yamamoto, T., Nishimura, T., Suzuki, T., and Tamon, H., *J. Non-Cryst. Solids*, 288, 46, 2001.
145. Kocklenberg, R., Mathieu, B., Blacher, S., Pirard, R., Pirard, J.P., Sobry, R., van den Bossche G., *J. Non-Cryst. Solids*, 225, 8, 1998.
146. Yamamoto, T., Nishimura, T., Suzuki, T., and Tamon, H., *Carbon*, 39, 2369, 2001.
147. Tamon, H., Ishizaka, H., Yamamoto, T., and Suzuki, T., *Carbon*, 38, 1099, 2000.
148. Tamon, H., Ishizaka, H., Yamamoto, T., and Suzuki, T., *Carbon*, 37, 2049, 1999.
149. Tani, T., *Composites A*, 30, 419, 1999.
150. Bellini, J.V., Morelli, M.R. and Kiminami, R.H.G.A., *Mater. Lett.*, 57, 3775, 2003.
151. Bellini, J.V., Morelli, M.R. and Kiminami, R.H.G.A., *Mater. Lett.*, 57, 3325, 2003.

152. Bellini, J.V., Morelli, M.R. and Kiminami, R.H.G.A., *J. Mater. Sci. Mater. Electr.*, 13, 485, 2002.
153. Ivanov, V.K., Baranov, A.N., Oleinikov, N.N., and Tretyakov, Y.D., *Russ. J. Inorg. Chem.*, 47, 1769, 2002.
154. Ivanov, V.K., Baranov, A.N., Oleinikov, N.N., and Tretyakov, Y.D., *Inorg. Mater.*, 38, 1224, 2002.
155. Ivanov, V.K., Oleinikov, N.N., and Tretyakov, Y.D., *Dokl. Chem.*, 386, 277, 2002.
156. Ivanov, V.K., Baranov, A.N., Mazo, G.N., Oleinikov, N.N., and Tretyakov, Y.D., *Russ. J. Inorg. Chem.*, 48, 304, 2003.
157. Liu, Z., Ooi, K., Kanoh, H., Tang, W., and Tomida, T., *Langmuir*, 16, 4154, 2000.
158. Nivoix, V., and Gillot, B., *Solid State Ionics*, 111, 17, 1998.
159. Rizzato, A.P., Broussous, L., Santilli, C.V., Pulcinelli, S.H., and Craievich, A.F., *J. Non-Cryst. Solids*, 284, 61, 2001.
160. Sameti, M., Bohr, G., Ravi Kumar, M.N.V., Kneuer, C., Bakowsky, U., Nacken, M., Schmidt, H., Lehr, C.-M., *Int. J. Pharm.*, 266, 51, 2003.
161. Roncari, E., Galassi, C., Ardizzone, S., and Bianchi, C.L., *Colloids and Surfaces A*, 117, 267, 1996.
162. Moritz, T., and Nagy, A., *J. Nanoparticle Res.*, 4, 439, 2002.
163. Blomquist, H.R., Process for preparing free-flowing particulate phase stabilized ammonium nitrate, U.S. Patent application 20,020,096,235, July 25, 2002.
164. Segawa, H., Kawano, T., and Yoshida, K., *Electron. Lett.*, 37, 1381, 2001.
165. Ma, D., Schadler, L.S., Siegel, R.W., and Hong, J.-I., *Appl. Phys. Lett.*, 83, 1839, 2003.
166. Sasaki, T., *Supramol. Sci.*, 5, 367, 1998.
167. Boiadjieva, T., Cappelletti, G., Ardizzone, S., Rondinini, S., and Vertova, A., *Phys. Chem. Chem. Phys.*, 5, 1689, 2003.
168. Izutsu, H., Nair, P.K., and Mizukami, F., *J. Mater. Chem.*, 7, 855, 1997.
169. Deng, H.M., Ding, J., Shi, Y., Liu, X.Y., and Wang, J., *J. Mater. Sci.*, 36, 3273, 2001.
170. Bogicevic, C., Laher-Lacour, F., Malibert, C., Dkhil, B., Menoret, C., Dammak, H., Giorgi, M.L., Kiat J.M., *Ferroelectrics*, 270, 57, 2002.
171. Martynenko, L.I., Shlyakhtin, O.A., and Charkin, D.O., *Inorg. Mater.*, 33, 489, 1997.
172. Martynenko, L.I., Shlyakhtin, O.A., Milovanov, S.V., Goreljskiy, S.I., and Charkin, D.O., *Inorg. Mater.*, 34, 487, 1998.
173. Rego-Monteiro, V.A., de Souza, E.F., de Azevedo, M.M.M., and Galembeck, F., *J. Colloid Interface Sci.*, 217, 237, 1999.
174. Nosaka, Y., Shibamoto, M., and Nishino, J., *J. Colloid Interface Sci.*, 251, 230, 2002.
175. Jiansirisomboon, S., MacKenzie, K.J.D., Roberts, S.G., and Grant, P.S., *J. Eur. Ceram. Soc.*, 23, 961, 2003.
176. Martin, H.-P., Ecke, R., and Muller, E., *J. Eur. Ceram. Soc.*, 18, 1737, 1998.
177. Tappan, B.C., and Brill, T.B., *Propel. Explos. Pyrotech.*, 28, 72, 2003.
178. Primo-Martin, V., and Jansen, M., *J. Solid State Chem.*, 157, 76, 2001.
179. Lobanov, M.V., Balagurov, A.M., Pomjakushin, V.J., Fischer, P., Gutmann, M., Abakumov, A.M., D'yachenko, O.G., Antipov, E.V., Lebedev, O.I., and Van Tendeloo, G., *Phys. Rev. B*, 61, 8941, 2000.
180. Malchus, M., and Jansen, M., *Z. Anorg. Allg. Chem.*, 624, 1846, 1998.

181. El-Himri, A., Sapina, F., Ibanez, R., and Beltran, A., *J. Mater. Chem.*, 10, 2537, 2000.
182. El-Himri, A., Cairols, M., Alconchel, S., Sapina, F., Ibanez, R., Beltran D., and Beltran A., *J. Mater. Chem.*, 9, 3167, 1999.
183. Alconchel, S., Sapina, F., Beltran, D., and Beltran, A., *J. Mater. Chem.*, 9, 749, 1999.
184. Langbein, H., Bremer, M., and Krabbes, I., *Solid State Ionics*, 101–103, 579, 1997.
185. Deliyanni, E.A., Bakoyannakis, D.N., Zouboulis, A.I., Matis, K.A., and Nalbandian, L., *Microporous Mesoporous Mater.*, 42, 49, 2001.
186. Bakoyannakis, D.N., Deliyanni, E.A., Zouboulis, A.I., Matis, K.A., Nalbandian, L., and Kehagias, T., *Microporous and Mesoporous Mater.*, 59, 35, 2003.
187. Bremer, M., and Langbein, H., *Solid State Sci.*, 1, 311, 1999.
188. El-Fadli, Z., Metni, M.R., Sapina, F., Martinez, E., Folgado, J.V., Beltran, D., and Beltran A., *J. Mater. Chem.*, 10, 437, 2000.
189. Novoselov, A.V., Zimina, G.V., Filaretov, A.A., Shlyakhtin, O.A., Komissarova, L.N., and Pajaczkowska, A., *Mater. Res. Bull.*, 36, 1789, 2001.
190. Trofimenko, N.E., Paulsen, J., Ullmann, H., and Muller, R., *Solid State Ionics*, 100, 183, 1997.
191. Langbein, H., Christen, S., and Bonsdorf, G., *Thermochim. Acta*, 327, 173, 1999.
192. Boerger, A., Dallmann, H., and Langbein, H., *Thermochim. Acta*, 387, 141, 2002.
193. McHale, J.M., Navrotsky, A., and Kirkpatrick, R.J., *Chem. Mater.*, 10, 1083, 1998.
194. Shlyakhtin, O.A., Vjunitsky, I.N., Oh, Y.-J., and Tretyakov, Y.D., *J. Mater. Chem.*, 9, 1223, 1999.

# 5 Environmentally Benign Approach to Synthesis of Titanium-Based Oxides by Use of Water-Soluble Titanium Complex

*Koji Tomita, Deepa Dey, Valery Petrykin, and Masato Kakihana*

## CONTENTS

# I. INTRODUCTION

Titanium is widely used in forms of titanium metal, as a simple oxide, nitride, carbide, or as a component of numerous important functional multicomponent oxides. The preparation of complex oxides containing titanium relies mostly on the conventional ceramic method and utilizes titanium oxide as one of the reagents. Miniaturization of electronic devices, wider application of titanium oxide as a photocatalyst as well as preparation of thin films and coatings requires application of solution based synthesis methods for materials preparation. In such a case titanium chloride, sulfate or alkoxides are frequently used as the source of titanium. For example, titanium tetraisopropoxide ($Ti(C_3H_7O)_4$) is a liquid soluble in organic solvents, that is commonly utilized as a titanium source for the synthesis of oxide films and powders of the multicomponent oxides.[1–10] However, because of its high sensitivity to moisture, hydrolysis of the alkoxide occurs rapidly on exposure to traces of water vapor in air, resulting in precipitation of titanium hydroxide (or hydrated titanium oxide; $TiO_2 \cdot nH_2O$). Moreover, the organic solvent and the alkoxides are inflammable and harmful, thus they should be handled in the dry inert atmosphere. The industrial process, in addition, is also restricted by the considerations of cost and environmental impact of the alkoxides and organic solvents. Although titanium chloride ($TiCl_4$) and titanium sulfate ($Ti(SO_4)_2$) are non-inflammable, their solutions should be strongly acidic in order to prevent hydrolysis, while presence of the residual chloride or sulfate may deteriorate the functional properties of the materials. Obviously from the viewpoint of handling, processing, cost and environmental considerations, it is desirable to have stable, noncorrosive, non-toxic solutions of titanium compounds in water, preferably close to the neutral pH. However, this is a very challenging task for the case of Ti(IV) ion. Very few compounds like $K_2[TiO(C_2O_4)_2]$[11–14] and "Ti-acetic acid" (alkoxyacetate complex) solution,[15–19] form a stable titanium complex in water.

One can easily summarize the requirements that should be fulfilled before the titanium-based ceramic materials can be obtained through an aqueous solution synthesis method. First, the precursor compound should possess good solubility in water, and preferably it should be stable over a wide pH range. In ideal case such compound should be a weighing form for titanium; however, from the practical considerations it is sufficient to have a stock solution stable for a reasonably long period of time. Second, the reagent should be non-toxic, relatively cheap and its impact on the environment should be small. Its composition and chemistry should be simple and the reactions with other cations that will be introduced to the system must be well-predictable. The tendency to form precipitates with many cations, like in the case of oxalate ions, must be avoided. Finally, from an industrial point of view, the overall process should be cost effective and environmentally benign.

A series of new compounds that satisfy these conditions has been isolated recently. Using different hydroxycarboxylic acids as chelating agents, titanium (IV) complexes that are stable in an aqueous solution have been obtained.[20–22]

Hydroxycarboxylic acids, which include citric acid, malic acid, lactic acid, etc., are benign to the environment and very convenient for the solution processing. Moreover, since these reagents can form stable complexes with other cations, they rarely yield a precipitate. For several complexes single crystals of well-defined composition suitable for the X-ray structural analysis were isolated. Thus, these water-soluble titanium complexes of hydroxycarboxylic acids are promising precursors for the synthesis of ceramics from an aqueous solution and their industrial utilization is expected in the future. In this chapter we decribe the method of synthesis, structural analysis, and stability of these complexes. The examples of multicomponent oxide materials preparation using these compounds are presented.

## II. SYNTHESIS OF WATER-SOLUBLE TITANIUM COMPLEXES

In this section, the method of synthesizing water-soluble titanium complexes using hydroxycarboxylic acid and the nature of the complexes are described. Metallic titanium powder reacts with hydrogen peroxide in the presence of ammonia as described in Equation 5.1, yielding a yellowish solution with dissolution of titanium powder in hydrogen peroxide:

$$Ti + 3H_2O_2 + NH_3 \rightarrow [Ti(OH)_3O_2] + H_2O + NH_4^+. \tag{5.1}$$

Peroxotitanium complex $[Ti(OH)_3O_2]$, obtained from this reaction, is unstable and decomposes to produce a precipitate of titanium hydroxide (Equation 5.2), even at room temperature:

$$[Ti(OH)_3O_2] + H_2O \rightarrow Ti(OH)_4 + O_2 + OH^-. \tag{5.2}$$

When an excess of hydrogen peroxide exists in the solution, titanium keeps dissolving in the solution in accordance with the reaction of Equation 5.3, which involves the reaction of Equation 5.2 as an intermediate:

$$Ti(OH)_4 + H_2O_2 \rightarrow [Ti(OH)_3O_2] + H_3O^+. \tag{5.3}$$

An excess of hydrogen peroxide helps in sustaining the peroxotitanium complex in the solution without the precipitation of titanium hydroxide. At this point, adding a reagent that can help to form a stable complex with titanium in an aqueous solution before titanium hydroxide precipitates out, producing a transparent aqueous solution, would be of immense importance. However, most of the organic ligands including EDTA fail to form a stable complex with Ti(IV) and precipitation of titanium hydroxide occurs.[23,24] Nevertheless, a series of reagents that belong to hydroxycarboxylic acids can form stable complexes with titanium in an aqueous solution. Citric acid, lactic acid, malic acid, tartaric acid, glycolic

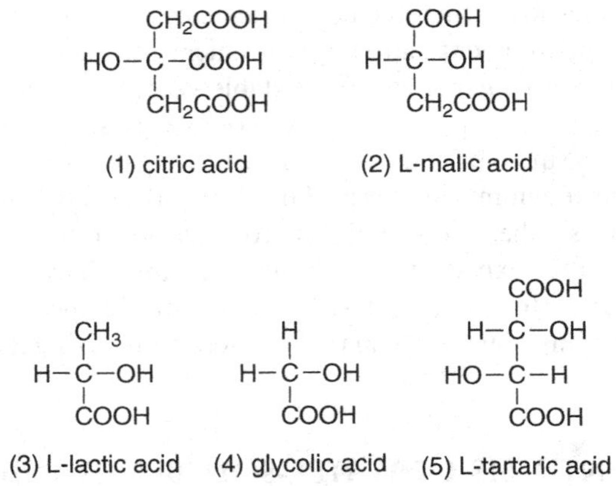

$$
\begin{array}{cc}
\text{CH}_2\text{COOH} & \text{COOH} \\
| & | \\
\text{HO}-\text{C}-\text{COOH} & \text{H}-\text{C}-\text{OH} \\
| & | \\
\text{CH}_2\text{COOH} & \text{CH}_2\text{COOH} \\
\\
\text{(1) citric acid} & \text{(2) L-malic acid}
\end{array}
$$

$$
\begin{array}{ccc}
 & & \text{COOH} \\
 & & | \\
\text{CH}_3 & \text{H} & \text{H}-\text{C}-\text{OH} \\
| & | & | \\
\text{H}-\text{C}-\text{OH} & \text{H}-\text{C}-\text{OH} & \text{HO}-\text{C}-\text{H} \\
| & | & | \\
\text{COOH} & \text{COOH} & \text{COOH} \\
\\
\text{(3) L-lactic acid} & \text{(4) glycolic acid} & \text{(5) L-tartaric acid}
\end{array}
$$

**FIGURE 5.1** Structures of various hydroxycarboxylic acids.

acid, etc., belong to the hydroxycarboxylic acids and have one or more hydroxyl groups and carboxyl groups, as shown in Figure 5.1. Dissociation of the acids occurs in an aqueous solution and the deprotonated hydroxyl groups and carboxyl groups coordinate titanium, which then leads to the formation of a stable complex in an aqueous solution (detailed structures of these complexes are described in the next section). For example, when using citric acid, it reacts as in Equation 5.4 and stable citratoperoxotitanium complex, $[Ti(C_6H_4O_7)O_2]^{-2}$, is formed:

$$[Ti(OH)_3O_2] + C_6H_8O_7 \rightarrow [Ti(C_6H_4O_7)O_2]^{-2} + 2H_2O + H_3O^+. \quad (5.4)$$

This complex is stable and does not decompose in the manner observed for peroxotitanium complex, and can exist in an aqueous solution.

All complexes are stable in acidic, neutral, and weakly alkaline solutions. The color is orange in acidic solution and changes to yellow in alkaline solution. In the strongly alkaline environment they are unstable and yield precipitation of titanium hydroxide. Heating above 100°C also results in the decomposition of the peroxocomplexes. During gradual evaporation, some of the complexes form single crystals of the corresponding ammonium salts (Figure 5.2). They can be further purified by recrystallization from water and used for single-crystal x-ray diffraction studies.

## III. STRUCTURE OF WATER-SOLUBLE TITANIUM COMPLEX

Some hydroxycarboxylic acid titanium complexes yield the single crystals of ammonium salt by concentrating the solution. To obtain a high-quality single

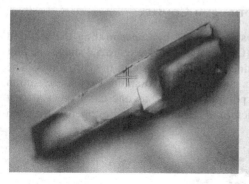

**FIGURE 5.2** Single crystal of water-soluble titanium complex ammonium salt: (left) citric acid complex; (right) lactic acid complex.

crystal, recrystallization must be strictly controlled by keeping in mind parameters such as the concentration of hydroxycarboxylic acid and ammonia, the temperature of the solution, and the rate of evaporation. Not all hydroxycarboxylic acid titanium complexes yield single crystals. In some cases, polycrystalline samples are obtained; in other cases, gelation and polymerization occur.

Structural analysis of the single crystal provides precise information about locations of elements (except for hydrogen), which makes it possible to discuss the architecture and stability of the complex. On the other hand, such information may not fully reflect the dominating form of the complex in the solution; neither does it provide an idea about the equilibrium state when the compound is dissolved in water. In this case, spectroscopic methods can be complementary to the x-ray diffraction experiment.

Thermogravimetric/differential thermal analysis (TG/DTA), infrared (IR) spectroscopy, Raman scattering, hydrogen nuclear magnetic resonance (HNMR), $^{13}$C-NMR, elemental analysis, etc., are also useful for structural analysis of the single crystal. TG/DTA reveals the amount of crystal water, approximate amount of elements, and the formula weight from the ash content. IR spectroscopy and Raman scatterings provide information about the nature of functional groups and presence of hydrogen in hydroxyl groups and carboxyl groups. NMR provides detailed chemical information about a neighborhood of hydrogen and carbon. These results and the content of elements from elemental analysis can determine the formula of the single crystal. Using this formula and the results of single-crystal x-ray diffraction, the location of each element is determined and the steric structure of the complex is revealed. For example, the results of structural analyses and the discussion of stability based on the results of citratoperoxotitanium complex and lactatotitanium complex are described below.

## A. STRUCTURE OF CITRATOPEROXOTITANIUM COMPLEX

The composition of the single crystal obtained by concentrating the solution stabilized by citric acid is $(NH_4)[Ti(C_6O_4O_7)O_2]\cdot 2H_2O$. Figure 5.3 shows the unit cell of the crystal structure determined by single-crystal x-ray diffraction. The compound is a tetranuclear complex containing four titanium atoms; each anion is separated and does not form a polymeric chain, as one can expect at first glance. Figure 5.4 presents the structure of the individual citratoperoxotitanium

**FIGURE 5.3** Unit cell of citratoperoxotitanium complex ammonium salt.

**FIGURE 5.4** Structure and labeling of citratoperoxotitanium complex.

complex anion and the labeling, indicating that the complex consists of two binuclear fragments interlinked through the bridging carboxylate group (C1–O1–O2) of one citrate ligand. The salient feature of the citrate ligand in this complex is its tridentate nature as a whole molecule. As a result, the deprotonated alcoholic oxygen atom (O3) and the two oxygen atoms (O4, O5) from the C2 and C3 carboxylate groups build up fused five- and six-member chelate rings (Ti–O4–C2–C4–O3 and Ti–O5–C3–C5–C4–O3, respectively), providing the maximum stability for the complex. The coordinated peroxo group (O6–O7) may play two important roles: first, it retards further polymerization of the tetrameric anion by occupying the free sites in the equatorial pentagonal plane that can also be active sites for nucleophilic attack and, consequently, hydrolysis; second, it provides a negative charge to the complex, thus making possible formation of the ammonium salt soluble in water. Further detailed procedures and crystallographic data are reported in Kakihana et al.[20]

## B. STRUCTURE OF LACTATOTITANIUM COMPLEX

The single crystal that is obtained from stabilizing the solution by L-lactic acid was analyzed and the composition was found to be $(NH_4)_2[Ti(C_3H_4O_3)_3]$. The structure of the complex determined by single-crystal x-ray diffraction is shown in Figure 5.5. Notice that titanium is coordinated by six oxygen atoms from three lactic acid molecules forming the stable five-member rings composed of Ti–O2–C3–C4–O5. The $TiO_6$ octahedron is slightly distorted with the O–Ti–O angle of 162.7° and the titanium atom is shifted from the center, giving the two

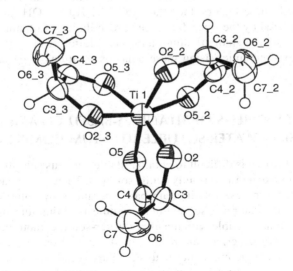

**FIGURE 5.5** Structure and labeling of lactatotitanium complex.

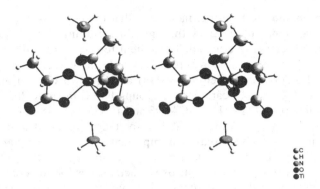

**FIGURE 5.6** Three-dimensional image of lactatotitanium complex.

sets of Ti–O interatomic distances of 2.071 Å and 1.848 Å. The lactic acid molecules preserved their chirality. Lactic acid acts as a bidentate ligand, and there are no potential free groups for oligomerization. At the same time, tight coordination creates spatial difficulties for nucleophilic attack—the first step in hydrolysis. Figure 5.6 shows the three-dimensional image of the complex. Details of the structural analyses of lactatotitanium complex ammonium are forthcoming.

In conclusion, to obtain a stable complex of titanium in aqueous solution, the following four factors should be considered: (1) the ligand should contain both a hydroxyl group and carboxyl group; (2) the ligand should coordinate titanium in such a way that it forms either a five- or six-member ring; (3) the coordination number of titanium must be six or seven; (4) the complex should be anionic. With these factors, the attack of Ti by $H_2O$ or $OH^-$ in an aqueous solution is hindered by the steric factor and precipitation of titanium hydroxide is prevented, that is, a stable titanium aqueous solution is obtained. Synthesis of titanium-based ceramics using these aqueous solutions is described in the next section.

## IV. SYNTHESIS OF TITANIUM-BASED CERAMICS USING WATER-SOLUBLE TITANIUM COMPLEX

As described above, it is difficult to obtain a stable aqueous solution containing titanium species except in a strongly acidic solution. Therefore organic solvents are used for the synthesis of titanium-based ceramics by solution methods. However, utilizing water as a solvent in the synthesis of titanium-based ceramics has now become possible with the use of water-soluble titanium complexes. In this section we describe our results in the synthesis of titanium-based ceramics from an aqueous solution using hydroxycarboxylic acid titanium complexes.

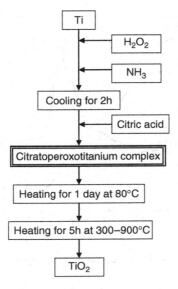

**FIGURE 5.7** Flowchart for synthesis of TiO$_2$ from an aqueous solution using citric acid as the complexing agent.

## A. TiO$_2$

Synthesis of TiO$_2$ is the basis for the synthesis of ceramics using water-soluble titanium complexes. Figure 5.7 shows the flowchart for the synthesis of TiO$_2$ using citratoperoxotitanium complex. Metallic titanium powder is favored as the source of titanium. Although TiCl$_4$ and Ti(SO$_4$)$_2$ are also available, these are not preferable because Cl$^-$ and SO$_4^{-2}$ have a tendency to remain in the products when these are used as a source of titanium. Metallic titanium is preferable in the form of powder, because the reaction rate of equation (1), forming peroxotitanium complex by adding titanium into the mixture of hydrogen peroxide and ammonia, is slow. If granular titanium is used, it does not dissolve completely and remains in the solution. This reaction should be carried out in an ice-cold condition because of the exothermic nature of the reaction. If there is no cooling, the temperature of the solution becomes high and decomposition of hydrogen peroxide is promoted, then the titanium does not dissolve. After the titanium dissolves, citric acid is added to the solution as a complexing agent. Because the ratio of the complex formed by titanium and citric acid is 1:1, one or more moles of citric acid are needed per mole of titanium. The amount is different for each complexing agent; that is, three or more moles of lactic acid and one or more mole of malic acid are needed per mole of titanium. This solution is heated at 80°C, and yellowish "gel" is obtained after the solvent is evaporated. This yellowish gel can dissolve in water. This

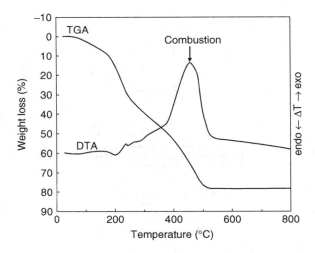

**FIGURE 5.8** TG/DTA curves of "gel" containing citratoperoxotitanium complex.

may indicate that the structure of the complex can be retained in the gel. The results of TG/DTA of this gel are shown in Figure 5.8.

There is weight loss due to the evaporation of water that remained in the gel above 100°C. Next hydrogen and nitrogen evaporate as $H_2O$ and $NH_3$, resulting in carbonization of the sample. After this, combustion of the carbonized sample occurs at 400 to 500°C and carbon is removed as $CO_2$, yielding $TiO_2$. This "gel" is heated for 5 h at 300 to 900°C; the x-ray diffraction (XRD) patterns of these heated samples are shown in Figure 5.9. The pattern of the sample heated to 300°C has a broad peak at low 2 Theta, indicating an amorphous phase. Crystallization of $TiO_2$ does not occur at this temperature because much of carbon remains in the sample, as seen in the results of TG/DTA. Although the pattern at 400°C is somewhat broad, it corresponds to that of anatase-type $TiO_2$. Therefore synthesis of $TiO_2$ by the present method is possible at temperatures above 400°C. Crystallization proceeds at 500°C, and the peaks of anatase-type $TiO_2$ become sharp. Anatase to rutile phase transition occurs partially at 600°C, with the existence of both phases at this temperature. Single-phase rutile-type $TiO_2$ is obtained at temperatures above 700°C. Anatase to rutile phase transition depends on the amount of impurities or additives in the $TiO_2$, as has been reported by various authors,[25-29] where the transition temperature increases with increases in the amount of the impurity. In the case where $Ti(SO_4)_2$ is used as the source of titanium for the synthesis of $TiO_2$, $SO_4^{-2}$ has a tendency to remain in $TiO_2$ and anatase to rutile transition does not occur until about 800°C. In contrast, anatase to rutile transition occurs at 600°C in the present method, indicating that the residual carbon in $TiO_2$ is easily removed at low temperatures.

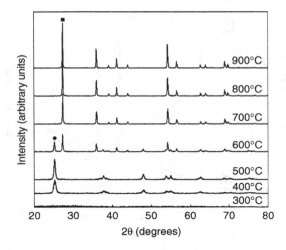

**FIGURE 5.9** X-ray diffraction patterns of $TiO_2$ synthesized from citratoperoxotitanium complex at different temperatures: circle denotes the position of the main peak of anatase-type $TiO_2$; square denotes the position of the main peak of rutile-type $TiO_2$.

## B. BaTiO₃

Complex oxides can also be synthesized from an aqueous solution using the water-soluble titanium complex. Synthesis of $BaTiO_3$ using citratoperoxotitanium complex is described below. Using the same method as in the synthesis of $TiO_2$, titanium powder is dissolved into a mixture of hydrogen peroxide and ammonia under ice-cold conditions, followed by addition of citric acid to obtain a stable aqueous solution. The amount of citric acid should be in excess, considering the coordination of barium. Next, a barium salt is added, maintaining the same molar amount as that of titanium. Among the numerous barium compounds, $Ba(CH_3COO)_2$ and $BaCO_3$ are preferable because of their solubility in the citric acid solution. After evaporation of the solvent, a yellowish "gel" is obtained and $BaTiO_3$ is synthesized by heating of the gel.

Figure 5.10 shows the results of TG/DTA of the gel. It is carbonized up to 400°C, and combustion occurs at about 450°C with the generation of heat. Compared to the synthesis of $TiO_2$, the amount of citric acid is large, therefore the amount of generated heat is also large. However, a small amount of carbon originating from the citric acid remains in the sample as complex carbonate of barium and titanium. This carbonate decomposes at temperatures above 600°C. $BaCO_3$ or complex carbonate of barium and titanium frequently appears as an impurity in the synthesis of $BaTiO_3$ by solution methods.[30-34] Therefore a decrease in the amount of carbonates leads to a reduction in the temperature and time of the heat treatment. In the present method, the amount of organics used for the synthesis is less than that for the method that uses titanium alkoxide and organic solvent. Consequently the amount of carbonates is decreased and the heat treatment temperature and time are reduced. The XRD patterns of the samples heated

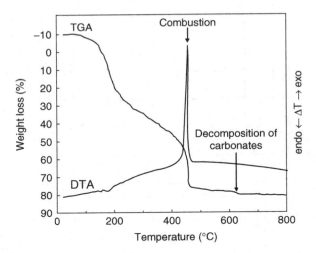

**FIGURE 5.10** TG/DTA curves of "gel" containing citratoperoxotitanium complex and barium.

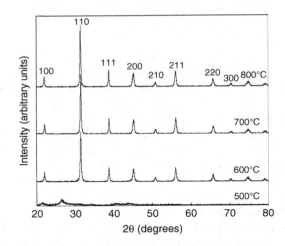

**FIGURE 5.11** X-ray diffraction patterns of $BaTiO_3$ synthesized from citratoperoxotitanium complex at different temperatures.

at 500 to 800°C for 5 h are shown in Figure 5.11. Consistent with the results of TG/DTA, carbonates remain in the sample heated up to 500°C and the peaks of $BaTiO_3$ do not appear. Carbonates are completely removed from the samples at temperatures above 600°C and single-phase $BaTiO_3$ is obtained.

## C. $BA_xSR_{1-x}TIO_3$

In the case of synthesis of complex oxide by the present method, all of the metallic elements contained in the product must be transferred to the solution. When

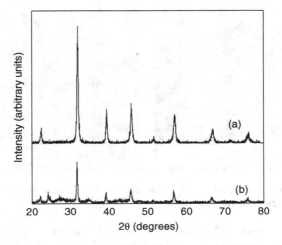

**FIGURE 5.12** X-ray diffraction patterns of $Ba_{0.5}Sr_{0.5}TiO_3$ synthesized from (a) lactatotitanium complex and (b) citratoperoxotitanium complex.

$Ba_xSr_{1-x}TiO_3$ is synthesized, barium and strontium carbonates or acetates are dissolved in the aqueous solution containing water-soluble titanium complex. Although strontium salts can be dissolved in water relatively easily (there are numerous water-soluble strontium salts), strontium forms precipitate with citric acid that is hardly soluble in water. Therefore to utilize citratoperoxotitanium complex for the synthesis of $Ba_xSr_{1-x}TiO_3$ is difficult. In such a case, the complexing agent should be changed.

As described above, titanium forms a stable complex with not only citric acid, but also lactic acid. Lactate ion does not form precipitate with strontium or barium. Therefore all of the titanium, barium, and strontium ions can be kept in an aqueous solution with use of lactic acid, that is, synthesis of $Ba_xSr_{1-x}TiO_3$ is possible. The ratio of the metallic elements can be arbitrarily changed, and thus all compositions of $Ba_xSr_{1-x}TiO_3$ can be synthesized.

Figure 5.12 shows XRD patterns for $Ba_{0.5}Sr_{0.5}TiO_3$ synthesized using citric acid or lactic acid as a complexing agent. The heat treatment temperature of both samples is the same—700°C—nevertheless, there are no impurities in the sample obtained with lactic acid, but in the case where citric acid was used, impurities that seem to be due to carbonate were detected. It appears that, in the case of citric acid, sufficient miscibility is not obtained due to the precipitation of strontium and citric acid. In contrast, high miscibility is retained in the case of lactic acid, yielding a single phase of $Ba_{0.5}Sr_{0.5}TiO_3$.

## D. SrTiO₃

$SrTiO_3$ is a very important industrial material because of its dielectric and photocatalytic properties.[35–43] In a photocatalytic reaction, the photogenerated electron and hole move to the surface of the photocatalyst and react with the molecules

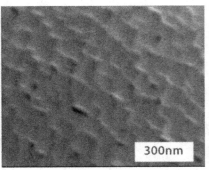

**FIGURE 5.13** SEM images of SrTiO$_3$ synthesized by (left) a solid state reaction at 1000°C and (right) the present method at 1000°C.

adsorbed on the surface. Photocatalytic activity increases with a decrease in the particle size because more of the photogenerated electrons and holes reach the surface.[44,45] Also, the activity increases with increasing specific surface area. On the other hand, promotion of electron-hole recombination by a defect or an impurity is a factor in decreasing activity. Thus high crystallinity, small particle size, and large specific surface area are necessary to obtain a photocatalyst of high activity.

In general, the heat treatment temperature in solution methods is lower compared to solid state reactions; consequently grain growth is retarded and the particle of the sample becomes small with an increase in surface area. Therefore one can expect an increase in photocatalytic activity by solution methods.

Scanning electron microscopy (SEM) images of SrTiO$_3$ synthesized using a lactatotitanium complex and solid state reaction are shown in Figure 5.13. The particle size on these images is about 200 nm in solid state reactions and about 90 nm in the present method after calcination at 1000°C. The particle size decreases as the heat treatment temperature is reduced. The specific surface areas of these samples are shown in Table 5.1. Compared to the sample for the solid state reaction, the specific surface area of the sample heated to 1000°C is two times larger, and the specific surface area increases as the heat treatment temperature is decreased; the maximum value is 13 times larger than that of the solid state reaction.

Films are synthesized by coating with a solution containing a stoichiometric amount of the constituent elements, followed by heat treatment of the coated substrate. An example of the synthesis of SrTiO$_3$ film on a glass substrate is described below. Two solutions of different concentrations were used for the coatings. Coatings were conducted by a spin coater. The coated substrates were heated for 2 h at 600°C. The x-ray diffraction patterns of the films are shown in Figure 5.14 and indicate formation of SrTiO$_3$ films on the substrates. SEM images of the cross section and the surface are shown in Figure 5.15. The thickness is about 1.3 μm (higher concentration solution, Figure 5.15a) and about 400 nm

**TABLE 5.1**
**Specific Surface Areas of SrTiO$_3$ Synthesized by the Present Method at Different Temperatures and Solid State Reactions**

| Heat-Treatment Temperature (°C) | Specific Surface Area (m²/g) |
|---|---|
| 500 | 15.13 |
| 600 | 55.42 |
| 700 | 38.64 |
| 800 | 23.00 |
| 900 | 17.38 |
| 1000 | 7.95 |
| 1000 (SSR) | 4.35 |

*Note:* SSR = solid state reaction.

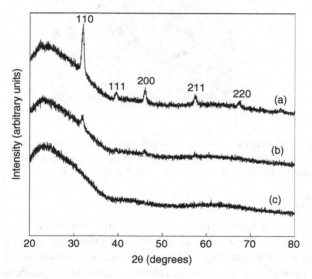

**FIGURE 5.14** X-ray diffraction patterns of SrTiO$_3$ films on glass plates synthesized from lactatotitanium complex aqueous solution: (a) denser solution; (b) dilute solution; (c) glass substrate.

(dilute solution, Figure 5.15b). The thickness can be changed by changing the concentration of the solution. However, cracks can form during the synthesis of thicker films (Figure 5.15c). Further details of this study are reported by Tomita and Kakihana.[46]

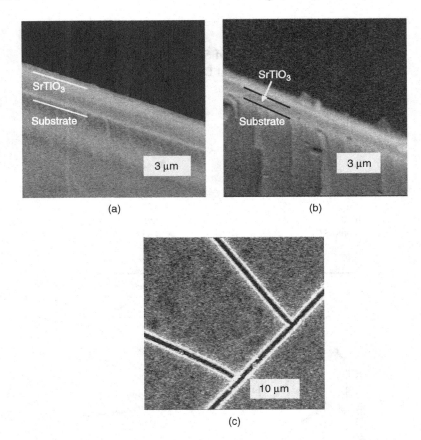

(a)                                      (b)

(c)

**FIGURE 5.15** SEM images of SrTiO$_3$ films on glass plates: (a) cross section of the film synthesized from dense solution; (b) cross section of the film synthesized from dilute solution; (c) surface of the film synthesized from dense solution.

## E.   Bi$_4$Ti$_3$O$_{12}$

Bi$_4$Ti$_3$O$_{12}$ with a layered perovskite structure (Aurivillius phase) is well-known ferroelectrics. BLSFs are one of the most promising materials for ferroelectric random access memory (FeRAM) due to their outstanding ferroelectric properties. Although bismuth ions are relatively stable in the acidic solutions, on dilution or upon increasing the pH partial hydrolysis occurs yielding insoluble compounds of BiO$^+$ ions. Therefore it is necessary to solubilize bismuth in water to synthesize Bi$_4$Ti$_3$O$_{12}$ from an aqueous solution. We tried to solubilize bismuth in water using a number of complexing agents and found that lactic acid forms a stable complex with bismuth and solubilizes bismuth in water. Using lactic acid, Bi$_4$Ti$_3$O$_{12}$ was synthesized from an aqueous solution.

Bi-lactic acid aqueous solution and lactatotitanium complex aqueous solution were stoichiometrically mixed (Bi:Ti = 4:3). After evaporation of the mixed

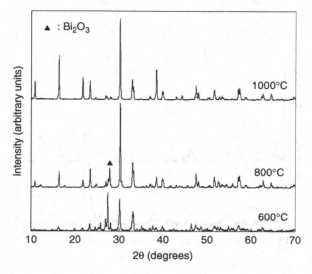

**FIGURE 5.16** X-ray diffraction patterns of $Bi_4Ti_3O_{12}$ synthesized from an aqueous solution using lactic acid at different temperatures.

solution, heat treatment was carried out for 5 hours at 600°C, 800°C, and 1000°C. XRD patterns of the heated samples at each temperature are shown in Figure 5.16. Although $Bi_4Ti_3O_{12}$ was detected as the main phase, the presence of $Bi_2O_3$ was observed as an impurity, and its amount decreased as the heat treatment temperature increased. In this case, however, Bi deficient phase is not detected in the sample heat treated at 1000°C. Therefore loss of bismuth is negligible in the synthesis of $Bi_4Ti_3O_{12}$ powder by the present method.

In contrast to powder synthesis, volatilization of bismuth oxide during heat treatment should be controlled in the synthesis of $Bi_4Ti_3O_{12}$ film. For example, a coating of $Bi_4Ti_3O_{12}$ from an aqueous solution on a silicon substrate sputtered with platinum for an electrode is described. The substrate was coated with the aqueous solution containing a stoichiometric ratio of bismuth and titanium, followed by drying and heating at 400°C for pyrolysis of lactic acid. This coating-drying-pyrolysis was repeated until the desired thickness has been achieved. The prepared films were crystallized by heating at 600°C and 800°C. The XRD patterns of these films are shown in Figure 5.17. $Bi_2Ti_2O_7$ detected in the sample heated at 800°C has a pyrochlore structure and does not have ferroelectric properties. This impurity is caused by volatilization of bismuth during the heat treatment. Therefore low-temperature heat treatment should be carried out in order to prevent volatilization of bismuth and obtain a single-phase film of $Bi_4Ti_3O_{12}$ by the present method. Figure 5.18 shows an SEM image of a cross section of the sample heated at 600°C. The thickness is about 230 nm. There are no cracks on the surface and packing density is high, therefore electric insulation, necessary for FeRAM, is good.

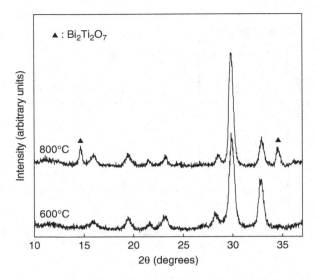

**FIGURE 5.17** X-ray diffraction patterns of $Bi_4Ti_3O_{12}$ films on silicon substrate synthesized from an aqueous solution using lactic acid at different temperatures.

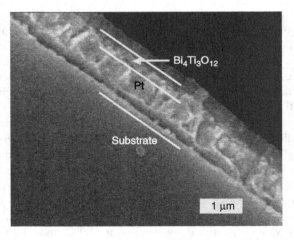

**FIGURE 5.18** SEM image of cross section of $Bi_4Ti_3O_{12}$ film on silicon substrate sputtered with platinum synthesized from an aqueous solution using lactic acid.

## F. "COMPOUND" PRECIPITATION METHOD

"Sol-gel" method described above is a process where the stoichiometric amount of the constituent metallic elements are dissolved in an aqueous solution and the solution is evaporated and heated to obtain the final product. In contrast, the "compound" precipitation method is where a unique compound containing metal ions in the desired stoichiometry forms a precipitate that can be isolated for

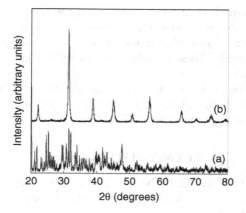

**FIGURE 5.19** X-ray diffraction patterns of (a) "compound" precipitation of citratoperoxotitanium complex with barium and (b) $BaTiO_3$ synthesized by "compound" precipitation.

further heat treatment. For example, in the case of $BaTiO_3$, a compound with the Ba:Ti ratio of 1:1 is filtered out and heat treated to obtain the product. In this method, since to form the "compound," precipitation containing metallic elements in the ratio of the product is a prerequisite, synthesis of ceramics containing equal or more than three kinds of metallic elements is difficult. This method should be distinguished from the conventional co-precipitation method when the insoluble salts of metals form fine powders of the individual compounds intimately mixed with each other during the rapid precipitation process. Thus, the "compound precipitation" method is superior to "sol-gel" and especially to co-precipitation method because it provides the homogeneity on the atomic level, which is never achieved by co-precipitation and which is exceptional even for "sol-gel" techniques.

Citratoperoxotitanium complex forms "compound" precipitation with barium and strontium by controlling the pH, temperature, and concentration of the solution. The ratios of metallic elements of both precipitates are 1:1, therefore $BaTiO_3$ and $SrTiO_3$ are synthesized by heat treatment. The color of both precipitates is orange. The XRD pattern of the "compound" precipitation with barium, shown in Figure 5.19a, indicates that this precipitate is not amorphous but crystalline. The composition of the precipitate was found to be $Ba[Ti(C_6H_4O_7)O_2]\cdot2H_2O$ by the results of TG/DTA and elemental analysis. This precipitate was heat treated for 5 h at 700°C to obtain $BaTiO_3$. The XRD pattern of the heat-treated sample is shown in Figure 5.19b. Single-phase $BaTiO_3$ is obtained, indicating that this precipitate can be used as a source of $BaTiO_3$. Further details of this study are reported in Reference 47.

## V. CONCLUSION

In this chapter, the merits of utilizing water as the solvent in the synthesis of ceramics, the synthesis and structure of water-soluble titanium complex, and examples of the synthesis of various ceramics using the complex and water solvent were described. A series of new water-soluble titanium complexes has been synthesized that makes possible the use of water as a solvent in synthesis of titanium-based ceramics. These complexes are easily synthesized, have low cost, are harmless and safe, consequently they are favorable materials from an industrial and environmental point of view. Also, these complexes are effective in improving the properties of the product, therefore they are interesting as a subject of research. Further study will lead to more practical applications.

## ACKNOWLEDGMENTS

We are grateful to Prof. Sasaki, Dr. Nakamura, Dr. Shiro, and Dr. Tada for single-crystal x-ray diffraction and NMR measurement, analysis and contributions of data, and helpful discussions.

## REFERENCES

1. Negishi, N. Takeuchi, K., and Ibusuki, T., *J. Sol-Gel Sci. Technol.*, 13, 691, 1998.
2. Kajihara, K., and Yao, T., *J. Sol-Gel Sci. Technol.*, 12, 193, 1998.
3. Nam, H.J., Itoh, K., and Murabayashi, M., *Electrochemistry*, 70, 429, 2002.
4. Shimooka, H., and Kuwabara, M., *J. Am. Ceram. Soc.*, 78, 2849, 1995.
5. Yamashita, Y., Yoshida, K., Kakihana, M., Uchida, S., and Sato, T., *Chem. Mater.*, 11, 61, 1999.
6. Veith, M., Mathur, S., Lecerf, N., Huch, V., Decker, T., Beck, H.P., Eiser, W., and Haberkorn R., *J. Sol-Gel Sci. Technol.*, 15, 145, 2000.
7. Kakihana, M., Okubo, T., Arima, M., Nakamura, Y., Yashima, M., and Yoshimura, M., *J. Sol-Gel Sci. Technol.*, 12, 95, 1998.
8. Smith, J.S., Dolloff, R.T., and Mazdiyasni, K.S., *J. Am. Ceram. Soc.*, 53, 91, 1979.
9. Varma, H.K., Pillai, P.K., Mani, T.V., Warrier, K.G.K., and Damodaran, A.D., *J. Am. Ceram. Soc.*, 77, 129, 1994.
10. Rho, Y.H., Kanamura, K., and Umegaki, T., *Chem. Lett.*, 12, 1322, 2001.
11. Stahler, A., *Chem. Ber.*, 37, 4405, 1904.
12. Stahler, A., *Chem. Ber.*, 38, 2619, 1905.
13. Pecsok, R.L., *J. Am. Chem. Soc.*, 73, 1304, 1951.
14. Eve, D.J., and Fowles, G.W.A., *J. Chem. Soc. A*, 9, 1183, 1966.
15. Phule, P.P., and Khairulla, F., *Mater. Res. Soc. Symp. Proc.*, 180, 527, 1990.
16. Varma, H.K., Pillai, P.K., Sreekumar, M.M., Warrier, K.G.K., and Damodaran, A.D., *Br. Ceram. Trans. J.*, 90, 189, 1991.
17. Kao, C.F., and Yang, W.D., *Mater. Trans. JIM*, 37, 142, 1996.
18. Yi, G., and Sayer, M., *J. Sol-Gel Sci. Technol.*, 6, 75, 1996.
19. Zheng, H., Liu, X., Meng, G., and Sorensen, O.T., *J. Mater. Sci. Mater. Electron.*, 12, 629, 2001.
20. Kakihana, M., Tada, M., Shiro, M., Petrykin, V., Osada, M., and Nakamura, Y., *Inorg. Chem.*, 40, 891, 2001.
21. Kakihana, M., Szanics, J., and Tada, M., *Bull. Korean Chem. Soc.*, 20, 893, 1999.

22. Tada, M., Yamashita, Y., Petrykin, V., Osada, M., Yoshida, K., and Kakihana, M., *Chem. Mater.*, 14, 2845, 2002.
23. Pecsok, R.L., and Maverick, E.F., *J. Am. Chem. Soc.*, 76, 358, 1953.
24. Sweetser, P.B., and Bricker, C.E., *Anal. Chem.*, 26, 196, 1954.
25. Rao, C.N.R., Turner, A., and Honing, J.M., *J. Phys. Chem. Solids*, 11, 173, 1959.
26. Yoganarasimhan, S.R., and Rao, C.N.R., *Trans. Faraday Soc.*, 58, 1579, 1962.
27. Sullivan, W.F., and Coleman, J.R., *J. Inorg. Nucl. Chem.*, 24, 645, 1962.
28. Hishita, S., Mutoh, I., Koumoto, K., and Yanagida, H., *Ceram. Int.*, 9, 61, 1983.
29. Yang, J., and Ferreira, J.M.F., *Mater. Lett.*, 36, 320, 1998.
30. Arima, M., Kakihana, M., Yashima, M., and Yoshimura, M., *Eur. J. Solid State Inorg. Chem.*, 32, 863, 1995.
31. Gopalakrishnamurthy, H.S., Subba Rao, M., and Narayanan Kutty, T.R., *J. Inorg. Nucl. Chem.*, 37, 891, 1975.
32. Hennings, D., Rosenstein, G., and Schreinemacher, H., *J. Eur. Ceram. Soc.*, 8, 107, 1991.
33. Hasenkox, U., Hoffmann, S., and Waser, R., *J. Sol-Gel Sci. Technol.*, 12, 67, 1998.
34. Wada, S., Chikamori, H., Noma, T., and Suzuki, T., *J. Mater. Sci. Lett.*, 19, 245, 2000.
35. Wagner, F.T., and Somorjai, G.A., Photocatalytic hydrogen production from water on Pt-free SrTiO₃ in alkali hydroxide solutions, *Nature*, 285, 559, 1980.
36. Domen, K., Naito, S., Soma, M., Onishi, T., and Tamaru, K., *J. Chem. Soc. Chem. Commun.*, 12, 543, 1980.
37. Carr, R.G., and Somorjai, G.A., Hydrogen production from photolysis of steam adsorbed onto platinized SrTiO₃, *Nature*, 290, 576, 1981.
38. Domen, K., Naito, S., Onishi, T., and Tamaru, K., *Chem. Phys. Lett.*, 92, 433, 1982.
39. Domen, K., Naito, S., Onishi, T., and Tamaru, K., *J. Phys. Chem.*, 86, 3657, 1982.
40. Li, Q.S., Domen, K., Naito, S., Onishi, T., and Tamaru, K., *Chem. Lett.*, 3, 321, 1983.
41. Baba, R., and Fujishima, A., *J. Electroanal. Chem.*, 213, 319, 1986.
42. Domen, K., Kudo, A., Onishi, T., Kosugi, N., and Kuroda, H., *J. Phys. Chem.*, 90, 292, 1986.
43. Domen, K., Kudo, A., and Onishi, T., *J. Catal.*, 102, 92, 1986.
44. Jing, L., Xu, Z., Sun, X., Shang, J., and Cai, W., *Appl. Surf. Sci.*, 180, 308, 2001.
45. Kudo, A., and Kato, H., *Chem. Phys. Lett.*, 331, 373, 2000.
46. Tomita, K., and Kakihana, M., *Trans. Mater. Res. Soc. Jpn.*, 28, 377, 2003.
47. Tada, M., Tomita, K., Petrykin, V., and Kakihana, M., *Solid State Ionics*, 151, 293, 2002.

# 6 Peroxoniobium-Mediated Route toward the Low-Temperature Synthesis of Alkali Metal Niobates Free from Organics and Chlorides

*Deepa Dey and Masato Kakihana*

## CONTENTS

## I. INTRODUCTION

Niobium oxides (niobates) are the topic of active research due to their burgeoning technological importance, particularly with regard to their successful use in a variety of applications in electronics and electro-optics, including dielectric ceramics,[1] piezoelectric ceramics,[2] and ferrites.[3] Along with these applications, the use of LiNbO$_3$ as a photorefractive material for nonvolatile volume holographic data storage systems[4] and its use as a substrate for the preparation of gigahertz-range surface acoustic wave (SAW) transducers[5] because of its high electromechanical coefficient, make it a potent material for technological advancement. On the other hand, the introduction of a small amount of doped metal ions such as barium, lithium, etc. in $NaNbO_3$ makes it a challenging material with piezoelectric, ferroelectric, pyroelectric, and electro-optic properties.[6,7] It has been observed that the chemical and microstructural homogeneity of alkali metal niobates seriously affect the electro-optical behavior, therefore the synthesis of ceramic powders with good stoichiometry and homogeneity is necessary to develop these materials. Their traditional method of preparation often leads to powders of poor compositional homogeneity, especially because of the easy volatilization of alkali metal, since high temperature is required for synthesis. In turn, a number of alternatives based on wet chemical methods for low-temperature synthesis of niobates are becoming increasingly important. These methods, including sol-gel, the polymerizable complex method, the amorphous complex method, etc. provide excellent techniques for the synthesis of highly pure multi-component oxides at a reduced temperature. However, there are certain weaknesses related to these methods, the most severe being the use of a huge amount of organics, sometimes reaching almost 80% of the total weight.[8] This causes concern due to difficulty in effectively removing large amounts of organic substances. Keeping these in mind, we thought of designing a method that would overcome the problem of organics, while keeping the advantages of the above methods.

The peroxide-based method is not common in the synthesis of ceramics. However, one can harness the important property of dissolution of inert metal oxides in hydrogen peroxide for synthesizing technologically important materials, thus providing a route where metal oxides can be used instead of other metal sources. Thus we have developed a method based on the peroxide route for the successful synthesis of $LiNbO_3$ and $NaNbO_3$ powders at reduced temperatures without the use of organics. The first step involves the dissolution of niobium pentoxide in hydrogen peroxide in the presence of ammonia, leading to the formation of niobium peroxo complex followed by its reaction with corresponding cations. The advantages of this method are (1) the use of $Nb_2O_5 \cdot 3H_2O$ as the niobium source instead of alkoxides or chlorides; (2) the water solubility of the niobium peroxide complex; (3) the use of no organic substances; (4) the simplicity of the procedure, thus eliminating time-consuming steps; (5) water, a convenient and environmentally compatible medium, is preferred to other toxic organic solvents; and (6) it is cost effective.

Moreover, the isolation of self-assembled $LiNbO_3$ powders using this route has added credibility to the methodology; they were otherwise prepared by templating colloidal crystals of polyelectrolyte-coated spheres.[8] The interest in $LiNbO_3$ inverse opals stems from the fact that they have a constant refractive index, but a spatially periodic second-order nonlinear susceptibility. Such nonlinear periodic structures allow for efficient quasi-phase-matched second-order harmonic generation, which could find applications where simultaneous conversion of multiple wavelengths is required.[9] Thus, in this chapter we will focus our attention on the peroxide method for the clean and low-temperature synthesis of $LiNbO_3$ and $NaNbO_3$ powders.

## II. EXPERIMENTAL SYNTHESIS OF $LiNbO_3$ AND $NaNbO_3$ POWDERS

The key to successful synthesis and isolation of transition metal peroxo complexes depends on the reaction conditions, such as temperature and the pH of the medium.[10] With a slight variation of the reaction pH, one can end up with compound of different stoichiometry. In the present case, for the dissolution of niobium pentoxide, we found a pH of 11 (vide experimental) to be conducive for precipitation and isolation of highly crystalline and stable niobium peroxo compound from aqueous solution having a composition $[Nb(O_2)_4]^{-3}$. The reaction can be summarized as shown in Equation 6.1. In most of the peroxo metal complexes reported so far, it has been observed that in order to stabilize a peroxo metal complex in solution and then isolate it in solid state requires the presence of an ancillary ligand. However, in the present case we could isolate the

$$Nb_2O_5 + 8H_2O_2 + 6NH_3 \rightarrow 2(NH_4)_3[Nb(O_2)_4] + 5H_2O \qquad (6.1)$$

peroxoniobate complex in the absence of ancillary ligand. Furthermore, in the absence of base or at a pH lower than 11, the dissolution of niobium pentoxide in virgin hydrogen peroxide was not favored. Thus, having achieved the successful synthesis of niobium peroxo compound, we employed this as the precursor for the synthesis of niobium-based ceramics.

The flowchart depicting the synthesis of $NaNbO_3$ by the peroxide method is shown in Figure 6.1. In a typical synthesis, 40 ml of 30% hydrogen peroxide was added to 2 g (6.25 mmol) of $Nb_2O_5 \cdot 3H_2O$, followed by the addition of 5 ml of aqueous ammonia (28%), maintaining the pH at 11. The solution was allowed to stand at room temperature for 30 min. Thereafter, a clear colorless solution characteristic of orthoperoxoniobate,[11] of the general formula $[Nb(O_2)_4]^{-3}$, was obtained. To confirm the presence of peroxoniobium compound, a portion of the clear solution was allowed to stand at room temperature for 12 h, which isolated white shiny crystals suitable for single-crystal x-ray diffraction measurement. The precise peroxide content was determined by redox titration involving potassium permanganate solution[12] in the presence of boric acid, suggesting the occurrence of four peroxide groups per $Nb^{5+}$ center. An aqueous solution of 0.66 g (6.22 mmol) of $Na_2CO_3$ (or $LiOH \cdot H_2O$) in 15 ml of distilled water was added to the peroxoniobate solution. The solution was allowed to stand at room temperature for 30 min and then was warmed to 70 to 80°C for 10 min, resulting in a vigorous evolution of gas. The exothermic nature of the reaction at this point is due to the evolution of active oxygen and of ammonia as reported[13] for tetraperoxoniobate, which decomposes on heating at 80 to 100°C. After a few seconds, white precipitate was formed. The precipitate was filtered, washed with water,

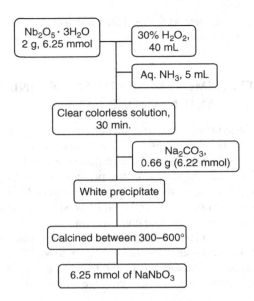

**FIGURE 6.1** Flowchart showing the synthesis of NaNbO$_3$.

and dried in a vacuum over concentrated H$_2$SO$_4$. The white powder thus obtained was subjected to heat treatment up to 600°C for 2 h. Thus the peroxide method is extraordinary simple, and since no organics are used, it is also extremely clean compared to previously reported methods.

## III. RESULTS AND DISCUSSION

The thermogravimetric profiles obtained for both NaNbO$_3$ precipitate (Figure 6.2) and LiNbO$_3$ precipitate were more or less similar. In the thermogram of NaNbO$_3$ precipitate, weight loss was observed in two steps. The first step, between 40 and 200°C, shows a weight decrease of 17.48% followed by another weight loss of 15.02% between 200 and 380°C. A broad endothermic depression at 71°C and a small sharp exothermic peak at 380°C can be seen in the differential thermal analysis (DTA) curve. The weight losses are most likely due to the dehydration and evaporation of water, which is consistent with the endothermic depression at 71°C. The DTA curve shows an exothermic peak between 390 and 410°C that can be attributed to NaNbO$_3$ crystallization. Similarly in the case of LiNbO$_3$, the thermogram shows two weight losses for a total of 32%, corresponding to the dehydration and evaporation of water molecules. The broad exothermic peak between 390 and 440°C was credited to the crystallization of LiNbO$_3$ powder.

The x-ray diffraction pattern of the precipitated powders and the powders calcined at different temperatures are shown in Figure 6.3 and Figure 6.4, respectively. The NaNbO$_3$ precipitate and powder calcined at 300°C/2 h showed broad reflections between 20 and 69°C, which are characteristic of low-temperature

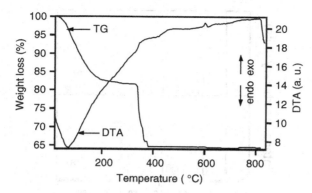

**FIGURE 6.2** TG/DTA curves for the precipitate obtained from $[Nb(O_2)_4]^{-3}$ and $Na^+$ solutions used as a precursor for synthesis of $NaNbO_3$ powder.

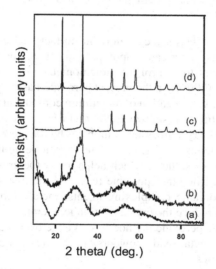

**FIGURE 6.3** X-ray diffraction profile for (a) $NaNbO_3$ precipitate and powders calcined for 2 h at (b) 300°C, (c) 400°C, and (d) 500°C.

formation of the $NaNbO_3$ phase; in fact, the initiation of crystallization was observed for the powder calcined at 300°C. During the course of calcinations, pure crystalline $NaNbO_3$ powder was attained at a very low temperature (400°C) (Figure 6.3c) without the presence of any intermediate phase. All diffraction peaks in the curve were assigned to the $NaNbO_3$ phase. Similarly, complete crystallization of $LiNbO_3$ powder occurred when its powder precursor was heat treated at 400°C for 2 h (Figure 6.4a). Lowering of the operational temperature is of particular importance in the synthesis of complex oxides with elements having significant vapor pressure, even at relatively low temperatures.

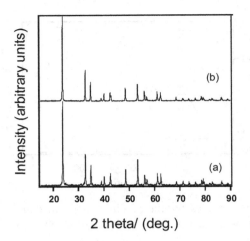

**FIGURE 6.4** X-ray diffraction pattern for the LiNbO₃ powders calcined for 2 h at (a) 400°C and (b) 600°C.

The synthesis of LiNbO₃ is a case in point. In fact, the severe loss of lithium components due to its significant vapor pressure at temperatures higher than 600°C makes it difficult to control the stoichiometry of complex oxides.[14] The operative temperature (400°C) in the present peroxide-based synthesis of LiNbO₃ is low enough to avoid volatilization of the lithium component during calcinations and high enough to fully crystallize LiNbO₃. In comparison to the other soft solution-based method, crystallization of niobates by the present method has been obtained at very low temperatures, one of the reason being the efficient mixing of the ions in the peroxo solution which helps lower the crystallization temperature down to 400°C. The other possible reason is the total absence of organic substances, such as organic complexing agents, metal alkoxides, and organic solvents, which helps reduce the crystallization temperature and also reduces the reaction time. Thus the peroxide method has an edge over the other soft solution methods in obtaining metal oxide powders at relatively lower temperatures and being free from organics.

The crystalline particle size for NaNbO₃ (400°C) and LiNbO₃ (400°C) determined by the x-ray diffraction pattern using the Debye-Scherrer equation showed values of 29 and 25 nm, indicating the small particle size. Small particle size is often related with high specific surface area. This point was probed for the NaNbO₃ powders calcined at 400 and 500°C, as we could get high Brunauer-Emmet-Teller (BET) specific surface areas of 18.72 and 24.60 m²/g. For the powder calcined at 400°C, we observed a slight decrease in the value, which might be due to the presence of a small amount of amorphous powder, as observed in the x-ray diffraction. For the NaNbO₃ powders calcined at 400°C and 500°C, the equivalent diameter ($D_{BET}$) calculated by using the relation

$$D_{BET} = 6/\rho_T S_{BET},$$

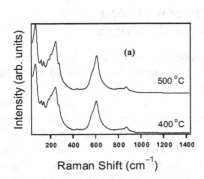

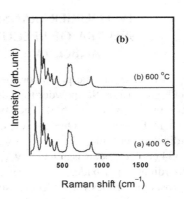

**FIGURE 6.5** Raman spectra of (a) NaNbO$_3$ powders calcined at 400°C and 500°C and (b) LiNbO$_3$ powders calcined at 400°C and 600°C.

where $\rho_T$ is the theoretical density (4.575 g/cm$^3$) of NaNbO$_3$, was found to be 70.05 and 53.30 nm.

The LiNbO$_3$ and NaNbO$_3$ powders prepared by the peroxide route were further characterized by Raman spectroscopy (Figure 6.5). Moreover, in recent years Raman spectroscopy is being used not only for identifying the pure phase of the material concerned, but it also helps in detecting the presence of impurities in the processed powders. For instance, the soft solution methods, such as the polymerizable and amorphous complex methods, use excess of organics, and several authors have claimed the synthesis of crystalline powders by these methods at low temperatures based only on x-ray diffraction evidence, often neglecting the possible presence of residual carbon. However, Camargo et al.[15,16] have shown using Raman spectroscopy that although the crystallization occurs at lower temperatures in these methods, the removal of residual carbon requires a temperature that is higher then the crystallization temperature obtained by x-ray diffraction. The Raman spectra of these x-ray crystallized products with residual carbon are broad and cannot be assigned with conformity to a particular phase. Only on further heating at higher temperatures does one get the Raman spectra that can be used for identification of a particular phase. Thus one can infer from this observation that by using the methods that employ excess organics, although crystallization may be obtained at a lower temperature, the particulates of carbon remain until the samples are further processed at higher temperatures. The Raman spectra obtained for the LiNbO$_3$ and NaNbO$_3$ powders obtained by the peroxide-based method calcined at 400°C were characteristic of well-known LiNbO$_3$ and NaNbO$_3$ patterns and showed no extra peak due to any impurities. Thus Raman spectroscopy, in addition to identifying the pure phase, also confirms the fact that the peroxide method provides the purist final product.

## IV. FOURIER TRANSFORM INFRARED (FTIR) SPECTRA OF PRECURSOR SOLIDS AND THE ANbO$_3$ (A = Li OR Na) POWDERS

To gain insight into the precursor (peroxoniobium complex) → intermediate → ANbO$_3$ (A = Li or Na) powder transition, FTIR spectra were recorded at various processing steps (Figure 6.6). The spectrum of the initial peroxide precursor is shown in Figure 6.6a. One can clearly observe the distinctive features of the coordinated peroxide. The most significant bands, observed at 813 cm, 581 cm, and 534 cm have been assigned to $v_{(O-O)}(v_1)$, $v_{(Nb-O2)}(v_3)$, and $v_{(Nb-O2)}(v_2)$ modes[17,18] of coordinated peroxide, respectively. The O-O and Nb-O$_2$ bands are important spectroscopic probes for molecular structure determination and are amenable to infrared and Raman spectroscopic studies. The observed positions of $v_{(O-O)}$ and $v_{(Nb-O2)}$ modes correspond to those expected for a triangularly bidentate (C$_{2v}$) bonded peroxide ligand to the Nb$^{5+}$ center.

Addition of LiOH·H$_2$O (or Na$_2$CO$_3$) solution to the niobium peroxo solution produces a clear colorless solution with lots of effervescence. The solution remains clear when it is allowed to stand at room temperature for several hours. However, when the solution is warmed on a water bath at 80 to 90°C for 10 min, the reaction becomes exothermic with further evolution of gas. After a few seconds, some white compound precipitates out. The FTIR for this compound is shown in Figure 6.6b. The infrared (IR) features obtained for the white powder were completely different from those of the peroxide precursor, with a broad band being observed between 700 cm and 400 cm in the case of lithium salt (Figure 6.6b) and a wide band at 540 cm with a shoulder at 770 cm in the case of sodium, with additional bands at about 3400 cm and 1600 cm due to coordinated water. The FTIR spectra for the lithium and sodium powders heated at 400°C showed IR features that are characteristic of LiNbO$_3$ and NaNbO$_3$. Thus

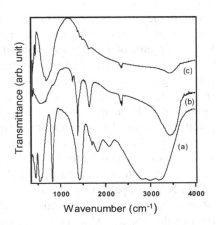

**FIGURE 6.6** FTIR spectra of (a) [Nb(O$_2$)$_4$]$^{-3}$, (b) intermediate, and (c) LiNbO$_3$ powder.

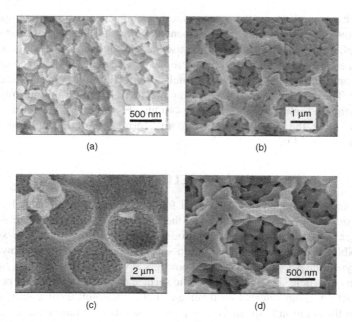

**FIGURE 6.7** Scanning electron micrographs of (a) LiNbO₃ powder calcined at 500°C and (b–d) self-assembled inverse opals of LiNbO₃ powders calcined at 600°C

an increase in the reaction temperature above ambient favors the formation of inorganic material free from peroxide moiety. Postsynthesis heating of the powder to 400°C results in removal of water molecules and crystallization of LiNbO₃ and NaNbO₃. This is a remarkable result in view of the fact that the maximum temperature used in the peroxo process is only 400°C, whereas the so-called amorphous citrate method required 650°C and 550°C to obtain truly homogeneous and transparent samples.

All these results convincingly indicate the success of the peroxide route over other reported methods in lowering the crystallization temperature for LiNbO₃ and NaNbO₃ down to 400°C and isolating powders free of organics. Further, lowering of the crystallization temperature to 400°C may be associated with an improved level of mixing of cations in the powder precursor derived from the newly developed aqueous solution route. Overall, the procedure developed in the present study permits the preparation of pure and homogeneous LiNbO₃ and NaNbO₃ at reduced temperatures.

The scanning electron micrographs (depicted in Figure 6.7a) for LiNbO₃ powder calcined at 500°C showed a fine, uniform particle size distribution and homogeneous microstructure. The average particle size determined from the scanning electron micrograph images was found to be about 100 nm.

Interestingly, the powder calcined at 600°C showed the formation self-assembled inverse opals of LiNbO₃ (Figure 6.7b–d). Usually the reported methods for the preparation of LiNbO₃ inverse opals use colloidal crystals as templates,[8] but by the present method we have been able to obtain them using no template at all. Moreover, these materials are of immense significance because of their importance in photonics,[19] where they can be used as photonic crystals for the manipulation of electromagnetic waves, for example, by localizing light and controlling spontaneous emissions.[20,21] The channels connecting the pores are clearly observed. The average pore size was found to be 1.45 μm and the average size of the channels connecting the pores is 440 nm (Figure 6.7b). However, we are not clear about the mechanism behind the formation of the self-assembled macroporous $LiNbO_3$.

## V. CONCLUSION

Here we have developed a new method for the synthesis of niobates based on the peroxide route, which employs neither the use of organics nor chlorides, unlike the other soft solution methods; in fact, it uses a water-soluble $(NH_4)_3[Nb(O_2)_4]$ complex as the niobium source. Thus, this method allows us to obtain crystalline $NaNbO_3$ and $LiNbO_3$ powders at very low temperatures (400°C) with a small particle size and high specific surface area. The presence of peroxide oxygen acting as an internal oxidizer and better dissolution of corresponding cations in the solution helps in lowering the crystallization temperature. In addition, we observed the formation of self-assembled inverse opals of $LiNbO_3$ without the use of any templates. Thus we have been able to successfully unravel a new facet of metal-peroxo chemistry in obtaining industrially important $NaNbO_3$ and $LiNbO_3$ powders in an environmentally benign way, which opens a new path in the synthesis of other materials following this pathway.

## ACKNOWLEDGMENTS

We thank K. Tomita and K. Yoshioka for their assistance with electron microscopy and Raman measurements.

## REFERENCES

1. Nami, R.B., Capacitor manufacturers face challenges in the next century, *Ceram. Ind.*, December, 28, 1997.
2. Moulsoun, A.J., and Herbert, J.M., *Electroceramics*, Chapman & Hall, London, 1990.
3. Sugimoto, M., *J. Am. Ceram. Soc.*, 82, 269, 1999.
4. Hesselink, L., Orlov, S.S., Liu, A., Akella, A., Lande, D., and Neurgaonkar, R.R., Photorefractive materials for nonvolatile volume holographic data storage, *Science*, 282, 1089, 1998.
5. Takagaki, Y., Wiebicke, E., Kostial, H., and Ploog, K.H., *Nanotechnology*, 13, 15, 2002.
6. Pardo, L., Duran-Martin, P., Mercurio, J.P., Nibou, L., and Jimenez, B., *J. Phys. Chem. Solids*, 58, 1335, 1997.

7. Konieczny, K., *Mater. Sci. Eng. B*, 60, 124, 1999.
8. Wang D., and Caruso, F., *Adv. Mater.*, 15, 205, 2003.
9. Broderick, N.G.R., Ross, G.W., Offerhause, H.L., Richardson, D.J., and Hanna, D.C., *Phys. Rev. Lett.*, 84, 4345, 2000.
10. Connor, J.A., and Ebsworth, E.A.V., *Adv. Inorg. Chem. Radiochem.*, 6, 279, 1964.
11. Clark, R.J.H., and Brown, D., *The Chemistry of Vanadium, Niobium and Tantalum*, Pergamon Press, New York, 598, 1973.
12. Vogel, A.I., *A Textbook of Quantitative Inorganic Analysis*, Longmans, Green and Co., New York, 295, 1962.
13. Selezneva, K.I., and Nisel'son, L.A., *Russian J. Inorg. Chem.*, 13, 45, 1968.
14. Hultgren, R.R., Orr, R.L., Anderson, P.D., Kelly, K.K., *Selected Values of Thermodynamic Properties of Metals and Alloys*, John Wiley & Sons, New York, 153, 1963.
15. Camargo, E.R., Popa, M., and Kakihana, M., *Chem. Mater.*, 14, 2365, 2002.
16. Camargo, E.R., and Kakihana, M., *Solid State Ionics*, 151, 413, 2002.
17. Griffith, W.P., and Wickens, T.D., *J. Chem. Soc.* A, 2, 397, 1968.
18. Griffith, W.P., *J. Chem. Soc.*, Nov., 5345, 1963.
19. Krauss, T.F., and de la Rue, R.M., *Prog. Quant. Electron.*, 23, 51, 1999.
20. Yablonovitch, E., *Phys. Rev. Lett.*, 58, 2059, 1987.
21. John, S., *Phys. Rev. Lett.*, 58, 2086, 1987.

# 7 Synthesis and Modification of Submicron Barium Titanate Powders

*Burtrand I. Lee, Xinyu Wang, D.H. Yoon,*
*Prerak Badheka, Lai Qi, and Li-Qiong Wang*

## CONTENTS

As miniaturization with increasing functionality and acceleration of clock speed of electronic devices continues, the desired characteristics and processability of the starting powders for passive components become more of an issue. For the capacitor material barium titanate ($BaTiO_3$; BT), one processing issue is the chemical stability of BT in water as the slip medium. Aqueous processing of BT is considered to be a replacement for the current state-of-the-art nonaqueous processing technology for obvious reasons. A greater volumetric efficiency of passive components, and multilayer ceramic capacitors (MLCCs) in particular, points toward thinner dielectric layers or embedded design. All of these require a smaller particle size of the raw BT powders with closely controlled particle morphology, high dispersibility, and better dielectric properties. Thus nanocrystalline BT particles have been prepared by ambient-condition sol processes starting from soluble precursors of barium and titanium under controlled experimental variables. The resulting particles of different sizes were characterized for morphology and crystal phases.

Commercial submicron BT particles were modified for the study of the preferred crystal phase relationship to the corresponding dielectric properties. The results reveal that the crystal phase of the BT particles in the nanometer size range relates to the impurities incorporated in the BT crystal lattice. The responsible impurity has been identified as hydroxyls. BT is considered to meet the demands for current and future capacitor applications by modifying submicron BT particles.

## I. INTRODUCTION

Barium titanate, owing to its excellent dielectric and ferroelectric properties, is the most widely used material for electronic capacitors. BT-based dielectrics have dominated the ceramic capacitor industry since the 1950s, representing 80 to 90% of the business.[1,2] BT-based MLCCs are processed beginning with fine BT powders dispersed in a liquid medium and tape-cast followed by electroding, laminating, dicing, and sintering. Traditionally BT powder is prepared by a solid state reaction of fine barium carbonate ($BaCO_3$) and titanium dioxide ($TiO_2$) powders at high temperatures of around 1000°C followed by milling.[3,4]

With the rapid miniaturization of electronic devices, the downsizing of MLCCs has been accelerating. As a result, it is expected that the thickness of each dielectric layer in MLCCs will become less than 1 μm. Consequently the particle size of BT powders will continue to decrease to a few tenths of a nanometer. In the past few decades, extensive studies have been conducted to produce nanosized BT powders with narrow particle size distributions, controlled morphology, and high purity. BT nanocrystals have been synthesized by various techniques: the hydrothermal method,[5–24] the sol-gel process,[25–30] low-temperature aqueous synthesis (LTAS),[31–33] low-temperature direct synthesis (LTDS),[34,35] combustion synthesis,[36] oxalate coprecipitation,[37,38] microwave heating,[39] and the microemulsion process.[40] Among these methods, wet chemical synthesis, such as hydrothermal and LTAS, involves reactions in strongly alkaline aqueous media. These provide a promising method for producing nanocrystalline BT with narrow particle size distributions, high purity, and controlled morphology of the resulting particles at a relatively low temperature under mild pressure.

However, in ferroelectric fine particles, ferroelectricity decreases with decreasing particle and grain size, and disappears below a certain critical size, the so-called the size effect in ferroelectrics.[41–47] The preferred tetragonal phase of BT may be unstable at room temperature for a crystallite size below a certain size at which the stable phase is cubic. Therefore the size effect in ferroelectric materials such as BT can be considered to be one of the most important phenomena of interest to the industry as well as the scientific community. Some researchers have estimated the critical size to be around 30 to 100 nm in fine BT particles.[48–50] The preferred tetragonal phase of BT can be identified by x-ray diffraction (XRD). Venigalla[51] showed that the size limit distinguishing the tetragonal phase from the cubic phase is approximately 200 nm by using XRD peak splitting of (002) and (200) at $2\theta = \sim 45$. Below this size, the BT particles appear to be cubic phase under XRD examination. We define this as tetragonality or the crystallographic c/a ratio of the BT crystal lattice. We have reported that the phase change from cubic to tetragonal, depending on the crystallite size, is a gradual transition with the increase in c/a ratio. Hence there is no one critical size dividing the phases[20] exhibiting a range of degree of tetragonality.

Despite the well-known chemical instability of BT in water,[52,53] aqueous processing of BT needs to replace the current state-of-the-art nonaqueous processing technology for the obvious reasons. The instability of BT in water, which affects the exhibited dispersion in an adverse manner, is caused by the preferential leaching of barium cations from the BT lattice. This behavior affects the dispersibility and the barium/titanium compositional ratio in the final product.[54,55] In order to use a water-based slip system with greater reliability, one may introduce a passivating agent layer (PAL) onto the BT particle surface to reduce $Ba^{2+}$ leaching. A PAL is also used as a dispersant to prevent the particles from coagulating in the water medium.[55]

Redispersible spherical BT nanoparticles with a narrow size distribution and a lower hydroxyl content were produced by our novel method, which we call the ambient condition sol (ACS) process, to produce pure crystalline nanometer-sized BT.[56–58] A controlled particle size with a uniform distribution in the 50 to 100 nm size range from different refluxing media (e.g., water, butanol, toluene) was rapidly produced. During the process, a polymeric surface modifier can be added to protect the surface from agglomeration and to promote redispersibility after drying. The ACS process is similar to hydrothermal synthesis, but without applying pressure.

Modern electronics in advanced electronic packaging applications, such as integral decoupling capacitors and tunable radiofrequency (RF) filters, will require embedded capacitors providing superior performance. Conventionally, discrete components are mounted onto a printed circuit board (PCB) or interconnected substrate. This results in higher parasitic capacitance, lower reliability, and large attachment area requirements.[59] Growing demand for faster clock speeds, miniaturization, and higher volumetric efficiency in MLCCs is leading the trend toward embedded passive components in system-on-packaging (SOP).[60–62] In one of the efforts, high dielectric constant (K) BT particles of approximately 0.2 μm were dispersed in the PCB polymer matrix (e.g., epoxy resin). This composite structure yielded the highest K less than about 150 at as high as 85 vol% ceramic content, with poor adhesion to metals and mechanical properties of the PCB, and dispersion difficulty of the ceramic particles in the polymer matrix.[63,64] The reasons for the unimpressive results are thought to be (1) the BT powder was predominantly cubic phase (i.e., low tetragonality), (2) the imperfect dispersion of the particles in the matrix, and (3) the particle size was too large, requiring high ceramic loading. Some commercially advertised embedded capacitors[59] are low K and the K needs to be increased.

**TABLE 7.1**
**List of BT Samples Prepared by the ACS Process in a Water Medium**

| Sample ID[a] | Ba/Ti | Mineralizer | pH | Reaction Time (h) |
|---|---|---|---|---|
| BT1 | 1.5:1 | KOH | 12.0 | 20 |
| BT2 | 1.5:1 | KOH | 14.0 | 20 |
| BT3 | 1.5:1 | KOH | 14.2 | 20 |
| BT4 | 1.5:1 | KOH | 14.0 | 6 |
| BT5 | 3:1 | KOH | 14.0 | 20 |
| BT6 | 1.5:1 | KOH | 14.0 | 20 |
| BT7 | 1.5:1 | KOH | 14.0 | 20 |
| BT8 | 1.5:1 | TMAH | 14.0 | 20 |
| BT9 | 1.5:1 | KOH | 14.0 | 20 |

[a] BT1–5 and 8: $BaCl_2 + TiCl_4$ precursors; BT6: $Ti(OC_2H_5)_4$ as titanium source; BT7: $Ba(OOCCH_3)_2$ as barium source; BT9: with 2 mg APA/ml.
TMAH = tetramethylammonium hydroxide.

## II. SUBMICRON BT PARTICLE SYNTHESIS VIA THE ACS PROCESS

A somewhat simpler method than the hydrothermal process to produce submicron-size BT powders is a novel technique called the ACS process.[56–58] We examine here some of the chemical experimental variables affecting particle morphology in the process. Here we define the ambient condition as near room temperature and pressure. To study the influence of experimental parameters on the properties of final BT products, a series of BT samples have been prepared as described in Table 7.1.

Dynamic light scattering (DLS) (model BI-9000AT, Brookhaven Instruments Corp., Holtsville, NY) was used to estimate the particle size and particle size distribution of the resulting BT powders. The samples were diluted with deionized water and ultrasonicated for 15 min just before the size analysis. Room temperature XRD (Scintag PAD V using CuKα with $\lambda = 0.15406$ nm) was used for crystalline phase identification and determination of the crystallite size of the powder. The crystallite size of $BaTiO_3$ powders was calculated by the Scherrer equation:

$$d_x = \frac{0.94\lambda}{\beta \cos\theta} \tag{7.1}$$

where $d_x$ is the crystallite size, $\lambda$ is the x-ray wavelength, $\beta$ is the full-width at half-maximum (FWHM), and $\theta$ is the diffraction angle. The (200) peak was used to calculate the crystallite size.

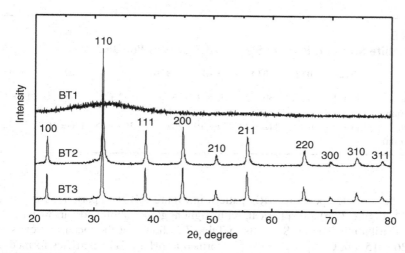

**FIGURE 7.1** Room temperature XRD patterns of BaTiO$_3$ samples prepared at pH 12 (BT1), 14 (BT2), and 14.2 (BT3).

The morphology of the BT powders was examined by scanning electron microscopy (SEM) and transmission electron microscopy (TEM). A small amount of the BT powders were pressed on a carbon tape, which attached to the brass sample stub for SEM. To prepare TEM samples, a tiny amount of the BT particles was dispersed in isopropanol by grinding in an agate mortar before placing it on a copper grid.

## A. EFFECT OF SOLUTION PH

To study the impact of the initial solution pH on particle morphology and crystal structure in the ACS process, BT gels were formed at pH 12, 14, and 14.2 for BT1, BT2, and BT3, respectively. The room temperature XRD patterns in Figure 7.1 indicate that the pH needs to be greater than 12 for BT powders to be crystalline. At pH 14.0 and 14.2, the powders were crystallized into cubic phase BT. It can be concluded that alkalinity plays an important role in the crystallization of BT in the ACS process. According to the thermodynamic model[65] for the hydrothermal process, phase-pure BT can only be obtained at a pH higher than 13.5.

The diffraction peaks of BT3 from pH 14.2 are slightly broader than BT2 from pH 14.0, indicating a smaller crystallite size of BT3. Calculated from the FWHM of (200) peak using equation (1), the crystallite size of BT3 is indeed smaller (42.6 ± 1.8 nm) than that of BT2 (51.5 ± 1.5 nm), as shown in Table 7.2. At a higher pH, a greater supersaturation by the increased solubility of the precursor gel leads to higher nucleation rate than growth rate.[9]

Figure 7.2 shows the scanning electron micrographs of the BT powders synthesized under different pHs and barium/titanium ratios. No individual particles could be seen in the chalky mass of BT1 gelled at pH 12 (Figure 7.2a), agreeing

**TABLE 7.2**
**Crystallite Size and Particle Size of ACS BaTiO$_3$ Powders**

|                       | BT2       | BT3      | BT4       | BT5      | BT6       | BT7       | BT8       | BT9      |
|-----------------------|-----------|----------|-----------|----------|-----------|-----------|-----------|----------|
| Crystallite size (nm) | 52 ± 2    | 43 ± 2   | 50 ± 2    | 42 ± 2   | 50 ± 2    | 53 ± 2    | <10       | 47 ± 1   |
| Particle size (nm)    | 126 ± 15  | 91 ± 18  | 112 ± 20  | 79 ± 10  | 207 ± 23  | 164 ± 38  | 146 ± 16  | 82 ± 13  |

with the room temperature XRD pattern in Figure 7.1. BT2 particles formed at pH 14 (Figure 7.2b) are spherical, with slight agglomeration and narrow particle size distribution. The DLS results in Table 7.2 show that the mean particle size is 126 ± 15 nm. Compared with BT2 formed at pH 14, BT3 particles formed at pH 14.2 (Figure 7.2c) are more agglomerated and have a smaller particle size of 91 ± 18 nm. These results show that a higher pH of the reaction medium leads to a smaller BT particle size for the reasons given above for the crystallite size.

**FIGURE 7.2** Scanning electron micrographs of BaTiO$_3$ powders: (a) BT1 (pH 12, Ba/Ti = 1.5); (b) BT2 (pH 14, Ba/Ti = 1.5); (c) BT3 (pH 14.2, Ba/Ti = 1.5); (d) BT5 (pH 14, Ba/Ti = 3). The scale bar = 1μm.

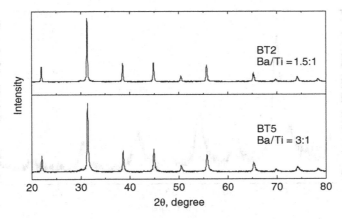

**FIGURE 7.3** Room temperature XRD patterns of $BaTiO_3$ powders prepared with different initial barium/titanium ratios.

## B. Effect of the Initial Barium/Titanium Ratio

The influence of the barium/titanium ratio in the starting reaction on the particle size under the same pH, temperature, and time conditions is shown in Figure 7.2b, Figure 7.2c, and Table 7.2 for BT2 and BT5. They show smaller particle sizes for higher barium/titanium ratios. Figure 7.3 shows the cubic phase structure with symmetry Pm3m for both barium/titanium ratios of 1.5 and 3. One can also notice the increased degree of agglomeration of BT5 particles with higher barium/titanium ratios. The DLS results show that the mean particle size of BT5 is $79 \pm 10$ nm. Similar observations were reported by Wada et al.[35] for BT powders prepared by their LTDS method. At a fixed reaction temperature, increasing the barium/titanium ratio resulted in smaller crystallite size (37.0 nm when barium/titanium = 5 and 12.9 nm when barium/titanium > 35). Calculating from the FWHM of (200) peak by equation (1), the crystallize size of the BT5 sample (barium/titanium = 3) is $42 \pm 2$ nm, slightly smaller than that of BT2 from barium/titanium = 1.5.

## C. Effect of Mineralizer

Since the precursor BT gel in the ACS process is obtained by chemically reacting and precipitating the starting barium and titanium chemicals in a highly basic solution, could the inorganic base potassium hydroxide (KOH) be replaced by an organic base, and in turn can some of the morphology changes in the final products be expected? Tetramethylammonium hydroxide (TMAH) was tried in place of KOH to adjust the pH of the precursor solution to 14.0 in BT8. The room temperature XRD pattern, as shown in Figure 7.4, of the BT8 powders shows a cubic phase, similar to the other BTs. However, significant XRD peak broadening indicates either poor crystallinity or very small crystallite size less than 10 nm.

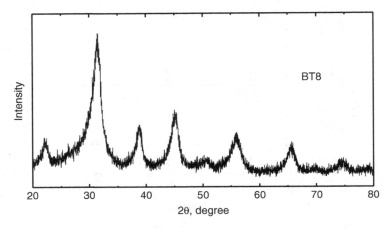

**FIGURE 7.4** Room temperature XRD pattern of BT8 (with TMAH) powder showing significant peak broadening.

All BT8 particles are spherical, with a softer appearance (Figure 7.5a) as compared to those particles prepared with KOH. Moreover, each individual BT8 particle is found to comprise many much smaller particles at a closer observation by TEM (Figure 7.5b). Thus the broad peaks room temperature XRD pattern in Figure 7.4 must be caused by the very small crystallite size of less than 10 nm.

Similar to KOH as a strong base, TMAH undergoes complete dissociation in water:

$$N(CH_3)_4OH \rightarrow N(CH_3)_4^+ + OH^-. \tag{7.2}$$

While the precursor gels are forming via chemical precipitation, some of the quaternary methyl ammonium cations are adsorbed onto the cluster of precursor gels. Obviously $N(CH_3)_4^+$ ion is much larger than $K^+$ and can stabilize and inhibit

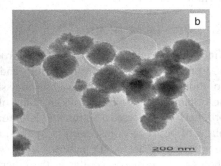

**FIGURE 7.5** (a) Scanning electron and (b) transmission electron micrographs of BT8 powders.

further growth of the nuclei. Since the particles are so small (<10 nm by XRD), they tend to aggregate into larger particles. The DLS results show the particle size of BT8 to be 146 ± 16 nm, indicating the aggregates of tiny crystals are difficult to break with 15 min of ultrasonication. The high weight loss (13.2 wt%) at 800°C in the thermal weight loss (TGA) curve (not shown) confirms that there is a large amount of organic or volatile species adsorbed on BT8 powder.

## D. Effect of Polymer Additive

In nanoparticle BT synthesis by the ACS process, BT nanoparticle nuclei are formed from water-soluble chemical precursor gel, but the growth takes place in another liquid medium under refluxing. A BT sample labeled as BT9 has been made in the presence of ammonium polyacrylate (APA) with a mean molecular weight of 6000 g/mol at a concentration of 0.002 g/ml in water refluxing medium. All other experimental parameters, such as the pH, barium/titanium ratio, and reaction time were the same as for BT2.

The scanning electron micrograph of the particles (Figure 7.6) reveals near-spherical BT powders. The redispersibility of the particles was tested by drying the powder in a vacuum oven at 70°C followed by dispersing the dried powder in water with 10 min of ultrasonication. Particle size measurements were then made by the DLS technique. The redispersibility of the dried nanoparticle BT in terms of the particle size was compared with the particle size before drying. The mean particle sizes differ by less than 10%, as shown in Figure 7.7. The XRD results show pure cubic phase for this BT as well. The crystallite size of BT9 calculated by FWHM of (200) peak is 47 ± 1 nm. Moreover, BT9 particles are smaller than those of BT2, suggesting that the particle growth is inhibited by the adsorbed APA molecule. The procedure with APA has been shown to be effective for drying wet BT particles of nanometer size without an adverse effect on the particle morphology, crystallinity, etc.

**FIGURE 7.6** Redispersible ACS BT (BT9) prepared with APA.

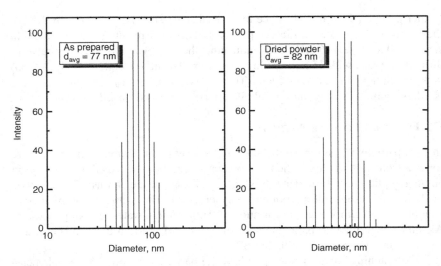

**FIGURE 7.7** DLS particle size distribution of ACS BT dispersed in water before and after drying.

## III. BARIUM ION LEACHING IN WATER

Cabot's hydrothermal BT-08 powder (Cabot Performance Materials, Boyertown, PA), having a specific surface area of 8.5 $m^2/g$, a barium/titanium ratio of 0.998, and a median particle size of 0.24 μm was used for the kinetic study of $Ba^{2+}$ leaching from BT particles. Different PALs at 0.5 wt% loading were added to the BT slurry to observe the effect of a PAL on the kinetics of $Ba^{2+}$ leaching at different aging times up to 96 h at room temperature. The organic PALs used are listed in Table 7.3.

Ba$^{2+}$ leaching, monitored by the ethylenediaminetetraacetic acid (EDTA) titration method[54] at pH 8 and room temperature as a function of aging time, is quite high in the first few minutes of aging. Blanco-Lopez et al.[66] also reported $Ba^{2+}$ ion leaching in water occurring quickly and little $Ba^{2+}$ ion concentration change after several hours. The effects of organic adsorbates, that is, PAL or

---

**TABLE 7.3**
**List of Organic Passivation Agents and Their Structural Names**

KD-6: copolymer of methyl methacrylate backbone with polyethylene glycol side chains
APA: ammonium polyacrylate
PAsA-Na: polyaspartic acid-sodium salt
Oxalic acid: $(COOH)_2 \cdot 2H_2O$
Citric acid: $HOCCOOH(CH_2COOH)_2 \cdot 2H_2O$
PAMPA-MA: copolymer of 2-acylamido-2-methyl propanesulfonic acid with maleic anhydride

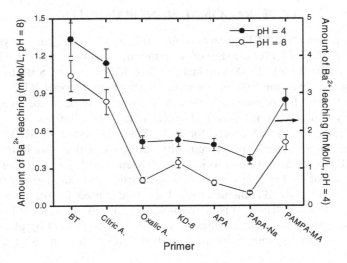

**FIGURE 7.8** Effect of organic passivating agents on $Ba^{2+}$ leaching at pH 4 and 8 after 2 days of aging.

primer, on the passivation of $Ba^{2+}$ ions leaching at pH 4 and 8 for 2 days are presented in Figure 7.8. After mixing the BT powder with 0.5 wt% PAL or primer, this slurry was dried instantly at 120°C for 12 h to make a PAL-coated BT powder. The number of $Ba^{2+}$ ions leaching are clearly reduced with PALs and showed three to five times less leaching than without a PAL.

For KD-6, shown to be an effective PAL, Figure 7.9 demonstrates the effect of $Ba^{2+}$ ion in the slip as shown by SEM microstructures of BT slips with and without an excess $Ba^{2+}$ ion added as barium hydroxide (BaOH). A poorer microstructure with increased porosity is the result due to the presence of $Ba^{2+}$ ions in the slip. Therefore it is desirable to keep the $Ba^{2+}$ concentration to a minimum in the slip medium. This can be achieved by introducing a PAL.

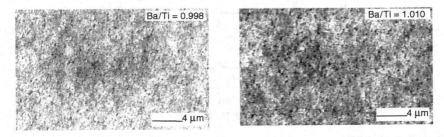

**FIGURE 7.9** BT powder dispersed in water with KD-6 without additional $Ba^{2+}$ ions (left) and with 1 mol% $Ba^{2+}$ addition (right).

## IV. TETRAGONAL NANOPARTICLES

The room temperature XRD patterns of ACS BT showed cubic phase crystalline BT formed at 20 min after the start of the refluxing reaction. However, a Raman spectrum and dark-field TEM examinations revealed that the ACS BT had some tetragonality as was the case in our redispersible hydrothermal BT.[20,67] These XRD pure cubic BT powders of 100 nm in particle diameter coated with an organic layer may be useful as embedded capacitors for enhanced dispersibility in the polymer matrix of choice for PCB. Since the synthesized BT in the nanometer range exhibits low tetragonality due to the critical size limit for the tetragonal phase,[51] it is desired to increase the tetragonality in this size range. The question of XRD tetragonality discussed earlier was examined by treating an XRD cubic phase BT chemically.[66] The powder used for this experiment was the same BT-08 having a mean particle size of 240 nm, since it exhibited XRD cubic phase. Here the chemical treatment was carried out by placing the cubic BT powder in dimethylforamide at 170°C for 6 to 24 h under a sealed condition.

Figure 7.10 shows the XRD patterns for BT-08 powder before and after chemical treatment. It is clear that the cubic phase BT transformed to tetragonal as shown by the doublet peaks at $2\theta = 45°$ for (002) and (200), as shown in Figure 7.11. We used the greater peak separation corresponding to a higher tetragonality or higher c/a ratio. Did the increased tetragonality come by the growth in crystallite size?

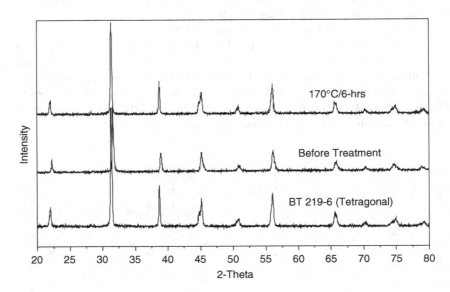

**FIGURE 7.10** XRD patterns of BT-08 powders before and after chemical treatment. BT 219-6 (obtained from Ferro Electronic Materials Corp., Penyan, NY), which is a commercially known tetragonal BT, is shown as a reference.

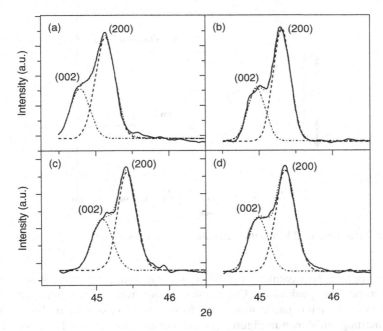

**FIGURE 7.11** XRD peaks of BT-08 for (002) and (200) planes after chemical treatment: (a) 170°C/6 h, (b) 170°C/12 h, (c) 200°C/4 h, and (d) 200°C/8 h.

Figure 7.12 shows the particle morphology before and after chemical treatment. The SEM particle morphology appeared to be unchanged. The particle size determined by DLS and Brunauer-Emmet-Teller (BET) surface area measurements for the chemically treated BT-08 revealed that the particle size and the surface area decreased slightly by the chemical treatment. The differential scanning calorimetry (DSC) in Figure 7.13 for the treated BT shows the Curie transition peak due to the tetragonal to cubic phase transition. How did this

**FIGURE 7.12** Scanning electron micrographs of BT-08 before chemical treatment (a) and after treatment (b) for 6 h at 170°C.

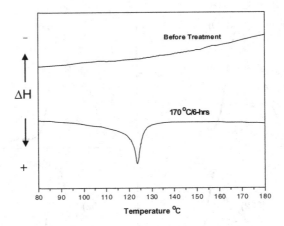

**FIGURE 7.13** DSC of BT-08 before and after chemical treatment.

happen? Before the treatment, the BT was cubic and hence there was no $\Delta H$ endothermic DSC peak at the Curie point arising from the tetragonal to cubic transition as the temperature was raised from room temperature to 180°C. The TGA thermogram shown in Figure 7.14 exhibits an approximate 1% weight loss due to water or hydroxyl, mostly from the BT crystal lattice in the temperature range of 300 to 400°C before the treatment and essentially no weight loss after the treatment.

The Fourier transform infrared spectroscopy (FTIR) spectra in Figure 7.15 are for BT-08 before and after chemical treatment for different durations. The

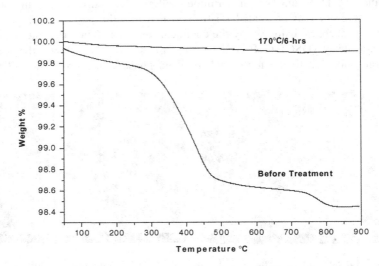

**FIGURE 7.14** TGA of BT-08 powder before and after chemical treatment.

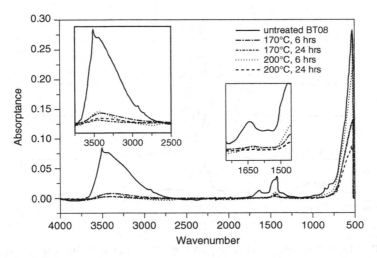

**FIGURE 7.15** FTIR spectra of BT-08 after chemical treatment.

results support the enhanced tetragonality related to the loss of water or hydroxyl (OH) from the BT as shown by the reduction in the OH stretching band at approximately 3400 cm$^{-1}$. However, the OH examination by FTIR cannot distinguish if the origin of the OH is from the BT surface or from the BT crystal lattice. We carried out a solid state nuclear magnetic resonance (NMR) spectroscopy study to distinguish the two different origins of OH. Solid state $^{1}$H magic angle spinning NMR experiments were carried out with a Chemagnetics spectrometer (magnetic field 11.74 T, resonance frequency 499.9 MHz) using a variable-temperature double-resonance probe. Samples were loaded into 5-mm zirconia PENCIL rotors and spun at 10 kHz. Single-pulse Bloch-decay spectra were collected at room temperature with a 4.0 μsec (90°) $^{1}$H pulse, a spectrum width of 100 kHz, and a repetition delay of 10 sec. The number of transients was 48 and the $^{1}$H NMR chemical shifts were referenced to solid tetrakis(trimethylsilyl)silane ($C_{12}Si_5H_{36}$; TTMS) at 0 ppm. TTMS was also used as an external standard for quantification of the number of protons by taking the spectra under identical conditions.

Two sharp resonances are observed at 5.0 ppm and 5.8 ppm referenced to TTMS. As shown in Figure 7.16, the NMR peak ratios of two different protons, that is, surface OH (5.0 ppm) to lattice OH (5.8 ppm), decreased after the treatment. By deconvoluting these two resonances, we obtained the area ratios of these two peaks, the normalized peak areas for 5.8 ppm resonance (normalized to the sample weight), and the moles of protons per gram of sample based on the TTMS proton NMR spectra. The results, within 10 to 20% uncertainty, arising from the deconvolution and background subtraction are given in Table 7.4. The peak ratio of $OH_{lattice}/OH_{surface}$ decreased close to one-half after 24 h of treatment of the BT-08 sample. Therefore the cubic to tetragonal transition may have been

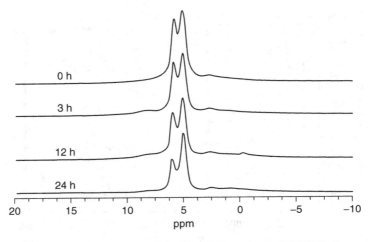

**FIGURE 7.16** A 500 MHz solid state proton NMR spectra of BT-08 before and after chemical treatment.

**TABLE 7.4**
**High-Resolution Solid State NMR of BT after Chemical Treatment**

| | Treatment (h) | | | |
| --- | --- | --- | --- | --- |
| Parameters | 0 | 3 | 12 | 24 |
| Ratio (5.8 ppm/5.0 ppm) | 0.70 | 0.60 | 0.52 | 0.39 |
| Normalized peak area/g (5.8 ppm) | 240.9 | 223.3 | 209.6 | 152.1 |
| Moles of protons/g of sample | $4.6 \times 10^4$ | $4.3 \times 10^4$ | $4.0 \times 10^4$ | $2.9 \times 10^4$ |

promoted by the elimination of the lattice protons, as suggested by Hennings et al.,[15] without the growth in grain size.

## V. TETRAGONALITY AND K

Can the dielectric constant (K) of BT with a particle size of less than 200 nm be increased? As stated earlier in the background section, the K values need to be higher for embedded capacitor applications. How does the tetragonality in terms of the c/a ratio of the BT unit cell relate to the K of the particles? The K values of BT-08 powders treated chemically were measured for the test specimens of BT-08 powder consolidated by polyvinyl butyral (PVB) binder with silver electrodes and by being placed under a Hewlett-Packard LCR meter. Our preliminary

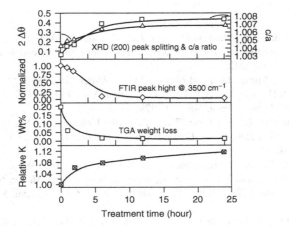

**FIGURE 7.17** Compilation of the changes in BT-08 by chemical treatment at 170°C for various lengths of time.

results show a substantial increase in the value with the treatment corresponding to the property of tetragonal phase BT as compared with the cubic phase (Figure 7.12a). Figure 7.17 is a compilation of several sets of the data resulting from the chemical treatment on BT-08. It shows the relative K increase as the tetragonality increase. It also shows the approximate 0.4% increase in the c/a ratio along with a reduction in TGA weight loss by the chemical treatment. Based on the combination of the TGA, FTIR, XRD, and solid state NMR, the positive change of the tetragonality and the corresponding K of BT is attributed to the loss of the lattice OH in the BT crystal lattice.

## VI. CONCLUSIONS

Surface coating by organic passivation agents is effective in reducing $Ba^{2+}$ ion leaching. BT nanocrystals and redispersible particles can be formed in different liquid media at near-ambient conditions. Higher pH and barium/titanium ratios in the synthesis favor smaller BT particle sizes. These nanocrystals were identified as metastable cubic phase BT by XRD with some tetragonality. Tetragonality can be increased by chemical treatment of cubic phase BT. These modifications are expected to be useful for the new embedded capacitors.

## ACKNOWLEDGMENTS

This work was sponsored in part by the U.S. National Science Foundation under grant no. DMR-9731769, by the Division of Materials Science, Office of Basic Energy Sciences, U.S. Department of Energy at Oak Ridge National Laboratory and at Pacific Northwest National Laboratory.

## REFERENCES

1. Henderson, I., and Staut, R., *Ceram. Ind.*, December, 30, 1996.
2. Mann, L., presented at a graduate seminar, Clemson University, Clemson, SC, September 26, 2002.
3. Bauger, A., Mutin, J.C., and Niepce, J.C., *J. Mater. Sci.*, 18, 3041, 1983.
4. Chu, M.S.H., and Rae, A.W.I.M., *Am. Ceram. Soc. Bull.*, 74, 69, 1995.
5. Hertl, W., *J. Am. Ceram. Soc.*, 71, 879, 1988.
6. Vivekanandan, R., and Kutty, T.R.N., *Powder Technol.*, 57, 181, 1989.
7. Hennings, D., Rosenstein, G., and Schreinemacher, H., *J. Eur. Ceram. Soc.*, 8, 107, 1991.
8. Hennings, D., and Schreinemacher, S., *J. Eur. Ceram. Soc.*, 9, 41, 1992.
9. Dutta, P.K., Asiaie, R., Akbar, S.A., and Zhu, W.D., *Chem. Mater.*, 6, 1542, 1994.
10. Xia, C.T., Shi, E.W., Zhong, W.Z., and Guo, J.K., *J. Eur. Ceram. Soc.*, 15, 1171, 1995.
11. Eckert, J.O., Jr., Houston, C.C.H., Gersten, B.L., Lencka, M.M., and Riman, R.E., *J. Am. Ceram. Soc.*, 79, 2929, 1996.
12. Asiaie, R., Zhu, W.D., Akbar, S.A., and Dutta, P.K., *Chem. Mater.*, 8, 226, 1996.
13. Shi, E.W., Xia, C.T., Zhong, W.Z., Wang, B.G., and Feng, C.D., *J. Am. Ceram. Soc.*, 80, 1567, 1997.
14. Zhu, W.D., Akbar, S.A., Asiaie, R., and Dutta, P.K., *Jpn. J. Appl. Phys.*, 36, 214, 1997.
15. Hennings, D., Metzmacher, C., and Schreinemacher, B.S., *J. Am. Ceram. Soc.*, 84, 179, 2001.
16. Urek, S., and Drofenik, M., *J. Eur. Ceram. Soc.*, 18, 279, 1998.
17. Pinceloup, P., Courtois, C., Leriche, A., and Thierry, B., *J. Am. Ceram. Soc.*, 82, 3049, 1999.
18. Clark, I.J., Takeuchi, T., Ohtori, N., and Sinclair, D., *J. Mater. Chem.*, 9, 83, 1999.
19. MacLaren, I., and Ponton, C.B., *J. Eur. Ceram. Soc.*, 20, 1267, 2000.
20. Lu, S.W., Lee, B.I., Wang, Z.L., and Samuels, W.D., Hydrothermal synthesis and structural characterization of $BaTiO_3$ nanocrystals, *J. Crystl. Growth*, 219, 269, 2000.
21. Hu, M.Z., Miller, G.A., Payzant, E.A., and Rawn, C.J., *J. Mater. Sci.*, 35, 2927, 2000.
22. Hu, M.Z., Kurian, V., Payzant, E.A., Rawn, C.J., and Hunt, R.D., *Powder Technol.*, 110, 2, 2000.
23. Xu, H.R., Gao, L., and Guo, J.K., *J. Eur. Ceram. Soc.*, 22, 1163, 2002.
24. Ciftci, E., Rahaman, M.N., and Shumsky, M., *J. Mater. Sci.*, 36, 4875, 2001.
25. Shimooka, H., and Kuwabara, M., *J. Am. Ceram. Soc.*, 79, 2983, 1996.
26. Frey, M.H., and Payne, D.H., *Chem. Mater.*, 7, 123, 1995.
27. Matsuda, H., Kuwabara, M., Yamada, K., Shimooka, H., and Takahashi, S., *J. Am. Ceram. Soc.*, 81, 3010, 1998.
28. Lee, B.I., and Zhang, J.P., *Thin Solid Films*, 388, 107, 2001.
29. Matsuda, H., Kobayashi, N., Kobayashi, T., Miyazawa, K., and Kuwabara, M., *J. Non-Crystl. Solids*, 271, 162, 2000.
30. Beck, H.P., Eiser, W., and Haberkorn, R., *J. Eur. Ceram. Soc.*, 21, 687, 2001.
31. Nanni, P., Leoni, M., Buscaglia, V., and Alipandi, G., *J. Eur. Ceram. Soc.*, 14, 85, 1996.
32. Leoni, M., Viviani, M., Nanni, P., and Buscaglia, V., *J. Mater. Sci. Lett.*, 15, 1302, 1996.

33. Viviani, M., Lemaitre, J., Buscaglia, M.T., and Nanni, P., *J. Eur. Ceram. Soc.*, 20, 315, 2000.
34. Wada, S., Chikamori, H., Noma, T., and Suzuki, T., *J. Mater. Sci.*, 35, 4857, 2000.
35. Wada, S., Tsurumi, T., Chikamori, H., Noma, T., and Suzuki, T., *J. Crystl. Growth*, 229, 433, 2001.
36. Anuradha, T.V., Ranganathan, S., Mimani, T., and Patil, K.C., Combustion synthesis of nanostructured barium titanate, *Scripta Mater.*, 44, 2237, 2001.
37. Stockenhuber, M., Mayer, H., and Lercher, J.A., *J. Am. Ceram. Soc.*, 76, 1185, 1993.
38. Gallagher, P.K., and Schrey, F., *J. Am. Ceram. Soc.*, 46, 567, 1963.
39. Ma, Y., Vileno, E., Suib, S.L., and Dutta, P.K., *Chem. Mater.*, 9, 3023, 1997.
40. Wang, J., Fang, J., Ng, S.C., Gan, L.M., Chew, C.H., Wang, X.B., and Shen, Z.X, *J. Am. Ceram. Soc.*, 82, 873, 1999.
41. Kinoshita, K., and Yamaji, A., *J. Appl. Phys.*, 45, 371, 1976.
42. Arlt, G., Hennings, D., and De With, G., *J. Appl. Phys.*, 58, 1619, 1985.
43. Ishikawa, K., Yoshikawa, K., and Okada, N., *Phys. Rev. B*, 37, 5852, 1988.
44. Uchino, K., Sadanaga, E., and Hirose, T., *J. Am. Ceram. Soc.*, 72, 1555, 1989.
45. Frey, M.H., and Payne, D.A., *Phys. Rev. B*, 54, 3158, 1996.
46. Wada, S., Suzuki, T., and Noma, T., *J. Ceram. Soc. Jpn.*, 104, 383, 1996.
47. McCauley, D., Newnham, R.E., and Randall, C.A., *J. Am. Ceram. Soc.*, 81, 979, 1998.
48. Hsiang, H.-I., and Yen, F.-S., *J. Am. Ceram. Soc.*, 79, 1053, 1996.
49. Xu, H., Gao, L., and Guo, J., *J. Eur. Ceram. Soc.*, 22, 1163, 2002.
50. Wada, S., Yasuno, H., Hoshina, T., Nam, S.-M., Kakemoto, H., and Tsurumi, T., Dielectric properties of nm-sized barium titanate crystallites with various particle size, presented at the 105th Annual Meeting of the American Ceramic Society, Nashville, TN, April 26–29, 2003.
51. Venigalla, S., Hydrothermal $BaTiO_3$-based dielectric for BME X7R/X5R MLCC Applications, Center for Dielectric Studies, Spring Meeting, Pennsylvania State University, April 22–23, 2002.
52. Wang, X., and Lee, B.I., Dispersion of barium titanate with polyaspartic acid in aqueous media, *Colloids Surf.*, 202, 71, 2002.
53. Wang, X., Lee, B.I., Lu, S., and Mann, L., Dispersion and aging behavior of barium titanate in water, *Mater. Res. Bull.*, 35, 2555, 2000.
54. Yoon, D., and Lee, B.I., *J. Mater. Sci. Mater. Electron.*, 14, 165, 2003.
55. Lee, B.I., Wang, X., Yoon, D.H., and Hu, M., Aqueous processing of barium titanate powders, *J. Ceram. Proc. Res.*, 4 17, 2003.
56. Wang, X., Lee, B.I., Hu, M., and Payzant, A., *J. Mater. Sci., Mater. Electron.*, 14, 557, 2003.
57. Lee, B.I., Wang, X., and Hu, M., Nanocrystal barium titanate via low temperature ambient pressure conditions, U.S. Patent application in process, June 2002.
58. Wang, X., Lee, B.I., Hu, M., and Payzant, A., *J. Mater. Sci. Lett.*, 22, 557, 2003.
59. Gould Electronics, TCC embedded planar capacitor technology, available at www.gould.com.
60. Reuss, R.H., and Chalamala, B.R., Microelectronics packaging and integration, *MRS Bull.*, 28, 11, 2003.
61. Tumala, R., Chahal, P., and Bhattachrya, S., Recent advances in integral passives at PRC, presented at the International Microelectronics and Packaging Society 35th Nordic Conference, Stockholm, Sweden, September 1998.

62. *Proceedings of the 51st Electronic Components and Technology Conference*, IEEE, Piscataway, NJ, 2001.
63. Windlass, H., Raj, P.M., Balaraman, D., Bhattacharya, S.K., and Tummala, R.R., "Colloidal processing of polymer ceramic nanocomposites for integral capacitors," in *Proceedings of the International Symposium on Advanced Packing Materials: Processes, Properties, and Interfaces*, IEEE, Piscataway, NJ, 2001.
64. Rao, Y., Ogitani, S., Kohl, P., and Wong, C.P., High dielectric constant polymer-ceramic composite for embedded capacitor application, in *Proceedings of the International Symposium on Advanced Packing Materials: Processes, Properties, and Interfaces*, IEEE, Piscataway, NJ, 2000.
65. Lencka, M.M., and Riman, R.E., *Chem. Mater.*, 5, 61, 1993.
66. Blanco-Lopez, M.C., Rand, B., and Riley, F.L., *J. Eur. Ceram. Soc.*, 17, 281, 1997.
67. Lu, S., Lee, B.I., and Wang, Z., Synthesis of redispersible nano-sized barium titanate powders by hydrothermal method, in *Ceramic Transactions*, vol. 106, *Dielectric Materials & Devices*, edited by K.M. Nair and A.R. Bhalla, American Ceramic Society, Westerville, OH, 2000.
68. Badheka, P., Magadala, V., Lee, B.I., and Bhaduri, S., Tetragonality and dielectric constant of submicron sized BT powders, Paper AM-S19-30-2003, presented at the American Ceramic Society annual meeting, Nashville, TN, April 30, 2003.

# 8 Magnetic Particles: Synthesis and Characterization

*Masataka Ozaki*

## CONTENTS

## I. INTRODUCTION

Ever since iron powders were employed for magnetic recording systems shortly before World War II, large amounts of magnetic particles have been produced for magnetic recording media and extensive efforts have been made to produce better magnetic particles.[1-5] Originally the use of magnetic particles was limited to audio tapes, but they are now employed in a variety of magnetic recording systems in the form of tapes, cards, and flexible and rigid disks. Magnetic particles are of importance not only in industrial technology, but also in our environment and in the functions of some biosystems, as well as being of scientific interest.[6,7] Recently magnetic particles have found wide applications in biological and medical diagnostic fields.[8] Attempts have also been made to use magnetic particles in biotechnology, such as in the labeling and separation of cells, purification of deoxyribonucleic acid (DNA), etc.[9] Techniques are being developed in order to produce new magnetic particles and to introduce new functions to particulate materials by incorporating magnetic material (e.g., coating magnetic particles on nonmagnetic materials, coating nonmagnetic particles on magnetic materials, using combinations of enzymes on the surfaces of magnetic particles, etc.).[10–13]

In using fine magnetic particles, particle size is the most important parameter, as well as other qualities such as crystallinity and composition, since the magnetic properties of the particles are strongly influenced by them. Therefore it is advantageous if the particles formed are uniform in their size and shape (i.e., monodispersed or having a narrow size distribution in a desired size range).

Colloidal particles with a narrow size distribution have been pursued by colloid chemists for a long time. The work of LaMer and Dinegar[14] gave us some basic principles for the formation of monodispersed particles. According to the theory, to produce colloidal particles having a narrow size distribution, the nucleation and growth process must be carefully controlled. It should also be noted that many experimental results contradict this theory. For example, many monodispersed particles are formed by the association of subunits of small particles. Privman et al.[15] proposed a theory for the growth of monodispersed particles in solutions in order to explain the experimental results.

The conditions necessary for the formation of magnetic particles are essentially the same as those for nonmagnetic particles, but some special precautions are necessary because of the strong magnetic interactions among the particles. In producing monodispersed particles, the essential parameters are (1) separation of the nucleation process from the growing process, (2) protection of particles from aggregation, (3) controlled supply of precursor material, and (4) the temperature and pH of the solution. These parameters are intimately tied to each other, and sometimes they are difficult to separate. Thus the concentration of the starting material, reaction temperature, and pH of the solution must be optimized. Dispersion reagents such as surfactants and polymers must be chosen carefully. A continuous supply of precursor material is sometimes provided by controlled decomposition of another material. Recently a new method for controlling size and shape by dissolution was developed.[16]

Since the field related to magnetic particles is broad, this article is written from a colloid chemistry point of view. Readers should refer to the articles by Privman et al.,[15] Sugimoto,[17] and Matijević[18] for information on the formation of monodispersed particles.

## II. SIZE OF MAGNETIC PARTICLES AND MAGNETIC PROPERTIES

When a magnetic material is placed in a magnetic field, it is magnetized, as shown in Figure 8.1, with the change in the applied magnetic field strength. Bulk material

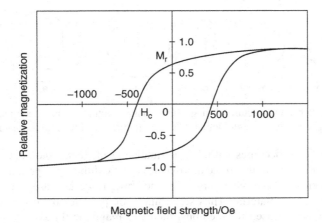

**FIGURE 8.1** Magnetization curve: $M_r$, residual magnetization; $H_c$, coercive force.

or powder is not magnetized originally unless it is exposed to a magnetic field, since the original directions of the magnetic dipoles in magnetic domains of bulk material or in powder are random. Thus the magnetization of magnetic solid is brought about by the orientation of dipoles. As the magnetic field increases, the degree of orientation of the dipole increases and saturation magnetization is attained at the perfect orientation of the dipoles. Even though the magnetic field is decreased to zero, the material will have a certain magnetization, the so-called residual or remnant magnetization, designated by $M_r$. A magnetic field of opposite direction is required to bring the magnetization to zero. This magnetic field is called the coercive force and is frequently designated by $H_c$. Coercive force is essential for magnetic recording media, as is remnant magnetization. If we remove the magnetic field at the coercive force, the magnetization of the material remains to some extent. A negative magnetic field slightly stronger than the coercive force is required to bring the magnetization to zero. This magnetic field is called the remnant coercive force.

As the size of the particle decreases, the number of the magnetic domains in the particle decreases, changing the domain structure from multidomain to single domain. The domain structures in particles are shown schematically in Figure 8.2. Generally, the coercive force of multidomain particles is smaller than single-domain particles, since the rotation of the magnetic moment in the former particles occurs easily from domain walls. The coercive force of a single-domain particle is determined by the so-called magnetocrystalline anisotropy together with the shape anisotropy. The coercive force due to the shape anisotropy increases with an increase in the aspect ratio if the particle size remains the same. Therefore elongated single domain particles are preferentially employed for magnetic recording media. As we will see later in this chapter, the particles available for magnetic media must be single-domain particles having high saturation magnetization and proper coercive force.

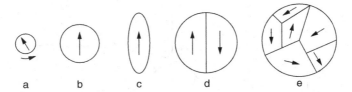

**FIGURE 8.2** Schematic picture of magnetic domain structures of magnetic particles: (a) superparamagnetic; (b, c) single-domain; (d) two-domain; and (e) multidomain particle.

If the particle becomes sufficiently small, the magnetic moment shows no preferential orientation due to the thermal agitation exhibiting superparamagnetic property. Such particles have very small coercive force and are not useful for magnetic recording material. Thus small magnetic particles having a superparamagnetic nature are used for the constituents of magnetic fluids.[19,20]

In a stable dispersion of fine magnetic particles, particles are free to rotate. The magnetic field coercive force applied to a magnetic particle having a permanent magnetic moment, m, will cause orientation along the direction of the field. The orientation may be disturbed by thermal agitation. The degree of the orientation can be described by the Langevin equation:

$$m_{av} = m(\coth \alpha - 1/\alpha),\qquad(8.1)$$

where $m_{av}$ is the average magnetic moment and $\alpha = mH/kT$, k being the Boltzmann constant and $T$ being the absolute temperature. Therefore the magnetization of a stable dispersion of magnetic particles can be given by

$$M = nm_{av},\qquad(8.2)$$

where $n$ is the number of the particles per volume. Thus the dispersion of magnetic particles will show a magnetization curve similar to that of paramagnetic material.

There is a strong interaction between magnetic particles, which is far reaching compared to the interaction due to van der Waals force. The attractive magnetic energy, $U_A^{Mag}$, between two magnetic dipoles is shown in Equation 8.3:[21]

$$U_A^{Mag} = 1/4 \ \pi\mu_0 r^3[m_1 m_2 - 3(m_1 + r/r)(m_2 + r/r)],\qquad(8.3)$$

where $m_1$ and $m_2$ are the magnetic dipole moments of the particles, $r$ is the vector joining the centers of the two dipoles, and $\mu_0$ is the magnetic permeability of the vacuum. The magnetic dipole moment, assuming a single-domain particle, is expressed as $m = I_0 v$, with $I_0$ being the magnetization per unit volume of the particle and v its volume. When two dipoles, $m_1$ and $m_2$, are oriented in the same direction on the same line, Equation 8.3 can be written as

$$U_0 = -m_1 m_2 / 2\pi\mu_0 r^3.\qquad(8.4)$$

$U_0$ is the minimal potential energy of the system. The magnetic interaction energy increases with the second power of its volume, that is, proportional to the sixth power of their radius. Therefore the magnetic interaction energy increases rapidly as the size of the particle increases. The interaction between particles of strong magnetic material such as magnetite or maghemite is difficult to obtain unless the particle size is small, which leads to superparamagnetic particles.

The magnetic interaction energy between superparamagnetic particles will be reduced since the dipole moment in such particles has no preference for a particular orientation due to free rotation of the magnetic moment by the thermal agitation. The magnetic interaction energy increases if the interaction works between multiple particles. These multiple interactions become important in a concentrated dispersion.[22] We must also consider that the interaction between magnetic particles is reduced when the interparticle distance becomes comparable to or smaller than the particle size. It should be noted that the magnetic interaction is reduced to 1/10 to 1/20 of the interaction calculated by Equation 8.4.[23]

In addition to the mutual interaction, when magnetic particles are placed in a gradient of a magnetic field, they are forced to move, which makes it easy to either collect them in a desired place or to readily suspend them. Therefore they can be used for extraction and purification of biomaterials, diagnosis, and for the production of magnetic fluids.

## III. MAGNETIC PARTICLES FOR MAGNETIC RECORDING MEDIA

Magnetic particles are widely employed for recording media, not only for audio and video recording systems, but also for a variety of computer media in the form of tapes, flexible and rigid disks, and cards, although recently some magnetic recording systems are being replaced by digital video disk (DVD) systems. Extensive research is thus being conducted to produce magnetic particles that have better recording performance.

Particulate magnetic recording materials are required to have (1) high coercivity, (2) large saturation magnetization, (3) good dispersibility and orientability, (4) narrow distributions in size and shape, and (5) high chemical and mechanical strength.[1–5,24,25] The magnetic particles must have large saturation magnetization and proper coercive force so the magnetic recording media has enough output signal strength and reliability. However, there are not many materials that can satisfy all of these demands.

After finding that elongated maghemite ($\gamma$-$Fe_2O_3$) particles showed superior magnetic properties for recording performance, maghemite particles attracted a lot of interest from recording tape manufacturers. Maghemite particles were the most popular material until cobalt-modified maghemite appeared. Iron particles are employed for digital video cassette tapes, still camera disks, and high-quality audio tapes.[26] Much attention has been focused on fine barium ferrite particles for perpendicular recording materials, as well as for use in videotapes to improve their recording properties.[27,28]

## A. ACICULAR $\gamma$-Fe$_2$O$_3$ PARTICLES

Since the crystal habit of $\gamma$-Fe$_2$O$_3$ is cubic, it cannot be formed in elongated shape directly. Therefore the particles are prepared from other compounds by solid phase transformation. Though $\gamma$-Fe$_2$O$_3$ can be prepared from either $\beta$-FeOOH or $\gamma$-FeOOH, most commercially available particles are produced from $\alpha$-FeOOH (goethite). The original shape and size of $\alpha$-FeOOH particles are kept almost the same throughout the dehydration and reduction process, followed by oxidation. For high-density recording, particles must be small, and particles are made to have a large aspect ratio to achieve a large coercive force. Therefore the size and shape of the starting $\alpha$-FeOOH particles must be carefully controlled. However, the conditions and mechanism for the formation of oxyhydroxides were not well known until Kiyama and Takada[29] revealed them after extensive studies. Today, goethite particles having desired sizes and shapes are available.

The seed particles of goethite are prepared by the reaction

$$Fe^{2+} + OH^- + 1/2O_2 \rightarrow \alpha\text{-FeOOH}.$$

The seed particles are then grown to desired sizes in the presence of iron and ferrous ions. The goethite particles are chemically dehydrated by heating, reduced by hydrogen into magnetite, and finally oxidized into maghemite through the reactions,

$$\alpha - FeOOH \xrightarrow[\sim 300°C]{\text{heat}} \alpha - Fe_2O_3 \xrightarrow[300-400°C]{H_2} Fe_3O_4 \xrightarrow[\sim 250°C]{O_2} \gamma - Fe_2O_3$$

The temperatures and the length of the reduction and reoxidation time are optimized and carefully controlled. The $\gamma$-Fe$_2$O$_3$ particles on the market are 0.2 to 0.5 $\mu$m in length, with an aspect ratio of about 10. Typical $\gamma$-Fe$_2$O$_3$ particles are shown in Figure 8.3.

When $\alpha$-FeOOH particles are used for the starting material, it is very difficult to avoid pore formation completely in the dehydration process, which necessitates heating at high temperatures to decrease pores. It was expected that nonporous particles could be formed if elongated hematite particles were used for the starting material. The first successful method for the production of elongated hematite particles was developed by Matsumoto et al.[30] using hydrothermal reactions. The maghemite particles prepared from the hematite particles were poreless and were more ellipsoidal than other magnetic particles. The good magnetic properties of the solids could be accounted for by the fact that the magnetic field lines in the particles were more parallel than in other particles.[31] Maghemite particles having good magnetic properties could also be produced from spindle-type hematite particles prepared by forced hydrolysis of ferric chloride solutions in the presence of small amounts of phosphate ions.[32] Arndt[33] obtained maghemite particles of improved quality from acicular hematite particles prepared by doping tin in the hydrothermal transformation of precipitated Fe(OH)$_3$.

**FIGURE 8.3** Transmission electron micrograph of typical magnetic particles, cobalt-modified maghemite particles, utilized for VHS videotapes, $H_c$ = 705 Oe. (Courtesy of Dr. Horiishi, Toda Kogyo Corp., Hiroshima, Japan.)

## B. COBALT MODIFIED γ-FE$_2$O$_3$ PARTICLES

It was very difficult to increase the coercive force of $\gamma$-Fe$_2$O$_3$ beyond 500 Oe. According to the theory of magnetism, a coercive force of $\gamma$-Fe$_2$O$_3$ particles larger than 500 Oe is obtainable, but the highest coercive force reported is 450 Oe to 470 Oe.[34] Magnetic particles having a greater coercive force were required for high-density recording media. It was well known that $\gamma$-Fe$_2$O$_3$ particles containing cobalt had a greater coercive force. However, the temperature dependency of the coercive force of the solids was large, since the coercive force of such $\gamma$-Fe$_2$O$_3$ particles was due to magnetocrystalline anisotropy rather than shape anisotropy. Great improvements of magnetic properties in coercive force were achieved by forming a thin layer of cobalt oxide compounds on acicular $\gamma$-Fe$_2$O$_3$ particles.[35] The modification of $\gamma$-Fe$_2$O$_3$ particles was made by immersing the particles in a solution containing Co$^{2+}$ and Fe$^{2+}$ followed by addition of alkaline solution. Today, most home videotapes and audio tapes of high bias use these particles.

## C. IRON PARTICLES

Since iron has a large saturation magnetization of 216 emu/g, it must be one of the best materials for magnetic recording media. Although iron particles were the first particulate materials employed for tapes, they were not used after the introduction of maghemite particles. This was partly because of the difficulty in overcoming corrosion with small iron particles. It did not take long before iron powders again attracted the interest of tape manufacturers as a material for high-density recording systems.

Although iron particles can be prepared by a variety of methods, most of the particles on the market are produced from goethite through dehydration and reduction processes.[26,36] Therefore most processes for iron particle production are the same as the production processes for $\gamma$-$Fe_2O_3$ particles; that is, dehydrated goethite particles are reduced into iron rather than magnetite and the surfaces of the particles are carefully oxidized to prevent further oxidation by forming a thin oxide layer.[37] The saturation magnetization of the powder is 60 to 70% that of pure iron and the coercive force is greater than 1000 Oe. Special procedures, such as the addition of silicone oil or silver or cobalt compounds during the reduction process, are effective in preventing mutual sintering of particles, leading to the powders improved magnetic properties.[38] Elongated iron particles are superior particulate materials for magnetic recording media, but the problem associated with corrosion has not been overcome completely for long storage times. Chen et al.[39] reported that iron particles of 70 nm in length with an aspect ratio of about 10 and having a coercive force of 1500 Oe to 2000 Oe can be prepared by reducing $\beta$-$FeOOH$ particles suspended in liquid crystals using $NaBH_4$ as a reducing reagent at 273 K.

## D. Chromium Dioxide Particles

Chromium dioxide ($CrO_2$) particles were developed as a high-quality magnetic recording material[40] and were used until cobalt modified $\gamma$-$Fe_2O_3$ particles appeared on the market. Commercially available $CrO_2$ is produced by oxidation of $Cr_2O_3$ under high temperature and high pressure. The particles produced are single crystals and acicular in shape, with large aspect ratios and good magnetic properties. However, $CrO_2$ particles are not widely used today because of their high production cost and their toxicity. Although these particles are not used much today, the technical term "$CrO_2$ position" was used for a while for the bias position, along with the "High" or "Metal" position for audio recorders and tapes.

## E. Barium Ferrite Particles

In the ordinary recording system, the magnetic medium is magnetized along the direction of motion of the magnetic medium, which is called the longitudinal recording method. In contrast to this method, Iwasaki[41] and Sugata et al.[42] proposed a different type of recording method, called the perpendicular recording method, for high-density recording systems, in which the magnetic medium is magnetized perpendicular to the direction of motion of the magnetic medium. Therefore the axis of the magnetic particles for this medium must be oriented perpendicular to the supporting film. Platelet barium ferrite particles seemed to be a suitable material for this purpose, as it has an axis perpendicular to the flat surface. Barium ferrite has been widely used as a precursor material for permanent magnets. However, the coercive force of the particles was too large for magnetic medium. Kubo et al.[43] developed new plate-like barium ferrite particles with the proper coercive force by using the glass crystallization method in solid phase reactions. Acicular barium ferrite particles doped with $Co^{2+}$ and $Ti^{4+}$ were prepared using $\alpha$-FeOOH precursors as the starting material by a conventional sintering method. The produced $BaCo_xTi_xFe_{12-2x}O_{19}$ consisted of some grains with their c-axis along the short axis of the acicular particle and having a coercive force of approximately 1500 Oe, which is suitable for perpendicular recording media.[44]

Elongated magnetic particles are employed for the ordinary longitudinal recording system. It was suggested by Lemke[45] that the vertical components of the recorded signals play an important role as the recording density increases or as the recording wavelength becomes shorter. Based on the same principle, videotapes having improved recording properties were made by mixing small amounts of barium ferrite particles to modify cobalt particles.[27]

## IV. COERCIVE FORCE AND DISPERSIBILITY OF MAGNETIC PARTICLES FOR RECORDING MEDIA

Coercive force is an extrinsic property that is influenced by many factors, such as size, shape, and packing density. Maghemite particles prepared directly from precipitated spindle-type hematite particles show a tendency toward an increase in coercive force with a decrease in particle size and an increase in aspect ratio.[32] However, the quantitative relationship between the coercive force and the size and/or aspect ratio is still unknown.

It is most desirable if measurements of the coercive force are carried out for individual particles, since particulate material contains particles of different sizes, although they may be highly monodispersed in size and shape. A novel method was developed to measure the rotation of magnetization of individual particles by Knowels,[46] who applied a pulse magnetic field to a particle dispersed in a viscous liquid and observed under a microscope the rotation of the solid. From the obtained remnant coercive force, he explained the rotation of the magnetization in the particle using the fanning reversal mechanism. On the other hand, Aharoni[47] showed that the rotation mechanism could be explained by the curling model using reported experimental data of individual particles. Measurements of the coercive force of individual particles were also made using Lorenz microscopy and atomic force microscopy (AFM) with maghemite powders prepared from directly precipitated spindle-type hematite particles.[48,49] It was suggested that the reversal of particle magnetization could be explained by curling over a narrow range near 0 degrees, and by coherent rotation at larger angles. It was also suggested that the dynamics of the reversal occurs via a complex path, and that a complex theoretical approach would be required to provide a correct description of thermally activated magnetization reversal, even in a single-domain magnetic particle.

Magnetic particles are embedded in plastic binder for use as magnetic recording media. Therefore particles must have good dispersibility in organic solvents to have large squareness and smooth-coated surfaces. However, it is especially difficult to disperse magnetic particles, as compared to nonmagnetic particles, due to strong magnetic interactions. Usually the surface of the magnetic particle is coated with a polymer to prevent aggregation. Inoue et al.[50] investigated the effect of an epoxy resin adsorbed layer on the stability of $\gamma$-$Fe_2O_3$ particles dispersed in organic solvents and found that although both the height of the maximum and the depth of the secondary minimum in the total potential energy of the colloidal interaction strongly affected the dispersion stability, the former was more effective than the latter. The experimentally obtained surface roughness of the tape film produced by using polymer resin adsorbed magnetic particles increased with the depth of the calculated total potential energy minimum rather than with the decrease in the height of the maximum energy.

Homola and Lorenz[51] showed that the recording performance of rigid disks was considerably improved by using magnetic particles coated with small colloidal silica particles. The dispersion property of the coated particles was enhanced by the control of the separation distance between the magnetic particles, resulting in good particle orientability and leading to improved recording performance.

# V. NANOMAGNETIC PARTICLES AND MAGNETIC FLUIDS

Historically dispersions of small magnetic particles having a superparamagnetic nature were used for the observation of magnetic domains. Recently nanoparticles have been intensively studied because of their unique physical properties, which cannot be achieved by bulk materials.

When a concentrated suspension of magnetic particles is placed in the gradient of a magnetic field, forces act on the particles and magnetic interaction between these is enhanced. The magnetic field attracts particles and the dispersing liquid moves together with the particles. Therefore the liquid behaves as if it is magnetized. Such a behavior is typical of a stable dispersion containing magnetic particles. In the 1960s, "magnetic fluid" or "ferro-fluid" was introduced as a new material. Since then, it has been used in a variety of applications.[52,53] Today, magnetic fluids are employed in many industrial technologies. For example, lubricating oil in which magnetic particles are dispersed works as a good sealing material when it is suspended by a magnetic field. The magnetic particles remain in the magnetic field together with the oil, which works as a sealing material. Today, magnetic fluid is used to seal computer disk units. The constituent of the magnetic fluid is the concentrated dispersion of superparamagnetic particles protected with surfactants in an appropriate liquid.

Since the discovery of magnetic fluids, techniques for the production of a variety of magnetic fluids, including water-based and oil-based magnetic fluids, have been developed. The first magnetic fluid was prepared by milling magnetic material in a nonpolar organic liquid in the presence of oleic acid. Concentrated dispersions of magnetic particles were also independently prepared by Satoh et al.[54] using precipitated magnetite particles. According to their method, the fine magnetite particles were obtained by adding alkaline solution into an aqueous solution containing $Fe^{2+}$ and $Fe^{3+}$ ions following the overall reaction:

$$2Fe^{3+} + Fe^{2+} + 8OH^- \rightarrow Fe_3O_4 + 4H_2O. \tag{8.5}$$

The magnetite particles were covered with oleic acid and were filtered, followed by dispersing in an organic liquid to form a stable dispersion of magnetic particles. Urea and urotropin, which on decomposition cause a controlled release of OH ions, have been employed to precipitate magnetite.[55,56]

Fine metal particles such as cobalt and iron can be prepared from metal carbonyl compounds. Thomas[57] succeeded in forming particles of cobalt with sizes ranging from 2 to 30 nm and having a narrow size distribution by decomposing cobalt carbonyl compounds in the presence of suitable surface active reagents.[57] By the same procedure, iron and nickel particles were prepared. Papirer et al.[58] studied the decomposition of a toluene solution of $Co_2(CO)_8$ in the presence of a surface active reagent and found that at least two factors are responsible for the formation of particles having an extremely narrow size distribution: the division of the system into microreactors and a diffusion-controlled growth mechanism of the individual particles.

Gobe et al.[59] prepared magnetite particles using the microemulsion method. The procedure was the mixing of water/isooctane or water/cyclohexane microemulsion with aqueous $FeCl_3$ and aqueous $NH_3$, followed by addition of aqueous $FeCl_2$ with vigorous stirring. In this procedure, aerosol OT was necessary as a surfactant in order to solubilize adequate amounts of $FeCl_2$ in the hydrocarbon used. Extensive studies were also made by Pileni[60] and Petit et al.[62] using reverse micelles as microreactor systems for the production of nanoparticles, including magnetic ones. Figure 8.4 shows an electron micrograph of nanometer-scale cobalt particles so obtained, which are well crystallized and of very narrow size distribution.

Nickel nanoparticles with a saturation magnetization of 22 emu/g were prepared by hydrazine reduction of nickel chloride in ethylene glycol without protective agent.[62] Nanometer-size iron oxide particles were prepared by using unilamellar vesicles.[63] Adding alkaline solution to vesicles containing intravesicular solutions of $Fe^{2+}$, $Fe^{3+}$, and $Fe^{2+}/Fe^{3+}$ resulted in the formation of membrane-bound discrete particles of goethite, magnetite, and ferrihydrite. The particles were very small, in the range of 1.5 to 12 nm. These results, together with particle formation in microemulsions, are not only of interest in colloid

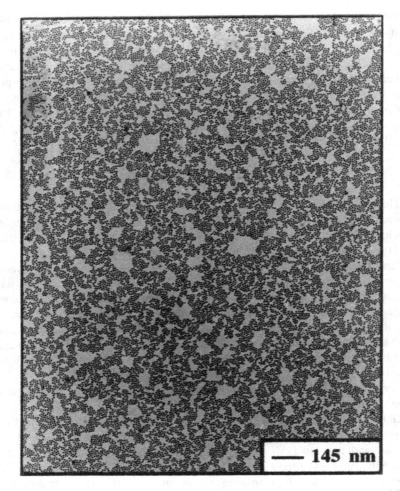

**FIGURE 8.4** Electron micrograph of nanometer size cobalt particles. (Courtesy of Prof. Pileni, University of Marie P. Curie, Paris, France).

chemistry, but also have significance for mineralization in biosystems, such as magnetotactic bacteria, where particles are formed within enclosed organic compartments. $\gamma$-Fe$_2$O$_3$ nanoparticles with a variety of shapes were prepared by controlling the chemical microenvironment of the reaction system under gamma irradiation.[64]

According to Tang et al.,[65] MnFeO$_4$ particles of relatively small size (5 to 25 nm) can be obtained through the reaction

$$MnCl_3 + 2FeCl_3 + xNaOH \rightarrow MnFeO_4 + (x - 8)NaOH + 4H_2O + 8NaCl$$

for $x > 8([Me]/[OH])$. Whereas $Mn_xFe_{3-x}O_4$ ($0.2 < x < 0.7$) particles of larger size, up to 180 nm, are produced from ferrous salts, [Me] is the concentration of metal ions. They found that in either case, the particle size appeared to be a unique function of the ratio of metal ion concentration to hydroxide concentration. A variety of ferrites can be obtained by similar reactions.

Small metal particles can also be obtained by vacuum evaporation in low-pressure inert gas.[66] Magnetic particles of metals such as iron, cobalt, nickel, and alloys of these metals can be prepared by this method. Although the amounts obtainable by this method are limited, the particles are clean compared to particles precipitated from solutions. They are mainly used for studies of the physical properties of fine particles.

## VI. METAL PARTICLES BY THE POLYOL PROCESS

In the polyol process, powdered inorganic compounds such as $Co(OH)_2$ or $Ni(OH)_2$ are suspended in a liquid polyol such as ethylene glycol.[67] The suspension is then heated to the boiling point or near the boiling point of the polyol under continuous stirring. A complete reduction of these compounds can be achieved within a few hours. In this reaction, the polyol acts as a solvent for the starting compound, subsequently, for example, ethylene glycol reduces the cobalt(II) or nickel(II) species in the liquid phase to metallic states in which nucleation and growth of the particles occurs. Metal particles obtained by the polyol process are composed of equiaxed particles with a mean size of 1 to 10 μm and having narrow size distributions.

## VII. COMPOSITE MAGNETIC PARTICLES

Attempts have been made to improve the essential properties of magnetic materials and to introduce new magnetic functions to particles by coating them with nonmagnetic or magnetic material. Iron particles covered with polystyrene were used for making the core of a high-frequency transformer.[68] Ishikawa et al. coated large polymer particles with a thin magnetic film using the ferrite plating method developed by Abe and Tamura.[69] The particles thus obtained were found to be useful as toners and carriers in copying machines. Silica particles can easily be dispersed in aqueous solutions over a wide pH range. Therefore surface coating by silica promotes the dispersibility of iron oxides in aqueous solutions.[70] The coating also reduces magnetic interactions, enhancing dispersibility. Various methods for coating particles have been developed by Ohmori and Matijević[10] and others, including silica coating on spindle-type iron particle cores. Small silica particles having cores of magnetic materials were obtained by depositing a silica layer from sodium silicate on freshly

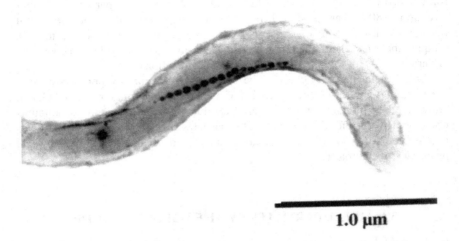

1.0 μm

**FIGURE 8.5** Electron micrograph of a magnetotactic bacteria. (Courtesy of Prof. Matsunaga, Tokyo University of Agriculture and Engineering, Tokyo, Japan.)

prepared magnetite particles, followed by forming a silica layer from tetramethoxysilane (TMS)-ethanol solution,[11] where the second coating proceeded through polymerization on the particle surface.

Hirano et al.[71] reported that spherical carbon particles containing highly dispersed cementite (Fe$_3$C) particles were formed by heating copolymers of divinylbenzene and vinylferrocene in a high-pressure bomb at 125 MPa and 650°C. The cementite particles thus formed could be transformed into α-Fe by further heating at 850°C for 6 h. When the same copolymers were heated with water, magnetite particles were formed. A scanning electron micrograph of the spherical carbon particles containing dispersed magnetite particles is shown in Figure 8.5. By using this technique, spherical carbon particles containing metal particles such as cobalt or cobalt-alloys can also be prepared.

The so-called dry mixing or dry blending method has been used for the modification of particles in powder technology.[72] In this method, surface modification of coarse particles is carried out by mixing fine particles and coarse particles with a ceramic mortar or with a centrifugal rotating mixer. This procedure can be used for the production of a variety of composite magnetic particles.

Inada et al.[12] succeeded in combining enzyme on synthesized magnetic particles. Thermosensitive magnetic particles that are useful for separation of enzyme and biomaterials have been reported.[73] Furthermore, since magnetic polymeric microsphere particles were developed for labeling and separation of biocells,[74] studies into the applications of magnetic particles in the biology and medical fields are under way.[8,9]

## VIII. MAGNETIC PARTICLES IN BIOSYSTEMS

It is known that magnetic particles can be found in the bodies of some animals.[6] It is also believed that certain animals have the ability to detect magnetic fields.[75] In 1975, Blakemore[76] found that some bacteria have magnetic particles in their cells and navigate along geomagnetic fields using these magnetic particles. The magnetic particles obtained from magnetotactic bacteria were confirmed as crystallized magnetite particles, with sizes ranging from 50 to 100 nm.[77] The particles found in some bacteria are cubic-like, have a narrow size distribution, and are aligned in single or multiple chains, more or less parallel to the axis of the cell. The magnetic moment of each particle is small, but the total magnetic moments of the aligned particles is large enough to allow orientation along the geomagnetic field. It is believed that magnetotactic bacteria navigate using these magnets as a direction detector toward north in the northern hemisphere and toward south in the southern hemisphere. The magnetite particles isolated from the bacteria showed similar magnetic properties to those of synthesized particles, and the aligned magnetic particles in their bodies were found to be a good model of the chain of spheres theory for the rotation of the magnetic moment of a magnetic particle.[78] It was also found that the surfaces of the magnetic particles isolated from magnetotactic bacteria were covered with a strong organic membrane. Matsunaga and Kamiya[13] succeeded in immobilizing glucose oxidase and uricase on organic membranes attached to magnetic particles. The enzymes bonded on the particles were more active than enzymes combined on synthesized magnetic particles. Such particles could easily be separated from reactant solution. Introductions of magnetic particles into bioparticles such as blood cells and microphages have also been successful. Such small bioparticles carrying magnetic particles can easily be moved to desired places by applying a magnetic field. Figure 8.5 shows an electron micrograph of a magnetotactic bacterium.

Ferritin is a protein consisting of 24 polypeptide subunits arranged in a roughly spherical cage within which magnetic particles called magnetoferritin is encased. It is interesting for scientific purposes because of its very narrow size distribution, with an average diameter of 10 nm.[79,80]

## IX. HEMATITE PARTICLES: WEAKLY MAGNETIZED PARTICLES

Although hematite has long been known as a material having a weak magnetic property called parasitic ferromagnetism,[81] little attention was paid to hematite dispersion as a magnetic colloid until a loose and reversible agglomeration of hematite dispersion was reported.[82,83] The spontaneous magnetization of the solid is about 0.2 emu/g, which is less than 1/200 of magnetite. Hematite particles can be dispersed in an aqueous solution without being strongly aggregated because of its low magnetization.

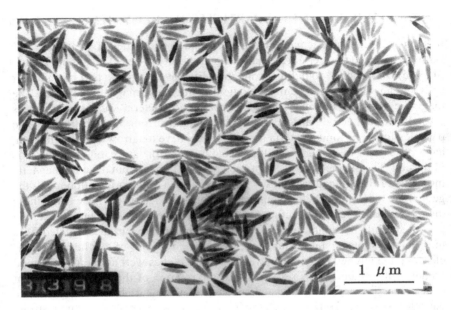

**FIGURE 8.6** Electron micrograph of monodispersed spindle-type hematite particles prepared by hydrothermal reaction of FeCl$_3$ solution in the presence of small amounts of phosphate ions.

Hematite particles having narrow size distributions can be prepared in the form of spheres, cubes, spindles, or platelets. Even dumbbell-type particles can be formed. Matijević and Scheiner[84] succeeded in preparing monodispersed spherical hematite particles by forced hydrolysis of ferric chloride solutions. Since then, monodispersed hematite particles have been widely used for studies of colloidal dispersions. Spindle-type hematite particles of extremely narrow size distribution, as shown in Figure 8.6, were obtained by hydrothermal reactions in the presence of small amounts of phosphate ions.[85] The aspect ratio of the spindle-type particles can be controlled by the concentration of phosphate anions. High-resolution electron microscopy revealed that ellipsoidal hematite particles are composed of an ordered aggregation of smaller ellipsoids. However, the electron diffraction on a single particle showed that the mosaic structure behaves as a single crystal.[86] Figure 8.7 shows the electron diffraction pattern of a single particle obtained by selected area electron diffraction. Uniform spindle-type hematite particles with nanometer size ranges were also obtained by forced hydrolysis of iron(III) perchlorate in the presence of urea and phosphate ions.[87]

Elongated hematite particles were also produced in basic conditions by aging freshly precipitated ferric hydroxides at temperatures of 100°C to 200°C in the presence of small amounts of organic phosphonic acid or hydroxycarboxylic acid

**FIGURE 8.7** Electron diffraction pattern of a spindle-type hematite particle.

compounds as crystal growth control reagents. These elongated hematite particles were used for the production of maghemite particles.[30] Similar spindle-type hematite particles were developed by Sugimoto et al.[88] using a method similar to the so-called sol-gel method. They called the new method the gel-sol method.

The lengths of the elongated particles can be shortened by dissolution in hydrochloric acid or oxalic acid.[16] It is interesting to note that only the length becomes shorter by dissolution. This makes it easy to obtain hematite particles having the desired aspect ratio. This technique may also be applied to other particles having anisotropic shapes.

Cubic and platelet hematite particles are produced by transformation from other particles in aqueous solutions. In this procedure, preformed particles are recrystallized from other precipitates into their final forms. Large cubic-like hematite particles were produced through conversion of previously generated $\alpha$-FeOOH in acidic solutions of HCl at 373 K. Plate-like hematite particles were formed from $\alpha$-FeOOH or Fe(OH)$_3$ under strong basic conditions at elevated temperatures.[89,90] There was a critical temperature for each alkaline concentration in the hydrothermal transformation.[89] The platelets produced were single crystals with (001) flat planes.[91]

As stated above dispersions of hematite particles show some interesting magnetic behaviors.[82,83] Chantrell et al.[92] simulated the structure of the agglomerates of magnetic particles, formed in a magnetic fluid in the presence of and in the absence of an external magnetic field, using the Monte Carlo method.

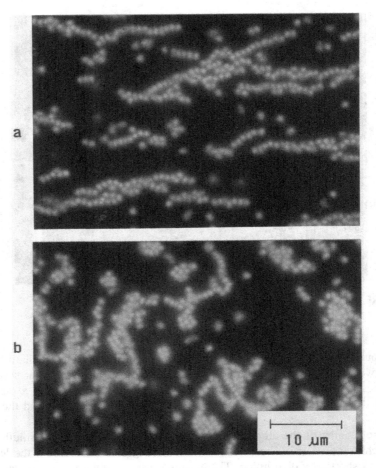

**FIGURE 8.8** Photomicrographs of chain-like agglomerates: (a) chain-like agglomerates of hematite particles in the presence of the geomagnetic field; (b) in the absence of the field.

According to their simulations, the agglomerates were straight under an external field, but were randomly oriented when the field was absent.[92] Similar structures of agglomerates were observed using an aqueous suspension of coarse spindle-like hematite particles of about 1.5 μm long. Figure 8.8 shows the agglomerates of the hematite particles observed in the presence and absence of an external magnetic field. Figure 8.9 shows the highly ordered structure of the spindle-like particles. Chantrell et al.[92] also suggested the formation of loop-like agglomerates in the absence of an external field. Similar loop formations were also visually observed in a dispersion of coarse elongated hematite particles.[93]

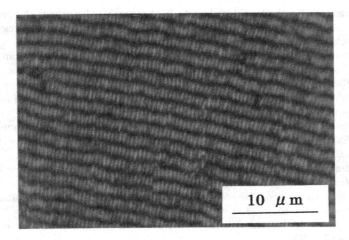

**FIGURE 8.9** Photomicrograph of the highly ordered structure observed in a spindle-like hematite suspension.

## REFERENCES

1. Abe, N., Ejiri, K., and Hanai, K., Advanced technology for magnetic tapes, *J. Magn. Soc. Jpn.*, 26, 1090, 2002.
2. Bate, G., Magnetic recording materials since 1975, *J. Magn. Magn. Mater.*, 100, 413, 1991.
3. Hibst, H., Magnetic pigment for recording information, *J. Magn. Magn. Mater.*, 74, 193, 1988.
4. Bate, G., Present and future of magnetic recording media, in *Ferrites: Proceedings of the ICF3*, H. Watanabe, S. Iida, and M. Sugimoto, Eds., Center for Academic Publications, Tokyo, Japan, 1981.
5. Imaoka, Y., Takada, K., Hamabata, T., and Maruta, F., Advances in magnetic recording media from maghemite and chromium dioxide to cobalt adsorbed gamma ferric oxide, 516, 1981.
6. Kirschvink, J.L., and Gould, J.L., Biogenic magnetite as a basis for magnetic field detection in animals, *Biosystems*, 13, 181, 1981.
7. Dormann, J.L., and Fiorani, D., Eds., *Magnetic Properties of Fine Particles*, North-Holland, Amsterdam, 1992.
8. Hafeli, U., Schutt, W., Teller, J., and Zuborowski, M., Eds., *Scientific and Clinical Applications of Magnetic Carriers*, Plenum Press, New York, 1997.
9. Tajima, H., System for automated extraction using magnetic particles, *J. Magn. Soc. Jpn.*, 22, 1010, 1998.
10. Ohmori, M., and Matijević, E., Preparation and properties of uniform coated inorganic colloidal particles 8: Silica on iron, *J. Colloid Interface Sci.*, 160, 288, 1993.
11. Philipse, A.P., Bruggen, M.P.B., and Pathmamanoharan, C., Magnetic silica dispersions: preparation and stability of surface modified silica particles with a magnetic core, *Langmuir*, 10, 92, 1994.

12. Inada, Y., Takahashi, K., Yoshimoto, T., Kodera, Y., Matsushima, A., and Saito, Y., Application of PEG-enzyme and magnetite-PEG-enzyme conjugates for biotechnical process, *Trends Biotech.*, 6, 131, 1988.

13. Matsunaga, T., and Kamiya, S., Use of magnetic particles isolated from magnetotactic bacteria for enzyme immobilization, *Appl. Microbiol. Biotechnol.*, 26, 328, 1987.

14. LaMer, V.K., and Dinegar, R.H., Theory, production and mechanism of formation of monodispersed hydrosols, *J. Am. Chem. Soc.*, 72, 4847, 1950.

15. Privman, V., Goia, D.V., Park, J., and Matijević, E., Mechanism of formation of monodispersed colloids by aggregation of nanosize precursors, *J. Colloid Interface Sci.*, 213, 36, 1999.

16. Ozaki, M., Sanada, J., Hori, H., and Hori, M., *ITE Lett. Batteries New Technol. Med.*, 3, 472, 2002.

17. Sugimoto, T., Preparation of monodispersed colloidal particles, *Adv. Colloid Interface Sci.*, 28, 65, 1987.

18. Matijević, E., Preparation and properties of uniform size colloids, *Chem. Mater.*, 5, 412, 1993.

19. Scholten, P.C., How magnetic can a magnetic fluid be?, *J. Magn. Magn. Mater.*, 39, 99, 1983.

20. Vékás, L., Rasa, M., and Bica, D., Physical properties of magnetic fluids and nanoparticles from magnetic and magneto-rheological measurements, *J. Colloid Interface Sci.*, 231, 247, 2000.

21. Chikazumi, S., *Physics of Magnetism*, Wiley, New York, 1965.

22. Scholten, P.C., and Tjaden, D.L.A., Mutual attraction of superparamagnetic particles, *J. Colloid Interface Sci.*, 73, 254, 1980.

23. Okada, I., Ozaki, M., and Matijević, E., Magnetic interactions between platelet-type colloidal particles, *J. Colloid Interface Sci.*, 142, 251, 1991.

24. Bate, G., Oxides for magnetic recording, in *Magnetic Oxides, Part 1*, D.J. Craik, Ed., Wiley, New York, 1975.

25. Morrish, A.H., Morphology and physical properties of gamma ferric oxide, in *Crystals: Growth and Applications*, vol. 2, H.C. Freyhard, Ed., Springer, Berlin, 1979.

26. Chubachi, R., and Tamagawa, N., Characteristics and application of metal tape, *IEEE Trans. Magn.*, 20, 45, 1984.

27. Fugiwara, T., Isshiki, M., Koike, Y., and Oguchi, T., Recording performances of Ba-ferrite coated perpendicular magnetic tapes, *IEEE Trans. Magn.*, 18, 1200, 1982.

28. Yashiro, T., Kikuchi, Y., Matsubayashi, Y., and Morizumi, H., The effects of barium ferrite particles added to VHS tapes, *IEEE Trans. Magn.*, 23, 100, 1987.

29. Kiyama,. K., and Takada, T., Iron compounds formed by the aerial oxidation of ferrous salt solutions, *Bull. Chem. Soc. Jpn.*, 45, 1923, 1972.

30. Matsumoto, M., Koga, T., Fukai, K., and Nakatani, S., Production of acicular ferric oxide, U.S. Patent 4,202,871, 1980.

31. Corradi, A.R., Andress, S.J., French, J.E. Bottoni, G., Candoflo, D., Cecchetti, A., and Masoli, F., Magnetic properties of new (NP) hydrothermal particles, *IEEE Trans. Magn.*, 20, 33, 1984.

32. Ozaki, M., and Matijević, E., Preparation and magnetic properties of monodispersed spindle-type $\gamma$-$Fe_2O_3$ particles, *J. Colloid Interface Sci.*, 107, 199, 1985.

33. Arndt, V., $\gamma\text{-Fe}_2\text{O}_3$ of improved quality from direct synthesis of acicular $\alpha\text{-Fe}_2\text{O}_3$, *IEEE Trans. Magn.*, 24, 1796, 1988.

34. Yada, Y., Miyamoto, S., and Kawagoe, H., A new high $H_c$ gamma ferric oxide exhibiting coercive force as high as 450–470 Oe, *IEEE Trans. Magn.*, 24, 1973.

35. Umeki, S., Saitoh, S., and Imaoka, Y., A new high coercive magnetic particle for recording tape, *IEEE Trans. Magn.*, 10, 655, 1974.

36. Asada, S., Preparation of fine acicular iron particles with high coercivity by reduction method, *Nippon Kagaku Kaishi*, 1985, 22, 1985.

37. Asada, S., Surface stabilization of fine acicular iron particles for magnetic recording media, *Nippon Kagaku Kaishi*, 1984, 1372, 1984.

38. van der Giessen, A.A., and Klomp, C.J., The preparation of iron powders consisting of submicroscopic elongated particles by pseudomorphic reduction of iron oxides, *IEEE Trans. Magn.*, 5, 317, 1969.

39. Chen, M., Tang, B., and Nikles, D.E., Preparation of iron nanoparticles by reduction of $\beta\text{-FeOOH}$ particles, *IEEE Trans. Magn.*, 34, 1141, 1998.

40. Chen, H.Y., Hiller, D.M., Hudson, J.E., and Westenbroek, C.J.A., Advances in properties and manufacturing of chromium dioxide, *IEEE Trans. Magn.*, 20, 24, 1984.

41. Iwasaki, S., Perpendicular magnetic recording, *IEEE Trans. Magn.*, 16, 71, 1980.

42. Sugata, N., Maekawa, M., Ohta, Y., Okinaka, K., and Nagai., N., Advances in fine magnetic particles for high density recording, *IEEE Trans. Magn.*, 31, 2854, 1995.

43. Kubo, O., Ido, T., and Hidehira, Y., Barium ferrite super-fine particles produced by glass crystallization method, *Toshiba Rev.*, 43, 897, 1988.

44. Kakizaki, K., and Hiratuka, N., Magnetic properties and crystal structures of acicular barium ferrite particles, *J. Magn. Soc. Jpn.*, 22(suppl. S1), 129, 1998.

45. Lemke, J.U., Ultra high density recording with new heads and tapes, *IEEE Trans. Magn.*, 15, 1561, 1979.

46. Knowels, J.E., Magnetic properties of individual acicular particles, *IEEE. Trans. Magn.*, 17, 3008, 1981.

47. Aharoni, A., Angular dependence of nucleation field in magnetic recording media, *IEEE Trans. Magn.*, 22, 149, 1986.

48. Salling, C., Schultz, S., McFadyen, I., and Ozaki, M., Measuring the coercivity of individual sub-micron ferromagnetic particles by Lorentz microscopy, *IEEE Trans. Magn.*, 27, 5184, 1991.

49. Lederman, M., Schultz S., and Ozaki, M., Investigation of the dynamics of the magnetization reversal in individual single-domain ferromagnetic particles. *Phys. Rev. Lett.*, 73, 1986, 1994.

50. Inoue, H., Fukke, H., and Katsumoto, H., Effect of polymer adsorbed layer on magnetic particle dispersion, *IEEE Trans. Magn.*, 26, 75, 1990.

51. Homola, A.M., and Lorenz, M.R., Novel magnetic dispersions using silica stabilized particles, *IEEE Trans. Magn.*, 22, 716, 1986.

52. Rosensweig, R.E., Magnetic fluids, *Sci. Am.*, 247, 136, 1982.

53. Taketomi, S., and Chikazumi, S., *Zisei Ryutai [Magnetic Fluids]*, Nikkan Kogyo Shinbun, Tokyo, 1988.

54. Satoh, T., Higuchi, S., and Shmoiizaka, J., Dispersion property of a magnetite colloid in cyclohexane and van der Waals energy, *Abstracts of the 19th National Meeting of the Japan Chemical Society*, 1966.

55. Šarić, A., Musić, S., Nomura, K., and Popović, S., Influence of urotropin on the preparation of iron oxides from $FeCl_3$ solutions, *Croatia Chem. Acta*, 71, 1019, 1998.

56.  Matsuda, K., and Kamiya, I., Formation of magnetite by application of hydrolysis of urea, *Nippon Kagaku Kaishi*, 1983, 23, 1983.

57.  Thomas, J.R., Preparation and magnetic properties of colloidal cobalt particles, *J. Appl. Phys.*, 37, 2914, 1966.

58.  Papirer, E., Horny, P., Balard, H., Anthore, H., Petipas, C., and Martinet, A., The preparation of ferrofluid by decomposition of dicobalt octacarbonyl, *J. Colloid Interface Sci.*, 94, 220, 1983.

59.  Gobe, M., Konno, K., Kandori, K., and Kitahara, A., Preparation and characterization of monodispersed magnetite sols in W/O microemulsion, *J. Colloid Interface Sci.*, 93, 293, 1983.

60.  Pileni, M.P., Nanosized particles made in colloidal assemblies, *Langmuir*, 13, 3266, 1997.

61.  Petit, C., Taleb, A., and Pileni, M.P., Cobalt nanosized particles organized in a 2D superlattice: synthesis, characterization and magnetic properties, *J. Phys. Chem. B*, 103, 1805, 1999.

62.  Wu, S.-H., and Cher, D.-H., Synthesis and characterization of nickel nanoparticles by hydrazine reduction in ethylene glycol, *J. Colloid Interface Sci.*, 259, 282, 2003.

63.  Mann, S., and Hannington, J.P., Formation of iron oxide in unilamellar vesicles, *J. Colloid Interface Sci.*, 122, 26, 1988.

64.  Xu, F., Zhang, X., Xie, Y., Tian, X., and Li, Y., Morphology control of $\gamma$-Fe$_2$O$_3$ nanocrystals via PEG polymer and accounts of its Mössbauer study, *J. Colloid Interface Sci.*, 260, 160, 2003.

65.  Tang, Z.X., Sorensen, C.M., Klabunde, K.J., and Hadjipanayis, G.C., Preparation of manganese ferrite fine particles from aqueous solution, *J. Colloid Interface Sci.*, 145, 38, 1991.

66.  Granqvist, C.G., and Buhrman, R.A., Ultra fine metal particles, *J. Appl. Phys.*, 47, 2200, 1976.

67.  Fievet, F., Lagier, J.P., and Figlarz, M., Preparing of metal powders in micrometer and submicrometer sizes by the polyol process, *MRS Bull.*, 5, 29, 1989.

68.  Ochiai, K., Horie, H., Kamohara, H., and Morita, M., An encapsulation process for magnetic metal powder and its application to powder core manufacturing, *Nippon Kagaku Kaishi*, 1987, 233, 1987.

69.  Abe, M., and Tamura, Y., Ferrite plating in aqueous solution: new technique for preparing magnetic thin film, *J. Appl. Phys.*, 55, 2614, 1984.

70.  Furusawa, K., Muramatsu, H., and Majima, T., Characterization of silica-coated hematite and application to the formation of composite particles including egg yolk PC liposomes, *J. Colloid Interface Sci.*, 264, 95, 2003.

71.  Hirano, S., Yogo, T., Suzuki, H., and Naka, S., Synthesis of iron dispersed carbons by pressure pyrolysis of divinylbenzene-vinylferrocene copolymer, *J. Mater. Sci.*, 18, 2811, 1983.

72.  Ukita, K., Kuroda, M., Honda, H., and Koishi, M., Characterization of powder-coated microsponge prepared by dry impact blending method, *Chem. Pharm. Bull.*, 37, 3367, 1989.

73.  Molday, R.S., Yen, S.P.S., and Rembaum, A., Application of magnetic microspheres in labelling and separation of cells, *Nature*, 268, 437, 1977.

74.  Kondo, A., and Fukuda, H., Preparation of thermo-sensitive magnetic hydrogel microspheres for antibody and application to enzyme immobilization, *J. Ferment. Bioeng.*, 41, 99, 1994.

75.  Kirschvink, J.L., Homing in on vertebrates, *Nature*, 390, 339, 1997.

76. Blakemore, R.P., Magnetotactic bacteria, *Science*, 190, 377, 1975.
77. Masuda, T., Endo, J., Osakabe, N., Tonomura, A., and Arii, T., Morphology and structure of biogenic magnetite particles, *Nature*, 302, 411, 1983.
78. Moskowitz, B.M., Frankel, R.B., Flanders, P.J., Blakemore, R.P., and Schwartz, B.B., Magnetic properties of magnetotactic bacteria, *J. Magn. Magn. Mater.*, 73, 273, 1988.
79. Dickson, D.P.E., Nanostructured magnetism in living systems, *J. Magn. Magn. Mater.*, 203, 46, 1998.
80. Roshko, R.M., and Moskowitz, B.M., A Preisach analysis of magneto-ferritin, *J. Magn. Magn. Mater.*, 177–181, 1461, 1998.
81. Dunlop, D.J., Magnetic properties of fine-hematite, *Ann. Geophysique*, 27, 269, 1971.
82. Ozaki, M., Suzuki, H., Takahashi, K., and Matijević, E., Reversible ordered agglomeration of hematite particles due to weak magnetic interactions, *J. Colloid Interface Sci.*, 113, 76, 1986.
83. Ozaki, M., Egami, T., Sugiyama, N., and Matijević, E., Agglomeration in colloidal hematite particles due to weak magnetic interactions, *J. Colloid Interface Sci.*, 126, 212, 1988.
84. Matijević, E., and Scheiner, P., Ferric hydrous oxide sols. III. Preparation of uniform particles by hydrolysis of Fe(III)-chloride, -nitrate, -perchlorate solutions, *J. Colloid Interface Sci.*, 63, 509, 1978.
85. Ozaki, M., Kratohvil, S., and Matijević, E., Formation of monodispersed spindle-type hematite particles, *J. Colloid Interface Sci.*, 102, 146, 1984.
86. Ocaña, C., Morales, M.P., and Serna, C.J., The growth mechanism of ellipsoidal particles in solution, *J. Colloid Interface Sci.*, 171, 85, 1995.
87. Ocaña, C., Morales, M.P., and Serna, C.J., Homogeneous precipitation of uniform $\alpha$-$Fe_2O_3$ particles from ion salts solutions in the presence of urea, *J. Colloid Interface Sci.*, 212, 317, 1999.
88. Sugimoto, T., Wang, Y., Itoh, H., and Muramatsu, A., Systematic control of sizes, shape and internal structure of monodispersed $\alpha$-$Fe_2O_3$ particles, *Colloids Surf. A*, 134, 265, 1998.
89. Nobuoka, S., and Ado, K., Studies on thin iron oxide platelets: formation of $\alpha$-iron oxide by hydrothermal reactions, *Shikizai*, 60, 265, 1987.
90. Ozaki, M., Ookoshi, N., and Matijević, E., Preparation and magnetic properties of uniform hematite platelets, *J. Colloid Interface Sci.*, 137, 546, 1990.
91. Shindo, D., Lee, B.T., Waseda, Y., Muramatsu, A., and Sugimoto, T., Crystallography of platelet-type hematite particles by electron microscopy, *Mater Trans. JIM*, 34, 580, 1993.
92. Chantrell, R.W., Bradbury, A., Poppelwell, J., and Charles, S.W., Particle cluster configuration in magnetic fluids, *J. Phys. D Appl. Phys.*, 13, L119, 1980.
93. Ozaki, M., Sanada, J., and Isobe, A., Direct observation of loop-like agglomerations of spindle-type hematite particles due to weak magnetic interactions, *Colloids Surf.*, 109, 117, 1996.

# 9 Synthesis and Surface Modification of Zinc Sulfide-Based Phosphors

*Lai Qi, Burtrand I. Lee, David Morton, and Eric Forsythe*

## CONTENTS

## I. INTRODUCTION

Phosphors are solid materials that can absorb a certain kind of energy and convert it into electromagnetic radiation that is usually within the infrared (IR), visible, and ultraviolet (UV) ranges. This radiation process is called luminescence. Luminescence can be excited by different types of energy. Photoluminescence (PL) is excited by electromagnetic radiation (often UV light). Cathodoluminescence (CL) is excited by energetic electron beams. Electroluminescence is excited by electric fields. Chemiluminescence is excited by chemical energy released from reactions. X-ray luminescence is excited by high-energy x-rays, and so on.

Numerous inorganic materials have been found or synthesized to luminesce. And thousands of them had been used as phosphors in illumination and display devices. Although some phosphors have demonstrated excellent optical properties to meet current requirements, the research on phosphors continues because the current phosphors are far from perfect. Industry and researchers continue to look for either new phosphors or methods to modify the current phosphors to make them more energy efficient, higher in brightness, and give them better low-voltage properties.[1] The commonly used phosphors for display purposes and their synthesis methods are listed in Table 9.1.

Since its invention by Karl Ferdinand Braun in 1897,[2] the cathode-ray tube (CRT) has become the most widely adopted device for information display, such as conventional televisions (TVs), computer monitors, and projection TVs. Although several other techniques have demonstrated the potential to partially replace CRTs, they are either still being researched to overcome technical shortcomings or are struggling with manufacturing cost problems. Such display devices include electroluminescence (EL), field-emission displays (FEDs), light-emitting diodes (LEDs), and plasma displays. Liquid crystal displays (LCDs) have become popular in recent years due to breakthroughs in manufacturing techniques.[3] However, LCDs cannot completely replace CRTs because of their innate drawbacks (i.e., viewing angle and brightness problems).[4]

Among all the CRT phosphors, the one worth mentioning is the zinc sulfide (ZnS) phosphor family. It includes ZnS:Cu,Al (green), ZnS:Ag,Cl (blue), and ZnS:Mn (orange), which are representative for their excellent properties in their respective color domain. ZnS is a semiconductor with a band gap of 3.85 eV, which makes it a suitable material for accommodating a variety of dopants. The dopant ions do not change the ZnS structure or form separate phases. Under controlled conditions, the doped ions distribute homogeneously within the ZnS matrix, forming numerous lattice "defects" due to their different atomic size, electronegativity, or valence. Those "defects" create energy levels within the ZnS band gap, like forming some steps where electrons can be excited to or have a rest when they are dropping back. When the electrons are transmitted back and forth between bands or levels, the energy difference between the "steps" is emitted as electromagnetic radiation. With the careful selection of dopants, the wavelength of the radiation from most phosphors is within the visual range (400 nm to 700 nm). A schematic of semiconductor-type luminescence by electron transitions is shown in Figure 9.1.

## II. GENERAL SYNTHESIS TECHNOLOGY

Zinc sulfide possesses either a cubic or hexagonal structure at room temperature, depending on the synthesis conditions. The synthesis methods for the ZnS family phosphors include the solid state reaction and homogeneous coprecipitation. In the solid state reaction, phosphor-grade ZnS is prepared first through rigorous purification processes.[15] Then the ZnS powder is well mixed with a solution of dopant compounds. After the solvent is evaporated and the dopant ions cling to the ZnS particle surface, this phosphor precursor is sent to the furnace, where it is annealed in a controlled atmosphere at a certain temperature. Powders are usually washed with diluted acid and water after annealing to remove the residual dopants and possible contaminants.

**TABLE 9.1**
**Composition and Optical Properties of Commonly Used CRT Phosphors**

| Composition | EIA Symbol | Emission Color | Peak Wavelength (nm) | 1/10 Delay Time[a] | Applications | Ref. |
|---|---|---|---|---|---|---|
| ZnS:Ag | P11 | Blue | 460 | MS | Lamps, projection tube | 5 |
| $Y_2SiO_5$:$Ce^{3+}$ | P47 | Violet-blue | 400 | VS | Photographic applications | 6 |
| $CaWO_4$:W | P5 | Blue | 420 | MS | Lamps | 7 |
| $Zn_2SiO_4$:Ti | P52 | Violet-blue | 400 | MS | Photographic applications | 8 |
| (Zn,Cd)S:Cu,Al | — | Yellowish green | 560 | M | Graphic display | 5 |
| ZnS:Ag + ZnS:Cu,Al | B | White | — | MS | Lamps, backlights | 9 |
| ZnO:Zn | P24 | Bluish green | 510 | S | Flying spot, EL, CRTs | 5 |
| ZnS:Cu | P31 | Green | 520 | MS | Oscilloscopes, lamps | 5 |
| $Zn_2SiO_4$:$Mn^{2+}$ | P1 | Yellowish green | 525 | M | Oscilloscopes, radar | 5 |
| ZnS:Cu,Al | P22 | Green | 550 | M | Graphic display | 10 |
| $Y_3Al_5O_{12}$:$Tb^{3+}$ | P46 | Yellowish green | 530 | VS | Flying spot | 11 |
| $Y_2O_2S$:$Eu^{3+}$ | P54 | Red | 606 | M | Lamps | 12 |
| $Y_2O_3$:$Eu^{3+}$ | P56 | Red | 611 | M | Projection tube | 13 |
| $YVO_4$:$Eu^{3+}$ | P49 | Reddish orange | 619 | M | Graphic display | 14 |

*Note:* EIA, Electronics Industry Alliance.

[a] VS, 1 μsec or less; S, 1–10 μsec; MS, 10 μsec to 1 msec; M, 1 msec to 100 msec; L, 100 msec to 1 sec; VL, 1 sec or more.

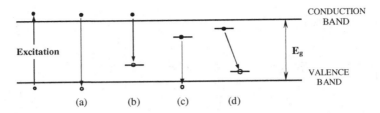

**FIGURE 9.1** Schematic of emissive transitions of electrons in semiconductors: (a) band-band emission; (b) a free electron recombines with a trapped hole; (c) a trapped electron recombines with a free hole; (d) donor-acceptor pair emission.

In the chemical coprecipitation method, zinc and dopant ions are atomically mixed in a homogeneous solution, and then coprecipitated at the same time by addition of sulfide anions from such agents as $H_2S$ gas,[16] thioacetamide,[17] $Na_2S$,[18] and $(NH_4)_2S$.[10] An annealing treatment is necessary for the coprecipitated ZnS phosphors to achieve a homogeneous distribution of dopants and to crystallize the precipitate into the expected crystal structure. But this annealing treatment can be carried out at a lower temperature and for a shorter time compared with those in the solid state reaction. Lower annealing temperatures and shorter times are helpful to reduce particle growth and agglomeration. Compared with the solid state reaction, the coprecipitated ZnS phosphors have a smaller size and narrower size distribution. However, the processing costs prevent this technique from being used in large-scale production. Thus the solid state reaction route is still being used in the industry.

ZnS-based phosphors can also be made in the form of thin films by either physical or chemical deposition. However, these techniques are quite different and are beyond the scope of this chapter.

## III. SURFACE TREATMENT AND COATING TECHNOLOGY

Most phosphors are made as powders 1 to 10 μm in diameter before being applied to CRT production. Surface treatment and surface coating are necessary for some phosphors before the screening step, especially for the sulfide-based phosphors because of their susceptibility to chemical degradation during extended electron bombardment. Accumulated knowledge of surface treatments and coatings on phosphor particles has been protected by patents.

Surface treatments and coatings on phosphors are usually carried out for the following reasons: protection of phosphors from degradation in brightness and chromaticity during a lifetime of service; contrast improvement by pigment coating; surface cleaning of residues and defects; and improvement of screening characteristics.

Possible chemical and mechanical damage to ZnS phosphors may be caused during the slurry process, baking step, and a lifetime of service. Currently known coatings are $SiO_2$,[19,20] metal oxide,[21,22] and organic coating.[23] Igarashi et al.[24] reported that the degradation of ZnS-based phosphors is reduced by 40% after ZnO coating. Wagner et al.[25] reported a coating-enhanced brightness of 40 to 60% in ZnS phosphors by $SiO_2$.

For comfortable viewing by the human eye, displays should have a contrast as high as possible to the background. That is why we use white chalk on blackboard, where the reflective handwriting obtains a high contrast to the absorptive background. In contrast, we use color markers on a white board, where the characters are absorptive but the background is highly reflective. Thus TVs and monitors normally have a dark screen, because we want the pixels excited by electron beams to be emissive, while the unexcited ones are completely dark. Although phosphors are normally transparent to visible light, the particle-coated phosphor screen appears white due to the numerous highly reflective particle surfaces. This white screen is uncomfortable and tiresome for human eyes to watch. So coatings on phosphor particles are usually adopted in manufacturing. Although a black coating is the best choice for contrast purposes, blue and red phosphors are used because of the consideration of brightness. Moreover, blue and red coatings render better color chromaticity. However, pigment coating for green phosphors is usually unnecessary because ZnS-based phosphors themselves are green, and an appropriate green pigment is not available.[26]

Other than the considerations mentioned above, current CRT phosphors have to be modified before extending their service to future applications in EL and FE devices, such as improved luminescent efficiency at low voltage and solving the charging problem at low excitation voltage.[1,27]

## A. Luminescent Enhancement of ZnS:Cu by $BaTiO_3$ and $SrTiO_3$ Coating

The concept of applying dielectric coating to phosphor particles originated from the concept of increasing the actual voltage on phosphors by the capacitance effect in EL displays.[28] In a typical design of thin-film electroluminescence (TFEL) and AC powder EL, the phosphor layer is sandwiched between a pair of insulating layers. The insulating layers should have a high breakdown electric field strength, high dielectric constant (K), and good adhesion to both the phosphor and electrode layers. However, high K materials usually do not show high electrical breakdown strength, as reported by Fujita et al.[29] Suitable insulating materials include $SiO_2$, $Al_2O_3$, $TiO_2$, $Si_3N_4$, $BaTiO_3$, $SrTiO_3$, etc. $BaTiO_3$ is a promising candidate among all the choices because it not only has a very high dielectric constant, but it also has a variety of synthesis methods at low temperatures.[30]

Using a $BaTiO_3$ coating on phosphor particles has several advantages over coating on phosphor layers. First, the sandwich structure is not necessary, and is replaced by one phosphor-$BaTiO_3$ mixture layer, thus the manufacturing process is simplified. Second, it provides better luminescence characteristics due to the surface modification effect.[22] Third, phosphor particles are better sealed and protected from electrical breakdown and moisture penetration through pinholes in the insulating layers of the sandwich structure.[31]

The coating of $BaTiO_3$ on ZnS:Cu starts with the preparation of a $BaTiO_3$ solution (BT sol) of barium acetate and titanium lactate. The stoichiometrically prepared BT sol is a clear acidic solution. ZnS:Cu phosphor particles are then dispersed in the BT sol under ultrasonication for 30 minutes before filtration or centrifugation. Annealing of the powders in borosilicate vials is carried out by placing them in an alumina crucible packed with activated carbon granules, which provides a slightly reducing atmosphere during annealing to protect the phosphors. The crucible is covered with an alumina cover and annealed in air at a temperature of 500 to 600°C for 2 h. A continuous coating will be formed on the ZnS:Cu particle surface. Another way to prepare the $BaTiO_3$ coating is to use barium and titanium alkoxides, as described by Nakamura et al.[28] A similar procedure can be applied to prepare $SrTiO_3$ coatings.

It has been observed that the dielectric constant of capacitors in a multilayer form (with silver as the electrodes) is somewhat higher than those in a disk form.[32] Also, an enhanced dielectric constant has been reported with the addition of silver in the perovskite structure[33] due to the improved sintering density and enlarged actual electrode areas. It is worth investigating the effect of this K enhancement of coatings on the luminescent intensity of phosphors.

A $BaTiO_3$ coating with silver addition can be prepared by modifying the $BaTiO_3$ route mentioned above. During the preparation of clear BT sol, silver nitrate or silver acetate is added together with a large volume of triethanol amine. The weight ratio of silver to $BaTiO_3$ ranges from 0.1 to 10 wt%. The following coating operation is the same as that for the pure $BaTiO_3$ coating. Figure 9.2 is the PL excitation spectra of the as-received $BaTiO_3$-coated, $SrTiO_3$-coated, and $BaTiO_3$/Ag-coated green phosphor ZnS:Cu (Sarnoff Corp., Princeton, NJ).

Both $BaTiO_3$ and $SrTiO_3$ coatings show enhanced PL excitation properties, possibly due to the surface modification on phosphor particles. Nevertheless, $BaTiO_3$ coating with silver addition shows an inferior result compared with the noncoated sample. The excitation intensity of the $BaTiO_3$/Ag-coated sample in the UV range (250 to 400 nm) is lower by almost an order of magnitude than that of the pure $BaTiO_3$-coated sample. And this intensity continues to go down to near zero beyond 450 nm. This intensity reduction is possibly caused by the gray color of the $BaTiO_3$/Ag coating. The silver cations are reduced to metal silver and form very fine dispersed particles attaching to the $BaTiO_3$ particles at temperatures of 200 to 400°C, which gives the dark color. Figure 9.3 is the PL emission spectra of the $BaTiO_3$-coated, $SrTiO_3$-coated, and $BaTiO_3$/Ag-coated ZnS:Cu phosphor. Both coatings of $BaTiO_3$ and $SrTiO_3$ improve the emission intensity, while the coating of $BaTiO_3$/Ag lowers emission intensity, which agrees with the observation in the PL excitation spectra. Other than the coating effect, another contribution to this PL improvement by the $BaTiO_3$ coating is possibly

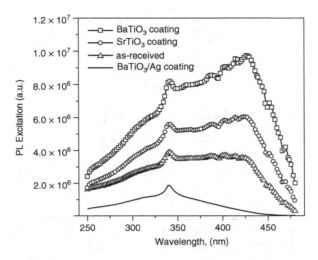

**FIGURE 9.2** PL excitation spectra of ZnS:Cu after coating with different materials.

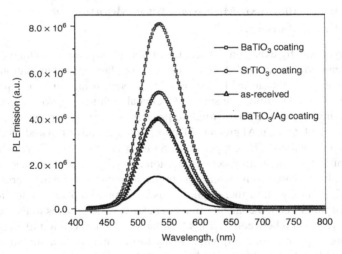

**FIGURE 9.3** PL emission spectra of ZnS:Cu after coating with different materials.

from the acidic media in BT sol, which slightly etched the particle surface during the dispersion and removed defects.

It is believed that this effect of enhanced PL may be exhibited to a greater extent in EL performance, since BaTiO$_3$ is an excellent dielectric material. However, ZnS:Cu phosphor for CL cannot be used for EL directly because of the low doping concentration. Typical doping concentrations in EL ZnS phosphors should be about an order higher than in their CL counterparts.

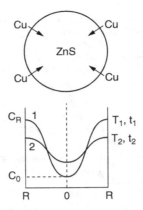

**FIGURE 9.4** Schematic demonstration of copper distribution within a ZnS particle. Curves 1 and 2 represent two different distributions under different annealing conditions. However, curves 1 and 2 have the same average concentration.

## B. Surface Etching and Microstructural Modeling of ZnS:Cu,Al Phosphor

A dead layer usually exists on the surface of phosphor particles, which degrades the phosphors due to crystal defects, residues, oxidation, contamination, etc. A surface treatment with an acid wash is widely used in industry to improve the quality of oxide and sulfide phosphors. Chemical etching on phosphors can also be used to investigate the microstructure properties of the phosphor particles.[34,35]

Commercial ZnS:Cu,Al green CRT phosphor is usually prepared by the solid state reaction method. The copper concentration is one of the key factors that directly influences the luminescence emission intensity. The luminescence intensity normally increases with increasing copper concentration up to a certain value and then decreases with further doping. This is a common phenomenon for most doped phosphors[36] called concentration quenching, which relates to the nonemissive cross-relaxation between dopants.[37] In the solid state reaction of ZnS:Cu,Al, the dopants (copper, aluminum) will actually form concentration gradients during annealing rather than uniform distributions. Any variation of the annealing conditions (i.e., temperature or time) will change the gradient shape and thus change the luminescence intensity, while the average concentration may remain the same, as shown in Figure 9.4. Therefore the average concentration may not truly relate to the luminescent properties.

A novel method of investigating the efficiency of copper doping concentration by etching has been recently reported.[35] The commercial ZnS:Cu,Al phosphor (P22-GN4, Kasei Optonix, Ltd., Kanagawa, Japan) was deeply etched by hydrochloric acid. The diameter of the etched particles ranged from 7.2 μm to 100 nm, as shown in Figure 9.5.

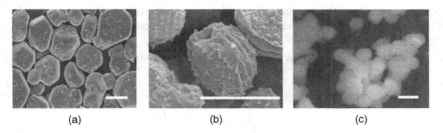

**FIGURE 9.5** Scanning electron micrographs of differently etched ZnS:Cu,Al particles: (a) as received; (b) etched 60 seconds; (c) etched 14 min. The scale bars represent 5 μm, 5 μm and 100 nm, respectively.

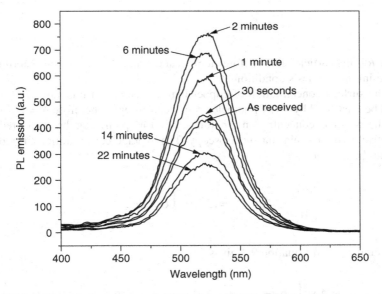

**FIGURE 9.6** PL emission spectra of differently etched ZnS:Cu,Al phosphors.

The PL emission intensity of differently etched particles varied according to etching time, as shown in Figure 9.6. The reason for this variation is the radial removing of portions along the concentration gradient of the dopants.

Based on the model in Figure 9.4, the diffusion of copper in a spherical ZnS particle is governed by Fick's second law:[38]

$$\frac{\partial C}{\partial t} = D \times \left( \frac{\partial^2 C}{\partial r^2} + \frac{2}{r} \times \frac{\partial C}{\partial r} \right),$$ (9.1)

where $C$ and $D$ are the concentration and diffusion coefficients, respectively. The diffusion coefficient could reasonably be treated as a constant for the low copper concentration.

A general solution to the Fick equation in polar coordinates is

$$C_{r,t} = \frac{\alpha}{2\sqrt{\pi(Dt)^3}} \exp\left(-\frac{r^2}{4Dt}\right), \qquad (9.2)$$

where $\alpha$ is the mass of copper deposited on the particle surface before annealing and $t$ is the annealing time. For the commercial samples, $\alpha$ and $t$ have fixed values because they are already made. Thus the solution equation can be simplified as

$$C_r = A\exp[-(R-r)^2 / B] \quad , \qquad (9.3)$$

where $R$ is the particle radius before etching, and $A$ and $B$ are constants determined by the initial synthesis conditions.

The surface concentration of copper before etching (in this case, 179 ppm) could be measured by x-ray photoelectron spectroscopy. Therefore $A = 179$ ppm and the average concentration (in this case, 140 ppm) can be measured by inductively coupled plasma spectroscopy. The average concentration of copper, $C_m$, can be derived as

$$C_m = (3/R^3)\int_0^R C_r r^2 dr \ . \qquad (9.4)$$

Replacing $C_r$ with Equation 9.3 gives

$$C_m = \frac{3AB}{2R^2}\left[(\frac{R\sqrt{\pi}}{\sqrt{B}} + \frac{\sqrt{B\pi}}{2R})erf(\frac{R}{\sqrt{B}}) + \exp(\frac{-R^2}{B}) - 1\right] = 140\,ppm \ , \quad (9.5)$$

where $B$ (in this case, 4.2 $\mu m^2$) is solved for by iterative calculation.

The calculated gradient curve, $C_r = 179\exp[(3.6\ r)^2/4.2]$, is shown as curve (a) in Figure 9.7. Curve (b) is the calculated average concentration of etched particles, based on the gradient in curve (a), which agrees with the measurement. Curve (c) is the polynomial simulation of the measured PL relative intensities. A 140% increase of the PL intensity was observed from the particles with 1 $\mu m$ of the top surface etched away.

The luminescence intensity should be a function of a number of factors:

$$I_{em} = f(I_{ex}, M, Xtal, P, C_r...), \qquad (9.6)$$

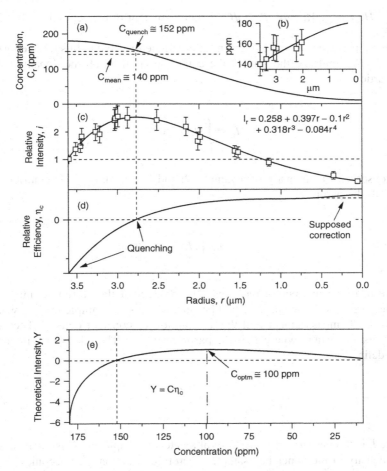

**FIGURE 9.7** Plot of functions (abscissa: particle radius from particle surface to the center): (a) the calculated copper concentration gradient ($C_{quench}$ is the quenching concentration, $C_{mean}$ is the mean concentration); (b) the calculated mean concentration of etched particles based on the gradient in curve (a); (c) the measured relative PL intensity and the polynomially fitted curve; (d) the calculated relative CCE curve; (e) the theoretical output intensity curve ($C_{optm}$ is the optimized concentration).

where $I_{em}$, $I_{ex}$, $M$, $Xtal$, $P$, and $C_r$ are the emission intensity, excitation intensity, sample loading mass, crystallinity, purity, and doping concentration, respectively. For differently etched samples, all the variables, except $C_r$ were kept the same because $I_{ex}$ and $M$ were experimentally set to be constant. *Xtal* and $P$ were supposed to have no change after etching. The only variation was the dopants distribution, $C_r$. So Equation 9.6 can be rewritten as

$$I_{em} = H_c \times C_r, \tag{9.7}$$

where $H_c = f(I_{ex}, M, Xtal, P, C_r)/C_r$ is the coefficient of concentration efficiency (CCE).

For a particle with radial concentration gradient $C_r$ and radius $r$, the emission intensity $I_r$ should be the sum of the contributions from all the dopants within the particle:

$$I_{r'} = \int_0^{r'} H_c C_r 4\pi r^2 dr \ .$$  (9.8)

The observed PL intensity of a sample should be the sum of the contributions from all the particles:

$$I_{r'} = n_{r'} \int_0^{r'} H_c C_r 4\pi r^2 dr \ ,$$  (9.9)

where $n_r$ is the number of particles. The product of the particle number and single-particle volume was constant because of the same sample loading weight for each PL measurement and the assumption of constant density. Therefore $n_r \times V_r$ is a constant, where $V_r$ is the particle volume. The relative PL intensity, $i_r$, is defined as

$$i_{r'} = \frac{I_{r'}}{I_R} \ ,$$  (9.10)

where $I_r$ is the measured PL intensity of an etched sample and $I_R$ is the measured PL intensity of the nonetched sample. Therefore Equation 9.10 becomes

$$i_{r'} = \beta r'^{-3} \times \int_0^{r'} H_c C_r 4\pi r^2 dr,$$  (9.11)

where

$$\beta = R^3 \times \left( \int_0^R H_c C_r 4\pi r^2 dr \right)^{-1}$$

is a constant.

To separate $H_c$, Equation 9.4 was differentiated by radius, giving

$$H_c = \frac{1}{\beta}(\frac{di_{r'}}{dr'} \times \frac{r'}{3C_{r'}} + \frac{i_{r'}}{C_{r'}}) .$$
(9.12)

As defined, $H_c$ should give absolute values. However, in practical cases, it is impossible to calculate. Thus the relative CCE, $_c$, is defined as $_c = H_c \infty$ .
Therefore

$$\eta_c = \frac{di_{r'}}{dr'} \times \frac{r'}{3C_{r'}} + \frac{i_{r'}}{C_{r'}} .$$
(9.13)

The derivative $di_r/dr$ could be derived from the fitting equation shown in curve (c) of Figure 9.7. Thus Equation 9.13 can be calculated, which is shown as curve (d).

In Figure 9.7, the copper concentration, PL intensity, and doping efficiency are related to each other through the abscissa of particle radius. The relative CCE becomes zero at the concentration that gives the highest PL intensity (152 ppm) because of the fact that any further doping beyond this point decreases the total intensity, which indicates a negative CCE. The concentration quenching starts from this point because increased cross-relaxation will cause more light to be absorbed than emitted in those overdoped portions. As the concentration decreases, the efficiency increases gradually to a limit. The explanation is, as the doping concentration decreases, the distance between dopants increases. The chance of nonradiative relaxation between dopants decreases, so the CCE increases. The highest efficiency should be at the zero concentration, where the "cross-relaxation" is null. The instrumental errors became significant when the radius approached zero. A correction of the CCE curve at radius zero is suggested (dotted line).

Based on a homogeneous distribution model, another function, the theoretical output intensity ($Y$), is defined as

$$Y = C \times \eta_c ,$$
(9.14)

which is shown as curve $e$ in Figure 9.7. The optimum doping concentration, giving the highest luminescent output, is about 100 ppm.

Investigation and modeling of the ZnS:Cu,Al microstructure found that although the comprehensive properties had been optimized, commercial ZnS:Cu,Al particles still have a 1 μm overdoped surface layer. Luminescence intensity was significantly improved when this layer was etched away.

## ACKNOWLEDGMENT

Partial funding for phosphor surface modification was provided by the U.S. Army Research Office under contract no. DAAD 19-02-D-001.

## REFERENCES

1. Shea, L.E., Low-voltage cathodoluminescent phosphors, *Electrochem. Soc. Interface*, 7, 34, 1998.
2. Woodcock, S., and Leyland, J.D., *Displays*, July, 69, 1979.
3. Jun, S., Advances in large size LCD-TV: the technical challenges in 40 inch LCD for digital TV, *Society for Information Display Conference Record*, International Display Research Conference, 2002.
4. Petrov, V.F., Liquid crystals for advanced display applications, *Proc. SPIE*, 2408, 84, 1995.
5. Leverenz, H.W., *Final Report on Research and Development Leading to New and Improved Radar Indicators*, PB25481, Office Bub. Board, Washington, D.C., 1945.
6. Bosze, E.J., Hirata, G.A., and McKittrick, J., Investigation of the chromaticity of blue emitting yttrium silicate, *Mater. Res. Soc. Symp. Proc.*, 560, 15, 1999.
7. Min, K.-W., Mho, S.-I., and Yeo, I.-H., Electrochemical fabrication of luminescent $CaWO_4$ and $CaWO_4$:Pb films on W substrates with anodic potential pulses, *J. Electrochem. Soc.*, 146, 3128, 1999.
8. Miyata, T., Minami, T., Saikai, K., and Takata, S., $Zn_2SIO_4$ as a host material for phosphor-emitting layers of TFEL devices, *J. Lumin.*, 60–61, 926, 1994.
9. Ozawa, L., *Application of Cathodoluminescence to Display Devices*, Kodansha, Tokyo, 1994.
10. Qi, L., Lee, B.I., Kim, J.M., Jang, J.E., and Choe, J.Y., Synthesis and characterization of ZnS:Cu,Al phosphor prepared by a chemical solution method, *J. Lumin.*, 104, 261, 2003.
11. Scholl, M.S., and Trimmier, J.R., Luminescence of YAG:Tm, Tb, *J. Electrochem. Soc.*, 133, 643, 1986.
12. Tseng, Y.H., Chiou, B.S., Peng, C.C., and Ozawa, L., Spectral properties of $Eu^{3+}$-activated yttrium oxysulfide red phosphor, *Thin Solid Film*, 330, 173, 1998.
13. Hao, J.H., Studenikin, S.A., and Cocivera, M., Blue, green and red cathodoluminescence of $Y_2O_3$ phosphor films prepared by spray pyrolysis, *J. Lumin.*, 93, 313, 2001.
14. Kandarakis, I., Cavouras, D., Kanellopoulos, E., Nomicos, C.D., and Panayiotakis, G.S., Image quality evaluation of $YVO_4$:Eu phosphor screens for use in X-ray medical imaging detectors, *Radiat. Meas.*, 29, 481, 1998.
15. Leverenz, H.W., *Introduction to Luminescence of Solids*, John Wiley & Sons, New York, 1950.
16. Gallagher, D., Heady, W.E., Racz, J.M., and Bhargava, R.N., Homogeneous precipitation of doped zinc sulfide nanocrystals for photonic applications, *J. Mater. Res.*, 10, 870, 1995.
17. Vacassy, R., Scholz, S.M., Dutta, J., Plummer, C.J.G., Houriet, R., and Hofmann, H., Synthesis of controlled spherical zinc sulfide particles by precipitation from homogeneous solutions, *J. Am. Ceram. Soc.*, 81, 2699, 1998.

18. Lu, S.W., Lee, B.I., Wang, Z.L., Tong, W., Wagner, B.K., Park, W., and Summers, C.J., Synthesis and photoluminescence enhancement of $Mn^{2+}$-doped ZnS nanocrystals, *J. Lumin.*, 92, 73, 2001.

19. Villalobos, G.R., Bayya, S.S., Sanghera, J.S., Miklos, R.E., Kung, F., and Aggarwal, I.D., Protective silica coating on zinc-sulfide-based phosphor particles, *J. Am. Ceram. Soc.*, 85, 2128, 2002.

20. Abrams, B.L., Roos, W., Holloway, P.H., and Swart, H.C., Electron beam-induced degradation of zinc sulfide-based phosphors, *Surf. Sci.*, 451, 174, 2000.

21. Choi, H.-H., Ollinger, M., and Singh, R.K., Enhanced cathodoluminescent properties of ZnO encapsulated ZnS:Ag phosphors using an electrochemical deposition coating, *Appl. Phys. Lett.*, 82, 2494, 2003.

22. Sawada, M., Oobayashi, S., Yamaguchi, K., Takemura, H., Nakamura, M., Momose, K., and Saka, H., Characteristics of light emission lifetime of electroluminescent phosphor encapsulated by titanium-silicon-oxide film, *Jpn. J. Appl. Phys.*, 41, 3885, 2002.

23. Chen, S., and Liu, W., Characterization and antiwear ability of non-coated ZnS nanoparticles and DDP-coated ZnS nanoparticles, *Mater. Res. Bull.*, 36, 137, 2001.

24. Igarashi, T., Kusunoki, T., Ohno, K., Isobe, T., and Senna, M., Degradation proof modification of ZnS-based phosphors with ZnO nanoparticles, *Mater. Res. Bull.*, 36, 1317, 2001.

25. Wagner, B.K., Russell, G., Yasuda, K., Summers, C.J., Do, Y.R., Yang, H.G., and Park, W., Thin $SiO_2$ coating on ZnS phosphors for improved low-voltage cathodoluminescence properties, *J. Mater. Res.*, 15, 2288, 2000.

26. Shionoya, S., and Yen, W.M., *Phosphor Handbook*, CRC Press, Boca Raton, FL, 1999.

27. Kominami, H., Nakamura, T., Sowa, K., Nakanishi, Y., Hatanaka, Y., and Shimaoka, G., Low voltage cathodoluminescent properties of phosphors coated with $In_2O_3$ by sol-gel method, *Appl. Surf. Sci.*, 113–114, 519, 1997.

28. Nakamura, T., Kamiya, M., Watanabe, H., and Nakanishi, Y., Preparation and characterization of zinc sulfide phosphors coated with barium titanate using sol-gel method, *J. Electrochem. Soc.*, 142, 949, 1995.

29. Fujita, Y., Kuwata, J., Nishikawa, M., Tohda, T., Matuoka, T., Abe, A., and Nitta, T., *Proc. Soc. Inf. Disp.*, 25, 177, 1984.

30. Lee, B.I., and Zhang, J., Dielectric thick film deposition by particle coating method, *Mater. Res. Bull.*, 36, 1065, 2001.

31. Yan, S., Maeda, H., Hayashi, J.-I., Kusakabe, K., Morooka, S., and Okubo, T., Low-temperature plasma coating of electroluminescence particles with silicon nitride film, *J. Mater. Sci.*, 28, 1829, 1993.

32. Chu, M.S.H., and Hodgkins, C.E., in *Advances in Ceramics*, vol. 19, *Multilayer Ceramic Devices*, J.B. Blum and W. Roger Cannon, Eds., American Ceramics Society, Westerville, OH, 1986.

33. Hwang, H.J., Nagai, T., Ohji, T., Sando, M., Toriyama, M., and Niihara, K., Curie temperature anomaly in lead zirconate titanate/silver composites, *J. Am. Ceram. Soc.*, 81, 709, 1998.

34. Ozawa, L., *Cathodoluminescence, Theory and Applications*, VCH, Weinheim, Germany, 1990.

35. Qi, L., Lee, B.I., Gu, X.J., Grujicic, M., Samuels, W.D., and Exarhos, G.J., Concentration efficiency of doping in phosphors: investigation of the copper- and aluminum-doped zinc sulfide, *Appl. Phys. Lett.*, 83, 4945, 2003.

36. Danielson, E., Golden, J.H., McFarland, E.W., Reaves, C.M., Weinberg, W.H., and Wu, X.D., A combinatorial approach to the discovery and optimization of luminescent materials, *Nature*, 389, 944, 1997.

37. Maruyama, T., Yamada, H., Mochizuki, T., Akimoto, K., and Yagi, E., Quenching mechanism of luminescence in Sm-doped ZnS, *J. Crystl. Growth*, 214/215, 954, 2000.

38. Crank, J., *The Mathematics of Diffusion*, Oxford University Press, Fair Lawn, NJ, 1975.

# 10 Characterization of Fine Dry Powders

*Hendrik K. Kammler and Lutz Mädler*

## CONTENTS

# I. INTRODUCTION

Physical and chemical characterization of dry powders is essential in manufacturing of high-performance ceramic materials. The characterization enables process control and custom product design. However, a large number of powder properties must be defined to understand this relationship completely. Therefore a large variety of characterization methods have been developed. In this chapter, selected methods for dry powder analysis are described, starting with their operating principals, types of data obtained, limitations, and examples. The examples are mainly taken from the dry manufacture of ceramic powders by means of aerosol flame technology.[1] Therefore the main focus of this chapter is based on particle characterization and sizing techniques of dry powders. Particle characteristics such as primary particle size, agglomerate size, mass-fractal dimension, particle size distribution, particle shape, phase composition, surface characteristics, porosity, electronic structure, and the applicable size range for successful operation are opposed to the analysis techniques. The latter are discussed in terms of imaging techniques, spectroscopic analysis, interaction with gases and liquids, and the behavior of particles in applied force fields (Table 10.1). Thus the techniques are explored and summarized with respect to their mode of operation and primary physical meaning of the output parameters. Particle diameters are distinctly derived from the corresponding measured moments of the particle analysis (see the Appendix).

Dry powders with nanoscale structures are typically characterized by microscopic means (transmission electron microscopy [TEM]/scanning electron microscopy [SEM]) for their shape, size, and structure. Diffraction, interference, and scattering techniques (x-ray diffraction [XRD], small angle x-ray scattering [SAXS], and light scattering) give information about size, structure, crystallinity, and phase composition, while absorption and emission spectroscopic methods (ultraviolet-visible [UV-vis], Fourier transform infrared spectroscopy [FTIR], x-ray fluorescence [XRF], x-ray photoelectron spectroscopy [XPS], Auger spectroscopy, or extended x-ray absorption fine structure [EXAFS]) provide information on phase and surface composition as well as the electronic structure of the sample. The interaction with gases and liquids without reaction is used in adsorption/desorption techniques that are widely used to measure surface area and pore sizes. Surface characteristics can be determined as in temperature programmed reduction/ temperature programmed oxidation (TPR/TPO) or thermogravimetric analysis (TGA), for example. Particle sizes can be further characterized by applying inertial force fields during sieving and sedimentation or electrical fields for determining the mobility of particles.

**TABLE 10.1**
**Overview of Common Dry Powder Characterization Techniques Discussed in This Chapter in Terms of Particle Properties**

| | Imaging Techniques | | Spectroscopic Analysis — Diffraction and Scattering | | | Spectroscopic Analysis — Absorption/Emission | | | Interaction with Gases and Liquids — Without Reactions | | Interaction with Gases and Liquids — With Reactions | | Behavior in Applied Force Fields — Inertial Fields | | Behavior in Applied Force Fields — Electrical Fields |
|---|---|---|---|---|---|---|---|---|---|---|---|---|---|---|---|
| | TEM | SEM | XRD | SAXS | Light Scattering | XRF, XPS, Auger, EXAFS | UV-Vis | FTIR | Adsorption Desorption BET (SSA) | Hg Porosimetry | TPR/ TPO | TGA | Sedimentation | Sieving | DMA |
| $d_{prime}$ | X | X | X | X | (X) | | | | X | | | | | | |
| $d_{aggl}$ | X | X | | X | X | | | | | | | | X | X | X |
| $D_f$ | X | | | X | X | | | | | | | | X | | (X) |
| Size distribution | X | X | | X | | | | | | | | | | X | X |
| Shape | X | X | | | | | | | (X) | | | | | | |
| Phase composition | | | X | | | X | | | | | | | | | |
| Surface composition and characterization | X | | | | | X | | X | | | X | X | | | |
| Porosity | | | | | | | | | X | X | | | | | |
| Electronic structure | | | | | | | X | | | | | | | | |
| Size deter. range | 5 nm – 1 mm | 500 nm – 1 mm | 4 – 60 nm | 0.5 nm – 1 mm | 30 nm – 900 mm | – | – | – | 3 nm – 1 mm | 3 nm – 1 mm | – | – | 20 nm – 100 mm | 30 μm – 125 mm | 5 nm – 500 nm |

## II. IMAGING TECHNIQUES

### A. TRANSMISSION ELECTRON MICROSCOPY

Transmission electron microscopy can be used to examine particles that are too small for investigation with optical microscopes. TEM provides a powerful method for determining particle shapes as well as their size and degree of agglomeration. It uses the electrons the same way that optical microscopes use light (photons); the particles under investigation in an electron beam absorb and scatter electrons to produce a two-dimensional image.[2] In TEM, electrons are generated by thermoionic emission from a heated tungsten filament and are focused by magnetic lenses, which serve the same function as optical lenses. However, magnetic lenses have the advantage that their focal length can be adjusted by controlling the current through them. The interior of an electron microscope must be under high vacuum (less than $10^{-7}$ atm) to prevent scattering of the electron beam by air molecules.

For sample preparation, typically very fine carbon-coated copper meshes 3 mm in diameter (here called TEM grids) are used. To form a high-contrast picture, their carbon film has to be sufficiently thin compared to the particles, causing only slight attenuation in comparison with the particles, which extensively scatter and absorb the electron beam. Very small amounts of powder are placed on the TEM grids by dipping the grid in the powder, preparing a dilute solution with an evaporating fluid, or by depositing the particles in situ by thermophoresis or electrostatic forces. A rule of thumb for coverage of the TEM grid is that about 10% or less should be covered with particles.[3]

The insert of Figure 10.1 shows part of a TEM grid that was used for *in situ* (thermophoretic) particle sampling[3] in a premixed flame aerosol reactor making 11 g/h $TiO_2$ nanoparticles 0.5 cm above the burner.[4] By evaluating several TEM pictures from the same grid or from other grids sampled at the same location, here the size of 761 primary particles was evaluated, resulting in a firm particle size distribution (Figure 10.1b). The moment obtained from TEM counting of regular-shaped particles is the $M_{1,0}$, which results in the average diameter, $d_{1,0}$ (see the appendix for details). From Figure 10.1a, it appears that the $d_{1,0}$, $d_{1,2}$, and the geometric standard deviation, $\sigma_g$ (characterizing the width of the log-normal particle size distribution), reach almost asymptotic values when counting more than 500 particles. From a similar particle size evaluation at different heights above the burner, experimental data for the particle growth can be obtained that can be compared to particle growth models. Both average particle size and particle size distribution[4,5] can potentially be predicted by this method.

In the case of small particles on the surface of a larger matrix, adsorption/desorption analysis (e.g., Brunauer-Emmit-Teller [BET]; see Section IV, "Interaction with Gases and Liquids") is limited and can give only an average particle size. High-resolution TEM can be used to determine the particle size distribution of these small particles, as demonstrated by Mädler et al.[6] To confirm XRD data, they counted and classified small gold particles on larger $SiO_2$ support particles

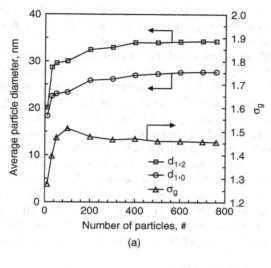

(a)

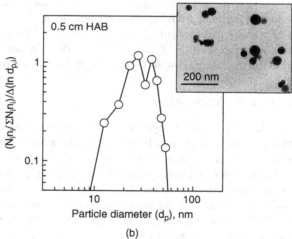

(b)

**FIGURE 10.1** (a) Effect of the number of particles counted on the average mean primary particle diameter, $d_{1,0}$ (circles), and Sauter mean primary particle diameter, $d_{1,2}$ (squares), as well as the geometric standard deviation (that is, a measure of the width of the size distribution, triangles). (b) The corresponding primary particle size distribution and a TEM picture of the investigated titania nanoparticles collected with a thermophoretic sampler directly from a premixed $TiO_2$ flame at a height of 0.5 cm above the burner. (Courtesy of H.K. Kammler and S.E. Pratsinis.)

from high-resolution TEM images. The count mean diameter, $d_{1,0}$, of the gold particle size distribution was 9 nm, with a standard deviation of 2.3 nm, while the mass mean diameter was 10.6 nm and the count geometric standard deviation was 1.27 from 190 particles. They found that the measured size distribution of gold particles was narrower than that of the self-preserving distribution for particles made by coagulation, supporting further that the latter was not a dominant particle growth mechanism, as with titania or silica.[1]

The electron diffraction pattern observed with a transmission electron microscope shows whether the particles are polycrystalline (ring patterns), a single crystal (spot patterns), or amorphous. Furthermore, high-resolution TEM can visualize the lattices of particles. Tani et al.[7] produced 2:1 $ZnO:SiO_2$ mixtures of nanoparticles by flame spray pyrolysis, where each primary particle was a nanocomposite in which very fine ZnO crystals of 1 to 3 nm in diameter were dispersed in the amorphous $SiO_2$ phase. From high-resolution TEM they determined a lattice distances of 0.163 nm, which was in agreement with the distance (0.162 nm) of the (110) plane in hexagonal ZnO.[7] High-resolution TEM can also help to detect different materials such as a carbon layer on the surface of $TiO_2$ nanoparticles[8] (shown in Figure 10.2a). Combining TEM with electron spectroscopic imaging (ESI) can identify the elements in the sample. In this example, the carbon is on the surface of the titania, as shown in the elemental maps of carbon and titanium (Figure 10.2b).

Koylu et al.[9] used extensive TEM studies to derive the mass fractal dimension, $D_f$, of flame-made soot particles from Equation A1 (see the appendix) by extracting the number of primary particles per agglomerate, $N$, the primary particle size, $d_p$, and the radius of gyration of the agglomerate, $R_g$. The constant fractal prefactor, $A$, as well as the fractal dimension, $D_f$, were then determined by regression.

## B.  SCANNING ELECTRON MICROSCOPY

In contrast to TEM, SEM creates images with a three-dimensional appearance created by secondary electrons emitted from the sample surface. The secondary electrons are emitted from atoms interacting with the main electron beam focused on the sample. The number of secondary electrons depends on the sample topography and atomic composition, where valleys emit few electrons compared with higher points. In order to increase the sample conductivity and therefore avoid charging nonconducting powder, samples are typically sputtered with a gold, tungsten, or platinum film 2 to 5 nm thick.

# III. SPECTROSCOPIC ANALYSIS (INTERACTION WITH ELECTROMAGNETIC WAVES WITH ENERGIES FROM $10^4$ TO $10^{-2}$ eV)

## A.  DIFFRACTION AND SCATTERING

### 1.  X-Ray Diffraction

X-ray diffraction is one of the most widely used dry powder probing methods. The energies of the employed electromagnetic waves are several thousand electron volts. In most cases the x-ray source is the characteristic K$\alpha$ radiation that

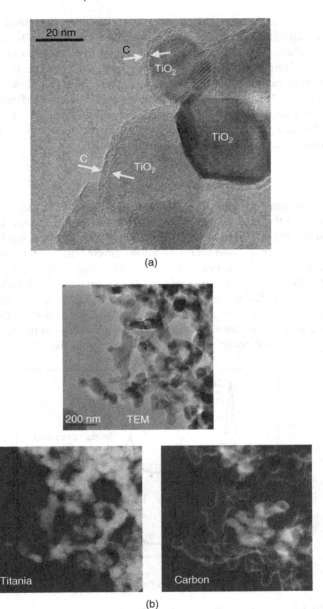

**FIGURE 10.2** (a) High-resolution TEM pictures of carbon-coated $TiO_2$ particles with 15 wt% carbon. The carbon layer (indicated with the arrows) can be clearly observed on top of the titania particles. (b) Elemental analysis of titanium and carbon can detect the domains of the respective elements as demonstrated here for a $TiO_2$ powder with 52 wt% carbon. Clearly the carbon shells on top of the $TiO_2$ particles can be observed along with separate pure carbon particles. (Courtesy of H.K. Kammler and S.E. Pratsinis.)

originates when high-energy electrons interact with a copper target (8.05 keV, 0.154 nm). Therefore the radiation can penetrate solids and interact with their internal structure. The spherical scattered x-rays from each atom of the solid combine, resulting in diffraction effects. If the powder sample is composed of fine, crystalline randomly oriented particles, a certain crystal plane is by chance at the right angle θ with the incident beam for constructive interference. A set of those planes (0.1 to 10% of the irradiated crystallites) generate a signal on the detector. The angle at which the detector receives this interference signal is also θ, and gives the corresponding lattice spacing by Bragg's law:

$$n \lambda = 2d \sin\Theta; \; n = 1, 2, \ldots, \tag{10.1}$$

where $n$ is an integer called the order of reflectance, $\lambda$ is the wavelength of the x-ray, $d$ is the lattice spacing, and $\Theta$ is the angle between the incident beam and the normal to the reflecting lattice plane. The spacing between the lattices is characteristic and therefore allows phase and crystal structure identification as well as atomic composition. As an example, Figure 10.3 compares the XRD patterns of two samples of $Ce_{0.7}Zr_{0.3}O_2$ prepared with different liquid carriers using the flame spray pyrolysis method.[10] The XRD pattern clearly shows the difference in crystal structure and homogeneity of the powders. Depending on the precursor solution, either a product containing a ceria-rich (Figure 10.3, bottom trace, left peak) and a zirconia-like phase (right peak)

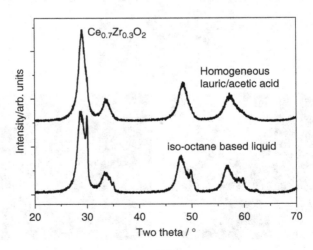

**FIGURE 10.3** Powder x-ray patterns of ceria-zirconia made by using two different liquid carriers for the precursors. The isooctane-based carrier composition forms a product containing a ceria-rich (Figure 10.2, left peak, bottom trace) and a zirconia-like phase (right peak). In contrast, application of the lauric/acetic acid-based carrier solution results in a single mixed oxide phase (top trace) (From Stark, W.J., Mädler, L., Maciejewski, M., Pratsinis, S.E., and Baiker, A., Flame synthesis of nanocrystalline ceria-zirconia: effect of carrier liquid, *Chem. Commun.*, 5, 588, 2003.)

forms or a single mixed oxide phase (top trace) can be obtained. Generally the change in lattice parameters (e.g., of fluorite-type oxides) due to the formation of solid solutions can be predicted and the composition evaluated.[11] A homogeneous solid solution of ceria/zirconia mixed oxides is especially important for the current generation of three-way catalysts for automotive exhaust gas treatment, which relies on ceria-zirconia as a dynamic oxygen source for the conversion of remainders from incomplete combustion and the removal of NO.

For a polycrystalline powder sample, the diffraction theory predicts that the signal will be within an exceedingly small range of the diffraction angle, θ. In practice, such sharpness will never be observed because of the combined effects of instrumental and physical factors. Therefore the final diffraction line profile results from the convolution of a number of independent contributing shapes. The main components can be divided into the instrumental contributions and spectral distribution and the intrinsic profile.[12] Instrumental contributions are mainly the flat specimen, specimen transparency, axial divergence of incident beam, and receiving slit. The spectral distribution is caused by the inherent width and asymmetry of the spectral profile of the x-ray source. Both effects can be eliminated by applying the fundamental parameter approach.[13] The intrinsic profile results from the powder crystallites themselves. In addition, there are two principle sample effects, which broaden the profile shape. One is the microstrain of the sample and the other is the crystallite size. If the crystallite size becomes smaller than about 1 μm, the long-range order is interrupted, resulting in an incomplete destructive interference in the scattering direction where the scattered x-rays are out of phase. The crystal size can be calculated using the Scherrer relation:[14]

$$L = \frac{K\lambda}{\beta \cos\Theta} , \qquad (10.2)$$

where $L$ is a measure of the dimension of the crystallite in the direction perpendicular to the reflecting plane, $\beta$ is the breadth of the pure diffraction profile on the $2\Theta$ scale in radians, and K is a constant approximately equal to unity and related to both the crystallite shape and to the way in which $L$ and $\beta$ are defined. Scherrer's original derivation was based on the assumption of a Gaussian shape profile and small cubic crystals of uniform size, in which case K = 0.89.[15] For nonuniform crystal sizes, Delhez et al.[16] point out that the integral breath in reciprocal space results from integration over the entire volume of the sample:

$$(\beta \cos\Theta / \lambda)^{-1} = d_{1,3} = \frac{1}{V} \iiint t \, dx \, dy \, dz , \qquad (10.3)$$

where $V$ is the volume of the domain, $t$ is the thickness measured through the points $x$, $y$, $z$ in the direction parallel to the diffraction vector. For example, if the crystal domains are spheres they will have a diameter of $4/3 d_{1,3}$ for all diffraction planes.[17] For other shapes, $d_{1,3}$ depends on direction and therefore on the diffraction plane. The dimension of the crystallite $L$ is in fact a ratio of two moments of the distribution, which can be written for a cube:

$$\frac{\int x^4 q_0(x)\, dx}{\int x^3 q_0(x)\, dx} = \frac{M_{4,0}}{M_{3,0}} = M_{1,3} \Rightarrow \int x\, q_3(x)\, dx = d_{1,3}, \qquad (10.4)$$

where $q_0(x)$ and $q_3(x)$ are the number and volume distributions, respectively (see the appendix).[18] Very often the integral apparent crystallite size is calculated for a projected area average, choosing the coordination system so that one axis is perpendicular to the reflecting planes.[19–21] In this way one obtains

$$\frac{\int x^2 q_2(x)\, dx}{\int x\, q_2(x)\, dx} = \frac{M_{2,2}}{M_{1,2}} = M_{1,3} \Rightarrow \int x\, q_3(x)\, dx = d_{1,3}, \qquad (10.5)$$

which results again in the moment of the average volume of the distribution. As an example, when comparing the obtained average crystal size with electron micrographs, the moments of the distribution have to be converted. This was also pointed out by Solliard.[22] The strength of the crystal size determination is illustrated in Figure 10.4, analyzing an inhomogeneous powder. In this study, one-third of the $CeO_2$ powder mass consists of large crystals (average size 155 nm) and two-thirds of smaller ones (average size 8 nm).[23] The average crystal sizes were obtained on the basis of the fundamental parameter approach and Rietveld analysis.[24] The analysis was carried out by regressing the measured XRD pattern with the crystalline data of cubic ceria, assuming that the background was a linear function. Since the integral of the corresponding XRD signals is directly related to particle mass, an estimate of the powder composition can be made. Another example is shown in Figure 10.5. Here, an XRD pattern of 4 wt% gold on titania and its reconstruction using the structural information of anatase, rutile, and gold is shown.[6] The reconstructed pattern resulted in a crystal size of 16.1 nm, while the evaluated gold content (3.9 wt%) of the powder was in good agreement with the nominal content of 4 wt%. Figure 10.6 shows an example of a 2:1 $ZnO:SiO_2$ powder[7] where the XRD pattern was described well as the sum of the linear background, the crystalline ZnO, and the broad peak attributed to an amorphous phase (e.g., $SiO_2$) by the fundamental parameter approach and the Rietveld method. In this example, the XRD could resolve structures less than 2 nm.

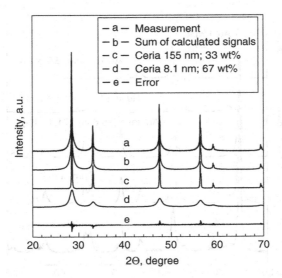

**FIGURE 10.4** X-ray diffraction pattern of FSP-made ceria with strong bimodal crystal size prepared from Ce(Ac)$_3$ in pure acetic acid. One-third of the powder mass consists of large crystals (average size 155 nm) (GOF = 1.38). The measured XRD was fitted with two cubic ceria modes (PDF 81-792). (From Mädler, L., Stark, W.J., and Pratsinis, S.E., Flame-made ceria nanoparticles, *J. Mater. Res.*, 17, 1356, 2002.)

## 2. Small-Angle X-Ray Scattering

Small-angle x-ray scattering requires a special detector arrangement to measure at small and very small angles (0.0002° to 6°). In this technique, the intensity of the scattered photons, $I(q)$, is monitored as a function of the absolute value of the momentum transfer vector, $q$, for an elastic scattering event:

$$q = \frac{4 \cdot \pi \cdot \sin\left(\frac{\Theta}{2}\right)}{\lambda} , \qquad (10.6)$$

where $\lambda$ is the wavelength of the employed irradiation and $\Theta$ is the scattering angle.[25] Laboratory sources typically operate under vacuum using K$\alpha$ x-ray sources similar to those described in the previous section, while for ambient air measurements and scattering at larger angles, collimated photons, longer coherence lengths, and higher energy densities are needed. This is achieved by using synchrotron radiation (e.g., the advanced photon source at the Argonne National Laboratories, Argonne, IL).

Ultra small-angle x-ray scattering (USAXS) is capable of determining all characteristics of agglomerate and primary particles from a single scattering experiment: mass fractal dimension, $D_f$, radii of gyration (primary particles and

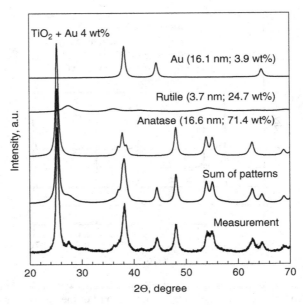

**FIGURE 10.5** The 4 wt% gold on titania XRD pattern and its reconstruction using the structural information of anatase, rutile, and gold. The reconstructed pattern resulted in a crystal size of 16.1 nm. The evaluated gold content (3.9 wt%) of the powder is in good agreement with the nominal content of the prepared solution (4 wt% gold). (From Mädler, L., Stark, W.J., and Pratsinis, S.E., Simultaneous deposition of Au nanoparticles during flame synthesis of TiO$_2$ and SiO$_2$, *J. Mater. Res.*, 18, 115, 2003.)

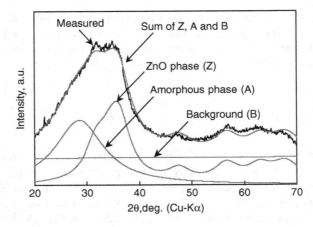

**FIGURE 10.6** Result of the fitting using the fundamental parameter approach in ZnO/SiO$_2$ powder (2:1). The black and gray lines correspond to the measured and calculated XRD patterns, respectively. (From Tani, T., Mädler, L., and Pratsinis, S.E., Synthesis of zinc oxide/silica composite nanoparticles by flame spray pyrolysis, *J. Mater. Sci.*, 37, 4627, 2002.)

agglomerates), and the powder volume:surface ratio, $d_{V/S}$, using the unified fitting equation.[26–28] The latter diameter can be derived from the scattering curve:[29]

$$d_{V/S} = \frac{6 \cdot V}{S} = \frac{6 \cdot Q}{\pi \cdot B_1} \,, \tag{10.7}$$

where the Porod invariant, $Q$, is the integral of the scattering curve and reflects the scattering power,

$$Q = \int_0^\infty q^2 \, I(q) \, dq \tag{10.8}$$

and the Porod constant, $B_1$, can be determined from Porod's law:[30]

$$I(q) = B_1 \, q^{-4} \,. \tag{10.9}$$

The $d_{V/S}$ is the same ratio of moments that is measured by nitrogen adsorption (BET; see Section IV, "Interaction with Gases and Liquids," and the appendix):

$$\frac{M_{3,0}}{M_{2,0}} = \frac{\int x^3 \cdot q_0(x) dx}{\int x^2 \cdot q_0(x) dx} = \frac{1}{M_{-1,3}} = M_{1,2} \;\Rightarrow\; d_{1,2} = \int x \cdot q_2(x) dx \tag{10.10}$$

The mass-fractal dimension of the ramified agglomerates is determined from the slope of the weak power law decay in between the power law regimes that follow Porod's law (Equation 10.9):

$$I(q) = B_2 q^{-D_f} \,, \tag{10.11}$$

where $D_f$ is the mass fractal dimension and $B_2$ is a power law prefactor:[27]

$$B_2 = \frac{G_2 \cdot D_f}{R_{g2}^{D_f}} \Gamma \frac{D_f}{2} \,, \tag{10.12}$$

where $\Gamma$ is the gamma function, $R_{g2}$ is the radius of gyration of the agglomerates, and $G_2$ is determined from Guinier's law:[31]

$$I(q) = G_2 \cdot \exp\left\{ -\frac{q^2 R_{g2}^2}{3} \right\} \,. \tag{10.13}$$

The measurement error from SAXS can be estimated by the propagated error in the scattering data, which is generally very low. The USAXS measurement can be normalized by the Porod invariant $Q$ or the absolute intensity can be used if the sample thickness and packing density are known. The Porod analysis relies on the particle surface being smooth and sharp, an assumption of Porod's law (Equation 10.10). This assumption can be verified through observation of a scattering regime with a negative slope of 4 in the log-log plots. This is clearly a good assumption as experimentally shown for many oxide nanoparticles.[29] The USAXS also relies on the accuracy of the global unified function in separating the primary particle structure from the other structures present in the scattering curve, that is, the mass-fractal agglomerate structure and the structure of "soft" agglomerates (aggregates). It is difficult to quantify the error involved in this separation. However, it is expected that this error would be higher for mass-fractal samples. Since there is no indication of a difference between the results for mass-fractal (agglomerated) and non-mass-fractal (nonagglomerated) samples,[29] it would seem that the global unified fit is performing adequately in this regard.

Small-angle x-ray scattering is able to distinguish between agglomerated and nonagglomerated primary particles. In the latter case, the mass-fractal regime that is the intermediate power law dependence between the two Porod regimes (power laws with a negative slope of 4) vanishes as a result of the correlation between the particles. This is clearly observed in Figure 10.7 for nonagglomerated (upper scattering pattern) and agglomerated $SiO_2$ (lower scattering pattern) made by flame spray pyrolysis.[32]

Agglomerates with an almost monodispersed size are manufactured by Tokuyama Soda (Tokyo, Japan) and are marketed as Spherosil. These particles consist of smaller subunits, as indicated in the TEM shown in Figure 10.8a,b. Figure 10.8c represents the scattering pattern of this powder, which clearly shows the humps of the developing sphere function, indicating a very narrow size distribution. Determining the radius of gyration of the subunits (embedded primary particles) and the size of the almost spherical clusters of primary particles (large conglomerate particles in Figure 10.8b) with the global unified fitting equation,[26–28] corresponding diameters of the equivalent sphere of 2 nm and 115 nm are derived, respectively. The Sauter mean particle diameter (Equation 10.7) of the clusters is 80 nm, which is less than that derived from the equivalent sphere, but it is in very good agreement with the TEM observations and consistent with Kammler et al.[29]

## 3. Light Scattering

Light scattering is normally performed in the visible range of the electromagnetic spectrum ranging from about 1 eV to 10 eV. When a light beam is illuminating a particle cloud it will be absorbed or scattered, or both, depending on the wavelength of the light and the optical properties of the particles, such as the size, shape, and refractive indices of the particle and the surrounding medium.

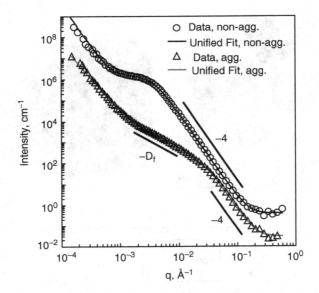

**FIGURE 10.7** Scattering curves recorded by USAXS of nonagglomerated (upper pattern) and agglomerated (lower pattern) flame-made silica nanoparticles. The use of third-generation synchrotron radiation enables the detection over more than four orders of magnitude in $q$, resulting in information about primary particle and agglomerate size. The mass fractal dimension, $D_f$, can be determined from the slope of the intermediate power law dependence as indicated. The latter is absent for the upper pattern indicating nonagglomerated particles as the correlation between the particles vanishes. This pattern was shifted by one order of magnitude in intensity for clarity. Furthermore, primary particle size, agglomerate size, and $D_f$ can be determined with the global unified fitting Equation 10.29. (Courtesy of H.K. Kammler and G. Beaucage.)

The net result of absorption and scattering is extinction, while the light that is not extinct is said to be transmitted. In terms of the extinction efficiency, $Q_e$, which is defined as the ratio of the radiant power scattered and absorbed by a particle to the radiant power geometrically incident on the particle (proportional to the cross-sectional area). Therefore the extinction efficiency per cross-sectional area of the particle is the sum of its scattering and absorption efficiency per cross-sectional area, $Q_s$ and $Q_a$, respectively:[2]

$$Q_e = Q_s + Q_a. \tag{10.14}$$

Both the scattering and absorption characteristics of particles can be described by the complex refractive index $m = n - ik$, where the real part describes the scattering and the imaginary part the absorption characteristics, respectively. In general, the light scattering processes can be completely described by the Lorentz-Mie theory. However, due to its complexity, a number of approximations have been developed that are valid in certain size ranges. The ranges are determined

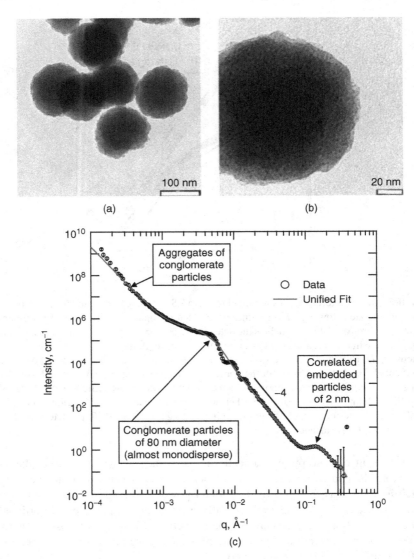

**FIGURE 10.8** (a,b) TEM pictures of Spherosil (Tokuyama Soda, Tokyo, Japan) that consists of very small primary particles (around 2 nm) and almost spherical conglomerates of these. (c) Shows corresponding USAXS scattering pattern. At large $q$ (approximately 0.1 Å), the primary particles can be clearly identified, while the humps of the developing sphere function, typical for monodispersed systems, is clearly observed for the conglomerate particles at $q \sim 0.07$ Å. The slope of $-4$ indicates a smooth primary particle surface. Using the global unified fitting equation,[26,27] the diameters of the primary and conglomerate particles can be determined. (Courtesy of H.K. Kammler, G. Beaucage, and S.E. Pratsinis.)

by the relation between the particle diameter, $d_p$, and the wavelength of the light, $\lambda$. A dimensionless parameter $\alpha = \pi d_p/\lambda$ has been defined. For $\alpha \ll 1$, the oscillating electric field of the light induces an oscillating dipole in the particle, causing symmetrical scattering—the so-called Rayleigh scattering. In this regime the scattering efficiency per cross-sectional area $Q_s$ is

$$Q_s = \frac{8}{3} \left( \frac{\pi d_p}{\lambda} \right)^4 \text{Re} \left( \frac{m^2 - 1}{m^2 + 2} \right)^2 \propto \alpha^4, \qquad (10.15)$$

and therefore the total scattered intensity is proportional to $d_p^6$. The absorption coefficient per cross-sectional area, $Q_a$, is

$$Q_a = -4 \left( \frac{\pi d_p}{\lambda} \right) \text{Im} \left( \frac{m^2 - 1}{m^2 + 2} \right) \propto \alpha, \qquad (10.16)$$

and therefore is proportional to $d_p^3$, and thus to the volume of the particle. The angular distribution in the Rayleigh regime depends on the polarization. There are a variety of instruments that use the light scattering in the regime of $\alpha \leq 1$, especially when sizing aerosols.[33] Most instruments rely on a calibration due to the complex effect of the refractive index and particle shape. Another type of instrument applies the scattering of particles for counting (optical particle counter) and time of flight determination.

Scattering from large particles, and therefore for $\alpha \gg 1$, can be treated with simple geometric optics, where the extinction coefficient per cross-sectional area, $Q_e$, is constant. In this case, the light hitting the particle led to reflection, refraction, and absorption, while the light passing on the edge gives rise to scattering.[34] The scattered intensity is proportional to $d_p^2$. The scattering of light on a particle leads to diffraction in the same way as one observes it at a single slit. When parallel light waves are used, this phenomenon is called Fraunhofer diffraction. In this case, a characteristic diffraction pattern can be observed in the forward direction of the scattering event. The relative shape of this pattern depends on the particle size and shape and not on the refractive index of the particle. Therefore it is often used in particle sizing instruments, which rely on a known uniform particle shape and the assumption of a certain particle size distribution. Unfortunately most of the instrument manufacturers do not disclose the exact transfer function for data deconvolution and regression routines used in their instruments. Nevertheless, it is the most common technique for particle sizing ($\alpha \gg 1$), which is also called angular light scattering, and has been accepted by the international particle sizing community.[35] The sizing range of such instruments is from about 1 μm to several hundreds of micrometers. The fractal dimension can also be measured with such a system.[36]

For particles of a diameter on the order of the wavelength of incoming light ($\alpha \approx 1$), the oscillations are too large for Rayleigh scattering because the uniformity of the electromagnetic field over the particle is not given. There is no simple relation between the extinction efficiency and the particle size, as well as other properties such as refractive index and particle shape.[34] A very good description of the theory and applicable computer codes is given by Bohren and Huffman.[37] This regime ($\alpha \approx 1$) of light scattering is not commonly used for direct particle sizing because of detailed information on particle shape and refractive index are needed. However, changes in particle systems (e.g., particle growth) can be observed and monitored very well.

## B. ABSORPTION/EMISSION

### 1. High-Energy Spectroscopy

Absorption and emission in the high-energy spectrum of several kiloelectron volts are very common in dry powder analysis, especially in catalysis.[21] The high energies are supplied either by bombarding the sample with electrons or x-ray photons. When exciting the sample with primary electrons, very sensitive surface analysis can be performed using information of the emitted photons. This method is known as x-ray fluorescence (XRF), which gives a fingerprint of the atoms on the surface (Mosley's law). At the same time, an Auger transition by the emission of secondary electrons is possible. These emitted Auger electrons carry information from the surface of the particles and can be used to identify elemental compositions of the surface (Auger spectroscopy, AES). However, not only electrons are applied to study absorption effects of particles, but also high-energy x-rays are used. When the atoms absorb such photons with energies above the ionization energy, the excess energy will be imparted to the ejected electron in the form of kinetic energy. By knowing the source frequency and measuring the kinetic energies of the ejected electrons, the ionization energy of an electron can be determined and the binding energy can be evaluated. This method is known as x-ray photoelectron spectroscopy (XPS). It not only reveals binding energies but also the elemental composition and oxidation stages. When irradiating the samples with energies just above the absorption edge, an electron can be emitted and can be scattered by its neighboring atoms. This scattering results in interferences, which carry information on inter atomic distances and atom concentrations in the vicinity of the absorbing atom. This method is often called extended x-ray absorption of fine structures (EXAFS).

### 2. UV-Vis Spectroscopy

Most of the UV-vis measurements are focused on the electronic state of the substance and its photophysical behavior, such as absorption, fluorescence, and phosphorescence. The mathematical-physical basis of light absorption measurements in the UV-vis region with energies from about 1 to 100 eV   (10 to 1000 nm) is Beer's law:

$$\left[ \frac{I}{I_0} \right]_\lambda = \exp(-\varepsilon_\lambda \, c \, \ell) = \tau_\lambda = 1 - A_\lambda \, , \qquad (10.17)$$

where $I_0$ is the intensity of the monochromatic light entering the sample and $I$ is the intensity of the light emerging from the sample; $c$ is the concentration of the light-absorbing substance; $\ell$ is the path length through the sample, $\varepsilon_\lambda$ is the decadic extinction coefficient, $\tau_\lambda$ is the transmission, and $A_\lambda$ is the absorption. For a constant extinction coefficient and path length, this method can be used to determine the relative concentration or the absolute concentration of the sample after calibration. However, the extinction coefficient is an important characteristic of the sample. Its functional correlation with the wavelength ($\varepsilon_\lambda = f(\lambda)$) is called the absorption spectrum.[38]

In spectroscopy, the sum of transmittance, reflectance, and absorbance is always equal to unity. Beer's law is valid for cases where scattering and reflection are negligible. In this case, the absorbance can be directly related to the transmittance, which holds true for these molecularly dispersed systems. However, measuring the reflectance of a sample, the transmittance becomes zero and the absorbance can be directly related to the reflectance. Here, the light is reflected diffusively on the surface and inside a powder sample. This problem was first discussed by Kubelka and Munk.[39] In order to decrease the signal:noise ratio, the following parameter was introduced to present the diffuse reflectance data, the so-called Kubelka-Munk function, $F(R)$:

$$F(R_\infty) = \frac{(1 - R_\infty)^2}{2\,R_\infty} = \frac{K}{S}\;, \qquad (10.18)$$

where the square of the absorptance is divided by two times the reflectance, $R$, which describes the diffuse reflecting power at a penetration depth of infinity. These data can also be expressed in terms of the absorption coefficient, $K$, and the scattering coefficient, $S$. The theory implies an infinitely thick sample, where the background reflectance is zero. Since the reflecting power is the ratio of absorption and the scattering coefficient, and does not depend on their absolute values, it can be easily measured.[40] Since the diffuse reflecting power cannot be measured accurately in conventional instruments, $R$ is always related to a white standard as a reference and therefore obtained as a relative value: $R'_\infty = R_{sample}/R_{standard}$.[38] The so-called dilution method implies that the absolute reflecting power of the standard has to be known, especially where MgO, $SiO_2$, $MgSO_4$, and $BaSO_4$ are used. Measurements of concentrations can also be achieved with this method, and a very detailed discussion on reflectance spectroscopy can be found elsewhere.[40] An example is given in Figure 10.9, which shows absorption spectra of ZnO crystallites of different sizes as determined by XRD. The bulk optical absorption spectra were recorded in the Kubelka-Munk mode while, prior to absorption measurements, the as-prepared particles were homogeneously diluted 10 times with barium sulfate. A blueshift of the absorption spectrum is observed with decreasing ZnO size, indicating its quantum size effect. Samples with $d_{XRD} > 12$ nm start to absorb in the blue for smaller ZnO crystals, the absorption shifts into the UV region, resulting in a white powder that is

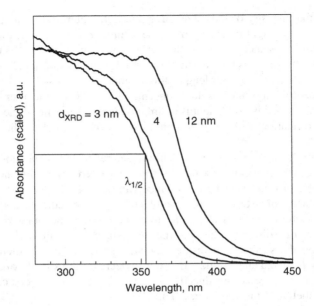

**FIGURE 10.9** Optical absorbance spectra of as-prepared flame-made ZnO nanoparticles showing the blueshift associated with decreasing crystallite size. Coprecipitation with amorphous silica was used to control the crystallite size. (From Mädler, L., Stark, W.J., and Pratsinis, S.E., Rapid synthesis of stable ZnO quantum dots, *J. Appl. Phys.*, 92, 6537, 2002.)

attractive for applications in which transparency is of importance, such as polymer stabilizers or photoinitiation of polymerization.[41]

### 3. Fourier Transformed Infrared Spectroscopy

Fourier transformed infrared (FTIR) spectroscopy is the most common form of vibrational spectroscopy. Applied energies range from 0.01 to 1 eV (100 to 10,000 cm$^{-1}$). Infrared spectroscopy probes the transition between vibrational energy levels, which occur when photons in that energy range (or frequency, $\nu$) are absorbed by the surface molecules of the particle surface. There are several fundamental vibrations for linear and nonlinear molecules. In general, the vibrational frequencies increase with increasing bond strength and with decreasing mass of vibrating atoms:[21]

$$\nu = \frac{1}{2\pi}\sqrt{\frac{k}{\mu}} \, , \tag{10.19}$$

with

$$\frac{1}{\mu} = \frac{1}{m_1} + \frac{1}{m_2} \, , \tag{10.20}$$

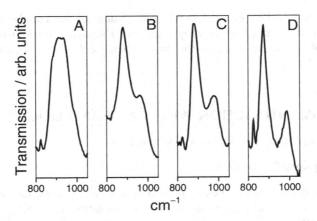

**FIGURE 10.10** DRIFT spectra for silica containing (a) no titanium, (b) 0.9 wt% $TiO_2$, (c) 1.3 wt% $TiO_2$, and (d) 3.2 wt% $TiO_2$. The 960 $cm^{-1}$ absorption (Ti-O-Si) grows for increasing titania contents and correlates qualitatively with catalytic activity. (From Stark, W.J., Pratsinis, S.E., and Baiker, A., Flame made titania/silica epoxidation catalysts, *J. Catal.*, 203, 516, 2001.)

where $k$ is the force constant of the bond, $\mu$ is the reduced mass, and $m_i$ is the mass of the vibrating atoms. When measuring dry powders, scattering on the sample surface influences the measurement. The solution to this problem is equivalent to the diffuse reflectance UV-vis spectroscopy. The application of the Kubelka-Munk theory (see Section III.B.2, "UV-Vis Spectroscopy," for details) is very common in FTIR analysis. The method applied is often called diffuse reflectance Fourier transformed infrared spectroscopy (DRIFTS). For DRIFTS measurements, KBr is often used as the standard and dilution compound. Figure 10.10 shows a DRIFTS measurement of dehydrated titania/silica powders.[42] Spectra were recorded on an FTIR instrument (Perkin Elmer, Model 2000) containing a diffuse reflectance unit and a controlled environmental chamber equipped with two ZnS windows. Prior to measurements, the powders were outgassed in an argon stream (Pangas, 99.999%; 5 ml/min) for 1 h at room temperature. Dehydration was performed under flowing argon at 573 K for 1 h. Pure silica (Figure 10.10a) showed no absorption signal around 960 nm. A sample containing 0.9 wt% titania (Figure 10.10b) gave rise to a weak signal at 960 nm, indicating the presence of some titanium-oxygen-silicon units. The same band appeared more pronounced in catalyst containing 1.3 wt% titania (Figure 10.10c). Particles containing 3.2 wt% titania (Figure 10.10d) produced a broad signal at 960 nm, indicating a high density of titanium-oxygen-silicon units in the powder.

## 4.  Ultrasound Methods

Acoustics attained recognition in the field of colloidal science very recently. Ultrasound wavelengths are in the range from 10 $\mu$m to 1 mm. The advantage of the method is that ultrasound can propagate through concentrated suspensions and allows colloidal measurements without dilution. Its main applications are in the fields of particle sizing, rheology, and electrokinetics in solution. A recent overview of ultrasound methods for characterizing colloids is given by Dukhin and Goetz.[43]

# IV. INTERACTION WITH GASES AND LIQUIDS

## A. WITHOUT REACTION/PHYSICAL INTERACTION

### 1. Adsorption/Desorption

Physical adsorption of nitrogen or helium at low and constant temperatures on the particle surface (at the boiling point of liquid nitrogen, 77 K) is one of the most frequently used methods for determining the surface area and porosity of dry powders. The technique is relatively simple. It measures the uptake of gas (adsorptive) at increasing partial pressure over the sample (adsorbent) in the adsorption experiment and the release of the adsorbed gas at decreasing partial pressure in the desorption experiment. This results in full isotherms that are shown for different powders in Figure 10.11. The type I isotherm (following the IUPAC classification) is characteristic for materials encountering micropores (pores smaller than 2 nm). The steep rise at low relative pressure indicates filling of the micropores, while the plateau at higher values shows the absence of multilayer adsorption of the external surface, which is often very small for microporous materials. The type II isotherm describes nonporous or macroporous (pores larger than 50 nm) samples. Thus point B indicates the start of the linear section of the isotherm, which is considered to be the stage in the adsorption process where the monolayer is completed and the multilayer is about to begin. Type III isotherms are rarely seen and are characteristic for strong adsorbate-adsorbate interactions. Type IV isotherms are characteristic of mesoporous (pores ranging in size from 2 to 50 nm) materials and show a hysteresis loop that is typical for capillary condensation in mesopores. It shows the characteristic point B at lower relative pressures as well. Type V isotherms are similar to type III isotherms, but mesopores are present. Type VI isotherms may be seen with ideal multilayer adsorption on nonporous surfaces, where the step height represents the monolayer capacity for adsorbed layers.[44,45]

Full adsorption and desorption isotherms (Figure 10.12) can give information about pore structure and pore size distribution. Strobel et al.[46] measured adsorption/desorption isotherms of a typical flame-made powder (5 wt% platinum on $Al_2O_3$) and a commercially available catalyst (E4759) for enantioselective hydrogenation (Figure 10.12a). They determined pore size distributions from the desorption branch of the full isotherm. Both materials possessed similar specific surface areas. However, the flame-made powder contained virtually only macropores, which originate from particle agglomeration, whereas the commercial standard catalyst E4759 is a mesoporous material (Figure 10.12b). The flame-made catalyst showed much improved activity compared to the commonly used commercial E4759 catalyst.

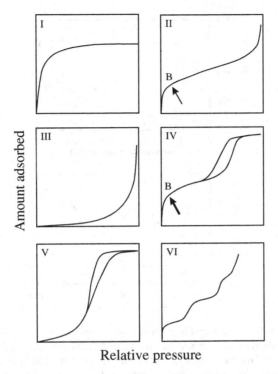

Relative pressure

**FIGURE 10.11** IUPAC classification of physisorption isotherms. (Following Schoofs, T., Surface area analysis of finely divided and porous solids by gas adsorption measurements, in *Particle and Surface Characterisation Methods*, R.H. Müller and W. Mehnert, Eds., Medpharm GmbH Scientific, Stuttgart, 1997.)

This improvement in activity is traced to the open and highly accessible structure of flame-made nanoparticle agglomerates and therefore decreased mass transfer limitations compared to the conventional porous catalysts.

## 2. Specific Surface Area, BET Analysis

To determine the surface area of dry powders, it is only necessary to record the first part of the adsorption branch, reducing the experimental time significantly (to less than 0.5 h). When increasing the partial pressure of the adsorbate over the sample, a monolayer of adsorptive builds up, while with increasing relative pressure, multilayer adsorption occurs. Brunauer et al.[47] derived a relation from gas-kinetic and statistical models on how this monolayer coverage can be determined from the mentioned experiment, which nowadays is often called BET-isotherm:

$$\frac{p_r}{V_A \cdot (1 - p_r)} = \frac{1}{V_M \cdot C} + \frac{C-1}{V_M \cdot C} p_r, \qquad (10.21)$$

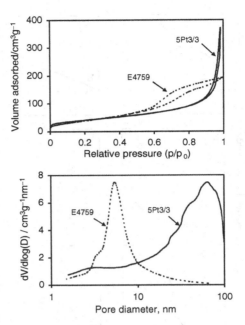

**FIGURE 10.12** Nitrogen adsorption isotherms and corresponding pore size distributions of a typical FSP-derived powder and the commercial reference catalyst E4759. Note that the particles of the flame-made catalyst are virtually nonporous. The indicated macropores originate from interstitial voids of the agglomerated particles. (From Strobel, R., Stark, W.J., Madler, L., Pratsinis, S.E., and Baiker, A., Flame-made platinum/alumina: structural properties and catalytic behaviour in enantioselective hydrogenation, *J. Catal.*, 213, 296, 2003.)

where $p_r = p/p_0$ is the relative pressure. $p$ and $V_A$ are the equilibrium pressure and adsorbed volume, respectively, and $p_0$ is the equilibrium vapor pressure at analysis temperature, $V_M$ is the volume that is needed for monomolecular coverage of the surface. Furthermore, $C$ is defined as

$$C = A \cdot \exp\left(\frac{E_1 - E_L}{R \cdot T}\right), \tag{10.22}$$

where $E_1$ is the energy of adsorption of the first layer and $E_L$ is the energy of liquefaction of the adsorptive, $A$ is a kinetic factor, $R$ is the universal gas constant, and $T$ is the absolute temperature.[45]

A known mass of degassed (free of water and volatile compounds) powder with a total surface of more than 2 m² is used for analysis. By adding defined doses of the analysis gas and measuring the equilibrium pressure, gas adsorption isotherms can be obtained (Equation 10.21). From this measurement the mass-specific surface area (SSA) of the particles is obtained. Specific surface areas can range up to several hundred square meters per gram in the case of highly porous materials or very fine particles. The average primary particle size can be obtained, assuming that the particles are spherical, nonporous, smooth, and monodispersed:

$$d_{BET} = \frac{6}{\rho \cdot SSA} \,, \tag{10.23}$$

where $\rho$ is the bulk density of the investigated material. From Equation 10.23 it becomes apparent that the $d_{BET}$ describes the volume:surface diameter ratio, thus the ratio of the third to the second moment of the number-based size distribution (see the Appendix for details) is

$$\frac{M_{3,0}}{M_{2,0}} = \frac{\int x^3 \cdot q_0(x)dx}{\int x^2 \cdot q_0(x)dx} = \frac{1}{M_{-1,3}} = M_{1,2} \quad \Rightarrow \quad d_{1,2} = \int x \cdot q_2(x)dx \tag{10.24}$$

The standard BET method requires three or more points in the relative pressure range ($p/p_0$) of 0.05 to 0.3, which is typically the region of a linear relationship between $p_r$ and $p_r/[V(1 - p_r)]$ (Equation 10.21). This must be verified by fitting a regression line through the data, and the regression coefficient should be 0.999 or better when using five or six equally spaced measurement points. The BET theory can only be applied straightforward for powders that follow IUPAC type II or type IV isotherms (Figure 10.11). Good indications that the BET theory is not valid are negative C-values or values greater than 300. The latter indicates very strong attraction for the surface or preferential adsorption.[45] For flame-made nanoparticles, which are typically are nonporous (type II isotherm), the C-value is usually between 80 and 120. In general, BET analysis may be insensitive to rough surfaces or ink-bottle-shaped pores, where the nitrogen molecules are too large to cover the whole particle surface area or to penetrate into internal pores. Furthermore, internal (closed) pores cannot be detected by this adsorption method. The fundamental assumption in BET theory is that the forces active in gas condensation are also responsible for the binding energy in multilayer adsorption. The error in the gas adsorption measurement is also associated with the accuracy of the weight of the sample before or after nitrogen adsorption.

## 3. Mercury Porosimetry

This technique utilizes mercury intrusion to obtain pore size distributions in powders and solids.[48] A sample is placed into the instrument, where mercury, under controlled pressure, is forced into the pores of the material. As the pressure is increased, more mercury enters the pores. The accessible pore size corresponding to a given pressure is obtained from the Washburn equation,

$$d = \frac{-4 \cdot \gamma \cdot \cos\theta}{p} , \qquad (10.25)$$

which relates the applied pressure, $p$, to the intruded diameter, $d$, of a cylindrical pore together with the surface tension, $\gamma$, and the contact angle, $\theta$. Thus mercury porosimetry is useful for determining the pore size distribution in particles. It can be applied for pores ranging from 3 nm to 360 μm.[45]

## B. WITH REACTION/CHEMICAL INTERACTION

### 1. Temperature Programmed Reaction

Temperature programmed reaction (TPR) methods monitor a chemical reaction while the temperature is increased linearly with time. This method is often used in catalysis, where interpretation is straightforward most of the time. However, the data reduction to activation energies and preexponential factors of the Arrhenius equation is a difficult task.[21] Often the reactions are grouped into reduction (TPR) and oxidation (TPO), while the uptake or the release of an oxidant or reducing compound is monitored.

### 2. Thermogravimetric Analysis

This technique measures the change in the weight of a sample as it is heated at a known rate.[49] The primary use of thermogravimetric analysis (TGA) for materials produced by aerosol processes is for determination of the extent of reaction and for detection of adsorbed species such as water vapor. The off-gases of the reaction or desorption can be monitored with mass spectrometry or FTIR. TGA determines the identity of the species released at a certain temperature. Using the sample weight loss rate, kinetic parameters can be determined.[50,51] TGA and differential thermal analysis (DTA, see below) typically require 50 mg samples. TGA can be used as a simple and fast method for determining carbon content[52] as well as surface OH groups[53] of oxide nanopowders made by flame aerosol and sol-gel processes. In the latter study it was shown that it is possible to distinguish between physically adsorbed and chemically bound water and to rapidly determine the OH surface density even of small powder samples (less than 0.2 g) after calibrating the TGA with LiAlH$_4$ titration data. The high accuracy of the OH surface density determination by TGA is confirmed further with additional LiAlH$_4$ titration data of silica powders and by comparison with the specifications of commercially available silica Aerosil and titania P25 powders.

## 3. Differential Thermal Analysis

Differential thermal analysis (DTA) measures the amount of heat released or absorbed by a sample as it is heated at a known rate.[49] When the enthalpy change is determined, the method is called differential scanning calorimetry (DSC). The presence of exothermic or endothermic processes at certain temperatures provides information about the nature of phase changes and chemical reactions occurring in the material as it is heated. DTA can often be used as a sensitive method for establishing the presence or absence of secondary phases in samples if these phases undergo phase transformations at known temperatures.[33]

## V. BEHAVIOR IN APPLIED FORCE FIELDS

### A. INERTIAL FIELDS

### 1. Sedimentation

In colloidal systems the sedimentation principle uses the fact that particles of different mass have different velocities in a fluid at rest when the inertial force acting on the particle is equal to the difference between its weight minus buoyancy and the drag force of the medium:

$$m_p \frac{du}{dt} = V_p(\rho_p - \rho_M)\, g - c_W\, \frac{\pi}{4}\, d_p^2\, \frac{\rho_M}{2}\, u^2 , \qquad (10.26)$$

where $m_p$ is the mass of the particles, $V_p$ is their volume, $g$ is the gravitational constant, $d_p$ is the particle diameter, and $\rho_p$ and $\rho_M$ are the particle and medium density, respectively. Furthermore, $c_W$ is the drag coefficient. In case of nonspherical particles, a shape factor has to be applied. The drag coefficient depends on the particle Reynolds number, $Re_p$. For $Re_p < 1$, $c_W = 24/Re$ (Stokes regime), and with this the particle size is directly related to the migration velocity, $u_{st}$:

$$u_{st} = \frac{\Delta\rho\, g}{18\, \mu}\, d_p^2 . \qquad (10.27)$$

The methods of sedimentation analysis are grouped according to the following fundamental criteria:[54] measurement and suspension arrangement (e.g., gravitational or centrifugal force field, incremental or cumulative methods) and type of measuring principle (e.g., gravimetric determination, absorption of electromagnetic radiation by the sedimenting solids, or backscattering of radiation). The methods are applicable for particle sizes larger than 50 μm, while the size range of the measurement can be extended to larger than 0.1 μm if a centrifugal force field is used. It is worth noting that the sedimentation principles result in mass-related particle size distributions ($q_3(x)$). However, when suspending the particles in the medium, care must be taken to break loose agglomerates and to measure at the isoelectric point where the particles carry a net zero charge in the given solvent. Practical guidelines are given by Jillavenkatesa et al.[55]

## 2.  Sieving

Particle size analysis by sieving is one of the oldest methods, and can be applied for particles from about 125 mm down to about 90 μm; when using a liquid medium (wet sieving), the smallest size is about 10 μm. In the coarse range of larger than 1 mm, sieving practically has a monopoly.[54] The main problem associated with sieving is not the actual measurement (weighing to individual fractions), but rather is the cut size at which separation takes place. The cut size depends not only on the sieving apparatus, but also on the sieving medium, sieving time, and many other parameters (e.g., acicular particle, powder aggregation, or particle fracture). It should be noted that sieving gives directly the particle mass distribution ($q_3(x)$). Practical guidelines are given by Jillavenkatesa et al.[55]

## B.  ELECTRICAL FIELDS

## 1.  Differential Mobility Analyzer

Particles smaller than about 100 nm in diameter are difficult to measure optically, and inertial or gravitational separation in gases is only possible at pressures well below atmospheric. However, when applying electrostatic forces to such particles, detection and sizing are possible.[34] Most particles are readily charged during generation or a charge can be applied by passing them through an ionized gas. If the charge distribution is known, the size distribution can be obtained by mobility analyzing techniques. In this case, the mechanical and electrical mobilities of the particles are compared when the charged particles are in an electric field relative to the gas motion. From the resulting migration velocity, $u$, one can determine the electrical particle mobility, $Z_p$, and therefore the particle mobility equivalent diameter in the Stokes regime, $d_{mob}$, according to

$$\frac{n_e\, e\, C_c(d_{mob})}{3\pi\ \mu\ d_{mob}} = \frac{u}{E} = Z_p\ , \tag{10.28}$$

where $n_e$ is the number of charges, $e$ is the elementary unit of charge, $E$ is the electrical field, $\mu$ is the gas viscosity, and $C_c$ is a correction term accounting for noncontinuum effects. This method is very valuable, especially for dry agglomerated powders suspended in air, since it directly measures the radius of gyration of agglomerated particles when their fractal dimension is larger than 2. This method can be applied to obtain particle size distributions based on number ($q_0(x)$) when the particles are fractioned according to their migration velocity.[56] After the classification step, the particles are detected by nucleation particle counters or electrometers. Recently the instrument has been used to fractionate particles in order to obtain almost monodispersed particles, which is especially important in electronic applications.[57]

## 2. Electrophoresis

The physical principle underlying differential mobility analyzer (DMA) also enables the measurement of the potential, which gives information on the charge distribution of dispersed particle in liquids and their boundary layers (electrochemical double layer). Particles in a colloidal suspension usually carry an electrical charge due to their surface chemistry. Sometimes the surfaces of the particles contain chemical groups that can ionize. In other cases there may be deliberately added chemical compounds that preferentially adsorb on the particle surface to generate the charge. In such suspensions, the migration velocity, $u$, of the particles in the electric filed can usually be observed optically (see also light scattering section), which is directly proportional to the $\zeta$ potential (Smoluchowski):

$$u = \frac{\varepsilon E}{6 \pi \mu} \zeta ,  \tag{10.29}$$

where $\varepsilon$ is the dielectric constant of the solution. This equation is valid for nondeformable, nonconductive, and spherical particles at low concentrations, which do not lead to a breakthrough of the electrochemical double layer.

## VI. SUMMARY

Particle characterization and sizing techniques with the main focus on dry powder analysis were explored and summarized with respect to their mode of operation and primary physical meaning of the output parameters. Thereby, particle diameters are distinctly derived from the respective moments obtained from the corresponding particle analysis techniques. Dry powders with nanoscale structures are typically characterized by microscopic means (TEM/SEM) for their shape, size, and structure. Diffraction and scattering techniques (XRD, SAXS, and light scattering) provide information about size, structure, crystallinity, and phase composition, while absorption and emission spectroscopic methods (UV-Vis, FTIR, XRF, XPF, Auger spectroscopy, and EXAFS) provide hints about phase and surface composition, as well as the electronic structure of the sample. The interaction with gases and liquids is used in adsorption/desorption techniques that measure surface area, pore sizes, and surface characteristics when reactions are involved, as in TGA. Larger ranges of particle sizes can be characterized by applying force fields in sieving, sedimentation, and sizing by the particle's mobility.

## VII. APPENDIX

Particle size is one of the main characteristics of dry powders. Most of the dry powders are composed of primary particles, which are held together by various forces (e.g., van der Waals and electrical forces). The formed agglomerates are characterized by the force strength between the primary particles. If the primary particles form sinter necks, they are often referred as "hard" agglomerates or simply as agglomerates (breakage of these hard agglomerates typically involves significant power or energy input), while "soft" agglomerates (aggregates) can be broken down to individual particles (primary particles) with significantly less energy input. The aggregate strength depends on the process for their manufacture and the material system. The structure of the agglomerates is characterized by its fractal dimension, where the mass ($m_{agg}$) of an agglomerate is related to its size, $d$, by a power law:

$$m_{agg} \propto d^{D_f} , \qquad (A1)$$

where the exponent, $D_f$, is the fractal dimension. The mass of such an agglomerate is directly related to the number of primary particles, $N_p$, in that agglomerate. Therefore the number of primary particles, as expressed by Friedlander,[58] is

$$N_p = A \cdot \left( \frac{2R_g}{d_p} \right)^{D_f} , \qquad (A2)$$

where $A$ is a constant, $R_g$ is the radius of gyration (directly proportional to the periphery agglomerate diameter) of the agglomerate, and $d_p$ is the primary particle diameter. However, when determining primary particle diameters or agglomerate diameters, the particle size distribution (PSD) should always be considered. This is very important, especially when size averages are obtained with a certain measuring technique. Therefore special care has been taken to evaluate the given size averages or moments of the distribution from the measurement techniques discussed in the text. The moment $M_{k,r}$ of a uniform cubic particle collective (or PSD) is defined as

$$M_{k,r} = \int_0^{\infty} x^k \cdot q_r(x) \, dx , \qquad (A3)$$

where $x$ is the variable of integration (length scale or size), $q_r(x)$ is the frequency distribution (size distribution), and $r$ is the quantity type (0 = number, 1 = length, 2 = area, 3 = volume, mass). From every momentum a representative length scale or size can be derived from the following equation:

$$x_{k,r} = (M_{k,r})^{1/k} , \qquad (A4)$$

and moments can be rearranged giving the following rule:

$$M_{k,q} = \frac{M_{k+q-r,r}}{M_{q-r,r}} \cdot \qquad (A5)$$

For log-normal particle size distributions, simple algebraic relations for the conversion between different moments can be obtained.[2]

# REFERENCES

1. Pratsinis, S.E., Flame aerosol synthesis of ceramic powders, *Prog. Energy Combust. Sci.*, 24, 197, 1998.
2. Hinds, W.C., *Aerosol Technology*, John Wiley & Sons, New York, 1999.
3. Dobbins, R.A., and Megaridis, C.M., Morphology of flame-generated soot as determined by thermophoretic sampling, *Langmuir*, 3, 254, 1987.
4. Kammler, H.K., Jossen, R., Morrison, P.W., Jr., Pratsinis, S.E., and Beaucage, G., The effect of external electric fields during flame synthesis of titania, *Powder Technol.*, 310, 135–136, 2003.
5. Tsantilis, S., Kammler, H.K., and Pratsinis, S.E., Population balance modeling of flame synthesis of titania nanoparticles, *Chem. Eng. Sci.*, 57, 2139, 2002.
6. Mädler, L., Stark, W.J., and Pratsinis, S.E., Simultaneous deposition of Au nanoparticles during flame synthesis of $TiO_2$ and $SiO_2$, *J. Mater. Res.*, 18, 115, 2003.
7. Tani, T., Mädler, L., and Pratsinis, S.E., Synthesis of zinc oxide/silica composite nanoparticles by flame spray pyrolysis, *J. Mater. Sci.*, 37, 4627, 2002.
8. Kammler, H.K., and Pratsinis, S.E., Carbon-coated titania nanoparticles: continuous, one-step flame-synthesis, *J. Mater. Res.*, 18, 2670, 2003.
9. Koylu, U.O., Xing, Y.C., and Rosner, D.E., Fractal morphology analysis of combustion-generated aggregates using angular light scattering and electron microscope images, *Langmuir*, 11, 4848, 1995.
10. Stark, W.J., Mädler, L., Maciejewski, M., Pratsinis, S.E., and Baiker, A., Flame synthesis of nanocrystalline ceria-zirconia: effect of carrier liquid, *Chem. Commun.*, 5, 588, 2003.
11. Kim, D.J., Lattice-parameters, ionic conductivities, and solubility limits in fluorite-structure Hf-4+O2, Zr-4+O2, Ce-4+O2, Th-4+O2, V-4+O2 oxide solid-solutions, *J. Am. Ceram. Soc.*, 72, 1415, 1989.
12. Snyder, R.L., Analytical profile fitting of x-ray powder diffraction profiles in Rietveld analysis, in *The Rietveld Method*, R.A. Young, Ed., Oxford University Press, Oxford, 1993.
13. Cheary, R.W., and Coelho, A., A fundamental parameters approach to x-ray line-profile fitting, *J. Appl. Crystallogr.*, 25, 109, 1992.
14. Scherrer, P., Bestimmung der Grösse und der inneren Struktur von Kolloidteilchen mittels Röntgenstrahlen, *Nachr. Ges. Wiss. Göttingen*, 98, 1918.
15. Klug, H.P., and Alexander, L.E., *X-ray Diffraction Procedures: For Polycrystalline and Amorphous Materials*, John Wiley & Sons, New York, 1974.

16. Delhez, R., de Keijser, T.H., Langford, J.I., Louer, D., Mittemeijer, E.J., and Sonneveld, E.J., Crystal imperfection broadening and peak shape in Rietveld method, in *The Rietveld Method*, R.A. Young, Ed., Oxford University Press, Oxford, 1993.

17. Wilson, A.J.C., *X-ray Optics*, John Wiley & Sons, New York, 1962.

18. Langford, J.I., Louer, D., and Scardi, P., Effect of a crystallite size distribution on x-ray diffraction line profiles and whole-powder-pattern fitting, *J. Appl. Crystallogr.*, 33, 964, 2000.

19. Cohen, J.B., X-ray-diffraction studies of catalysts, *Ultramicroscopy*, 34, 41, 1990.

20. Langford, J.I., and Wilson, A.J.C., Scherrer after 60 years—survey and some new results in determination of crystallite size, *J. Appl. Crystallogr.*, 11, 102, 1978.

21. Niemantsverdriet, J.W., *Spectroscopy in Catalysis. An Introduction*, Wiley-VCH Verlag, Weinheim, Germany, 2000.

22. Solliard, C., Structure and strain of the crystalline lattice of small gold and platinum particles, *Surf. Sci.*, 106, 58, 1981.

23. Mädler, L., Stark, W.J., and Pratsinis, S.E., Flame-made ceria nanoparticles, *J. Mater. Res.*, 17, 1356, 2002.

24. Young, R.A., *The Rietveld Method*, Oxford University Press, Oxford, 1993.

25. Roe, R.-J., *Methods of X-ray and Neutron Scattering in Polymer Science*, Oxford University Press, New York, 2000.

26. Beaucage, G., Approximations leading to a unified exponential power-law approach to small-angle scattering, *J. Appl. Crystallogr.*, 28, 717, 1995.

27. Beaucage, G., Small-angle scattering from polymeric mass fractals of arbitrary mass-fractal dimension, *J. Appl. Crystallogr.*, 29, 134, 1996.

28. Beaucage, G., and Schaefer, D.W., Structural studies of complex-systems using small-angle scattering—a unified Guinier power-law approach, *J. Non-Crystl. Solids*, 172, 797, 1994.

29. Kammler, H.K., Beaucage, G., Mueller, R., and Pratsinis, S.E., Structure of flame-made silica nanoparticles by ultra small-angle x-ray scattering, *Langmuir*, 20, 1915, 2004.

30. Porod, G., Small angle x-ray scattering, in *Small Angle X-ray Scattering, Part II*, O. Glatter and O. Kratky, Eds., Academic Press, London, 1982.

31. Guinier, A., and Fournet, G., *Small Angle Scattering of X-rays*, John Wiley & Sons, New York, 1955.

32. Mueller, R., Mädler, L., and Pratsinis, S.E., Nanoparticle synthesis at high production rates by flame spray pyrolysis, *Chem. Eng. Sci.*, 58, 1696, 2003.

33. Baron, P.A., and Willeke, K., *Aerosol Measurement*, John Wiley & Sons, New York, 2001.

34. Flagan, R.C., Electrical techniques, in *Aerosol Measurement*, P.A. Baron and K. Willeke, Eds., John Wiley & Sons, New York, 2001.

35. Xu, R., *Particle Characterization: Light Scattering Methods*, Kluwer Academic, Dordrecht, The Netherlands, 2000.

36. Bushell, G., Amal, R., and Raper, J., The effect of polydispersity in primary particle size on measurement of the fractal dimension of aggregates, *Part. Syst. Charact.*, 15, 3, 1998.

37. Bohren, C.F., and Huffman, D.R., *Absorption and Scattering of Light by Small Particles*, John Wiley & Sons, New York, 1998.

38. Perkampus, H.H., *UV-vis Spectroscopy and Its Applications*, Springer Verlag, New York, 1992.

39. Kubelka, P., and Munk, F., Ein Beitrag zur Optik der Farbanstriche, *Zeitschr. Techn. Physik*, 12, 593, 1931.
40. Kortüm, G., *Reflectance Spectroscopy: Principles, Methods, Applications*, Springer Verlag, New York, 1969.
41. Mädler, L., Stark, W.J., and Pratsinis, S.E., Rapid synthesis of stable ZnO quantum dots, *J. Appl. Phys.*, 92, 6537, 2002.
42. Stark, W.J., Pratsinis, S.E., and Baiker, A., Flame made titania/silica epoxidation catalysts, *J. Catal.*, 203, 516, 2001.
43. Dukhin, A.S., and Goetz, P.J., *Ultrasound for Characterizing Colloids*, Elsevier, Amsterdam, 2002.
44. Schoofs, T., Surface area analysis of finely divided and porous solids by gas adsorption measurements, in *Particle and Surface Characterisation Methods*, R.H. Müller and W. Mehnert, Eds., Medpharm GmbH Scientific, Stuttgart, 1997.
45. Webb, P.A., and Orr, C., *Analytical Methods in Fine Particle Technology*, Micromeritics Instrument Corp., Norcross, GA, 1997.
46. Strobel, R., Stark, W.J., Madler, L., Pratsinis, S.E., and Baiker, A., Flame-made platinum/alumina: structural properties and catalytic behaviour in enantioselective hydrogenation, *J. Catal.*, 213, 296, 2003.
47. Brunauer, S., Emmett, P.H., and Teller, E., Adsorption of gases in multimolecular layers, *J. Am. Chem. Soc.*, 60, 309, 1938.
48. Reed, J.S., *Principles of Ceramic Processing*, John Wiley & Sons, New York, 1995.
49. Hench, L.L., and Gould, R.W., *Characterization of Ceramics*, Marcel Decker, New York, 1971.
50. Biswas, P., Li, X.M., and Pratsinis, S.E., Optical wave-guide preform fabrication—silica formation and growth in a high-temperature aerosol reactor, *J. Appl. Phys.*, 65, 2445, 1989.
51. Biswas, P., Zhou, D., Zitkovsky, I., Blue, C., and Boolchand, P., Superconducting powders generated by an aerosol process, *Mater. Lett.*, 8, 233, 1989.
52. Kammler, H.K., Mueller, R., Senn, O., and Pratsinis, S.E., Synthesis of silica-carbon particles in a turbulent $H_2$-air flame aerosol reactor, *AIChE J.*, 47, 1533, 2001.
53. Mueller, R., Kammler, H.K., Wegner, K., and Pratsinis, S.E., The OH-surface density of $SiO_2$ and $TiO_2$ by thermogravimetric analysis, *Langmuir*, 19, 160, 2003.
54. Bernhardt, C., *Particle Size Analysis, Classification and Sedimentation Methods*, Chapman & Hall, London, 1994.
55. Jillavenkatesa, A., Dapkunas, S.J., and Lum, L.-S.H., *Particle Size Characterization*, National Institute of Standards Technology Washington, D.C., 2001.
56. Knutson, E.O., and Whitby, K.T., Aerosol classification by electric mobility: apparatus, theory, and applications, *J. Aerosol Sci.*, 6, 443, 1975.
57. Kennedy, M.K., Kruis, F.E., Fissan, H., Mehta, B.R., Stappert, S., and Dumpich, G., Tailored nanoparticle films from monosized tin oxide nanocrystals: particle synthesis, film formation, and size-dependent gas-sensing properties, *J. Appl. Phys.*, 93, 551, 2003.
58. Friedlander, S.K., *Smoke, Dust, and Haze: Fundamentals of Aerosol Dynamics*, Oxford University Press, New York, 2000.

# Section II

## Powder Processing at Nanoscale

# 11 Theory and Applications of Colloidal Processing

*Wolfgang Sigmund, Georgios Pyrgiotakis, and Amit Daga*

## CONTENTS

## NOTATIONS

The following is a list of symbols and notations used in text. Symbols not included in this list are assumed constants.

| Symbol | Meaning |
|--------|---------|
| P | Gas pressure |
| V | Volume |
| $\alpha$, $\beta$ | Van der Waals Constants |
| $k_B$ | Boltzmann constant |
| T | Temperature |
| W(r) | Energy |
| R | Distance from the center of the mass |
| $\alpha_0$ | Atomic radius |
| d | Separation distance |
| R. $\alpha$ | Radius |
| $\rho$ | Density |
| A | Hamaker constant |
| $A_{132}$ | Hamaker constant between material 1 and 2 in medium 3 |
| L | Chain length |
| $\sigma$ | Monomer length |
| $\varepsilon_i$, i=1,2,3 | Dielectric permittivity |
| $\varepsilon_0$ | Dielectric permittivity of vacuum |
| v | Frequency |
| $n_i$, i=1,2,3 | Index of refraction |
| $C_{IR}$, $C_{UV}$ | Constants |
| $v_{IR}$, $v_{UV}$ | Infrared and ultraviolet major absorption peaks |
| $e_{el}$ | Electron charge |
| $\Psi(x)$ | Surface potential |
| $\Psi_\delta$, $\Psi_0$ | Surface potential at the surface |
| $\Psi_\zeta$, $\zeta$ potential | Zeta potential |
| z | Ionic strength |
| $\kappa$ | Inverse Debye length |
| $\rho_i$ | Ion concentration |
| $z_i$ | Electrolyte valence |
| J | Number of collision per unit time |
| $t_{1/2}$ | Time for 50% coagulation |
| n | Numbers of particles |
| D | Diffusion coefficients |
| $D_{eff}$ | Effective diffusion coefficients |
| $\eta$ | Viscosity |
| P | Probability |
| $E_b$ | Energy barrier |
| W | Stability ratio |
| V(d) | Energy |
| $\chi$ | Flory-Huggins parameter |
| $\overline{\phi}_p$ | Average polymer volume fraction |
| $\delta$ | Polymer layer thickness |
| $\overline{MW}$ | Average molecular weight |
| $\rho_p$ | Polymer density |
| $\tau$ | Shear stress |
| $\dot{\gamma}$ | Shear rate |

| | |
|---|---|
| $G$ | Shear modulus |
| $\phi$ | Solids loading |
| $K$ | Hydrodynamic factor |
| $\phi_{max}$ | Maximum solids loading |
| $\phi_{eff}$ | Effective solids loading |
| $\eta_r$ | Relative viscosity |
| $\phi_O$ | Initial volume fraction of particles |
| $\phi_g$ | Gelation threshold |
| $G_0$ | Constant |
| $S$ | Structural parameter |
| $\theta$-temperature | Temperature where segment-segment interaction is equal to segment-solvent interaction |

Due to the high melting temperatures and the inherent brittle nature of ceramic materials 90% of the ceramics on the world market are processed via powder-metallurgical routes. Furthermore, advanced ceramics require beneficiation steps to optimize particle mixing and particle packing. This is usually done in liquids; therefore particles need to be wetted by the liquid and then dispersed, yielding slurry. For pressing techniques, slurries are typically spray dried to form granules. Alternatively, it is possible to use these slurries directly in either drained casting, where the liquid is removed by suction into pores, leaving the touching particles behind, or by direct casting, where the entire slurry system is transferred into a gel. This chapter introduces the fundamental physics and chemistry for colloidal processing and demonstrates their application in direct casting techniques.

The reader will learn about the origins of attractive forces between particles and how to overcome these forces, yielding a dispersed system. Furthermore, it is shown how these interparticle forces relate to the flow behavior of the slurry system, and finally, how this knowledge can be applied to current direct casting techniques.

Before we go into detail it is important to note that due to the interdisciplinary nature of this field, certain words have different meanings to physicists, surface scientist, and ceramists, especially the word force. Today's physics knows only four forces; two forces of very short range that are responsible for acting on neutrons, protons, electrons, and other subatomic particles; and two forces of longer range—electromagnetic and gravitational force. This chapter deals with variations in electromagnetic force only. Due to the specific shape of the force-distance curves, surface scientists give new names to interactions that are all based on electromagnetic force.

## I. COLLOIDAL PROCESSING THEORY

### A. VAN DER WAALS FORCES

The behavior of particles larger than 100 μm in diameter is mainly controlled by gravity, that is, they do not tend to stick. This changes dramatically for particles

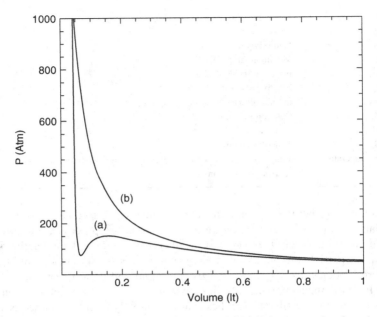

**FIGURE 11.1** Description of (a) van der Waals and (b) ideal gas equation for water. For low energy (low pressure, high volume) the equations are similar, for high energy though the difference is apparent.

less than 100 μm in diameter. Surface forces govern them. These forces can be a million times or more stronger than the gravitational pull and therefore these particles have a tendency to stick. As mentioned above, even though all surface forces are of electrodynamic origin, several interactions have specific names. First, we look into the most important and ubiquitous force responsible for sticking, the so-called van der Waals force. The origin of the theory on van der Waals forces dates back to the late 18th century. At that time the ideal gas equation was used to characterize and predict the behavior of every molecular or atomic gas. This approximation seemed to work well for monatomic gases like helium, xenon, and neon, as well as for simple diatomic gasses like $Cl_2$ and $H_2$. It seemed to fail though for compound gases with polar behavior, like $H_2O$ (Figure 11.1). In 1876 Johannes D. van der Waals ascribed this behavior to possible intermolecular interaction (despite the fact that the model of matter consisting of atoms was not accepted at that time) and suggested a modification to the well-established ideal gas equation known today as the van der Waals equation of state:[1]

$$(P - \alpha/V^2)(V - \beta) = k_B T \tag{11.1}$$

Today, collectively those interactions are called van der Waals interactions. Electrostatic (coulomb) and hydrogen-bonding forces, however, are excluded from this group basically due to the nature of the force.

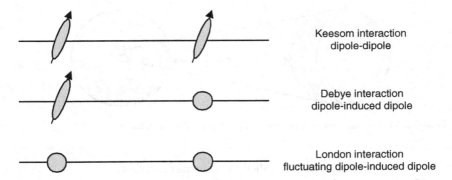

**FIGURE 11.2** The three major types of van der Waals interactions.

The origin of the van der Waals forces is known to be electromagnetic, and more specifically, dipole interaction. All gases, despite the fact that they are neutral, might have a nonhomogeneous charge distribution (e.g., in water, the shared electrons are pulled closer to the oxygen than the hydrogen) and with the proper molecule structure (in water, the two hydrogen atoms and the oxygen atom form a 109.47° angle) can result in a dipole. Those dipoles produce vibrating electromagnetic waves and they interact with each other by inducing attractive or repulsive forces. By 1930 the dipole-dipole interaction had already been studied by classical physics and the interaction energy had been calculated to be

$$W(r) = -C/r^6. \tag{11.2}$$

In the 1930s, Eisenschitz and London[2] did analytical calculations with the perturbation theory in quantum mechanics and estimated the interaction energy between two induced dipoles as

$$W(r) = -\frac{3}{4}\alpha_0^2 I \frac{1}{(4\pi\varepsilon_0)r^6}, \tag{11.3}$$

which is very similar to the classic expression, with the only difference being the numerical prefactor. The fluctuating dipole-induced dipole interaction is the strongest of the van der Waals interactions and is called a London interaction, but there are also dipole-dipole and dipole-induced dipole interactions that are called Keesom and Debye interactions, respectively (Figure 11.2).

In 1937 Hamaker had the idea of expanding the concept of the van der Waals forces from atoms and molecules to solid bodies.[3] He assumed that each atom in body 1 interacts with all atoms in body 2, and with a method known as pairwise summation (Figure 11.3), found an expression for the interaction between two spheres of radius $R_1$ and $R_2$:

$$W(d) = -\frac{A}{6d}\left(\frac{R_1 R_2}{R_1 + R_2}\right), \tag{11.4}$$

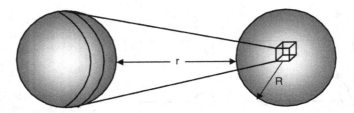

**FIGURE 11.3** Principles of pairwise summation used by Hamaker.

where $A$ is a material constant, called the Hamaker constant.

This theory can be extended to several other geometries and many kinds of interaction can be predicted (Table 11.1).

The major disadvantage of this microscopic approach theory was the fact that Hamaker knowingly neglected the interaction between atoms within the same solid, which is not correct, since the motion of electrons in a solid can be influenced by other electrons in the same solid. So a modification to the Hamaker theory came from Lifshitz in 1956 and is known as the Lifshitz or macroscopic theory.[4] Lifshitz ignored the atoms completely; he assumed continuum bodies with specific dielectric properties. Since both van der Waals forces and the dielectric properties are related with the dipoles in the solids, he correlated those two quantities and derived expressions for the Hamaker constant based on the dielectric response of the material. The detailed derivations are beyond the scope of this book and readers are referred to other publications. The final expression that Lifshitz derived is

$$A_{132} = \frac{3}{2} k_B T \sum_{n=0}^{\infty} \left[ \frac{\varepsilon_1(i\nu_n) - \varepsilon_3(i\nu_n)}{\varepsilon_1(i\nu_n) + \varepsilon_3(i\nu_n)} \right] \left[ \frac{\varepsilon_2(i\nu_n) - \varepsilon_3(i\nu_n)}{\varepsilon_2(i\nu_n) + \varepsilon_3(i\nu_n)} \right], \quad (11.5)$$

where the pointers 1, 2, and 3 refer to material 1 and 2 interacting in medium 3, and $\varepsilon(\nu)$ is the dielectric permittivity of the material in frequency $\nu$. The expression, although it looks simple, is not easily calculated because there are complications in finding the dielectric permittivity in every frequency and is even more difficult at frequencies of zero and infinity. A simplification of this expression is to assume that the main contribution is coming from the ultraviolet (UV) and infrared (IR) part of the spectrum. Under this approach, the expression of the dielectric permittivity can be written as

$$\varepsilon(\nu) = 1 + \frac{C_{IR}}{1 - i\nu/\nu_{IR}} + \frac{C_{UV}}{1 - i\nu/\nu_{UV}}. \quad (11.6)$$

More approximations can be made from this point to further simplify the calculations, but all of them require experimental inputs at some point about

**TABLE 11.1**
**Van der Waals Forces for Different Geometries**

| Two atoms | Two spheres |
|---|---|
| ○　　○ |  |
| $W(d) = -\dfrac{C}{d^6}$ | $W(D) = -\dfrac{A}{6d}\left(\dfrac{R_1 R_2}{R_1 + R_2}\right)$ |

| Atom surface | Sphere-surface |
|---|---|
| ○　▮ | |
| $W(D) = -\dfrac{\pi C \rho}{6d^3}$ | $W(D) = -\dfrac{AR}{6d}$ |

| Parallel chain molecules | Surface-surface |
|---|---|
|  |  |
| $W(d) = \dfrac{3\pi A L}{8\sigma^2 d^5}$ | $W(D) = -\dfrac{A}{12\pi d^2}$, per unit area |

the optical properties of the material. If the optical data of the materials are known, then the calculation of the Hamaker constant can be completed. There is an important aspect to be noted in the Lifshitz theory in that the Hamaker constant can be positive or negative[5,6] depending on the materials, and sometimes even zero; something that has critical importance for colloidal stability, as will be explained later. This can be seen easier in the Tabor-Winterton

**TABLE 11.2**
**Isoelectric Point and Hamaker Constants in Vacuum and Water for Various Materials[106]**

| Material | Hamaker Constant in Vacuum ($10^{-20}$ J) | Hamaker Constant in Water ($10^{-20}$ J) | Isoelectric Point (pH) |
|---|---|---|---|
| $TiO_2$ | 15.3 | 5.35 | 4–6 |
| $SiO_2$ (amorphous) | 6.5 | 0.46 | 2–3 |
| $\alpha$-$Al_2O_3$ | 15.2 | 3.67 | 8–9 |
| $Si_3N_4$ | 16.7 | 4.85 | 9 |
| $BaTiO_3$ | 18 | 8 | 5–6 |
| $ZnO$ | 9.21 | 1.89 | 9 |

approximation, which, starting from the Lifshitz theory, simplifies the Hamaker constant:[7]

$$A_{132} = \frac{3}{4}k_B T \left(\frac{\varepsilon_1 - \varepsilon_3}{\varepsilon_1 + \varepsilon_3}\right)\left(\frac{\varepsilon_1 - \varepsilon_3}{\varepsilon_1 + \varepsilon_3}\right)$$
$$+ \frac{3h\nu_e}{8\sqrt{2}} \frac{\left(n_1^2 - n_3^2\right)\left(n_2^2 - n_3^2\right)}{\left(n_1^2 + n_3^2\right)^{1/2}\left(n_2^2 + n_3^2\right)^{1/2}\left\{\left(n_1^2 + n_3^2\right)^{1/2} + \left(n_2^2 + n_3^2\right)^{1/2}\right\}} \tag{11.7}$$

In the case where the medium and the solids have the same index of refraction ($n_1 = n_2$ and $\varepsilon_1 = \varepsilon_2$), the Hamaker constant $A_{132}$ goes to zero. This method is known as index matching and is used to minimize or eliminate the effect of the Hamaker constant. Table 11.2 lists Hamaker constants for several materials. Figure 11.4 demonstrates the van der Waals forces for the system of alumina-alumina in water for three different particle radii.

As mentioned earlier, the van der Waals interaction occurs due to the synchronization of dipoles. The electromagnetic field travels with the speed of light, so time is required for dipole synchronization. For long distances, the dipoles fail to synchronize properly so the effect is not as strong as predicted by the theory. This phenomenon is known as the retardation of van der Waals force or Casimir-Polder[8] correction and starts to be noticeable at distances greater than 5 nm. Similar decay of the van der Waals interaction can be observed in the case of two particles in a polar medium. In this case the free charges of the medium are screening the dipoles. This phenomenon is known as screened van der Waals force and is important in small distances, about half the Debye length, $\kappa^{-1}$, the meaning of which is explained later in this chapter.

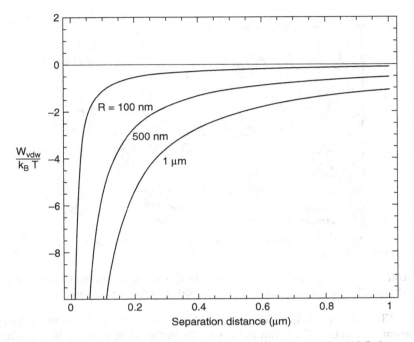

**FIGURE 11.4** Van der Waals energy for three sizes of spherical particles ($Al_2O_3$ in water). R = radius of the spherical particle. The ordinate is normalized by a Boltzmann factor to make for easy comparison to Brownian energy ($E_{Br} = 2/3\ k_B\ T$).

## B. ELECTRIC DOUBLE LAYER FORCES

For the processing of ceramics in liquids, it is important to introduce repulsive forces to overcome attractive van der Waals forces. One type of force is the so-called electric double layer (EDL) force. Some books refer to this force as electrostatic force. To avoid confusion, the term EDL force is used throughout this chapter to clearly show that the physics of particles in liquids strongly differs from particles in air, where electrostatic forces apply that follow Coulombs law. This section describes the chemistry in the development of surface charges on particles and the physics equation that governs the forces.

A perfect stoichiometric crystal with complete absence of defects is considered neutral, which implies that the surface will be neutral too, thus there is no net surface charge. This is true for diamond. As soon as the surface is oxidized, hydroxyl and other groups form that develop a charge as soon as it is immersed in a protic liquid. For any material, there are four major sources of this charge: ionization or dissolution of surface groups, specific ion adsorption, ion exchange, and solution of specific ions out of the surface.

In 1901 Stern[9] assumed that when particles are submerged in an electrolyte containing solvent, there is a second layer of charges with opposite sign (counter ions) formed on top of the first layer.[10,11] Thus he called the system electric double

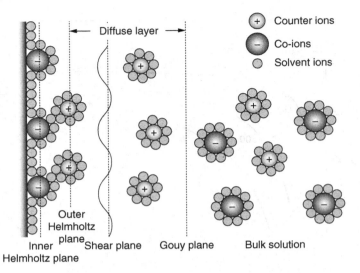

**FIGURE 11.5** Electric double layer (EDL). In this example: the surface selectively adsorbs negative ions and a second layer is formed from the counter ions.

layer (EDL) (Figure 11.5). Later Gouy[12] and Chapman[13] modified the Stern theory by assuming that the EDL is formed between the surface charge and a diffuse (buffer) layer that extends to a certain distance from the surface (a few nanometers).[14] The EDL is creating a potential known as surface potential, and decays away from the surface. The general solution of this potential is given from the combination of two equations, the Poisson equation for the potential decay and the Boltzmann equation describing the statistical distribution of ions. The Poisson-Boltzmann equation,

$$\frac{d^2\Psi(x)}{dx^2} = -\left(\frac{ze\rho_0}{\varepsilon\varepsilon_0}\right)e^{-ze\Psi/K_BT} , \qquad (11.8)$$

does not have an analytical solution, but can be solved numerically for a specific case. A simplification of this is the Debye-Hückel[15] equation, which assumes simple exponential decay:

$$\Psi(x) = \Psi_0 e^{-\kappa x} , \qquad (11.9)$$

where $\Psi(x)$ is the potential at distance $x$, $\Psi_0$ is the value at the surface, and $\kappa$ is the Debye parameter. $1/\kappa$ is known as the Debye length, which physically gives the range of the field and is given as a function of the electrolyte concentration ($\rho$) and valence ($z$) (see Table 11.3 for examples):

$$\frac{1}{\kappa} = \left[\frac{\varepsilon_0\varepsilon k_B T}{e_{el}^2 \sum_i \rho_i z_i^2}\right]^{1/2} . \qquad (11.10)$$

**TABLE 11.3**
**The Debye Length as a Function of the Concentration for Different Types of Electrolyte**

| $1/\kappa$ (nm) | Electrolyte Valence |
|---|---|
| $\dfrac{0.304}{\sqrt{[AX]}}$ | For 1:1 electrolyte (NaCl, KCl) |
| $\dfrac{0.176}{\sqrt{[AX_2]}}$ | For 2:1 electrolyte (CaCl$_2$, Na$_2$SO$_4$) |
| $\dfrac{0.152}{\sqrt{[A^{+2}X^{-2}]}}$ | For 2:2 electrolyte (CaCO$_3$, MgSO$_4$) |

A: refers to the counter ion.

The measurement of this surface potential ($\Psi_\delta$ or $\Psi_0$) is impossible due to the hydrodynamic behavior of the system that generates a thin layer of attached liquid around the particles. However, there is a plane where the shear starts (shear plane), and at this plane the surface potential can be measured and the value is known as the zeta potential ($\Psi_\zeta$). Besides the indifferent counter- and co-ions in solution, there are also so-called potential determining ions (chemists call them adsorbing ions). For most systems these are H$^+$ and OH$^-$ ions that can adsorb directly on the particle surface and alter the $\zeta$-potential. There is a pH value for which the $\zeta$ potential becomes zero and is called the isoelectric point (IEP), as shown in Figure 11.6.

Increasing the concentration of an indifferent monovalent electrolyte in a solution will decrease the value of $\zeta$ potential and the Debye length, and eventually the EDL will collapse, but it will not change the IEP. For divalent electrolytes, however, and for high concentrations, the sign of the $\zeta$ potential can be reversed and the IEP is shifted. So, for example, if a gold surface, which is positively charged, is submerged in a Na$_2$SO$_4$ solution, the SO$_4^{-2}$ ions will adsorb on the surface, and if the concentration is enough, all the surface sites are covered. In this case there will be an excess of negative charge that will inverse the sign of $\zeta$-potential. If this happens, the ions are referred to as potential determining or specifically adsorbing.

The surface charge is important for the colloidal stability because the induced entropic forces can separate or coagulate the particle suspension. For two particles of radius $R_1$ and $R_2$, the interaction energy is[16]

$$W(d) = 64\pi\varepsilon_0\varepsilon_r\gamma_1\gamma_2\left(\frac{R_1 R_2}{R_1 + R_2}\right)\left(\frac{k_B T}{ve}\right)^2 e^{-\kappa d}, \qquad (11.11)$$

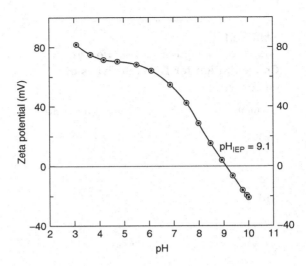

**FIGURE 11.6** Zeta potential as a function of pH for alumina ($Al_2O_3$).

where

$$\gamma_i = \tanh\left(ve\Psi^i_\delta/4k_BT\right)$$

$\varepsilon_0$ is the vacuum dielectric permittivity, $\varepsilon_r$ is the medium dielectric constant, $e$ is the elementary charge, and $v$ is the number of electrons. For same-material particles, the force will always be repulsive, but the range will differ depending on the electrolyte concentration (Figure 11.7). For particles of different material or different surface chemistry that are in a medium at a pH that is between the two IEPs, then the $\zeta$- potentials will have different signs, and thus the force will be attractive. For any other pH value, the force will be repulsive, but the magnitude will depend on the pH. It must be noted here that for very short distances (less than 10 nm), the surface charge interaction is very strong and it forces the charges to be redistributed on the surfaces, and while the separation distance approaches zero, the surface charge density becomes a function of the distance. This phenomenon is known as charge regulation[17] and reduces the effective EDL stabilization.

## C. Colloidal Stability

In 1942 Derjaguin and Landau[18] and Verwey and Overbeek[19] (DLVO) suggested that the total energy of a colloidal system is the summation of van der Waals (vd W) and EDL interactions. For the case of spherical particles of radius $R$, this has the form

$$W(d) = \underbrace{64\pi\varepsilon_0\varepsilon_r\gamma_1\gamma_2\left(\frac{R_1R_2}{R_1+R_2}\right)\left(\frac{k_BT}{ve}\right)^2 e^{-\kappa d}}_{EDL} - \frac{A}{6D}\left(\frac{R_1R_2}{R_1+R_2}\right). \qquad (11.12)$$

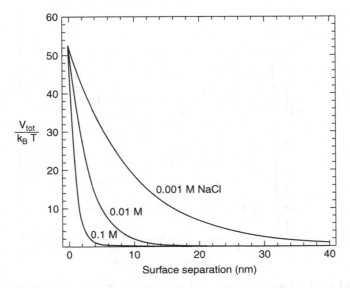

**FIGURE 11.7** Electric double layer interaction between two spheres of 100 and 500 nm, for three different values of monovalent electrolyte (0.1 M, 0.01 M and 0.001 M), which results in three different values for the EDL thickness (0.96 nm, 3.04 nm, and 9.61 nm, respectively).

In most cases of colloidal suspensions, the attraction part comes from van der Waals forces and repulsion comes from EDL forces. A graph of these interactions is shown in Figure 11.8, where, for specific parameters, a repulsive barrier is created. In order for the particles to coagulate they must overcome this barrier and reach the primary minimum. For small barriers, simple mechanical energy, like vibration, is enough to cause coagulation, but for larger barriers, addition of electrolyte or a shift in pH toward the IEP is usually used to collapse the EDL and destabilize the suspension. As shown in Figure 11.8, just before the barrier there is a small, but significant minimum where the system is stable and still at a significant separation distance. This is known as the secondary minimum, and at this point the system is weakly flocculated and only small energies are required to move it from this position.

## D. COAGULATION KINETICS

For the case where the system has no barrier, the primary minimum will cause two particles to coagulate every time they collide. This is known as fast coagulation, and since the sticking coefficient is almost one, the coagulation kinetics are governed by diffusion. According to the von Smoluchowski[20–22] theory of fast coagulation, the rate at which primary particles disappear is

$$\frac{dn}{dt} = -8r_c \, Dn^2 \qquad (11.13)$$

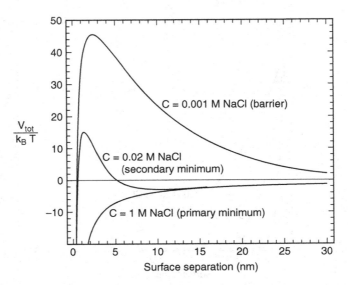

**FIGURE 11.8** DLVO interaction for $Al_2O_3$ in water with three different electrolyte concentrations. For low electrolyte concentration the effective barrier can stabilize the particles at distance $\kappa^{-1}$ (~9 nm). For larger concentrations a secondary minimum can cause weak coagulation of the suspension at the secondary minimum. Even higher concentrations can completely eliminate any repulsion (fast coagulation).

and the number of collisions per unit time is

$$J = 4Dr_c n_0 \qquad (11.14)$$

and the diffusion coefficient, $D$, is

$$D = \frac{K_B T}{6\pi\eta r_c} \qquad (11.15)$$

where $n$ is the number of primary particles (which have not yet coagulated), $n_0$ is their number at $t = 0$, $r_c = 2R$, and $\eta$ is the viscosity of the system. An important parameter to characterize the coagulation kinetics is the time required for 50% coagulation and is calculated from the previous equation to be

$$t_{1/2} = \frac{1}{8\pi r_c D n_0} \cdot \qquad (11.16)$$

Typical values calculated from the previous equations are ~0.1 to 1 s. In reality, coagulation times are longer. This can be attributed to the fact that for large distances the solvent acts as a continuum medium, so the opposition of the medium to the particle movement is incorporated in the equations with the viscosity coefficient. However, for smaller separation distances the solvent cannot be considered as a continuum any more. This results in a reduced diffusion coefficient for smaller distances. The analytical solution to this problem is very difficult; however, there are some empirical equations that can be used. The effective diffusion coefficient as a function of the separation distance $r$ is[23]

$$D_{eff} = D \frac{6x^2 - 20x + 16}{6x^2 - 11x} , \qquad (11.17)$$

where $x = d/R$. Even with this correction there is some deviation from the actual time, because the theory is neglecting the presence of particle clusters, although they are constantly forming via the collisions.

In the case of energy barriers, the previous theory is not applicable anymore, since besides diffusion, the coagulation depends on the barrier threshold, too. The coagulation rate is expected to be less, and so it is called *slow coagulation*. The probability for the particles to overcome the barrier depends on the energy barrier ($E_b$) and the temperature according to the Boltzmann distribution,

$$P \propto e^{-E_b/k_B T} . \qquad (11.18)$$

So a reasonable modification of the disappearing rate equation, for the case of the slow coagulation will be

$$\dot{n} = \frac{dn}{dt} = -8\pi r_c D n^2 e^{-E_b/k_B T} . \qquad (11.19)$$

The ratio between the two rates is called the stability ratio, $W$, and is

$$W = \frac{\dot{n}_{fast}}{\dot{n}_{slow}} = e^{-E_b/k_B T} \qquad (11.20)$$

Still that theory is not fully satisfying because it does not include the particle-particle interactions, which in this case are long range ($\sim\kappa^{-1}$). A more accurate expression for the stability ratio for this case is

$$W = r_c \int_{d=2R}^{d=\infty} \left(\frac{1}{d}\right) e^{W(d)/k_B T} \partial d , \qquad (11.21)$$

which for the case of an EDL stabilized suspension can be approximated with

$$W = \left(\frac{1}{\kappa r_c}\right) e^{E_b/k_B T} . \qquad (11.22)$$

Figure 11.9 shows a stability diagram for silicon nitride powder in water.

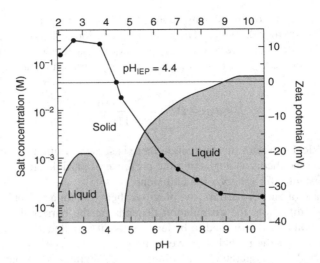

**FIGURE 11.9** Stability diagram for $Si_3N_4$ particles as produced from calculations (IEP 4.4) assuming 90% probability of coagulation for solid formation. The particle size is assumed to be 10 nm and the system is at room temperature. Also shown superposed is the zeta potential for the same particles.

### E. STERIC AND ELECTROSTERIC FORCES

As seen in Figure 11.4, van der Waals forces are size and material dependent, so there are cases where the EDL barrier is too low to stabilize the colloid, especially when there is only a small potential (less than 15 mV) and high ionic strength involved. For those cases, other techniques, such as steric[24] and electrosteric[25] stabilization, have been developed. For steric stabilization, selected organic macromolecules or polymers are added to the suspension. These molecules are designed to adsorb on the particle surface. In order to be a good dispersant the polymer has to adsorb on the surface and have good solubility in the dispersing medium. There are three main categories of theories that have been developed for polymers on surfaces for colloidal stabilization: mean field theories,[26–31] scaling theories,[32–34] and Monte Carlo theories.[35–38] These theories have all been successfully applied to describe the impact on colloidal stability. It has been shown that mean field theories work best for polymers at the θ-temperature, Monte Carlo for poor solubility and scaling theory for well-soluble polymers. Even though some theories, such as scaling theory, have been developed for endgrafted polymers on surfaces, they can be applied for adsorbed polymer layers. Since this theory applies to well-soluble polymers, that is, the polymer segment-solvent interaction is more favorable than the segment-segment interaction, a few quantitative formulas will be presented that apply to ceramic processing.

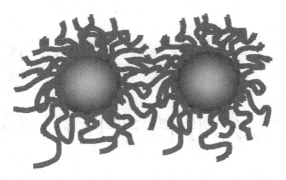

**FIGURE 11.10** Steric stabilization for adsorbed homopolymer molecules.

The interaction of two adsorbed polymer layers in close proximity depends on the polymer solubility (polymer segment-solvent interaction), the surface coverage (adsorption density), as well as the chain length. At low polymer concentrations and with high chain length, one polymer chain can adsorb on more than one particle, and thus flocculate the system. This is referred to as bridging flocculation. At high polymer concentrations, that is, full polymer surface coverage of adsorption sites on one particle, the polymer can form two conformations: mushroom or brush.[39-41] Figure 11.10 shows an adsorbed homopolymer.

When polymer-coated surfaces come together the layers are forced to penetrate into each other. The increase in segment density removes freedom of movement of the polymer chains, which results in a sudden drop in entropy. This is not a physically favorable process, so the particles cannot come closer than roughly twice the layer thickness. If the layer thickness is larger than the range of the van der Waals interaction, then it will act as an effective barrier, by preventing the coagulation, called steric stabilization. For layer thickness $\delta$ and separation distance $2\delta > d > \delta$, the repulsive interaction is due to the layer mixing opposition and is equal to[42]

$$V_{steric}^{mix}(d) = \frac{32\pi a k_B T \bar{\phi}_p (0.5 - \chi)}{5 v \delta^4} \left( \delta - \frac{d}{2} \right)^6 \qquad (11.23)$$

where $v$ is the molar volume, $\chi$ is the Flory-Huggins parameter, and $\bar{\phi}_p$ is the average volume fraction of segments in the adsorbed layer. At smaller separation distances, in addition to the mixing constraint, there is the elastic behavior of the layers, so the interaction is the summation of the two contributions:[23]

$$V_{steric}(d) = V_{steric}^{mix}(d) + V_{steric}^{elastic}(d) \qquad (11.24)$$

where

$$V_{steric}^{mix}(d) = \frac{4\pi\alpha\delta^2 k_B T \bar{\phi}_p (0.5-\chi)}{v}\left(\frac{d}{2\delta} - \frac{1}{4} - \ln\frac{d}{\delta}\right) \quad (11.25)$$

$$V_{steric}^{mix}(d) = \frac{2\pi\alpha\delta^2 k_B T \bar{\phi}_p \rho_p}{MW}\left\{\frac{d}{\delta}\ln\left[\frac{d}{\delta}\left(\frac{3-d/\delta}{2}\right)^2\right] - 6\ln\left(\frac{3-d/\delta}{2}\right) + 3\left(1-\frac{d}{\delta}\right)\right\}$$

$$(11.26)$$

There are five kinds of steric stabilizers:

- Homopolymers: Simple polymer chains that can adsorb on the surface and entropic repulsion stabilizes the particles. Branched polymers usually adopt more complicated conformations which results in a further entropy increase and more effective stabilization.
- Diblock copolymers: Where one block is strongly adsorbed on the surface and the other is extended into the solvent.
- Comb-like copolymers: A combination of two polymers, one that strongly adsorbs on the surface (train configuration) and acts as an anchored backbone for the other polymer to attach and extend into the solvent.
- Surfactants: Short chains consisting of an anchoring functional group and a short organic chain, usually hydrophobic or hydrophilic.
- Polyelectrolytes: Polymers that contain at least one ionizable group (i.e., carboxylic acid) such as polyacrylic acid (PAA).

In the last case, the repulsion is even greater due to the strong long-range EDL repulsion between the charged molecules, known as electrosteric stabilization.[43] In the case of high ionic strength, the polymer chains are neutral so they act like a polymer, and depending on the conditions they can have brush, pancake, or mushroom conformation. At low ionic strength, however, the charge is not screened and the electrostatic interaction among the chains pushes them to extend away from the surface, to brush conformation, or to adsorb flat in the pancake conformation if other interactions are stronger than the charge repulsion.

## F.  Depletion

The origin of the depletion forces is attributed to a concentration gradient that may exist in the solution from the region around the particles to the region between the particles.[44,45] This happens only for large particles (approximately 500 nm) and at the presence of a solute, known as depletant, usually polymer or nanoparticles (<<50 nm). At low concentrations the depletant molecules are excluded in the region between the particles, so the concentration is decreased

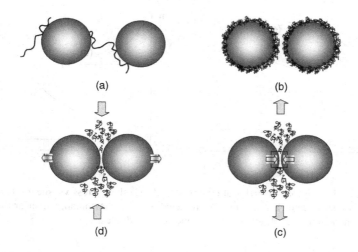

**FIGURE 11.11** Different interaction with homopolymer molecules. Cases (a) and (b) the molecule adsorbs on the particle and results flocculation, at low concentration and steric stabilization at high concentrations respectively. In cases (c) and (d) does not adsorb and results depletion flocculation, at low concentration, and depletion stabilization at high concentration respectively.

compared to the bulk solution. This gradient is pushing the particles together and the system is flocculated. At higher concentration, the solvent molecules are excluded and the depletant molecules are in higher concentration at the narrow region between the particles, and thus depletion stabilization occurs. For low polymer concentrations, flocculation occurs, and for higher concentrations, stabilization occurs, similar to the case of the steric repulsion as shown in Figure 11.11. In the depletion case, the flocculation to stabilization transition is possible, however, in the case of the steric forces, redispersion is not possible with a simple increase in the concentration.

## G. Rheology of Ceramic Slurries

Ceramic processing typically requires that a high solids loading low-viscosity slurry be prepared. The rheological behavior of such complex fluids is still very challenging to predict quantitatively, but general simple rules have been developed that guide the processing.[46] Complex fluids have a shear dependent flow behavior, as shown in Figure 11.12.

Flow behavior also depends on the solids loading, which is the volume percent of the particles in the solution. The flow of a system can be categorized into two types: viscous and viscoelastic. Viscous solutions are solutions where viscosity ($\eta$) depends on the shear stress ($\tau$) and the shear rate ($\dot{\gamma}$):

$$\tau = \eta\dot{\gamma}. \tag{11.27}$$

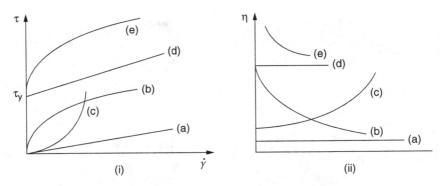

**FIGURE 11.12** Viscous behavior of complex fluids; (i) shear stress vs. shear rate and (ii) viscosity vs. shear rate. The notation for the curves is (a) Newtonian, (b) shear thinning, (c) shear thickening, (d) Bingham plastic, and (e) pseudoplastic.

For Newtonian fluids, the viscosity does not change with the applied stress, curve (a) in Figure 11.12. There is also the case where the viscosity decreases with increasing shear rate, and those fluids are called shear thinning (Figure 11.12, curve (b)). Sometimes the particle network can create an internal stress, as seen in curves (d) and (e). Finally, there is the case of shear thickening, where the viscosity of the fluid increases with increasing shear rate (c).[47]

Complex fluids are viscoelastic when the fluid still maintains internal stress after the external shear stress has ceased. The internal stress decays with time; the time required for the fluid to recover to the initial state is called the relaxation time. For this case the shear modulus ($G^*$) is a complex number:

$$G^* = G' + iG'' = G^* \cos \vartheta + G^* \sin \vartheta \qquad (11.28)$$

where $\vartheta$ is the phase between the applied strain and the stress in the fluid. The $G'$ and $G''$ can be calculated with the Maxwell model for viscoelastic materials. If the strain immediately results then the fluid behaves like a solid, meaning that it stores all the energy and the phase is zero. For the other extreme case where the phase is 90 then the fluid behaves like a viscous solution and the energy is dissipated.[48]

Several models have been developed to describe these phenomena quantitatively, the main difference being the interaction potential between the particles. There are two major approaches: the hard sphere and the soft sphere. The hard sphere assumes that the only interaction between particles is a strong repulsion at the point of contact. The soft sphere is more realistic and assumes a potential with a barrier and a primary minimum like in DLVO theory (Figure 11.8).

Hard sphere systems are characterized by viscous flow and for low solids loading (less than 5%) they can be described as Newtonian fluid. At higher loadings, cluster formation takes place and the fluid can acquire shear thinning or thickening behavior. The viscosity and solids loading are correlated with the

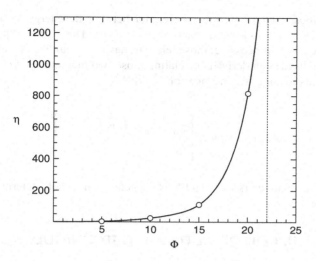

**FIGURE 11.13** Viscosity vs. volume solids loading. Krieger-Dougherty relation. For small loading (<5%) it is linear as predicted by Einstein.

Krieger-Dougherty relation:[49]

$$\eta = \eta_0 \left( 1 - \frac{\phi}{\phi_{max}} \right)^{-K\phi_{max}} \tag{11.29}$$

where $K$ is the hydrodynamic shape factor and $\phi_{max}$ is the maximum volume loading (Figure 11.13). For low solids loading, where there is no Brownian interaction and no hydrodynamic phenomena,[50] it approaches a linear relation, similar to the Einstein relation:[51]

$$\eta = \eta_0 (1 + 2.5\phi) \tag{11.30}$$

This is obvious in Figure 11.13, where for low solids loading the dependence can be assumed to be linear with a good approximation. Most ceramic colloids are characterized as soft sphere since the total energy of the slurry has a barrier and the primary minimum. The distance for the barrier from the surface is given by the Debye length for EDL stabilized systems and the thickness of the adsorbed layer for the sterically stabilized system. Particles with radius $R$, stabilized at distance $\delta$ behave as particles with a radius of about $(R + \delta)$. The layers can be accounted for in the solids loading by using the effective solids loading, which for small $\delta/R$ is[52]

$$\phi_{eff} = \phi \left( 1 + \frac{\delta}{R} \right)^3, \tag{11.31}$$

and then the slurry can be treated as slurry with the Krieger-Dougherty relation.

It is obvious that here is a direct correlation between the energy barrier created by the DLVO theory and the viscosity of the system. Due to the importance of viscosity in ceramic processing, those relations have been studied already. Empirical relations have been derived correlating those two magnitudes with temperature and solids loading dependence for $Al_2O_3$:[53]

$$\eta_r(\phi_0, T) = 20 \left[ \frac{\phi_0}{\phi_g} e^{-\frac{E_b}{kT} + 1.62 \left( 1.48 + \frac{T-90}{90} \right)} - 1 \right]^{2.4}, \qquad (11.32)$$

where $\eta_r$ is the relative viscosity and $\phi_g$ is the gelation threshold, a term explained later in the chapter.

## II. COLLOIDAL FORMING TECHNIQUES

### A. INTRODUCTION

The first section of this chapter has dealt with the theoretical aspects of colloidal processing of ceramics. Colloidially fabricated ceramics have several advantages over dry powder processed ceramics. These advantages are obtained by the wet processing route used in colloidal ceramic processing, which can produce ceramics with optimal particle packing and a homogenous microstructure. Therefore high-quality ceramics can be expected from colloidal processing methods.[54] A reduction in maximum flow size and an increased reliability has been demonstrated for colloidally processed $Si_3N_4$ ceramics.[55]

Colloidal processing techniques can be divided into two categories: drain casting and direct casting. Drain casting is widely used in the ceramic industry and includes techniques such as slip casting, pressure casting, and centrifugal casting. These techniques use a porous mold, which by capillary forces induces a liquid-solid separation to produce a solid green body. The capillary forces cause stress gradients and can lead to nonuniform densities of the green bodies and also give rise to limitations in producing complex-shaped thin-walled components. In addition, liquid flow will affect the suspension microstructure, leading to orientation of elongated particles such as whiskers or fibers. These difficulties can result in a degradation of properties and a decrease in reliability of the final product.[54] Therefore research has been focused on developing other alternatives for colloidal ceramic processing where ceramic bodies can be prepared directly from slurry without liquid removal. Such direct casting techniques have the advantages of colloidal processing while overcoming the problems of drain casting.

Direct casting can be classified into two categories—physical gelation or chemical gelation—on the basis of fundamental physical and chemical principles of the dispersing mechanism and gelation reaction. Physical gelation refers to a change in colloidal properties that cause a percolating particle network to form, gelling the entire slurry. Gelation is called chemical when the dispersing medium changes its properties due to a chemical reaction, typically causing an immobilization of the liquid or chemically consuming the liquid (see HAS process). Physical gelation techniques include temperature-induced forming (TIF), temperature-induced gelation (TIG), and direct coagulation casting (DCC). Chemical gelation techniques include hydrolysis-assisted solidification (HAS), freeze casting, and gel casting.[54,56]

## B. PREDICTION OF GEL STRENGTH

The interparticle bonding of the ceramic body can be defined as gel strength. The gel strength of ceramic green bodies is critical in the fabrication of crack-free demoldable ceramic bodies. If gel strength is too low, the ceramic particles of the body will not stay connected, and therefore during the demolding and handling procedures, the ceramic part will crumble. One way to predict the gel strength of ceramic bodies is to apply the use of percolation theory in ceramic colloidal systems. The basic idea of percolation theory is the existence of a sharp transition at which the long-range connectivity of the system disappears or goes in the opposite direction. This transition occurs abruptly when some generalized density in the system reaches a critical value (percolation threshold). This is the reason that there is no generalized theory on the gel strength, but individual works for specific systems.[57-60] Two different approaches for two different systems are

$$G = G_0 \left( \frac{\phi}{\phi_g} - 1 \right)^s \qquad (11.33)$$

$$G = G_0 (\phi - \phi_g)^s , \qquad (11.34)$$

where $s$ is constant, depending on the microstructure of the network, and $\phi_g$ is the gelation threshold. According to Equation 11.33 and Equation 11.34 ideal castable slurry should have a low gelation threshold and the highest possible solids loading (Figure 11.14). This volume fraction threshold should vary with the molecular weight of the bridging agent (chemical gelation) because of the increasing possibility of the interactions between dispersant chains and particles. In the absence of a bridging agent (physical gelation) the network is held together by van der Waals interactions. Optimizing the system in this case relies on increasing the effectiveness and the range of the force by varying the Hamaker constant, by changing solvent (Equation 11.5 and Equation 11.7), or the size of the particles (Equation 11.4).

## C. CHEMICAL GELATION TECHNIQUES: GELCASTING

The general principle of gelcasting is to have a three-dimensional polymer network surrounding a suspension of ceramic particles. The ceramic slurry contains

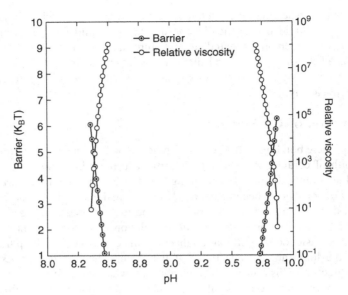

**FIGURE 11.14** Viscosity and separation barrier vs. the pH. The calculation is done with the use of Equation 11.34 for alumina particles of radius 100 nm and $10^{-3}$ M salt concentration. The maximum of the barrier and the minimum of the viscosity are at pH = 9.1. The data about the zeta potential as a function of pH are obtained from Figure 11.6.

particles that are dispersed by EDL or steric means typically using a dispersant for maximum solids loading. High solids loading is important for dense ceramics since it minimizes shrinkage during sintering. The polymer network is essentially a binder that is either polymerized or gelled. Through an initiation process, the polymer network will form and hold the ceramic particles together to form a demoldable dense green body, which is in the shape of a nonporous mold cavity. Molds can be fabricated out of metal or plastic materials and complex shapes can be attained.[61,62]

The gelcasting process was first developed in the 1960s for hard metals[63] and further developed for ceramic materials at the Oak Ridge National Laboratory (Oak Ridge, TN).[64] Figure 11.15 describes the gelcasting process. This variation of gelcasting involves a mixture of dispersed ceramic slurry with a monomer solution, which is poured into a mold and then polymerized. The monomer solution originally developed was acrylamide. The neurotoxic acrylamide monomer was then replaced with methacrylamide, which is less toxic. The mixture is then milled and later a polymerization initiator is added. A catalyst may be added to speed up the polymerization. The polymerization will permanently gel the liquid containing the trapped particles and retain the shape of the mold. After casting and gelation, the ceramic part can then be demolded and dried, leaving a compact powder with an open pore network. The drying time is typically the longest step in the gelcasting process. If drying occurs too quickly, warpage and cracking may occur. Humidity environments are controlled during drying to

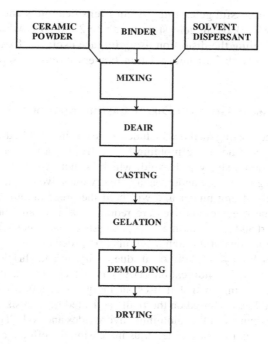

**FIGURE 11.15** Gelcasting process flowchart.

minimize cracking and warping. The green body can then undergo standard binder burnout and sintering procedures.[58]

Since then, different alternative techniques of gelcasting have been explored. The main emphasis has been to develop gelcasting processes with different binders that polymerize or gel through either a chemical reaction process[65–70] or a thermal process (heating or cooling). Researchers have successfully demonstrated that naturally occurring materials such as agarose,[71,72] agar,[73,74] carrageenan,[75,76] and gelatin[77,78] can be used as a gelcasting binder through a cooling mechanism. These materials are of interest since they are naturally occurring, inexpensive, and nontoxic. Other materials such as albumin have been used as a gelling binder that gels upon heating.[79–81]

Gelcasting has several advantages over other traditional processes such as slip casting. Gelcasting typically has a more uniform powder packing and is a 50% faster process than slip casting. Green strengths obtained from gelcasting are quite high, which is advantageous in green machining and handling. In addition, reusable nonporous metal or plastic molds can be used, which can provide an economic cost advantage.[56,58]

There are still several technical challenges with gelcasting. Gelcast components can be difficult to demold and may adhere to the mold surface. A gelcasting system requires a suitable combination of mold materials and demolding release agents to ensure part removal. This selection often can be the most delicate part of developing a successful gelcasting system. Densification is another challenge with gelcasting. Shrinkage in gelcast parts can be 20 to 30%, which can lead to distortions. Controlling the distortion using furnace cycles or fixtures can be an issue. Also, typically the organic gelation binders require a separate burnout step.[4,82,83]

## D. OTHER CHEMICAL GELATION DIRECT CASTING METHODS

Several other direct casting methods include freeze casting,[84–87] hydrolysis-assisted solidification,[88–92] and adiabatic molding.[93] The freeze casting process involves the casting of ceramic slurry in a mold, which is then frozen and undergoes a sublimation drying procedure under vacuum to dry the solvent. The frozen solvent acts as a relatively strong binder and will hold the green ceramic part together. Therefore minimal organic additives are required, which can lead to a higher-purity ceramic part and elimination during processing of an organic binder burnout step. Removing the solvent through a sublimation procedure eliminates cracking and warping, which may normally occur due to drying and shrinkage.

Hydrolysis-assisted solidification (HAS) is a technique that is somewhat similar to freeze casting in that it uses an inorganic reaction to consume the dispersing media. HAS is based on thermally activated hydrolysis of AlN powder in an aqueous medium, yielding aluminum hydroxides and $NH_3$. The process gels a castable slurry by consuming water, thus increasing the effective solids loading and by undergoing a pH increase due to the formation of ammonia during hydrolysis of AlN. Both of these increase the viscosity of the slurry, yielding the formation of a solid ceramic part.

High green strengths can be achieved since the HAS process precipitates aluminum hydroxides. The high strength allows the production of complex shapes such as thin-walled shapes and gives low shrinkage on drying. One drawback of the HAS process is that the hydrolysis reaction yields $Al_2O_3$ in the sintered part, thus limiting its applications to systems where this is not a concern.

Adiabatic molding is a novel water-based process that involves the phase transformation of water from solid to liquid at $-10°C$ by a change in pressure. Examples of the materials used in this process are $Al_2O_3$ powder, polyvinyl alcohol, and water. These ingredients are mixed to make a dough, which is then placed in the form of a pellet into the top of a forming die. The die is then equilibrated to $-10°C$, and later, as pressure is applied to the pellet during a pressing operation, the ice is melted and a part is formed in the die. As the pressure is released, the water solidifies and the part is ejected out of the die.

## E. DIRECT COAGULATION CASTING

The DCC process involves an EDL-stabilized ceramic slurry that is destabilized by a time-delayed reaction involving enzyme catalysts.[94–99] The enzymes trigger chemical reactions that can be used to either increase the ionic salt content at a constant pH or shift the pH to the IEP. An increase in the salt concentration will cause a compression of the EDL surrounding the ceramic particles, resulting in controlled coagulation. By shifting the pH of the ceramic slurry to the IEP, a controlled coagulation is achieved. Figure 11.9 illustrates this in a colloidal stability diagram by changing the pH or salt concentration. The rheological properties of a concentrated $Al_2O_3$ suspension can be transformed from a dispersed, fluid state into a coagulated, rigid state.

The ceramic slurry used in DCC consists of a high solids loaded suspension of ceramic particles such as $Al_2O_3$ with an enzyme catalyst. An example of an enzyme-catalyzed reaction is shown in Equation 11.36,

$$(NH_2)_2CO + 2H_2O \xrightarrow{urease} 2NH_4^+ + CO_3^{-2} \qquad (11.35)$$

which states that the decomposition of urea by urease yields diammonium carbonate with a buffer regime at pH = 9. The reaction kinetics depend on conditions such as temperature, pH, and urease (catalyst) concentration. The reaction is inhibited because of enzyme adsorption on $Al_2O_3$ for low solids loading suspensions and, to a greater degree, because of hindered diffusion for concentrated suspensions. The reaction will start after the last ingredient is added to the slurry, and the slip must be cast. The process requires several minutes for consolidation and temperature can be used to control this time to a certain extent. Alkaline swellable thickeners can be used to increase the elasticity and strength of the wet green bodies.[93] Figure 11.16 describes a processing flowchart for DCC.

Advantages of DCC are that it can be used to fabricate complex-shaped ceramic components with good mechanical strengths at low costs. DCC allows for casting at room temperature and can use inexpensive reusable metal or plastic molds. This technique is also independent of the size and wall thickness of the component. In addition, since small amounts of organic compounds (dispersants and enzymes) are used, typically a separate burnout step is not required.[93,94]

## F. TEMPERATURE-INDUCED FORMING AND TEMPERATURE-INDUCED GELATION

Temperature-induced forming (TIF) involves an EDL-stabilized slurry that is flocculated by temperature control.[53,100–104] TIF utilizes the temperature dependency of the solubility to gel the system. EDL-stabilized slurries can be made at room temperature, which at 80°C, the increased ionic strength will cause flocculation. The process can be enhanced by polymer bridging (Figure 11.11a). At room temperature, TIF makes use of a low molecular weight dispersant, which is absorbed and induces a high surface charge for stabilization. With increasing temperatures, TIF uses the increasing dissolution of the ceramic powder to exchange the lower molecular weight dispersant with a higher molecular weight polymer. Controlling the amount of the high molecular weight polymer below the level necessary for stabilization will yield gelation via bridging flocculation.[99] Normally, stronger gelation will result by using a high molecular weight polymer.

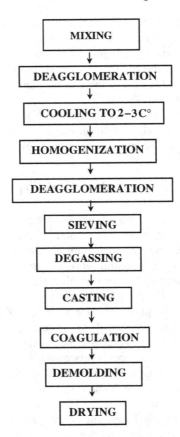

**FIGURE 11.16** DCC process flowchart.

After drying, the precipitation of dissolved materials at particle contacts increases the green strength and aids in sintering.[100]

Figure 11.17 describes the TIF gelation mechanism. Components of this slurry are ceramic powder ($Al_2O_3$), solvent (water), dispersant (citric acid), and high molecular weight polymer (polyacrylic acid). At room temperature, the slurry is initially EDL stabilized and negatively charged polyacrylic acid does not adsorb on the surface of the $Al_2O_3$ because of full surface coverage of the citrate anions. At elevated temperature, the increasing solubility of $Al_2O_3$ in water weakens the EDL by salt increase. In addition, desorption of citrate anions from the $Al_2O_3$ surface will also occur, resulting in adsorption sites for polyacrylic acid chains. Controlling the amount of polyacrylic acid will ensure low surface coverage and cause bridging flocculation.[102]

Temperature-induced forming has several advantages when compared to conventional slip casting techniques. TIF utilizes a gelation technique for consolidation, and in principle will produce green ceramic parts with less stress when compared

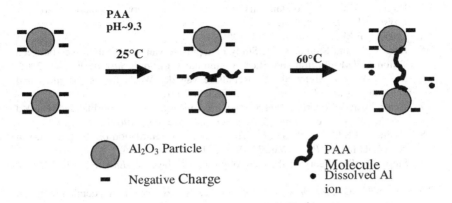

**FIGURE 11.17** Proposed Gelation Mechanism for TIF.

with drain casting methods. In addition, TIF uses only temperature to control the rheology of the slurries. The bridging flocculation gelation mechanism of TIF is a physical gelation process, which indicates that the degree of gelation can be modified by altering processing parameters such as temperature and time. This adjustability allows for building ceramic parts layer by layer without delamination issues.[100,102]

Temperature-induced gelation involves flocculation of sterically stabilized slurries with changes in temperature.[105] An example of a TIG slurry includes the following components: $Al_2O_3$ and $SiC\text{-}Al_2O_3$ ceramic whisker composite, pentanol solvent, and a dispersant—an amphiphilic polymer (Hypermer KD3). As the solvency decreases to a critical level, the adsorbed polymer layer will collapse, which will induce flocculation when the van der Waals forces overcome the remaining steric repulsive forces. The solvency can be decreased by a temperature change or addition of a nonsolvent. At room temperature, this slurry will transform from a dispersed state to a flocculated state.

## III. CONCLUSION

The understanding of the underlying fundamental physics and chemistry of processing continually advances. This chapter introduced fundamental concepts of physics and showed how they apply to colloidal processing. The understanding helped to improve many colloidal forming techniques including slip casting and pressure slip casting, which is widely used in industry, and allowed the development of novel forming methods like HAS, DCC, and TIF.

## REFERENCES

1. van der Waals, J.D., The equation of state for gases and liquids, Nobel Lecture, December 12, 1910.
2. Eisenschitz, R. and London, F., The relation between the van der Waals forces and the homeopolar valence forces, *Zeits. Physik.*, 60, 491, 1930.

3. Hamaker, H.C., The London van der Waals attraction between spherical particles, *Physica IV*, 10, 1048, 1937.
4. Lifshitz, E.M., *Sov. Phys.*, 2, 73, 1956.
5. Lee, S.-W. and Sigmund, W., Study of repulsive van der Waals forces between Teflon AFTM thin film and silica or alumina, *J. Colloids Surf. A*, 204, 43, 2002.
6. Lee, S.-W. and Sigmund, W., Repulsive van der Waals forces for silica and alumina, *J. Colloid Interface Sci.*, 243, 365, 2001.
7. Tabor, D. and Winterton, R.H.S., The direct measurement of normal and retarded van der Waals forces, *Proc. R. Soc. Series A*, 312, 435, 1969.
8. Casimir, H.B.G. and Polder, D., The influence of retardation on the London-van der Waals forces, *Phys. Rev.*, 73, 360, 1948.
9. Stern, O., The theory of the electrolytic double layer, *Z. Elektrochem.*, 30, 508, 1924.
10. Israelachvili, J., *Intermolecular and Surface Forces*, 2nd ed., Academic Press, San Diego, 214, 221–223, 1992.
11. Helmholtz, H.L.F., *Ann. Physik.*, 7, 337, 1879.
12. Gouy, G., About the electric charge on the surface of an electrolyte, *J. Phys. A*, 9, 457, 1910.
13. Chapman, D., A contribution to the theory of electrocapillarity, *Phil. Mag.*, 25, 475, 1913.
14. Israelachvili, J., *Intermolecular and Surface Forces*, 2nd ed., Academic Press, San Diego, 239, 1992.
15. Debye, P. and Huckel, E., The interionic attraction theory of deviations from ideal behavior in solution, *Z. Phys.*, 24, 185, 1923.
16. Bhattacharjee, S., Elimelech, M., and Borkovecb, M., DLVO interaction between colloidal particles: beyond Derjaguin's approximation, *Creatica Chem. Acta*, 71, 883, 1998.
17. Israelachvili, J., *Intermolecular and Surface Forces*, 2nd ed., Academic Press, San Diego, 230, 235–236, 243, 1992.
18. Derjaguin, B.V. and Landau, L.D., Theory of stability of highly charged lyophobic sols and adhesion of highly charged particles in solutions of electrolytes, *Acta Physicochim. URSS*, 14, 633, 1941.
19. Verwey, E.J.W. and Overbeek, J.T.G., *Theory of Stability of Lyophobic Colloids*, Elsevier, Amsterdam, 1948.
20. von Smoluchowski, M., *Phys. Z.*, 17, 557, 1916.
21. von Smoluchowski M., *Phys. Z.* 17, 585, 1916.
22. von Smoluchowski M., *Z. Phys. Chem. (Lpz)*, 92, 129, 1917.
23. Myers, D., *Surfaces, Interfaces, and Colloids: Principles and Applications*, John Wiley & Sons, New York, 1999.
24. Lange, F.F., Powder processing science and technology for increased reliability, *J. Am. Ceram. Soc.*, 72, 3, 1989.
25. Biggs, S. and Healy, T.W, Electrosteric stabilization of colloidal zirconia with low molecular weight polyacrylic acid, *J. Chem. Soc. Faraday Trans.*, 90, 3415, 1994.
26. Schultz, R.C. and Flory, J.P., *J. Polymer Sci.*, 15, 231, 1955.
27. Dondos, A. and Benoit, H., *Macromol. Chem.*, 133, 119, 1970.
28. Brochard, F. and de Gennes, P.G., *Ferroelectrics*, 30, 33, 1980.
29. Magda, J.J., Fredrickson, G.H., Larson, R.G., and Helfand, E., Macromolecules, 21, 726, 1988.
30. Vilgis, T.A., Sans, A., and Jannink, G., *J. Phys. II France*, 3, 1779, 1993.

31. Stapper, M. and Vilgis, T.A., *Europhys. Lett.*, 1, 7, 1998.

32. Vilgis, T.A., Polymer theory: path integrals and scaling, *Phys. Rep.*, 336, 167, 2000.

33. de Gennes, P.G., *Phys. Lett. A*, 38, 339, 1972.

34. de Gennes, P.G., *Scaling Concepts in Polymer Physics*, Cornell University Press, Ithaca, NY, 1979.

35. Pedersen, J.S., Laso, M., and Schurtenberger, P., A Monte Carlo study of excluded volume effects in worm-like micelles, *Phys. Rev. E*, 54, R5917, 1996.

36. Jerke, G., Pedersen, J.S., Egelhaaf, S.U., and Schurtenberger, P., Flexibility of charged and uncharged polymer-like micelles, *Langmuir*, 14, 6013, 1998.

37. Pedersen, J.S. and Schurtenberger, P., Static properties of semidilute solutions of polystyrene: a comparison of Monte Carlo simulations and small-angle neutron scattering results, *Europhys. Lett.*, 45, 666, 1999.

38. Cannavacciuolo, L., Sommer, C., Pedersen, J.S., and Schurtenberger, P., Size, flexibility, and scattering functions of semiflexible polyelectrolytes with excluded volume effects: Monte Carlo simulations and neutron scattering experiments, *Phys. Rev. E*, 62, 5409, 2000.

39. Bird, R.B., Curtiss, C.F., Armstrong, R.C., and Hassager, O., *Dynamic of Polymeric Liquids*, Vol. 2, *Kinetic Theory*, 2nd ed., John Wiley & Sons, New York, 1987.

40. Doi, M. and Edwards, S.F., *The Theory of Polymer Dynamics*, Clarendon Press, Oxford, 1986.

41. Flory, P.J., *Statistical Mechanics of Chain Molecules*, Hanser Publisher, Munich, 1969.

42. Vincent, B., Edwards, J., Emmett, S., and Jones, A., Depletion flocculation in dispersions of sterically stabilized particles ("soft spheres"), *Colloids Surf.*, 18, 261, 1986.

43. Dobrynin, A.V., Colby, R.H., and Rubinstein, M., Scaling theory of polyelectrolyte solutions, *Macromolecules*, 28, 1859, 1995.

44. Mao, Y., Cates, M.E., and Lekkerkerker, H.N.W., Depletion force in colloidal systems, *Phys. A*, 222, 10, 1995.

45. Mao, Y., Cates, M.E., and Lekkerkerker, H.N.W., Theory of the depletion force due to rod-like polymers, *J. Chem. Phys.*, 106, 3721, 1997.

46. Bergstrom, L., Rheology of concentrated suspension, in *Surface and Colloidal Chemistry in Advanced Ceramic Processing*, R.J. Pugh and L. Berstrom, Eds., Marcel Dekker, New York, 1994.

47. Barnes, H.A., Hutton, J.F., and Walters, K., *An Introduction to Rheology*, Elsevier, Amsterdam, 1989.

48. Russell, W.B., Concentrated colloidal dispersions, *MRS Bull.*, 16, 27, 1991.

49. Krieger, I.M., Rheology of monodisperse lattices, *Adv. Colloid Interface Sci.*, 3, 111, 1972.

50. Brown, R., A brief account of microscopical observations made in the months of June, July, and August, 1827, on the particles contained in the pollen of plants; and on the general existence of active molecules in organic and inorganic bodies, *Phil. Mag.*, 4, 161, 1828.

51. Einstein, A., Über die von der molekularkinetischen Theorie der Wärme geforderte Bewegung von in ruhenden Flüssigkeiten suspendierten Teilchen, *Ann. Phys.*, 17, 549, 1905.

52. Ogden, A.L. and Lewis, J.A., Effect of nonadsorbed polymer on the stability of weakly flocculated nonaqueous suspensions, *Langmuir*, 12, 3413, 1996.

53. Yang, Y.-P. and Sigmund, W., Effect of particle volume fraction on the gelation behavior of the temperature induced forming (TIF) aqueous alumina suspensions, *J. Am. Ceram. Soc.*, 84, 2138, 2001.

54. Sigmund, W.M., Bell, N.S., and Bergstrom, L., Novel powder-processing methods for advanced ceramics, *J. Am. Ceram. Soc.*, 83, 1557, 2000.

55. Pujari, V.K., Tracey, D.M., Foley, M.R., Paille, N.I., Pelletier, P.J., Sales, L.C., Wilkens, C.A., and Yeckley, R.L., Reliable ceramics for advanced heat engines, *Am. Ceram. Soc. Bull.*, 74, 86, 1995.

56. Tari, G., Gelcasting ceramics: a review, *Am. Ceram. Soc. Bull.*, 82, 43, 2003.

57. Rueb, C.J. and Zukoski, C.F., Viscoelastic properties of colloidal gels, *J. Rheol.*, 41, 197, 1997.

58. Stauffer, D., *Introduction to Percolation Theory*, Taylor & Friends, London, 1985.

59. Yanez, J.A., Laarz, E., and Bergström, L., Viscoelastic properties of particle gels, *J. Colloid Interface Sci.*, 209, 162, 1999.

60. Yang, Y. and Sigmund, W.M., Expanded percolation theory model for the temperature induced forming (TIF) of alumina aqueous suspensions, *J. Eur. Ceram. Soc.*, 22, 1791, 2002.

61. Omatete, O.O., Janney, M.A., and Strehlow, R.A., Gelcasting—a new ceramic forming process, *Am. Ceram. Soc. Bull.*, 70, 1641, 1991.

62. Omatete, O.O., Janney, M.A., and Nunn, S.D., Gelcasting: from laboratory development toward industrial production, *J. Eur. Ceram. Soc.*, 17, 407, 1997.

63. Golibersuch, E.W., Method of making cemented carbide articles and articles produced thereby, Can. Patent no. 624,217, 1962.

64. Omatete, O.O. and Janney, M.A., Method for molding ceramic powders using a water-based gel casting, U.S. Patent no. 5,028,362, 1991, and U.S. Patent no. 5,145,908, 1992.

65. Janney, M.A., Omatete, O.O., Walls, C.A., Nunn, S.D., Ogle, R.J., and Westmoreland, G., Development of low-toxicity gelcasting systems, *J. Am. Ceram. Soc.*, 81, 581, 1998.

66. Morissette, S., Lewis, J., and Cesarano, J., Solid freeform fabrication of aqueous alumina-poly(vinyl alcohol) gelcasting suspensions, *J. Am. Ceram. Soc.*, 83, 2409, 2000.

67. Ma, J., Xie, Z., Miao, H. et al., Gelcasting of alumina ceramics in the mixed acrylamide and polyacrylamide systems, *J. Eur. Ceram. Soc.*, 23, 2273, 2003.

68. Cai, K., Huang, Y., and Yang, J., Gelcasting of alumina with low-toxicity HEMA system, *J. Inorg. Mater.*, 18, 343, 2003.

69. Ma, J., Xie, Z., Miao, H. et al., Elimination of surface spallation of alumina green bodies prepared by acrylamide-based gelcasting via poly(vinylpyrrolidone), *J. Am. Ceram. Soc.*, 86, 266, 2003.

70. Ma, J., Me, Z., Miao, H. et al., Gelcasting of ceramic suspension in acrylamide polyethylene glycol systems, *Ceram. Int.*, 28, 859, 2002.

71. Ewais, E., and Ahmed, Y., Consolidation of silicon carbide in aqueous medium based on gelation of agarose, *Br. Ceram. Trans.*, 101, 255, 2002.

72. Millan, A., Baudin, C., Moreo, R. et al., Influence of gelling additives in the green properties of $Al_2O_3$ bodies obtained by aqueous gel casting, *Key Eng. Mater.*, 206, 413, 2002.

73. Olhero, S., Tari, G, Coimbra, MA, et al., Synergy of polysaccharide mixtures in gelcasting of alumina, *J. Eur. Ceram. Soc.*, 20, 423, 2000.

74. Millan, A., Moreno, R., and Nieto, M., Thermogelling polysaccharides for aqueous gelcasting—Part 1: a comparative study of gelling additives, *J. Eur. Ceram. Soc.*, 22, 2209, 2002.

75. Santacruz, I., Nieto, M., and Moreno, R., Rheological characterization of synergistic mixtures of carrageenan and locust bean gum for aqueous gelcasting of alumina, *J. Am. Ceram. Soc.*, 85, 2432, 2002.

76. Santacruz, I., Baudin, C., Nieto, M. et al., Improved green properties of gelcast alumina through multiple synergistic interaction of polysaccharides, *J. Eur. Ceram. Soc.*, 23, 1785, 2003.

77. Chen, Y., Xie, Z., Yang, J. et al., Alumina casting based on gelation of gelatine, *J. Eur. Ceram. Soc.*, 19, 271, 1999.

78. Vandeperre, L., De Wilde, A., and Luyten, J., Gelatin gelcasting of ceramic components, *J. Mater. Process Tech.*, 135, 312, 2003.

89. Schilling, C., Tomasik, P., Li, C. et al., Protein plasticizers for aqueous suspensions of micrometric- and nanometric-alumina powder, *Mater. Sci. Eng A Struct.*, 336, 219, 2002.

80. Garrn, I., Reetz, C., Brandes, N. et al., Clot-forming: the use of proteins as binders for producing ceramic foams, *J. Eur. Ceram. Soc.*, 24, 579, 2004.

81. Lyckfeldt, O., Brandt, J., and Lesca, S., Protein forming—a novel shaping technique for ceramics, *J. Eur. Ceram. Soc.*, 20, 2551, 2000.

82. Millan, A., Nieto, M., Baudin, C. et al., Thermogelling polysaccharides for aqueous gelcasting—Part II: influence of gelling additives on rheological properties and gelcasting of alumina, *J. Eur. Ceram. Soc.*, 22, 2217, 2002.

83. Zhou, L., Huang, Y., and Xie, Z., Gelcasting of concentrated aqueous silicon carbide suspension, *J. Eur. Ceram. Soc.*, 20, 85, 2000.

84. Novich, B., Sundback, C., and Adams, R., Quickset™ injection molding of high-performance ceramics; *Ceram. Trans.*, 26, 157, 1992.

85. Sofie, S., and Dogan, F., Freeze casting of aqueous alumina slurries with glycerol, *J. Am. Ceram. Soc.*, 84, 1459, 2001.

86. Jones, R., Near net shape ceramics by freeze casting, *Ind. Ceram.*, 20, 117, 2000.

87. Statham, M., Hammett, E., Harris, B. et al., Net-shape manufacture of low-cost ceramic shapes by freeze-gelation, *J. Sol-Gel Sci. Technol.*, 13, 171, 1998.

88. Kosmac, T., Novak, S., and Sajko, M., Hydrolysis-assisted solidification (HAS): a new setting concept for ceramic net-shaping, *J. Eur. Ceram. Soc.*, 17, 427, 1997.

89. Novak, S., Kosmac, T., Krnel, K. et al., Principles of the hydrolysis assisted solidification (HAS) process for forming ceramic bodies from aqueous suspension, *J. Eur. Ceram. Soc.*, 22, 289, 2002.

90. Kosmac, T., Novak, S., and Krnel, K., Hydrolysis assisted solidification process and its use in ceramic wet forming, *Z. Metallkd.*, 92, 150, 2001.

91. Kosmac, T., The potential of the hydrolysis assisted solidification (HAS) process for wet forming of engineering ceramics, *Key Eng. Mater.*, 2, 357, 1999.

92. Krnel, K., and Kosmac, T., Use of hydrolysis-assisted solidification (HAS) in the formation of $Si_3N_4$ ceramics, *Mater. Sci. Forum*, 413, 75, 2003.

93. King, A., and Keswani, S., Adiabatic molding of ceramics, *Am. Ceram. Soc. Bull.*, 73, 96, 1994.

94. Gauckler, L., Graule, T., and Baader, F., Ceramic forming using enzyme catalyzed reactions, *Mater. Chem. Phys.*, 61, 78, 1999.

95. Graule, T., Gauckler, L., and Baader, F., Direct coagulation casting, a new green shaping technique, Part I: processing principles, *Ind. Ceram.*, 16, 31, 1996.

96. Kosmac, T., Near-net-shape of engineering ceramics: potential and prospects of aqueous injection molding (AIM), in Proceedings of NATO ARW on Engineering Ceramics '96, G.N. Batini, M. Havier, and P. Sejgalik, Eds., Kluwer, New York, 1997.

97. Si, W., Graule, T., Baader, F. et al. Direct coagulation casting of silicon carbide components, *J. Am. Ceram. Soc.*, 82, 1129, 1999.

98. Balzer, B., Hruschka, M., and Gauckler, L., Coagulation kinetics and mechanical behavior of wet alumina green bodies produced via DCC, *J. Colloid Interface Sci.*, 216, 379, 1999.

99. Wei, L., Zhang, H., Jin, Y. et al., Rapid coagulation of silicon carbide slurry via direct coagulation casting, *Ceram. Int.*, 30, 411, 2004.

100. Bell, N., Wang, L., Sigmund, W., and Aldinger, F., Temperature-induced forming: application of bridging flocculation to near-net-shape part production of ceramics, *Z. Metallkd.*, 90, 388, 1999.

101. Aldinger, F., Sigmund, W., and Yanez, J., Formgebungsmethode für Keramiken und Metalle in wässrigen Systemen mittels Temperaturänderung, German Patent no. 19,751,696.3, 1998.

102. Wang, L., and Aldinger, F., Enhanced gelation by cations in temperature-induced forming, *Adv. Eng. Mater.*, 2, 821, 2000.

103. Yang, Y., and Sigmund, W., Preparation, characterization, and gelation of temperature-induced forming (TIF) alumina slurries, *J. Mater. Synth. Process.*, 9, 103, 2001.

104. Yang, Y., and Sigmund, W., Rheological properties and gelation threshold of temperature induced forming (TIF) alumina suspensions with variation in molecular weight of polyacrylic acid, *J. Mater. Synth. Process.*, 10, 249, 2002.

105. Bergström, L., and Sjostroum, E., Temperature-induced flocculation of concentrated ceramic suspensions: rheological properties, *J. Eur. Ceram. Soc.*, 19, 2117, 1999.

106. Bergström, L., Hamaker constants of inorganic materials, *Adv. Colloid Interface Sci.*, 70, 125, 1997.

# 12 Nano/microstructure and Property Control of Single and Multiphase Materials

*Philippe Colomban*

## CONTENTS

# I. INTRODUCTION

One of the key steps in the advancement of ceramic processing is the achievement of a special, controlled and reproducible micro(nano)structure. Reducing the dimension of a "meso/macroscopic" phase down to the nanometer scale produces moieties in which interfacial volumes dominate. This results in absolutely unique properties. Sol-gel routes, and more generally synthesis through liquid precursors, offer the possibility to control the structure at the scale of the precursor and its aggregates (1 to 100 nm). This allows the preparation of more homogeneous materials than the usual ball-milling process. Materials made with (or from) nanophases have received considerable attention in the last few years, but their characterization is not easy. A large variety of nanomaterials are now being developed, if not already commercially available, for applications taking advantage of their (1) optical properties (pigments in traditional ceramics and for the cosmetics industry, fluorescent markers, quantum dots, photonic crystals for multiplexing and switching in optical networks, quantum computer components, etc.); (2) mechanical properties (fibers, wear-resistant, anticorrosion, and cutting coatings, "nanopolishing" SiC, diamond, boron carbide powders, high impact strength nanocomposites, etc.); (3) magnetic properties (data storage, reading heads, giant magnetoresistance [GMR] materials, etc.); (4) high specific surface area (propulsion, filters, nanosensors, semiconductor nanowires, catalysts, etc.); (5) electrical properties (miniaturized silicon chips, single electron transistors, carbon nanotubes or even silicon nanotubes transistors, etc.); and (6) biocompatibility (*in vivo* drug delivery, diagnostic and monitoring devices, etc.). Most of these materials are ceramics synthesized through a route involving a liquid step. In fact, potters and ceramists have been using such nanoscience (clays form a "physical" gel; Figure 12.1a) for thousands of years in clay-based ceramics: most of the properties of traditional ceramics (plasticity, shapability, rheology, low sintering temperature, etc.) are directly related to the fact that natural clay—the raw material—is nanosized and the so-called sol-gel route only results from the transposition of the clay route to simpler oxide, carbide, etc. compositions.[1,2] In the last 10 centuries, nanosized ceramic pigment powders were also created to color enamels and glass.[3] Because of human eye sharpness, pigment must have a particle size close to 100 nm to give homogeneous coloration.

The challenge for ceramists is to achieve perfect control on related properties. This obviously requires correlating the parameters of the synthesis process with the resulting nano/microstructure. This chapter focuses on the ways to prepare multiphase materials and their main controlling parameters, on the precursors and their structure, and on the porosity, densification, and final mechanical properties. The goal here is not to give a complete overview, but to address the most important points and the ways to solve most of the problems with some examples. Particular attention will be given to nondestructive methods able to characterize the materials over different scales (from atomic distances to full size).

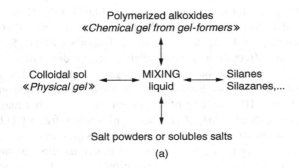

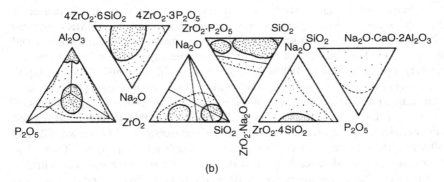

**FIGURE 12.1** (a) Schematic flowchart of hybrid routes. (b) Sol-gel forming compositions (alkoxide route) in the $SiO_2$-$Al_2O_3$-$ZrO_2$-$P_2O_5$, $4ZrO_2 \cdot 6SiO_2$-$Na_2O$-$4ZrO_2 \cdot 3P_2O_5$, $ZrO_2$-$Na_2O$-$SiO_2$, $ZrO_2$-$P_2O_5$, $ZrO_2 \cdot Na_2O$-$SiO_2$, $ZrO_2 \cdot 4SiO_2$-$Na_2O$-$P_2O_5$, and $SiO_2$-$P_2O_5$-$Na_2O \cdot CaO \cdot 2Al_2O_3$ phase diagrams. The representation corresponds to the final oxide composition. Solid line indicates the limit of optically clear gel by slow hydrolysis. Dashed line indicates the limit of easy synthesis of (translucent) monolithic gels. (Reprinted from Colomban, P., Gel technology in ceramics, glass-ceramics and ceramic-ceramic composites, *Ceram. Int.*, 15, 23, 1989. With permission from Elsevier.)

## II. PRECURSORS AND ANALYSIS METHODS

Precursors must have different properties:[4–14] (1) a high content of the final elements (mostly aluminum, silicon, zirconium, titanium, phosphorus), (2) a low content of health hazardous elements and elements that corrode the equipment (e.g., chlorine, sulfur), (3) a viscosity adapted to the process: low viscosity for preform infiltration, medium viscosity for spinning and coating, (4) a controlled precursor-ceramic transformation (bubbling is researched for foams but not for dense parts), (5) the ability to be mixed with other precursors or to be processed ("good" hydrolysis rate), and (6) low cost.

Regarding sol-gel routes, the most used reagents are those made with gel-former compositions: aluminium-s-butoxide $[Al(OC_4H_9)_3]$, tetraethoxysilane $[Si(OC_2H_5)_4]$ (and homologous methoxysilane), aluminum-silicon ester

$[(OC_4H_9)_2$-Al-O-Si$(OC_2H_5)_3]$, zirconium-i-propoxide $[Zr(OC_3H_7)_4]$, titanium butoxide $[Ti(OC_4H_9)_4]$ and propoxide $[Ti(OC_3H_7)_4]$, tributylborate $[B-(OC_4H_9)_3]$, and tributylphosphate $[P-(OC_4H_5)_3]$.[6–17] Figure 12.1b shows the basic $SiO_2$-$Al_2O_3$-$ZrO_2$-$P_2O_5$ quaternary diagram and some related ternary diagrams. The dotted regions correspond to composition ranges easily giving optically clear (high dot density) or translucent (low dot density) gels. The dashed line corresponds to the limit of easy synthesis. The handling of aluminum and some other reagents like germanium propoxide $[Ge(OC_3H_7)_4]$ requires a glove box free of $H_2O$ traces. Homemade preparations of very hygroscopic alkoxides guarantees their quality. However, in the preparation of multicomponent materials through the sol-gel route or by mixing organic precursors, determination of the metal element content is not obvious.[7] In commercially available reagents, the metal content often varies from batch to batch and a slight prehydrolysis is often observed. The degree of hydrolysis can be determined by nuclear magnetic resonance (NMR), infrared (IR) absorption, or Raman scattering.[14,16–18] This modifies the metal content and the viscosity. The viscosity of the alkoxides can be adjusted at values of approximately 1 poise by heating below 80°C for most of the reagents. Mixing of liquid precursors promotes the intimate combination of various precursors at the molecular scale and allows adjustment of the viscosity to void filling requirements.

One of the main interests of liquid alkoxides is that they exhibit a rather good ceramic yield: typically the alkoxide-to-ceramic conversion yield is between 20 and 30 wt%. Thermal treatments under various atmospheres ($NH_3$, $H_2S$, CO) lead to (oxy)-nitrides, -sulfides, -carbides, and derivatives.[1,19] However, carbides, nitrides, and their oxyderivatives are generally prepared by cross-linking and ceramization of specific precursors: polycarbosilanes, polysilazanes. Since the work of Yajima et al.,[20,21] polycarbosilane (PCS) is well known as a precursor of SiC materials.[22,23] The high viscosity of this precursor is convenient for fiber spinning, but prevents its use to fill a fiber preform already filled with a submicrometer powder. Unfortunately the oxygen-driven reticulation of the chains of the precursor, convenient for films and fiber, is not obvious inside a thick sample, which makes it difficult to control the stoichiometry in the whole. Polyvinylsilane (PVS) fulfills the specific requirements for the infiltration of a porous body:[22] (1) a thermostat behavior with a cross-linking temperature (200°C to 300°C) far enough from the beginning of pyrolysis (400°C) to avoid formation of bubbles during the cycle; (2) a relatively low molecular weight (MW < 1000), which ensures a sufficiently low viscosity (e.g., 0.3 poise at 130°C) to allow injection in the fiber preform at "low" temperatures; (3) a very high ceramic yield: 64% in weight; and (4) a silicon/carbon stoichiometry leading to air-stable materials. A variety of precursors of carbide, nitrides, and their mixture are now available for many applications. Last-generation precursors use $\gamma$ radiation to develop an intrinsic cross-linking between precursor chains instead of the extrinsic Si-O-Si bridging obtained by controlled oxidation.

Generally, the main questions that need an answer are:

Is the expected composition and structure achieved? Most of the materials prepared through liquid precursor routes are first obtained in an amorphous or metastable form. These forms are very similar to those obtained by quenching from a high-temperature melt, but usually retain hydrogen (protons, OH groups, and –CH branches).[24–26] In the liquid state the local structure is determined by geometry constraints. On the contrary, in the solid state the structure of materials equilibrated at a given temperature is dominated by long-range Coulombic interactions. Knowledge of the phase relationship diagram is mandatory and thermal expansion measurement (Figure 12.2a) is the simplest way to determine it.[4,25–28] Differential thermal analysis (DTA) and phase characterization methods (x-ray diffraction [XRD], Raman scattering, IR absorption) are useful methods to confirm the phase relationship.

Is the sample free of hydrogen? Sol-gel-prepared ceramics easily retain a few wt% of protons up to 1000°C or more, and these protons drive the sintering and crystallization mechanisms.[24,26] Rather similar amounts of hydrogen are retained in carbides/nitrides derived from polymeric precursors (e.g., in SiC fibers thermally treated at approximately 1200°C). This hydrogen can generate unwanted gas departure.

Is the density/porosity convenient for the final use? What is the optimum processing temperature? Is the material homogeneous? What are the differences between the skin and the bulk? Where are the second phases located?

Another question is how to select the method with the lower cost. The use of hybrid methods in which part of the elements is introduced from metal-organic precursors offers the best compromise (Figure 12.1a).

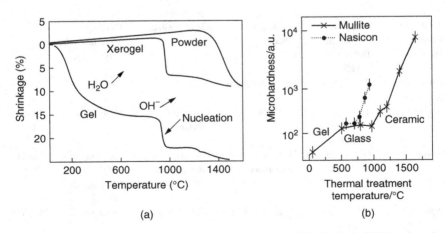

(a)  (b)

**FIGURE 12.2** (a) Comparison of the linear shrinkage of mullite ($3Al_2O_3 \cdot 2SiO_2$) micronic powder, 600°C thermally treated gel (xerogel) and pristine gel. (After Reference 87.) (b) Plot of the R.T. microhardness for mullite and Nasicon ($Na_3Zr_2Si_3PO_{12}$) as a function of thermal treatment. (After Reference 27.)

The final question is how to control the achieved materials. Methods and representative references are listed in Table 12.1. Two of these methods, the depth-sensing indentation and the micro-Raman scattering, were developed recently and will be addressed in the last part of this chapter.

**TABLE 12.1**
**Methods of Analysis for Nano/Microsized Materials**

| Sensitivity | Method | Information | Advantages | Drawbacks | Ref. |
|---|---|---|---|---|---|
| Bond length Structure | Dilatometry | Densification Phase transition Crystallization CTE Anharmonicity Sintering mechanisms | Analysis of the whole sample High sensitivity Easy atmosphere control | Generally destructive | 4, 25–28 |
| Bond length Structure Composition | DTA | Phase transformation /transition | Routine High sensitivity | | 4, 28 |
| Bond length Structure Symmetry Composition Geometry | Vibrational spectroscopy (IR, Raman) Microspectros copies | Structure Crystallization Phase transition Strain/stress Phase location | Minor phases can be studied Phase and properties mapping Properties predictions | Very different cross sections | 16, 51, 52, 89, 94, 107 |
| Bond length Structure Symmetry | X-ray, electron (neutron) Diffraction (TEM) | Structure Phases | Good database | Destructive Materials to be crystalline | 15 |
| Composition Geometry | SEM | Phase locations | | | 5 |
| Density | Thermal expansion/ shrinkage | Sintering | Routine technique | Destructive | 7, 36, 40 |
| Density | Gas adsorption | Porosity Reactivity | Pore size and distribution | Very long time of analysis Difficulty of drying | 30, 31 |
| Density | X-ray (neutron) scattering | Phase contrast | Whole sample | Unequivocal | 33, 38 |
| Mass Composition | TG | Gas evolution Oxidation | Routine | Destructive | 27, 31 |

## III. DENSIFICATION

We will take examples among silicates, aluminates, and zirconates prepared from the alkoxide sol-gel route that serve as the basic compounds for many applications. Figure 12.3 illustrates the structure change at the bond length scale when a gel/porous glass densifies by dehydroxylation-crystallization. The departure of some OH groups destabilizes the pore surface and gives rise to a high surface mobility. Rearrangement in the crystalline form induces densification.[24,26] Consequently the microhardness increases (Figure 12.2b).[27] This sketch illustrates that the densification and crystallization process is intimately related to the gel structure.

### A. GEL STRUCTURE AND COMPOSITION

Gelation of liquid oxide precursors like alkoxides ($M(OR)_n$) results from hydrolysis-polycondensation:

$$M(OR)_n + nH_2O \rightarrow M(OH)_n + nROH, \qquad (12.1)$$

and simultaneously, generally[1,2,8–11,29]

$$M(OH)_n \rightarrow n/2H_2O + nMO_{n/2}. \qquad (12.2)$$

After drying the resultant gel has a complex composition such as

$$MO_{n-x-\varepsilon}(OH)_{2x}(O\text{-}R)_\varepsilon mH_2O \qquad (12.3)$$

with $x \sim 0.1$ to $0.3$, $m = 3$ to $6$, and $\varepsilon \sim 0.01$.[30–33] These gels are made of small polymeric entities (diameter 0.5 nm to 5 nm), which are more or less aggregated and densely packed according to the process (hydrolysis-polycondensation rate, initial alkoxide-to-solvent and alkoxide-to-water ratios).[17,18,33] Small angle x-ray/neutron diffraction analysis and scanning electron microscopy (SEM)/transmission electron microscopy (TEM) images show that oxide gels are made of

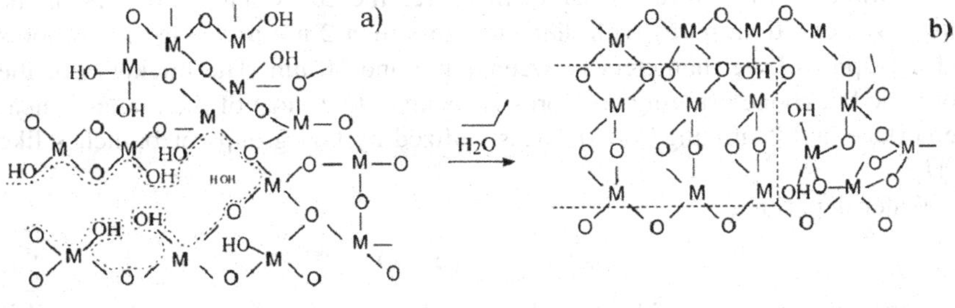

FIGURE 12.3 Sketch of the nucleation-densification process (b) involved with the dehydroxylation of a gel surface (a). (After Reference 27.)

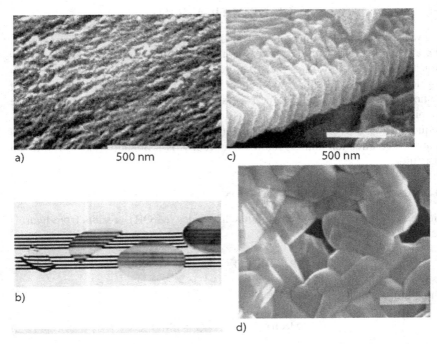

**FIGURE 12.4** Scanning electron micrography of optically/translucent NASICON gels (a-c) prepared by different gels routes. An example of fine grained optically clear ceramic (1400°C fired mullite) is shown in (d) (a-d, reprinted from *Ceramic Powder Science and Technology*, Messing, G., Ed., © 1989, with permission from The American Ceramic Society; (d) after "Sciences & Techniques", Céramiques: les progrès de la Chimie, © Ph. Colomban).

rather globular polymeric entities (diameter 1 nm to 2 nm), which are more or less densely packed according to the hydrolysis-polycondensation rate and the initial alkoxide-solvent-water ratios. An example is given in Figure 12.4: the size of the aggregated moieties is <20 nm. Dense packing of these aggregates is obtained by sedimentation (slow hydrolysis using an inert solvent) or by pressure (compaction of a gel powder prepared by rapid hydrolysis). Krypton, CO, argon, or $N_2$ Brunauer-Emmit-Teller (BET) measures the equivalent diameters of the pores. By definition, pores with diameters less than 2 nm are called micropores and mesopores have diameters between 2 nm and 50 nm. The thickness of the inorganic backbone bridging the pores is so small that most of the matter is near the surface and "dangling" bonds are stabilized by OH groups or branches like $-CH_2$.

Water departure,

$$2OH^- \rightarrow H_2O + O_{solid} \tag{12.4}$$

destroys the inorganic backbone and drives the crystallization-densification.[24–31] There is thus a relationship between the backbone composition, the hydroxyl content, and the meso/microposity (and hence between this porosity and the water content) and the sintering temperature. The nanostructure and the porosity (and hence the densification and many other properties) can therefore be tailored by the chemical parameters controlling the inorganic polymerization and the hydroxyl content: relative proportion of reagents, additions of complexing liquids, temperature, sonification, sintering atmosphere, and heating rate.[7,25,28,30–33] These remarks are general: for instance, the chain length of SiC (or C) polymeric precursor determines the silicon/carbon and carbon/hydrogen stoichiometry, because terminal branches contain more carbon and hydrogen.

Like many hydrates, gels present a unique challenge when densified under pressure at room temperature.[34–37] Due to the liquid (water generally) present at the boundary, viscous rearrangement under pressure allows densification and optimization of the contact between polymeric grains, approaching the "perfect" packing obtained by sedimentation: true sintering can be achieved under the combined action of the pressure and the liquid phase soaked in the pores (water generally, but also alcohol). Gel sintering is associated with an increase in optical clarity.[36] On heating, the grains formed retain a memory of the packing of primary moieties. The grain growth does not destroy this arrangement. Figure 12.4b shows gel pellets of Nasicon solid solutions prepared by different sol-gel routes. Translucent or even optically clear samples are easily obtained. The plasticity of the gel offers the possibility of facilitating the compaction of any kind of particle, especially those with an elongated shape (whiskers, platelets), by coating each particle with a gel film.[37] The gel coating lubricates the particle and promotes homogeneity of the applied pressure. Plasticity and shapability disappear with the loss of the water adsorbed at the "grain" surface, above ~200°C. Depending on the composition, the shrinkage rate then decreases up to 500 (zirconia) to 1000°C (aluminosilicates), where drastic shrinkage takes place. The latter results from the backbone rearrangement due to the departure of the last stabilizing OH groups (Figure 12.2 and Figure 12.3). Note that because of the ability of the gel to be deformed, a memory effect very similar to the well-known paste memory of kaolin-rich porcelain paste can be observed. An example is shown in Figure 12.5, with a compacted pellet of zirconia gel fired at 1200°C.

Gels transform more or less gradually into a meso/microporous glass.[24–26] A logarithmic plot of the weight loss vs. the inverse of the temperature allows the onset determination of water and hydroxyl departures, and hence the measurement

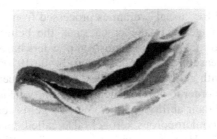

**FIGURE 12.5** Self-wrapped zirconia pellet after firing at 1200°C (diameter ~10 mm).

**TABLE 12.2**

**Gel Compositions, Specific Area and Hydroxyl Coverage Ratio for Some Heat-Treated Gels**

| Precursor | Thermal Treatment (°C) | Formula | Specific Surface Area (m²/g) | OH(nm²) |
|---|---|---|---|---|
| Aluminum butoxide | 300 | $Al_2O_{2.5}(OH)_{0.9}$ 0.4 $H_2O$ | 530 | 8.8 |
| | 600 | $Al_2O_{2.8}(OH)_{0.35}$ 0.3 $H_2O$ | 350 | 5.5 |
| Aluminum butoxide + silicon methoxide (1/1 molar) | 400 | $Si_2Al_2O_{6.5}(OH)$ 1.2 $H_2O$ | 680 | 3.5 |
| | 600 | $Si_2Al_2O_{6.7}(OH)_{0.6}$ 1.9 $H_2O$ | 630 | 2.2 |
| (1/2 molar) | 400 | $Si_2Al_2O_{6.7}(OH)_{0.65}$ 0.9 $H_2O$ | 515 | 3.1 |

After Vendange, V., and Colomban, P., Determination of the hydroxyl group content in gels and porous "glasses" issued of alkoxide hydrolysis by combined TGA and BET analysis, *J. Porous Mater.*, 3, 193, 1996.

of $x$ and $m$ values in the gel formula.[26,31] Measurement of the specific surface area by BET methods allows the calculation of the hydroxyl coverage ratio. Representative measured gel compositions are given in Table 12.2. The high value of the hydroxyl coverage ratio explains the high correlation between the hydroxyl group elimination and the nucleation-densification reaction.

Note that the high specific surface area of gels is not related to their apparent density, but to the intrinsic characteristics of the network. For instance, specific surface areas as high as ~1000 m².g⁻¹ are achieved for optically clear gels prepared by slow hydrolysis (their density is close to 1.5 to 2 g.cm⁻³),[31,33] but also for aerogels made through supercritical drying[38] and for gels obtained by oxyhydration of aluminum in the presence of mercury (the apparent density of these last materials is approximately 0.05 g.cm⁻³).[39]

## B. Sintering

It is well established for "usual" ceramics processed from crystalline powders that the denser the material in the green state, the better the densification. However, although sol-gel prepared nanosized powders have been available for many decades, their ease of use is more recent. In the 1970s, poor density was generally obtained with gel powders because many engineers overheated them before compaction in order to develop their crystallinity. This thermal treatment drastically decreased their ability to sinter. When a gel is converted by thermal treatment into a meso/microporous xerogel (also called "porous glass") above approximately 400°C, OH/$H_2O$ loss and squeezing of the $(Si-O)_n$ network occur. This first step in shrinkage (Figure 12.2) related to the gel densification was suppressed. Consequently the gel plasticity disappeared. The presence of some M-OH "dangling" bonds in place of M-O-M bridges stabilizes the surface of the porous network, which makes the densification sensitive to the atmosphere ($H_2$, $H_2O$, vacuum or air).[26,27] Densification and nucleation for refractory or glass-like compositions result from the short range structural rearrangement induced by surface dehydroxylation at a temperature of about 0.5 $T_m$ (melting temperature).[1,36] It can be compared to the 0.8°C $T_m$ value usually required for the sintering of a fine ceramic powder. There is thus a direct relationship between the hydroxyl content, the composition, and the meso/microporosity of a (xero)gel. A thermal treatment above the crystallization step leads to a powder which has lost its sinterability: the grain surface has been smoothed, which hinders the sintering at the temperature expected compared with ball-milled powders of the same grain size. The nanostructure and the sinterability can therefore be tailored by the chemical parameters controlling the inorganic polymerization and the hydroxyl content (relative proportions of reagents, addition of complexing liquids, temperature, sonification, sintering atmosphere, etc.). In the case of glass-forming compositions, a viscous sintering may occur while the dehydroxylation-densification proceeds.[1,40] In this case, bloating can be observed if gas departure ($H_2O$ from dehydroxylation, $CO_2$ from oxidation of unhydrolyzed organic branches) is hindered up to a viscous state, allowing the formation of bubbles.

Convenient sol-gel processing, compaction, and firing allow the synthesis of optically clear ceramics. Two examples are presented in Figure 12.6, a ferroelectric PLZT 65/35 ceramic prepared by hybrid sol-gel route[41] and a mullite matrix prepared using aluminum-silicon ester and silicon-methoxide.[23] Note the opaque skin around the transparent PLZT ceramics is related to PbO loss. When a partial pressure higher than the equilibrium pressure required to avoid any PbO excess

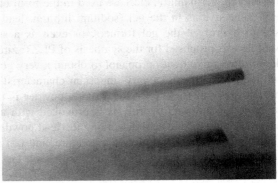

FIGURE 12.6 (a) Optically clear PLZT 65/35 ceramic sintered in PbO-enriched atmosphere (thickness ~5 mm). (b) Translucent mullite ceramic with SiC fibers (diameter 12 μm).

at the grain boundaries in the bulk is used, it gives a red shade to the yellow-colored optically clear ceramic.

## C. Multicomponent Materials

For a long time, the use of polymeric and sol-gel precursors was limited to simple compositions: $SiO_2$, $TiO_2$, $ZrO_2$, etc. The main difficulty in multicomponent and nonstoichiometric materials is to achieve the right composition. This requires the precise determination of the elemental composition of liquid precursors or reagents and to take into account any loss related to the various transformation rates, volatility, polycondensation, or even some polymerization.[1,2,6,7,42–45] For instance, aluminum alkoxides are very hygroscopic (instant hydrolysis with water traces), whereas tributylphosphate can stay months in humidity without hydrolysis. Some reagents are volatile (e.g., many methoxides). Consequently the hydrolysis of a mixture of reagents could give local heterogeneities. During the drying of gels containing elements that do not participate in the polymeric network (e.g., alkali ions), they migrate to the evaporation front and the gel (and container) surfaces have higher concentrations of these elements. This problem was first encountered (and solved) at the very beginning of the sol-gel history for the synthesis of PLZT (optically clear ferroelectric ceramics made of lanthanum lead zirconate-titanate)[41] and then NASICON (a solid solution of sodium zirconium phosphate-silicate solid electrolytes).[6,7] The proposed solutions were (1) to use reagents in which both elements link early [e.g., the aluminum-silicon ester $(OC_4H_9)_2\text{-}Al\text{-}O\text{-}Si(OC_2H_5)_3$ has intermediate behavior between pure aluminum butoxide and silicon ethoxide].[16–18,44–46] Controlled hydrolysis is thus very easy and catalysis is not required, as for the hydrolysis of silicon-alkoxides; (2) to use an inert reagent (propanol, hexane) as a solvent for the alkoxide mixture in order to control the hydrolysis by water diffusion;[6,7,45–52] (3) to increase the chain length of the most hygroscopic reagent in order to slow down its hydrolysis rate; (4) to use hybrid methods: only the gel formers (silicon, aluminum, zirconium, titanium, phosphorus, germanium, etc.) are used in the form of alkoxides, the elements to be present as ions in the gel (sodium, lithium, lead) being added in the water used to hydrolyze the gel formers, or even as a solid powder.[1,6,41,47] A good example was proposed for the synthesis of PLZT: zirconium and titanium alkoxides are first mixed with propanol to obtain a very good dispersion of zirconium and titanium elements, a very important characteristic for the final ferroelectric and optical properties.[41,53,54] Lead oxide powder is then mixed to the alkoxide alcoholic mixture and hydrolysis is provoked with an aqueous solution of lanthanum acetate (acetic acid dissolving the lead powder). Lead oxide gives rise to a liquid phase at relatively low temperature, and fast diffusion of $Pb^{2+}$ ions takes place before the melting temperature; with such fast diffusion of ions, any heterogeneity will rapidly disappear on heating. Thus, in hybrid routes, the use of sol-gel or polymeric precursors has to be limited to the elements with the lowest diffusion coefficients. In this way it is possible to prepare better materials at lower cost.

# IV. COMPOSITES AND MULTIPHASE MATERIALS

## A. PARTICULATE-REINFORCED MATERIALS

The "simplest" application of sol-gel materials is as abrasive grains. Abrasive particles can be dispersed into vitrified or resin-bonded grinding wheels, coated abrasive belts, sheets, and discs. In 1981, 3M Company introduced the first sol-gel abrasive particle, referred to as Cubitron, in the Regal-coated abrasive fiber disc product line.[55] The idea to use sol-gel processing originated from the processing route of ceramic fibers developed at 3M by Sowman[56] some years before. Commercially available boehmite (AlOOH) sol is derived from the hydrolysis of aluminum alkoxide, a by-product of the Ziegler process for the production of long-chain alcohols. When a metal salt is added to the boehmite sol, it provokes gelation and a stiff gel is formed. Afterwards this gel is dried, crushed, and screened to the appropriate size distribution, calcined, and then sintered at a temperature between 1200 and 1600°C. The resulting material is screened and ready to be used as an abrasive particle.

Nonoxide ceramics are finding increasing applications as high-speed cutting tools for metals.[57] The combination of sol-gel processing and reaction-sintering has been used to prepare dense nonoxide abrasive grit.[58] An $Al_2O_3$ sol was prepared by dispersing boehmite powder into $HNO_3$ containing water heated to ~80°C. The sol was seeded with $\alpha Al_2O_3$ crystallites. The abrasive grains were prepared from the dispersion of carbon black and $TiO_2$ in the seeded $Al_2O_3$ sol. Glycerol was added to prevent oxidation of the carbon black during calcination. The gel was dried, crushed, screened, and calcined at 1000°C and then thermally heated to 1400°C and then to 1900°C under a flowing $N_2$ atmosphere for the formation of TiN and AlN by carbothermal reduction. The final microstructure (AlN/TiN/AlON) is controlled by the amount of $TiO_2$ in the sol and by the carbon content. The great advantage of sol-gel abrasive particles is the ability to introduce chemical changes, which modify the alumina crystal structure and enable the optimization of grinding.

Particulate reinforcement (platelets and whiskers) is used to improve the toughness of small shaped pieces or to facilitate machining. This type of reinforcement is also used to increase the toughness of monolithic ceramics and of the matrix of fiber-reinforced composites. Figure 12.7a shows an example application, a dispersion of submicronic zironia in a mullite matrix in between woven SiC fibers. Another example application is the "machinable" sol-gel prepared mica-ceramic composite that can be cut by a conventional metallic saw[59] and is used as an insulating material for precision machines and as a substrate for electronic parts. The homogeneous dispersion of anisotropic particles in a powder and the compaction of the resulting mixture are always difficult because the applied load decreases with the number of contact points. The use of liquid aids (colloidal processing) counteracts this drawback, although imperfectly. Gel embedding offers a new route for the mixing of powders and particulate reinforcements due to the incorporation of a viscous substance, the reactivity of which can be tailored by compositional design.[37] Cold molding makes it possible to prepare shaped and crack-free pieces. The process is the following: powder and

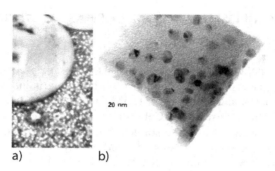

a)                    b)

**FIGURE 12.7** (a) Scanning electron micrograph (SEM) of a mullite matrix CMC reinforced with submicronic zirconia (fiber diameter 12 μm). (b) TEM of a cobalt dispersion in an aluminosilicate matrix synthesized by the infiltration-$H_2$ firing route.[87]

particles are dispersed together in an anhydrous solvent (e.g., propanol) before the addition of alkoxides. After mixing, pH-controlled water is poured in under vigorous mechanical stirring, which has to be maintained until a paste of thick consistency is obtained. The paste is then dried to give a flour-like powder. The powder is compacted and sintered. Gel embedding promotes a homogeneous load/pressure transfer and hence suppresses the densification hindrance induced by the platelet addition. For example, zirconia gel embedding of the mullite powder increases the microhardness by 30% and the toughness by 40%. The combination of zirconia gel embedding and platelets addition increases the $K_{IC}$ value to about 5 MPa.m$^{1/2}$.[37]

## B. Metal- and Ceramic-Ceramic Nanocomposites

Metal-ceramic microcomposites (cermets) have been prepared for a long time, especially by dispersion of nickel particles in a silica matrix. Metal nanocomposites (iron, nickel, tin, and copper particles in a silica or alumina matrix) have been prepared by different groups.[60-64] More complex materials with nanodispersion of pure or alloyed iron, cobalt, and nickel in aluminosilicate matrices have also been prepared.[65,66] The main methods can be classified as follows: (1) the mixing of the (sub)micronic metal powder within a liquid matrix precursor; (2) the mixing of the alkoxides with an aqueous solution of metal ions and controlled $H_2$ reduction on firing; and (3) the preparation of a porous host matrix impregnated by a concentrated solution of transition metal nitrates and controlled reduction. Homogenous dispersions are achieved at the submicronic scale (Figure 12.7b). About 5% of matter in volume (15% in weight) can be incorporated in one cycle of impregnation/firing. The metal content can be increased by successive impregnation-firing cycles (up to ~50% in volume). Combined infiltrations with different metal and alloy precursors can tailor the magnetic and electromagnetic properties, in particular the microwave absorption between 0.1 and 10 GHz.[66] These new materials can find applications at higher temperatures, as their magnetic properties are maintained up to 550°C.

The preparation of ceramic (oxide-oxide, carbide-carbide, etc.) nanocomposites is also rather old and many authors have reported some exceptional properties.[49,50] This field of study developed rapidly. It is clear that more work is needed in order to further understand the mechanics of nanocomposites.

## C.  ONE-DIMENSIONAL AND TWO-DIMENSIONAL FIBER-REINFORCED COMPOSITES

One of the main problems in the preparation of ceramic matrix composites (CMCs) is achieving a low open porosity in the matrix in order to protect the fiber from the environment and to optimize the mechanical strength and toughness. Fibrous fracture of CMCs is mandatory to get a stable and reliable strength value to be used for parts design. In the case of weavable fibers forming a yarn, the interfiber voids are a few microns or less in size, which makes infiltration of liquid or gaseous precursors mandatory.[12,23,67] In many cases, interyarn voids (1 to 100 mm$^3$) of textile preforms are accessible only through interfiber voids close to 1 to 5 μm. Chemical vapor infiltration (CVI), a method based on the infiltration of gaseous precursors, is more expensive because of the duration of the synthesis cycle and the need for specific tools for each cycle.[68,69] Furthermore, gaseous precursors are not well suited for the multicomponent oxide compositions required for functional composites. The presence of long-fiber geometrically invariant reinforcement inhibits the coherent shrinkage of the matrix. In the case of one-dimensional or two-dimensional (textile) reinforcements, this phenomenon can be solved by hot-pressing: the in-plane shrinkage can be counterbalanced by the thickness reduction if a viscous behavior is achieved during the hot-pressing.[67] A flowchart for this process is summarized in Figure 12.8.

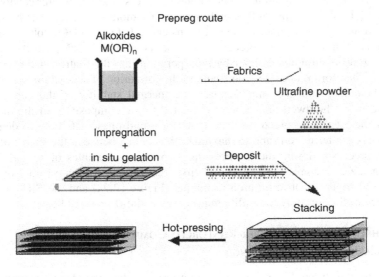

**FIGURE 12.8** Schematic of the sol-gel process for the fabrication of woven fabrics reinforced CMCs.[87]

Many attempts to prepare dense ceramic matrix composites by sol-gel methods have been made. A lot of them used mullite as the matrix material because of its excellent high-temperature strength and creep resistance, good chemical and thermal stability, low thermal expansion coefficient, and low permittivity. First, Qui and Pantano,[70] Pannhorst et al.,[71] and then Russell-Floyd et al.[72] demonstrated that sol-gel processing can be used. A group from Sheffield[73-75] pointed out that highly densified unidimensional continuous carbon fiber-reinforced mullite composites could be successfully prepared by a single-stage infiltration process, using Ludox® colloidal silica sol (Dupont AS40) and α-alumina powder (A1000G, Alcoa), followed by hot-pressing. The pH of the Ludox® sol was adjusted to 2 to 3 before dispersion of alumina powder.

Composites reinforced with woven (carbon, SiC, or oxide) fibers could be successfully prepared by a two-step process,[76-81] as illustrated in Figure 12.8. First the fibers of a two-dimensional woven fabric are impregnated with a liquid alkoxide mixture (for instance, a mixture of zirconium-i-propoxide, aluminum-silicon ester, tributylborate and germanium propoxide for mullite matrix composite), which fills in the voids between the fibers, *in situ* transforms into a gel by reaction with atmospheric moisture and is to be converted into a (glass-) ceramic by pyrolysis during the hot-pressing step. Before pyrolysis, a fine amorphous matrix precursor powder, suspended in chlorobenzene, is deposited onto the layers of woven fiber fabric (prepreg textile). This matrix precursor powder is prepared by rapid hydrolysis of the appropriate mixtures of alkoxides in propanol. The resulting gel is dried at about 750°C to drive most of the water and part of the hydroxyl groups out and thus reduce the subsequent shrinkage of polymeric oxide network. Then the doubly impregnated layers of textile fabric are stacked together in a graphite mold and hot-pressed at temperatures between 1000 and 1400°C. Below 300°C, gel viscosity promotes homogeneous pressure application. A liquid sintering aid (for instance, $B_2O_3$ or $GeO_2$) obtained by pyrolysis of appropriate gel precursors can be incorporated.[76-78] Then the liquid sintering aid is eliminated either by incorporation into the matrix or by volatilization. Therefore it does not contribute to the formation of second phases at the grain boundary and does not decrease the thermal stability of the composite significantly. The dwell temperature is related to the composition of the matrix and of the interphase precursor. The transient liquid phase acts both to densify the matrix (sintering aid) and to maximize the contact between the grains of the matrix precursor during the hot-pressing step. Some examples of sol-gel fiber coating are a zirconia deposit on SiC fiber reinforcing a glass-ceramic matrix processed by the melt-infiltration technique (Figure 12.9a) and the SiC-coating interface in SiC-reinforced mullite matrix composite (Figure 12.9b).

## D. THREE-DIMENSIONAL FIBER-REINFORCED COMPOSITES: NEAR NET-SHAPE SINTERING

For continuous three-dimensional fiber-reinforced bodies, the idea of using liquid ceramic precursors for the impregnation of the fibrous preform originates in the

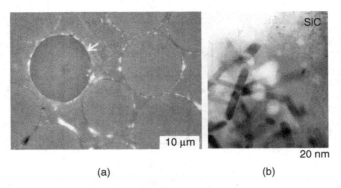

(a) (b)

**FIGURE 12.9** (a) SEM of a sol-gel $ZrO_2$ coated SiC fiber reinforced glass-ceramic. (b) TEM detail of the SiC-mullite coating interface; see the elongated mullite crystals (l = 5–10 nm).

preparation of carbon/carbon composites by pitch or phenolic resin infiltration.[23] To mimic this route, the slip-casting of a submicronic powder has been proposed,[82] but this led to highly porous samples (open porosity close to 35 to 40%) exhibiting very poor mechanical properties and low protection of the fibers against corrosive atmospheres. Also, the range between the consolidation temperature and the temperature at which the matrix shrinkage leads to matrix cracking, within the geometric invariant preform or fabric, is small. Due to the fact that the achievement of a zero-porosity matrix is unrealistic for thermostable three-dimensional reinforcements (fabrics, felts, etc.), we will consider methods for increasing the mechanical strength of a porous body. The first idea is to avoid cracks, maintaining a coherent matrix (maximization of the number of interparticulate bonds without shrinkage), the second one is to improve the strength of the interparticulate bonds and/or of the intergranular phase (maximization of the reaction between grains), and the last one is to introduce between a phase between the grains, which hinders the shortening of the distance between their centers.

The first requirement can be obtained by filling the voids between slip-cast grains with a ceramic precursor, which, after a thermal treatment, is transformed into a refractory phase. The resulting intergranular phase may form an inert barrier between slip-cast grains and thus put the shrinkage off to higher temperatures.[12,83] The mean size of alumina particle is 0.6 μm, which is allowed to pass through interfiber voids. Thermal treatment at 1000°C was carried out in order to initiate the consolidation of the matrix without developing significant shrinkage of alumina grains.

Figure 12.10 shows a schematic of the two-step infiltration process.[23] The first step consists of the preparation of a powder compact within the fiber preform by a routine slip-cast infiltration, the body being dried, and then strengthened by heating at a temperature close to the onset of shrinkage. The second step consists of the infiltration of the strengthened body by a liquid (polymeric) precursor that is first converted into a solid (a gel by reaction with water or diols if alkoxides are used[23,77,83,84]), or by the action of oxygen if polyvinylsilane (PVS) is used to

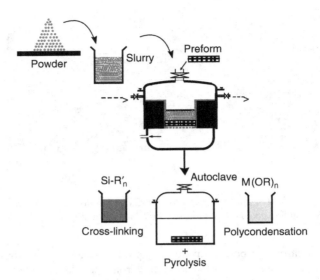

**FIGURE 12.10** Schematic of the oxide (alkoxides) or nonoxide (polymers) matrix precursors infiltration for fabricating three-dimensional reinforced composites. (After Reference 87.)

prepare SiC matrix,[85] and then, during pyrolysis, into a refractory intergranular phase. This second step can be repeated as many times as necessary to optimize the interparticle bonding.

As explained in the previous part, optimization of the microstructure might involve (1) an increase in ceramic yield by the pyrolysis of the precursor, (2) an increase in thermostability of the resultant phase, in order to limit the shrinkage of the refractory interphase and postpone the whole matrix shrinkage toward higher temperatures, or (3) the optimization of the fiber-matrix interface. For instance, various polymeric precursors can lead to inert, refractory, and nonreacting interphases: zirconium-i-propoxide as zirconia precursor, aluminum-butoxide as alumina precursor, lanthanum-alkoxide, and homologues as rare earth oxide precursors. Sometimes titanium-i-propoxide, tetraethoxysilane, and aluminum-silicon ester can be used as rutile, silica, and aluminosilicate precursors, respectively. These last three compounds, which slightly react with many oxides, might also be used to strengthen the interparticle bonds. It is very important, however, to simultaneously maintain a net-shape sintering behavior. The room-temperature and high-temperature flexural strength was increased after five (Figure 12.11) cycles of postinfiltration *in situ* hydrolysis-polycondensation and 1000°C heating, zirconium-i-propoxide being used for the first four cycles and aluminum-silicon ester for the last one.[12,83] Confirmation was given by tensile test, at room temperature. The increase in mechanical strength arises from the filling of the voids (porosity decreases), but also from the blocking of shrinkage due to the zirconia "inert" phase. The high stability of zirconia coatings has been used in many oxide-oxide composites. The addition

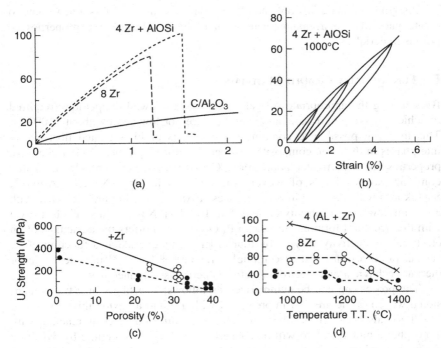

**FIGURE 12.11** Improvements of the mechanical properties of three-dimensional reinforced CMCs by hybrid infiltration routes: (a) R.T. flexural stress-strain plots for a three-dimensional carbon fiber reinforced composite before and after cycles of infiltration (comparison between eight cycles with zirconium propoxide and four cycles plus a last infiltration with aluminum-silicon ester; (b) plot of the mechanical strength as a function of the final open porosity for composites and matrix of equivalent porosity, before and after infiltration (Reprinted from Colomban, P. and Wey, M., Sol-gel control of the matrix net-shape sintering in 3D reinforced ceramic matrix composites, *J. Eur. Ceram. Soc.*, 17, 1475, 1997. With permission from Elsevier); (c) R.T. tensile behavior; (d) comparison of the R.T. mechanical strength after thermal treatments at various temperatures. (Reprinted from Colomban, P., Tailoring of the nano/microstructure of heterogeneous ceramics by sol-gel routes, *Ceram. Trans.*, 95, 243, 1998. With permission from The American Ceramic Society.)

of aluminosilicate precursor for the last infiltration strengthens the bridge between particles and decreases the open porosity. Tensile measurements show that the mechanical properties achieved through the oxide route are very similar to those of the usual SiC made by the CVI process, although the matrix Young's modulus is lower in the former (30 GPa instead of 75 GPa for the CVI SiC matrix). A rather pronounced hysteresis in loading-unloading cycles is observed. The same behavior is observed for composites prepared by combining SiC powder and PVS infiltration.[23,84]

Alloying the materials obtained from liquid precursors offers a new route to create particular micro/nanostructures in order to optimize the properties of porous materials.

## E. FUNCTIONALLY GRADED MATERIALS

By selecting the appropriate sol-gel precursors, the dwell temperature required to achieve maximum densification can be raised or lowered by about 100°C.[1,36] This makes it possible to combine several kinds of impregnated fiber/matrix interphases in the same composite in order to tailor its physical and/or chemical properties for a particular application. Composites have been fabricated that combine various kinds of woven fibers (SiC Nicalon®, Nextel®, Almax®, Saphikon®) with various kinds of matrices (LAS, mullite [a pure dielectric with a real microwave permittivity close to 5 at 10 GHz], Nasicon [a solid electrolyte with the structural formula $Na_{1+x}Zr_2Si_xP_{3x}O_{12}$ (0 x 3) offers the advantage of an electrical conductivity varying by four orders of magnitude as a function of $x$], celsian [a corrosion-resistant ceramic], zirconia)[6–10,76,78,79,86,87] in order to tailor thermomechanical and electromagnetic properties simultaneously.

The processing can be summarized as follows: The fibers of a two-dimensional woven fabric are first impregnated with a liquid alkoxide mixture, which fills the voids between the fibers, transforms *in situ* into a gel by reaction with atmospheric moisture, and will be converted into a (glass)-ceramic by pyrolysis during the hot-pressing of prepreg fabrics covered with the matrix powder precursors. Depending on both the matrix and the fibers, several different "interface precursors" can be used, either alone or in various combinations. Finally, the doubly impregnated, coated woven fabrics are stacked in a graphite mold and hot-pressed, typically between 950 and 1400°C. By selecting the appropriate sol-gel precursors, the dwell temperature required to achieve maximum densification can be raised or lowered by about 100°C. This makes it possible to combine several kinds of fiber-reinforced layers in the same composites in order to tailor, unidirectionally, the microwave absorption along the direction perpendicular to the body. Examples given in Figure 12.12 show (1) a composite consisting of a zirconia matrix and a mullite matrix, both reinforced by SiC Nicalon NLM 202 fibers (4 + 4 layers), and (2) NASICON matrix composite reinforced on one side by low-permittivity Nextel mullite woven fibers (4 layers) and on the other side by conducting Nicalon SiC woven fibers (2 layers).

# V. CHARACTERIZATION OF MULTIPHASE MATERIALS

One of the difficulties in the characterization of multiphase materials is the need to analyze the different phases at various scales: at the scale of each phase, at the scale of their association (interfaces and intrafaces), and at the scale of the parts. We will address two methods, which can be used very rapidly for this type

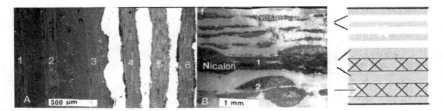

**FIGURE 12.12** Examples of FGM CMCs: (left) the combination of NLM Nicalon SiC fiber reinforced zirconia (in white) and mullite (in black) matrices; (right) a Nasicon matrix reinforced with mullite (Nextel) and SiC (NLMTM) fibers (see the sketch). (Reprinted from Colomban, P., Process for fabricating a ceramic matrix composite incorporating woven fibers and materials with different compositions and properties in the same composite, *Mater. Technol.*, 10, 89, 1995. With permission.)

of study. Both methods are associated with optical microscopes, which provides easy analysis of regions ranging in size from a few square centimeters to 1 $\mu m^2$.

## A. DEPTH-SENSING MICROINDENTATION

The measurement of local mechanical properties is an important step in understanding of the macroscopic behavior of multiphase materials. The indentation hardness test is probably the simplest method of measuring the mechanical properties of materials. Figure 12.2b shows the evolution of the microhardness as a function of the thermal treatment temperature of a Nasicon sample. The use of load-controlled depth-sensing hardness testers which operate in the (sub)micron range enables the study of each component of the composite more precisely.

Following the work of Loubet et al.,[88] Young's modulus ($E$) and Vickers' microhardness ($H_v$) can be extracted from the unloading part of load-displacement plots by considering the different contributions of the elastoplastic behavior of the indented materials. Figure 12.13a shows a schematic of an indentation test and Figure 12.13b gives the corresponding curves of loading and unloading versus in-depth penetration. For materials with high plasticity, the remnant penetration depth of indentation ($h_p$) is close to $h_e$, given by the intersect of the initial unloading step tangent with the x-axis; for plastic material $h_p = h_{max}$. The area delimited by loading and unloading curves is proportional to the work of indentation. On the other hand, the hysteresis is very small ($h_p \ll h_e < h_{max}$) for densely packed structures and high Young's modulus phases (e.g., SiC and sapphire). The slope of the straight line is directly related to Young's modulus:

$$\left(\frac{1-\nu^2}{E} + \frac{1-\nu_0^2}{E_0}\right)^{-1} = \frac{dF}{dh} 4\sqrt{\pi}(h_p \tan\theta)^{-1} \qquad (12.5)$$

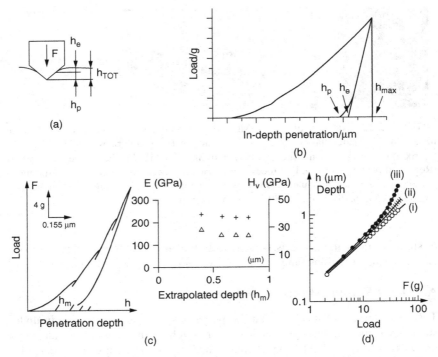

**FIGURE 12.13** (a) Sketch of the indenter penetration in a ceramic and (b) corresponding loading-unloading plot. (c) Example of the partial unloading from which the Young's modulus and microharness are extracted. (Reprinted from Colomban, P. et al., Sol-gel mullite matrix-SiC and -mullite 2D woven fabrics composites with or without zirconia containing interphase. Elaboration and properties, *J. Eur. Ceram. Soc.*, 16, 301, 1996. With permission from Elsevier.) (d) Comparison of the logarithmic plots for indented fibers in their matrix: a fiber sliding is observed for fibers (ii) and (iii).

where $dF/dh$ is the slope of the curve, $h_p$ is the plastic penetration, $\theta$ is the half angle of the Vickers' pyramid faces, $E$, $v$, $E_0$, $v_0$, being the Young's modulus and the Poisson's coefficients of the material and the indenter, respectively. The observation of noise and kinks on the loading/unloading curves is related to the formation of cracks. Young's modulus is measured for the constant regime observed for a set of indentation under different loads. Partial unloading (see Figure 12.13c, where three partial unloadings have been made before the load has been removed) is sufficient to extract the hardness, $H_c$, value used for the calculation.[37,78,89] Recent homemade or commercially available instruments allow the study of submicronic regions (nanoindenters). This method is well-suited to analyze the various phases and allow the study of the gel-ceramic transformation. Indentation of a fiber, its coating, or the matrix allows one to observe the evolution of the material as a function of processing and thermal aging. Note that if the applied load exceeds the threshold load $(F_s)$, the fiber slides within the matrix and a step becomes visible at the fiber periphery: after indentation, part of the fiber push-down is permanent (Figure 12.9a). This makes it possible to study the departure of the fiber/matrix mechanics from the linear behavior on logarithmic plots of the in-depth penetration as a function of the applied load. This gives evidence of the fiber sliding in its matrix (Figure 12.13d). The high strength and toughness of (glass)-ceramic matrix composites directly result from the low fiber-matrix bonding originating in the processing. This interphase acts as a "fuse," deflecting the matrix microcracks parallel to the fiber axis and thus avoiding the early failure of the fibers. This interphase is formed *in situ* during the hot-pressing as a result of the fiber-matrix chemical reaction or results from the deposit of a thin coating of carbon, BN, SiC, $ZrO_2$, etc.,[23,78,90,91] when nonreactive matrices are used. An important parameter to be controlled is the sign and the level of the residual mechanical stresses in the composite arising after processing from the mismatch of fiber (coating) and matrix (substrate) coefficients of thermal expansion. The sliding strength is usually measured on one-dimensional composites using instrumented or noninstrumented microhardness testers following different models.[92,93]

## B. Micro-Raman Spectrometry

Routine x-ray diffraction techniques fail to analyze nanosized/nanocrystalline materials: small size effect and short-range disorder are hardly distinguished and long counting times are required. On the contrary, Raman spectroscopy has demonstrated its capability to analyze low crystallinity and amorphous materials. Covalent-bonded materials are well-suited systems for Raman analysis. Their spectra are easily understood with the molecular model (the vibrational unit is built with the stronger bonds of the unit cell). However, their spectra do not change very much with "particle" size. On the contrary, because of long-range Coulomb interactions, spectra of ionocovalent structures are very sensitive to particle size, but can be understood with the molecular model. Thus Raman parameters (peak intensity, wave number, bandwidth, and shape) are very sensitive to any change of the chemical bonds: length, strength, geometry, etc. An example is given in Figure 12.14a with the plot of the wave number and bandwidth of the main Raman peak of disordered carbon (sp$^{2/3}$ hybridized C–C bond) for different SiC fibers (NLM™, Hi™, and Hi-S™ from Nippon Carbon; TM™, TE™, ZM™, and ZE™ from Ube Industry; and SA© from Sylramic) as a function of their thermal treatment in different temperatures and atmospheres.[94–97] Modifications induced by contact with a ceramic matrix or a coating can also be considered. Most of the properties controlled by the particle size (electrical conductivity, mechanical strength; e.g., in Figure 12.14b) are correlated to Raman parameters, which conversely can be used to predict these properties.

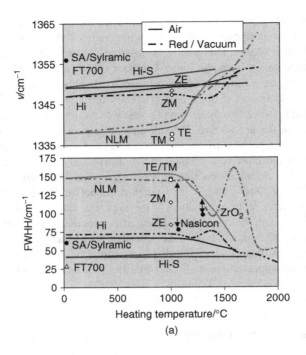

(a)

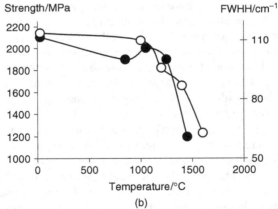

(b)

**FIGURE 12.14** (a) Plot of the wave number and bandwidth (full width at half height) of the $sp^{2/3}$ C–C raman peaks for different SiC fibers after various thermal treatment in air or in nonoxidizing atmospheres. (b) Comparison between ultimate tensile strength and $sp^{2/3}$ peak FWHH as a function of thermal treatment for SiC Hi Nicalon fibers. (Adapted from Colomban, P., Raman microscopy and imaging of ceramic fibers in CMCs and MMCs, *Ceramic Trans.*, 103, 517, 2000. With permission.)

## 1. The Raman Effect in Nanomaterials

First, Raman spectroscopy has an intrinsic "nanospecificity" since the probes are the (ionocovalent) bonds themselves. The study of extended domains includes nanodomains also (the volume probed by the spot determines how many). Using the molecular description, the vibrational atomic unit is the "molecule" found in the structure when considering only the strongest covalent bonds between metal cations and oxygen (or carbon, nitrogen, etc.) anions. Repetition of this vibrational unit must give the whole structure, like for the unit cell in crystallography. Stretching modes are characteristic of the chemical bond and allow for composition identification. Bending modes are more sensitive to the neighboring entities and hence to the short-range order. Vibrational and external modes, which correspond to the relative motions of the vibrational units, depend on the structure. Roughly speaking, Raman spectra are the projection of all the vibrations on the energy axis. Structures made of well-defined covalent vibrational units will give nice sharp peaks, whereas broad bands will indicate that different configurations and/or many electric or atomic defects exist in the material.

If the proportion of atoms belonging to the (near) surface region is significant (Figure 12.15a), the disorder related to the various atomic arrangements will contribute to Raman features.[94–102] We can say empirically that the chemical bonds located as far as 5 atomic distances (the approximately 1 nm short interaction length expected for covalent-bonded materials) or even 15 atomic distances (the long interaction length of approximately 3 nm expected for ionocovalent structures) from the skin atoms have a state different from those in the bulk. It is obvious that vibrational properties of surface atoms will dominate the spectra for a grain size of less than 20 nm (1 nm interaction length) or les than 50 nm (3 nm interaction length).

## 2. Correlation Between Raman Parameters and Grain Size

Many techniques for the preparation of nanosized materials (sol-gel, thermal treatment of polymeric precursor, electrochemical deposition, atomic layer deposition [ALD], etc.) lead to amorphous or low-crystallinity compounds by quenching of a liquid-state local structure or a very disordered state. At a given temperature, two phenomena can be at the origin of the broadening of the Raman spectrum: (1) the loss of periodicity because of the large contribution of surface atoms, and (2) a low crystallinity, that is to say, short-range disorder or bond distortion. In many cases the exact origin is not obvious and a comparison must be made with TEM.[102]

Different models have been used to derive the particle size from Raman spectra[90,96,99–105] As an example, we shall briefly explain the phonon confinement model (PCM). The scattering of one photon by $n$ phonons is governed by the momentum conservation. Only vibrations from the center of the Brillouin zone (BZC) should therefore be active in one phonon process (first-order Raman spectrum) and this is actually the case in large and flawless crystals, where

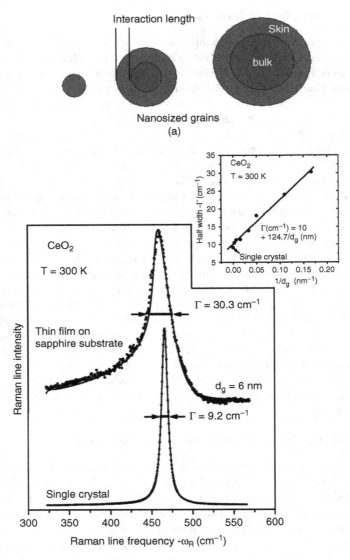

**FIGURE 12.15** (a) Comparison of the relative proportion of skin and bulk atoms in nanosized grains for a given interaction length. (Reprinted from Colomban, J., Raman analysis and "smart" imaging of nanophases and nanosized materials, *Spectroscopy Europe*, 15, 8, 2003. With permission from Wiley.) (b) Comparison of the Raman spectra of CeO$_2$ in a single crystal and in a nanocrystalline ceramic (6 nm grains). Note the plot of the half width as a function of the inverse grain size. (Reprinted from Kosacki, I. et al., Raman scattering and lattice defects in nanocrystalline CeO$_2$ thin films, *Solid State Ionics*, 149, 99, 2002. With permission from Elsevier.)

phonons virtually propagate to "infinity." However, short-range disorder, bond distortion, or nanocrystallinity can destroy periodicity and confine the phonons spatially. This introduces an uncertainty on $k$ (the selection rule is broken by crystalline imperfection), and because dispersion curves are not symmetrical with respect to BZC, the "sampling" of $k$-space provokes band shifting and asymmetry. In the first approximation, Raman line broadening can be described by the (linear) dependence of its half width upon the inverse grain size (Figure 12.15), as reported previously for many nanocrystalline materials including $CeO_2$, Si, Ge, GaAs, and diamond (see References 98, 99, and 100 and references therein).

### 3. Raman Images

Figure 12.16 illustrates how Raman microspectrometry can be used to image nanophase distribution in nanosized materials. We will present the case of SiC fibers for aerospace applications, prepared with different technologies. These

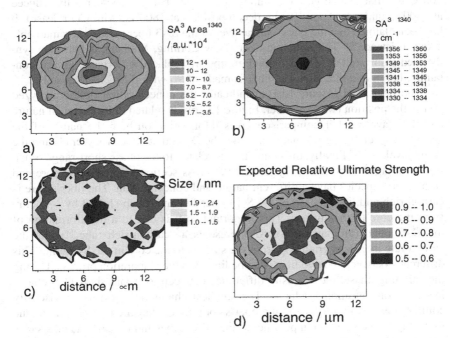

**FIGURE 12.16** Raman images of SA3 Ube Industries fiber for D band ($sp^{3/2}$) area (a) and wave number (b) ($25 \times 25$ spectra, $\lambda = 632$ nm, 0.5 mW, 120 sec/dot). The size distribution of carbon grains was calculated using the correlation between the coherent length and the Raman peaks ratio ID/IG. The relationship correlating Raman studies, TEM examination, and the ultimate strength measurements is used to anticipate the "relative" mechanical strength in the different regions of the fiber (d). (Reprinted from Colomban, P., Raman analysis and "smart" imaging of nanophases and nanosized materials, *Spectroscopy Eur.*, 15, 8, 2003. With permission from Wiley.)

fibers consist of nanosized more or less crystalline SiC grains, with a second phase of disordered carbons. Pure diamond and graphite have sharp peaks at 1331 cm$^{-1}$ and 1581 cm$^{-1}$, respectively. The two main bands of amorphous carbons are then assigned to diamond-like (D band) and graphite-like (G band) entities. Because the diamond cross section is much lower than that of graphite (approximately 10$^2$), a weak $C_{sp}^3$–$C_{sp}^3$ stretching mode is expected. In fact, given the small size of carbon moieties, the contribution of the chemical bonds located near their surface will be large. The D band can then be assigned to vibrations involving $C_{sp}^3$-mixed $C_{sp}^2$/$_{sp}^3$ bonds, hereafter called sp$^{2/3}$ (assignment of Raman peaks in disordered carbons has been discussed previously[90,94,97]). This band presents a strong resonant character (contrary to the sp$^3$ C mode in diamond), evidenced by a high dependence of the intensity and position on wavelength.[90]

The Tyranno SA3 fiber (UBE Industries Ltd., Ube, Yamaguchi, Japan) is synthesized from graft PCS where methyl groups are replaced by organic groups containing aluminum as a crystal growth inhibitor. Reticulation is obtained by 160°C thermal oxidation, the resulting oxycarbide being carbothermally reduced at 1400°C under inert gas, before decomposition at 1800°C (with CO release). It has been claimed that the glassy silica layer of the SA3 fiber presents a remarkable alkali resistance.[106] The Sylramic fiber (Dow Corning, Midland, MI) is also synthesized from graft PCS, but methyl groups are replaced by organic groups containing boron and titanium. The fiber is made of 96 wt% β-SiC nanocrystals.

The peak area (Figure 12.16a) indicates the concentration of the analyzed phase, the position (wave number, Figure 12.16b) is related to its chemical nature and the wave number comparison gives information on both the nature of the species (especially for resonant lines like the D band at about 1330 cm$^{-1}$) and the strain level.[90,94,107] Finally, the bandwidth is related to the phase short-range order. Raman images of SA3 and Sylramic fiber cross sections indicate that core and skin carbon species have a different nature and concentration. The fiber's core is carbon rich and, in the case of the SA3 fiber D band, is up-shifted by ~30 cm$^{-1}$ from the core to the surface. This important shift is due to a transformation of the C–C bond (aromatization). SA3 presents a "carbon-rich tube," while the carbon rate increases gradually from the surface to the core in the Sylramic. This difference between the two fibers can be linked to the elaboration process. During the spinning phase, the viscosity difference between long and short PCS chains leads to concentration gradients. The smallest chains having a lower viscosity could preferentially migrate to the fiber's core. Since they are rich in carbon, due to the high concentration in polymer terminations, they induce carbonated species concentration gradients. These heterogeneities have important consequences on the mechanics of each fiber.

## 4. Prediction of Material Properties from Raman Parameters

The relationship between grain size and material properties is known for a large number of materials. Once the relationship between Raman parameters and grain size is established, it becomes possible to correlate Raman parameters with the properties to be achieved in the different parts of the materials. This work was made for $CeO_2$, an electrolyte for high-temperature fuel cells and its electrical properties (e.g., the concentration of defects). Raman prediction was in agreement with electrochemical and thermodynamic measurements.[9,100,101] Prediction of the mechanical properties of SiC fibers from Raman spectra is also possible.

A model has been used to calculate carbon grain size distribution in the as-received SA3 cross section (Figure 12.16c). A ring of larger carbon species appears on the fiber's periphery, indicating a higher short-range order, which is consistent with the decrease in Raman peak widths. This size increase near the surface can be linked to the elaboration process, revealing a higher temperature in this area. Heat-treated fibers show a large increase in this ring. The observed size distribution indicates that crystal growth of the carbon moieties is obtained even deep in the fiber (a small area of low short-range order remains at the core). The relationship between Raman parameters and mechanical properties allows conversion of the Raman image in a predictive "mechanical" image. As an example, we present in Figure 12.16c the expected ultimate strength to be achieved for the different fiber regions. In this example, the highest values are expected from the fiber core. Other types of fibers exhibit homogeneous behavior or an inverse configuration. Note that if a map gives a didactic view, a line scan offers a better compromise between rapidity and quality of information. An example is shown in Figure 12.17 with a line scan (~18 μm) between two adjacent NLM SiC fibers in a mullite matrix composite (a $ZrO_2$-based coating has been applied onto the fibers).

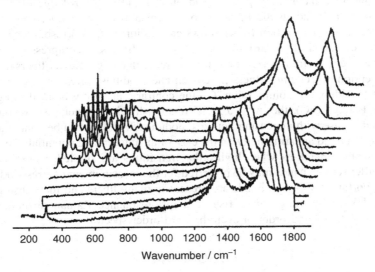

**FIGURE 12.17** Spectral in-line scan (2 μm step) of a composite cross section. Spectra were recorded from one NLM202 fiber to another, through the {aluminosilicate + ($ZrO_2$; $GeO_2$)} tailored interface and the mullite matrix. (Reprinted from Colomban, P., Raman microspectrometry and imaging of ceramic fibers in CMCs and MMCs, 103, 517, 2000. With permission from The American Ceramic Society.)

## 5. Residual Stress/Strain in Multiphase Materials

Fiber stress determination is of major importance for the modeling of composites and it is now well established that Raman microspectroscopy, with its main advantage being its nondestructive nature, makes it possible in composites.[107] First, Galiotis[108] and Young[109] demonstrated that Raman spectroscopy is an excellent method to follow the deformation of aramid and carbon fibers. This is a result of variation in the (stretching) vibrational wave number, as a consequence of the anharmonicity of the interatomic bonds. The relationship linking Raman wave number shifts ($\Delta\nu$) to the tensile strain ($\Delta\varepsilon$) is linear, $\Delta\nu = J\Delta\varepsilon$.

It makes it possible to characterize fibers, composites,[107–111] and films.[112,113] The only requirement for such studies is to use matrices sufficiently translucent to allow *in situ* recording of the fiber signals through the matrix (a few microns to a few tens of microns inside) and the spectrum of the matrix should not overlap with that of the fibers. The quantitative information on the fiber strain distribution comes from a calibration of the "wave number shift" of the peak against strain. During the last few years, micro-Raman spectroscopy has been increasingly used as a powerful technique to measure local residual stress in submicronic electronic devices. Such localized stresses can induce various types of structural defects modifying the electronic properties. As pointed out in Section III.A, knowledge of the stress concentrations in fiber-reinforced composites is of major importance. Since such stress concentrations cannot be measured easily, many analytical theories have been proposed to understand the micromechanics of composites using the microindentation push-in and push-out tests. Knowledge of Young's modulus ($E$), found in the literature or obtained from local microindentation measurements (see Section III.A) allows calculation of the corresponding stress distribution. A shift toward higher energies indicates a compressive strain, whereas a tensile strain corresponds to a decrease in energy. Kevlar fibers display an exceptional strength/stiffness in tension and modulus softening, followed by abrupt "yielding" in compression. On the contrary, in tightly bonded inorganic materials (e.g., SiC), the $|\Delta\nu| = J \times |\Delta\varepsilon|$ relationship is symmetrical. Absolute $J$ values decrease with increasing Young's modulus.[107,114] This can be related to the decrease in bond anharmonicity. In the case of an anharmonic potential, the wave number shift increases when high-order transitions are studied. Consequently high-order Raman bands may be preferred for the measurement of stress-induced shifts. Fortunately carbon bonds are present in many fibers and their absorption in the UV-visible range gives rise to a (pre)resonance Raman phenomenon, enhancing the second-order or even the third-order spectra.

## VI. SUMMARY

The examples reported in this chapter show that the new processing routes using liquid organometallic ceramic precursors allow the tailoring of multiphase and composite materials at the micro- and nanoscale, and hence the development of functionally graded materials. The stringent advantages are a better control of the matrix shrinkage during densification, the flexibility of the matrix composition from some mixtures of precursors, and a reduction in the cost and processing time in comparison with the CVI technology.

Depth-sensing microindentation allows the reliable determination of Young's modulus and Vickers' microhardness of the different phases made of grains larger than ~2 to 5 µm. Variations in chemical bonding and short-range order can be extracted from the intensity, wave number, and bandwidth of Raman peaks. The wave number shifts can be used to map the strain (and calculate the stress). Analysis of the bandwidth offers a tool to ascertain whether the wave number shift is strain induced or related to structural evolution of the Raman probe.

Raman spectroscopy provides a better understanding of the material evolution during the materials preparation and aging in operating atmospheres.

## ACKNOWLEDGMENTS

The author wishes to thank G. Gouadec, I. Kosacki, M. Parlier, E. Mouchon, E. Bruneton, S. Karlin, V. Vendange, M. Wey, J.L. Lagrange, and C. Courtemanche for their contributions.

## REFERENCES

1. Colomban, P., Gel technology in ceramics, glass-ceramics and ceramic-ceramic composites, *Ceram. Int.*, 15, 23, 1989.
2. Colomban, P., Chemical routes and sol-gel processes: the elaboration of ultrafine powders, *Ind. Ceram.*, 792, 186, 1985.
3. Pérez-Arantegui, J., Molera, J., Larrea, A., Pradell, T., Vendrell-Saz, M., Borgia, I., Brunetti, B.G., Cariati, F., Fermo, P., Mellini, M., Sgamellotti, A., Viti. C., Luster pottery from the thirteenth century to the sixteenth century: a nanostructured thin metallic film, *J. Am. Ceram. Soc.*, 84, 442, 2001.
4. Karlin, S., Colomban, P., Phase diagram, short range structure and amorphous phases in the $ZrO_2$-$GeO_2$(-$H_2O$) system, *J. Am. Ceram. Soc.*, 82, 735, 1999.
5. Segal, D., *Chemical Synthesis of Advanced Ceramic Materials*, Cambridge University Press, Cambridge, 1989.
6. Perthuis, H., and Colomban, P., Well densified NASICON-type ceramics elaborated using sol-gel process and sintering at low temperatures, *Mater. Res. Bull.*, 19, 621, 1984.
7. Perthuis, H., and Colomban, P., Sol-gel routes leading to NASICON ceramics, *Ceram. Int.*, 12, 39, 1986.
8. Mazdiyasni, K.S., Powder synthesis from metal-organic precursors, *Ceram. Int.*, 8, 42, 1982.
9. Yoldas, B.E., Effect of variation in polymerized oxides on sintering and crystalline transformations, *J. Am. Ceram. Soc.*, 65, 387, 1977.
10. Sakka, S., and Kamiya, K., Glasses from metal alcoholates, *J. Non-Cryst. Solids*, 42, 403, 1980.
11. Johnson, D.W., Jr., Sol-gel processing of ceramics and glass, *Am. Ceram. Soc. Bull.*, 64, 1587, 1985.

12. Colomban, P., and Wey, M., Sol-gel control of the matrix net-shape sintering in 3D reinforced ceramic matrix composites, *J. Eur. Ceram. Soc.*, 17, 1475, 1997.

13. Okamura, H., and Bowen, H.K., Preparation of alkoxides for the synthesis of ceramics, *Ceram. Int.*, 12, 161, 1986.

14. Bruneton, E., Bigarré, J., Michel, D., and Colomban, P., Heterogeneity, nucleation, shrinkage and bloating in sol-gel glass ceramics: the case of LAS compositions, *J. Mater. Sci.*, 32, 3541, 1997.

15. Klein, L.C., Ed., *Sol-Gel Technology*, Noyes Publications, Park Ridge, NJ, 1988.

16. Colomban, P., Raman studies of inorganic gels and of their sol-to-gel, gel-to-glass and glass-to-ceramic transformation, *J. Raman Spectrosc.*, 27, 747, 1996.

17. Pouxviel, J.C., Boilot, J.P., Dauger, A., and Wright, A., Gelation study of alumino-silicates by small-angle neutron scattering, *J. Non-Crystl. Solids*, 103, 331, 1988.

18. Pouxviel, J.C., Boilot, J.P., Smaihi, M., and Dauger, A., Structural study of alu-mino-silicate sols and gels by SAXS and SANS, *J. Non-Crystl. Solids*, 106, 147, 1988.

19. Poorteman, M., Descamps, P., Cambier, F., Plisnier, M., Canonne, V., and Des-camps, J.C., Silicon nitride/silicon carbide nanocomposite obtained by nitridation of SiC: fabrication and high temperature mechanical properties, *J. Eur. Ceram. Soc.*, 23, 2361, 2003.

20. Yajima, S., Okamura, K., Hayashi, J., and Omori, M., Development of a SiC fibre with high tensile strength, *Nature*, 261, 683, 1976.

21. Yajima, S., Okamura, K., Hayashi, J., and Omori, M., Synthesis of continuous SiC fibre with a high tensile strength, *J. Am. Ceram. Soc.*, 59, 324, 1976.

22. Noireaux, P., Jamet, J., Parlier, M., and Bacos, M.P., Polysilanes et leur procédé de préparation, French Patent no. FR 2642080, July 27, 1990.

23. Parlier, M., and Colomban, P., Composites à matrice céramique pour applications thermostructurales, *Recherche Aerospatiale*, 5–6, 457, 1996.

24. Colomban, P., and Vendange, V., Sintering of alumina and mullite prepared by slow hydrolysis of alkoxides: the role of the protonic species and of pore topology, *J. Non-Crystl. Solids*, 147, 245, 1992.

25. Bruneton, E., and Colomban, P., Influence of hydrolysis conditions on crystalli-zation, phase transitions and sintering of zirconias prepared by alkoxide hydrol-ysis, *J. Non-Crystl. Solids*, 147, 201, 1992.

26. Vendange, V., and Colomban, P., Densification mechanisms of alumina, alumino-silicates and aluminoborosilicates gels, glasses and ceramics, *J. Sol-Gel Sci. Tech-nol.*, 2, 407, 1994.

27. Colomban, P., Protonic defects and crystallization of Sol-Gel (Si,Ge) mullites and alumina, *Ceramics Today — Tomorrow's Ceramics*, P. Vincenzini, Ed., *Mater. Sci. Monogr.*, 66B, 599, 1991.

28. Colomban, P., Courret, H., Romain, F., Gouadec, G., and Michel, D., Sol-gel prepared pure and Li-doped hexacelsian polymorphs. An infrared, Raman and thermal expansion study of -phase stabilization by frozen short-range disorder, *J. Am. Ceram. Soc.*, 83, 2974, 2000.

29. Roy, R., Gel route to homogeneous glass preparation, *J. Am. Ceram. Soc.*, 52, 344, 1969.

30. Vendange, V., and Colomban, P., How to tailor the porous structure of alumina and aluminosilicate gels and glasses, *J. Mater. Res.*, 11, 518, 1996.

31. Vendange, V., and Colomban, P., Determination of the hydroxyl group content in gels and porous "glasses" issued of alkoxide hydrolysis by combined TGA and BET analysis, *J. Porous Mater.*, 3, 193, 1996.
32. Roy, D.M., and Roy, R., Synthesis and stability of minerals in the system MgO-$Al_2O_3$-$SiO_2$-$H_2O$, *Am. Miner.*, 40, 147, 1955.
33. Vendange, V., Colomban, P., and Larché, F., Pore size and liquid impregnation of microporous aluminosilicate gels and glasses, *Microporous Mater.*, 5, 389, 1996.
34. Howe, A.T., and Shilton, M., Densification of HUP solid electrolyte, *J. Solid State Chem.*, 34, 341, 1980.
35. Colomban, P., Ed., *Proton Conductors—Solids, Membranes and Gels—Materials and Devices*, Cambridge University Press, Cambridge, 1992.
36. Bouquin, O., Perthuis, H., and Colomban, P., Low temperature sintering and optimal physical properties: a challenge the NASICON ceramics case. *J. Mater. Sci. Lett.*, 4, 956, 1985.
37. Lagrange, J.L., and Colomban, P., Double particle reinforcement of ceramic matrix composites prepared through a sol-gel route, *Composites Sci. Technol.*, 58, 653, 1998.
38. Rigacci, A., Achard, P., Ehrburger-Dolle, F., and Pirard, R., Structural investigation in monolithic silica aerogels and thermal properties, *J. Non-Crystl. Solids*, 225, 260, 1998.
39. di Costanzo, T., Frappart, C., Mazerolles, L., Rouchaud, J.C., Fedoroff, M., Michel, D., Beauvy, M., Vignes, J.L., Fixation de divers polluants dans des alumines monolithiques poreuses, *Ann. Chim. Sci. Mater.*, 26, 67, 2001.
40. Sherer, G.W., Calas, S., and Sempéré, R., Densification kinetics and structural evolution during sintering of silica aerogel, *J. Non-Crystl. Solids*, 240, 118, 1998.
41. Colomban, P., Frittage de céramiques transparentes PLZT, *Ind. Ceram.*, 697, 531, 1976.
42. Lenfant, P., Plas, D., Ruffo, M., Boilot, J.-P., and Colomban, P., Céramiques d'alumine et de ferrite pour sonde à protons, *Mater. Res. Bull.*, 15, 1817, 1980.
43. Perthuis, H., and Colomban, P., $Li^+$ eucriptite superionic thick films, *J. Mater. Sci. Lett.*, 4, 344, 1985.
44. Perthuis, H., Velasco, G., and Colomban, P., $Na^+$ and $Li^+$ NASICON superionic conductor thick films, *Jpn. J. Appl. Phys.*, 23, 534, 1984.
45. Colomban, P., Boilot, J.-P., Polymètres inorganiques (xerogels et verres) dans les systèmes $M_2O$-$M'O_2SiO_2$-$P_2O_5X_2O_3$, *Rev. Chim. Miner.*, 22, 235, 1985.
46. Blanchard, N., Boilot, J.-P., Colomban, P., Pouxviel, J.-C., New glasses from metal-organic precursors: preparation and properties, *J. Non-Crystl. Solids*, 82, 205, 1986.
47. Colomban, P., Sol-gel synthesis and densification of NASICON powders, *Adv. Ceram.*, 21, 139, 1987.
48. Boilot, J.P., Gay, A., Colomban, P., Lejeune, M., Compositions solides amorphes et homogènes à base de dérivés métalliques, sous forme de gels polymérisés ou de verres, leur préparation et leur application, French Patent (CNRS) no. EN 83 06 934.
49. Colomban, P., and Mazerolles, L., Nanocomposites in mullite-$ZrO_2$ and mullite-$TiO_2$ systems synthesized through alkoxide hydrolysis gel routes. Microstructure and fractography, *J. Mater. Sci.*, 26, 3503, 1991.

50. Colomban, P., and Mazerolles, L., $SiO_2$-$AI_2O_3$ phase diagram and mullite nonstoichiometry of sol-gel prepared monoliths: influence on mechanical properties, *J. Mater. Sci. Lett.*, 2, 1077, 1990.

51. Colomban, P., Structure of oxide gels and glasses by IR and Raman scattering: I. Aluminas, *J. Mater. Sci.*, 24, 3002, 1989.

52. Colomban, P., Structure of oxide gels and glasses by IR and Raman scattering: II. Mullites, *J. Mater. Sci.*, 24, 3011, 1989.

53. Snow, G., Fabrication of transparent electronic PLZT ceramics by atmosphere sintering, *J. Am. Ceram. Soc.*, 56, 91, 1973.

54. Haertling, G.H., Ferroelectric ceramics: history and technology, *J. Am. Ceram. Soc.*, 82, 797, 1999.

55. Erikson, D.D., Wood, T.E., and Wood, W.P., Historical development of abrasive grain, Sol-gel processing symposium, *Ceram. Trans.*, 12, 95, 1998.

56. Sowman, H.G., U.S. Patent nos. 3,795,524, 3,709,706, 3,793,041, 3,916,584, 4,047,965, 4,125,406.

57. Whitney, E.D., and Vaidyanathan, P.N., Microstructural engineering of ceramic cutting tools, *Am. Ceram. Soc. Bull.*, 67, 1010, 1988.

58. Mathers, J.P., Forester, T.E., and Wood, W.P., Sol-gel preparation of non-oxide abrasives, *Am. Ceram. Soc. Bull.*, 68, 130, 1989.

59. Hamasaki, T., Eguchi, K., Koyanagi, K., Matsumoto, A., Utsunomiya, T., and Koba, K., Preparation and characterization of machinable mica/glass-ceramics by sol-gel process, *J. Am. Ceram. Soc.*, 71, 1120, 1988.

60. Hoffman, D., Roy, R., and Komarneni, S., Diphasic ceramic composites via a sol-gel method, *Mater. Lett.*, 2, 245, 1984.

61. Roy, R.A., and Roy, R., Diphasic xerogels: I. Ceramics metal composites, *Mater. Res. Bull.*, 19, 169, 1984.

62. Petrullat, J., Ray, S., Schubert, U., Guldner, G., Egger, C., and Breitscheidel, B., Preparation and processing of metal-ceramic composite materials, *J. Non-Crystl. Solids*, 147, 594, 1992.

63. Breval, E., Deng, Z., Chiou, S., and Pantano, C.G., Sol-gel prepared Ni-alumina composite materials. Part I, Microstructure and mechanical properties. *J. Mater. Sci.*, 27, 1464, 1992.

64. Stella, A., Cheyssac, P., De Silvestri, S., Kofman, R., Lanzani, G., Nisoli, M., Tognini, P., Self organized growth and ultrafast electrob dynamics in metallic nanoparticles, in *Nanophase and Nanocomposite Materials II*, S. Komarneni, J.C. Parker, and H.J. Wollenberger, Eds., *Mater. Res. Soc. Symp. Proc.*, 457, 155, 1997.

65. Vendange, V., and Colomban, P., Elaboration and thermal stability of alumina, alumino-silicate. Fe, Co, Ni magnetic nanocomposites prepared through a sol-gel route, *Mater. Sci. Eng. A*, 168, 199, 1993.

66. Colomban, P., and Vendange, V., Sol-gel routes towards magnetic nanocomposites with tailored microwave absorption, in *Nanophase and Nanocomposite Materials II*, S. Komarneni, J.C. Parker, and H.J. Wollenberger, eds., *Mater. Res. Soc. Symp. Proc.*, 457, 451, 1997.

67. Mouchon, E., and Colomban, P., Microwave absorbent: preparation, mechanical properties and RF/microwave conductivity of SiC (and/or mullite) fibres reinforced NASICON matrix composites, *J. Mater. Sci.*, 31, 323, 1996.

68. Naslain, R., Fiber-matrix interphases and interfaces in ceramic matrix composites processed by CVI, *Composite Interfaces*, 1, 253, 1993.

69. Droillard, B., Lamon, J., and Bourrat, X., Strong interfaces in CMCs-conditions for efficient multilayered interphases, *Mater. Res. Soc. Symp. Proc.*, 365, 371, 1995.

70. Qui, D., and Pantano, C.G., Sol-gel processing of carbon fiber-reinforced glass matrix, in *Ultrastructure Processing of Ceramics, Glasses and Composites*, J.D. Mackenzie and D.R. Ulrich, Eds., John Wiley & Sons, New York, 1987.

71. Panhorst, W., Spallek, M., Brueckner, R., Hegaler, H., Reich, C., Gratewohl, G., Meier, B., Spelmann, D.S., Fibre-reinforced glass ceramics fabricated by a novel process, *Ceram. Eng. Sci. Proc.*, 11, 947–63, 1990.

72. Russell-Floyd, R.S., Harris, B., Cooke, R.G., Laurie, J., Hammett, F.W., Jones, R.W., Wang, T., Application of sol-gel processing techniques for the manufacture of fiber-reinforced ceramics, *J. Am. Ceram. Soc.*, 76, 2635, 1993.

73. Guney, V., Jones, F.R., James, P.F., and Bailey, J.E., Alumina ceramic matrices for fibre composites prepared by modified sol-gel processing, Institute of Physics Publishing, London, 1990.

74. Wu, J., Chen, M., Jones, F.R., and James, P.F., Mullite matrix fibre reinforced composites by sol-gel processing, *Ceram. Trans.*, 46, 177, 1995.

75. Wu, J., Chen, M., Jones, F.R., and James, P.F., Characterization of sol-gel derived alumina-silica matrices for continuous fibre reinforced composites, *J. Eur. Ceram. Soc.*, 16, 619, 1996.

76. Colomban, P., Process for fabricating a ceramic matrix composite incorporating woven fibers and materials with different compositions and properties in the same composite, *Mater. Technol.*, 10, 89, 1995.

77. Colomban, P., Menet, M., Mouchon, E., Courtemanche, C., and Parlier, M., Composites céramique-céramique multicouches élabo-rés en utilisant un précurseur d'interface et un précurseur de matrice [Multilayer fiber-matrix ceramic composite material and process for its production], French Patent nos. (ONERA) FR 2672283 (Feb. 4, 1991) and EP 0 498 698 (Jan. 29, 1992), and U.S. Patent no. 07/830.904.

78. Colomban, P., Bruneton, E., Lagrange, J.L., Mouchon, E., Sol-gel mullite matrix-SiC and -mullite 2D woven fabrics composites with or without zirconia containing interphase. Elaboration and properties, *J. Eur. Ceram. Soc.*, 16, 301, 1996.

79. Colomban, P., and Lapous, N., New sol-gel matrices of chemically stable composites BAS, NAS and CAS, *Composites Sci. Technol.*, 56, 737, 1996.

80. Mouchon, E., and Colomban, P., Oxide ceramic matrix-oxide fibers woven fabric composites exhibiting dissipative fracture behavior, *Composites*, 26, 175, 1995.

81. Colomban, P., Composites ceramiques multiniredux ou l'interêt des méthods sol-gel, Proc. 8ᵉ Jouznées Nationales surles composites, JNC'8, Alix, O., Favre, J.P., and Ladevèze, P., Eds., AMAC, Paris, 1992, 73.

82. Jamet, J., Demange, D., and Loubeau, J., Procédé d'élaboration de composites, French Patent no. (ONERA) FR 2526785 (Nov. 18, 1985).

83. Colomban, P., Wey, M., and Parlier, M., Procédé d'élaboration d'un matériau céramique par infiltration d'un précurseur dans un support poreux céramique, French Patents (ONERA) FR 2713222 (June 9, 1995), EP 0.656.329 (Jan. 31, 1995), and 0656329 (June 17, 1998).

84. Touati, F., Gharbi, N., and Colomban, P., Structural evolution in polyolysed hybrid organic-inorganic alumina gels, *J. Mater. Sci.*, 35, 1565, 2000.

85. Parlier, M., and Ritti, M.H., State of the art and perspective for oxide-oxide composites, *Aerospace Sci. Technol.*, 7, 211, 2003.

86. Colomban, P., Sol-gel route to functional and hierarchical ceramid matrix composites, in Proceedings ICIM '96, ECSSM '96, Lyon '96, 3rd International Conference on Intelligent Materials, June 3–5, 1996, P.F. Gobin and J. Tatibouet, Eds., SPIE, Lyon, 1996.

87. Colomban, P., Sol-gel control of the micro/nanostructure of functional ceramic-ceramic and metal-ceramic composites, *J. Mater. Res.*, 13, 803, 1998.

88. Loubet, J.L., Georges, J.M., Marchesini, O., and Meille, G., Vickers indentation curves of magnesium oxide, *J. Tribol.*, 106, 43, 1984.

89. Colomban, P., Tailoring and control of the micro/nanostructure of functional (FGM) CMC's and MMC's, *J. Korean Ceram. Soc.*, 5, 55, 1999.

90. Gouadec, G., Colomban, P., and Bansal, N.P., Raman study of Hi-nicalon fiber reinforced celsian composites. Part I: Distribution and nanostructure of different phases, *J. Am. Ceram. Soc.*, 84, 1129, 2001.

91. Gouadec, G., Colomban, P., and Bansal, N.P., Raman study of Hi-nicalon fiber reinforced celsian composites. Part II: Residual stress in the fibers, *J. Am. Ceram. Soc.*, 84, 1136, 2001.

92. Hsueh, C.H., Interfacial debonding and pull-out stresses of fiber reinforced composites, *Mater. Sci. Eng. A*, A154, 125, 1992.

93. Marshall, D.B., Analysis of fiber debonding and sliding experiments in brittle matrix composites, *Acta Metal. Mater.*, 40, 427, 1992.

94. Colomban, P., Stress- and nanostructure-imaging of ceramic fibers and abradable thermal barrier coatings by Raman microspectrometry—state of the art and perspectives, Proceedings of the 24th Annual Cocoa Beach Conference and Exposition, Cocoa Beach, Florida, Jan. 23–28, 2000, T. Jessen and E. Ustundag, Eds., *Ceram. Eng. Sci. Proc.*, 21, 143, 2000.

95. Colomban, P., and Havel, M., Raman imaging of stress-induced phase transformation in transparent ZnSe ceramics and sapphire single crystal, *J. Raman Spectrosc.*, 33, 789, 2002.

96. Havel, M., and Colomban, P., Rayleigh and Raman image of the bulk/surface surface nanostructure of SiC based fibres, *Composite B Eng.*, 35B, 353, 2004.

97. Havel, M., and Colomban, P., Skin/bulk nanostructure and corrosion of SiC based fibres. A surface Rayleigh and Raman study, *J. Raman Spectrosc.*, 34, 786, 2003.

98. Colomban, P., Raman/Rayleigh study of nanophases, Proceedings of the 27th Annual Cocoa Beach Conference and Exposition, Cocoa Beach, Florida, January 26–31, 2003, W.H. Kriven and H.T. Lin, Eds., *Ceram. Eng. Sci. Proc.*, 24, 41, 2003.

99. Gouadec, G., and Colomban, P., Raman spectroscopy of nanomaterials: how do spectra relate to particle size and local mechanics?, Progress in crystal growth, in press, 2005.

100. Suzuki, T., Kosacki, I., Colomban, P., and Anderson, H.U., Electrical conductivity and lattice defects in nanocrystalline $CeO_2$ thin films, *J. Am. Ceram. Soc.*, 84, 2007, 2001.

101. Kosacki, I., Suzuki, T., Anderson, H., and Colomban, P., Raman scattering and lattice defects in nanocrystalline $CeO_2$ thin films, *Solid State Ionics*, 149, 99, 2002.

102. Surca-Vuk, A., Orel, B., Drazic, G., and Colomban, P., Vibrational spectroscopy and analytical electron microscopy studies of Fe-V-O and In-V-O thin films, in *Nanostructured Materials*, H. Hofman, Z. Rahman, U. Schubert, Eds., Springer, Wien, 2002.

103. Parayanthal, P., and Pollak, F.H., Raman scattering in alloy semiconductors: "spatial correlation" model, *Phys. Rev. Lett.*, 52, 1822, 1984.

104. Solin, S.A., and Caswell, N., Raman scattering from alkali graphite intercalation compounds, *J. Raman Spectrosc.*, 10, 129, 1981.

105. Duval, E., Far-infrared and Raman vibrational transitions of a solid sphere: selection rules, *Phys. Rev. B*, 46, 5795, 1992.

106. Ishikawa, T., Kohtoku, Y., Kumagawa, K., Yamamura, T., and Nagasawa, T., High-strength alkali-resistant sintered SiC fibre stable to 2200°C, *Nature*, 391, 773, 1998.

107. Colomban, P., Analysis of strain and stress in ceramic, polymer and metal matrix composites by Raman spectroscopy, *Adv. Eng. Mater.*, 4, 535, 2002.

108. Galiotis, C., Laser Raman spectroscopy, a new stress/strain measurement technique for the remote and on-line non-destructive inspection of fiber reinforce polymer composites, *Mater. Technol.*, 8, 203, 1993.

109. Young, R.J., Raman spectroscopy and mechanical properties, in *Characterization of Solid Polymers*, S.J. Spelles, Ed., Chapman & Hall, London, 1994.

110. Beyerlein, J., Amer, M.S., Schadler, L.S., and Phoenix, S.L., New methodology for determining *in situ* fiber, matrix and interface stresses in damaged multifiber composites, *Sci. Eng. Composite Mater.*, 7, 151, 1998.

111. Wu, J., and Colomban, P., Raman spectroscopy study on the stress distribution in the continuous fibre reinforced ceramic matrix composites, *J. Raman Spectrosc.*, 28, 523, 1997.

112. Mohrbacker, H., Van Acker, K., Blanpain, B., Van Houtte, P., and Celis, J.P., Comparative measurement of residual stress in diamond coatings by low-incident-beam-angle-diffraction and micro-Raman spectroscopy, *J. Mater. Res.*, 11, 1776, 1996.

113. de Wolf, I., Vanhellemont, J., Romano-Rodriguez, A., Norström, H., and Maes, H.E., Micro-Raman study of stress distribution in local isolation structure and correlation with transmission electron spectroscopy, *J. Appl. Phys.*, 71, 898, 1992.

114. Gouadec, G., and Colomban, P., Non-destructive mechanical characterization of SiC fibers by Raman spectroscopy, *J. Eur. Ceram. Soc.*, 21, 1249, 2001.

# 13 Nanocomposite Materials

*Sridhar Komarneni*

## CONTENTS

# I. INTRODUCTION

The term "nanocomposites" was first coined by Roy et al. sometime during the period 1982 to 1983 to describe the major conceptual redirection of the sol-gel process, that is, using the solution sol-gel (SSG) process to create maximally heterogeneous instead of homogeneous materials.[1-4] Di- and multiphasic nanoheterogeneous sol-gel materials were prepared and documented in 1984.[1-4] Nanocomposites should be clearly differentiated from "nanocrystalline" and "nanophase" materials, which refer to single phases in the nanometer range. "Nanocomposites" refers to composites of more than 1 Gibbsian solid phase where at least one dimension is in the nanometer range and typically all solid phases are in the 1 to 100 nm range. The solid phases can be amorphous, semicrystalline, or crystalline, or combinations thereof. They can be inorganic, organic, or both, and essentially of any composition. The "nanocomposite" theme has now been widely and accurately copied and used worldwide. Although the term "nanocomposites" was coined only recently, nanocomposites are pervasive throughout biological systems (e.g., plants and bones). Only very few man-made materials, such as intercalation compounds (e.g., graphite intercalation compounds, pillared clays, and clay-organic complexes) and entrapment-type compounds (e.g., zeolite-organic complexes) have dealt with this size of material. In the biological world, plants form nanocomposites with the accumulation of significant amounts of inorganic components such as silicon, calcium, aluminum, etc. at the tissue and cellular level to deal with the mechanical and biophysical demands of their survival. In the animal world, bones, teeth, and shells consist of nanocomposites of inorganic and organic materials to achieve several key properties. The objective of this chapter, however, is to review the work on man-made nanocomposite materials and to identify areas where further developments are likely to occur.

# II. DIFFERENT FAMILIES OF NANOCOMPOSITES

Although nanocomposites can be classified as other composites based on connectivity,[5] here I have identified several major families of nanocomposites based on their material function, physical and chemical differences, temperature of formation, etc. There are five major groups of nanocomposites at present: (1) sol-gel nanocomposites, which are composites made at low temperatures (less than 100°C), and these nanocomposite precursors can lead to homogeneous single crystalline phase ceramics or multiphasic crystalline ceramics with high-temperature heating; (2) intercalation-type nanocomposites, which can be prepared at low temperatures (less than 200°C) and lead to useful materials with heating to modest temperatures (less than 500°C); (3) entrapment-type nanocomposites, which can be prepared from three-dimensionally linked network structures such as zeolites which can also be synthesized at low temperatures (less than 250°C); (4) electroceramic nanocomposites, which can be prepared by mixing nanophases of ferroelectric, dielectric, superconducting, and ferroic materials in a polymer matrix at low temperatures (less than 200°C); and (5) structural ceramic nano-composites, which are prepared by traditional ceramic processing at very high temperatures (1000 to 1800°C). These five major categories can be subdivided further and are described below in detail.

## A. SOL-GEL NANOCOMPOSITES

The worldwide goal of all SSG work has been ultrahomogeneity, while our goal in this area changed in the early 1980s to the preparation of nanocomposites that exhibit ultraheterogeneity or nanoheterogeneity. This conceptual innovation of nanocomposite materials was a radically new direction for sol-gel research. The concept of diphasic ceramic-ceramic gels as a new class of materials with inter-esting potential was first introduced by Roy.[2] This new direction for sol-gel processing science is now well established in our laboratory and elsewhere.[3-37] The goal of ceramic materials processing via the nanocomposite ceramic gels is to exploit the thermodynamics of metastable materials and, in particular, to utilize the heat of reaction of the discrete phases and the advantages offered by epitaxy. Here the concepts are illustrated with the processing of densification of different types of sol-gel nanocomposites, leading to the crystallization and densification of mullite, alumina, and zircon ceramics which have numerous technological applications as infrared transmitting materials, refractory materials, substrate materials, and high-temperature structural materials. Readers should refer to the numerous publications cited above for a thorough understanding of these con-cepts. In addition to the above exploitation of the thermodynamics of sol-gel materials, one can also process sol-gel materials into various shapes such as films, fibers, and monoliths with designed pore sizes in which organic and inorganic second phases can be incorporated.

Sol-gel nanocomposites are further subdivided into six categories: (1) com-positionally different nanocomposites, (2) structurally different nanocomposites, (3) both compositionally and structurally different nanocomposites, (4) nanocom-posites of gels with precipitated phases, (5) nanocomposites of xerogels with metal phases, and (6) nanocomposites of inorganic gels and organic molecules.

## 1. Compositionally Different Sol-Gel Nanocomposites

These are very intimate mixtures composed of two or more solid phases that differ in composition and each with particle sizes of 10 to 20 nm. Solid phases of these dimensions produce "sols" when dispersed in a liquid. Two or more sols of different composition can be uniformly mixed and gelled to obtain composi-tionally different nanocomposites. Figure 13.1a shows the transmission electron microscopy (TEM) picture of a sol-gel nanocomposite of mullite composition consisting of spherical silica particles (20 nm) and rod-like alumina (boehmite) particles (approximately 7 nm). Such a uniform physical mixture can be distin-guished from a homogeneous sol-gel material which does not show any nonunifor-mity because it is mixed on an atomic scale (Figure 13.1b).[8,32] The compositionally

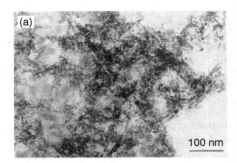
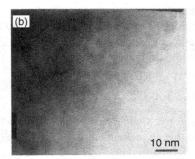

**FIGURE 13.1** Transmission electron micrographs of mullite composition gels: (a) nanocomposite and (b) homogeneous (single phase).

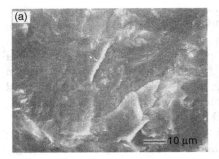
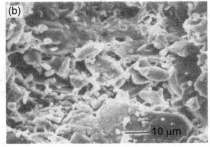

**FIGURE 13.2** Scanning electron micrographs of fracture surfaces of cordierite composition gels sintered at 1300°C for 2 h (a) nanocomposite and (b) homogeneous (single phase). (Reprinted with permission from Kazakos, A.M., Komarneni, S., and Roy, R., *J. Mater. Res.*, 5, 1095, 1990.)

different sol-gel nanocomposites have been shown to sinter to crystalline products in several compositional systems[8,23,26,27,37] with close to theoretical density at much lower temperatures than the homogeneous gels.[23,27] Figure 13.2 compares the scanning electron micrographs of sintered bodies of cordierite made from nanocomposites and homogeneous gels. This example clearly shows that the compositionally different sol-gel nanocomposites densify much better than the homogeneous gels. Similar results have been obtained in other compositional systems such as $Al_2O_3$-$SiO_2$, $SiO_2$-$MgO$, $Al_2O_3$-$TiO_2$, etc.[8,30,37] The enhanced densification of the nanocomposite gels made from two or more sols or nanophases may be attributed to the heat of reaction among the sols or nanophases. The nanocomposite gels store much higher metastable energy than the single-phase gels (Figure 13.3)[4] and thus the enhanced densification of the nanocomposites may be attributed to the additional energy provided during the exothermic heat of reaction. Another reason for enhanced densification in the nanocomposites appears to be due to the simultaneous densification and crystallization,

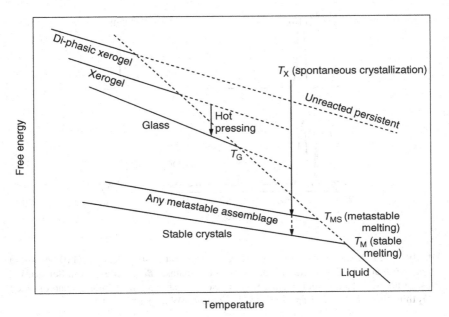

Temperature

**FIGURE 13.3** Representation of G-T relations among isoplethal phases. (Reprinted with permission of the American Ceramic Society, *J. Am. Ceram. Soc.*, 67, 468, 1984.)

unlike the homogeneous gels where crystallization precedes densification because of atomic-scale mixing. Once the crystallization of the equilibrium phase takes place, it is difficult to densify unless very high temperatures are used. The above concept of compositionally different sol-gel nanocomposites is generalizable and applicable to all oxide ceramics that can be used as structural or electroceramic materials. The lower processing temperatures are not only useful in conserving energy, but are also advantageous in preparing electroceramics and multilayer capacitors, which contain volatile elements and therefore need to be processed at low temperatures.

## 2. Structurally Different Sol-Gel Nanocomposites

These nanocomposites consist of two or more solid phases with the same composition but different structure. Examples include mixtures of ultrafine crystalline seeds in amorphous or semicrystalline xerogels. The gel materials are highly amenable for uniform distribution of ultrafine crystalline seeds. Using the system $Al_2O_3$, the effects of $\alpha$-$Al_2O_3$ seeds in lowering crystallization temperature through epitaxy[10–16,20,24–25] have been clearly demonstrated. Further evidence for epitaxy has been thoroughly demonstrated with sol-gel films on single crystal substrates,[31,35,36] where the gel crystallized into a single crystal-like film with the same orientation as the substrate. Figure 13.4 shows an x-ray diffractogram of a $TiO_2$ film on a $TiO_2$ single crystal (110) substrate with the same orientation and

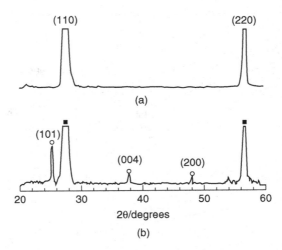

**FIGURE 13.4** X-ray diffractogram of a TiO$_2$ film on rutile single crystal (110) substrate (a) after 6 h at 900°C; (b) after 1 h at 650°C. O = anatase, ■ = rutile. (From Selvarj, U., Prasadarao, A.V., Komarneni, S., and Roy, R., Sol-gel fabrication of epitaxial and oriented TiO$_2$ thin films, *J. Am. Ceram. Soc.*, 75, 1167, 1992. With permission.)

is indistinguishable from the substrate below.[36] The role of solid state epitaxy in lowering the crystallization temperature is now well established in numerous compositional systems, including electroceramics. The resultant effects of this structural epitaxy on microstructure and sintering of alumina and other gels have already been reported. Using the structurally different nanocomposite sol-gels, it is now possible to crystallize some feldspar gels (and glasses; see below) which have been found previously to be impossible to crystallize. The effect of solid state epitaxy in compositional systems that have high-energy barriers is significant and can be exploited in the making of all types of ceramics. Further developments are likely to occur in electroceramics, where there are some useful phases, such as lead zinc niobate, which cannot be crystallized under ordinary pressures.

### 3. Both Compositionally and Structurally Different Sol-Gel Nanocomposites

These nanocomposites are a combination of the above two types of nanocomposites and consist of compositionally discrete phases with crystalline seeds of the equilibrium phase. This combination utilizes the heat of reaction of the compositionally discrete phases and the lowering of the energy barrier through epitaxial growth on the crystalline seeds. Using zircon[17] as the prototype model, we have shown that both the compositionally and structurally different sol-gel nanocomposites crystallize at a lower sintering temperature than either the compositionally or structurally different sol-gel nanocomposites. Table 13.1 shows the lowest temperatures at which zircon formed for different types of nanocomposites. It is obvious from Table 13.1 that both the compositionally

**TABLE 13.1**
**Lowest Temperature at which Zircon Formed in Different Mono-
and Nano-Composite Precursors[17]**

| Structural Diphasicity | Compositional Diphasicity | |
| --- | --- | --- |
| | No | Yes |
| No | 1325°C | 1175°C |
| Yes | 1100°C | 1075°C |

Reprinted with permission of Chapman & Hall.

and structurally different nanocomposite gels yielded zircon at the lowest temperature. Similar results were obtained for $ThO_2$-$SiO_2$ and $Al_2O_3$-MgO systems.[18,19] In the latter case, densification and microstructural studies of 93% $Al_2O_3$-7% MgO (abrasive grain composition by 3M Co.) gel seeded with both α-$Al_2O_3$ and $MgAl_2O_4$ seeds revealed that double seeding led to complete densification with very fine microstructure.[19]

## 4. Nanocomposites of Gels with Precipitated Phases

These are a type of ceramic-ceramic nanocomposite that are prepared by the growth of extremely fine crystalline or noncrystalline phases inside the pores of a premade gel (e.g., $SiO_2$) structure.[1,6,7] The growth of fine phases is accomplished by soaking the gel in metal salt solution and subsequent precipitation of the metal with selected anions. Such precipitation presents a vastly more versatile (with respect to composition of the matrix) approach to such nanocomposite materials than is possible with, say, precipitation out of a glass. The gel pores can be modified by liquid and gas deposition techniques.[38] This leads to modification of the chemical character and the effective pore size and gives rise to nanophases smaller than the size of the pores.[38] The nanocomposite materials made by this method extend the processing options for photochromic glasses and catalytic materials with interesting and improved transport, catalytic, and mechanical properties.

## 5. Nanocomposites of Xerogels with Metal Phases

The sol-gel process has been extended to the preparation of new diphasic xerogels, leading to new ceramic-metal nanocomposite materials.[3] These nanocomposites have been prepared by two methods (Figure 13.5). Xerogels of $Al_2O_3$, $SiO_2$, and $ZrO_2$ have been prepared as the matrices with copper, platinum, and nickel (5 to 50 nm) as the dispersed metal phases. Very finely dispersed metal particles (2 to 4 nm) have been deposited by liquid and gas deposition techniques in sol-gel membranes.[38] These materials are obviously important as catalysts and they can be optimized by manipulating the process parameters of the sol-gel methods.

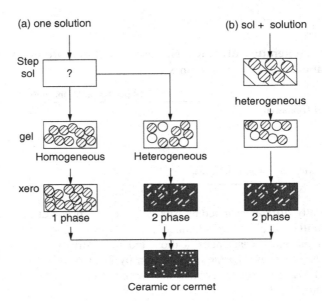

**FIGURE 13.5** Two preparation routes used in making ceramic-metal nanocomposites. Method (a): solution mixing of all components simultaneously. Method (b): uses a premade sol to which a further solution is added before gelation. (From Roy, R.A., and Roy, R., *Mater. Res. Bull.*, 19, 169, 1984. With permission.)

Iron/silica or alumina gel nanocomposites have been prepared and properties such as spin glass magnetic behavior, change in magnetic state with ammonia treatment, and iron magnetic moments[39–43] have been investigated. Sol-gel-derived glass-metal nanocomposites involving iron, nickel, and copper in silica glass matrix have been prepared, and electrical and optical properties have been studied.[44–46] Although the use of these nanocomposites in electrical, magnetic, and optical devices is a long shot, their future in catalysis appears to be bright.

## 6. Nanocomposites of Inorganic Gels and Organic Molecules (Dyes)

The sol-gel process is highly amenable for incorporating optically active organic molecules, such as laser dyes, in porous gel or glass-like matrices because the gels can be prepared at room temperature and the porosity can be controlled. One can incorporate the organic species, including polymers, during gelation[47] or the organic molecules can be introduced into the premade sol-gel matrices through diffusion. Various laser dyes, conducting and conjugated polymers, polymers that contain hydrogen bond acceptor groups, and photochromic molecules have been successfully incorporated into silica gels,[47,48] and these nanocomposites may have interesting optical, nonlinear optical, conducting, and photochromic properties, with potential applications in optical devices, laser materials, and chemical sensors. Studies in the last several years have demonstrated that the activity of the different molecules is not impaired by their incorporation in sol-gel matrices. Further studies in this area include interactions between the sol-gel matrix and the guest molecules for optimizing the properties of the resulting nanocomposites for various applications, and the engineering of the gel pore structure to incorporate selective chemical molecules, leading to the development of optically based chemical sensors. Further details about these nanocomposites can be obtained from an excellent review by Dunn and Zink.[48]

## B. INTERCALATION-TYPE NANOCOMPOSITES

Naturally occurring or synthetic crystals of layer structure, such as graphite and clay, can be intercalated with inorganic and organic species to generate bidimensional nanocomposites. The layered crystals are of two types: (1) those with an unbalanced charge on the layers and (2) those with neutral layers. The 2:1 clay minerals and hydrotalcites (anionic clays) belong to the first group, while the 1:1 clay minerals and graphite are examples of the second type. Table 13.2 gives numerous examples of the layered crystals[49] that can be utilized in the preparation of nanocomposites. Although there are numerous layered crystals, only a few, such as clays and graphite, have been extensively studied as intercalation compounds. There is a large amount of literature on the graphite intercalation compounds[50] and this subject will not be covered here. The main intercalation-type composites that will be described in detail are pillared clays, metal intercalated clays, and clay-organic composites.

## 1. Pillared Clays

Swelling clay minerals such as montmorillonite, hectorite, saponite, nontronite, and beidellite of the smectite family can be pillared in the interlayers by exchanging their interlayer cations with polymeric hydroxy cations followed by dehydration, which leads to ceramic oxide pillars (Figure 13.6). The silicate layers are permanently propped apart by the oxide pillars and zeolitic micropores are formed between the silicate layers. Pillared clays are the perfect example of a nanocomposite because the clay layer is approximately 1 nm and the oxide pillars are approximately 1 to 2 nm. Vaughan et al.[51] were the first to synthesize pillared clays with alumina, and subsequently several other oxides such as $ZrO_2$, $Cr_2O_3$, $TiO_2$, $Fe_2O_3$, $Bi_2O_3$ etc.[52–57] and mixed oxides such as $Al_2O_3$-$SiO_2$, $SiO_2$-$TiO_2$ and $SiO_2$-$Fe_2O_3$[58–60] have been introduced as pillars between the clay layers. Pillared clays are versatile microporous materials because the dimensions and the surface characteristics of the micropores can be designed by changing the size and composition of the pillaring species. In addition, they have large surface areas and high thermal stabilities. Therefore these have potential applications as catalysts, catalyst substrates, selective adsorbents,[60–66] and desiccants.[67–69] We have determined the water adsorption and desorption isotherms of several pillared clays for determining their suitability as desiccants for gas-fired cooling and

**TABLE 13.2**
**Layered Crystals[49]**

| Molecular Layered Crystals | Cation Exchangeable Layered Crystals |
| --- | --- |
| Element | Silicates |
|   Graphite |   Montmorillonite |
| Chalcogenides |   Vermiculite |
|   $MX_2$ ($TiS_2$, $NbSe_2$, $MoS_2$) |   Hectorite |
|   $MPX_3$ ($MnPS_3$, $FePSe_3$) | Phosphates |
|   $Ta_2S_2C$ |   $Zr(HPO_4)_2 \cdot nH_2O$ |
| Oxides |   $Ti(HPO_4)_2 \cdot nH_2O$ |
|   $MoO_3$, $V_2O_5$ |   $Na(UO_2PO_4) \cdot nH_2O$ |
| Oxyhalides | Titanates |
|   FeOCl, VOCl, CrOCl |   $Na_2Ti_3O_7$ |
|   ZrNCl |   $KTiNbO_5$ |
| Hydroxides |   $Rb_xMn_xTi_{2-x}O_4$ |
|   $Zn(OH)_2$, $Cu(OH)_2$ | Vanadates |
| Silicates |   $KV_3O_8$ |
|   Kaolinite, halloysite |   $K_3V_5O_{14}$ |
|   $H_2Si_2O_5$ |   $CaV_6O_{16} \cdot nH_2O$ |
|   $H_2Si_{14}O_{29} \cdot 5H_2O$ |   $Na(UO_2V_3O_9) \cdot nH_2O$ |
| Miscellaneous | Niobates |
|   $Ni(CN)_2$ |   $K_4Nb_6O_{17}$ |
|   $VOSO_4$, $VOPO_4$ |   $KNb_3O_8$ |
|   $WO_2Cl_2$ | Miscellaneous |
| |   $Na_2W_4O_{13}$ |
| |   $Na_2U_2O_7$ |
| |   $Mg_2Mo_2O_7$ |
| |   Hydrotalcites |

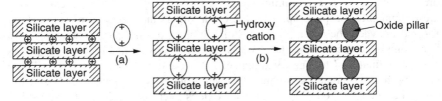

**FIGURE 13.6** Schematic illustration of pillaring process in bidimensional clay: (a) ion exchange with precursor cations and (b) conversion to oxide by calcination. (From Yamanaka, S., Design and synthesis of functional layered nanocomposites, *Am. Ceram. Soc. Bull.*, 70, 1056, 1991. With permission.).

dehumidification equipment. Figure 13.7 shows the adsorption isotherms of several pillared clays and the data show that these can serve as good desiccants.[69] Although a great deal of research has been done on the pillared clays, the pore size distribution is not well understood. One can estimate the pillar sizes from

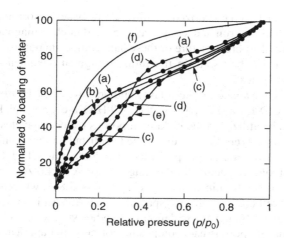

**FIGURE 13.7** Water adsorption isotherms of different smectites with alumina pillars: (a) nontronite treated with $NH_3$ and exchanged with $Ca^{2+}$, (b) original pillared nontronite, (c) original pillared saponite, (d) original pillared montmorillonite (kunimine), (e) original pillared hectorite (hectabrite AW), and (f) ideal isotherm shape for use in dehumidification and cooling equipment. (From Malla, P., and Komarneni, S., *Mem. Sci. Geol.*, 86, 59, 1990. With permission.)

**TABLE 13.3**
**Precursor Cations Used in Pillaring the Clay and Resulting Basal Spacings[63]**

| Pillar Oxide | Precursor | Basal Spacing/Å |
|---|---|---|
| $Al_2O_3$ | $[Al_{13}O_{14}(OH)_{24}]^{7+}$ | 17–19 |
| $ZrO_2$ | $[Zr_4(OH_{14})]^{2+}$ | 17–20 |
| $Fe_2O_3$ | $[Fe_3O(OCOCH_3)_6]^+$ | 17 |
| $Cr_2O_3$ | $[Cr_n(OH)_m]^{(3nm)+}$ | 17–21 |
| $Bi_2O_3$ | $[Bi_6(OH)_{12}]^{6+}$ | 16 |
| $Al_2O_3$-$SiO_2$ | $[Al_{13}O_4(OH)_{24n}]$-$[OSi(OH)_3]_n^{7+}$ | 17–19 |
| $TiO_2$ | Sol solution | 24–27 |
| $SiO_2$-$TiO_2$ | Sol solution | 40–50 |
| $SiO_2$-$Fe_2O_3$ | Sol solution | 40–100 |

the basal spacings (Table 13.3) by subtracting 9.6 Å for the thickness of the clay layer, but the size of the pores along the *a–b* direction is mostly unknown. It is imperative that more research be done to determine the pore sizes of these samples before they can find wide-ranging applications. The pore size along the *a–b* direction is expected to depend upon the charge density, the uniformity of the charge distribution, the size and type of polymeric cations used, etc. The pore structures are usually characterized by nitrogen or water adsorption treatments. Techniques such as neutron and x-ray scattering and nuclear magnetic resonance may help unravel the pore structure of these nanocomposite porous materials and this information will be useful in the future design of chemical sensors from theses nanocomposites.

In addition to the above cationic clays, anionic clays of the hydrotalcite group, $[Mg_2Al(OH)_6]^+Cl \cdot XH_2O$, and $[Al_2Li(OH)_6]^+Cl \cdot H_2O$ are potential layered compounds that can be utilized in the making of porous nanocomposites. The anionic species in the interlayers can be exchanged with bulkier inorganic or organic anions.[70–78] The hydrotalcite group of materials containing $Cl^-$, $NO_3^-$, $SO_4^{2-}$, or $CrO_4^{2-}$ have interlayer spacings in the range of 3.0 Å to 4.0 Å.[72–74] Substitution of $CO_3^{2-}$ gives the smallest interlayer spacing, that is, 2.8 Å, whereas substitution with $Fe(CN)_6^{3-}$ or $Fe(CN)_6^{4-}$ gives a spacing of 6.1 Å.[74] An interlayer spacing of 8.1 Å was obtained by the intercalation of napthol yellow $S^2$ in hydrotalcite.[78] Thus one can design porous nanocomposites for potential applications as adsorbents. For example, a $Co(CN)_6^{3-}$ exchanged sample shows[78] the following order of adsorption for several hydrocarbons: hexane $\approx$ 2 methylpentane $>>$ cyclohexane $>$ methylcyclohexane. In addition to the pore size, the surface properties can be modified through numerous substitutions.[70–78] Because these materials are not stable above about 300°C, all the applications are restricted to temperatures below this. Although the anionic species in the interlayers of these nanocomposites are not dehydrated, unlike pillared clays, because of their poor thermal stability, it is not impossible to create pillared anionic clays with low-temperature dehydration of the anionic species through future molecular engineering.

## 2. Metal Intercalated Clays

Expandable layer silicates such as montmorillonite can be converted to efficient heterogeneous catalysts by introducing catalytically active sites or guest species between the layers or on the external surfaces. Previous attempts to produce intercalated zero-valent transition-metal particles in layer silicates, by hydrogen reduction for example, have, however, failed: the layers tend to collapse, sometimes followed by deposition of metal particles on the external surfaces. Recently Malla et al.[79,80] described the successful intercalation of copper metal clusters of 4 to 5 Å in montmorillonite by in situ reduction of $Cu^{2+}$ ions using ethylene glycol. These metal-cluster intercalates were stable up to at least 500°C. The clusters prop the silicate layers apart, much as metal oxides do in pillared clays, and may thus be able to introduce unique catalytic product selectivity through a molecular sieving effect similar to that in cluster-loaded zeolites. As metal clusters of these dimensions behave very differently from the bulk metal, intercalates of this sort may prove to be versatile catalysts.

## 3. Clay-Organic Nanocomposites

These are often referred to as clay/organic complexes and have been around since at least biblical times.[81] During this period, clays were used to decolorize edible oils and clarify alcoholic beverages utilizing clay/organic reactions or complexation.[81] The interaction of clays with organic species has been implicated in the origin of life.[82] The structure of swelling clays, especially of the smectite group, is amenable to forming clay-organic nanocomposites because of the weak bonding (van der Waals and electrostatic forces) between the layers. The clay-organic intercalation compounds can be grouped into two main types: those that are formed by ion exchange of interlayer exchangeable cations and those that are formed by adsorption of polar organic molecules. Clay-organic complexes have been utilized in the past for surface area measurements, layer charge determination, soil stabilization, etc. Recently clay-organic nanocomposites have been proposed for several new applications in the materials field. Because of the nanoscale mixing of inorganic and organic components in the clay/organic nanocomposites, they can be utilized as precursors in the preparation of structural nonoxide ceramic materials such as $Si_3N_4$, SiC, AlN, Sialon, etc.[83–91] When the clay-organic nanocomposites are heated in an inert atmosphere, carbothermal reduction reactions take place, leading to the formation of nonoxide structural ceramic materials as follows:

$$3SiO_2 + 6C + 2N_2 \rightarrow Si_3N_4 + 6CO \qquad (13.1)$$

$$SiO_2 + 3C \rightarrow SiC + 2CO. \qquad (13.2)$$

Several naturally occurring clay minerals such as montmorillonite, kaolinite, pyrophyllite, and illite have been studied as raw materials in the synthesis of nonoxide ceramic materials,[83–91] but the montmorillonite-polyacrylonitrile complex was found to be the best because of the intimate mixing. An alumina pillared clay has been found to be a suitable precursor to mix with carbon in the synthesis of β-sialon by carbothermal reduction.[92] Sugahara et al.[84] used a magadiite $(Na_2Si_{14}O_{29}-nH_2O)-C_{12}H_{25}N(CH_3)_3^+$ nanocomposite and a physical mixture of magadiite with carbon in their carbothermal reduction reactions and found that the former yielded β-SiC, while the latter yielded $SiO_2$. Thus the reaction process in the nanocomposite is quite different because of the intimate mixing achieved on a nanoscale.

Clay-organic nanocomposites have also been proposed as low dielectric constant substrates.[93–96] These nanocomposites consist of a quasi-two-dimensional layered structure (fluorohectorite or other swelling clay) and an organic compound such as polyaniline, $n$-$C_6H_5NH_2$ intercalated between the layers by ion exchange. The ceramic (clay) layer imparts good mechanical and thermal stabilities, while the organic compound gives low relative permittivity and good processibility to the nanocomposite material. The present disadvantage with these nanocomposites is that they are hydrophilic, that is, they adsorb water, which can increase the relative permittivity. To alleviate this problem, future work needs to deal with the incorporation of hydrophobic polymers in the interlayers. With other applications in mind, clay-organic nanocomposites having hydrophobic characteristics have already been prepared by exchanging aminosilane or organic chrome complex for $Li^+$ or $Na^+$ from swollen clay gels.[97] These nanocomposite gels can be processed to form paper, board, film, fiber, and coatings, and the dried gel powders can be hot-pressed to give a body of cross-linked organic polycation-mica derivatives.

Intercalation of electroactive polymers such as polyaniline and polypyrrole in mica-type layered silicates leads to metal-insulator nanocomposites.[98,99] The conductivity of these nanocomposites in the form of films is highly anisotropic, with the in-plane conductivity $10^3$ to $10^5$ times higher than the conductivity in the direction perpendicular to the film. Conductive polymer/oxide bronze nanocomposites have been prepared by intercalating polythiophene in $V_2O_5$ layered phase, which is analogous to clays.[100] Studies of these composites are expected not only to provide a fundamental understanding of the conduction mechanism in the polymers, but also to lead to diverse electrical and optical properties.

Intercalation of ethylenediamine functionalized buckminsterfullerene in fluorohectorite clay has been achieved[101] and these nanocomposites may lead to microporous materials analogous to pillared clays with the elimination of the ligands by suitable heat treatment in oxygen. A new microporous tubular silicate-layered silicate (TSLS) nanocomposite has been synthesized by selective hydrolysis of γ-aminopropyl triethoxysilane from the external surfaces of imogolite, which led to its intercalation into layered silicate.[102] The nitrogen adsorption and $t$-plot analysis of this porous nanocomposite showed a bimodal pore structure that is attributed to intratube and intertube adsorption environments.[103] The TSLS nanocomposite has been found to be active for the acid-catalyzed dealkylation of cumene at 350°C, but this composite has been found to be less reactive than a conventional $Al_2O_3$ pillared montmorillonite.[103] Large surface area (approximately 900 $m^2$ $g^{-1}$) microporous materials have been prepared[104,105] by calcining nanocomposites of alkyltrimethylammonium-kanemite ($NaHSi_2O_5 \cdot 3H_2O$), the latter being a layered polysilicate. During the organic intercalation, the $SiO_2$ layers in the complexes condense to form three-dimensional $SiO_2$ networks. The calcined products of the complexes have micropores in the range of 2 to 4 nm (Figure 13.8), and such materials are expected to find applications in catalysis.

Clay-fluorescence dye nanocomposites have been prepared by ion exchange and the fluorescence properties of different dyes as affected by the inorganic crystal field of the clay have been extensively investigated.[106–110] Such confinement can lead not only to high thermal stability, but also to higher luminescence efficiency.[109–110]

## C. ENTRAPMENT-TYPE NANOCOMPOSITES

The entrapment-type nanocomposites can be prepared from zeolites and they are of two types: zeolite-inorganic and zeolite-organic. Zeolite crystals are three-dimensionally linked network structures of aluminosilicate, aluminophosphate (ALPO), and silicoaluminophosphate (SAPO) composition and are porous, the pores being in the range of 2.8 to 10 Å. Many of the highly siliceous, ALPO, and SAPO zeolites have been synthesized using organic templates such as tetrapropyl

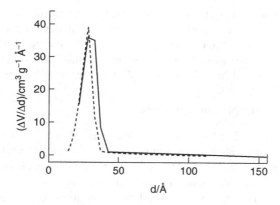

**FIGURE 13.8** Pore size distributions of (-) calcined product obtained from hexadecyltri-methylammonium-kanemite complex and (---) calcined product obtained from trimethyl-silylated derivative. (From Yanagisawa, T., Shimizu, T., Kuroda, K., and Kato, C., *Bull. Chem. Soc. Jpn.*, 63, 1535, 1990. With permission.)

ammonium tetratmethyl ammonium, di-n-propylamine, etc.[111–112] After synthesis, the organics are removed by different techniques, the main one being combustion, to get access to all the pore space.

## 1. Zeolite-Inorganic Nanocomposites

Fine metal clusters supported in zeolites are a good example of this type of nanocomposite and they possess unique catalytic properties, molecular selectivity, and polyfunctional activity.[113–115] A large volume of literature exists on the preparation and catalytic properties of metal clusters or aggregates dispersed in zeolites.[116,117] Various methods, such as ion exchange, evaporation, irradiation, thermal decomposition, particle beam, impregnation, adsorption, or deposition and coprecipitation, have been used to introduce metal ions or complexes that are then reduced to zero-valent metal forms by molecular or atomic hydrogen, ammonia, metal vapors, and various organic compounds. We have recently used[118] a new approach, that is, the polyol process to entrap nickel and copper metal clusters in zeolites. The interface chemistry associated with nanophase confinement and packaging and some features of three-dimensional surface confinement using zeolites and molecular sieves has recently been reviewed,[119] and silver sodalites have been touted as novel optically responsive nanocomposites.[120] Future studies may exploit the zeolite-inorganic nanocomposites for materials applications other than catalysis because of the great potential for nanodesigning.

## 2. Zeolite-Organic Nanocomposites

There has not been a great deal of work in utilizing the zeolite-organic nanocomposites directly in materials applications. Recently pyridine ($C_5H_{11}N$)-incorporated ZSM-39 and Dodecasil-3C zeolites have been synthesized[121] and these nanocomposites show an optical memory effect and interesting domain structure. The ability to rotate polar groups within unusual symmetries gives rise to field responses in molecular nanocomposite properties.[121] Future work in this area needs to emphasize the growth of large single crystals of zeolites with organic molecules incorporated in them for optical and other applications.

## D. ELECTROCERAMIC NANOCOMPOSITES

During the past two decades Newnham et al. have developed a large family of microcomposite materials with properties superior to those obtainable from single phases for use as electrochemical transducers, positive thermal coefficient (PTC) and negative thermal coefficient (NTC) thermistors, piezoresistors, and chemical sensors, and recently they have turned their attention to nanocomposites for electronic applications.[122,123] Recent advances in both information and charge storage in the electronics industry may be attributed to electroceramic nanocomposites and especially those that are based on ferroic materials because both the presence of domain walls and the ferroic transition are affected by the crystallite size. The size dependence of ferroic properties is shown in Figure 13.9.[122] Multi-domain effects accompanied by hysteresis take place in large crystallites. Reductions in size (Figure 13.9) lead to single-domain particles, and yet smaller sizes to destabilized ferroics with large property coefficients, and finally, to normal behavior as the particle size approaches the nano or atomic scale.[122] Because of the nanoscale, there are different quantum effects leading to variations in the energy states and electronic structures of their components. Other basic features of these nanocomposites are the remarkable modification of the electronic structure by

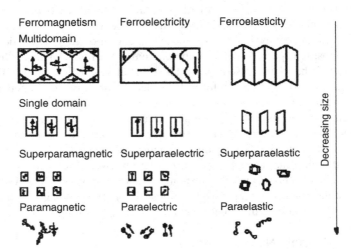

**FIGURE 13.9** Changes in the domain configurations of ferroics as a function of size. (From Newnham, R.E., McKinstry, S.E., and Ikawa, H., *Mater. Res. Soc. Symp. Proc.*, 175, 161, 1990. With permission.)

widespread interface interaction at the electronic level and the great variety of nanostructures ranging from high-level ordered three-dimensional periodic structures to a stoichiometrically dispersed medium of nanoparticles.[124] Electroceramic nanocomposites can be further classified into (a) magnetic, (b) ferroelectric, and (c) superconducting/ferroelectric, (d) dielectric, and (e) conducting, semiconducting, and insulating types of materials. Two excellent reviews have been published on magnetic, ferroelectric, and ferroelastic nanocomposites[122,123] and readers are advised to refer to these for more details.

Among the electroceramic nanocomposites, the magnetic nanocomposites consisting of small particles or ultrathin films have been investigated extensively. The use of nanoparticles in magnetic recording media can lead to smaller storage units where the information is stored at a greater density.[122] Nanoparticles are also used in ferrofluids, where the size of the crystallites is small enough to prevent settling in the fluid. Ferrofluids can be construed as solid-fluid nanocomposites and they have applications in noncontaminating seals, loudspeakers, ink jet printers, levitation systems for separating materials of different density, vibration dampers, engines for converting low-grade heat to usable energy, and devices to measure very small inclination angles.[122,125–128] A novel fabrication of a cobalt-chromium nanocomposite by radio frequency sputtering has led to very high storage density with the perpendicular recording technique.[129,130] Another innovative example is the processing of a 1-3 nanocomposite for uniform, high-density magnetic components.[130] In this work, an aluminum alloy substrate is first oxidized to achieve a regular network of honeycomb cells on the surface. The columnar pores which are formed during the oxidation are etched and then backfilled with iron to create a high density of magnetic elements with practical values for the coercive force.[122,125]

## 1. Ferroelectric Nanocomposites

These nanocomposites have not been explored to a significant extent because research has concentrated on microcomposites, which have been found to be extremely useful as pressure transducers, vibration dampers, and transducers. The need for optical transparency or low driving voltages may lead to the development of ferroelectric nanocomposites in the future.[122] A solid 0-3 ferroelectric nanocomposite has been recently prepared using ultrafine (approximately 20 nm) $PbTiO_3$ powders dispersed in a polymeric matrix. This type of composite may be useful in optical applications. Single-phase relaxer ferroelectrics exhibit compositional and order-disorder inhomogeneities on a nanometer scale.[122–136] These single phases on a macroscopic scale can be construed as "nanocomposites" if the definition of the nanocomposite term can be extended to include materials that show inhomogeneities in structure, composition, or properties on a nanoscale. For example, the $A(B'_{1/2}, B''_{1/2})O_3$ and $A(B'_{2/3}, B''_{1/3})O_3$ perovskites have been found to display microdomains (approximately 2 nm to 3 nm in size) of 1:1 ordering on the B sublattice dispersed in a disordered matrix.[134–136] Other types of ferroelectric nanocomposites can be prepared by mixing two or more nanophases. By mixing nanoparticles of a ferroelectric in an organic liquid, one can design ferroelectric fluids that are analogous to the ferromagnetic fluids discussed above. A ferromagnetic fluid prepared from ultrafine $BaTiO_3$ particles and an organic carrier liquid showed a maximum in the dielectric constant at the tetragonal-cubic phase transition of the perovskite phase.[122,137] By dispersing 10 nm particles of $BaTiO_3$ in a mixture of heptane and oleic acid, Bachman and Bamer[138] showed that the nanoparticles show permanent polar moments. These types of ferroelectric fluids may be useful as an alternative to liquid crystals in display panels.[122] Future research in ferroelectric nanocomposites may lead to applications in optical devices.

## 2. Superconducting Ferroelectric Nanocomposites

Thin-film heterostructures of $Bi_4Ti_3O_{12}/Bi_2Sr_2CuO_{6+x}$ have been grown on single crystals of $SrTiO_3$, $LaAlO_3$, and $MgAl_2O_4$ by pulsed laser deposition.[139] These films have been found to be ferroelectric and the thickness of the layers can be in the nanometer range if so desired. These thin films look promising for use as novel, lattice-matched, epitaxial ferroelectric film/electrode heterostructures in nonvolatile memory applications.

## 3. Dielectric Nanocomposites

There is a need for substrates with very low dielectric constants (less than 3) in very large scale integration (VLSI) devices so that acceptable limits of cross-talk, signal line impedance, and transmission delay are maintained.[140] Ceramic-polymer nanocomposites using fumed silica and polydimethyl siloxane have been fabricated to achieve low dielectric constant substrates.[141] Clay/organic nanocomposites are also proposed as low dielectric constant substrates (see above). Future studies in this area may try to take advantage of highly microporous, hydrophobic zeolites mixed in an organic or inorganic matrix to achieve very low dielectric constant substrates for electronic packaging.

## 4. Conducting/Semiconducting/Insulating Nanocomposites

Nanocomposites in the form of superlattice structures have been fabricated with metallic,[142] semiconductor,[143] and ceramic materials[144] for semiconductor-based devices.[145] The material is abruptly modulated with respect to composition and/or structure. Semiconductor superlattice devices are usually multiple quantum structures, in which nanometer-scale layers of a lower band gap material such as GaAs are sandwiched between layers of a larger band gap material such as GaAlAs.[145] Quantum effects such as enhanced carrier mobility (two-dimensional electron gas) and bound states in the optical absorption spectrum, and nonlinear optical effects, such as intensity-dependent refractive indices, have been observed in nanomodulated semiconductor multiple quantum wells.[146] Examples of devices based on these structures include fast optical switches, high electron mobility transistors, and quantum well lasers.[146] Room-temperature electrochemical deposition of nanomodulated (5 nm to 10 nm) ceramic superlattice thin films of $Ti_aPb_bO_c/Ti_dPb_eO_f$ has been recently reported.[146] The electrochemical method offers several advantages over vapor deposition methods such as molecular beam epitaxy for depositing nanomodulated materials with nearly square-wave modulation of composition and structure, because the low processing temperatures minimize interdiffusion.[146] It is hoped that these structures will show quantum electronic, optical, or optoelectronic effects as the modulation wavelength approaches the electron mean path.[146]

Other methods such as sputtering, electrodeposition, and chemical techniques[147–153] have also been used to prepare nanocomposites. Nanocomposites of molybdenum in aluminum matrix were prepared by sputtering techniques. Nanoscale particles of molybdenum were first produced by high-pressure sputtering at more than 100 mTorr in a thermal gradient and then they were embedded in an aluminum matrix by normal sputtering.[148] A new class of diamond-like nanocomposite materials was synthesized,[149] consisting of silicon-oxygen and transition metal networks imbedded in a diamond-like carbon matrix. In metal containing nanocomposites, the conductivity can be varied continuously over 18 orders of magnitude by varying the concentration of metal atoms. Conductivities as high as 104 S cm$^{-1}$ have been achieved with tungsten-containing films.[149] Clusters of metallic silver have been formed by electrodeposition within the surface of an oxide glass which was subjected to an alkali silver ion exchange process.[150] The clusters have a fractal structure and the fractal growth within the glass results in the formation of a glass-metal nanocomposite with a particle diameter of about 12 nm. A new approach in nanocomposite synthesis has been developed and it involves rapid condensation of metallic and nonmetallic species produced by laser-induced reactions.[151] These composite surface layers form by codeposition of fine amorphous silica fibers of 25 to 120 nm in diameter and a metal matrix where the fibers exist in the form of a random weave structure. A chemical technique has been used to produce metal-polymer nanocomposites.[152] In this method, poly(vinylpyridine)s were heated with copper formate in MeOH to 125°C and the resulting thermal decomposition of the complex initiated a redox reaction that reduced the $Cu^{2+}$ to copper metal and oxidized the formate to $CO_2$ and $H_2$, leading to solid copper-polymer nanocomposites containing up to 23 wt% copper. A review article by Hirai and Sasaki[153] deals with the *in situ* preparation methods, such as chemical vapor deposition, for different nanocomposites.

## E. STRUCTURAL CERAMIC NANOCOMPOSITES

Glass ceramics, which are well-known materials, constitute a type of ceramic nanocomposites with nanocrystals being embedded in the glassy phase. These will not be dealt with here. However, one recent breakthrough in glass ceramics is worth mentioning.[154] Albite glass, which has been thought to be impossible to crystallize, has been crystallized by seeding both gels and glasses with fine albite seeds, that is, the nanocomposite approach.[154] Recently Niihara and his colleagues[155–173] developed ceramic nanocomposites from oxide-nonoxide and

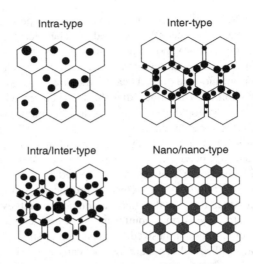

**FIGURE 13.10** Schematic illustration of ceramic nanocomposites. (From Niihara, K. and Nakahira, A., in *Ceramics: Toward the 21st Century*, Soga, N. and Kato, A., Eds., Ceramic Society of Japan, Tokyo, 1991, pp. 404-417. With permission.)

nonoxide-nonoxide mixtures and they have classified these into four categories: (a) intragranular, (b) intergranular, (c) both intra- and intergranular, and (d) nano/nanocomposites (Figure 13.10). In the intra- and intergranular nanocomposites, the nanosize particles are dispersed mainly within the matrix grains or at the grain boundaries of the matrix, respectively (Figure 13.10). The intragranular and intergranular nanocomposites and their combination showed tremendous improvement in mechanical properties such as hardness, strength, and creep and fatigue fracture resistances even at high temperatures compared to those of monophasic and microcomposites. The nano/nanocomposites were found to offer advantages in machinability and superplasticity.[173] Ceramic nanocomposites can be fabricated by chemical vapor deposition,[174–177] pressure-less sintering, HIPing, and hot-pressing.[155–173] The fabrication process of oxide-based nanocomposites is shown in Figure 13.11. Niihara and co-workers succeeded in preparing numerous ceramic nanocomposites, including $Al_2O_3/SiC$, $Al_2O_3/Si_3N_4$, $Al_2O_3/TiC$, mullite/SiC, $B_4C/SiC$, $B_4C/TiB_2$, SiC/amorphous SiC, $Si_3N_4/SiC$, and others. Table 13.4 shows the significant improvements in mechanical properties that can be achieved by the nanocomposite route for various compositional systems.[173] The improvement in mechanical properties is attributed to the nanosize dispersions.[173] Niihara and Nakahira[173] have also made hybrid composites with nanocomposites and microcomposites and found significant improvements in not only toughness, but also strength. Low-temperature methods have been used to fabricate ceramic nanocomposites, in addition to the above solid state method. A gel-based method has been used to prepare $SiC/Al_2O_3$ nanocomposites that also exhibited enhanced mechanical properties.[178] A novel epitaxial growth of

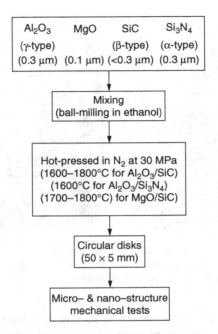

**FIGURE 13.11** The fabrication process of nanocomposites. (From Niihara, K. and Naka-hira, A., in *Ceramics: Toward the 21st Century*, Soga, N. and Kato, A., Eds., Ceramic Society of Japan, Tokyo, 1991, pp. 404-417. With permission.)

**TABLE 13.4**
**Improvement of Mechanical Properties Observed for the Ceramic Nanocomposites**

| Composite System | Toughness/MPa/m$^{1/2}$ | Strength/MPa | Maximum Operating Temperature[a]/°C |
|---|---|---|---|
| Al$_2$O$_3$-SiC | 3.5–4.8 | 350–1520 | 800–1200 |
| Al$_2$O$_3$-Si$_3$N$_4$ | 3.5–4.7 | 350–850 | 800–1300 |
| MgO-SiC | 1.2–4.5 | 340–700 | 600–1400 |
| Si$_3$N$_4$-SiC | 4.5–7.5 | 850–1550 | 1200–1400 |

[a] Maximum operating temperature at the high loads.

Reprinted with permission from Niihara, K., and Nakahira, A., in *Ceramics: Toward the 21st Century*, Soga, N. and Kato, A., Eds., Ceramic Society of Japan, Tokyo, 1991, pp. 404-417.

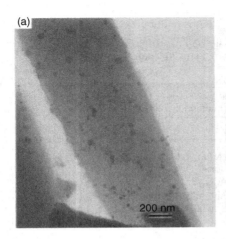 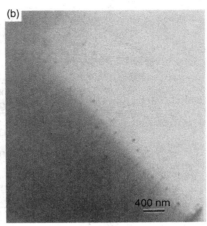

**FIGURE 13.12** Transmission electron micrographs of synthetic nanocomposite opal showing two different arrays of nanosized (7 to 50 nm) $ZrO_2$ balls in the void spaces of silica balls. (From Simonton, T.C., Roy, R., Komameni, S., and Breval, E., *J. Mater. Res.*, 1, 667, 1986. With permission.)

nickel (II) hydroxide on layer silicate followed by sintering led to metal-ceramic nanocomposites.[179] A proprietary, apparently low-temperature process has been used to produce synthetic opal that is a nanocomposite of amorphous silica and crystalline zirconia.[180] Synthetic opal invented by a French chemist, Gilson, is commercially available from Nakazumi Chemicals.[181] Unlike natural opal, synthetic opal revealed the presence of separate crystalline $ZrO_2$ balls that were nearly spherical and ranged in diameter from 7 to 50 nm among the 200 nm noncrystalline $SiO_2$ balls. The $ZrO_2$ balls were arranged in basically two types of patterns: hexagonal rings and nearly square grids (Figure 13.12a,b). This nanocomposite also showed significantly higher fracture toughness compared to the monophasic natural opal.[180] The above studies demonstrate that the nanocomposites are clearly superior to the monophasic or microcomposite alternatives in ceramics processing.

## III. CONCLUSION

The use of the nanocomposite approach in sol-gel, catalytic, optical, sensor, electroceramic, and structural ceramic materials is now well established. Sol-gel nanocomposites have been shown to lower crystallization temperatures and enhance the densification of ceramic materials, in general. Tailoring of nanocomposites with low-dimensional solids such as clays has led to novel microporous materials, which may find applications in catalytic, sensor, and laser materials. Nanodesigning in the electroceramics area has yielded improved dielectric, optical, optoelectronic, magnetic, quantum electronic, and superconducting properties. Structural ceramics processing through nanocomposites has resulted in significant improvements in mechanical properties such as hardness, strength, creep and fatigue fracture resistances, machinability, and superplasticity. Dramatic improvements in materials functions are expected in the future through judicious processing of nanocomposites.

## ACKNOWLEDGMENT

Reproduced with permission from the Royal Society of Chemistry. This article was originally published in *Journal of Materials Chemistry* 2:1219–1230 (1992) and the current version is slightly modified. Since this article was originally published, a great deal of research has been done on this topic and many proceedings have been published by the Materials Research Society, American Ceramics Society, and others, and readers are referred to these proceedings for the latest information.

## REFERENCES

1. Roy, R., Komarneni, S., and Roy, D.M., *Mater. Res. Soc. Symp. Proc.*, 32, 347, 1984.
2. Roy, R., Abstracts, presented at the Materials Research Society Annual Meeting, Boston, MA, 1982.
3. Roy, R.A., and Roy, R., *Mater. Res. Bull.*, 19, 169, 1984.
4. Hoffman, D.W., Roy, R., and Komarneni, S., *J. Am. Ceram. Soc.*, 67, 468, 1984.
5. Newnham, R.E., Skinner, D.P., and Cross, L.E., *Mater. Res. Bull.*, 13, 525, 1978.
6. Hoffman, D.W., Roy, R., and Komarneni, S., *Mater. Lett.*, 2, 245, 1984.
7. Hoffman, D.W., Komarneni, S., and Roy, R., *J. Mater. Sci. Lett.*, 3, 439, 1984.
8. Komarneni, S., Suwa, Y., and Roy, R., *J. Am. Ceram. Soc.*, 69, C-155, 1986.
9. Suwa, Y., Roy, R., and Komarneni, S., *J. Am. Ceram. Soc.*, 68, C-238, 1985.
10. Roy, R., Suwa, Y., and Komarneni, S., in *Science of Ceramic Chemical Processing*, L.L. Hench and D.R. Ulrich, Eds., John Wiley & Sons, New York, 1986.
11. Suwa, Y., Komarneni, S., and Roy, R., *J. Mater. Sci. Lett.*, 5, 21, 1986.
12. Suwa, Y., Roy, R., and Komarneni, S., *Mater. Sci. Eng.*, 83, 151, 1986.
13. Kumagai, M., and Messing, G.L., *J. Am. Ceram. Soc.*, 67, C-230, 1984.
14. Kumagai, M., and Messing, G.L., *J. Am. Ceram. Soc.*, 68, 500, 1985.
15. Messing, G.L., McArdle, J.L., and Shelleman, R.A., *Mater. Res. Soc. Symp. Proc.*, 73, 471, 1986.
16. Shelleman, R.A., Messing, G.L., and Kumagai, M., *J. Non-Crystl. Solids*, 82, 277, 1986.
17. Vilmin, G., Komarneni, S., and Roy, R., *J. Mater. Sci.*, 22, 3556, 1987.
18. Vilmin, G., Komarneni, S., and Roy, R., *J. Mater. Res.*, 2, 489, 1987.
19. Komarneni, S., Suwa, Y., and Roy, R., *J. Mater. Sci. Lett.*, 6, 525, 1987.
20. Yarbrough, W., and Roy, R., *J. Mater. Res.*, 2, 494, 1987.
21. Roy, R., *Mater. Sci. Res.*, 21, 25, 1987.
22. Roy, R., *Science*, 238, 1664, 1987.
23. Kazakos-Kijowski, A., Komarneni, S., and Roy, R., *Mater. Res. Soc. Symp. Proc.*, 121, 245, 1988.

24. McArdle, J.L., and Messing, G.L., Tietz, L.A., and Carter, C.B., *J. Am. Ceram. Soc.*, 72, 864, 1989.
25. Huling, J.C., and Messing, G.L., *J. Am. Ceram. Soc.*, 72, 1725, 1989.
26. Roy, R., and Komarneni, S., U.S. Patent no. 4,828,031, 1989.
27. Kazakos, A.M., Komarneni, S., and Roy, R., *J. Mater. Res.*, 5, 1095, 1990.
28. Ravindranathan, P., Komarneni, S., and Roy, R., *J. Am. Ceram. Soc.*, 12, 1024, 1990.
29. Pach, L., Roy, R., and Komarneni, S., *J. Mater. Res.*, 5, 278, 1990.
30. Kazakos, A., Komarneni, S., and Roy, R., *Mater. Lett.*, 9, 405, 1990.
31. Kazakos, A., Komarneni, S., and Roy, R., *Mater. Lett.*, 10, 75, 1990.
32. Komarneni, S., and Roy, R., *Ceram. Trans.*, 6, 209, 1990.
33. Roy, R., *Mater. Res. Soc. Symp. Proc.*, 175, 15, 1990.
34. Komameni, S., Kazakos, A.M., and Roy, R., U.S. Patent no. 5,030,592, 1991.
35. Selvaraj, U., Prasadarao, A.V., Komarneni, S., and Roy, R., *Mater. Lett.*, 12, 311, 1991.
36. Selvaraj, U., Prasadarao, A.V., Komarneni, S., and Roy, R., *J. Am. Ceram. Soc.*, 75,1167, 1992.
37. Prasadarao, A.V., Selvaraj, U., Komarneni, S., Bhalla, A.S., and Roy, R., *J. Am. Ceram. Soc.*, 75,1529, 1992.
38. Burggraaf, A.J., Keizer, K., and Van Hassel, B.A., *Solid State Ionics*, 32, 771, 1989.
39. Shull, R.D., and Ritter, J.J., *Mater. Res. Soc. Symp. Proc.*, 195, 435, 1990.
40. Shull, R.D., Ritter, J.J., and Swartzendruber, L.J., *J. Appl. Phys.*, 69(pt 2A), 5144, 1991.
41. Shull, R.D., Ritter, J.J, Shapiro, A.J., and Swartzendruber, L.J., *J. Appl. Phys.*, 67(pt. 2A), 4490, 1990.
42. Shull, R.D., Ritter, J.J., Shapiro, A.J., Swartzendruber, L.J., and Bennett, L.H., *Mater. Res. Soc. Symp. Proc.*, 132, 179, 1988.
43. Shull, R.D., Ritter, J.J., Shapiro, A.J., Swartzendruber, L.J., and Bennett, L.H., *Mater. Res. Soc. Symp. Proc.*, 206, 455, 1991.
44. Chatterjee, A., and Chakravorty, D., *J. Phys. D Appl. Phys.*, 23, 1097, 1990.
45. Chatterjee, A., and Chakravorty, D., *J. Phys. D Appl. Phys.*, 22, 1386, 1989.
46. Das, G.C., and Chakravorty, D., *Bull. Mater. Sci.*, 12, 449, 1989.
47. Prasad, P.N., *Mater. Res. Soc. Symp. Proc.*, 180, 741, 1990.
48. Dunn, B., and Zink, J.I., *J. Mater. Chem.*, 1, 903, 1991.
49. Yamanaka, S., Proceedings of the Seventh Seminar on Frontier Technology—Nano-Hybridization and Creation of New Functions, Feb. 7–10, 1989, Oiso, Japan.
50. Whittingham, M.S., and Jacobson, A.J., Eds., *Intercalation Chemistry*, Academic Press, New York, 1982.
51. Vaughan, D.W.E., Lussier, R.J., and Magee, J.S., U.S. Patent no. 4,176,090, 1974.
52. Yamanaka, S., and Brindley, G.W., *Clays Clay Miner.*, 27, 119, 1979.
53. Brindley, G.W., and Yamanaka, S., *Am. Mineral.*, 64, 830, 1979.
54. Pinnavaia, T.J., Tzou, M.S., and Landau, S.D., *J. Am. Chem. Soc.*, 107, 4783, 1985.
55. Sterte, S., *Clays Clay Miner.*, 34, 658, 1986.
56. Yamanaka, S., Doi, T., Sako, S., and Hattori, M., *Mater. Res. Bull.*, 19, 161, 1984.
57. Yamanaka, S., Yamashita, G., and Hattori, M., *Clays Clay Miner.*, 28, 281, 1980.
58. Sterte, J., and Shabtai, J., *Clays Clay Miner.*, 35, 429, 1987.
59. Yamanaka, S., Nishihara, T., and Hattori, M., *Mater. Res. Soc. Symp. Proc.*, 111, 283, 1988.

60. Yamanaka, S., Matsumoto, H., Okumura, F., Yoshikawa, M., and Hattori, M., Abstracts of the 28th Annual Meeting of the Basic Science Division of the Ceramic Society of Japan, Tokyo, Japan, 1990.
61. Pinnavaia, T.J., *Science*, 220, 365, 1983.
62. Yoneyama, H., Haga, S., and Yamanaka, S., *J. Phys. Chem.*, 22, 4833, 1989.
63. Yamanaka, S., *Am. Ceram. Soc. Bull.*, 70, 1056, 1991.
64. Zielke, R.C., and Pinnavaia, T.J., *Clays Clay Miner.*, 36, 403, 1988.
65. Nolan, T., Srinivasan, K.R., and Fogier, H.S., *Clays Clay Miner.*, 37, 487, 1989.
66. Srinivasan, K.R., Fogler, H.S., Gulari, E., Nolan, T., and Schultz, J.S., *Environ. Prog.*, 4, 239, 1985.
67. Malla, P.B., Yamanaka, S., and Komarneni, S., *Solid State Ionics*, 32/33, 354, 1989.
68. Yamanaka, S., Malla, P.B., and Komarneni, S., *J. Colloid Interface Sci.*, 134, 51, 1990.
69. Malla, P., and Komarneni, S., *Mem. Sci. Geol.*, 86, 59, 1990.
70. Miyata, S., *Clays Clay Miner.*, 23, 396, 1975.
71. Miyata, S., and Kumura, T., *Chem. Lett.*, 843, 1973.
72. Miyata, S., and Okada, A., *Clays Clay Miner.*, 25, 14, 1977.
73. Kikkawa, S., and Koizumi, M., *Mater. Res. Bull.*, 17, 191, 1982.
74. Cavalcanti, F.A.P., Schutz, A., and Biloen, P., in *Preparation of Catalysts IV*, Delmon, B., Grange, P., Jacobs, P.A., and Poncelet, G., Eds., Elsevier, Amsterdam, pp. 165–174.
75. Miyata, S., and Okada, A., *Clays Clay Miner.*, 25, 14, 1977.
76. Schutz, A., and Biloen, P., *J. Solid State Chem.*, 68, 360, 1987.
77. Idemura, S., Suzuki, E., and Ono, Y., *Clays Clay Miner.*, 37, 553, 1989.
78. Suzuki, E., Idemura, S., and Ono, Y., *Clays Clay Miner.*, 37, 173, 1989.
79. Malla, P.B., Ravindranathan, P., Komarneni, S., and Roy, R., Intercalation of copper metal clusters in montmorillonite, *Nature*, 351, 555, 1991.
80. Malla, P.B., Ravindranathan, P., Komarneni, S., Breval, E., and Roy, R., *J. Mater. Chem.*, 2, 559, 1992.
81. Theng, B.K.G., *The Chemistry of Clay-Organic Reactions*, John Wiley & Sons, New York, 1974, p. 343.
82. Jacks, G.V., *Soils Fertil.*, 26, 147, 1973.
83. Sugahara, Y., Kuroda, K., and Kato, C., *J. Am. Ceram. Soc.*, 67, C-247, 1984.
84. Sugahara, Y., Sugimoto, K.I., Yanagisawa, T., Nomizu, Y., Kuroda, K., and Kato, C., *Yogyo-Kyokai-Shi*, 95, 1, 1987.
85. Sugahara, Y., Kuroda, K., and Kato, C., *Clay Sci.*, 7, 17, 1987.
86. Sugahara, Y., Kuroda, K., and Kato, C., *J. Mater. Sci.*, 23, 3572, 1988.
87. Sugahara, Y., Kuroda, K., and Kato, C., *Ceram. Int.*, 14, 1, 1988.
88. Sugahara, Y., Sugimoto, K.I., Kuroda, K., and Kato, C., *J. Am. Ceram. Soc.*, 71, C-325, 1988.
89. Sugahara, Y., Miyamoto, J., Kuroda, K., and Kato, C., *Appl. Clay Sci.*, 4, 11, 1989.
90. Mostaghaci, H., Riley, F.L., and Torre, J.P., *Int. J. High Technol. Ceram.*, 4, 51, 1988.
91. Sugahara, Y., Miyamoto, J., Kuroda, K., and Kato, C., *Ceram. Int.*, 14, 163, 1988.
92. Hrabe, Z., Komarneni, S., Malla, P.B., Srikanth, V., and Roy, R., *J. Mater. Sci.*, 27, 4614, 1992.
93. Mehrotra, V., and Giannelis, E.P., *Mater. Res. Soc. Symp. Proc.*, 171, 39, 1990.
94. Mehrotra, V., Kwon, T., and Giannelis, E.P., *Mater. Res. Soc. Symp. Proc.*, 171, 167, 1990.

95. Giannelis, E.P., Mehrotra, V., and Russell, M.W., *Mater. Res. Soc. Symp. Proc.*, 171, 180, 1990.

96. Mehrotra, V., and Giannelis, E.P., *Solid State Commun.*, 77, 155, 1991.

97. Hoda, S.N., and Olszewski, A.R., U.S. Patent no. 4,454,237, 1984.

98. Mehrotra, V., and Giannelis, E.P., *Solid State Commun.*, 77, 155, 1991.

99. Mehrotra, V., and Giannelis, E.P., *Solid State Ionics*, 51, 1, 1992.

100. Kanatzidis, M.G., Wu, C.G., Marcy, H.O., DeGroot, D.C., and Kannewurf, C.R., *Chem. Mater.*, 2, 222, 1990.

101. Mehrotra, V., Giannelis, E.P., Ziolo, R.F., and Rogalskyj, P., *Chem. Mater.*, 4, 20, 1992.

102. Johnson, L.M., and Pinnavaia, T.J., *Langmuir*, 7, 2636, 1991.

103. Werpy, T.A., Michot, L.J., and Pinnavaia, T.J., *ACS Symp. Series*, 437, 119, 1990.

104. Yanagisawa, T., Shimizu, T., Kuroda, K., and Kato, C., *Bull. Chem. Soc. Jpn.*, 63, 988, 1990.

105. Yanagisawa, T., Shimizu, T., Kuroda, K., and Kato, C., *Bull. Chem. Soc. Jpn.*, 63, 1535, 1990.

106. Endo, T., Sato, T., and Shimada, M., *J. Phys. Chem. Solids*, 47, 799, 1986.

107. Endo, T., Nakada, N., Sato, T., and Shimada, M., *J. Phys. Chem. Solids*, 49, 1423, 1988.

108. Shimada, M., Proceedings of the Seventh Seminar on Frontier Technology—Nano-Hybridization and Creation of New Functions, Feb. 7–10, 1989, Oiso, Japan.

109. Endo, T., Nakada, N., Sato, T., and Shimada, M., *J. Phys. Chem. Solids*, 50, 133, 1989.

110. Thomas, J.M., in *Intercalation Chemistry*, Whittingham, M.S., and Jacobson, A.J., Eds., Academic Press, New York, 1982.

111. Wilson, S.T., Lok, B.M., Messina, C.A., Cannan, T.R., and Flanigen, E.M., *Adv. Chem. Ser.*, 104, 1146, 1982.

112. Barrer, R.M., *Hydrothermal Chemistry of Zeolites*, Academic Press, London, 1982.

113. Nicolaides, C.P., and Scurrell, M.S., in *Keynotes in Energy Related Catalysis*, Kaliaguine, S., Ed., Elsevier, New York, 1988.

114. Minachev, W.M., and Isakov, I.Y., *Am. Chem. Soc. Monogr.*, 171, 552, 1976.

115. Jacobs, P.A., in *Metal Clusters in Catalysis*, Gates, B.C., Guczi, L., and Knözinger, H., Eds., Elsevier, New York, 1986.

116. Gallezot, P., and Bergeret, G., in *Metal Microstructures in Zeolites*, Jacobs, P.A., Jaeger, N.I., Jiru, P., and Schulz-Ekloff, G., Eds., Elsevier, Amsterdam, 1982, p. 162.

117. Jaeger, N.I., Ryder, P., Schulz-Ekloff, G., in *Structure and Reactivity of Modified Zeolites*, Jacobs, P.A., Jaeger, N.I., Jiru, P., Kazansky, V.B., and Schulz-Ekloff, G., Eds., Elsevier, Amsterdam, 1984, p. 299.

118. Malla, P.B., Ravindranathan, P., Komarneni, S., Breval, E., and Roy, R., *Mater. Res. Soc. Symp. Proc.*, 233, 207, 1991.

119. Stucky, G.D., *Mater. Res. Soc. Symp. Proc.*, 206, 507, 1991.

120. Ozin, G.A., Stein, A., Godber, J.A., and Stucky, G.D., Report no. AD-A208208, National Technical Information Service, Springfield, VA, 1988, pp. 29.

121. Chae, H.K., Klemperer, W.G., Payne, D.A., and Suchicital, C.T.A., [abstract], Advanced Materials Science and Engineering Society Symposium on Hydrothermal Reactions, Tokyo, Japan, 1989, p. 16.

122. Newnham, R.E., McKinstry, S.E., and Ikawa, H., *Mater. Res. Soc. Symp. Proc.*, 175, 161, 1990.

123. Newnham, R.E., and Trolier-McKinstry, S.E., *Ceram. Trans.*, 8, 235, 1990.
124. Wen, L., and Huang, R., *C-MRS Int. Symp. Proc.*, 1, 117, 1991.
125. Tsuya, N., Saito, Y., Nakamura, H., Hayano, S., Furugohri, A., Ohta, K., Wakui, Y., and Tokushima, T., *J. Magn. Magn. Mater.*, 54, 1681, 1986.
126. Bacri, J.-C., Perzynski, R., and Salin, D., *Endeavor*, New Series, 12, 76, 1988.
127. Rosenweig, R.E., *Sci. Am.*, 247, 136, 1982.
128. Mehta, R.V., in *Thermomechanics of Magnetic Fluids*, Berkovsky, B., Ed., Hemisphere Publishing, Washington, DC, 1978.
129. Camras, M., *Magnetic Recording Handbook*, Van Nostrand Reinhold, New York, 1988.
130. White, R.M., *Sci. Am.*, 243, 138, 1980.
131. Lee, M., Halliyal, A., and Newnham, R.E., *Ferroelectrics*, 87, 71, 1988.
132. Cross, L.E., *Ferroelectrics*, 76, 241, 1987.
133. Smolensky, G.A., *J. Phys. Soc. Jpn.*, 28(suppl), 26, 1970.
134. Chen, J., Chan, H.M., and Harmer, M.P., *J. Am. Ceram. Soc.*, 72, 593, 1989.
135. Randall, C.A., Barber, D.J., Groves, P., and Whatmore, R.W., *J. Mater. Sci.*, 23, 3678, 1988.
136. Randall, C.A., Bhalla, A.S., and Cross, L.E., *Jpn. J. Appl. Phys.*, 29, 327, 1990.
137. Miller, D.V., Ph.D. Thesis, Pennsylvania State University, University Park, PA, 1991.
138. Bachmann, R., and Bamer, K., *Solid State Commun.*, 68, 865, 1988.
139. Ramesh, R., Inam, A., Chan, W.K., Wilkens, B., Myers, K., Remschnig, K., Hart, D. L. and Tarascon, J. M., *Science*, 252, 944, 1991.
140. Das, A., Ph.D. Thesis, Pennsylvania State University, University Park, PA, 1988.
141. Das, A., Srinivasan, T.T., and Newnham, R.E., *Mater. Res. Soc. Symp. Proc.*, 167, 165, 1990.
142. Falco, C.M., and Schuller, I.K., in *Synthetic Modulated Structures*, Chang, L.L. and Giessen, B.C., Eds., Academic Press, Orlando, FL, 1985, pp. 339–362.
143. Leavens, C.R., and Taylor, R., Eds., *Interfaces, Quantum Wells, and Superlattices*, Plenum, New York, 1988.
144. Terashima, T., and Bando, Y., *J. Appl. Phys.*, 56, 3445, 1984.
145. Mendez, E.E., and Von Klitzing, K., Eds., *Physics and Applications of Quantum Wells and Superlattices*, Plenum, New York, 1987.
146. Switzer, J.A., Shane, M.J., and Phillips, R.J., *Science*, 247, 444, 1990.
147. Chow, G.M., Patnaik, A., Schlesinger, T.E., Cammarata, R.C., Twigg, M.E., and Edelstein, A.S., *J. Mater. Res.*, 6, 737, 1991.
148. Chow, G.M., Holtz, R.L., Chien, C.L., and Edelstein, A.S., *Mater. Res. Soc. Symp. Proc.*, 195, 623, 1990.
149. Dorfmann, V.F., Skotheim, T.A., and Pypkin, B.N., *Proc. Electrochem. Soc.*, 91(8), 393, 1991.
150. Roy, S., and Chakravorty, D., *Appl. Phys. Lett.*, 59, 1415, 1991.
151. Chow, G.M., and Strutt, P.R., *High Temp. Sci.*, 27, 311, 1990.
152. Lyons, A.M., Nakahara, S., and Pearce, E.M., *Mater. Res. Soc. Symp. Proc.*, 132, 37, 1988.
153. Hirai, T., and Sasaki, M., *Mater. Sci. Monogr.*, 68, 541, 1991.
154. Selvaraj, U., Liu, C.L., Komameni, S., and Roy, R., *J. Am. Ceram. Soc.*, 74, 1378, 1991.
155. Uchiyama, T., Inoue, S., and Niihara, K., in *Silicon Carbide Ceramics*, Somiya, S. and Inomata, K., Eds., Uchida Rokakuho, Tokyo, 1989, pp. 193–200.

156. Sawaguchi, A., Toda, K., and Niihara, K., *J. Ceram. Soc. Jpn.*, 99, 523, 1991.
157. Niihara, K., and Nakahira, A., in *Proceedings of the MRS Meeting on Advanced Composites*, Plenum, New York, 1988.
158. Niihara, K., *J. Jpn. Soc. Powder Powder Metal.*, 37, 348, 1990.
159. Niihara, K., Nakahira, A., Ueda, H., and Sasaki, H., *Proc. 1st Jpn. Int. SAMPE Symp.*, 1120, 1989.
160. Niihara, K., Hirano, T., Nakahira, A., and Izaki, K., *Proc. MRS Meet. Adv. Mater.*, 107, 1988.
161. Niihara, K., Hirano, T., Nakahira, A., Suganuma, K., Izaki, K., and Kawakami, T., *J. Jpn. Soc. Powder Powder Metal.*, 36, 243, 1989.
162. Niihara, K., Izaki, K., and Kawakami, T., *J. Mater. Sci. Lett.*, 9, 598, 1990.
163. Niihara, K., Suganuma, K., and Izaki, K., *J. Mater. Sci.*, 9, 112, 1990.
164. Niihara, K., Izaki, K., and Nakahira, A., *J. Jpn. Soc. Powder Powder Metal.*, 37, 352, 1990.
165. Suganuma, K., Sasaki, G., Fujita, T., and Niihara, K., *J. Jpn. Soc. Powder Powder Metal.*, 38, 374, 1991.
166. Izaki, K., Nakahira, A., and Niihara, K., *J. Jpn. Soc. Powder Powder Metal.*, 38, 357, 1991.
167. Wakai, F., Kodama, Y., Sakaguti, S., Murayama, N., Izaki, K., and Niihara, K., *Nature*, 344, 421, 1990.
168. Niihara, K., and Nakahira, A., in *Advanced Structural Inorganic Composite*, Vincenzini, P., Ed., Elsevier Science, Trieste, Italy, 1990, pp. 637–664.
169. Niihara, K., and Nakahira, A., *Ann. Chim.*, 16, 479, 1991.
170. Niihara, K., and Nakahira, A., *Mater. Sci. Monogr.*, 68, 637, 1991.
171. Sawaguchi, A., Toda, K., and Niihara, K., *J. Am. Ceram. Soc.*, 74, 1142, 1991.
172. Niihara, K., Izaki, K., and Kawakami, T., *J. Mater. Sci. Lett.*, 10, 112, 1991.
173. Niihara, K., and Nakahira, A., in *Ceramics: Toward the 21st Century*, Soga, N. and Kato, A., Eds., Ceramic Society of Japan, Tokyo, 1991, pp. 404–417.
174. Niihara, K., and Hirai, T., *Ceramics*, 26, 598, 1986.
175. Wang, Y., Sasaki, M., and Hirai, T., *J. Mater. Sci.*, 26, 6618, 1991.
176. Wen, L., Gong, J., Yu, B., Huang, R., and Guo, L., *C-MRS Int. Symp. Proc.*, 4, 219, 1990.
177. Wang, Y., Sasaki, M., and Hirai, T., *J. Mater. Sci.*, 26, 5495, 1991.
178. Haaland, R.S., Lee, B.I., and Park, S.Y., *Ceram. Eng. Sci. Proc.*, 8, 872, 1987.
179. Ohtsuka, K., Koga, J., Tsunoda, M., Suda, M., and Ono, M., *J. Am. Ceram. Soc.*, 73, 1719, 1990.
180. Simonton, T.C., Roy, R., Komameni, S., and Breval, E., *J. Mater. Res.*, 1, 667, 1986.
181. Nakazumi, Y., Paper no. S-2, *27th Symposium on Synthetic Minerals*, 1982.

# 14 Molecular Engineering Route to Two Dimensional Heterostructural Nanohybrid Materials

*Jin-Ho Choy and Man Park*

## CONTENTS

Heterostructural nanohybrid materials are the group of nanohybrids[1-3] in which two distinct structural components are integrated into one chemical compound. In general, two-dimensional nanohybrids are those that contain the layered structure as one of their structural components. Diverse structural types can be combined into a layer structure to give rise to the interesting two-dimensional heterostructural nanohybrids such as layer-layer, particle-layer, and molecule-layer materials.[2,4,5] Because each component is regularly arranged and chemically interacts with each other in nanoscale and even in molecular scale, their nanohybrids exhibit the fascinating physical and chemical properties distinguished from each component and their physical mixture. It is widely recognized that hybridization leads to remarkable synergic effects such as enhancement of various stabilities, improvement of textural properties, increase in process ability, accomplishment of quantum size and/or confinement phenomena, etc.

Two-dimensional heterostructural nanohybrids have been prepared via lattice engineering by many different techniques. Although hybridization does leave individual layers intact, guest components lead to a significant change in their interlayer distance as well as their stacking mode. In particular, displacement of interlayer space is very sensitive to the size and shape of the guest components and the charge density (effective area per unit framework charge), which plays a key role in evaluating the degree of hybridization. Because the host layer component safely and stably accommodates guest components through electrostatic, covalent, and/or van der Waals bonding, hybridization can be facilitated by various reactions such as ion exchange, coprecipitation, reconstruction, and exfoliation-restacking, depending on the interaction between the host layer and guest components. Likewise, there are almost no limitations in hybridizing layer structure with various structural as well as functional components, which allows one to explore new potentials as well as to solve current problems in materials-related fields. This chapter describes the several types of two-dimensional heterostructural nanohybrids, comprising preparations, properties, and potentials.

# I. SWELLING MICA-TYPE CLAY MINERALS

Swelling mica-type clay minerals composed of octahedral and tetrahedral sheets are widely utilized in preparing two-dimensional heterostructural nanohybrids, most likely because of their natural ubiquity, high stability, swelling property, and wide applications.[4-10] In fact, smectite and vermiculite groups have been the preferred layer components to be hybridized with a variety of inorganic and organic components.

These clays have been hybridized with diverse structural types of components such as nanoparticles, clusters, complex compounds, polymers, molecules, and ions. Their potential applications are found in many fields as inorganic catalysts, adsorbents, ceramics, coatings, and even drug delivery carriers. Various preparation methods have been developed such as pillaring, intercalation, and delamination techniques. The representative examples include organic-clay hybrids,[11-13] metal oxide-pillared clays,[14-18] and bioclay hybrids.[19]

## A. STRUCTURAL PROPERTIES

The fundamental structure of unit crystal is based on the mica framework, as shown in Figure 14.1. The unit layer consists of one octahedral sheet sandwiched

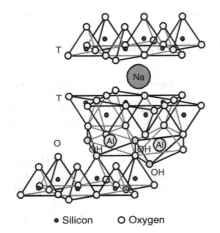

● Silicon    O Oxygen

**FIGURE 14.1** Crystal structure of 2:1 type aluminosilicate (montmorillonite).

between two tetrahedral sheets.[5–10] The cations in the tetrahedral sheet are typically $Si^{4+}$ and $Al^{3+}$, while those in the octahedral sheet are $Al^{3+}$, $Fe^{3+}$, $Mg^{2+}$, and $Fe^{2+}$. Because cations of both tetrahedral ($Si^{4+}$) and/or octahedral ($Al^{3+}$, $Fe^{3+}$, $Mg^{2+}$) sheets can be isomorphically substituted by less valent cations, the layer framework develops a permanent negative charge that is usually delocalized over the layer surface. The extent of the permanent charge is not affected by experimental conditions such as pH and ion activity. Thus negatively charged layers can accommodate an equivalent amount of exchangeable countercations in the interlayer space. Of course, additional negative charge could result from deprotonation of exposed hydroxyl groups in the deficient site of sheets and the cleaved edge and surface of the crystal. Unlike the permanent charge, the extent of this additional charge depends significantly on particle size as well as experimental conditions. Total cation exchange capacity is generally expressed in centiequivalents of exchangeable cations per kilogram of sample.[4,8]

The types of octahedral sheets, origin of the permanent charge, and charge density play a decisive role in classifying the species of mica-type clays, as indicated in Table 14.1. For example, the clays possessing a charge density per unit cell of 0.5 to 1.2 are classified as the smectite group, whereas those of 1.2 to 1.8 are classified as the vermiculite group. In the smectite group, montmorillonite has the isomorphous substitution of $Mg^{2+}$ for part of the $Al^{3+}$ in dioctahedral sheets, while hectorite has that of $Li^+$ for $Mg^{2+}$ in trioctahedral sheets. Although hybridization is affected by various inherent properties of the clays such as charge site, type of exchangeable cations, and crystallinity, swelling behavior and charge density are most important.[20–22]

## B. Organic-Clay Nanohybrids

Organic-clay nanohybrids can be roughly grouped into two different types by crystalline property, that is, lattice regularity.[10–13] Depending on the interaction

**TABLE 14.1**
**Classification of 2:1 Type Phyllosilicate Clay**

| Clay | Layer Type | Ideal Composition $M_{int}(M_{oct})(M_{tetra})O_1(OH)_m \cdot nH_2O$ |
|---|---|---|
| Pyrophyllite | Dioctahedral | $(Al_2)(Si_4)O_{10}(OH)_2$ |
| Talc | Trioctahedral | $(Mg_3)(Si_4)O_{10}(OH)_2$ |
| **Smectite group** | | |
| Montmorillonite | Dioctahedral | $M_x(Al_{2x}Mg_x)(Si_4)O_{10}(OH)_2 \cdot nH_2O$ |
| Beidellite | Dioctahedral | $M_x(Al_2)(Si_{4x}Al_x)O_{10}(OH)_2 \cdot nH_2O$ |
| Nontronite | Dioctahedral | $M_x(Fe_2^{3+})(Si_{4x}Al_x)O_{10}(OH)_2 \cdot nH_2O$ |
| Saponite | Trioctahedral | $M_x(Mg_3)(Si_{4x}Al_x)O_{10}(OH)_2 \cdot nH_2O$ |
| Hectorite | Trioctahedral | $M_x(Mg_{3x}Li_x)(Si_4)O_{10}(OH)_2 \cdot nH_2O$ |
| **Vermiculite group** | | |
| Vermiculite | Dioctahedral | $M_x(Al_{2y}Fe_4^{3+})(Si_{4x}Al_x)O_{10}(OH)_2 \cdot nH_2O$ |
| Vermiculite | Trioctahedral | $M_x(Mg_3)(Si_{4x}Al_x)O_{10}(OH)_2 \cdot nH_2O$ |

of guest organic components with negatively charged layers of the clays, guest and/or layer components could be arranged either in ordered or disordered form. In general, a highly ordered two-dimensional heterostructure of organic-clay layers results from intercalation of monomeric organics with cationic form. On the other hand, hybridization of delaminated clay layers with polymeric organic components can develop disordered polymer-clay hybrids, typically with house-of-card structures.

Organic cations can be easily intercalated and stabilized in the interlayer space of layered clays by the ion exchange reaction, which evolves into an ordered organic/inorganic nanohybrid.[12] The representative organic cations are the amphiphilic organoammonium ions that exhibit a number of molecular structures and functionalities. In the nanohybrid, organic cations develop a unique arrangement at the molecular level within subnanosized two-dimensional interlayer spaces. Their polar part with cationic charge is electrostatically interacted with the rigid inorganic layer of clays, while the flexible nonpolar part is projected out and stabilized by hydrophobic interaction. The arrangement of the nonpolar part depends on its size and shape, charge density of the layer, and solvent. Likewise, hybridization of clay with amphiphilic organics leads to the evolution of new potentials as well as a dramatic change in physical and chemical properties of both the organics and the clays.[11] Applications of these nanohybrids include control of permeability, coating for sensors, devices for optics and electronics, adsorbents, etc.[10]

The intercalation of organic polymers into the interlayer space has also been widely studied since hybridization may not only produce unique materials with high dispersive ability, but may also enhance polymeric properties by using nanoscale reinforcements compared to the conventional particulate-filled

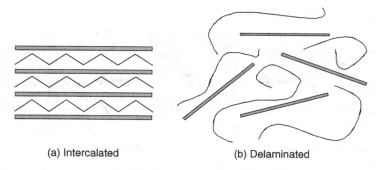

(a) Intercalated            (b) Delaminated

**FIGURE 14.2** Possible polymer/inorganic nanohybrid structures.

microcomposites.[23–25] Polymer-clay nanohybrids can be synthesized either in the ordered or disordered form (Figure 14.2).

The ordered lamella structure with a basal spacing of a few nanometers could be developed through intercalation of polymer chains into the interlayer space. On the other hand, exfoliation of clay layers and subsequent dispersion of delaminated layers in polymeric media result in disordered polymer/inorganic hybrids in which each exfoliated layer is irregularly interacted with polymer matrices. In addition, these polymer/inorganic nanocomposites can be also obtained by intercalation of monomers followed by *in situ* polymerization in the interlayer space. In particular, the disordered polymer/inorganic hybrids have been known to be effective in improving the performance properties of nanocomposite materials due to better phase homogeneity compared with that of intercalated polymer/inorganic hybrids.[26] Toyota researchers demonstrated that the disordered hybridization of delaminated clay layers with thermoplastic nylon-6 polymer matrix led to significantly improved thermal, mechanical, and flame-retardant properties.[27]

## C. INORGANIC-CLAY NANOHYBRIDS

Pillared clays are the representative particle layer nanohybrids that have a two-dimensional porous structure.[14–18] Pillared clays are usually mesomicroporous materials that exhibit intrinsic catalytic activity, large surface area, high thermal stability, and molecular sieving property, although their textural and catalytic properties are significantly affected by the characteristics of the pillars. Pillared clays are typically prepared by ion exchange reactions followed by stabilizing procedures. The ion exchange reaction facilitates intercalation of various cationic guest components such as metal chelates, metal polyoxycations, and positively charged colloidal particles into the interlayer space of clays.[14–18,28–32] Subsequent stabilizing procedures, usually calcination, allow the intercalated inorganic species to be converted into stable metal oxide clusters and nanoparticles within the interlayer space (Figure 14.3).

The most well-known examples are aluminum oxide pillared clays, which are stable in both oxidizing and reducing atmospheres.[14] $Al_2O_3$ pillared clays are

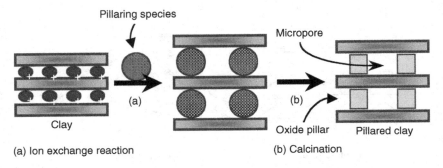

**FIGURE 14.3** Pillaring route to nanoporous inorganic media.

prepared by cation exchange reaction between the interlayer cations and aluminum-polyhydroxy cation ($[Al_{13}O_4(OH)_{24}(H_2O)_{12}]^{7+}$, Keggin-type ions). The intercalated aluminum-polyhydroxy cations are converted to $Al_2O_3$ by heating at 400°C in air. The $Al_2O_3$-mica-type clay nanohybrids have an interlayer spacing of 7 to 8 Å and a specific surface area of 300 m²/g. Their microporous structure exhibits high thermal stability. On the other hand, metal nanoparticle-clay nanohybrids are prepared either via *in situ* reduction of interlayered metal cations or via direct intercalation of metal nanoparticles whose surfaces are positively modified with organics.[28–32] In particular, surface modification has been widely applied to direct intercalation of various nanoparticles such as silica and titania nanoparticles (Figure 14.4).

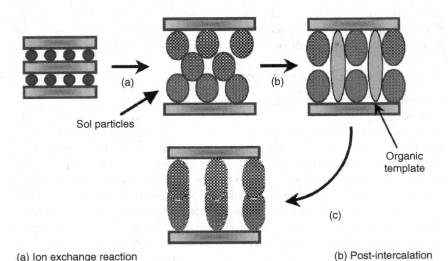

**FIGURE 14.4** Modification of interlayer pore structure by the postintercalation route.

In order to improve the textural properties of particle-clay nanohybrids, bulky organic cations are intercalated as a kind of template into particle-intercalated clays before stabilization procedures.[29,31] Intercalation of the organic cations results in the removal of some of the intercalated nanoparticles and/or in their rearrangement. Subsequent calcination leads to formation of additional pore space that is highly correlated to the geometry and size of the templates. This technique allows fine tuning of textural properties in the preparation of particle-clay nanohybrids. The clay nanohybrids intercalated with metals, oxides, and complexes have a broad range of applications. In particular, metal oxide particle-pillared clays have excellent potentials as catalysts, catalyst supports, selective adsorbents, etc.[14]

## D. DRUG-CLAY HYBRIDS

Biologically active organic molecules can also be hybridized with layered clays to evolve into bioclay hybrids. Recently they have attracted special attention because of their potential pharmaceutical applications.[19,33,34] It is generally conceived that natural layered clays, especially smectites, are stable, inert, and biocompatible enough to be employed as drug carriers. Furthermore, hybridization of drugs with clays offers fascinating features such as controlled and long-term release, an increase in water solubility, and even protective and targeted delivery. Likewise, drug-clay hybridization is utilized to increase drug efficiency as well as to reduce side effects.

Unique behaviors of intercalated drugs originate from the synergistic effects of organic-inorganic hybridization. In the hybrids, drug molecules are regularly arranged at the molecular level between two-dimensional inorganic layers. The stability and inertness of the inorganic layers impart remarkable stability as well as hydrophilic properties to drug molecules, while drug release is controlled mainly by layer charge. Furthermore, it is also possible for drugs to be delivered into specific target organs by surface modification of the clays. To date, only a few drugs have been intercalated into clays, including 5-fluorouracil, hydralazine hydrochloride, and diphenhydramine hydrochloride.[19] Although pesticide-clay hybrids for controlled release formulations were suggested a long time ago, drug-clay hybrids have been quite recently proposed for potential drug delivery systems. Thus the drug/inorganic nanohybrid materials may evolve into a useful drug delivery system.

Clay minerals have developed into organic/inorganic, inorganic/inorganic, bio/inorganic nanohybrids through the modification of their interlayer space. Thus, layered clay minerals are the excellent materials for producing new functionality and structure like various nanohybrids.

## II. LAYERED DOUBLE HYDROXIDES

Layered double hydroxides (LDHs) are a unique type of clay with anion exchange capacity. In nature, they are found in the carbonated hydrotalcite-like compounds that lack, in the strict sense, true anion exchange capacity due to an extremely high affinity for carbonate ion.[35,36] Fortunately LDHs are readily synthesized to have interlayered exchangeable anions such as nitrate, chloride, and sulfate. Like swelling layered clays, the interlayer distance of LDHs also depends exclusively on the properties of interlayered anions like size, shape, arrangement, and solvation capacity. Of course, intercalation of bulky interlayer anions results in a corresponding expansion of the interlayer space. As is expected because of their high affinity to carbonate ion, intercalated anions can be completely released by the anion exchange reaction with carbonate ion. Another property of LDHs is their structure-memory ability. Even after LDHs are thermally converted into the corresponding oxide forms, their rehydration results in the reconstruction of pristine LDHs. Likewise, LDHs exhibit excellent potentials as catalysts and catalyst precursors in a wide variety of base-catalyzed reactions,[37–40] as selective adsorbents for anionic substances,[41] and as fire retardants.[36]

Hybridization of LDHs with various guest components not only reinforces the potentials of both the LDHs and guest components, but also leads to a number of new properties. Currently, hybridization with bioactive substances for their safe, controlled, and targeted delivery is a hot issue, because most of the essential biomolecules are negatively charged. In addition, the acid lability and carbonate affinity of LDHs could be useful characteristics[36] for the safe and complete discharge of intercalated biomolecules. To date, a variety of LDH-derived hybrids have been developed, comprising organic-LDH, inorganic-LDH, and bio-LDH hybrids.

## A. Structural Properties

The general formula of LDHs is $[M^{2+}_{1-x}M^{3+}_{x}(OH)_2]^-(A^{m-})_{x/m}, nH_2O$, where the $M^{n+}$ are metal cations ($M^{2+} = Mg^{2+}, Zn^{2+}, Ni^{2+}, Ca^{2+}, Cu^{2+}, ..., M^{3+} = Al^{3+}, Fe^{3+}, ...$) and $A^m$ are interlayer anions ($A^{m-} = CO_3^{2-}, NO_3^-, SO_4^{2-}$, and other anionic species). The layer structure of LDHs is constructed with a stacking of brucite structure of $Mg(OH)_2$ in which $Mg(OH)_6$ octahedra are connected through edge sharing into two-dimensional sheets with a layer thickness of 4.8 Å. Some of the divalent cations in the brucite layer are substituted by trivalent cations such as $Al^{3+}$, which develop a permanent cationic layer charge. The interlayer space is occupied by charge-balancing anions that are typically bound to the layer through hydrogen bonding with water molecules[36] (Figure 14.5). Exchangeability of interlayered anions depends on their electrostatic interaction with the positively charged layer. Except for carbonate ion, most organic and inorganic anions are known to be exchangeable. Thus LDHs are widely applicable to various supramolecular structures or heterogeneous hybrid systems.

Layered double hydroxides are typically synthesized from coprecipitation of mixed metal cations by base titration either with or without hydrothermal treatment which usually enhances crystalline properties.[42,43] Because contamination with airborne carbonate ions frequently occurs during synthesis procedures, special caution is needed to prepare carbonate-free LDHs. On the other hand,

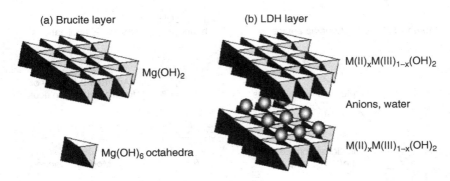

**FIGURE 14.5** The structure of LDHs: (a) brucite layer; (b) LDH layer.

two-dimensional heterostructured LDH hybrids can be prepared by three different techniques: coprecipitation, anion exchange, and reconstruction. Intensive attention has been given to the coprecipitation technique because it allows lattice engineering through fine tuning of the framework charge density, framework composition, anion properties, treatment condition, etc. Different preparation techniques accelerate the rapid expansion of hybrid systems from simple anion-based hybrids to complex and polymer-based hybrids and even to biosubstance-based hybrids.

## B. ORGANIC-LDH HYBRID SYSTEM

The introduction of organic molecules between the layers of LDHs creates the supramolecular structures that have potential applications in photoelectrical materials, organically modified electrodes, isomer separating matrixes, etc. To date, a variety of organic-LDH hybrids have been synthesized through hybridization with diverse organic components such as dye molecules, organometallic complexes, and even bulky C60 molecules.[44–50]

The most well-known organic-LDH hybrid system is the functionalized metal complex or organic molecule intercalated LDHs with advanced synergetic functions (Figure 14.6b). Mousty and Therius[44] hybridized LDHs with electroactive organic molecules such as m-nitrobenzene sulfonate, anthraquinone sulfonate, and 2,2'-azino-bis (3-ethylbenzothiaxoline-6-sulfonate) for enhancement of electrode function. Metal complex with large-ring ligand such as TSPP (5,10,15,20-tetra(4-sulfonatophenyl)porphyrin) was intercalated into $Mg_2Al(OH)_6$-LDH to make the organic-LDH hybrids with photochemical hole burning functions.[45] ZnTPPC ([tetrakis-(4-carboxyphenyl)porphyrinato]) was intercalated into LDH to induce a photochemical reaction in the particle edge.[46] Recently, hybridization selectivity of LDH is intensively studied for isomer separation. Lei et al.[47] studied the anionic exchange properties of $LiAl_2(OH)_6Cl$ LDH with various isomeric pyridinecarboxylate (2-PA, 3-PA, and 4-PA) and toluate (o-TA, m-TA, and p-TA) molecules for this purpose (Figure 14.6c).

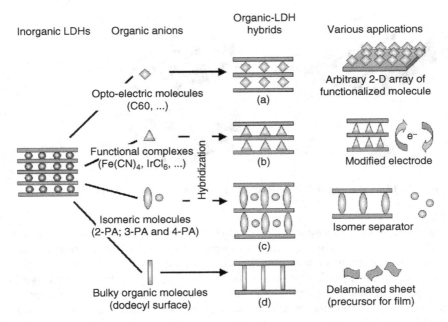

**Inorganic LDHs**　　**Organic anions**　　**Organic-LDH hybrids**　　**Various applications**

Opto-electric molecules
(C60, ...)

(a)

Arbitrary 2-D array of
functionalized molecule

Functional complexes
(Fe(CN)$_4$, IrCl$_6$, ...)

(b)

Modified electrode

Hybridization

Isomeric molecules
(2-PA; 3-PA and 4-PA)

(c)

Isomer separator

Bulky organic molecules
(dodecyl surface)

(d)

Delaminated sheet
(precursor for film)

**FIGURE 14.6** Synthetic route of various organic-LDH hybrid systems and their applications: (a) C60-LDH hybrid; (b) metal complex-LDH hybrid; (c) isomeric organic molecule-LDH hybrid; (d) bulky organic molecule-LDH hybrid and their delaminated product.

Recent hybridization of LDHs with bulky three-dimensional C60 molecules attracted special attention.[48] C60-LDH hybrids were synthesized by stepwise organic intercalation; bulky organic chain molecule such as dodecyl sulfate was intercalated into the LDH layers and then the C60 molecules were incorporated into the LDH layers (Figure 14.6a). C60 molecule regularly arranged within the two-dimensional interlayer space exhibited unique physicochemical properties that remained between those of the free molecules and the molecules trapped in the one-dimensional tunneling matrix. Nuclear magnetic resonance (NMR) studies confirmed that the C60 molecules trapped between the LDH layers could not freely rotate, unlike the powder-type three-dimensional C60 array.[48] This kind of hybrid system can expand the application range of optoelectrical materials.

The hybridization technique by delamination of LDH layers offers another interesting feature. Adachi-Pagano et al.[49] succeeded in delaminating LDH layers by exfoliating dodecyl sulfate intercalated LDH using various organic solvents. Similarly Hibino and Jones[50] reported delamination of glycine-LDH hybrid by foramide. The delaminated LDH colloidal suspension leads to remarkable expansion of the hybridization potentials of LDHs, especially to the ultrathin homogeneous LDH-based films as well as polymer-LDH hybrids. Likewise, delamination of LDH layers together with hybridization enable one to control structural and textural properties such as the layer stacking mode and porosity by a soft chemical route (Figure 14.6d).

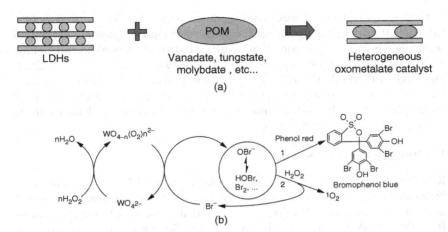

LDHs          +          POM          ➡          Heterogeneous
                  Vanadate, tungstate,                     oxometalate catalyst
                  molybdate , etc...

(a)

(b)

**FIGURE 14.7** Heterogeneous catalyst utilizing inorganic-LDH hybrid system. (a) The preparation of polyoxometalate-LDH hybrid and (b) the catalytic cycle in bromination with $WO_4^{2-}$ LDH. $H_2O_2$ binds to tungstate to form peroxotungstate at the surface of the LDH. Electrostatic attraction brings bromide to the surface, facilitating transfer of the activated oxygen atom from peroxotungstate to bromine. Reactions of two-electron-oxidized bromine species in solution: electrophilic bromination of phenol red into bromophenol blue (route 1), and bromide-assisted $1O_2$ generation from $H_2O_2$ (route 2). A preparative experiment of route 1 was run on a 250 ml scale, followed by work-up via extraction with $(n\text{-}C_6H_{13})4NCl$ to a $CH_2Cl_2$ layer. The 1H NMR spectrum of the bromophenol blue product (7.29 s,4H; 7.01 and 7.90 d,1H each, J 1/4 7:5 Hz; 7.41 and 7.50, t,1H each, J 1/4 7:5 Hz, TMS standard) is identical to that of authentic bromophenol blue. Figure 14.7b was cited from Itaya et al.[58]

## C. Inorganic-LDH Hybrid System

The main type of inorganic-LDH hybrid system comes from intercalation of oxometalates into the interlayer space of LDHs because of the high catalytic activity of both oxometalates and LDHs. Typical synthetic routes are similar to those of pillared clays, as mentioned before. After pristine nitrate/chloride-LDHs are anion exchanged with polyoxometalates, usually at elevated temperatures, the pillared polyoxometalates are stabilized within the interlayer space by calcination.

Recently Sels et al.[51] developed a novel tungstate-LDH hybrid as an oxidative bromination catalyst (Figure 14.7). They reported that the hybrid catalyst shows a 100-fold higher activity compared to that of conventional homogeneous catalyst. In addition, synthesis of nanoparticle-LDH hybrids has been attempted for catalytic applications. Although few, there are some reports of preparing porous LDH hybrids intercalated with metal nanoparticles by the exfoliation-restacking route. Figure 14.7 shows a schematic illustration of preparing heterogeneous LDH catalyst and the mechanism of the catalyst.

Intercalation of electronically active metal complexes has also been widely studied to enhance the reduction currents of electrodes.[52–58] The electrochemical responses of LDH hybrid electrodes modified with $[Fe(CN)_6]^{3-}$, $[Ru(CN)6]^{4-}$, $[Mo(CN)_8]^{4-}$, $[IrCl_6]^{2-}$, polyoxometalate ions, manganese porphyrins, and several organometals have been reported. Similarly, thin hybrid films of an amphiphilic anionic Ru(II) complex ($K_2wRu(CN)_4Lx$: Ls4,49-di-*n*-dodecylaminoyl-2,29-bipyridyl) with Mg-Al-Cl or Ni-Al-Cl LDHs were recently prepared through the adsorption of positively charged LDH particles by the floating film of $(wRu(CN)_4Lx)^{2-}$ at the air-water interface for potential utility in electrochemical sensing of biomolecular materials.

## D. Bio-LDH Hybrid System

Recent advances in the hybridization technique enable LDHs to accommodate biologically important substances such as drugs, genetic components, essential nutrients, and vitamins.[59–62] Current attention focused on bio-LDH hybrids comes from the remarkable biocompatibility of LDHs. In contrast to other inorganic matrices, not only is their layer framework labile to physiologic acid conditions, but also their positive layer charge selectively prefers abundant carbonate ions, which means the controlled and complete ion exchange of intercalated biosubstances. Likewise, LDHs hybridized with biosubstances have excellent potentials as safe, disposable, and even targeting carriers. Targeted delivery may be possible because bio-LDH hybrids can discharge their biosubstances to a specific organ through selective dissolution of the LDH layers.

The concept of the bio-LDH hybrid system was presented by Choy et al.,[59] who reported the successful hybridization of LDH with DNA strands by the ion exchange route. Since typical DNA molecules form strong complexes with trivalent metals such as $Al^{3+}$, the DNA-LDH hybrid could be attained only by the ion exchange route (Figure 14.8c). DNA-LDH hybrids are of great importance for two reasons: the successful hybridization of biomolecule and inorganic compounds, and the protective and release-controlling role of inorganic LDHs against rapid degradation and inadvertent release of unstable biosubstances. Intercalated DNA strands can endure strong enzymatic degradation by DNase I due to protective LDHs. Also, the DNA strands can be retrieved from the hybrid by simple acid treatment without any deterioration. Thus LDHs can function as a reservoir as well as a carrier for DNA strands (i.e., for genetic information). Biomolecules such as adenosine triphosphate (ATP) have also been successfully hybridized with LDHs.[60] When an ATP-LDH hybrid was tested for cellular uptake, the ATP-LDH hybrid showed a 25-fold higher uptake rate compared to ATP alone, which clearly supports that the delivery efficiency of ATP can be greatly enhanced by hybridization with LDHs (Figure 14.8d). Normally, biomolecules or bioactive organic molecules barely penetrate the cell membrane, mainly due to the selective permeability of cell membranes. However, bio-LDH hybrids can penetrate them rather easily.

Evidence on the role of LDH as an excellent biomolecule carrier has also been obtained for the antisense-LDH hybrid.[61,62] Therapeutic application of antisense is one of the newest technologies in biochemistry for repairing damaged

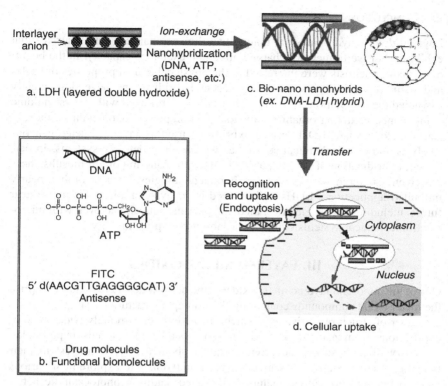

a. LDH (layered double hydroxide)

c. Bio-nano nanohybrids
(*ex. DNA-LDH hybrid*)

DNA

ATP

FITC
5′ d(AACGTTGAGGGGCAT) 3′
Antisense

Drug molecules
b. Functional biomolecules

d. Cellular uptake

**FIGURE 14.8** A schematic diagram of bio-LDH hybrid: (a) pristine LDH host; (b) biomolecules; (c) bio-LDH hybrid for biomolecule reservoir or carrier; (d) cellular uptake mechanism of bio-LDH hybrid for biomolecule carrier.

DNA sequences. Choy et al.[61] reported that the antisense-LDH hybrid exhibited strikingly high activity against the growth of cancer cells such as NIH3T3 and HL-60. Its application efficiency is far greater than that of antisense alone.

Bio-LDH hybrids have also been applied successfully in the pharmaceutical and cosmetics industries. A significant increase in drug efficiency was achieved by hybridization of LDH with the anticancer drug methotrexate (MTX). It was also suggested that the drug efficiency increased by protective delivery and controlled release of the intercalated MTX. Although the exact mechanism of drug delivery by LDH remains to be elucidated, the increase in drug efficiency could significantly contribute to alleviation of the serious side effects of drugs, in particular, anticancer drugs. Another type of bio-LDH hybrid is the vitamin-LDH hybrids found in cosmetics. Ascorbic acid, highly unstable vitamin C, was safely stabilized within the interlayer space of zinc-based hydroxide salt,[34] an LDH. The intercalated ascorbic acid was found to be slowly and steadily taken up by the skin.

## E. CONCLUSION

Layered double hydroxides are unique two-dimensional inorganic matrices that exhibit a positive framework charge with anion exchange capacity. In the beginning, many scientists were interested in the physicochemical properties of LDHs and their applicability as catalysts. Recent hybridizations have dramatically expanded the applications of LDHs. They can be hybridized with diverse anionic components, including organics, organometal complexes, oxometalates, biosubstances, etc. To date, LDH-derived hybrids are used as advanced heterogeneous catalysts and electrochemical sensors as well as drug delivery systems. In particular, hybridization of biocompatible LDH with many important biosubstances has given rise to remarkable advances in biotechnologies. Because of their nearly unlimited applications, LDHs will be used in many scientific fields in the near future, including optics, electrics, medical, pharmaceutical, and even with microelectromechanical systems (MEMS) or lab-on-a-chip technology.

## III. LAYERED METAL OXIDES

One of the important layered metal oxides is the perovskites. Perovskites comprise the metal oxide compounds containing the principal structure $CaTiO_3$. There are a number of structural derivatives which give rise to an extremely wide range of applications, from insulating ceramics to superconducting materials. In particular, discovery of their superconductivity brought about a revived interest in their potential.[63] Thus extensive research has been undertaken to explore their various physicochemical properties, including Brönsted acidity,[64] photocatalytic activity,[65-67] ionic conductivity,[68,69] intercalation behavior,[70-72] luminescence,[73] and magnetism.[74]

Perovskites are also an excellent inorganic component to be hybridized with other functional components such as inorganic and organic molecules, clusters, nanoparticles, and layers.[75-77] Of course, their physicochemical properties have been remarkably improved by hybridization. Furthermore, hybridization makes it possible to regularly combine two different properties at molecular level, such as plastic-magnetic,[78] superconducting-conducting,[79] and semiconducting-semiconducting.[76] For example, their hybridization with organic components can provide very useful properties by coupling together characteristics of both organic (e.g., plasticity, efficient luminescence) and inorganic compounds (e.g., magnetism, electrical mobility) at molecular level.[80] Only several representative two-dimensional hetrostructured hybrids will be described here, as well as with the structural properties of perovskites.

## A. STRUCTURAL PROPERTIES

Representative structures of two-dimensional layered perovskite can be described as a combination of a perovskite slab and an intervening layer. Three series of perovskites are well known, based on the intervening layer: Ruddlesden-Popper,

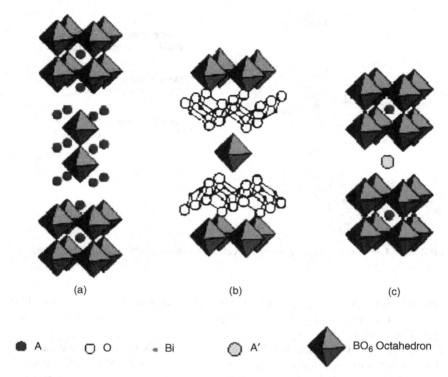

(a)                              (b)                              (c)

● A        ▢ O        ◂ Bi        ⬡ A'        ◆ BO$_6$ Octahedron

**FIGURE 14.9** Ideal tetragonal perovskite-like structures of the members of the (a) Ruddlesden-Popper, (b) Aurivillius, and (c) Dion-Jacobson series.

with an alkaline-earth metal-oxygen-rock salt layer;[81,82] Aurivillius phase, with a $Bi_2O_2$ layer as the intervening layer;[83–85] and Dion-Jacobson, with an alkali metal-oxygen layer[86] (see Figure 14.9 and Figure 14.10).

The Ruddlesden-Popper phase has the general formula $A'_2[A_{n1}B_nO_{3n+1}]$, where $[A_{n-1}B_nO_{3n+1}]$ donates perovskite-like slabs of $n$ octahedra in thickness, formed by slicing the perovskite structure along one of the cubic directions, and A indicates the interleaved cations. $Sr_2TiO_4$, $Sr_3Ti_2O_7$, $Sr_4Ti_3O_{10}$, and $A'_2[Ln_2Ti_3O_{10}]$ (A' = K, Rb; Ln = lanthanide) are included in this series.

The Dion-Jacobson series is represented by the general formula $A'[A_{n-1}B_nO_{3n+1}]$ (A' = alkali metal, A = alkaline-earth metal, B = Ti, Nb, and Ta). As shown in Figure 14.10, cation B is remarkably displaced along the $c$-axis from the center of the $BO_6$ octahedra. Accordingly, (i) the B-O$_{center}$ bond (bond between the B cation and the corner sharing oxygen along the $c$-axis) is relatively weak [R(B-O$_{center}$) ~ 2.2 Å], whereas (ii) the B-O$_{apex}$ bond (bond between the B cation and the oxygen sharing with the A cation) is strong with a highly covalent character [R(B-O$_{apex}$) ~ 1.8 Å]. This is because the alkali metal cations exhibit lower polarizing power than the transition metal ion, so alkali metal forms a less covalent bond with ligands. Similarly, the Aurivillius series is described by the

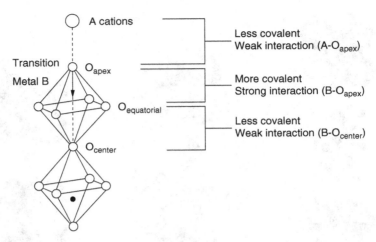

**FIGURE 14.10** Local structure around B cations in Dion-Jacobson-type layered perovskite, $AA'_{n-1}B_nO_{3n+1}$.

formula $Bi_2O_2[A_{n-1}B_nO_{3n+1}]$, consisting of alternate stacking of a fluorite-like $(Bi_2O_2)^{2+}$ unit and a perovskite layer. $Bi_4Ti_3O_{12}$, $Bi_3TiNbO_9$, and $SrBi_2Ta_2O_9$ are typical examples.

Anisotropy of chemical bonding in layered perovskites results in expansion through solvation and ion exchange, and even exfoliation of two-dimensional layers. When the individual structural elements are highly anisotropic, delamination occurs to form colloidal solutions that can be stabilized by the appropriate choice of cation/solvent combination or by the addition of surfactants.[87]

Layer modification, typically acidification, is generally required for reliable exfoliation of layered oxides with high layer charges. The acidic oxides still have strong interlayer bonding due to hydrogen bonding interactions, but these can be reduced by intercalation of suitable basic molecules. Partial intercalation of a primary amine can facilitate exfoliation by disrupting the hydrogen bonding interactions. In a related approach, introduction of hydrophilic groups into interlayer species also promotes exfoliation in water or solvents with high dielectric constants. Recently, electrostatic repulsion mediated by zwitter ions has also been used to exfoliate the layered oxides.[88]

Exfoliation is also facilitated by mechanical methods such as the application of high-intensity ultrasound. High-intensity ultrasound is known to increase the rate of intercalation reactions.[89] However, in recent studies on the exfoliation of the misfit layer compounds $PbNb_2S_5$ and $SmNb_2S_5$, some degree of dispersion was obtained by ultrasonic treatment in specific solvents such as ethanol and isopropanol.[90] The solvents were not intercalated, but acted to stabilize the dispersions. The dispersions contained a substantial fraction of particles with fewer than 10 layers, but only a small fraction was exfoliated. Along with the ion exchange reaction under relatively severe conditions compared with other layered clay materials, exfoliation plays an important role in hybridizing perovskites with heterostructured guest components through restacking.

## B. INORGANIC-LAYERED METAL OXIDE HYBRIDS

Intercalation of nanoparticles into the interlayer space of layered oxides is typically facilitated by the restacking-exfoliation technique, whereas molecular components are introduced by layer expansion mediated through the ion exchange reaction.[91] Intensive studies have been carried out to synthesize highly porous inorganic-inorganic hybrids with perovskites because many perovskites possess excellent photocatalytic activity.[92–94] Typical examples are the semiconductive-semiconductive hybrids such as anatase-layered titanate[76] and CdS-MoS$_2$ hybrids.[95] Intercalation of CdS into layered semiconductive materials has attracted special attention due to its unique features.[96] CdS has a sufficiently negative flat-band potential and effective absorption property in the visible region. However, it is not stable in aqueous solutions under irradiation to undergo electrochemical dissolution. Its hybrids with ultraviolet (UV)-sensitive perovskites could not only efficiently adsorb a widened spectrum in the UV-visible region, but also stabilize the interlayered nanosized CdS by protective perovskite layers. In these coupled semiconductor nanohybrids, the improvement of efficiency is easily expected as the result of a vectorial transfer of electrons and holes from one semiconductor to another. Furthermore, their texture can be significantly altered to exhibit high porosity with greatly increased surface area. CdS-perovskite hybrids have been synthesized either by exfoliation of perovskite layers and subsequent restacking of delaminated layers together with CdS nanoparticles or by introduction of Cd$^{2+}$ into the interlayer space of perovskite by the ion exchange reaction and subsequent synthesis of CdS nanoparticles by *in situ* reaction with H$_2$S (Figure 14.11).

On the other hand, inorganic molecules or clusters can also be intercalated into layered perovskites. In particular, halide-superconductor hybrids have been intensively studied to modulate electric conductivity by modification of their crystallographic structure because the Bi$_2$Sr$_2$Ca$_{n-1}$Cu$_n$O$_x$-type ($n$ = 1, 2, 3) high-temperature superconductors exhibit strong anisotropic physical properties, where electrical conductivity is metal-like in the (a,b) plane, whereas semiconductivity is observed along the $c$-axis. As a representative example, the ionic conductivities of several fast ionic conductors are compared with that of the AgI-Bi2212 hybrid, which shows a high ionic conductivity of $10^{-2}$ to $10^{-1}$ $\Omega$.[97]

## C. ORGANIC-LAYERED METAL OXIDE HYBRIDS

Organic-perovskite hybrids play an important role in the fabrication of various nanodevices as well as in elucidating their fundamental properties.[98] Intercalation of organic components into layered perovskites can be facilitated by various techniques such as ion exchange[99] and the exfoliation-restacking method.[100] In addition, the electrochemical intercalation method is also favorable for intercalation of organic molecules into transition metal oxides.[101] However, except for

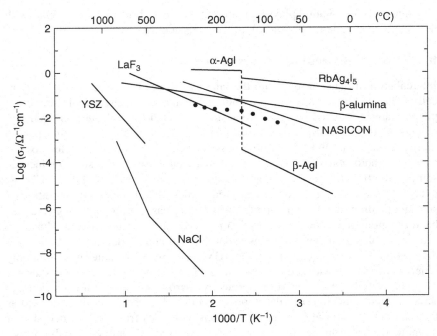

**FIGURE 14.11** Ionic conductivity of AgIBi2212 (shown as the solid circles) compared to the previously published data for the well-known solid ionic conductors AgI, $RbAg_4I_5$, $Na^+$-β-alumina ($NaAl_{11}O_{17}$), YSZ ($ZrO_2$-8% $Y_2O_3$), NASICON ($Na_{1+x}Zr_2P_{3-x}SiOxO_{12}$, $0 \leq x \leq 3$), $LaF_3$, and NaCl.

the exfoliation-restacking method, these procedures usually require drastic conditions or multistep processes because electron transfer from guests to metal oxides is believed to be necessary for reaction.

One of main thrusts in organic-perovskite hybrids has been in the preparation of stable delaminated perovskite sols as the building blocks of nanocomposite materials. A well-known process for exfoliation is to weaken the attractive interaction of the layers through intercalation of bulky organic components.[102] A representative example is exfoliation of bismuth-based cuprate superconductors (Figure 14.12).[103]

Another interesting process employs zwitteric molecules that bring about electrostatic repulsion depending on the pH.[88] A perovskite, $HCa_2Nb_3O_{10}$, was successfully exfoliated through its hybridization with aminoundecanoic acid. The delaminated organic-perovskite hybrid sols have been proved to be excellent precursors for industrially important ceramic nanodevices such as semiconductive and superconductive thin films, sensors, wave guides, photocatalysts, etc.[104–108]

Organic-perovskite hybrids that can be easily synthesized through ion exchange reactions can provide useful properties by themselves. Organic ammonium-perovskite hybrids self-organize a quantum well structure where a

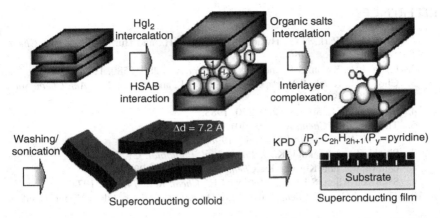

**FIGURE 14.12** Schematic illustration of the synthesis of the organic salt intercalate and superconducting film.

two-dimensional semiconductor layer and an organic ammonium $RNH_3$ layer are alternately assembled. As a result of a strong quantum confinement effect due to their low-dimensional semiconductor structure, they form a stable exciton with large binding energy of several hundred million electron volts and exhibit attractive optical properties due to the exciton, such as efficient photoluminescence, electroluminescence, and nonlinear optical effects. Also, the hybrid $(C_nH_{2n+1}NH_3)_2MnCl_4$ family is known as a good example of a quasi-two-dimensional Heisenberg antiferromagnet.[109]

## IV. CONCLUSIONS

In this chapter, chemically well-defined nanohybrid systems with inorganic/inorganic, organic/inorganic, and bio/inorganic heterostructures have been demonstrated, along with synthetic strategies such as intercalation, and exfoliation and successive restacking. The biomolecules can be intercalated into LDHs via the ion exchange reaction to construct bioinorganic nanohybrids. And it is demonstrated that inorganic supramolecules, such as LDHs with nanometer size, can play an important role as reservoirs for biomolecules and as delivery carriers for genes and drugs. A new class of inorganic/inorganic nanohybrids with high photocatalytic properties and with mixed conducting properties can be realized via intercalation or exfoliation-restacking. Organic/inorganic heterostructures with high-temperature superconductivity can be synthesized by hybridizing metal halides or organic salts with bismuth-based cuprates. These superconducting compounds are believed to be promising precursor materials for superconducting nanoparticles, thin and thick films, and wires. Therefore, hybridization of each heterogeneous material could lead to new uses as well as improvements in applications.

## REFERENCES

1. Choy, J.H., Park, N.G., Hwang, S.J., Kim, D.H., and Hur, N.H., *J. Am. Chem. Soc.*, 116, 11564, 1994.
2. Choy, J.H., Kwon, S.J., and Park, G.S., *Science*, 280, 1589, 1998
3. Choy, J.H., Kwak, S.Y., Park, J.S., Jeong, Y.J., and Portier, J., *J. Am. Chem. Soc.*, 121, 1399, 1999.
4. Pinnavaia, T.J., *Science*, 220, 220, 1983.
5. Thomas, J.K., *Chem. Rev.*, 93, 301, 1993.
6. Vaccari, A., *Catal. Today*, 41, 53, 1998.
7. Shichi, T., and Takagi, K., *J. Photochem. Photobiol.* C1, 113, 2000.
8. Ogawa, M., and Kuroda, K., *Chem. Rev.*, 95, 399, 1995.
9. Swartzen-Allen, S.L., and Matijevic, E., *Chem. Rev.*, 74, 385, 1974.
10. Theng, K.G., *The Chemistry of Clay-Organic Reactions*, Adam Hilger, London, 1974.
11. Choy, J.H., Kwak, S.Y., Han, Y.S., and Kim, B.H., *Mater. Lett.*, 33, 143, 1997.
12. Ogawa, M., and Kuroda, K., *Bull. Chem. Soc. Jpn.*, 70, 2593, 1997.
13. Carrad, K.A., *Appl. Clay Sci.*, 17, 1, 2000.
14. Ohtsuka, K., *Chem. Mater.*, 9, 2030, 1997.
15. Brindley, G.W., and Sempels, R.E., *Clay Miner.*, 25, 229, 1977.
16. Lahav, N., Shani, U., and Shabtai, J., *Clays Clay Miner.*, 26, 107, 1978.
17. Vaughan, D.E.W., Lussier, R.J., and Magee, J.S., Jr., U.S. Patent no. 4,176,090, 1979.
18. Yamanaka, S., and Brindley, G.W., *Clays Clay Miner.*, 27, 119, 1979.
19. Lin, F.H., Lee, Y.H., Jian, C.H., Wong, J.M., Shieh, M.J., and Wang, C.Y., *Biomaterials*, 23, 1981, 2002.
20. Fowden, L., Barrer, R.M., and Tinker, P.B., Eds., *Clay Minerals: Their Structure, Behaviour and Use*, Royal Society, London, 1984.
21. Newman, A.C.D., Ed., *The Chemistry of Clays and Clay Minerals*, Monograph 6 of the Mineralogical Society, Longman, London, 1987.
22. Wilson, M.J., *A Handbook of Determinative Methods in Clay Mineralogy*, Chapman & Hall, New York, 1987.
23. Fournaris, K.G., Karakassides, M.A., Petridis, D., and Yiannakopoulou, Y., *Chem. Mater.*, 11, 2372, 1999.
24. Wang, Z., and Pinnavaia, T.J., *Chem. Mater.*, 10, 1820, 1998.
25. Giannelis, E.P., *Adv. Mater.*, 8, 29, 1996.
26. Kim, C.E., Choy, J.H., and Hyung, K.W., *Tech. Inf. Ceram. Ind.* (Korean Ceramics Industry Association), 8, 6, 1983.
27. Usuki, A., Kawasumi, M., Kojima, Y., Okada, A., Kurauchi, T., and Kamingaito, O., *J. Mater. Res.*, 8, 1174, 1993.
28. Yamanaka, S., Nishihara, T., Hattori, M., and Suzuki, Y., *Mater. Chem. Phys.*, 17, 87, 1987.
29. Han, Y.S., Matsumoto, H., and Yamanaka, S., *Chem. Mater.*, 9, 2013, 1997.
30. Endo, T., Mortland, M.M., and Pinnavaia, T.J., *Clays Clay Miner.*, 28, 105, 1980.
31. Han, Y.S., and Choy, J.H., *J. Mater. Chem.*, 8, 1459, 1998.
32. Choy, J.H., Park, J.H., and Yoon, J.B., *J. Phys. Chem.*, 102, 5991, 1998.
33. Carretero, M.I., *Appl. Clay Sci.*, 21, 155, 2002.
34. Yang, J.-H., Lee, S.-Y., Han, Y.-S., Park, K.-C., and Choy, J.-H., *Bull. Korean Chem. Soc.*, 24, 499, 2003.

35. Bröcker, F.J., and Kainer, L., German Patent no. 2,024,282 (1970) to BASF AG and UK Patent no. 1,342,020 (1971) to BASF AG.

36. Cavani, F., Trifirò, F., and Vaccari, A., *Catal. Today*, 11, 173, 1991.

37. Dumitriu, E., Hulea, V., Chelaru, C., Catrinescu, C., Tichit, D., and Durand, R., *Appl. Catal. A*, 178, 145, 1999.

38. Velu, S., and Swamy, C.S., *Appl. Catal.*, 119, 241, 1994.

39. Ueno, S., Yamaguchi, K., Yoshida, K., Ebitani, K., and Kaneda, K., *J. Chem. Soc. Chem. Commun.*, 295, 1998.

40. Medina, F., Tichit, D., Coq, B., Vaccari, A., and Thy Dung, N., *J. Catal.*, 167, 142, 1997.

41. Serrano, J., Bertin, V., and Bulbulian, S., *Langmuir*, 16, 3355, 2000.

42. Roym, D.M., and Osborn, E.F., *Am. J. Sci.*, 251, 337, 1953.

43. Pausch, I., Lohse, H.H., Schürmann, K., and Allmann, R., *Clays Clay Miner.*, 34, 507, 1986.

44. Mousty, C., and Therias, S., *J. Electroanal. Chem.*, 374, 63, 1994.

45. Sakodam, K., Kominami, K., and Iwamoto, M., *Jpn. J. Appl. Phys.*, 27, L1304, 1988.

46. Robins, D.S., and Dutta, P.K., *Langmuir*, 12, 402, 1996.

47. Lei, L., Prasad, R., Jayan, V., and O'Hare, D., *J. Mater. Chem.*, 11, 3276, 2001.

48. Tseng, W.Y., Lin, J.T., Mou, C.Y., Cheng, S., Liu, S.B., Chu, P., and Liu, H.W., *J. Am. Chem. Soc.*, 118, 4411, 1996.

49. Adachi-Pagano, M., Forano, C., and Besse, J.-P., *Chem. Commun.*, 91, 2000.

50. Hibino, T., and Jones, W., *J. Mater. Chem.*, 11, 1321, 2001.

51. Sels, B., de Vos, D., Buntinx, M., Pierard, F., Kirsch-De Mesmaeker, A., and Jacobs, P., *Nature*, 400, 855, 1999.

52. Itaya, K., Chang, H.C., and Uchida, I., *Inorg. Chem.*, 26, 624, 1987.

53. Mousty, C., Therias, S., Forano, C., and Besse, J., *J. Electroanal. Chem.*, 347, 63, 1994.

54. Therias, S., and Mousty, C., *Langmuir*, 12, 4914, 1996.

55. Keita, B., Belhouri, A., and Nadjo, L., *J. Electroanal. Chem.*, 355, 235, 1993.

56. Keita, B., Belhouri, A., and Nadjo, L., *J. Electroanal. Chem.*, 314, 345, 1991.

57. Idemura, S., Suzuki, E., and Ono, Y., *Clays Clay Miner.*, 37, 553, 1989.

58. Itaya, K., Chang, H.C., and Uchida, I., *Inorg. Chem.*, 26, 624, 1987.

59. Choy, J.H., Kwak, S.Y., and Park, J.S., *J. Am. Chem. Soc.*, 121, 1399, 1999.

60. Choy, J.H., Kwak, S.Y., Park, J.S., and Jeong, Y.J., *J. Mater. Chem.*, 11, 1671, 2001.

61. Choy, J.H., Kwak, S.Y., Jeong, Y.J., and Park, J.S., *Angew. Chem. Int. Ed.*, 39, 4041, 2000.

62. Kwak, S.Y., Jeong, Y.J., Park, J.S., and Choy, J.H., *Solid State Ionics*, 151, 229, 2002.

63. Bednorz, J.G., and Müller, K.A., *Z. Phys. B*, 64, 189, 1986.

64. Gopalakrishnan, J., Uma, S., and Bhat, V., *Chem. Mater.*, 5, 132, 1993.

65. Domen, K., Yoshimura, J., Sekine, T., Tanaka, A., and Onishi, T., *Catal. Lett.*, 4, 339, 1990.

66. Yoshimura, J., Ebina, Y., Kondo, J., Domen, K., and Tanaka, A., *J. Phys. Chem.*, 97, 1970, 1993.

67. Domen, K., Yoshimura, J., Sekine, T., Tanaka, A., and Onishi, T., *Catal. Lett.*, 4, 339, 1990.

68. Sato, M., Watanabe, J., and Uematsu, K., *J. Solid State Chem.*, 107, 460, 1993.

69. Richard, M., Brohan, L., and Tournaux, M., *J. Solid State Chem.*, 112, 345, 1994.

70. Jacobson, A.J., Johnson, J.W., and Lewandowski, J.T., *Inorg. Chem.*, 24, 3729, 1985.
71. Uma, S., and Gopalakrishnan, J., *J. Solid State Chem.*, 102, 332, 1993.
72. Mohan Ram, R.A., and Clearfield, A., *J. Solid State Chem.*, 94, 45, 1991.
73. Kudo, A., *Chem. Mater.*, 9, 664, 1997.
74. Anderson, P.W., *Science*, 235, 1196, 1987.
75. Ebina, Y., Sasaki, T., Harada, M., and Watanabe, M., *Chem. Mater.*, 14, 4390, 2002.
76. Choy, J.H., Lee, C.H., Jung, H., Kim, H., and Boo, H., *Chem. Mater.*, 14, 2486, 2002.
77. Wang, L., Sasaki, T., Ebina, Y., Kurashima, K., and Watanabe, M., *Chem. Mater.*, 14, 4827, 2002.
78. Wang, Y., Teng, X., Wang, J., and Yang, H., *Nano Lett.*, 3, 789, 2003.
79. Choy, J.H., Hwang, S.J., Kim, Y.I., and Kwon, S.J., *Solid State Ionics*, 108, 17, 1998.
80. Marti, A.A., and Colon, J.L., *Inorg. Chem.*, 42, 2830, 2003.
81. Ruddlesden, S., and Popper, P., *Acta Crystallogr.*, 10, 538, 1957.
82. Ruddlesden, S., and Popper, P., *Acta Crystallogr.*, 11, 54, 1958.
83. Aurivillius, B., *Ark. Kemi.*, 1, 463, 1949.
84. Aurivillius, B., *Ark. Kemi.*, 1, 499, 1949.
85. Aurivillius, B., *Ark. Kemi.*, 1, 519, 1950.
86. Dion, M., Ganne, M., and Tournaux, M., *Mater. Res. Bull.*, 16, 1429, 1981.
87. Schaak, R.E., and Mallouk, T.E., *Chem. Commun.*, 706, 2002.
88. Han, Y.S., Park, I., and Choy, J.H., *J. Mater. Chem.*, 11, 1277, 2001.
89. Chatakhondu, K., Green, L.M.H., Mingos, D.M.P., Reynolds, J., *J. Chem. Soc. Chem. Comm.*, 900, 1987.
90. Bonneau, P., Mansot, J.L., and Rouxel, J., *Mater. Res. Bull.*, 28, 757, 1993.
91. Han, Y.S., Yamanaka, S., and Choy, J.H., *J. Solid State Chem.*, 144, 45, 1999.
92. Ebina, Y., Tanaka, A., Kondo, J.N., and Domen, K., *Chem. Mater.*, 8, 2534, 1996.
93. Kooli, F., Sasaki, T., Rives, V., and Watanabe, M., *J. Mater. Chem.*, 10, 497, 2000.
94. Abe, R., Shinohara, K., Tanaka, A., Hara, M., Kondo, J.N., and Domen, K., 9, 2179, 1997.
95. Lee, J.K., Lee, W., Yoon, T.J., Park, G.S., and Choy, J.H., *J. Mater. Chem.*, 12, 614, 2002.
96. Shangguan, W., and Yoshida, A., *J. Phys. Chem. B*, 106, 12227, 2002.
97. Choy, J.H., Kim, Y.I., Hwang, S.J., and Huong, P.V., *J. Phys. Chem. B*, 104, 7273, 2000.
98. Keller, S.W., Kim, H.-N., and Mallouk, T.E., *J. Am. Chem. Soc.*, 116, 8817, 1994.
99. Sasaki, T., Izumi, F., and Watanabe, M., *Chem. Mater.*, 8, 777, 1996.
100. Sukpirom, N., and Lerner, M.M., *Chem. Mater.*, 13, 2179, 2001.
101. Sugimoto, W., Terabayashi, O., Murakami, Y., and Takasu, Y., *J. Mater. Chem.*, 12, 3814, 2002.
102. Sasaki, T., Watanabe, M., Hashizume, H., Yamada, H., and Nakazawa, H., *J. Am. Chem. Soc.*, 118, 8329, 1996.
103. Choy, J.H., Kwon, S.J., Hwang, S.H., and Jang, E.S., *Mater. Res. Soc. Bull.*, 25, 32, 2000.
104. Kagan, C.R., Mitzi, D.B., and Dimitrakopoulos, C.D., *Science*, 286, 945, 1999.
105. Mitzi, D.B., Dimitrakopoulos, C.D., and Kosbar, L., *Chem. Mater.*, 13, 3728, 2001.
106. Hayakawa, T., Imaizumi, D., and Nogami, M., *J. Mater. Res.*, 15, 530, 2000.

107. Kang, S.-J., Choi, S.-K., Lee, H.-J., Lee, J.-H., and Kim, H.K., *Nonlinear Opt.*, 15, 181, 1996.
108. Furube, A., Shiozawa, T., Ishikawa, A., Wada, A., Domen, K., and Hirose, C., *J. Phys. Chem. B*, 106, 3065, 2002.
109. Era, M., Hattori, T., Taira, T., and Tsutsui, T., *Chem. Mater.*, 9, 8, 1997.

# 15 Nanoceramic Particulates for Chemical Mechanical Planarization in the Ultra Large Scale Integration Fabrication Process

*Ungyu Paik, Sang Kyun Kim, Takeo Katoh, and Jea Gun Park*

## CONTENTS

## I. INTRODUCTION

The semiconductor industry has been rapidly developing, achieving 1-gigabyte (GB) dynamic random access memory (DRAM) early in the 21st century. As semiconductor chip size is scaled to submicron dimensions and additional levels are added to multilevel interconnection schemes, the required degree of planarization is increased. Though numerous planarization processes exist, including resist etch back, spin on glass (SOG), boro-phosphate-silicate glass (BPSG) reflow, and so on, chemical mechanical planarization (CMP) is currently the only available technique used in wafer polishing that achieves line widths of 0.25 μm or less as well as multilevel interconnections for dynamic memory and microprocessor applications.[1-4]

The CMP process is composed of a chemical effect from nanosize ceramic particles and a physical effect from the pressed pad. Pads and slurries are the consumables of a CMP process. The polishing pads consist of polyurethane. Generally two types of pads (hard and soft types) are simultaneously used in the CMP process. A hard pad gives better local (within die) planarity, but a soft pad gives better uniformity of material removal across the entire wafer. A hard pad is mounted onto a softer pad to form a stacked pad. Figure 15.1 shows a stacked CMP pad; the hard pad is the Rodel IC-1000 and the soft pad is the Rodel Suba-IV.[5]

Nanoceramic particulates have been employed for the CMP process. The interactions of the ceramic particulates with organic additives, pad materials, and the wafer during the CMP process should be considered from the viewpoint of colloidal science. The surfaces to be polished involve silicon oxide and metals,

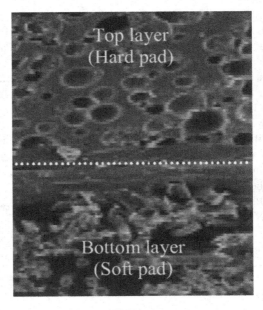

**FIGURE 15.1** The stacked pad for the CMP process.

including tungsten, aluminum, and copper.[6,7] Nanoceramic particulates are selected for the purpose of polishing. The chemistry of each slurry has its optimal composition for planarizing film. The interlayered dielectric (ILD) CMP process for planarization of silicon oxide uses nanoscale fumed silica slurry, which is well dispersed in a basic pH range. The shallow trench isolation (STI) CMP process, which is used to produce a newly introduced transistor by the technique of ultra large scale integration (ULSI), uses high-selectivity ceria slurry.[8,9] Metal CMP for planarizing metal film uses a slurry containing nanosize alumina or silica particles to polish aluminum or tungsten metal film with the additive, controlling oxidation of the film.[10] Metal for multilevel interconnection is substituted with copper as the processing speed of ULSI has increased, and nanosize colloidal silica particles are used to prevent scratches on the soft copper surface.[11] Nanoceramic particulate slurry for the CMP process contains various chemical additives. Adequate slurry for planarization of a wafer surface can be made using the additives mentioned above. The chemical composition of the slurry can be classified as follows:

Abrasive (nanosize ceramic particles): mechanically removes by physical and chemical reactions between the wafer surface and the pad.

Dispersant: increases the dispersion stability of the polishing particles in the slurry.

Surfactant: blends the chemical fluids thoroughly and consistently in the slurry, and controls the polishing rate of each film by selectively adsorbing on a designated film to achieve selective polishing.

Buffer solution: prevents the slurry from causing pH shock in the process and sustains suspension pH at a fixed value.

Bulk solution: induces a certain chemical reaction.

The CMP slurry should be well dispersed to planarize a large area of the wafer surface. Each additive should be used carefully, as they may affect the reactivity of the materials and the dispersion stability of the slurry. This chapter presents the physicochemical characteristics of nanoceramic particulates and an overview of ILD and STI CMP and the relationship between the characteristics of the nanoceramic particulates and CMP performance.

## II. ILD CMP AND THE STABILITY OF NANOSIZE FUMED SILICA PARTICLES

The ILD CMP process has been used to polish plasma-enhanced tetraethylorthosilicate (PETEOS) or high-density plasma chemical vapor deposition (HDPCVD) film on deposited silicon wafer. Figure 15.2 shows the ILD CMP process. The stacking of additional layers on top of one another produces a more and more rugged topography. Between each layer, the dielectric is deposited as an insulating material. To obtain a multilevel interconnection, the surface of the wafer must be

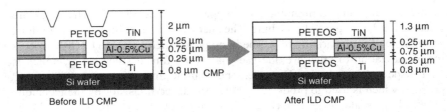

**FIGURE 15.2** A schematic of the ILD CMP process.

planarized to prevent the topography roughness from growing with each level. In order to effectively planarize the dielectric inserted between the metal wiring for multilayered wiring, ILD CMP is employed. The dispersion stability of nanosize fumed silica slurry is directly related to the polishing rate (removal rate), the surface scratch, and the uniformity (within wafer nonuniformity) of the wafer surface across the entire wafer. Controlling the dispersion stability of nanosize fumed silica slurry is a key parameter in ILD CMP process.

## A. BEHAVIOR OF CONCENTRATED NANOSIZE FUMED SILICA HYDROSOLS

Nanosize inorganic particles (i.e., less than 100 nm) are gradually being incorporated into a broad range of advanced devices and applications; some examples include silicon wafer polishing, planarization for semiconductor manufacturing (the CMP process), electronic packages, ultra-thin-film optical devices, advanced fuel cell catalysts, molecular conductors, and biochips.[12–19] Recent evidence[20] has indicated that classical colloid principles might not fully explain the complex behavior of concentrated nanosols.

According to Derjaguin-Landau-Verwey-Overbeek (DLVO) theory,[21] a cornerstone of modern colloid science, two types of forces exist between colloidal particles suspended in a dielectric medium: electrostatic forces, which result from an unscreened surface charge on the particle, and London-van der Waals attractive forces, which are universal in nature. The colloidal stability and rheology of oxide suspensions, in the absence of steric additives, can be largely understood by combining these two forces (assumption of additivity).

There are several reports[22–24] of the unique stability of nanosize silica hydrosols near the isoelectric point (IEP). The Canberra group[25] discovered experimentally the existence of short-range forces that play an important role in the interaction process and which must be added to those forces already accounted for by the original DLVO theory. These short-range interactions are referred to as structural forces.[26–28] Structural forces might explain some particular aspects of the stability behavior of silica nanosols, but they are insufficient to account for the apparent cooperative effects of solids loading and electrostatic forces found in the present study. Contrary to suspensions based on colloidal-size (100 to 1000 nm) silica[29] and other inorganic oxides[30] reported in the literature, we found that the rheological behavior of concentrated electrostatically stabilized silica nanosols is counter intuitive with regards to the predictions based on a standard interpretation of DLVO theory. Despite the high surface charge density electrokinetic potential at pH 8, nanosize fumed silica particles not only showed unstable rheological behavior that would normally indicate an unstable or aggregated suspension (i.e., pseudoplastic-high viscosity), but the rheology did not have the expected dependence on ionic strength. In this study, experimental measurements, DLVO calculations, and simple geometric considerations were used to understand the influence of solids loading and the electrical double-layer on the rheological behavior of concentrated silica (20 nm) nanosols and to compare their behavior with that of much larger silica microspheres, as well as like-size nano-alumina under similar conditions.

## 1. Electrokinetic Behavior of Nanosize Silica Hydrosols

By changing the pH, one can alter the magnitude (and sign) of the zeta potential ($\zeta$), while the addition of an inert electrolyte will affect both the magnitude of $\zeta$ and the electrical double-layer thickness. Thus both pH and electrolyte concentration will directly impact colloidal stability in an electrostatically stabilized system. Figure 15.3 compares $\zeta$ and viscosity (at a shear rate of 26.4 s$^{-1}$) as a function of pH for the nanosize fumed silica and silica microsphere suspensions. The average primary particle sizes were 20 nm for the nanosize fumed silica and 500 nm for the silica microspheres. Even at a solids concentration of 20%, the silica microspheres exhibit a fairly constant and low viscosity across the entire pH range, whereas nanosize fumed silica exhibits a strong pH dependence at a volume fraction of 13.2%, with an increase in viscosity near pH 7 in excess of 300%. Figure 15.4 shows the effect of inert electrolyte concentration on viscosity as a function of shear rate for highly charged 13.2% nanosize fumed silica at pH 8.

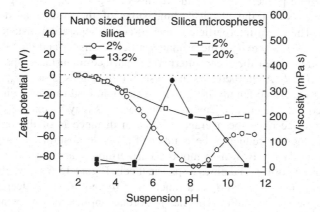

**FIGURE 15.3** The relationship between $\zeta$ potential (open) and viscosity (filled) for silica suspensions as a function of suspension pH: nanosize A90 vs. Geltech microspheres (G). Viscosity was determined at a shear rate of 26.4/sec. Particle volume fraction given in percent.

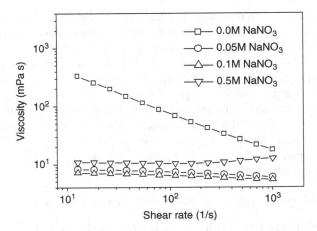

**FIGURE 15.4** The effect of electrolyte concentration on the viscosity of 13.2% A90 silica at pH as a function of shear rate.

Figure 15.3 indicates that for the silica microspheres, the $\zeta$ potential and viscosity both follow the expected behavior predicted by the classical DLVO theory. On the other hand, the nanosize fumed silica exhibits a discrepancy between the expectation of DLVO theory and the experimental results; that is, as the $\zeta$ of the nanosize fumed silica increases, viscosity sharply increases. Hence factors such as particle crowding, particle ordering, and electroviscous effects will also impact viscosity, in addition to aggregate or network formation.

## 2. Geometric Considerations

In order to more properly analyze the results of Figure 15.3 and Figure 15.4, it helps to first lay out the physical dimensions of the system as depicted in Figure 15.5. The mean interparticle center-to-center separation distance ($d_{c2c}$) is defined as $d_p/\Phi^{1/3}$, where $d_p$ is the primary particle diameter and $\Phi$ is the particle volume fraction. Then the mean interparticle surface-to-surface separation distance ($d_s$) is $d_{c2c}-d_p$. As $\Phi$ increases, the system dimensions, $d_s$ and $d_p$, eventually become of comparable length ($d_s/d_p \sim 1$), which can lead to constrained motion and excluded volume effects. That is, other particles may be excluded from the interparticle space once the average separation distance is of the order of the particle size, thereby reducing the number of possible positions each particle is able to sample during Brownian motion. Furthermore, each particle with a surrounding volume of liquid defines a spherical cell. Figure 15.6 shows the average cell radius, $r_{cell} = d_{c2c}/2$, and $d_s$ as a function of $\Phi$ and $d_p$. As $d_p$ decreases or $\Phi$ increases, $d_s$ becomes smaller. This has important implications for nanosize particles, and helps explain why it is so difficult to obtain low-viscosity concentrated nanosols in aqueous systems. This explanation may not be immediately obvious, since the critical $\Phi$ corresponding to $d_s/d_p = 1$ occurs at about 13%, irrespective of particle size.

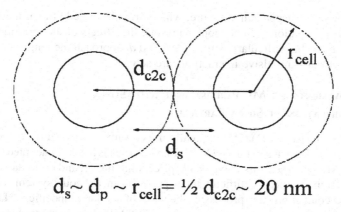

$$d_s \sim d_p \sim r_{cell} = \frac{1}{2} d_{c2c} \sim 20 \text{ nm}$$

**FIGURE 15.5** Diagram illustrating the relationship between average interparticle surface-to-surface separation distance, $d_s$, and other system dimensions, for a particle diameter $d_p$ = 20 nm and $\Phi$ = 13.2%.

However, the distance over which hydrodynamic and electrostatic forces act in solution is more or less independent of particle size at first approximation. As a result, when the average separation distance between particles is rather large, these forces dissipate before they can influence neighboring particles. As a result, particle motion is independent and the rheological behavior is Newtonian, so long as the particles remain stable and do not aggregate. On the other hand, as the average separation distance is reduced, these forces begin to influence the nearest neighbors, and the motion of nearby particles becomes coupled. Coupling leads

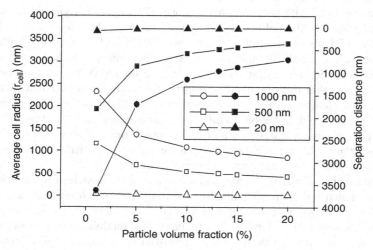

**FIGURE 15.6** Calculated average cell radius (circles) and surface-to-surface separation distance (triangles) as a function of particle volume fraction and particle size for silica.

to an increase in suspension structure, which provides an additional mechanism for viscous dissipation.[31] In aqueous nanosols, the effects of electrostatic forces on structure can be particularly strong as $d_p$ and $d_s$ approach the length scale over which short-range repulsive interactions are active.

## B. IMPROVEMENT OF CMP PERFORMANCE WITH SURFACE MODIFICATION OF SILICA PARTICLES

Interlayer dielectric (ILD) CMP typically uses a fumed silica slurry dispersed in an aqueous medium at a pH near 11.[32-34] Fumed silica is a widely adapted abrasive for ILD CMP because of its inexpensive price, high purity, and colloidal stability. However, fumed silica is difficult to disperse in an aqueous system, and it is difficult to control powder processing because of the large specific surface area of $90 \pm 15$ m²/g, making it very reactive. ILD CMP slurry was prepared at pH 11 to accelerate the chemical attack on the deposited PETEOS film on the wafer surface. But silica particles dispersed in aqueous media are partially dissolved at pH 11.[35-37] Consequently the removal rate decreased and microscratches were generated on the wafer surface due to agglomeration of silica particles as surface potentials decreased.[38-41]

As mentioned above, the dispersion stability of the slurry is directly related with CMP performance, removal rate, within-wafer nonuniformity (WIWNU), which is defined as the standard deviation divided by the average of remaining thickness after CMP, microscratching, and the remaining particle on the wafer. To avoid poor CMP performance, the dispersion stability of the slurry must be controlled by preventing silicon ion dissolution. Surface modification of the silica particle was produced by addition of an organic additive. Without surface modification, the amount of silicon dissolution was $1.370 \pm 0.002$ mol/L, while surfaces modified with poly(vinylpyrrolidone) (PVP) polymer yielded a dissolution of $0.070 \pm 0.001$ mol/L, almost 20 times less than the unmodified surface.

Table 15.1 shows the removal rate and WIWNU of silica slurry with and without surface modification.[42] In comparing the results with and without the modification, the removal rate is similar, but the final WIWNU of the modified slurry is better than that of unmodified slurry. The removal rate and final WIWNU results were closely correlated to the surface potential, rheological behavior, and large-particle size distribution. In effect, the surface modification strongly influenced the suspension stability, and hence the properties of wafer uniformity. The microscratches and remaining particles on the silicon wafer with and without the surface-modified slurry are shown in Figure 15.7.[42]

The number of microscratches and remaining silica particles for the modified slurry is much less than for unmodified slurry. PVP, which modifies the silica particles and plays a preventive role in dissolving silicon ions, is thought to improve the suspension stability. Due to the surface modification, microscratches on the silicon wafer are decreased, as improved suspension stability prevented the undesirable agglomeration. In addition, as the reactivity of silicon ion with the silicon wafer is much higher than that of silica particles, the number of

**TABLE 15.1**
**Removal Rate and WIWNU with and without Modification[39]**

| Wafer Number | Removal Rate (Å/min) | | WIWNU (%) | |
|---|---|---|---|---|
| | With Modification | Without Modification | With Modification | Without Modification |
| 1 | 2822 | 2873 | 3.57 | 8.09 |
| 2 | 2705 | 2767 | 3.75 | 8.62 |
| 3 | 2791 | 2813 | 3.96 | 9.43 |

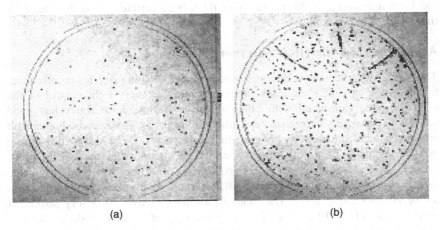

(a)                                    (b)

**FIGURE 15.7** Analysis of remaining silica particles (>0.189 μm) on silicon wafers after post-CMP cleaning: (a) modified slurry; (b) unmodified slurry.[39]

particles stuck on the wafer surface decreased because of the reduction in silicon ion dissolution in the case of the slurry modified with PVP.

## III. STI CMP AND THE PHYSICOCHEMIAL CHARACTERISTICS OF CERIA SLURRY

Shallow trench isolation (STI) is a relatively new technique that is replacing local oxidation of silicon (LOCOS) for the manufacture of 64 MB semiconductor devices with a line width of less than 0.25 μm.[43–45] Figure 15.8 shows the STI CMP process. The STI process is defined as (i) making a shallow trench to isolate active device regions physically, (ii) depositing silicon nitride on oxide films as a stopping layer, and (iii) depositing oxide films on the trench. Generally the STI process has the relative capability, compared to the LOCOS process, not only to deposit dielectrics to fill trenches isolating the active region at low temperature, but also to prevent bird's beak and dimensional limitations.[46,47] The key ingredients

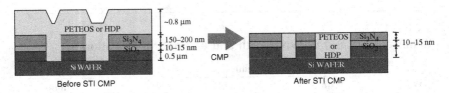

**FIGURE 15.8** A schematic of the STI CMP process.

to a successful STI process are the achievement of well-dispersed abrasive ceramic particles having high oxide-to-nitride selectivity and producing few microscratches on the wafer. Silica slurry had been conventionally used in the STI CMP process, but ceria (CeO$_2$) slurry with a high oxide-to-nitride selectivity has been introduced as the thickness of the silicon nitride film is decreased by the design rule restriction.[48]

## A. REQUIREMENT FOR HIGH-SELECTIVITY SLURRY

The STI CMP processes with conventional oxide polishing slurries require reactive ion etching (RIE), etchback preplanarization, or very tight control of the CMP process. Compared with other abrasive slurries, ceria slurry has good selectivity between silicon oxide and silicon nitride.[49–53] There are both chemical and mechanical interactions between the ceria particles and wafer film during polishing. Nitride film is mainly affected by the chemical factors and the properties of the CeO$_2$ abrasive particle. Ceria slurries offer improved oxide-to-nitride selectivity for planarizing the trench fill material, while utilizing the nitride film as the polishing stop layer. The oxide-to-nitride selectivity is a very important factor in the STI CMP process. It can significantly affect CMP-induced defects, such as erosion or dishing, and also is important for endpoint detection. Figure 15.9 shows the effect of overpolishing on device characteristics. Overpolishing due to nitride erosion or oxide dishing may cause degradation of device properties.

If the effective gate length ($L_{eff}$) decreases, subthreshold drain current ($I_D$) may increase, as shown by Equation 15.1:[54]

$$I_D = \frac{\mu_n Z C_i}{L}\left[(V_G - V_T)V_D - \frac{1}{2}V_D^2\right] \qquad (15.1)$$

where
  $I_D$ : drain current
  $\mu_n$ : electron mobility
  $Z$ : channel width
  $C_i$ : insulator capacitance per unit area
  $L$ : channel length
  $V_G$ : gate voltage
  $V_T$ : threshold voltage
  $V_D$ : drain voltage

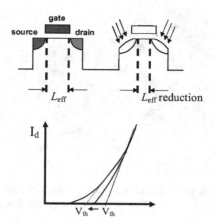

**FIGURE 15.9** The effect of overpolishing on the characteristics of the device: (a) the decrease in gate length by overpolishing; (b) the shift of threshold voltage in the device.

and the threshold voltage ($V_{th}$) shifts as shown by Equation 15.2:[55]

$$\Delta V_T = \frac{qN_{sub}d_{max}r_j\left(\left[\sqrt{\left\{1+\dfrac{2W_s}{r_j}\right\}}-1\right]+\left[\sqrt{\left\{1+\dfrac{2W_D}{r_j}\right\}}-1\right]\right)}{2C_{ox}L} \tag{15.2}$$

where

$V_T$ : threshold voltage

$d_{max}$ : the maximum width of the depletion region in the channel

$r_j$ : junction depth of the source/drain region

$N_{sub}$ : concentration of the substrate

$C_{ox}$ : capacitance of gate oxide

$L$ : channel length

$W_s, W_D$ : width of source and drain

In general, two mechanisms can be applied to improve the selectivity between PETEOS and $Si_3N_4$ during polishing of the pattern wafer. One is chemical control using a surfactant to reduce the removal rate of $Si_3N_4$; the other is mechanical control by improving the physical properties of ceria particles to enhance the removal rate of PETEOS.

## B. Synthesis of Ceria Particles and CMP Performance

Ceria particles are considered to be one of the best glass/$SiO_2$ polishing abrasives. This is suggested to be because of the reaction between ceria and $SiO_2$ film, which results in the formation of a chemical "tooth" between the silica surface and the ceria particles, and induced localized strain in the glass with particle

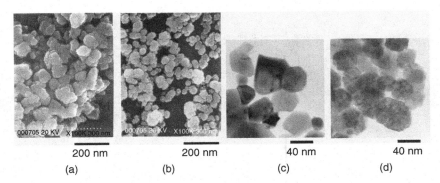

200 nm      200 nm      40 nm      40 nm

(a)      (b)      (c)      (d)

**FIGURE 15.10** Scanning electron and transmission electron micrographs of $CeO_2$ particles: (a) SEM (magnification ×100,000); (b) TEM (magnification ×300,000).[51]

movement.[56] As a consequence, the Si–O–Ce bonds can be rapidly removed by the mechanical force generated by a pressed pad and abrasive particles. This physicochemical reaction leads to the high removal rate of $SiO_2$ film by ceria particles. The physicochemical properties of ceria particles, such as crystallinity, particle roughness, and morphology, depend on the synthesis methods of cerium oxide. In this study, the influence of the ceria particle synthesis method on PETEOS and CVD nitride film removal rates are presented.

## 1. Physical Properties of Ceria Particles

The solid state displacement reaction method[57,58] and wet chemical precipitation method[59] were employed for synthesizing the ceria powders, and thus the ceria properties showed different features in several experiments. Figure 15.10 shows the morphology of the ceria particles observed with high-resolution scanning electron microscopy (SEM; S900, Hitachi, Japan) and transmission electron microscopy (TEM; JEM-2010, JEOL, Japan).[60] In the figure, the ceria particles have a polyhedral shape. Both of the powders have nearly the same size. The primary particle size is approximately 40 nm. However, the difference in crystal shape of the ceria particles was found on TEM analysis.

Figure 15.11 shows x-ray diffraction (XRD) profiles of ceria powders produced by precipitation.[60] The XRD data for the synthesized particles show characteristics of $CeO_2$ with a typical fluorite structure. Since the starting cerium salt was $Ce(NO_3)_3$, it required the oxidation of $Ce^{3+}$ to $Ce^{4+}$ in the solution. In this system, there is a possible cause for this oxidation. According to the Lewis definition of acids and bases, $Ce^{3+}$ is a Lewis base and $Ce^{4+}$ is a Lewis acid. Basic solution therefore favors $Ce^{4+}$ compared with $Ce^{3+}$. The crystallite size was calculated from the Scherrer formula:[61]

$$D = 0.9\lambda/(\beta\cos\theta),$$

where $\lambda$ is the wavelength of the x-rays, $\theta$ is the diffraction angle, and $\beta$ is the half width. The average crystallite size of $CeO_2$ calculated by the Scherrer equation

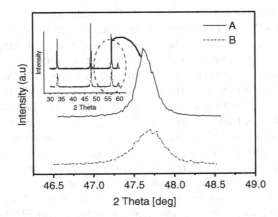

**FIGURE 15.11** X-ray diffraction pattern of CeO$_2$ powders synthesized by the precipitation method.[51]

from the XRD line broadening was 46 nm for powder A and 34 nm for powder B. The crystallite size increases as the calcined temperature increases.

## 2. STI CMP Performance with Ceria Slurries

Figure 15.12 shows the results of the CMP field evaluation.[60] The average PETEOS removal rate of slurry A was 2883 Å/min and of slurry B was 672 Å/min, as shown in Figure 15.12a. The WIWNU shows that ceria slurry B (0.7%) is better than ceria slurry A (1.9%). The average nitride removal rate of slurry A was 51 Å/min and of slurry B was 44 Å/min, as shown in Figure 15.12b. Thus the oxide-to-nitride selectivity is 56 for ceria slurry A and 15 for ceria slurry B. CMP field evaluation of ceria slurries having different crystallinities showed that slurry A had better crystallinity and smaller pore size and exhibited a higher removal rate of PETEOS than slurry B. Ceria slurry A showed a higher removal

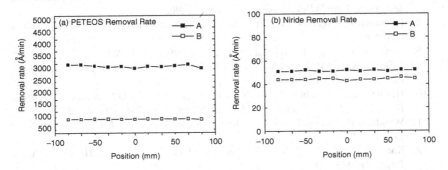

**FIGURE 15.12** The result of CMP field evaluation: (a) profiles of PETEOS removal rate; (b) profiles of nitride removal rate.[51]

rate and better planarization than slurry B. The oxide removal rate can be influenced by two CMP processing parameters: mechanical grinding and chemical interaction.[62] These mechanisms play simultaneous roles in polishing. Concerning the chemical interaction between PETEOS and ceria slurry, it was reported that Si–O–Ce bonding on the surface is a dominant mechanism.[62] During polishing the PETEOS film, the $SiO_2$ surface is first reacted with $CeO_2$ particles and chemically bonding Si-O-Ce is formed on the surface. Then mechanical tearing of the Si-O-Si bonds leads to removal of $SiO_2$ or $Si(OH)_4$ as monomer lumps. The lumps are then released from the $CeO_2$ particles downstream.[62]

Highly crystallized ceria particles have a great tendency to form a bond between cerium and silicon, increasing the oxide removal rate.[63] The ceria particles in slurry B are unlikely to have less oxidation during the wet chemical precipitation and these have less hard and less rigid surface. Therefore slurry B grains might have little effect in physically polishing with PETEOS film. The ceria particles in slurry A are almost fully crystalline on the surface after calcination at 800°C. The $CeO_2$ surface of slurry A supplies more potential sites to react and bond between cerium and silicon than slurry B. Therefore slurry A would interact with the oxide layer much more easily and hence slurry A has a higher removal rate of PETEOS.

## REFERENCES

1. Lin, C.F., Tseng, W.T., and Feng, M.S., *J. Electrochem. Soc.*, 146, 1984, 1999.
2. Sivaram, S., Bath, H., Leggeti, R., Maury, A., Monnig, K., and Tolles, R., *Solid State Technol.*, 35, 87, 1992.
3. Stiegerwald, J.M., Murarka, S.P., and Gutmann, R.J., *Chemical Mechanical Planarization of Microelectronic Materials*, John Wiley & Sons, New York, 1997.
4. Homma, Y., Furusawa, T., Morishima, H., and Sato, H., *Solid-State Electron.*, 41, 1005, 1997.
5. Wolf, S., in *Silicon Processing for the VLSI Era*, vol. 4, Lattice Press, Sunset Beach, CA, 2002, chap. 8.
6. Steigerwald, J., Muraraka, S., and Gutmann, R., in *CMP of Microelectronic Materials*, John Wiley & Sons, New York, 1996.
7. Coppetta, J., Rogers, C., Phillipossian, A., and Kaufman, F., *Proc. 2nd CMP-MIC*, 307, 1997.
8. Nojo, H., Kodera, M., and Nakata, K., *IEEE*, 96, 349, 1996.
9. Paik, U., Kim, J.P., Lee, T.W., Jung, Y.S., Park, J.G., and Hackley, V.A., *J. Korean Phys. Soc.*, 39, 197, 2001.
10. Hernandez, J., Wrschka, P., Hsu, Y., Kuan, T.-S., Oehrlein, G.S., Sun, H.J., Hansen, D.A., King, J., and Fury, M.A., *J. Electrochem. Soc.*, 146, 4647, 1999.
11. Wrschka, P., Hernandez, J., Oehrlein, G.S., and King, J., *J. Electrochem. Soc.*, 147, 706, 2000.
12. Katoh, T., Park, J.G., Lee, W.M., Jeon, H., Paik, U., and Suga, H., *Jpn. J. Appl. Phys.*, 41, 443, 2002.
13. Yang, J., Mei, S., and Ferreira, J.M.F., *J. Am. Ceram. Soc.*, 83, 1361, 2000.
14. Cross, L.E., *Ferroelectrics*, 76, 241, 1987.

15. Rozman, M. and Drofenik, M., *J. Am. Ceram. Soc.*, 81, 1757, 1998.
16. McCormick, P.G., Tsuzuki, T., Robinson, J.S., and Ding, J., *Adv. Mater.*, 13, 1008, 2001.
17. Singh, R.K., Lee, S.-M., Choi, K.-S., Basim, G.B., Choi, W., Chen, Z., and Moudgil, B.M., *J. Mater. Res. Bull.*, 27, 752, 2002.
18. Xia, B., Lenggoro, I.W., and Okuyama, K., *Chem. Mater.*, 14, 2623, 2002.
19. Beecroft, L. and Ober, C.K., *Chem. Mater.*, 9, 1302, 1997.
20. Raghavan, S.R., Walls, H.J., and Khan, S.A., *Langmuir*, 16, 7920, 2000.
21. Hiemenz, P.C. and Rajagopalan, R., *Principles of Colloid and Surface Chemistry*, Marcel Dekker, New York, 1997, chap. 13.
22. Healy, T.W., in *The Colloid Chemistry of Silica*, H.E. Bergna, Ed., American Chemical Society, Washington, DC, 1994, p. 147.
23. Depasse, J. and Watillon, A., *J. Colloid Interface. Sci.*, 33, 430, 1970.
24. Milonijic, S.K., *Colloids Surf.*, 63, 113, 1992.
25. Mahanty, J. and Ninham, B.W., *Dispersion Forces*, Academic Press, New York, 1979.
26. Israelachvili, J.N., *Intermolecular and Surface Forces*, Academic Press, San Diego, CA, 1987.
27. Lewis, J.A., *J. Am. Ceram. Soc.*, 83, 2341, 2000.
28. Horn, R.G., *J. Am. Ceram. Soc.*, 73, 1117, 1990.
29. Zaman, A.A., Moudgil, B.M., Fricke, A.L., and El-Shall, H., *J. Rheol.*, 40, 1191, 1996.
30. Zhou, Z., Scales, P.J., and Boger, D.V., *Chem. Eng. Sci.*, 56, 1290, 2001.
31. Russel, W.B., *J. Rheol.*, 24, 287, 1980.
32. Pye, J.T., Fry, H.W., and Schaffer, W.J., *Solid State Technol.*, 65, 1995.
33. Palla, B.J. and Shah, D.O., *J. Colloid Interface Sci.*, 102, 223, 2000.
34. Ali, I., Roy, S.R., and Shin, G., *Solid State Technol.*, 63, 1994.
35. Paik, U., Kim, J.P., Lee, T.W., Jung, Y.G., Park, J.G., and Hackley, V.A., *J. Korean Phys. Soc.*, 39, 201, 2001.
36. Paik, U., Hackley, V.A., and Lee, H.W., *J. Am. Ceram. Soc.*, 82, 833, 1999.
37. Hackley, V.A. and Malghan, S.G., *J. Mater. Sci.*, 29, 4420, 1994.
38. Palla, B.J., and Shah, D.O., *IEEE/CPMT International Electronics Manufacturing Technology Symposium*, 1999.
39. Sjöberg, S., *J. Non-Crystl. Solids*, 196, 51, 1996.
40. Iler, R.K., *The Chemistry of Silica*, John Wiley & Sons, New York, 1979.
41. Paik, U., Hackley, V.A., Choi, S.C., and Jung, Y.G., *Colloids Surf.*, A135, 77, 1998.
42. Kim, J.P., Paik, U., Jung, Y.G., Katoh, T., and Park, J.-G., *Jpn. J. Appl. Phys.*, 41, 4509, 2002.
43. Itoh, A., Imai, M., and Arimoto, Y., *Jpn. J. Appl. Phys.*, 37, 1697, 1998.
44. Boyd, J.M. and Ellul, J.P., *J. Electrochem. Soc.*, 143, 3718, 1996.
45. Lin, C.F., Tseng, W.T., Feng, M.S., and Wang, Y.L., *Thin Solid Films*, 347, 248, 1999.
46. Cheng, J.Y., Lei, T.F., Chao, T.S., Yen, D.L.W., Jin, B.J., and Lin, C.J., *J. Electrochem. Soc.*, 144, 315, 1997.
47. Smekaiin, K., *Solid State Technol.*, 40, 187, 1997.
48. Laparra, O. and Weling, M., in *Proceedings of the International Symposium on Chemical Mechanical Planarization II*, San Diego, 1998, Electrochemical Society, Pennington, NJ, 1998.

49. Kim, J.P., Jung, Y.S., Yeo, J.G., Paik, U., Park, J.G., and Hackley, V.A., *J. Korean Phys. Soc.*, 39, 197, 2001.
50. Katoh, T., Kang, H.G., Park, J.G., and Paik, U., *Jpn. J. Appl. Phys.*, 42, 1150, 2002.
51. Park, J.G., Katoh, T., Lee, W.M., Jeon, H., and Paik, U., *Jpn. J. Appl. Phys.*, 42, 5420, 2003.
52. Katoh, T., Kim, S.J., Paik, U., and Park, J.G., *Jpn. J. Appl. Phys.*, 42, 5430, 2003.
53. Kim, S.K., Lee, S., Paik, U., Katoh, T., and Park, J.G., *J. Mater. Res.*, 18, 2163, 2003.
54. Streetman, B.G. and Banerjee, S., *Solid State Electronic Devices*, 5th ed., Prentice Hall, Englewood Cliffs, NJ, 1999.
55. Wolf, S., *Silicon Processing*, vol. 3, Lattice Press, California, 1994.
56. Cook, L.M., *J. Non-Crystl. Solids*, 120, 152, 1990.
57. Hussein, G.A.M., *J. Anal. Appl. Pyrolysis*, 37, 111, 1996.
58. Djuricic, B. and Pickering, S., *J. Eur. Ceram. Soc.*, 19, 1925, 1999.
59. Fierro, J.L.G., Mendioroz, S., and Olivan, A.M., *J. Colloid Interface Sci.*, 100, 303, 1984.
60. Kim, S.K., Paik, U., Oh, S.G., Katoh, T., and Park, J.G., *Jpn. J. Appl. Phys.*, 42, 1227, 2003.
61. Maca, K., Trunec, M., and Cihlar, J., *Ceram. Int.*, 28, 337, 2002.
62. Hoshino, T., Kurata, Y., Terasaki, Y., and Susa, K., *J. Non-Crystl. Solids*, 283, 129, 2001.
63. Kim, J.Y., Kim, S.K., Paik, U., Katoh, T., and Park, J.G., *J. Korean Phys. Soc.*, 41, 413, 2002.

# Section III

Sol-Gel Processing

# 16 Chemical Control of Defect Formation During Spin-Coating of Sol-Gels

*Dunbar P. Birnie, III*

## CONTENTS

Spin-coating is a simple process for rapidly depositing thin coatings onto relatively flat substrates. The substrate to be covered is held by some rotatable fixture (often using a vacuum to clamp the substrate in place) and the coating solution is deposited onto the surface; the action of spinning causes the solution to spread out and form a very uniform coating of the chosen material on the surface of the substrate. Unfortunately there are many defects that can form during the deposition process that can detract from the desired coating uniformity. Some defects can be avoided simply by installing careful clean-room practices: good chemical cleaning of substrates and good air filtration to prevent particulates. Other defects arise as a direct result of solvent evaporation induced by the air-flow field above the rapidly spinning substrate. These defects will be the focus of this chapter.

The physics of the substrate rotation after the solution is dispensed leads to a fluid flow condition where the rotational accelerations are exactly balanced by the viscous drag felt within the solution at each stratum level. This flow condition was first described by Emslie, Bonner, and Peck (hereafter referred to as EBP)[1] and their article has been the foundation for many more recent studies. For most solutions, another important balance is established between viscous outward radial flow of the solution on the surface of the substrate and the evaporation of

**411**

solvent from the coating solution. Meyerhofer treated this dual-action process by splitting the spin-coating run into two stages—one controlled only by viscous flow and the other controlled only by evaporation.[2] With this approach he was able to predict the final coating thickness, $h_f$, in terms of several key solution parameters, according to:

$$h_f = x\left(\frac{e}{2(1-x)K}\right)^{1/3} ,$$ (16.1)

where $e$ and $K$ are the evaporation and flow constants, defined below, and $x$ is the solids content of the solution. The evaporation and flow constants are defined, respectively, as

$$e = C\sqrt{\omega}$$ (16.2)

and

$$K = \frac{\rho\omega^2}{3\eta} ,$$ (16.3)

where, $\omega$ is the rotation rate, $\rho$ is the solution's density, $\eta$ is its viscosity, and $C$ is a proportionality constant that depends on whether airflow above the surface is laminar or turbulent, and on the diffusivity of solvent molecules in air (since it is basically limited by the diffusion of the evaporating molecules through the aerodynamic boundary layer above the surface of the wafer during spinning). In prior work we have been able to measure $e$ and $K$ using a laser interferometric process,[3–5] and when several spin speeds are tested, the constant, $C$, can be determined as well.[3,5]

From a practical point of view, the absolute evaporation rate of the solvent during spinning is very important. For example, at a spin speed of 2000 rpm, the evaporation rate of ethanol is about 2 μm/s, while that of tetrahydrofuran (THF) is about 10 μm/s.[5] The fluid dynamics of radial flow for typical solutions show that the evaporation starts to dominate thinning for fluid layers that are in the neighborhood of 5 μm thick. So one can see that the evaporation rates are very substantial in comparison to the overall fluid thickness. The key difficulty is that because of the geometric constraints of the situation, all of the solvent must be removed through the top surface and some concentration gradients must exist within the coating solution to deliver the solvent to the surface, where it can be removed by the rapid airflow over the spinning wafer. These concentration gradients—and associated changes in concentration with time—are critical to understanding some of the defect modes that are observed in spin-coated sol-gel coatings. Two important defect modes are discussed in the next sections.

## I. STRIATION DEFECTS

One commonly occurring and prevalent class of coating thickness defects is "striations." These defects are radial ridges and thickness undulations that point directly along the flow direction during the spin-coating process. Figure 16.1 shows an optical micrograph of striations that occurred in a sol-gel-derived lead zirconate titanate (PZT) coating.[6] The radial flow direction was close to the horizontal direction in this picture. Brightness differences are due to interference effects and represent moderate thickness modulations with a nominal periodicity of about 100 μm, which is reasonably typical of striation defects. This ridge structure does not occur near the center of the wafer where the radial outflow is slower. Instead, a pattern results that is reminiscent of the shapes seen in buoyant Benard convection.[7] Figure 16.2 shows the pattern found at the center of the same wafer imaged in Figure 16.1. It is interesting that the thickness modulation wavelength is similar to that found in the surrounding wafer region where radial striation ridges are observed and that the striation spacing is essentially independent of radius.[8]

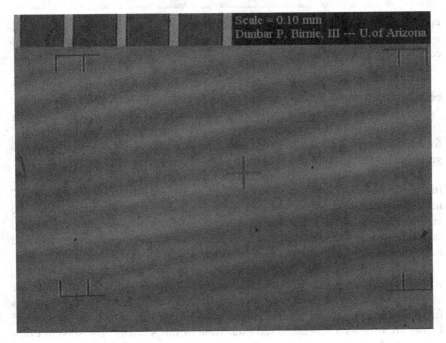

**FIGURE 16.1** Optical micrograph of typical striation defects, in this case found in a Pb(ZrTi)O₃ sol-gel coating on silicon. Color variations result from coating thickness differences. The radial outflow direction is aligned with the ridge features (nearly left to right in this picture).

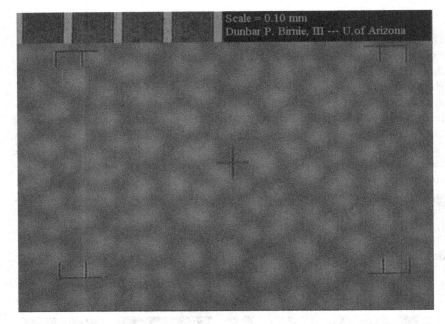

**FIGURE 16.2** Optical micrograph of cellular structure found at the center of the same wafer in Figure 16.1. The center region has flow velocities less than those needed to stretch cells into striation features.

Evaporation of solvent from the surface of the solution during spinning has been demonstrated to be the cause of these defects.[9,10] Daniels et al.[9] noticed that "when evaporation was completely eliminated by spinning the wafer in a totally closed, rotating chamber of 10 mm × 100 mm, striations were eliminated completely" for their photoresist layers. A similar observation was made more recently for sol-gel solutions.[10]

Daniels et al.[9] also demonstrated that surface tension plays a controlling role in the formation of striations. They added unspecified "surface leveling agents" into their solutions, which reduced the magnitude of the striations. That the cellular pattern found at the wafer center is so similar in appearance to Benard cells is quite interesting as well, since surface tension was ultimately demonstrated to be the cause of the patterns that he was observing a century ago.[11,12]

The association with thermocapillary convection gives an indication that when the surface tension rises—for any reason—then the coating may be unstable to convective fluid motions, and striation defects can form. This was the basis for proposing a solvent mixing strategy to reduce the effects of striations.[6] Figure 16.3 shows a schematic representation of this strategy. The solvent evaporation depletes the surface layer of solvent to some extent. This can have a strong effect on surface tension, depending on the solution components. If the physically higher spots experience somewhat faster evaporation (and therefore stronger composition variation) then the situation will be unstable if these high spots have

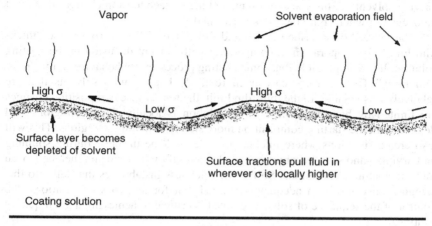

**FIGURE 16.3** Schematic representation of the mechanism by which striations develop during spin coating. Evaporation at the surface depletes the near-surface region, pushing the surface tension value up slightly. Random fluctuations in local surface tension cause lateral fluid motion. This motion causes the local surface tension differences to be amplified, creating thickness undulations that grow with time.

surface tension values that are higher than those found at nearby troughs in the coating. Then the bumps grow by winning a tug-of-war with their weaker neighbors. This instability is amplified by the fact that the depletion "skin" becomes thicker in those areas where the surface tension is pulling in and the "skin" becomes thinner in those areas where the surface tension is too weak and the solution gets stretched out. This is important because the depth of the skin layer will influence the top surface composition because of the competition between solvent removal rate from the surface and replenishment from within the coating solution. For a typical ethanolic sol-gel (where there is water in the solution for the hydrolysis of precursors), this surface layer will be enriched in water and this will force the surface tension to increase during evaporation, creating a situation that is unstable, and striations can develop.

The solvent mixing strategy[6] takes advantage of solvent properties by adjusting the recipe so that the most volatile solvent (and therefore the one depleted first during spin-coating) is one with a relatively high specific surface tension. This creates a situation where the depletion of this solvent will force the surface tension lower, and this makes any surface tension differences contribute to faster leveling of any perturbations that might exist.

We have published a demonstration of this strategy for an isopropanol-based aluminum-titanate sol-gel system.[13] The selected addition of ethyl acetate was effective at preventing striation formation for the aluminum-titanium recipe. More recently we have collected surface tension data for solutions with systematically varying solvent mixtures and substantiated the surface tension changes that must be occurring when spin-coating is performed.[14]

The instability described above and illustrated in Figure 16.3 is the "thick-film limit" for evaporation in competition with solvent diffusion in the coating solution. It is also possible that some coating processes are behaving at the "thin-film limit." This alternate case is more likely happen with solvents that are relatively less volatile. In the thin-film limit, the top-surface composition changes are limited by how deep the coating solution is locally. Thicker regions will deviate from the starting composition more slowly than thinner regions. This will also create situations where thickness ripples will be unstable during spinning and drying—and such instabilities can also amplify with time and then be frozen into the coating as striations. So far, the thickness instabilities that fall into this category have not been adequately studied, but for all types of striations, the control of the sequence of solvent removal is critical to achieving a good quality coating.

## II. SKIN-TEARING DEFECTS

As was seen above, many sol-gel solutions may experience solvent evaporation that is fast enough to force the creation of a "skin" on the top of the coating solution. Since evaporation is happening from the very beginning of the process, the skin may form even when very rapid fluid outflows are still being experienced. The enrichment of the precursors at the top surface can cause gelation of this skin, which will make it more resistant to stretching than the underlying fluid and can retard radial fluid motions as well as inhibit further solvent removal through the layer. It is even possible for the stretching forces to be large enough that fracture or tearing of the skin occurs. These surface ruptures create wispy locations of substantially thinner coating or small randomly located linear rips, but they do not typically result in complete penetration through the coating because the surface skin is still riding on a layer of more fluid solvent at the solution/substrate interface. The occurrence of these defects depends to a large degree on the volatility of key solvents used in the sol-gel process, with less volatile solvents being less likely to result in surface ruptures.

Figure 16.4 shows a particularly bad skin rupture defect and the catastrophic chain reaction that has occurred and left its imprint in the final coating thickness variations.[15] A black arrow indicates the radial flow direction for the coating fluid during spinning. The skin rupture was initiated at the defect point closest to the center of rotation. Once the skin ruptured, it was no longer retarded in its motion by the very high viscosity fluid in the skin itself. So fluid flow continuity was no longer required at the point of rupture. Thus the fluid was able to move out faster than before and overtake the regularly flowing fluid that was downstream and build into the droplet shapes that are evident in the lower part of the figure (four droplet paths can be seen in Figure 16.4). This suggests that contrary to the velocity profile found by EBP, there will be a retrograde shape with lower radial

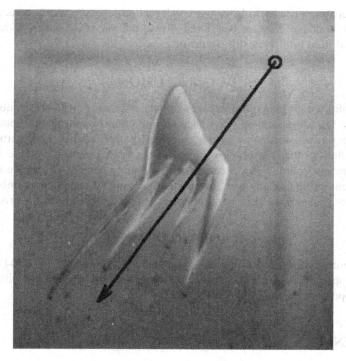

**FIGURE 16.4** Optical micrograph of a sol-gel coating on glass with a rupture defect. The center of wafer rotation is located within the black circle in the upper right corner. The black arrow shows the directly radial direction of fluid flow that occurred during spinning (figure shows an area approximately 20 mm × 20 mm).

velocities within the skin itself, but with faster flow rates deeper within the interior of the coating solution where the viscosity is still lower.

Prevention of these defects can be achieved by delaying the formation of the skin on the surface until fluid flow has slowed further. This can be done by reducing the evaporation rate of the solvent (by changing to a higher boiling point solvent, for example, or by spinning in a solvent-saturated environment as Daniels et al.[9] found for striation prevention). Further study of these tearing defects is needed.

## III. DISCUSSION

Interestingly, it is also possible for both types of defects to be formed by the addition of vapor from the surroundings during spinning. Water plays an important role in many sol-gel solutions and is usually also readily present as humidity in standard laboratory environments. This water vapor can enter into coating solutions just as easily as it can be removed—it just requires diffusion through the vapor boundary layer, but now in the opposite direction, from the ambient air into the coating. The hydrolysis and condensation reactions that are enabled by (and accelerated by) water can force the formation of a surface skin while substantial fluid motion is still occurring.

## IV. CONCLUSION

High-quality coatings require excellent thickness uniformity across substantial distances. Since evaporation plays such an important role in the development of the coating, it is critical that evaporation rates be identical over all parts of the wafer. Solutions that have high vapor pressure solvent constituents are intrinsically more susceptible to evaporation-related coating defects. Great care must be used when optimizing any new formulation to avoid surface tension imbalances and ensure that surface skin gelation does not occur too early in the process.

## ACKNOWLEDGMENT

The support of the National Science Foundation (grant no. EEC-0203504) is very greatly appreciated. Some of the work reviewed here was carried out with earlier NSF support under grant no. DMR-9802334.

## REFERENCES

1. Emslie, A.G., Bonner, F.T., and Peck, C.G., Flow of a viscous liquid on a rotating disk, *J. Appl. Phys.*, 29, 858, 1958.
2. Meyerhofer, D., Characteristics of resist films produced by spinning, *J. Appl. Phys.*, 49, 3993, 1978.
3. Birnie, D.P., III, and Manley, Combined flow and evaporation of fluid on a spinning disk, *Phys. Fluids*, 9, 870, 1997.
4. Birnie, D.P., III, Combined flow and evaporation during spin coating of complex solutions, *J. Non-Cryst. Solids*, 218, 174, 1997.
5. Haas, D.E., Quijada, J.N., Picone, S.J., and Birnie, D.P., III, Effect of solvent evaporation rate on "skin" formation during spin coating of complex solutions, in *Sol-Gel Optics V*, B.S. Dunn, E.J.A. Pope, H.K. Schmidt, and M. Yamane, Eds., *Proc. SPIE*, 3943, 280, 2000.
6. Birnie, D.P., III, Rational solvent selection strategies to combat striation formation during spin coating of thin films, *J. Mater. Res.*, 16, 1145, 2001.
7. Benard, H., Les Tourbillons Cellulaires dans une Nappe Liquide, *Rev. Gen. Sci. Pures Appl. Bull. Assoc. Franc. Avan. Sci.*, 11, 1261, 1900.
8. Haas, D.E., Birnie, D.P., III, Zecchino, M.J., and Figueroa, J.T., The effect of radial position and spin-speed on striation spacing in SOG coatings, *J. Mater. Sci. Lett.*, 20, 1763, 2001.
9. Daniels, B.K., Szmanda, C.R., Templeton, M.K., and Trefonas, P., III, Surface tension effects in microlithography—striations, in *Advances in Resist Technology and Processing III, Proc. SPIE*, Willson, C.G., Ed., 631, 192, 1986.
10. Du, X.M., Orignac, X., and Almeida, R.M., Striation-free, spin-coated sol-gel optical films, *J. Am. Ceram. Soc.*, 78, 2254, 1995.

11. Block, M.J., Surface tension as the cause of Benard cells and surface deformation in a liquid film, *Nature*, 178, 650, 1956.
12. Pearson, J.R.A., On convection cells induced by surface tension, *J. Fluid Mech.*, 4, 489, 1958.
13. Taylor, D.J., and Birnie, D.P., III, Striation prevention by targeted formulation adjustment: aluminum titanate sol-gel coatings, *Chem. Mater.*, 14, 1488, 2002.
14. Birnie, D.P., III, Kaz, D.M., and Taylor, D.J., Surface tension evolution during early stages of drying and correlation with defect reduction during spin-on of sol-gel coatings, Unpublished results.
15. Birnie, D.P., III, Surface skin development and rupture during sol-gel spin-coating, submitted to *J. Sol-Gel Sci. Technol.*, 31, 225, 2004.

# 17 Preparation and Properties of SiO$_2$ Thin Films by the Sol-Gel Method Using Photoirradiation and Its Application to Surface Coating for Display

*Tomoji Ohishi*

## CONTENTS

Functional ceramic thin films have been used in a variety of electronic devices. In particular, SiO$_2$ thin films are widely used for device passivation and protection of magnetic and optical disks.[1-3] The synthesis of SiO$_2$ thin films can be conveniently divided into physical routes, such as sputtering and physical vapor deposition (PVD), and chemical routes, such as chemical vapor deposition (CVD) and pyrolysis from precursor film. These methods are well suited to small scale, small area depositions; large area depositions are difficult because of the technological problems involved in scale-up.

On the other hand, sol-gel methods are recognized as a simple technology.[4-6] Because these processes are liquid phase processes, multicomponent and large area films can easily be made. A frequently encountered problem in organometallic sol-gel processing is the tendency for organic compounds to remain in the film after processing. Their removal typically requires heat treatment up to several hundred degrees centigrade.

A novel thin-film fabrication route using a sol-gel method together with photoirradiation curing has been investigated.[7-12] This process is an effective method for preparing ceramic thin films at low temperature. Previously the electrical properties of $Ta_2O_5$ films prepared using this novel low-temperature method were found to be similar to those of $Ta_2O_5$ films prepared by sputtering and CVD.[7,8] The preparation and properties of the $SiO_2$ thin films fabricated by this new process have been reported.[9,11] Physical properties such as film thickness, hydrogen fluoride (HF) etch rate, and hardness were studied as a function of process conditions, in particular, thermal heat treatment applied in the last step.

This chapter describes preparation of $SiO_2$ thin film by the sol-gel method using photoirradiation and the effect of heat treatment on the molecular structure of the $SiO_2$ in the film examined using Raman and nuclear magnetic resonance (NMR) spectroscopy. The relationships between the molecular characteristics and physical properties such as hardness and etch rate are derived, and applications for the new process are described.

## I. FILM FORMATION

The $SiO_2$ films were prepared according to the synthetic procedure outlined in Figure 17.1. Tetraethoxysilane (TEOS), water, nitric acid, and ethanol (1:12:0.26:45 by molar ratio) were mixed at room temperature and then magnetically stirred overnight. Drops of the mixed solution were deposited onto quartz substrates and silicon wafers while spinning at 2000 rpm. Substrates with precursor thin films were cured by placing them in an oven at a preset temperature (80 to 500°) for 30 min. This type was called heated films. For the second type, substrates with precursor thin films were placed onto a hotplate heated to a preset temperature (80 to 200°C). Thermal equilibrium was reached after 10 min of heating, then the radiation was begun. Ultraviolet radiation of two wavelengths, 254 and 184 nm (12 and 2.4 mW/cm²), was shown simultaneously on the films with a low-pressure mercury lamp for 10 min. Then the radiation was stopped, but heating was continued for 5 more min. This type was called irradiated films.

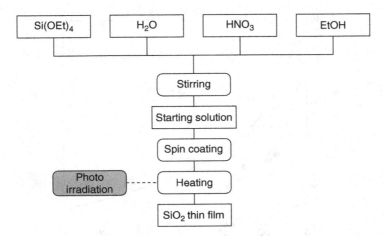

**FIGURE 17.1** Synthetic procedure for $SiO_2$ films by the sol-gel method using photoirradiation.

## II. MICROHARDNESS AND HF ETCH RATE

The thicknesses of the films as a function of heat treatment are plotted in Figure 17.2. In Figure 17.3 and Figure 17.4, the etch rate and microhardness are plotted as a function of heating temperature. The thickness of the irradiated film is reduced at lower temperatures compared to that of the heated films. In general,

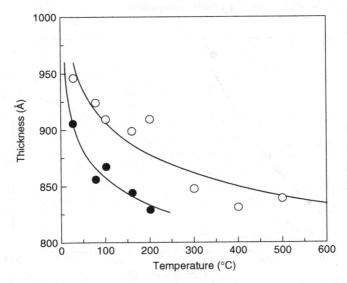

**FIGURE 17.2** Thermal treatment dependence of film thickness: ○ heated film, ● irradiated film.

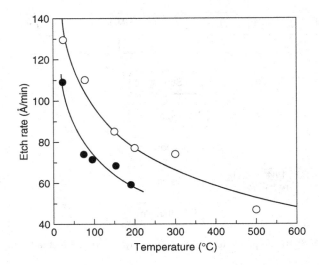

**FIGURE 17.3** Thermal treatment dependence of etch rate: ○ heated film, ● irradiated film.

densification of the film is constrained by the substrate in the planer direction and shrinkage is limited to the thickness direction. This indicates that densification occurs at a lower temperature with photoirradiation. The microhardness and etch rate of the irradiated films are higher and slower, respectively, than those of the heated films at the same temperature. This means that hard and dense films are obtained at low temperatures by photoirradiation.

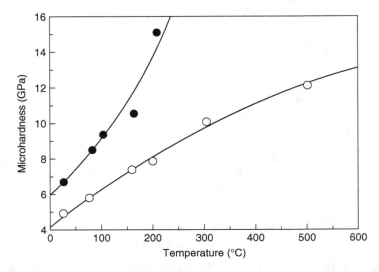

**FIGURE 17.4** Thermal treatment dependence of microhardness: ○ heated film, ● irradiated film.

# III. MOLECULAR STRUCTURE ANALYSES
# OF SIO$_2$ THIN FILMS

## A. RAMAN SPECTROSCOPY

According to the bond angle, SiO$_2$ bonds can be separated into linear-type stable bonds with a bond angle of 180° and folded angular-type unstable bonds anywhere between 120° and 180°.[13] From the precursor state to the final fully densified state, SiO$_2$ bonds tend to change from the folded angular type to the linear type. In Raman spectroscopy, nonlinear Si–O–Si bonds typically produce a vibrational peak at around 1050 cm$^{-1}$. As the number of folded nonlinear Si–O–Si bonds decreases with increasing heat treatment temperature, the 1050 cm$^{-1}$ peak intensity is expected to decrease. The Raman spectra as a function of temperature for heated and irradiated films are shown in Figure 17.5. For the latter, the 1050 cm$^{-1}$ peak (peak a), asymmetric vibration, disappears at 80°C, while it is still strong at 80°C for the former. This suggests that radiation promotes unfolding and straightening of the SiO$_2$ bonds. The 1050 cm$^{-1}$ peak for the heated films disappears only after further heat treatment (160°C). Peak intensity (Si–O–Si) at around 810 cm$^{-1}$ (peak b), symmetric vibration, increases with heat treatment temperature

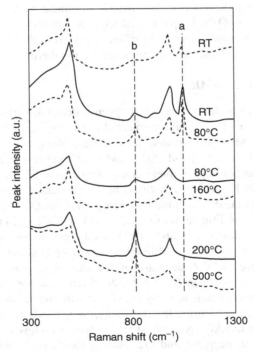

**FIGURE 17.5** Thermal treatment change of Raman spectra: ... heated film; — irradiated film. (a) Si-O-Si asymmetric stretch mode. (b) Si-O-Si symmetric stretch mode.

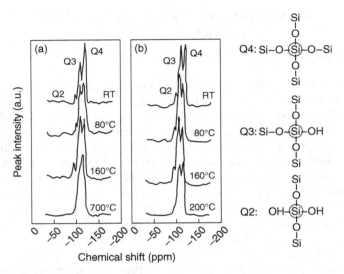

**FIGURE 17.6** Thermal treatment change of $^{29}$Si solid state NMR spectra.

up to 200°C. This indicates that the unstable nonlinear Si–O–Si is cleaved to yield stable linear Si–O–Si by photoirradiation or heat treatment. Peak b shifts to the higher wave number with thermal treatment from 807 to 815 cm$^{-1}$. This suggests that Si–O–Si bonds are strengthened by the thermal treatment.

## B. $^{29}$Si SOLID STATE NMR

Figure 17.6 shows $^{29}$Si solid state NMR spectra. Peaks at −90, −100, and −110 ppm corresponds to silicon atoms connected to two oxygen-silicon bonds, two oxygen-hydrogen bonds (designated the Q2 state), three oxygen-silicon bonds and one oxygen-hydrogen bond (Q3), and four oxygen-silicon bonds (Q4), respectively.[14] Changes in peak intensity as a function of heat treatment temperature are shown in Figure 17.7. With no heat treatment, the as-deposited film has Q3 at 34%, Q4 at 65%, and the remaining 1% from Q2. With irradiation at room temperature, the film shows Q2 at 5%, Q3 at 45%, and Q4 at 50%. As a function of temperature, Q4 exhibits a U-shaped dependence for both the heat-treated and irradiated samples, with minima occurring at about 200 and 100°C, respectively (Figure 17.7). A decrease of Q4 and an increase of Q2 and Q3 suggest an initial increase in Si–OH bonds. Such behavior, while seemingly inconsistent with intuition, is actually consistent with the unfolding behavior of the SiO$_2$ ring structure.[4] During the unfolding, SiO$_2$ bonds are cleaved and the cleaved bonds are quickly hydrolyzed in the wet gel films containing water, which causes Q4 to decrease and Q2 and Q3 to increase. With further heat treatment, dehydrolyzation dominates and Q4 turns upward, with corresponding decreases in Q2 and Q3.

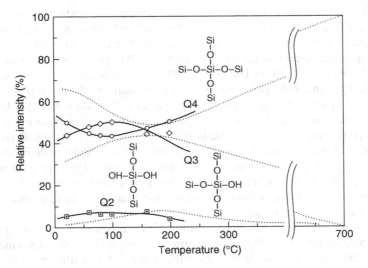

**FIGURE 17.7** Thermal treatment dependence of Q2, Q3, Q4: …: heated film; —: irradiated film.

With irradiated films, the starting Q4 is ~15% less than the heated film. This suggests that irradiation enhances the cleavage and unfolding of the SiO₂ ring structure, which is consistent with the Raman spectra observation.

## C. MECHANISM

The results of Raman and NMR spectroscopy suggest that the same reaction occurs in the photoirradiated film and the heated film. Figure 17.8 shows a mechanism for

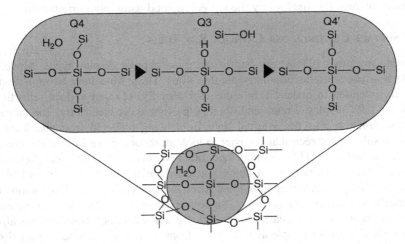

**FIGURE 17.8** Rearrangement of SiO₂ due to photoirradiation or thermal treatment.

the rearrangement of $SiO_2$ films due to photoirradiation or heat treatment. This model indicates that the molecular structures of $SiO_2$ films are rearranged and stabilized by photoirradiation and heat treatment. The Q4 in the film prepared at low temperatures forms unstable folded nonlinear $SiO_2$ species, which are obtained by the hydrolysis of TEOS. The unstable Si–O–Si bonds are hydrolyzed to yield Q3 bonding with OH by photoirradiation or heat treatment. Then, on treatment at higher temperature, Q3 rearranges to the stable linear Si–O–Si bond (Q4). In spite of the decrease in the Q4 fraction in all silicon atoms by low-temperature treatment, film properties are improved. It is due to the fact that decreasing of Q4 with unstable structure and increasing of Q4 with stable structure occurred simultaneously, as proved by the shift of the symmetric vibration peak to the higher wave number in Raman spectra. Photoirradiation enhances the cleavage and unfolding of the $SiO_2$ ring structure to promote densification of the thin film, resulting in greater microhardness and slower etch rates.

As shown previously, a technique for thin film preparation at low temperatures has been developed in which photoirradiation is combined with the sol-gel method. This has been used to produce $SiO_2$ thin films. Harder, denser films were prepared at lower temperatures than by conventional heat treatment. The hardness and HF etch rate were related to the molecular structure of the thin films. Photoirradiation cleaved Si–O–Si bonds so that unfolding of the $SiO_2$ ring structure was enhanced at low temperatures and densification of the thin film was promoted.

## IV. APPLICATION TO SURFACE COATING FOR DISPLAY

Modern displays have a variety of surface coatings, such as antireflection film, antistatic film, insulator film, and protection film.[15–17] $SiO_2$ thin film is widely used as a functional surface coating film for displays. This chapter describes surface coatings for displays by the sol-gel method using photoirradiation.

### A. SURFACE COATINGS FOR CATHODE RAY TUBES

Cathode ray tubes (CRTs) are one of the most popular display types. Modern CRTs now use antireflection/antistatic thin films. These films have two functions that are intended to make CRTs more user friendly: (1) they reduce reflection from the CRT display screen, and (2) they prevent static electricity. The antireflection function prevents external light from being reflected from the screen, which results in a screen that is easier to look at, reducing eye fatigue for visual display terminal (VDT) operators and other viewers. The antistatic function prevents static electricity from being generated on the CRT surface, and thus eliminates the problem of an individual receiving an electrical shock upon touching the display screen. The function also prevents dust from adhering to the screen and thus makes the screen easier to see. Finally, antireflection/antistatic films also reduce the danger for people near the CRT from the harmful effects of radiation leaking from the display unit.

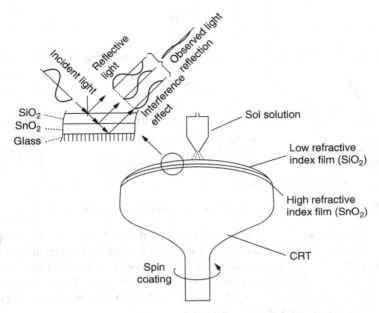

**FIGURE 17.9** Principle of the antireflective/antistatic thin film for CRT.

These films have been produced by sputtering, chemical vapor deposition, and vacuum evaporation. The sol-gel method has also received some attention as a way to prepare these films because of its low cost. But these film preparation methods have disadvantages, such as a need for vacuum equipment and high-temperature heat treatment, and difficulties in coating large areas.

This chapter reports on a newly developed, simple, low-temperature film formation technique for preparing antireflection/antistatic films. This process uses the sol-gel method and photoirradiation.

## B. ANTIREFLECTION/ANTISTATIC THIN FILMS FOR CRTs

Figure 17.9 shows the principle of the antireflection/antistatic thin films formed on the panel surface. There is a layer of film having a high refractive index (SnO$_2$) and a layer of film having a low refractive index (SiO$_2$). Light reflection is reduced since they are mutually compensated by the interference effect on all areas of the interface between these layers of film.

Reflection characteristics can be controlled by controlling the thickness and the refractive index of the film layers. SiO$_2$ is selected for the low refractive index film and SnO$_2$ containing Sb$_2$O$_3$ is selected for high refractive index films. SnO$_2$ containing Sb$_2$O$_3$ also serves as an effective antistatic film because it has good electrical conductivity.

These films are prepared by the sol-gel method using heat treatment. The SiO$_2$ film is formed using a SiO$_2$ sol solution and the SnO$_2$ film is formed using

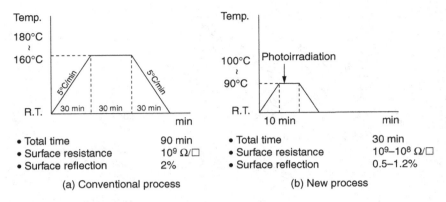

**FIGURE 17.10** Comparison of the conventional process and the new process.

a $SnO_2$ sol solution. The $SnO_2$ sol solution and the $SiO_2$ sol solution are spin-coated in turn on the CRT panel surface and then heat treated. Since this process is the last stage in the manufacture of the CRT, it should be done at a temperature as low as possible. Heat treatments at high temperatures are dangerous because CRTs are high-vacuum devices. In addition, high temperatures can also degrade the performance of the CRTs electron gun.

By applying the newly developed technique to this manufacturing process, we succeeded in producing a low-temperature preparation, within a short time, for high-performance antireflective/antistatic films.[10] Figure 17.10 compares the new process and a conventional process. The new process makes it possible to reduce treatment temperatures by almost 50% and treatment time to approximately 33% of conventional levels. Moreover, the photoirradiated films have better performance than conventional films in terms of better surface resistance and lower surface reflection.

## C. FORMATION OF ANTIREFLECTION/ANTISTATIC THIN FILMS

Figure 17.11 shows a flowchart describing the preparation of antireflection/antistatic thin films for CRTs. The antireflection/antistatic thin films were spin-coated on the CRT panel surface. A $SnO_2$ sol solution containing $Sb_2O_3$ was coated on the panel surface and dried at 45°C. Then the $SiO_2$ sol solution was coated and dried. The double-layered thin film was photoirradiated with 254 and 184 nm wavelength light for 10 min at the same time. The photo intensities were changed from 12 to 40 mW/cm². The substrate temperatures were changed from room temperature to 160°C during photoirradiation.

## D. CHARACTERISTICS OF ANTIREFLECTION/ANTISTATIC THIN FILMS

Figure 17.12 shows the photo intensities during the photoirradiation stage as well as the changes in surface resistance in relation to the changes in temperature. Photoirradiation that is performed with the substrate temperature between 90 and

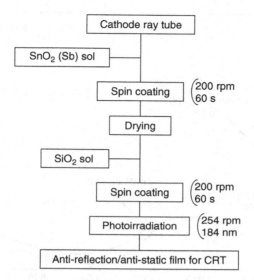

**FIGURE 17.11** Preparation flowchart for the antireflective/antistatic thin film.

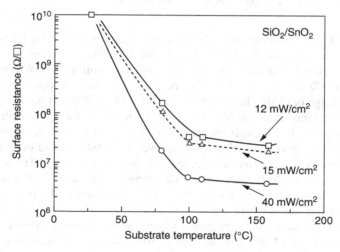

**FIGURE 17.12** Relationship between surface resistance and photo intensity.

100°C produces a film that has sufficiently low levels of surface electrical resistance. The antireflection/antistatic film prepared at a 40 mW/cm$^2$ photo intensity and 100°C has a low surface resistance of $4 \times 10^6$ $\Omega/\square$. The stronger the photo intensity is, the lower the surface electrical resistance becomes. We assume, furthermore, that the stronger the photo intensity is, the more extensively SnO$_2$ film shrinks, with the result that the SnO$_2$ film becomes dense.

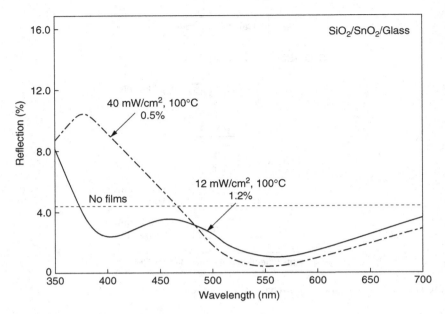

**FIGURE 17.13** Reflection curves of the irradiated films.

Figure 17.13 shows the reflection rate curves for antireflection/antistatic films. The solid line indicates the film prepared with a 12 mW/cm$^2$ photo intensity at 100°C; the dotted line indicates the film prepared with a 40 mW/cm$^2$ photo intensity at 100°C. For CRTs that have no such film, the reflection rate is about 4% in the range of visible light (350 to 700 nm). The CRTs that have antireflective/antistatic films show a reflection rate of 0.5 to 1.2% at a wavelength of 550 nm. Furthermore, the stronger the photo intensity used to prepare the film, the lower the reflection obtained. Since human eyes are most sensitive to light in the 550 nm range, it is necessary to lower the reflection in this region. The reflection curve for the antireflective/antistatic film prepared with a 15 mW/cm$^2$ photo intensity (data not shown) is almost the same as that for the film prepared with a 12 mW/cm$^2$ photo intensity.

Figure 17.14 shows the appearance of CRT surface coating prepared with a photo intensity of 12 mW/cm$^2$ at 100°C. Antireflective/antistatic thin film is prepared on half the area of the CRT panel surface, indicating good antireflection effect.

In the film prepared with 12 mW/cm$^2$ photo intensity at 100°C substrate temperature, the refractive index of the $SnO_2$ layer was 1.65 and that of the $SiO_2$ layer was 1.41. The refractive indexes of $SnO_2$ and $SiO_2$ were 1.75 and 1.43, respectively, in the film prepared with 40 mW/cm$^2$ at the same temperature. The values of the refractive index of the film obtained by photoirradiation are larger than those for the film without photoirradiation ($SnO_2$: 1.50; $SiO_2$: 1.36). The densification of the film occurs by photoirradiation.

**FIGURE 17.14** Photograph of the appearance of a CRT with an antireflective/antistatic thin film.

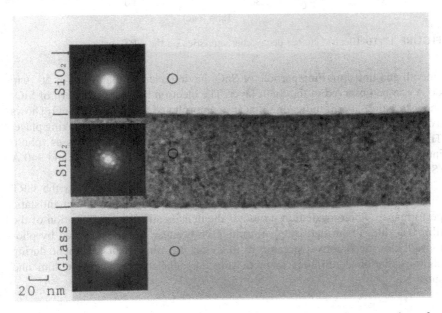

**FIGURE 17.15** Transmission electron microscopy photograph of a cross-section of a SiO$_2$/SnO$_2$ thin film.

Figure 17.15 shows a transmission electron microscopy photograph of a cross section of the antireflective/antistatic film (photo intensity 12 mW/cm$^2$, substrate temperature 100°C). The SiO$_2$ and SnO$_2$ films were prepared on a glass substrate and have thicknesses of 700 Å and 1000 Å, respectively. These films are very

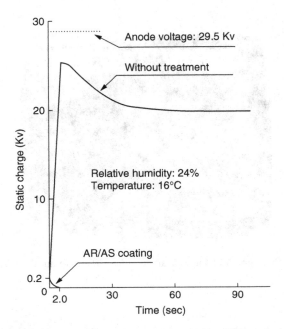

**FIGURE 17.16** Electric charge decay characteristics on the CRT surface.

smooth and uniform. Fine particle of $SnO_2$ having average diameters of 50 Å to 60 Å can be observed in the $SnO_2$ layer. The electron diffraction pattern of $SiO_2$ shows a halo pattern, indicating that it is amorphous, while that of $SnO_2$ shows a diffraction grating pattern, indicating that crystallization is partially taking place. The mechanism of crystallization of $SnO_2$ is not clear. Film thicknesses (photo intensity 40 mW/cm², substrate temperature 100°C) are 670 Å ($SiO_2$) and 940 Å ($SnO_2$).

Figure 17.16 shows the electric charge decay characteristics on the CRT surface. The CRT with the antireflective/antistatic coating has excellent antistatic performance. Photoirradiation promotes the densification and purification of the film. The film is purified by photoirradiation because ozone generated by photoirradiation decomposes organic residues on the film. Heat treatment during photoirradiation contributes to increasing the adhesion between the film and substrate.

## V. CONCLUSION

A technique for thin-film preparation at low temperatures has been developed in which photoirradiation is combined with the sol-gel method. This has been used to obtain $SiO_2$ thin films and surface coatings for display. Harder, denser films were prepared at lower temperatures than by conventional heat treatment. Antireflection/antistatic thin films prepared by this new process have low surface reflection and low surface electrical resistance. This method provides a significant savings in energy, since the treatment temperature and the treatment time are reduced compared to those of the conventional thin-film preparation methods.

## REFERENCES

1. Schere, M., Schmit, J., Lats, R., and Schanz, M., *J. Vac. Sci. Technol.*, A10, 1772, 1992.
2. Yamada, Y., Uyama, H., Watanabe, S., and Nozoye, H., *Appl. Opt.*, 38, 6638, 1999.
3. Matsumoto, M., *Mater. Stage*, 1, 65, 2001.
4. Brinker, C.J., and Schere, G.W., *Sol-Gel Science: The Physics and Chemistry of Sol-Gel Science*, Academic Press, San Diego, 1990, p. 573.
5. Yamamoto, Y., Kamiya, K., and Sakka, S., *J. Ceram. Soc. Jpn.*, 90, 328, 1982.
6. Ogiwara, S., and Kinugawa, K., 90, 157, 1982.
7. Ohishi, T., Maekawa, S., and Katoh, A., *J. Non-Crystl. Solids*, 147, 493, 1992.
8. Ohishi, T., and Katoh, A., *Br. Ceram. Trans.*, 92, 79, 1993.
9. Maekawa, S., and Ohishi, T., *J. Non-Crystl. Solids*, 169, 207, 1994.
10. Ohishi, T., Ishikawa, T., and Kamoto, D., *J. Sol-Gel Sci. Technol.*, 8, 511, 1997.
11. Maekawa, S., Okude, K., and Ohishi, T., *J. Sol-Gel Sci. Technol.*, 2, 497, 1994.
12. Imai, H., Hirashima, H., Awazu, K., and Onuki, H., *J. Ceram. Soc. Jpn.*, 102, 1094, 1994.
13. Lisovskii, I.P., Litovchenko, V.G., Lozinskii, V.G., and Steblovskii, G.I., *Thin Solid Films*, 213, 164, 1922.
14. Yasumori, A., *New Glass*, 4, 21, 1990.
15. Hayama, H., Aoyama, T., Utsumi, T., Miura, Y., Suzuki, A., and Ishiyama, K., *Natl. Tech. Rep.*, 40, 90, 1994.
16. Uyama, H., Tomioka, N., Takahashi, T., Harada, T., Yasuda, Y., Kobayashi, H., and Watanabe, H., *Proceedings of IDW '98*, 1998.
17. Itou, T., Onodera, M., and Matsuda, H., *SID '95 Digest*, 1, 25, 1995.

# Section IV

## Ceramics Via Polymers

# 18 Organosilicon Polymers as Precursors for Ceramics

*Markus Weinmann*

## CONTENTS

## I. INTRODUCTION

Thermolysis of organometallic polymers in controlled atmosphere and heat treatment conditions is a comparatively simple and inexpensive process for producing oxide and nonoxide ceramics. It provides a means for controlling and adjusting microstructure design and the shape of ceramic components which often cannot be achieved using "classical" techniques such as melting, sintering, or chemical vapor deposition (CVD). Figure 18.1 shows a flow diagram of the individual steps involved in the preparation of ceramics by polymer pyrolysis.[1]

Suitable polymers (suitable in this regard means that the polymer exhibits adjustable rheology, latent reactivity, and controllable pyrolytic degradation) are synthesized according to well-developed procedures from appropriate monomers

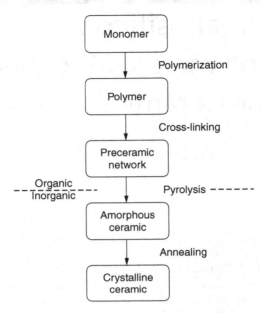

**FIGURE 18.1** Flow diagram representing the individual processing steps involved in the synthesis of precursor-derived ceramics.[1]

and subsequently cross-linked at moderate temperature, thereby transforming into an infusible preceramic network. The transformation into an inorganic material, that is, the ceramization, is performed by a sequence of thermally induced condensation reactions. In general, an amorphous metastable material forms, which by an additional heat treatment at higher temperatures can be crystallized into thermodynamically stable ceramic phases. Remarkably, precursor thermolysis allows for easy control of elemental composition, chemical homogeneity of the derived ceramics, and architecture on an atomic scale. The general idea behind the process concept is to generate preferred structural features in the organometallic precursors and to subsequently transform the precursors into ceramics with retention of the specially designed building blocks, requiring controllable condensation steps during the heat treatment.

Most frequently, organosilicon polymers are used as precursors for ceramics. They possess a versatile chemistry, which rapidly developed within the last few decades. As a consequence of being mostly reaction-controlled precursor thermolysis is especially valuable for the production of nonoxide ceramics, which possess a high degree of covalent bonding. Conventionally, low diffusion capability of such materials—even at very high temperatures—is overridden by the application of sintering aids. However, such additives degrade the otherwise unique high-temperature properties of nonoxide ceramics, such as chemical and mechanical stability. In contrast, thermolysis of appropriate organometallic polymers provides a means for realizing the production of nonoxide ceramics, thereby retaining their excellent properties without compromises.

Ceramics obtained from polymeric precursors are usually amorphous. Since substantial thermal activation is required for nucleation and crystallization, precursor-derived ceramics (PDCs) frequently remain amorphous or nanocrystalline up to rather high temperatures. For example, crystallization of a number of quaternary Si-B-C-N ceramics is retarded even up to 1800°C, resulting in excellent thermomechanical properties. Nevertheless, crystalline materials are of great interest because their microstructure formation can be controlled during devitrification, providing a means for stabilizing nanosized morphologies.

The excellent high-temperature properties of the ceramic materials strongly depend on the molecular structure and composition of the polymeric precursors. This chapter reviews the fundamentals of synthetic approaches to silicon-based nonoxide preceramic polymers and briefly discusses their processing.

## II. POLYMER SYNTHESIS

Preparative organosilicon chemistry offers manifold possibilities for the synthesis of precursors for nonoxide ceramics (Scheme 18.1). The focus has been on the synthesis of polymers such as polysilanes **A**, polysilazanes **B**, polycarbosilanes

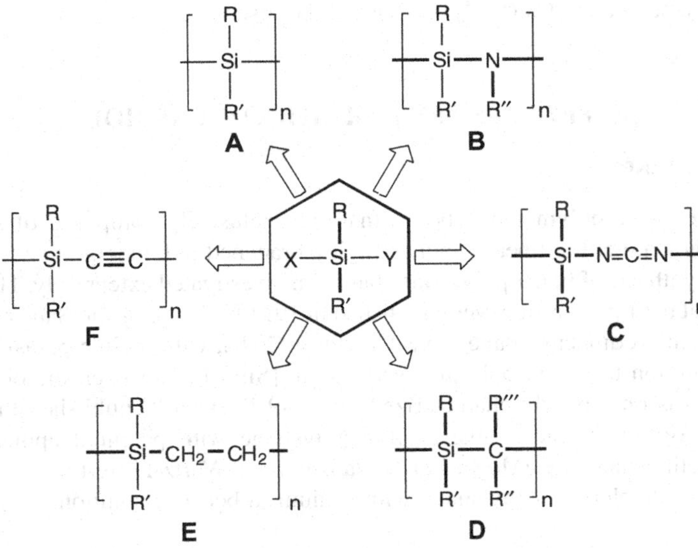

**SCHEME 18.1** Preparation of non-oxide organosilicon polymers as precursors for ceramics. X and Y are reactive substituents in the monomers, which are essential for polymerization and/or cross-linking reactions. In contrast, R and R' are singly bonded pendent or functional groups, which do not necessarily attribute to polymerization or cross-linking reactions.

**D-F** and polysilylcarbodiimides (PSCs) **C**. Moreover, syntheses of polysilanes with multi-element sequences were achieved.

The monomeric organosilicon compounds have reactive sites Si-X and/or Si-Y such as Si-Cl, Si-H, Si-N, Si-C=C, or Si-C≡C which enable polymerization by elimination, substitution (metathesis), or addition reactions. Usually, chlorosilanes (X, Y = Cl) are used as starting compounds because of their commercial availability, low cost, and well-developed chemistry.

Pendent groups R (R = singly bonded organic ligand) influence physical chemical properties, for example, solubility, viscosity, etc., to a significant extent. Moreover, they allow for further chemical modification by, for example, synthesis of derivatives, that is, incorporation of additional elements, attachment of functional groups, or by cross-linking. The latter issue is of major importance since a high cross-linking density is responsible for retaining structural features in the organometallic during the polymer-to-ceramic conversion. It avoids depolymerization reactions and subsequent volatilization of low-molecular weight species and thus extensive mass loss during ceramization. Cross-linking may be induced thermally or by surface modification, assuming that the polymers exhibit latent reactivity.

In the following, the synthesis of different types of organosilicon polymers as potential precursors for ceramics is highlighted topologically, starting from polysilanes with Si-Si linkages, followed by polycarbosilanes. Subsequently polysilazanes with Si-N building blocks and polymers with multi-element sequences, such as PSCs (Si–N=C=N), will be discussed.

## III. PRECURSORS FOR SILICON CARBIDE

### A. POLYSILANES

Polysilanes have polymer backbones that are exclusively composed of silicon atoms. They are either linear chain or highly branched polymers.

The synthesis of linear polysilanes has been investigated extensively. The first attempts at their preparation were published in 1921 by Kipping and Sands.[2] They reacted dichlorodiphenylsilane, $Ph_2SiCl_2$ (Ph = $C_6H_5$), with molten potassium in xylene solution to obtain poly(diphenylsilane), $[SiPh_2]_n$. However, the obtained polymer was only poorly characterized. In 1949 Burkhardt[3] published the first clear description of the synthesis of a polysilane with potential application. Poly(dimethylsilane), $[SiMe_2]_n$, was obtained in a Wurtz-Fittig-type coupling reaction[4] of dichlorodimethylsilane with sodium in benzene solution.

$$Cl-\underset{\underset{CH_3}{|}}{\overset{\overset{CH_3}{|}}{Si}}-Cl \quad \xrightarrow[- NaCl]{Na, benzene} \quad \left[ \underset{\underset{CH_3}{|}}{\overset{\overset{CH_3}{|}}{Si}} \right]_n$$

(18.1)

Poly(dimethylsilane) is an intractable white solid that is insoluble and decomposes without melting when heated to 250°C to form a polycarbosilane. It will be highlighted below that it represents the first precursor for the production of refractory SiC fibers.

In contrast to linear polysilanes, there was little progress made in the synthesis of highly branched polysilanes. This was because of incomplete reduction when using Wurtz-type reactions along with rearrangements resulting in complex product mixtures. As reported by Bianconi et al.,[5] these problems were overcome by using high-intensity ultrasound-promoted reactions between n-hexyltrichlorosilane, $n\text{-}C_6H_{13}SiCl_3$, and Na/K alloy. Under such conditions, "virtually homogeneous" reduction of alkyl silicon trichlorides with liquid Na/K emulsions could be achieved. In contrast to polysilanes bearing methyl groups or phenyl groups at the silicon atoms, n-hexylsilyne is soluble in organic solvents.

$$(C_6H_{13})SiCl_3 \quad \xrightarrow[-\text{ NaCl/KCl}]{\substack{\text{Na/K} \\ \text{[Ultrasound]}}} \quad \left[ \begin{array}{c} C_6H_{13} \\ | \\ \text{Si} \\ | \end{array} \right] \tag{18.2}$$

A remarkable reaction leading to polysilanes in which silicon atoms possess fourfold silicon coordination was published by Maxka et al.[6] Tetrakis(chlorodimethylsilyl)silane, which was obtained from $SiCl_4$, Li, $HMe_2SiCl$, and $CCl_4$[7,8] was coupled using lithium metal in tetrahydrofuran (THF) solution. $[(SiMe_2)_4Si]_n$ was obtained as a waxy orange solid with a yield of approximately 90%.

$$\begin{array}{c} \text{SiMe}_2\text{Cl} \\ | \\ \text{ClMe}_2\text{Si}-\text{Si}-\text{SiMe}_2\text{Cl} \\ | \\ \text{SiMe}_2\text{Cl} \end{array} \quad \xrightarrow[-\text{ LiCl}]{\text{Li}} \quad \left[ \begin{array}{cc} \text{Me}_2 & \text{Me}_2 \\ \text{Si} & \text{Si} \\ & \text{Si} \\ \text{Si} & \text{Si} \\ \text{Me}_2 & \text{Me}_2 \end{array} \right] \tag{18.3}$$

A crucial subject of polysilanes is their very often unsatisfactory ceramic yield. This is because of depolymerization during heat treatment and volatilization of low-weight molecular species. In the worst case, depolymerization results in a total breakdown of the precursor and 100% weight loss. For this reason, the presence of latent reactive sites in the precursors is required. Latent reactivity is provided by functional groups. For example, Si-H units can be used for hydrosilylation or dehydrocoupling of silanes (such reactions will be described below). Yet it is difficult to retain Si-H groups during reductive coupling of chlorosilanes; they undergo uncontrolled side reactions. Si-Cl moieties that allow for further modification and cross-linking by metathesis reactions are difficult to retain during the purification of polysilanes, which is in general performed in aqueous solution to remove alkali metal chloride and residual alkali metal. Under such conditions, remaining Si-Cl units are hydrolyzed to Si-OH moieties. Oxygen incorporated by this procedure is not released during thermolysis. It results in oxygen-contaminated silicon carbide ceramics which, compared to oxygen-free composites, possess unsatisfactory high-temperature properties. Nevertheless, Si-Cl units can be generated subsequently, after the reductive coupling, for example, by replacing phenyl groups in $[SiPhR]_n$ (R = alkyl, aryl) with chlorine atoms. For this purpose, poly(phenylsilane) was dissolved and treated with $HCl/AlCl_3$.[9] The attachment of vinyl groups to polysilane skeletons as reactive sites for further cross-linking was also described.[10] Such moieties allow for polyaddition by olefin polymerization or hydrosilylation.[11]

Because of the low solubility of most polysilanes in combination with expandable processing, that is, purification including the above mentioned hydrolysis steps, product yields and purity are often unsatisfactory if obtained by reductive coupling of chlorosilanes. A further negative aspect of this approach is the limited functional group tolerance. Accordingly, only the synthesis of polysilanes with inert pendant groups such as alkyl or aryl can be adequately achieved.

Alternatively, polysilanes can be obtained by transition metal-promoted dehydrocoupling reactions of hydridosilanes. In contrast to reductive coupling of chlorosilanes, no solid by-products form. The only by-product is hydrogen. Thus purification steps are not required. It was shown by several research groups that group 4 metallocenes (Equation 18.4) are uniquely active catalysts for the formation of Si-Si bonds by this procedure.

$$H-\underset{\underset{H}{|}}{\overset{\overset{R}{|}}{Si}}-H \quad \xrightarrow[-H_2]{\overset{R}{\underset{R}{M \leftarrow |||}}} \quad \left[ \overset{\overset{R}{|}}{\underset{\underset{H}{|}}{Si}} \right]_n$$

$$M = Ti, Zr$$

$$(18.4)$$

For example, Chang and Corey[12] used $\eta^2$-alkynyl titanocene or zirconocene complexes as catalysts and also provided a conclusive mechanism (Scheme 18.2). Initially the side-on $\eta^2$-coordinated alkynyl ligand is released from the original catalyst, thus forming a 14 valence electron (VE) metallocene species. This species oxidatively adds the silane (the transition metal inserts into the Si-H bond), whereby a 16 VE hydridosilyl metallocene forms. It is supposed that through hydrogen elimination, a metallocene-silanediyl complex forms, which undergoes rapid hydrosilylation with external $RSiH_3$. A subsequent hydride shift results in a rearrangement of the disilylmetallocene along with a transition metal-silicon bond cleavage and formation of a hydridosilyl metallocene. Repetitive hydrogen elimination and hydrosilylation result in the formation of the polymer.

Harrod[13] investigated the mechanism of $Cp_2TiMe_2$-catalyzed dehydrocoupling of silanes in detail. He found that the process occurs in two steps. A rapid autocatalytic reaction in which the catalytic species $Cp_2Ti$ forms is followed by a conproportionation of $Cp_2Ti$ and $Cp_2TiMe_2$ to yield $[Cp_2TiMe]_2$. The latter dissociates to $Cp_2TiMe$. $\sigma$-bond metathesis between Si-H and $Cp_2Ti$-Me gives

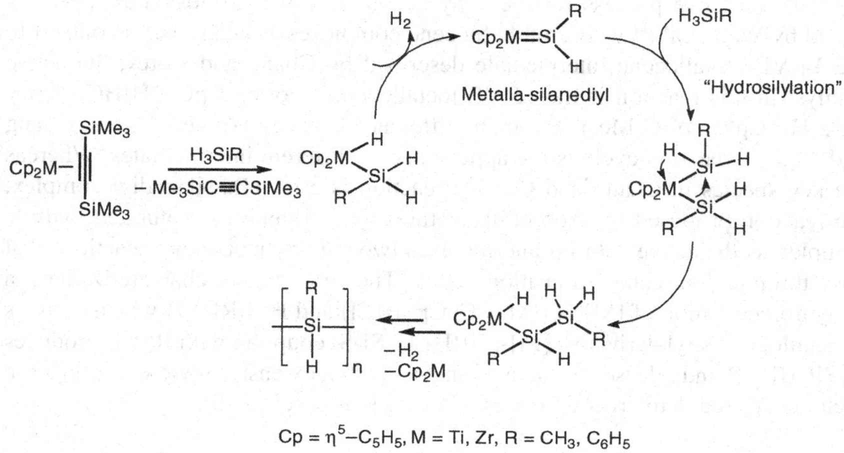

$Cp = \eta^5-C_5H_5$, M = Ti, Zr, R = $CH_3$, $C_6H_5$

**SCHEME 18.2** Proposed mechanism for the formation of polysilanes by metallocene-catalyzed dehydrocoupling of alkylsilanes, $RSiH_3$, via metallocene-silanediyl intermediates.[12]

Si-Me and $Cp_2Ti$-H, which by reaction with $Cp_2Ti$-Me result in the elimination of methane and the formation of two equivalents of $Cp_2Ti$. The second stage (Scheme 18.3) is (i) a formation of a dinuclear $Ti^{III}$ complex through reaction of $Cp_2Ti$ with Si-H, which then converts (via reactions ii and iii) into the dimer $[Cp_2TiSiR_2H]_2$. Though the precise manner of the transformation of $[Cp_2TiSiR_2H]_2$ into $Cp_2TiH$ and $Cp_2TiSiR_2SiR_2H$ (iv) is not understood in detail, it is the key to understanding the Si-Si linking and the polymerization of hydridosilanes using dimethyltitanocene as a catalyst.

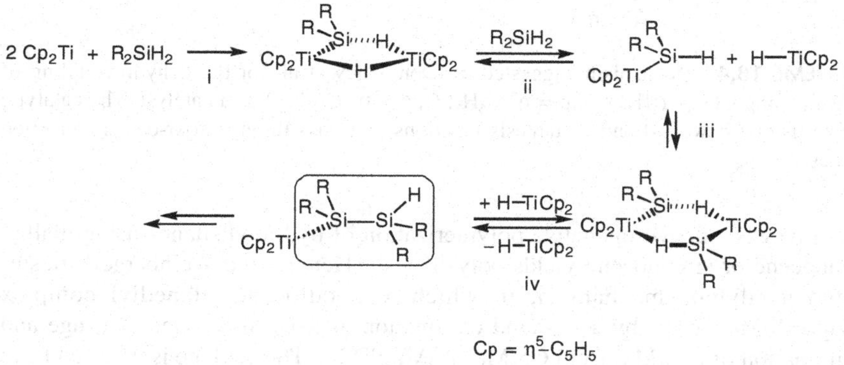

$Cp = \eta^5-C_5H_5$

**SCHEME 18.3** Mechanism of titanocene-catalyzed dehydrocoupling of hydridosilanes as proposed by Harrod.[13]

An alternative process for the dehydrocoupling of hydridosilanes was published by Woo et al.,[14] who used hafnocene complexes as catalysts. In contrast to the 14 VE metallocene intermediate described by Chang and Corey, the active catalyst in this reaction is the 16 VE metallocene chloride CpCp*HfHCl (Cp = $\eta^5$-$C_5H_5$; Cp* = $\eta^5$-$C_5Me_5$). The main difference between Woo et al.'s and Chang and Corey's reaction cycle is the appearance of different intermediates. Whereas the key species in Chang and Corey's reaction is a metalla-silanediyl complex, the reaction proposed by Woo et al. features a coordinative unsaturated hydride complex as the active catalyst and involves two σ-bond metathesis reactions that pass through four-center transition states. The first step is characterized by a dehydrometallation of [M]-H ([M = CpCp*HfCl) and H[SiRH]$_n$H which delivers a metallocene-silyl derivative [M]-[SiRH]$_n$H. Si-Si coupling with RSiH$_3$ produces H[SiRH]$_{n+1}$H and releases the active catalyst. For an extensive review on transition metal-catalyzed dehydrocoupling of silanes see Reference 15.

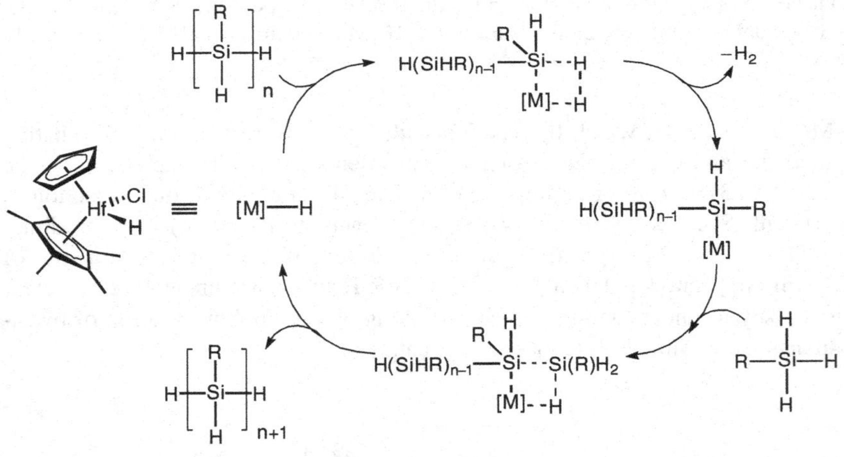

**SCHEME 18.4** Mechanism suggested by Don Tilley et al. for the dehydrocoupling of RSiH$_3$ using CpCp*HfHCl (Cp = $\eta^5$-$C_5H_5$; Cp* = $\eta^5$-$C_5Me_5$;) as a catalyst. The catalytic cycle involves two σ-bond metathesis reactions that pass through four-center transition states.[14]

Likewise, dehydrogenative polymerization of hydridodisilanes using dialkyl titanocene or zirconocene yields polysilanes.[16] Hengge and Weinberger[17] postulated a silylene mechanism, in which a metallocene silanediyl complex $Cp_2M=SiMe_2$ forms by a β*-bond elimination, that is, Si-Si bond cleavage and elimination of R-SiMe$_3$ from $Cp_2M(R)SiMe_2SiMe_3$. The silylene is released from $Cp_2M=SiMe_2$ and inserts either into a Si-H or Si-Si bond of a di- or oligosilane. In contrast to dehydrocoupling of monosilanes, branched oligomers and polymers are obtained.

An unusual dehydrocoupling was reported by Kimata et al.,[18] who performed Si-Si coupling from Si-H units electrochemically. Di- and trihydridomonosilanes were electrolyzed under constant-current conditions in $Bu_4NBF_4$/DME (DME = 1,2-dimethoxyethane) to yield the low-molecular weight oligomers $[SiR^1R^2]_n$ ($R^1R^2$ = (Me)(Ph), (H)(Ph), (H)(Hex); $n$ = 3–5).

Catalytic (Lewis base-induced) disproportionation of tetrachlorodimethyldisilane or dichlorotetramethylsilane is another efficient process to generate oligomeric silanes without formation of solid by-products.[19,20] The inexpensive starting compounds are obtained as a by-product during the synthesis of chlorosilanes in the Müller-Rochow process.[21] Depending on the molecular structure, that is, the number of chlorine atoms per silicon atom, mixtures of linear or branched polymers form.

$$ClMe_2Si\text{—}SiMe_2Cl \xrightarrow{\text{[Base]}} Me_2SiCl_2 + ClMe_2Si\overset{\overset{\displaystyle Me}{|}}{\underset{\underset{\displaystyle Me}{|}}{Si}}\text{—}SiMe_2Cl + Cl_2MeSi\overset{\overset{\displaystyle Me\ \ Me}{|\ \ \ |}}{\underset{\underset{\displaystyle Me\ \ Me}{|\ \ \ |}}{Si\text{—}Si}}\text{—}SiMeCl_2$$

+ further linear oligomers

$$(18.5)$$

For example, disproportionation of $Cl_2MeSi\text{-}SiMeCl_2$ delivered at least five structurally different oligomers, which could be assigned unequivocally by $^{29}Si$ nuclear magnetic resonance (NMR) spectroscopy. Silicon nuclei with $SiCCl_2Si$ (terminating groups), $SiCClSi_2$ ("linear" units), and $SiCSi_3$ (junctions) were detected. Against this, disproportionation of $ClMe_2Si\text{-}SiMe_2Cl$ exclusively delivered linear polysilanes.

## B. POLYCARBOSILANES

Polycarbosilanes have received much attention as precursors for high modulus silicon carbide and silicon carbide/carbon composites. In contrast to polysilanes, polycarbosilanes have polymer backbones, which are composed of silicon and carbon atoms. Alternating Si-C arrangements and polymers with multiple carbon sequences (Si-$C_n$) are known. Such multiple carbon sequences are typically alkyl $(CH_2)_n$, alkenyl CH=CH, alkynyl C≡C, or aryl $C_6H_4$ groups. Various types of reactions are known that release polycarbosilanes. The most prominent types are Grignard reactions or other metathesis reactions involving carbo anions and silicon halides or pseudohalides. Moreover, hydrosilylation reactions, for example, addition of Si-H units to olefins, dehydrocoupling involving C-H and N-H units, as well as ring opening polymerization (ROP), are well established procedures for the synthesis of polycarbosilanes.

Technically the most important access to polycarbosilanes is the Yajima process,[22] in which polydimethylsilane $[SiMe_2]_n$ (compare previous section) is

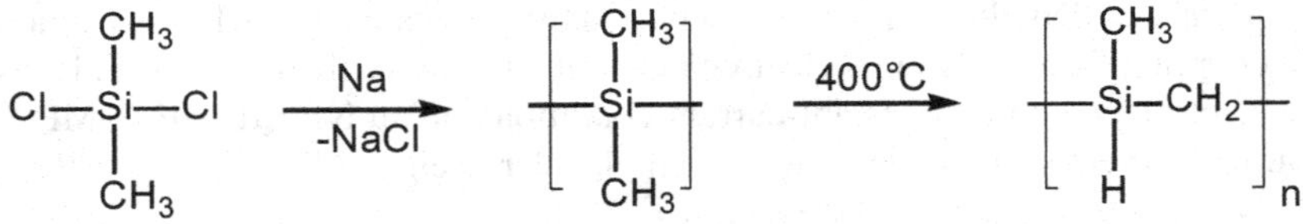

**SCHEME 18.5**  Synthesis of poly(dimethylsilylene) and thermal (Kumada-) rearrangement to poly(methylsilylene-methylene).[22]

thermally rearranged into poly(methylsilylene-methylene) $[HSi(CH_3)-CH_2]_n$. In contrast to $[SiMe_2]_n$, which is intractable and insoluble, $[HSi(CH_3)-CH_2]_n$ is a processable precursor. It can be cast or spun. Silicon carbide fibers obtained from this precursor are available under the trade name Nicalon™ or Hi-Nicalon™, depending on the curing procedure applied prior to thermolysis.

The rearrangement in the Yajima process includes a methylene migration from a pendant methyl group in $[SiMe_2]_n$ into the polymer backbone. The so-called Kumada reaction takes place at 400°C. It is initiated by homolytic Si-Si bond cleavage (step 1, Scheme 18.5) and formation of silicon radicals. Hydrogen radical migration (step 2) of a pendant methyl group to the silyl radical releases a methylene radical, which subsequently inserts into a Si-Si bond (step 3) to generate another silicon radical. Hydrogen radical migration releases the constituting polycarbosilane $[HSi(CH_3)-CH_2]_n$ group (step 4).

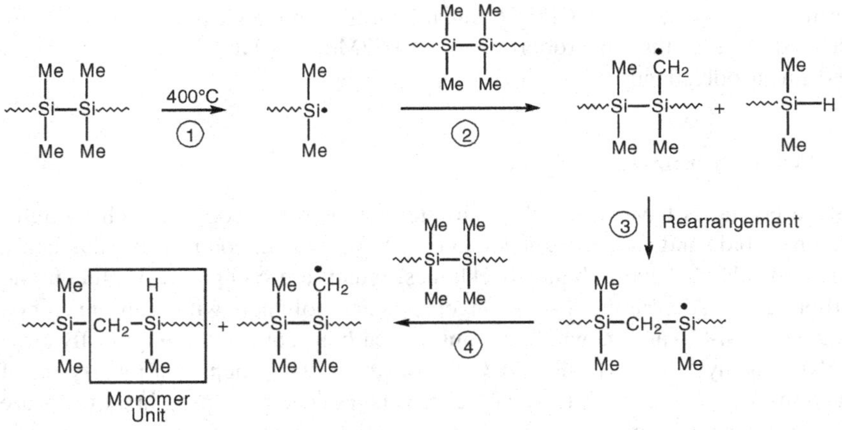

**SCHEME 18.6**  Mechanism of the Kumada-rearrangement.

As was previously mentioned, alternative synthetic routes to polycarbosilanes are metathesis reactions of carbo anions and silicon halides. For example, Wurtz-like condensation of dichlorosilanes and dibromoethane with sodium metal yields $[SiH_2CH_2]_n$.[23]

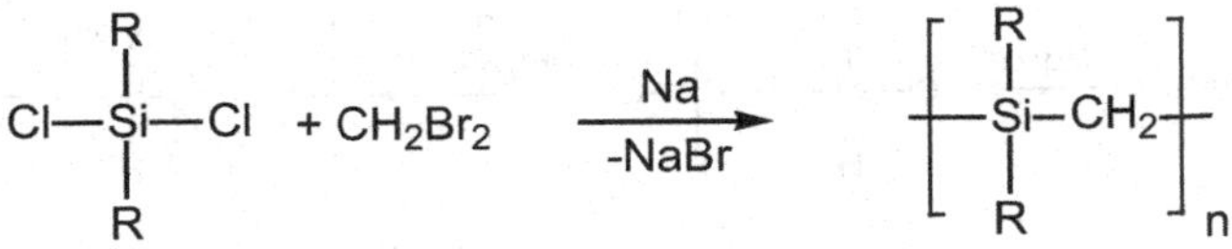

$$\underset{R}{\overset{R}{Cl-Si-Cl}} + CH_2Br_2 \xrightarrow[-NaBr]{Na} \left[ \underset{R}{\overset{R}{Si-CH_2}} \right]_n$$

$$(18.6)$$

A disadvantage of this type of dehalocoupling is low selectivity. Among preferred Si-C units, formation of undesired side products such as $[SiR_2]_n$ and $[CH_2]_n$ is observed. Nevertheless, derivatives with R = Ph ($C_6H_5$) were further reacted with HCl/AlCl$_3$ and LiAlH$_4$ to release a polycarbosilane $[SiH_2CH_2]_n$, also referred to as polysilaethylene because of its structural analogy with ethylene. Thermolysis of this precursor was accompanied by considerable depolymerization, and as a consequence, ceramic yields (~40%) were low. Polysilaethylene obtained by ROP of $[SiH_2CH_2]_n$, in contrast, delivered ceramics with a yield of 75%.[24] The reason for this difference remains unclear.

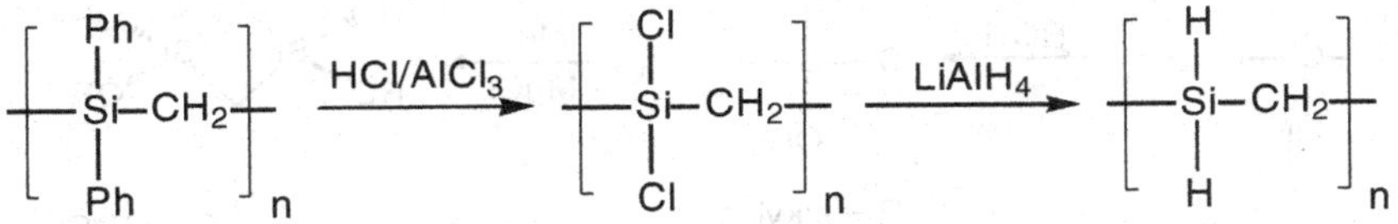

$$\left[ \underset{Ph}{\overset{Ph}{Si-CH_2}} \right]_n \xrightarrow{HCl/AlCl_3} \left[ \underset{Cl}{\overset{Cl}{Si-CH_2}} \right]_n \xrightarrow{LiAlH_4} \left[ \underset{H}{\overset{H}{Si-CH_2}} \right]_n$$

**SCHEME 18.7**  Synthesis of poly(silylenemethylene) from poly(diphenylsilylenemethylene)

The most thoroughly investigated type of dehalogenative coupling reactions of halosilanes and halohydrocarbons are Grignard reactions, that is, coupling of C-MgBr with Si-Cl units. Analogous to the above-described coupling of dichlorosilanes and dibromoethane, Hemida et al.[25] reacted mixtures of dichloromethane and methyldichlorosilane with Mg/Zn. The obtained ClH(H$_3$C)Si-CH$_2$-Si(CH$_3$)HCl was coupled with ($C_6H_5$)(CH$_3$)SiCl$_2$ using sodium metal. Subsequent treatment with HCl/AlCl$_3$ and LiAlH$_4$ delivered a polycarbosilane, which yielded nearly stoichiometric SiC ceramics.

Polycarbosilanes with alternating Si-C units can be obtained from chloromethyl-chlorosilanes, for example, Cl$_3$Si-CH$_2$Cl, and magnesium in diethylether solution. Under such conditions, Cl$_3$Si-CH$_2$Cl releases Cl$_3$Si-CH$_2$MgCl that immediately reacts intermolecularly with Si-Cl moieties. Following treatment of as-synthesized poly(dichlorosilylene-methylene), $[SiCl_2CH_2]_n$ with LiAlH$_4$ delivers polysilaethylene. Thermolysis of $[SiH_2CH_2]_n$ obtained on this reaction pathway produces near-stoichiometric silicon carbide.

A remarkable drawback in the synthesis of polysilaethylene using the Grignard approach in diethylether is the very long reaction time. As described by Wu and Interrante[26] this disadvantage can be overcome when changing the solvent toward more polar tetrahydrofuran. Unfortunately, selectivity decreases in this case and product mixtures are obtained that result from mono, double, or triple

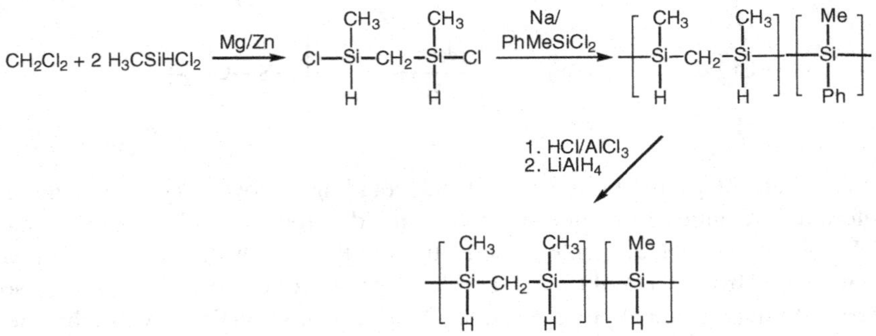

**SCHEME 18.8** Synthesis of a polycarbosilane to nearly stoichiometric SiC ceramics. The initial step is a Grignard coupling of dichloromethane and methyldichlorosilane.[25]

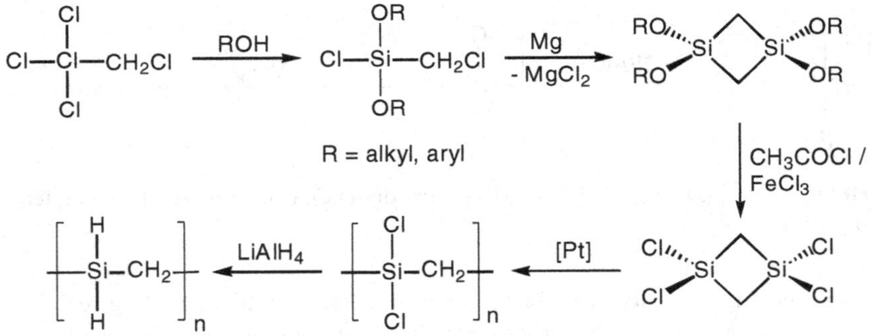

**SCHEME 18.9** Synthesis of polysilaethylene by Pt-catalyzed ring opening polymerization (ROP) of 1,3-disila-1,1,3,3-tetrachloro cyclobutane.[26]

metathesis. In contrast, single, and thus selective, metathesis reactions are observed if two of the three silicon-bonded chlorine atoms are initially replaced in alcoholysis reactions with alkyloxy or aryloxy groups:

Grignard reactions of $ClSi(OR)_2CH_2Cl$ yield 1,3-disilatetraalkyloxy- or aryloxy cyclobutane, which can easily be transformed into the respective tetrachloro-modified species by treatment with a mixture of $CH_3COCl$ and $FeCl_3$. Subsequent ROP and treatment with $LiAlH_4$ releases $[SiH_2CH_2]_n$ in a higher overall yield than described above. $[SiH_2CH_2]_n$ has a melting point of about 25°C and is readily soluble in common organic solvents. It gives near-stoichiometric silicon carbide upon thermolysis.[26]

Hydrosilylation of unsaturated hydrocarbons is the most essential procedure for creating Si-C bonds. It is an addition of Si-H units across π-systems such as C=C double or C≡C triple bonds and may occur on two different reaction modes. Radical hydrosilylation is initiated with ultraviolet (UV) light, γ-irradiation, or radical starters such as azobis(isobutyronitrile) (AIBN) or peroxides. However, there are some drawbacks which have to be taken into account, such as competition of radical-induced olefin polymerization with hydrosilylation. Such problems can be overridden by using suitable transition metal catalysts. The first work in this field was published by Speier, who explored a mixture of $H_2PtCl_6$/$^iPrOH$ to be catalytically efficient.[27] Remarkably, the catalyst has a high group tolerance and allows the introduction of functionalized silanes such as chlorosilanes, alkoxysilanes, or aminosilanes to olefins or acetylenes.

If a vinyl group is directly attached to the silicon atom bearing a hydrogen atom, hydrosilylation gives polycarbosilanes with $Si-CH_2-CH_2$ backbones. The procedure itself is straightforward: vinylsilane and catalyst are mixed at room temperature and subsequently heated to 200°C. Since solvents are not required, this method can be successfully applied for the preparation of bulk materials. Usually the addition occurs regioselectively, with silicon adding "end-on" to the terminal carbon atom and stereoselectively with retention at silicon.

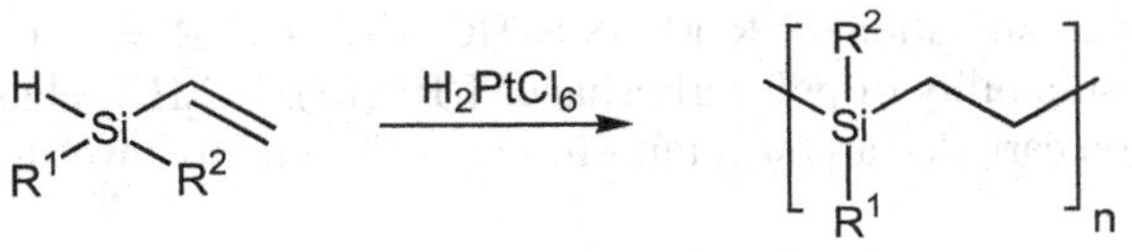

$$R^1, R^2 = H, Cl, alkyl, aryl, NR_2 \tag{18.7}$$

The mechanism of platinum-catalyzed hydrosilylation was studied in detail by Chalk and Harrod (Scheme 18.10).[28] The catalytically active species $[PtCl_4]^{2-}$ is obtained by a reduction of $[PtCl_6]^{2-}$ with isopropanol. One of the Cl ligands is then replaced with the olefin (A), whereby a 16 VE (VE = valence electron) $\eta^2$-olefin complex forms. Subsequent oxidative addition of external

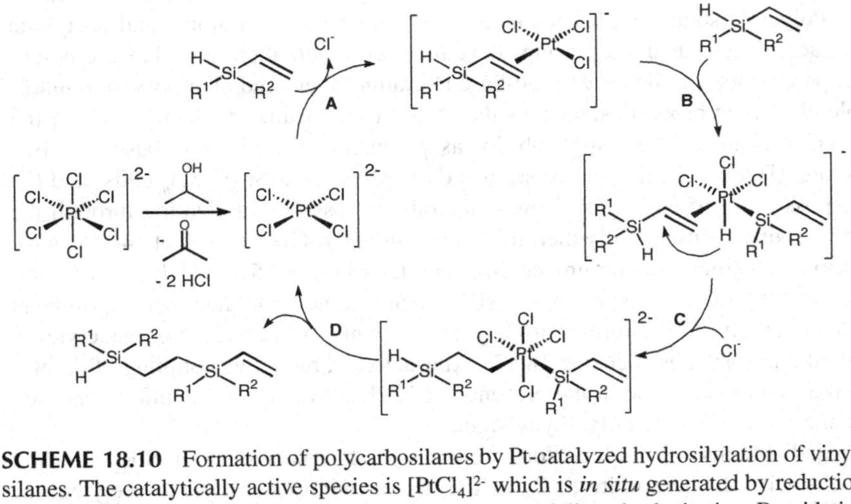

**SCHEME 18.10** Formation of polycarbosilanes by Pt-catalyzed hydrosilylation of vinylsilanes. The catalytically active species is $[PtCl_4]^{2-}$ which is *in situ* generated by reduction of $H_2PtCl_6$ with $^iPrOH$. Characteristic reaction steps are: A ligand substitution, B oxidative addition of Si-H, C H-migration, and D reductive elimination of the carbosilane.[28]

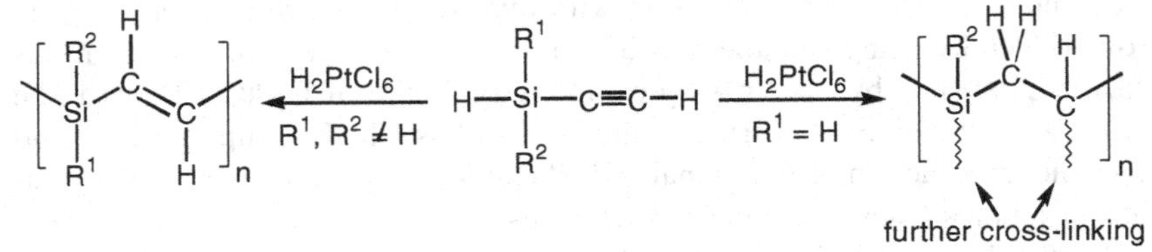

further cross-linking

**SCHEME 18.11**  Synthesis of polycarbosilanes by hydrosilylation of alkynylsilanes.[30,31]

$R^1R^1SiH(CH=CH_2)$ (B) yields an 18 VE species, which rearranges from $\pi$ to $\sigma$ complex by hydrogen migration from platinum to the coordinated olefin and Pt-C bond formation (C). The vacant coordination position that forms is occupied by an external Cl ligand. Reductive elimination of the carbosilane and regeneration of $[PtCl_4]^{2-}$ complete the cycle.

Synthesis by hydrosilylation of $(R^1)(R^2)SiH(HC=CH_2)$ ($R^1$, $R^2$ = H, Cl, alkyl, aryl, $NR_2$) and the suitability of polycarbosilanes $[(R^1)(R^2)Si-C_2H_4]_n$ obtained as precursors for silicon carbide/carbon composites were thoroughly investigated by Corriu et al.[29]

Similarly, $[PtCl_4]^{2-}$ catalyzed hydrosilylation of diorganoethinylsilanes $(R^1)(R^2)SiH(C\equiv CH)$ delivers linear poly(silylenevinylene) with an Si–C=C backbone.[30] The olefin units are reactive sites that allow for further modification or cross-linking, for example, by hydrosilylation. Facile cross-linking is indeed observed, if $R^1$ or $R^2$ are hydrogen atoms. In this case, double hydrosilylation of the C≡C units occurs and the carbon triple bonds in the starting compounds (the C=C units in the initially obtained polycarbosilanes) are transformed into aliphatic building blocks.[31] The increased cross-linking density is directly reflected in their physicochemical properties: polycarbosilanes received from $(R)SiH_2(C\equiv CH)$ are highly cross-linked, glass-like materials with high glass transition temperatures.

Polycarbosilanes with backbones consisting of silicon atoms and acetylene or diacetylene (butadiyne) groups have received much attention. They are potential precursors for silicon carbide-based ceramic fibers and also possess remarkable electronic properties, such as the ability to delocalize $\pi$-electrons along the polymer chain.[32] Their applicability as precursors for fibers is based on two factors: (i) ideal rheological properties due to the linear $Si-(C\equiv C)_n$ units, and (ii) the possibility of undergoing low-temperature cross-polymerization through the C≡C groups. Consequently thermolysis of $[Si(CH_3)_2-C\equiv C-C\equiv C]_n$ at 1400°C gave carbon-rich silicon-containing ceramics with a yield of 85%.[33-35] The most common procedure for the synthesis of silylene acetylene[36] and diacetylene polymers is the reaction of dilithium or di-Grignard reagents[37] of acetylene or diacetylenes with dichlorosilanes (Scheme 18.12). Alternatively, reductive coupling of dichlorosilanes and acetylene in the presence of Na/K alloy in THF solution was used for the preparation of polyalkynylsilanes.[38]

Ijadi-Maghsoodi and Barton[39] reported that replacement of $SiR_2$ building blocks with disilane units $Si_2R_4$ in the above mentioned silylene acetylene or diacetylene polymers reduces the amount of graphitic carbon in the resulting

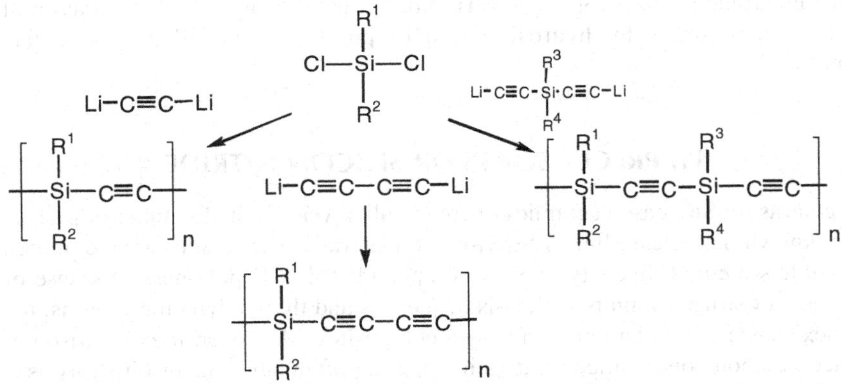

**SCHEME 18.12** Synthesis of silylene acetylene ("polyalkynylsilanes", left and right) or diacetylene polymers ("polybutadiynylsilanes", middle) by metathesis of dichlorosilanes and dilithium reagents of acetylene, silylenediacetylene and butadiyne, respectively.[36–41]

ceramic materials. For example, the C:SiC ratio in ceramics obtained from $[SiMe_2C\equiv C-C\equiv C]_n$ was 7:4 (wt%), whereas this ratio in $[(SiMe_2)_2C\equiv C-C\equiv C]_n$-derived ceramics was 1:1. Synthesis of $[(SiMe_2)_2C\equiv C-C\equiv C]_n$ was performed using a treatment of $ClMe_2SiSiMe_2Cl$ with $LiC\equiv C-C\equiv CLi$.[42] Likewise, copolymers $[(SiR_2)_2C\equiv C(SiR_2)_2C\equiv C]_n$ were obtained from $ClR_2SiSiR_2Cl$ and $LiC\equiv C(SiR_2)_2C\equiv CLi$.[43] A straightforward procedure for the preparation of such copolymeres is an anionic ROP of 1,2,5,6-tetrasilacycloocta-3,7-diynes, developed by Ishikawa et al.[44]

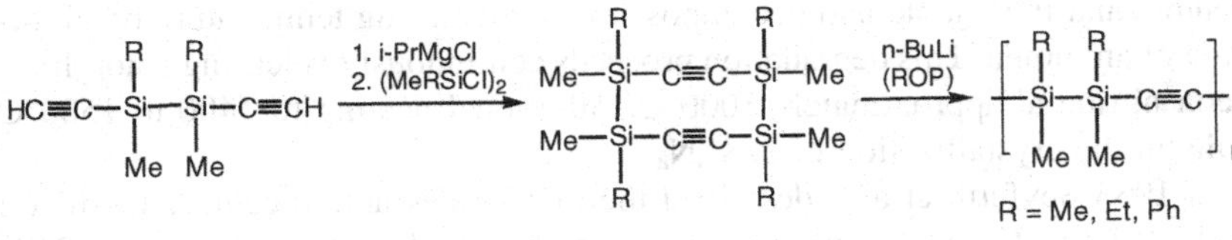

**SCHEME 18.13** Synthesis of disilylene acetylene polymers by anionic ring-opening polymerization (ROP) of 1,2,5,6-tetrasilacycloocta-3,7-diynes.[44]

This procedure starts with 1,2-diethynyldisilanes, which are deprotonated using alkyl magnesium chlorides. Subsequent treatment of the di-Grignard reagents with 1,2-dichlorodisilanes gives 1,2,5,6-tetrasilacycloocta-3,7-diynes, which can by recrystallized from benzene. ROP (Scheme 18.13, step 2) was carried out with 1.7 mol% of n-BuLi as a catalyst in THF solution.

In addition, alkynyl groups not only inhibit thermally inducted depolymerization, they can also be functionalized by, for example, hydroboration, hydrosilylation, or by attaching transition metal complex fragments. Accordingly, $C\equiv C$ units provide an access to multicomponent ceramics. For example, Corriu et al.[45] reported that $[SiR_2C\equiv C-C\equiv C]_n$ are precursors with interpenetrating networks to silicon carbide/metal carbide (M = Ti, Nb, Ta) nano composites. Preparation of branched polymers by hydrosilylation of poly(butadiynylsilanes) was also reported.[46]

## IV. PRECURSORS FOR SILICON NITRIDE

In contrast to SiC-based ceramics, there is only a very limited number of precursors known that release binary $Si_3N_4$-based materials. The reasons are the limited possible structural diversity of Si-N compounds (Si-N-H polymers), because of the lack of structural units with N-N sequences and thermodynamics, that is, the nonexistence of silicon nitride/nitrogen composites. Moreover, in comparison to silicon-carbon bonds, silicon-nitrogen bonds are often labile against hydrolysis.[47]

The first mention of a silicon/nitrogen polymer appeared in 1885 when Schutzenberger and Colson reported ammonolysis[48] of $SiCl_4$.[49] Polymerization was accompanied with precipitation of ammonium chloride, which was removed subsequently by filtration. An improved synthesis and a more detailed characterization were performed by Glemser and Naumann 75 years later.[50]

$$\underset{\underset{Cl}{|}}{\overset{\overset{Cl}{|}}{Cl-Si-Cl}} \quad \xrightarrow[- NH_4Cl]{NH_3} \quad [Si(NH)_2]_n$$

$$(18.8)$$

$[Si(NH)_2]_n$, also referred to as silicon diimide, is a colorless solid. It is insoluble and intractable and decomposes with increasing temperature by elimination of ammonia. This degradation proceeds continuously, releasing amorphous silicon nitride at approximately 1000°C. Additional heating to 1400 to 1500°C results in the crystallization of $\alpha$-$Si_3N_4$.

In 1983 Seyferth et al.[51] described the first processable precursor to silicon nitride. $[SiH_2NH]_n$, which was first described by Stock and Somieski in 1921,[52] was obtained from dichlorosilane, $H_2SiCl_2$, and ammonia as a mixture of linear oligomers and cyclomers (omitted in Scheme 18.14). $[SiH_2NH]_n$ ages rapidly by loosing hydrogen, thereby gradually increasing its viscosity.

To obtain high ceramic yields, further cross-linking of the as-obtained polymers by transition metal-catalyzed dehydrocoupling (see below) was required. Thermolysis to 1050°C of the cross-linked precursors released $\alpha$-$Si_3N_4$/$\beta$-$Si_3N_4$/$\alpha$-Si with a yield of 70%. "Free" silicon, however, worsens high-temperature properties. It is therefore desirable to remove it by performing thermolysis in an ammonia atmosphere, thereby transforming silicon into $Si_3N_4$. Unfortunately this procedure predominantly influences surface areas.

A critical issue is the use of $H_2SiCl_2$ in the technical scale. It is a highly flammable gas, which can disproportionate with formation of $SiH_4$ and $SiCl_4$. A manipulation, which allows for safer handling, was developed at Tonen Company.

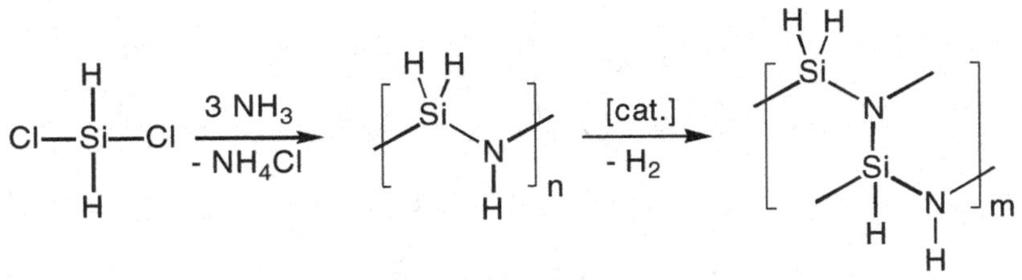

$$\text{Cl}-\underset{\underset{\text{H}}{|}}{\overset{\overset{\text{H}}{|}}{\text{Si}}}-\text{Cl} \xrightarrow[-\text{NH}_4\text{Cl}]{3\ \text{NH}_3} \left[\underset{\underset{\text{H}}{|}}{\overset{\overset{\text{H}\ \ \text{H}}{\diagdown\ /}}{\text{Si}}}-\text{N}\right]_n \xrightarrow[-\text{H}_2]{[\text{cat.}]} \left[\cdots\right]_m$$

**SCHEME 18.14** Synthesis of processable precursors to phase-pure silicon nitride by ammonolysis of dichlorosilane and subsequent cross-linking ([cat.] = catalyst).[51]

Prior to ammonolysis, $H_2SiCl_2$ is modified by reaction with pyridine, which results in the formation of $H_2SiCl_2 \cdot (NC_5H_5)_2$.[53]

Precursors that release stoichiometric silicon nitride in an inert gas atmosphere and which therefore avoid segregation of silicon metal can be obtained by coammonolysis of dichlorosilane and trichlorosilane.[54,55] Such polymers exhibit nitrogen:silicon ratios greater than 4:3. Compared to $[SiH_2NH]_n$ copolymers $[SiH_2NH]_m[SiH(NH)_{1.5}]_n$ are more highly cross-linked, but in contrast to $[Si(NH)_2]_n$, they are processable.

Attempts to synthesize phase-pure silicon nitride by thermolysis of ternary Si-C-N-H polymers have also been published. Narsavage et al.[56] reported on the synthesis and thermal conversion of tetrakis(ethylamino)silane, $Si(NHC_2H_5)_4$, in which silicon is exclusively bonded to nitrogen. The authors expected silicon nitride formation due to a volatilization of the nitrogen-bonded ethyl groups during thermolysis. However, elimination of the ethyl groups did not occur quantitatively, and consequently carbon was not removed completely during the heat treatment.

# V. PRECURSORS FOR SILICON NITRIDE/SILICON CARBIDE COMPOSITES

The thermodynamics of the above-elucidated SiC/C and $Si_3N_4$/Si composites are determined by the decomposition of silicon carbide and silicon nitride, respectively, into their elements. The chemistry of ternary Si-C-N composites is more complex. If producing Si-C-N ceramics for applications at elevated temperature, reactions between carbon and silicon nitride have to be considered. Figure 18.2, which exhibits a ternary phase diagram valid up to 1484°C (1 bar $N_2$) displays the situation. The only stable crystalline phases under these conditions are silicon carbide and silicon nitride. Ceramics with compositions in the three-phase field SiC/$Si_3N_4$/N are unknown (this is a consequence of the thermal instability of C-N bonds). Although composites within the three-phase field SiC/$Si_3N_4$/Si are thermodynamically stable even above 1500°C, such materials are rare. The reasons are difficulties in the synthesis of the required precursors and silicon melting above 1414°C. The latter aspect is of relevance, since liquid silicon dramatically worsens the mechanical properties of the derived ceramics.

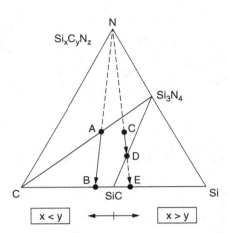

**FIGURE 18.2** Si-C-N phase diagram (T < 1484°C; 1 bar $N_2$). A: Ceramics obtained from VT50 (Hoechst AG, Germany); C: Ceramics obtained from NCP200 (Nichimen Corporation, Japan). Depending on the Si:C ratio, decomposition proceeds in either one or two steps.[60]

In general, compositions of Si-C-N ceramics $Si_xC_yN_z$ obtained from polymeric precursors are located either on the tie line $Si_3N_4$-C if obtained from nitrogen-rich polymers[57] or in the three-phase field $SiC/Si_3N_4/C$. At temperatures greater than 1484°C, such materials decompose due to a carbothermal reaction of Si-N units with "free" carbon. Accordingly, SiC forms and molecular nitrogen is released. Depending on the concentration of silicon ($x$) and carbon ($y$), SiC/C ($x < y$) or $SiC/Si_3N_4$ ($x > y$) will form. The latter decomposes into SiC/Si and $N_2$ if temperatures exceed 1841°C. These aspects are illustrated for two ceramics **A** and **C** obtained by thermolysis of commercial polymers VT50[58] and NCP200.[59] **A**, which contains more carbon than silicon, decomposes in a one-step reaction with formation of SiC/C (**B**), whereas decomposition of **C** proceeds in two steps via $SiC/Si_3N_4$ (**D**) to SiC/Si (**E**).

The onset of carbothermal reaction of Si-N units and decomposition of $Si_3N_4$ into the elements depends on the nitrogen partial pressure. Increasing $N_2$ pressure results in increased reaction temperatures, and vice versa. The quantitative context is given in the partial pressure diagrams shown in Figure 18.3.[60]

There are several structurally different types or polymers that are suitable precursors for ternary Si-C-N ceramics. By far the most investigated precursors are polysilazanes of the general type $[Si(R^1)(R^2)N(R^3)]_n$ ($R^1$, $R^2$, $R^3$ = H, alkyl, aryl, alkenyl, etc.). In contrast to the limited number of starting compounds, $H_xSiCl_{(4-x)}$ ($x$ = 0–3) as the silicon source and $NH_3$ or $H_2N$-$NH_2$ as the nitrogen source for synthesis of polysilazanes as precursors for binary Si-N ceramics, the chemistry of polycarbosilazanes, that is, carbon-containing or modified polysilazanes, is very multifaceted. The attachment of various organic groups to the silicon atoms allows adjustment of their physicochemical properties, to control their thermolysis chemistry, and also to influence materials properties. The first

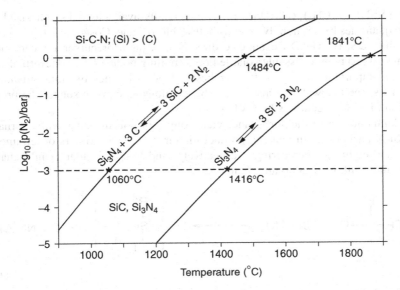

**FIGURE 18.3** Partial pressure diagram for Si-C-N composites ([Si] > [C]).[61] Presuming 1 bar $N_2$ atmosphere, carbothermal reaction of $Si_3N_4$ occurs at 1484°C. Increasing the nitrogen partial pressure to 10 bar shifts this temperature to 1700°C, whereas decreasing the nitrogen pressure to 1 mbar ($10^{-3}$ bar) should result in the degradation of Si-N units at approximately 1060°C. In 1 bar nitrogen, remaining $Si_3N_4$ dissociates into the elements at 1841°C. In $10^{-3}$ bar $N_2$, this value is shifted to 1416°C.

publications on polyorganosilazanes (not oligosilazanes, which are usually obtained as mixtures of cyclic dimers, trimers, and tetramers[62,63]) appeared in 1964 by Krüger and Rochow.[64] The method used most frequently for the synthesis of carbosilazanes and polycarbosilazanes is an ammonolysis of organosilicon chlorides. Reaction of monochlorosilanes $R_3SiCl$ delivers monomeric species.

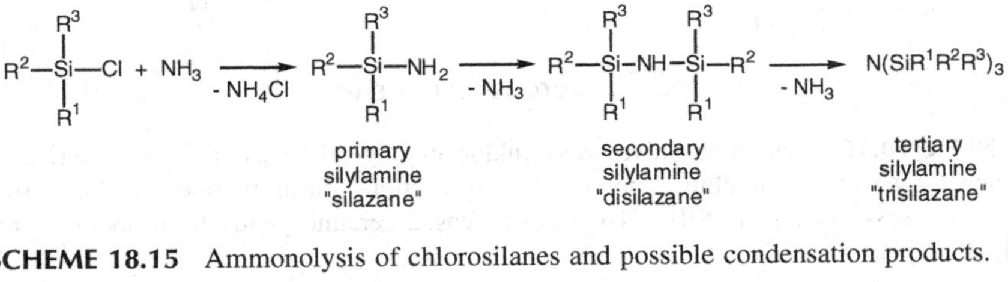

$$R^2{-}\underset{\underset{R^1}{|}}{\overset{\overset{R^3}{|}}{Si}}{-}Cl + NH_3 \xrightarrow[-NH_4Cl]{} R^2{-}\underset{\underset{R^1}{|}}{\overset{\overset{R^3}{|}}{Si}}{-}NH_2 \xrightarrow[-NH_3]{} R^2{-}\underset{\underset{R^1}{|}}{\overset{\overset{R^3}{|}}{Si}}{-}NH{-}\underset{\underset{R^1}{|}}{\overset{\overset{R^3}{|}}{Si}}{-}R^2 \xrightarrow[-NH_3]{} N(SiR^1R^2R^3)_3$$

|  | primary silylamine "silazane" | secondary silylamine "disilazane" | tertiary silylamine "trisilazane" |

**SCHEME 18.15** Ammonolysis of chlorosilanes and possible condensation products.

"Small" silicon-bonded substituents such as hydrogen or methyl result in disilazanes, $R_3Si$-NH-$SiR_3$. At higher temperatures they undergo transamination and condensation, thereby forming tertiary silylamines bearing $Si_3N$ motifs. Sterically demanding substituents R, such as isopropyl, isobutyl, and mesityl, inhibit this condensation. Consequently, primary silylamines of the type $R_3SiNH_2$ are obtained.[65,66] An industrially important monomeric silazane is hexamethyldisilazane,

$(H_3C)_3Si-NH-Si(CH_3)_3$ (HMDS). It is commercially available and synthesized in large quantities by ammonolysis of trimethylchlorosilane. Costs per kilogram are less than US$10. HMDS cannot be directly used as a precursor for ceramics because it is a nonprocessable volatile liquid (boiling point 125°C). Nevertheless, it is an important source for the synthesis of polysilazanes by transamination reactions (see below)[67] and also a valuable single-source precursor for the preparation of Si-C-N coatings by CVD processes.[68]

Ammonolysis of dichlorosilanes yields oligo- or polysilazanes with alternating Si-N backbones. In general, product mixtures of cyclic trimers or tetramers (six- and eight-membered rings, respectively) and low molecular weight chain molecules with M < 2000 are obtained.

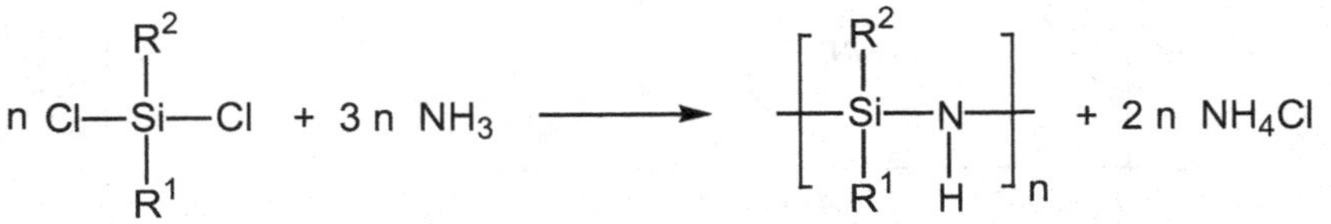

$$n \; Cl-\underset{\underset{R^1}{|}}{\overset{\overset{R^2}{|}}{Si}}-Cl \;+\; 3\,n \; NH_3 \;\longrightarrow\; \left[\begin{array}{c} R^2 \\ | \\ -Si-N- \\ | \quad | \\ R^1 \;\; H \end{array}\right]_n \;+\; 2\,n \; NH_4Cl$$

$$(18.9)$$

Because of the volatilization (small rings) or depolymerization followed by evaporation of low molecular weight species (chain molecules), oligo- or polysilazanes very often have unsatisfactory ceramic yields. Both volatilization and depolymerization can be efficiently avoided by cross-linking the polymers prior to thermolysis. Such cross-linking can be performed using different approaches. For example, Seyferth and Wiseman[69] published a method for increasing ceramic yields of silazanes bearing Si-H and N-H units:

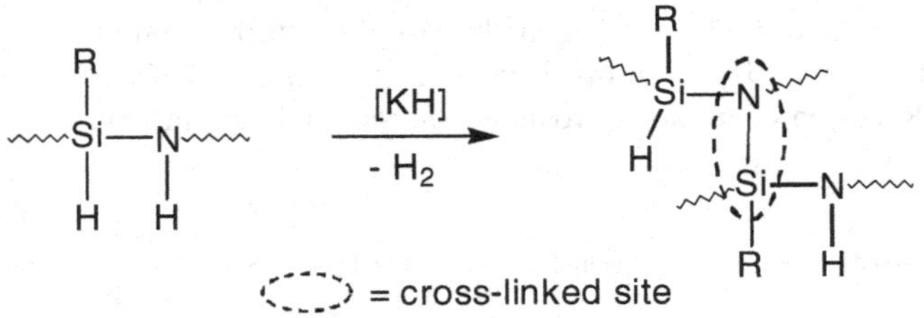

$$\begin{array}{c}\text{---} \end{array} = \text{cross-linked site}$$

**SCHEME 18.16** Dehydrogenative cross-linking of oligosilazanes. This step increases molecular weight and inhibits depolymerization reactions during thermolysis. Dehydrogenative cross-linking of $[SiH(CH_3)-NH]_n$ increased ceramic yields from about 30 to 80%.[69]

Seyferth and Wiseman reacted as-obtained silazanes with catalytic amounts of potassium hydride, KH. They observed dehydrocoupling of Si-H with N-H units and proposed a mechanism involving silylene-imine Si=N motifs as key species which rapidly add Si-H intermolecularly in hydrosilylation-type reactions. There has been, until now, no experimental proof for the proposed mechanism. Alternatively, a mechanism excluding silylene-imine formation was suggested for the n-butyl lithium-catalyzed (n-BuLi) dehydrocoupling of Si-H and N-H units in the synthesis of boron-modified silazanes.[70] It was stated that n-BuLi deprotonates N-H units. The resulting highly nucleophilic amides then replace silicon-bonded hydrides, thereby forming new Si-N bonds (the cross-linking units). Hydrides, which are released, again deprotonate N-H units, thus closing the catalytic cycle. For details consider Scheme 18.30.

Depolymerization reactions can also be avoided when attaching reactive groups to the polymer backbone, which increase the cross-linking density of the polymeric precursor at ambient temperature. In this regard, the attachment of vinyl groups $HC=CH_2$ was studied in much detail (see below).[71] According to Scheme 18.17, vinyl-substituted polysilazanes are best synthesized by ammonolysis of chlorovinylsilanes $(H_2C=CH)Si(R)Cl_2$ (R = H, $CH_3$).

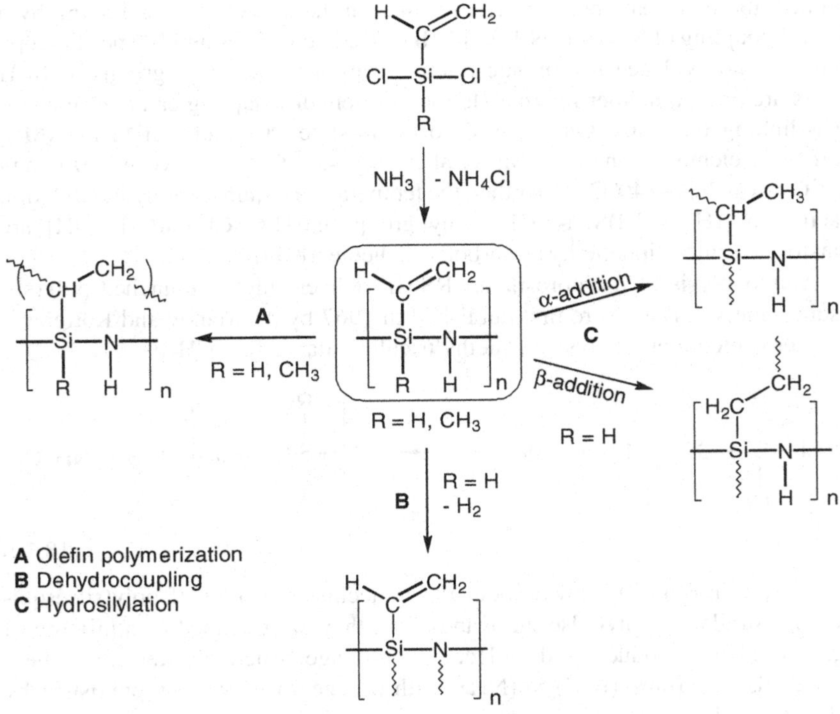

**A** Olefin polymerization
**B** Dehydrocoupling
**C** Hydrosilylation

**SCHEME 18.17**  Cross-linking chemistry of polyvinylsilazanes. *Trans*-amination is not considered.

The nature of the silicon-bonded group R and reaction conditions determine the cross-linking chemistry. If R is a substituent other than hydrogen, cross-linking can only occur via polymerization (**A**) of the vinyl groups. Hybridization of the carbon atoms of the olefin unit thereby changes from $sp^2$ to $sp^3$. The progress in this reaction can thus be easily monitored using $^{13}C$ NMR spectroscopy. The situation gets more complex if R = H. In this case, not only olefin polymerization but also dehydrocoupling (**B**) as well as hydrosilylation reactions (**C**) may occur. Though in the latter case β-addition is preferred, addition of silicon to the α-carbon atom may also occur, thus forming a mixture of two different regioisomers.

Chong Kwet Yive et al.[72] investigated in detail the thermal cross-linking of oligovinylsilazane $[(H_2C=CH)SiH-NH]_n$ (OVS) depending on the reaction conditions applied using solid state NMR and Fourier transform infrared (FTIR) spectroscopy. Heated in neat form, OVS leads to insoluble infusible solids within a few hours. The authors claimed that both NMR spectroscopy and elemental analysis point to the fact that hydrosilylation (**C**) is preferred. However, it does not proceed quantitatively and there is still a large number of C=C units detectable in the cross-linked precursors, even after heating to 110°C for 2 h. Hydrosilylation could be accelerated by addition of chloroplatinic acid (Speier's catalyst). In contrast, thermal treatment with KH resulted in the expected cross-linking by a dehydrocoupling of Si-H and N-H units (**B**). From both NMR and IR spectroscopy there was no evidence for anionic polymerization of the vinyl groups. If Si-H motifs are absent, neither hydrosilylation nor dehydrocoupling can contribute to cross-linking reactions. On the basis of solid-state $^{13}C$- and $^{29}Si$-MAS NMR, FTIR, and elemental analysis, Bill et al.[73] suggested that at between 250°C and 350°C—which is ~140°C higher than the temperature required for hydrosilylation reactions of $[(H_2C=CH)SiH-NH]_n$—vinyl groups in $[(H_2C=CH)Si(CH_3)-NH]_n$ are transformed into aliphatic hydrocarbons (Scheme 18.17A).

Ammonolysis of trichlorosilanes $RSiCl_3$ delivers highly branched poly(silsesquiazane)s.[74] They were first published in 1967 by Andrianov and Kotrelev,[75] who performed ammonolysis of methyltrichlorosilane (R = $CH_3$):

$$n\ Cl\!-\!\underset{\underset{Cl}{|}}{\overset{\overset{R}{|}}{Si}}\!-\!Cl\ +\ 4.5\,n\ NH_3 \longrightarrow \left[\underset{}{\overset{\overset{R}{|}}{Si}}(NH)_{1.5}\right]_n +\ 3\,n\ NH_4Cl$$

$$(18.10)$$

There is not much known about the molecular structure of poly(silsesquiazane)s. Similar to poly(silsesquioxane)s,[76–78] they are obtained as a mixture of highly branched products and molecules with cage structures. Recently a hexameric silsesquiazane $(H_3C)_6Si_6(NH)_9$ with a cage structure was published by Räke et al.[79] It was obtained from $NaNH_2$ and $(H_3C)SiCl_3$ in n-hexane solution at −78°C. The results of single-crystal x-ray diffraction displayed a cage structure in which two six-membered ring systems $[(H_3C)Si(NH)]_3$ with chair conformation are linked via the silicon centers each with an NH unit.

Due to their three-dimensional architecture, additional cross-linking of poly(silsesquiazane)s prior to thermolysis is not required. Nevertheless, there are two drawbacks that limit their applicability as preceramic polymers. First, the difficult workup, that is, removal of the couple product $NH_4Cl$ from the polymer is complicated and time intensive. Second, polysilsesquiazanes are difficult to process. Once the solvent is removed after synthesis, the polymers are in general insoluble, infusible, and intractable. Consequently it is difficult to obtain shapes other than simple bulk materials.

Copolymers are accessible by coammonolysis of chlorosilanes. This allows for specifically tuning chemical and physical chemical properties such as solubility, rheology, and (latent) reactivity. Tuning of these parameters using single source precursors is usually difficult. For example, coammonolysis of dichlorosilanes and trichlorosilanes releases copolymers $[Si(R^1)(R^2)-NH]_n[Si(R^3)(NH)_{1.5}]_m$. They can be processed like polysilazanes $[Si(R^1)(R^2)-NH]_n$, but have higher ceramic yields. It is also possible to replace ammonia with primary amines $RNH_2$ (R = Me, Et,...). For a detailed overview on such polymers see Reference 80.

From Figure 18.2 and Figure 18.3, it is evident that the overall composition of the Si-C-N ceramics determines their high-temperature chemistry. Almost all Si-C-N-H polymers published so far, deliver compositions on the tie line $Si_3N_4$-C or compositions within the three-phase field $Si_3N_4/SiC/C$. All these materials suffer from degradation by a carbothermal reaction of Si-N when temperatures exceed 1500°C in a $N_2$ atmosphere. Recently precursors for ceramics without "free" carbon were published, which do not undergo carbothermal reactions. The intention was to adjust the chemical composition of the ceramic materials already in the polymer and to design the precursors in a way that hydrogen is the only volatile thermolysis product. For a first investigation, a $SiC/Si_3N_4$ ceramic ($Si_4N_4C$) was considered. Based on the assumption that $[SiH(CH_3)NH]_n$ gives $Si_3N_4/SiC/3C$ and $[SiH_2NH]_n$ gives $Si_3N_4/Si$ (presuming only $H_2$ elimination occurs during thermolysis in both cases), a copolymer consisting of three monomer units of $SiH_2NH$ and one monomer unit of $SiH(CH_3)NH$ was synthesized: [81]

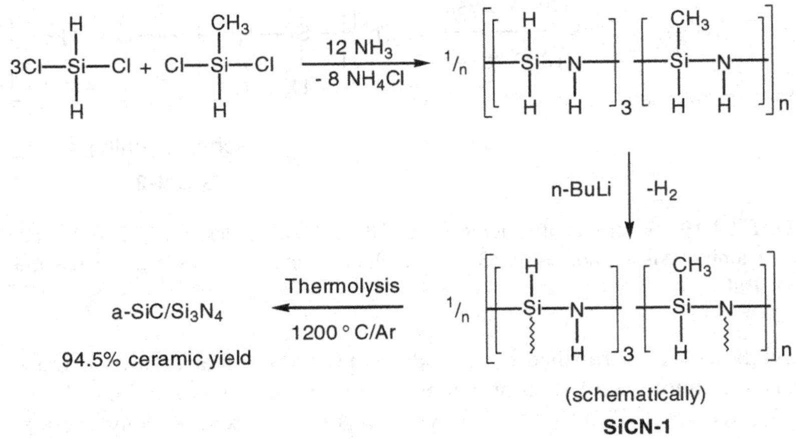

**SCHEME 18.18** Synthesis of $SiC/Si_3N_4$ ceramics without "free" carbon. **SiCN-1** is obtained by co-ammonolysis of dichlorosilane and methyldichlorosilane in a 3:1 molar ratio and subsequent cross-linking using n-BuLi. Thermolysis delivers a-$SiC/Si_3N_4$ in 94.5% yield.[81]

Synthesis was performed by coammonolysis of three equivalents of dichlorosilane and one equivalent of methyldichlorosilane. As-obtained $[(SiH_2NH)_3(SiH(CH_3)NH)]_n$ is a low-viscosity liquid that was further cross-linked in THF solution by catalytic amounts of n-BuLi. **SiCN-1** is a hard glass-like material that is insoluble. Thermogravimetric analysis (TGA) showed a two-step decomposition between 250 and 750°C, the weight loss was 5.5%. This value exactly corresponded to the hydrogen content of the polymer. Mass spectrometric investigations of the thermolysis gases indicated that predominately elimination of molecular hydrogen occurred. Ammonia and methane were released only in minor amounts (less than 1% compared to $H_2$). The x-ray amorphous ceramic residue is a bright gray material that has the desired overall composition $Si_4N_4C$.

Very recently, synthesis of a second precursor to a-$Si_4N_4C$ (Scheme 18.19) was published. It is structurally distinguished from the above-described precursor in that the methyl group is shifted from silicon to nitrogen.[82,83]

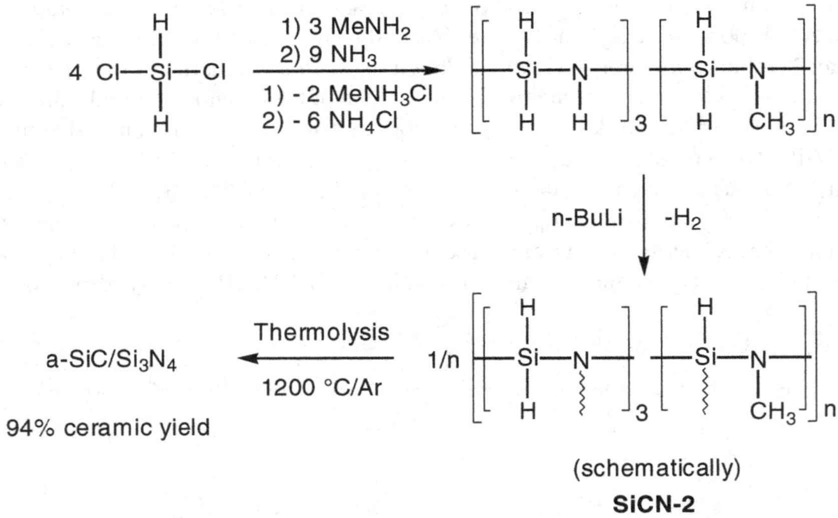

**SCHEME 18.19** Synthesis of stoichiometric SiC/$Si_3N_4$ ceramics. **SiCN-2** is obtained by stepwise aminolysis/ammonolysis (3:1) of dichlorosilane and subsequent cross-linking using n-BuLi.

Synthesis was performed by a modified procedure, that is, a two-step aminolysis/ammonolysis of dichlorosilane. Prior to thermolysis, which delivered a-$Si_4N_4C$ with a yield of 94%, the polymer was cross-linked by dehydrocoupling in the presence of catalytic amounts of n-BuLi. The ceramization progression of **SiCN-2** was similar to that of **SiCN-1**. A two-step decomposition was observed between 250 and 800°C, and hydrogen was the only gaseous product. The detected amounts of methylamine and methane were negligible. High-temperature TGA clearly points to the fact that degradation below 1500°C did not take place.

Very recently SiC-rich composites $(2SiC/Si_3N_4)$ without free carbon were obtained by thermolysis of a copolymer $[(SiH_2NH)_3(SiH(CH_2)_2SiHNH)]_n$.[82–84] The latter was obtained according to Scheme 18.18 by coammonolysis of dichlorosilane and 1,4-dichloro-1,4-disilabutane, $ClH_2Si(CH_2)_2SiH_2Cl$, in a 3:1 ratio. XRD of annealed ceramics showed that, due to the increasing amount of silicon carbide, the onset of crystallization could be shifted toward higher temperatures.[85]

Advanced synthetic procedures, including dehydrocoupling reactions of hydridosilanes and ammonia or amines, or redistribution reactions of hexamethyldisilazane (HMDS) with chlorosilanes avoid formation of solid couple products during polymerization. For example, ruthenium-catalyzed dehydrocoupling of silanes with ammonia or amines was published by Blum, Laine, and others.[86–88] They found that $Ru_3(CO)_{12}$ promotes such reactions efficiently.

$$R^1, R^2 = alkyl \tag{18.11}$$

Since the only by-product is hydrogen, the purification of the precursors can be performed by evaporating all volatile components from the reaction mixture. The catalyst used remains in the precursor.[89] An important advantage over ammonolysis reactions of chlorosilanes is the possibility of synthesizing highly cross-linked insoluble polymeric precursors, which are usually difficult to separate from solid by-products.

Liu and Harrod[90] reported on dehydrocoupling of ammonia and silanes catalyzed by dimethyltitanocene. Tertiary silanes, for example, $Ph_2SiMeH$, were transformed into disilazanes. Yet no polysilazanes could be obtained when reacting $PhSiH_3$ under similar conditions. Homodehydrocoupling (which proceeds with Si-Si linking; compare Scheme 18.3) effectively competed with amination reactions, indicating that the products obtained by ammonolysis of $PhSiH_3$ are polyaminosilanes $[SiPh(NH_2)]_n$ rather than polysilazanes $[SiPhH-NH]_n$.

Redistribution (*trans*-silylation) reactions of HMDS with dichlorosilanes is another pathway that delivers polysilazanes without formation of solid by-products. According to Mooser et al.,[67] such reactions can be performed without added catalyst.

$$\tag{18.12}$$

The synthetic procedure is simple: HMDS and chlorosilane are mixed at room temperature either in solution or in neat form using excess HMDS. The process takes advantage of the low boiling point of $Me_3SiCl$, which can be removed from the mixture continuously at ambient temperature, thus shifting the equilibrium of the reaction to the product side.

Previously it was shown that ROP of 1,3-disilacyclobutane is an efficient method for the synthesis of polycarbosilanes (Scheme 18.9). Likewise, ROP of 1,3-disiladiazetanes, cyclo-$(SiR_2\text{-}NR')_2$, delivers linear high-molecular weight polysilazanes. Synthesis of 1,3-disiladiazetanes from bis(amido)silanes $[R_2Si(NRLi)_2]$ and dichlorosilanes $[R_2SiCl_2]$ was reviewed in detail by Fink in 1966.[91]

Ring opening polymerization proceeds in the presence of suitable catalysts. Seyferth et al.[92] published research on organoalkali-catalyzed ROP of cyclo-$(SiMe_2\text{-}NMe)_2$. Depending on the solvent (THF, hexane) and the catalyst used (MeLi, n-BuLi, n-BuLi/KOt-Bu, KH), polymers with a maximum molecular weight of about 3000 g/mol were obtained. Thermolysis of $[SiMe_2NMe]_n$, however, delivered ceramics only in extremely poor yields (<10%), most probably because of the lack of functional sites in the precursor, which enable cross-linking reactions and thus avoid depolymerization during thermolysis.

Soum and others[93–97] studied ROP of 1,3-disiladiazetanes using strong acidic catalysts such as triflic acid methylester, $F_3CSO_3CH_3$, as well as basic catalysts such as MeLi:

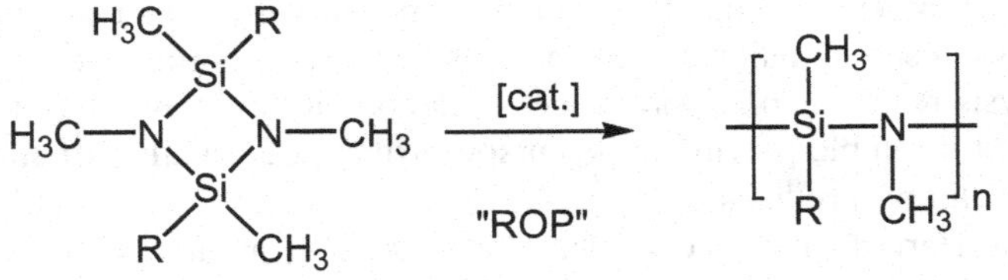

$$(18.13)$$

Four-membered ring systems, which have higher tension, react faster and more readily than cyclotrisilazanes, independent whether basic or acidic conditions are applied. Polysilazanes with molecular weights of up to 18,000 using acidic conditions and up to 100,000 using basic conditions together with narrow molar mass distribution could be obtained by this method. ROP allows for the synthesis of copolymers, if starting from cyclodisilazanes with differently substituted silicon atoms.

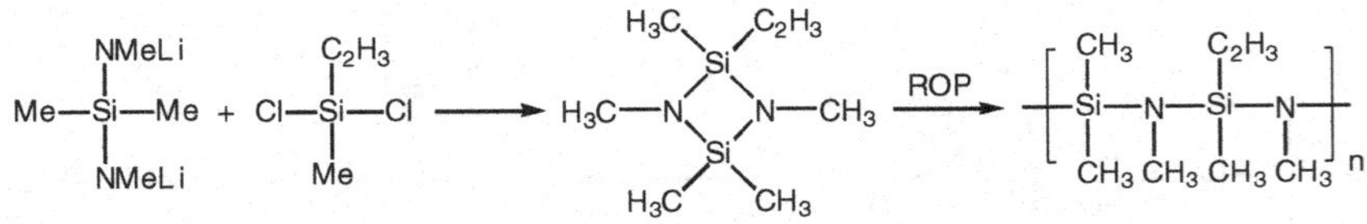

**SCHEME 18.20** Synthesis of copolymeric silazanes by ROP of cyclodisilazanes

This is an important issue with respect to obtaining preceramic polymers with tunable latent reactivity. For example, the attachment of vinyl groups would allow for cross-linking reactions by olefin polymerization (compare Scheme 18.17A). Replacement of SiMe$_2$ with SiHMe units introduces further cross-linking capabilities, such as hydrosilylation or dehydrocoupling, which are required for achieving high polymer-to-ceramic conversion yields.

Ring opening polymerization also allows for the synthesis of polycarbosilazanes with alternating SiR$_2$-CR$_2$-SiR$_2$-NR units, which otherwise are very difficult to obtain. For this purpose, cyclocarbosilazanes which have been obtained by reaction of bis(chlorodimethylsilyl)methane, R$_2$C(SiMe$_2$Cl)$_2$[98] with Li$_2$NR[99] were polymerized using acidic conditions. Applying basic conditions (organolithium initiators) resulted in side reactions at the methylene group.

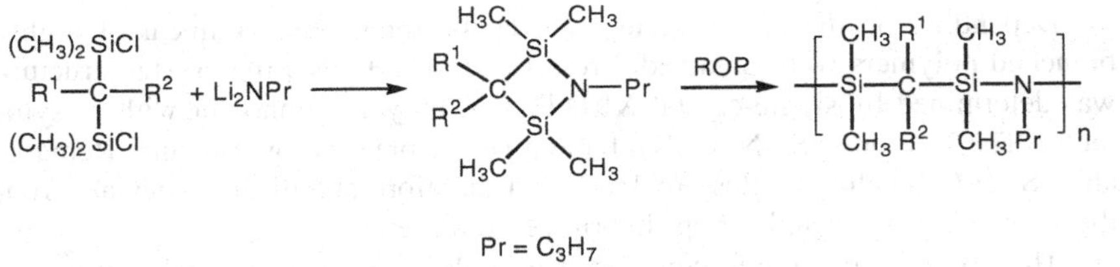

Pr = C$_3$H$_7$

**SCHEME 18.21** Synthesis of polycarbosilazanes with alternating SiR$_2$-CR$_2$-SiR$_2$-NR units by ring-opening polymerization of aza-1,3-disilethane.

Besides polysilazanes, PSCs are intensively investigated preceramic polymers for ternary Si-C-N materials. Their polymer backbone is composed of alternating Si–N=C=N units. Bis(silyl)carbodiimides, R$_3$Si–N=C=N–SiR$_3$ (R = alkyl) have been known since the early 1960s.[100–102] However, similar to low-molecular weight silazanes, they evaporate with heat treatment and are therefore not suitable as precursors for ceramics. PSCs were first obtained by Pump and Rochow[103] in 1964 by metathesis reactions of dichlorosilanes and disilvercyanamide.

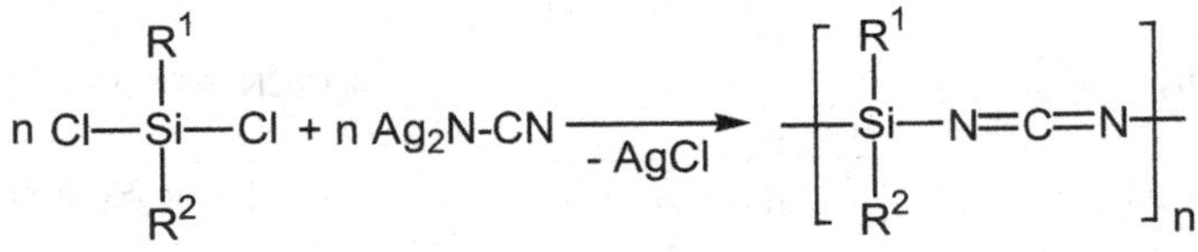

$$(18.14)$$

A cheaper approach to PSCs by means of a redistribution reaction of bis(trialkylsilyl)carbodiimides and dichlorosilanes or trichlorosilanes was described soon after in a patent by Klebe and Murray.[104] Similar to the synthesis of polysilazanes by transsilylation of dichlorosilanes or trichlorosilanes with HMDS (Equation 18.12), the procedure takes advantage of cheap and commercially available starting compounds, simple processing (i.e., no solid by-product formation), and high polymer yields.

$$n \; Cl\!-\!\underset{\underset{R^2}{|}}{\overset{\overset{R^1}{|}}{Si}}\!-\!Cl \; + \; n \; R_3Si\!-\!N\!=\!C\!=\!N\!-\!SiR_3 \; \xrightarrow[-\; R_3SiCl]{} \; \left[\!\underset{\underset{R^2}{|}}{\overset{\overset{R^1}{|}}{Si}}\!-\!N\!=\!C\!=\!N\!\right]_n$$

$$R = alkyl$$

(18.15)

Twenty years after the appearance of Klebe and Murray's patent,[104] the applicability of PSCs as precursors for Si-C-N ceramics was discovered. Riedel et al.[105] found that analogous to alkoxysilanes, pyridine-catalyzed reactions of chlorosilanes and bis(trimethylsilyl)carbodiimide possess typical sol-gel characteristics (Figure 18.4).[106]

Depending on the chlorosilane used, cyclic monomers or linear or highly branched polymers were obtained. From $(SiMe_2\text{–}N\!=\!C\!=\!N)_4$ the crystal structure was determined by single-crystal XRD.[107] It is a cyclic monomer with $C_i$ symmetry. The 16 atoms $(Si\text{–}N\!=\!C\!=\!N)_4$ build a nearly planar ring structure. Remarkably, Si-N-C bonding angles are 160°, and therefore significantly deviate from the 120° that is expected for $sp^2$-hybridized nitrogen.

The polymer-to-ceramic conversion was also studied thoroughly by means of solid state NMR and FTIR spectroscopy,[108] especially considering the structure and composition of thermolysis intermediates. Investigations by Riedel et al.[109] resulted in the discovery of the first crystalline phases in the Si-C-N system. They were obtained during thermolysis of $[Si(NCN)_2]_n$ (in terms of the principle of pseudo-chalcogenides,[110] an analogue of $SiO_2$), which was synthesized from $SiCl_4$ and $Me_3Si\text{-}NCN\text{-}SiMe_3$. The phase composition of the two crystalline phases **A** and **B** lie on the tie line $C_3N_4\text{-}Si_3N_4$ (Figure 18.5). **A** ($\beta$-$SiC_2N_4$ space group Pn3m)

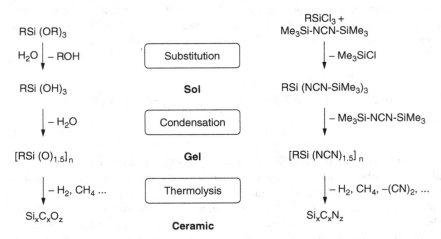

**FIGURE 18.4** Analogy of classical sol-gel processes of trialcoxysilanes (left) and nonoxide sol-gel processes of trichlorosilanes and bis(trimethylsilyl)carbodiimide.[106]

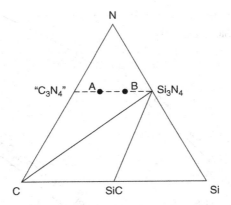

**FIGURE 18.5** Si-C-N phase diagram. A ($\beta$-SiC$_2$N$_4$) and B (Si$_2$CN$_4$) are the first crystalline ternary phases in the Si-C-N system. They were discovered by *in situ* x-ray diffraction during heating of [Si(NCN)$_2$]$_n$ at ~400 and ~920°C, respectively.[109]

crystallizes at temperatures greater than 400°C. At temperatures greater than 920°C, it decomposes with the loss of cyanogene and nitrogen to form **B** (Si$_2$CN$_4$, space group Aba2).

Recent investigations on PSCs focus on modifying their molecular structure in order to obtain precursors with pendant reactive substituents, such as hydrogen atoms or vinyl groups. The introduction of reactive substituents was intended to enable a facile modification with boron to allow for the production of quaternary Si-B-C-N ceramics.

## VI. PRECURSORS TO QUATERNARY SI-B-C-N CERAMICS

Takamizawa et al.[111] reported for the first time on quaternary precursor-derived Si-B-C-N ceramics. The goal was to produce ceramic fibers with high tensile strengths. Their approach was based on physical mixtures of polymers, that is, polymer blends of polydimethylsilane, [SiMe$_2$]$_n$, and B-trimethyl-N-triphenyl-borazine, [B(CH$_3$)-N(C$_6$H$_5$)]$_3$, or block copolymers of [SiMe$_2$]$_n$, [SiPh$_2$]$_n$, and borazine derivatives. Because of an inhomogeneous element distribution in the precursors, morphologically very inhomogeneous ceramic fibers without extraordinary high-temperature mechanical properties were obtained.

In the 1990s it was recognized that selected precursor-derived Si-B-C-N ceramics exhibit surprising high-temperature stability, oxidation resistance, and also superior high-temperature mechanical properties. Intensive studies in this field began in the U.S. and Europe, and many investigations in this field are still in progress. The reason for the interest is based on two factors: (i) the challenging properties of the ceramics and their many possible applications, and (ii) the fact that Si-B-C-N ceramics are kinetically stabilized and not thermally stable in a thermodynamic sense, as is evident from Figure 18.6.

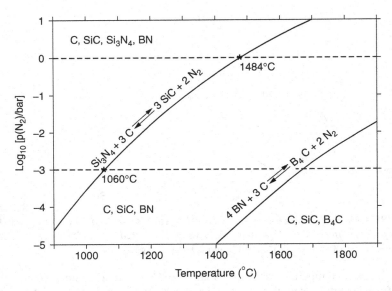

**FIGURE 18.6** Calculated partial pressure diagram for Si-B-C-N ceramics. The calculations predict that $SiC/Si_3N_4/BN/C$ composites behave thermally like boron-free $SiC/Si_3N_4/C$ materials (cf. Figure 18.3). Boron-containing phases (here BN) do not directly influence the thermal stability of $Si_3N_4$.[61]

## A. BORAZINE-BASED PRECURSORS FOR SI-B-C-N CERAMICS

Precursors to Si-B-C-N ceramics are generally divided into two groups: (i) those based on borazine or its derivatives as the boron-nitrogen source, and (ii) those without borazine units in the polymeric framework. The first borazine-based Si-B-C-N polymers where published by Nöth[112] in 1961, long before their potential as preceramics was recognized. He investigated the reaction of $BCl_3$ or $RBCl_2$ with $R_3SiNR_2$ and HMDS and observed that chloroboranes readily reacted with $R_3Si-NR_2$ under Si-N bond cleavage and B-N bond formation. In 1990, Seyferth et al.[113] reported the first polymeric Si-B-C-N precursor, obtained by a dehydrocoupling of borane dimethylsulfide, $BH_3 \cdot SMe_2$ and cyclotri(methylsilazane), $(SiMeH-NH)_3$. The mechanism presented in Scheme 18.22 is speculative. It is supposed that initial dehydrocoupling results in the formation of $N-BH_2$ units. The boron atoms are still Lewis acidic, therefore stabilization by intermolecular B-N coordination is expected to occur. Finally, a rearrangement involving cleavage of the cyclic silazane motifs along with a $\beta$-hydride shift from boron to silicon releases the borazine-based Si-B-C-N precursor.

Syntheses of Si-B-C-N polymers in which borazine units are directly attached to polysilazanes were investigated in detail by Sneddon et al.[114–116] The principle was to join borazine or its derivatives to oligomeric or polymeric silazanes by dehydrocoupling or dehydrosilylative (elimination of $HSiR_3$) coupling reactions. For example, hydridopolysilazane was treated with borazine at temperatures

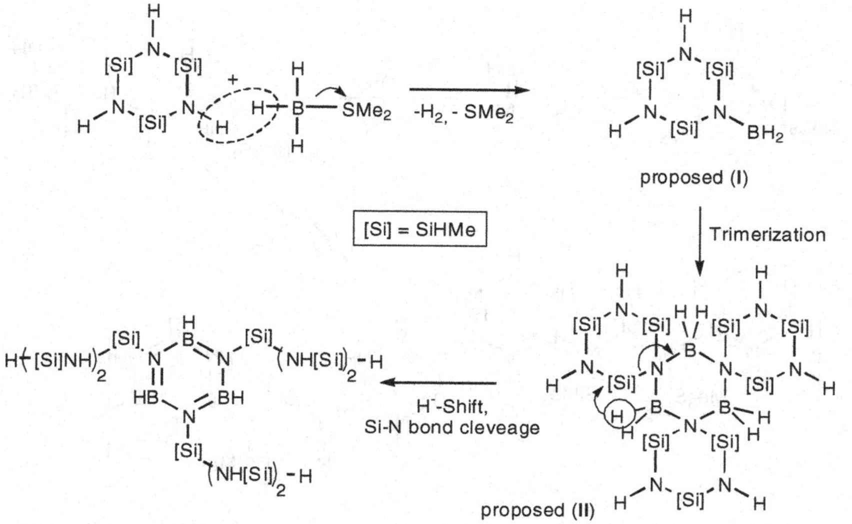

**SCHEME 18.22**  Synthesis of oligosilazane-functionalized borazine by dehydrocoupling of borane dimethylsulfide, $BH_3 \cdot SMe_2$ and cyclotri(methylsilazane), $(SiMeH-NH)_3$.[113] Intermediates (I) and (II) are proposed. (I) forms by dehydrocoupling of B-H and N-H and stabilizes by trimerization ($\rightarrow$ II). Si-N bond cleavage and $\beta$-hydride shift result in a borazine-based Si-B-C-N precursor.

between 50 and 90°C. The reaction produced polysilazanes with pendant borazine groups under retention of the polysilazane backbone.[114]

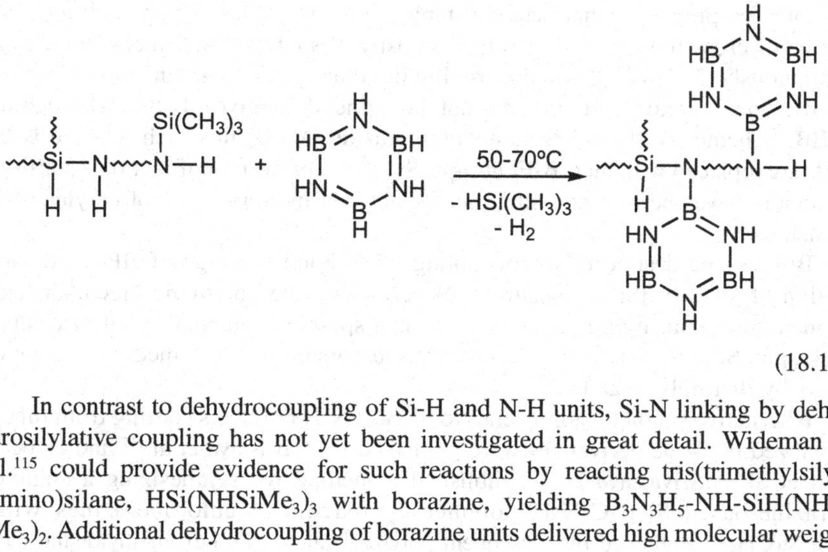

(18.16)

In contrast to dehydrocoupling of Si-H and N-H units, Si-N linking by dehydrosilylative coupling has not yet been investigated in great detail. Wideman et al.[115] could provide evidence for such reactions by reacting tris(trimethylsilylamino)silane, $HSi(NHSiMe_3)_3$ with borazine, yielding $B_3N_3H_5$-NH-SiH(NHSiMe_3)_2$. Additional dehydrocoupling of borazine units delivered high molecular weight

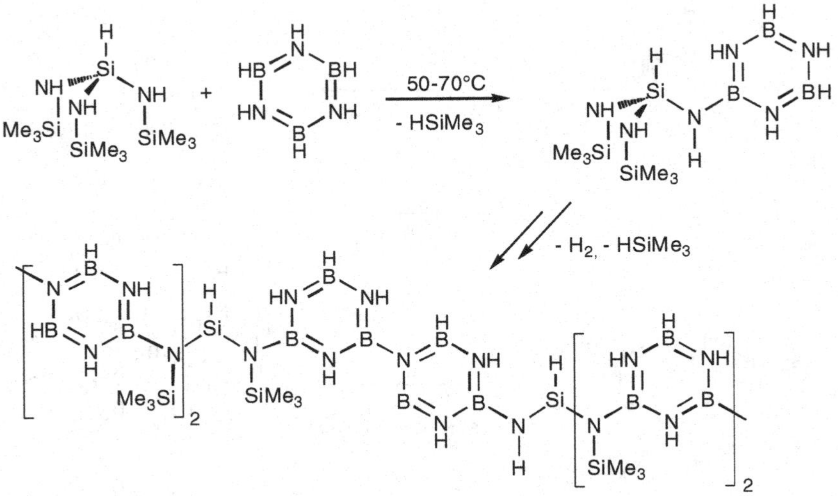

**SCHEME 18.23** Synthesis of Si-B-C-N-H polymers by dehydrocoupling and dehydrosilylative coupling of borazine and tris(trimethylsilylamino)silane, $HSi(NHSiMe_3)_3$.[115]

polymers possessing borazine and borazylene motifs, as depicted in Scheme 18.23. Ceramic yields of these preceramics were between 38 and 42%, depending on the silazane to borazine stoichiometry applied.

For the fabrication of ceramic fibers, polymers with linear molecular structures are desired. Polysilazanes with pendant borazine rings (Equation 18.16) carry both N-H and B-H units, which may react further. They are therefore not suitable for processes that require stable melt viscosities. Upon melting, the degree of cross-linking, and thus the viscosity of such polymer melts, increases continuously. To avoid thermal cross-linking during melt spinning, Wideman et al. designed mono- and difunctional borazine derivatives B-diethylborazine (DEB; Scheme 18.24) and B-monoethylborazine (MEB) in which reactive B-H units are replaced with inert B-Et groups. Synthesis of the monofunctional starting compounds was performed using metal-catalyzed hydroboration of ethylene by borazine.[116]

Besides the desired dehydrocoupling, Si-N bond cleavage of HPZ and formation of DEB-NHSiMe₃ occurred. Nevertheless, the polymeric precursor had an ideal glass-transition temperature for melt spinning and stable melt viscosity. Melt spun Si-B-C-N-H green fibers were subsequently transformed into ceramic fibers by thermolysis at 1400°C.

Alternative synthetic approaches to Si-B-C-N-H polymers starting from functionalized borazine derivatives were published by Srivastava et al.[117] and Haberecht et al.[118] Srivastava et al. published a method for synthesizing a soluble borazine-based Si-B-C-N-H polymer by reacting B-chloroborazines with $LiSi(Si(CH_3)_3)_3$ followed by subsequent polymerization with hexamethyldisilazane.

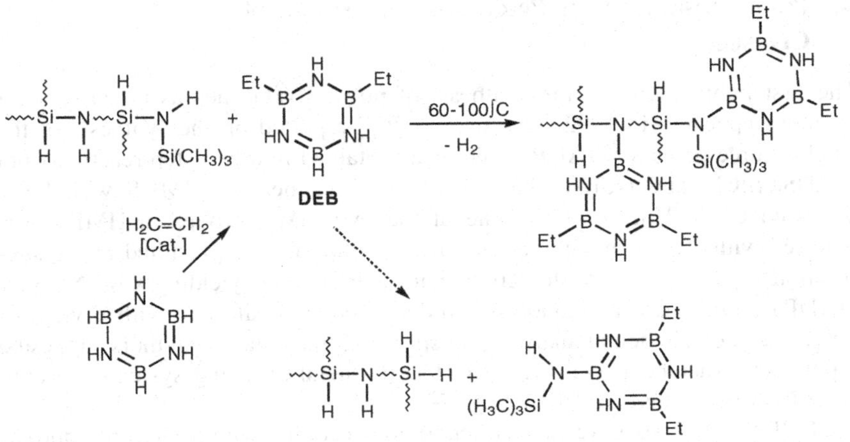

**SCHEME 18.24**  Dehydrocoupling of mono-functional B-diethylborazine DEB and perhydridopolysilazane. As a side reaction (indicated by the dotted arrow), Si-N bond cleavage of the silazane backbone and formation of DEB-NHSiMe₃ occurs.[116]

Thermolysis at 1400°C released ceramics with an overall composition of $Si_{0.29}BC_{0.12}N_{1.28}$. XRD investigations indicated a crystallization of h-BN, whereas formation of crystalline SiC or $Si_3N_4$ was observed only to a minor extent.

Haberecht et al.[118,119] obtained Si-B-C-N-H polymers from B-tris(trichloro-silylvinyl)borazine, $B_3N_3H_3(HC=CHSiCl_3)_3$, which was synthesized by a hydro-silylation of B-triethynylborazine, $B_3N_3H_3(C≡CH)_3$ with trichlorosilane. Amino-lysis of $B_3N_3H_3(HC=CHSiCl_3)_3$ using methylamine led to a soluble polysilazane with borazine units bridged by HC=CHSiNMe units, whereas hydrogenation with $LiAlH_4$ released a single-site molecular precursor (Scheme 18.25).[120] The latter delivered ceramics with a yield of 94%; hydrogen was the only volatile thermolysis product.

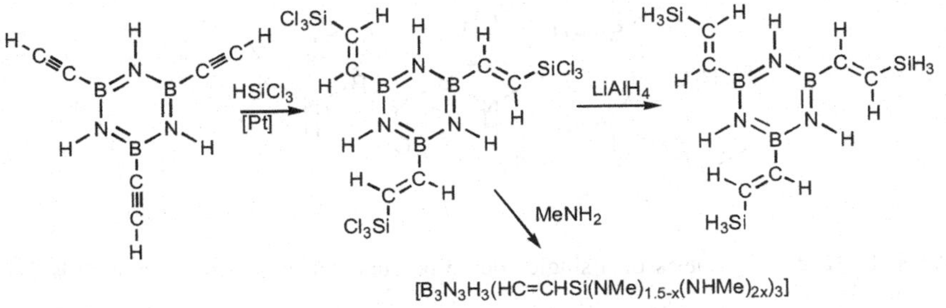

$[B_3N_3H_3(HC=CHSi(NMe)_{1.5-x}(NHMe)_{2x})_3]$

**SCHEME 18.25**  Synthesis of borazine-based Si-B-C-N-H polymers by hydrosilylation of B-triethynylborazine with trichlorosilane and subsequent hydrogenation with $LiAlH_4$ or aminolysis with methylamine, respectively.[119,120]

## B. POLYBOROSILAZANES AS PRECURSORS FOR SI-B-C-N CERAMICS

The first publications on the synthesis of polyborosilazanes as precursors for ceramics appeared in 1992. Jansen et al.[121,122] reported on the synthesis of the single-source precursor TADB, which was obtained in a two-step reaction from HMDS, SiCl$_4$, and BCl$_3$ (Scheme 18.26).[121] Treatment of HMDS with SiCl$_4$ produced Cl$_3$SiNHSiMe$_3$. Only one of the two SiMe$_3$ groups in HMDS was replaced with a SiCl$_3$ moiety. Subsequent addition of BCl$_3$ resulted in the substitution of the second SiMe$_3$ group with a BCl$_2$ unit, yielding Cl$_3$SiNHBCl$_2$ (TADB). Subsequent ammonolysis produced polyborosilazane with a yield of 80%. Based on elemental analysis and spectroscopic data, the authors suggested a polymer structure in which Si$_3$(NCH$_3$)$_3$ six-membered ring systems are connected via HN-B and N(CH$_3$)B units.[122]

Bulk thermolysis gave a black amorphous ceramic with a yield of approximately 50%. In a nitrogen atmosphere, the amorphous state was retained to 1900°C. No formation of any crystalline phase was observed below this temperature. The polymer could successfully be melt spun, and after curing in HSiCl$_3$, pyrolyzed to Si-B-C-N ceramic fibers.[123] Heating to 1500°C released 15 μm diameter ceramic SiBN$_3$C fibers with a yield of 55%. Tensile strength and Young's modulus were 3 GPa (at room temperature) and 300 GPa, respectively. Exposure to oxygen at 1500°C resulted in the formation of a 1.7 μm Si-C-O layer that acted as a protective layer, thus inhibiting further oxidation and protecting the fiber from full degradation.

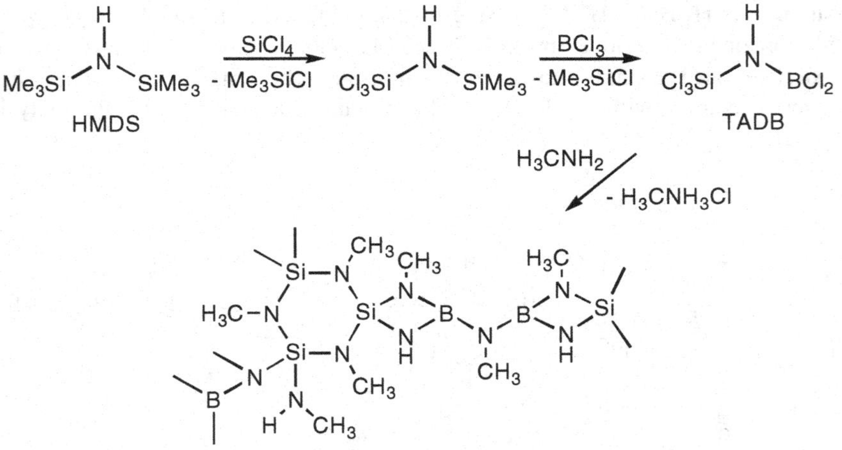

**SCHEME 18.26** Synthesis of a single source precursor to high-temperature Si-B-C-N ceramics.[122]

Riedel et al.[124,125] and Kienzle[126] published the synthesis of a polyborosilazane by ammonolysis of tris(dichloromethylsilyl-ethylene)borane, B(C$_2$H$_4$Si(CH$_3$)Cl$_2$)$_3$

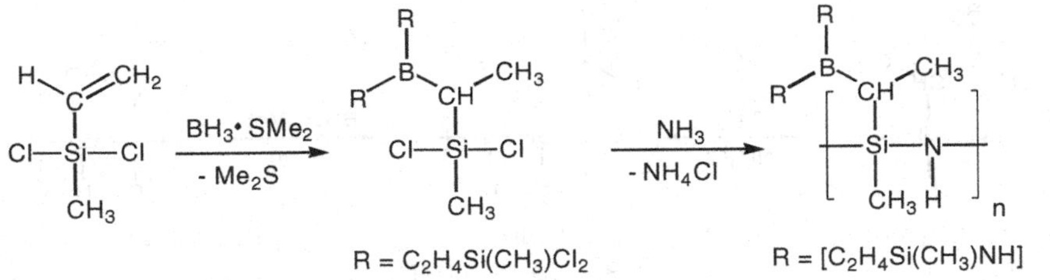

**SCHEME 18.27** Synthesis of C-B-C bridged polysilazane T(2)1 as precursor to high-temperature Si-B-C-N ceramics by ammonolysis of tris(dichloromethylsilyl-ethylene)borane.[124-126]

($C_2H_4$ = $CHCH_3$, $CH_2CH_2$). The latter was synthesized by hydroboration of dichloromethylvinylsilane, $(H_2C=CH)Si(CH_3)Cl_2$, according to a procedure described by Jones and Myers.[127]

Hydroboration of $(H_2C=CH)Si(CH_3)Cl_2$ occurs quantitatively, but as described previously, not regioselectively.[128] Instead, the boryl group may add in each of the three successive hydroboration steps at either the $\alpha$- or $\beta$-position of the vinyl function. The different regio and stereo isomers that form were not isolated prior to ammonolysis. Rather, the as-obtained product mixture was reacted without further purification with ammonia. The boron-modified polyborosilazane T(2)1, $[B(C_2H_4Si(CH_3)NH]_n$, was obtained with a yield of about 85%. It is composed of silazane chains, which are cross-linked via C-B-C bridges. Thermolysis produced $Si_3BC_{4.2}N_2$ ceramics with a yield of 50%. As-obtained ceramics possess excellent thermal stability; crystallization is retarded to approximately 1750°C, and thermally induced degradation with loss of gaseous species is not observed at temperatures below 1950°C.

Performing hydroboration of $(H_2C=CH)Si(CH_3)Cl_2$ using $H_2BCl\cdot SMe_2$ or $HBCl_2\cdot SMe_2$ results in $ClB(C_2H_4Si(CH_3)Cl_2)_2$ and $Cl_2B(C_2H_4Si(CH_3)Cl_2)$, respectively, which after ammonolysis release more highly cross-linked nitrogen-rich polyborosilazanes.[129] Therefore ceramic yields at 56% and 76%, respectively, are higher than that of T(2)1. Moreover, the boron:silicon ratio increases from 1:3 in T(2)1 to 1:2 to 1:1.

Alternatively derivatives of T(2)1 possessing additional cross-linking motifs instead of the chemically inert silicon-bonded methyl group were synthesized. Starting from $B(C_2H_4SiCl_3)_3$ or $B(C_2H_4SiHCl_2)_3$, which are accessible by hydroboration of vinyltrichlorosilane or vinyldichlorosilane, respectively, synthesis of highly cross-linked boron-modified polysilsesquiazane $[B(C_2H_4Si(NH)_{1.5}]_n$ and boron-modified polysilazane $[B(C_2H_4SiHNH]_n$ was achieved (M; Scheme 18.28).[130] As a result of the higher cross-linking density and latent reactivity, which is provided by N-H and/or Si-H units, ceramic yields (82% and 85%) were significantly higher than for T(2)1. A significant disadvantage in the synthesis of these precursors, however, was the difficult and time-intensive processing, that is, the separation of ammonium chloride, which was a result of the

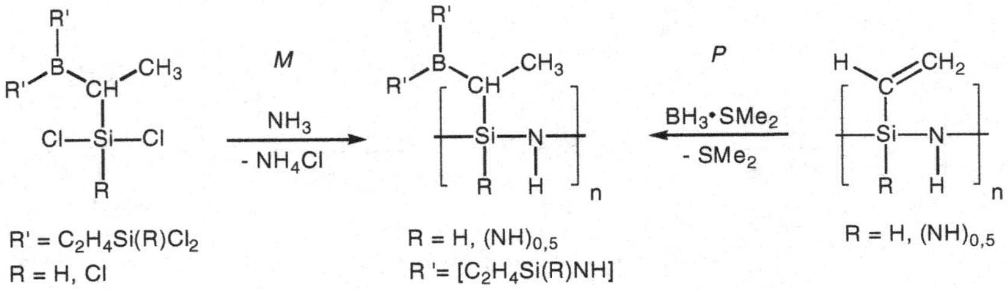

**SCHEME 18.28** Synthesis of boron-modified polysilazanes by ammonolysis of tris(chlorosilylethylene)boranes (*M*, monomer route) and by hydroboration of vinyl-substituted polysilazanes (*P*, polymer route).[125,130,131]

low solubility of the polymers and which caused extremely poor polymer yields. Processing can be facilitated and polymer yields can be increased considerably by switching the reaction sequence (P; Scheme 18.28).[131]

Accordingly, vinylchlorosilanes were initially treated with ammonia and the obtained polyvinylsilazanes [(H$_2$C=CH)SiRNH]$_n$ were subsequently reacted with Me$_2$S·BH$_3$ to give the corresponding polyborosilazanes with a yield of 100%. Purification was performed by removing the solvent and by-product dimethylylsulfane in a high vacuum.

Recently, complementary procedures for the synthesis of polyborosilazanes, which also avoid formation of solid by-products, were published. For example, dehydrocoupling of tris(hydridosilyl-ethylene)boranes, B(C$_2$H$_4$SiRH$_2$)$_3$, with ammonia or methylamine delivered various Si-B-C-N-H polymers as potential precursors for high-temperature ceramics.[132,133] The starting compounds B(C$_2$H$_4$SiRH$_2$)$_3$ (R = H, CH$_3$; Scheme 18.29) were obtained on two different reaction pathways, starting from vinylchlorosilanes (H$_2$C=CH)SiRCl$_2$. In the first route, tris(chlorosilyl-ethylene)boranes obtained from (H$_2$C=CH)SiRCl$_2$ and Me$_2$S·BH$_3$ were reacted with LiAlH$_4$ in diethylether solution. Hydroboration of hydridovinylsilanes (H$_2$C=CH)SiRH$_2$ (R = H, CH$_3$), which can be obtained from

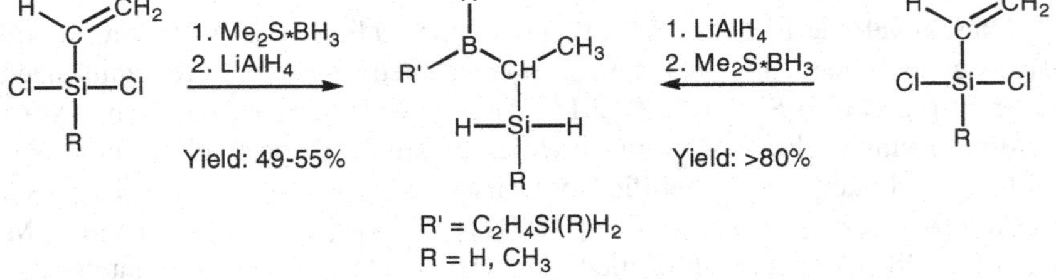

**SCHEME 18.29** Synthesis of tris(hydridosilyl-ethylene)boranes on different reaction pathways. Caution: Tris(hydridosilyl-ethylene)boranes are highly reactive and may explode spontaneously upon exposure to air!

($H_2C=CH$)$SiRCl_2$ and $LiAlH_4$ (Caution!), represents an alternative route that produces $B(C_2H_4SiRH_2)_3$ in much higher yields.

Ammonolysis of $B(C_2H_4SiRH_2)_3$ (Equation 18.17; $R'' = H$) does not occur directly and requires a catalyst. Similar to a method described by Seyferth et al.,[69] who used potassium hydride for the cross-linking of silazanes (cf. Scheme 18.16), 1 mol% of n-butyl lithium in n-hexane was applied. Silane and catalyst were dissolved in toluene/THF and the mixture was heated to 70°C. To avoid the loss of ammonia, the reaction vessel was equipped with a reflux condenser, which was cooled to −78°C (i-PrOH/CO₂). Purification of the precursors was performed by removing all volatile components in a high vacuum.

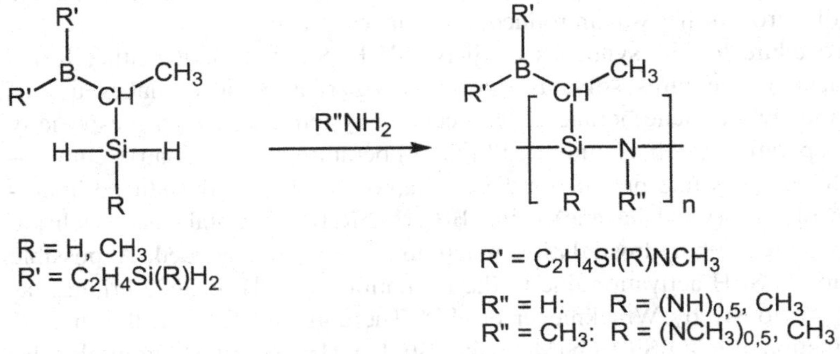

$R = H, CH_3$
$R' = C_2H_4Si(R)H_2$

$R' = C_2H_4Si(R)NCH_3$

$R'' = H: \quad R = (NH)_{0.5}, CH_3$
$R'' = CH_3: \quad R = (NCH_3)_{0.5}, CH_3$ (18.17)

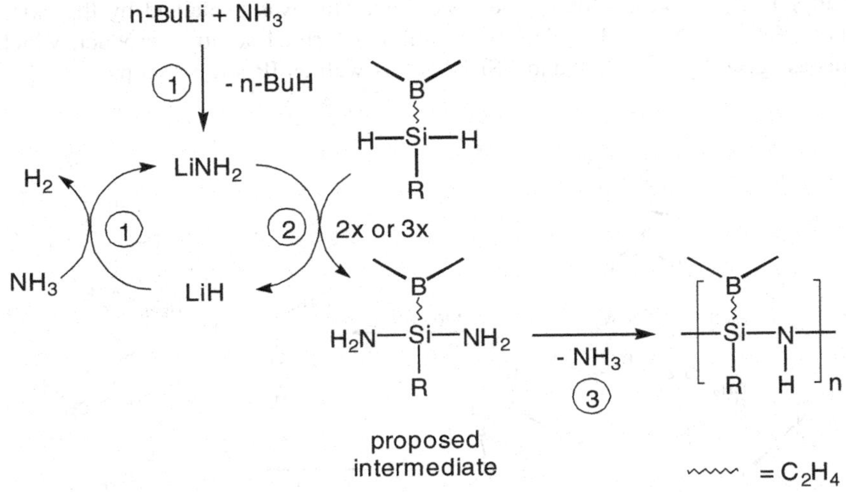

proposed
intermediate

$\rightsquigarrow = C_2H_4$

**SCHEME 18.30** Proposed mechanism of the n-BuLi catalyzed dehydrocoupling of tris(hydridosilyl-ethylene)boranes and ammonia. (1) deprotonation with formation of an amide, (2) substitution of silicon-bonded hydride with amide, (3) polymerization through condensation.

A possible mechanism for the base-catalyzed dehydrocoupling of $B(C_2H_4SiRH_2)_3$ was given (Scheme 18.30). Initially ammonia is deprotonated by n-BuLi. In the following, the nucleophilic amide replaces a silicon-bonded hydride, which itself deprotonates ammonia. Molecular hydrogen forms that evaporates. The cycle continues until all silicon-bonded hydrogen atoms are replaced with $NH_2$ units. Under the reaction conditions applied, $Si-NH_2$ units are not stable. They condense with elimination of ammonia, forming polysilazanes. The proposed mechanism was supported by the observation that more nucleophilic methylamine reacts without added catalyst. Nevertheless, aminolysis was also performed according to the procedure applied for ammonolysis and the influence of catalyst on the structure and chemical composition, that is, the amount of nitrogen that was introduced, was investigated.

A procedure for the synthesis of Si-B-C-N-H polymers that neither forms by-products nor requires solvents during polymerization also makes use of tris(hydridosilyl-ethylene)boranes. The precursor system was designed especially for the preparation of fiber-reinforced high-temperature ceramic matrix composites (CMCs). Glass-like precursors were obtained by thermally induced hydrosilylation of oligo(vinylsilazane) using $B(C_2H_4SiRH_2)_3$.[134] Catalysts, which are usually required for hydrosilylation reactions,[135] were not needed. A possible explanation is Si-H activation due to the formation of $Si-H-B$ 3c2e bridges, as reported previously by Wrackmeyer et al.[136] The authors observed that in 1,1-allyl borination of $HMe_2SiC{\equiv}CSiMe_3$ using $B(CH_2CH{=}CH_2)_3$, fast intramolecular hydrosilylation of one of the boron-bonded allyl groups with formation of 3,4-diallyl-1-sila-4-bora-cyclohept-2-en occurred. This was explained by the existence of a 3c2e Si-H-B bridge in the initially formed addition product, which was assigned by $^1H$-, $^{13}C$-, and $^{29}Si$-NMR, as well as IR spectroscopy.

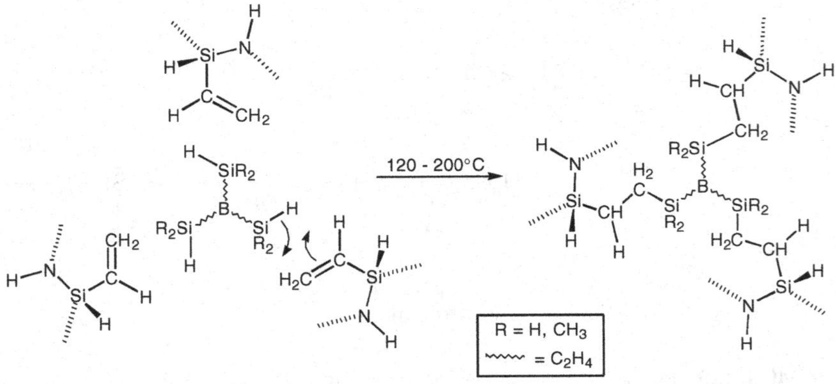

**SCHEME 18.31** Synthesis of polyborosilazanes through thermal induced hydrosilylation of oligovinylsilazane and tris(hydridosilylethyl)boranes (schematic representation).

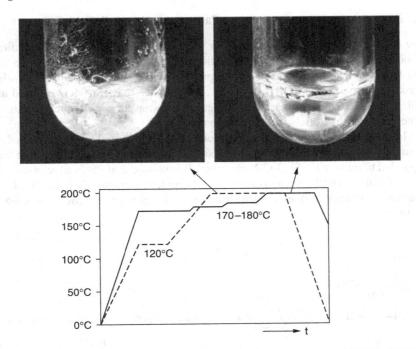

**FIGURE 18.7** Influence of the heating rate during the cross-linking of $[(H_2C=CH)SiH-NH]_n$ and $B(C_2H_4Si(CH_3)H_2)_3$ on the homogeneity of the resulting polymer.[137]

Performing the cross-linking reaction occurred straightforward. Nevertheless, the sensitivity of the starting compounds to oxygen and moisture required inert gas conditions. Prior to heating to approximately 200°C, $[(H_2C=CH)SiH-NH]_n$ and $B(C_2H_4SiRH_2)_3$ were thoroughly mixed and degassed. To avoid formation of bubbles and other inhomogeneities, gradual heating was applied.[137] The need for such stepwise heating is evident from Figure 18.7.

Originally the mixture was heated at 50°C/h to 120°C, and after a dwell time of 60 min, was heated at 30°C/h to 200°C. Using this heating program, only foamed polymers were obtained (Figure 18.7, left). By a modification of the heating process in which the mixture was heated to approximately 170°C in 2 h, followed by stepwise heating to 180°C, and then to 200°C, foaming could be avoided and dense, glass-like polymers were obtained (Figure 18.7, right). The progress of cross-linking was monitored by IR spectroscopy. Heating was continued until $\nu(=C-H)$ and $\nu(C=C)$ absorptions at ~3050 cm$^{-1}$ and ~1600 cm$^{1}$, respectively, vanished.

First tests have proven the silane/silazane mixture to be suitable for the preparation of fiber-matrix composites using the resin transfer molding (RTM) process.[138] The basics of this process are low viscosity (fluidity) of the starting compounds to allow for sufficient penetration, adequate vapor pressure, appropriate chemistry (no by-products), high reactivity during cross-linking, and high thermolysis conversion yields/low shrinkage during thermolysis.

## C. Polyborosilylcarbodiimides as Precursors for Si-B-C-N Ceramics

One of the very important processes for the formation of glasses and ceramics is the sol-gel process. The first publications, by Ebelmen,[139] appeared in 1844. The focus until today has been on the preparation of silicate ceramics. Mostly alkoxysilanes such as $Si(OR)_4$ (R = $CH_3$, $CH_2CH_3$, ...) or $RSi(OR)_3$ are used as molecular precursors. Remarkably, sol-gel processing allows for the formation of ceramic bulk materials, fibers, and coatings.[140]

In the previous section it was mentioned that Riedel et al.[105–107] developed a nonoxide sol-gel process for ternary Si-C-N ceramics. This concept can be expanded to quaternary Si-B-C-N ceramics. Boron-modified PSCs were obtained in quantitative yields from tris(chlorosilyl-ethylene)boranes, $B(C_2H_4SiRCl_2)_3$, which were reacted with excess bis(trimethylsilyl)carbodiimide, $Me_3Si-N=C=N-SiMe_3$.[141]

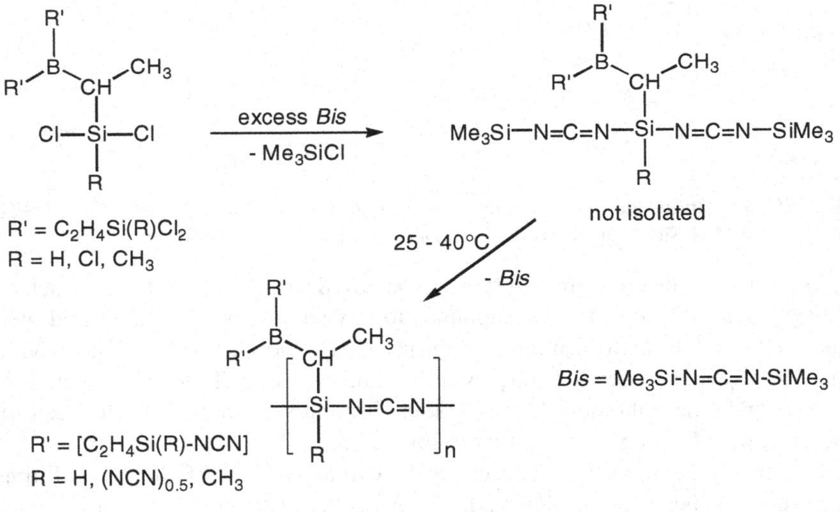

**SCHEME 18.32** Synthesis of boron-modified polysilylcarbodiimides from tris(chloro-silyl-ethylene)boranes and bis(trimethylsilyl)carbodiimide (*Bis*) under sol-gel conditions.[141]

Depending on the silicon-bonded substituents R in the starting compound, the viscosity of the solution increased considerably within 30 to 60 min (R = H, Cl) or within 24 h (R = $CH_3$). The obtained gels aged readily and excess *Bis* as well as the couple product, $Me_3SiCl$, which separated from the polymer, were distilled off under reduced pressure. The whole process is inexpensive, since excess bis and $Me_3SiCl$ can be recycled. Time-intensive processing steps are not required.

High-temperature investigations by TGA and XRD of annealed samples clearly showed that ceramics obtained from boron-modified PSCs do not possess satisfactory high-temperature properties. Thermal degradation was observed at approximately 1500°C. Preliminary studies on boron-modified polysilazanes with different nitrogen contents suggested that the decomposition of Si-B-C-N ceramics is strongly dependent on the amount of nitrogen bonded in the materials.[132] In contrast, thermal stability is not necessarily a direct function of the molecular structure. However, simply reducing the nitrogen content by performing the reaction with excess chlorosilanes is not possible, since the released polymers possess undesired large amounts of chlorine which lead to uncontrollable reactions during the polymer-to-ceramic conversion.

A procedure that allows for the synthesis of PSCs with lower (i.e., adjustable) nitrogen content is a dehydrocoupling of tris(hydridosilyl-ethylene)boranes, $B(C_2H_4SiRH_2)_3$, with cyan amide, $H_2N-CN$. The starting compounds were mixed in THF and refluxed for 12 h. Catalysts were not required. Dehydrocoupling started immediately and was recognized by strong $H_2$ evolution. When hydrogen evolution diminished, the solvent was removed in a vacuum and the precursors were obtained with a yield of 100%. Purification steps were not needed.[142]

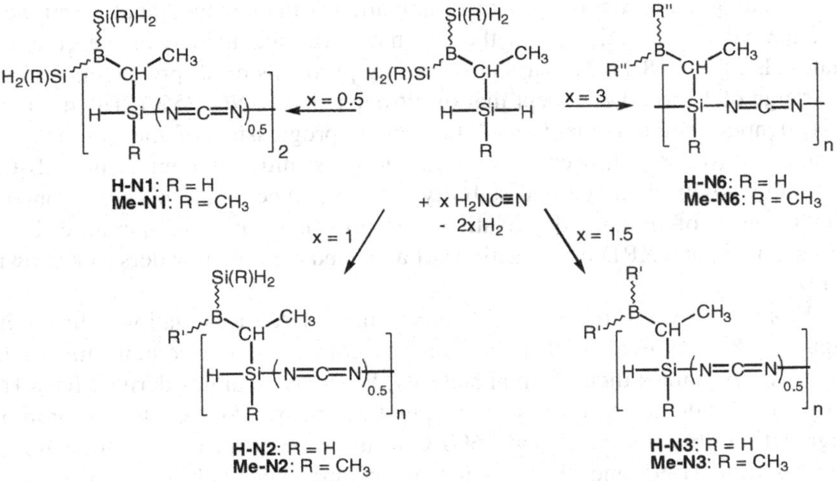

**SCHEME 18.33** Synthesis of B-modified PSCs by dehydrocoupling of $B(C_2H_4SiRH_2)_3$ with different amounts of $H_2N-CN$. Reactions were performed in boiling thf. Wave lines symbolize $C_2H_4$ units ($C_2H_4$ = CHCH$_3$, CH$_2$CH$_2$); R' = SiH(R)(NCN)$_{0.5}$ and R" = Si(R)NCN. Chemical compositions (idealized) ranging from ($C_{6.5}H_{20}NSi_3B$) (H-N1) to ($C_9H_{15}N_6Si_3B$) (H-N6) and ($C_{9.5}H_{26}NSi_3B$) (Me-N1) to ($C_{12}H_{21}N_6Si_3B$) (Me-N6) could be achieved by this procedure.[83,142]

**TABLE 18.1**

**Chemical Composition of B-Modified PSCs and Derived Ceramics**

| | Polymer[a] | Ceramic[a] | Ceramic Yield[b] |
|---|---|---|---|
| **H-N2** | $Si_3BC_7N_{2.3}H_{17}$ | $Si_3B_{1.2}C_{6.9}N_3$ | 84% |
| **H-N3** | $Si_3BC_7N_{3.2}H_{18}$ | $Si_3B_{1.2}C_{6.7}N_{3.2}$ | 84% |
| **H-N3.5** | $Si_3BC_{8.2}N_{3.7}H_{19}$ | $Si_3BC_{6.7}N_{3.6}$ | 83% |
| **H-N4** | $Si_3BC_{8.5}N_{3.9}H_{19}$ | $Si_3BC_{6.9}N_{4.1}$ | 80% |
| **H-N5** | $Si_3B_{0.9}C_{8.9}N_{4.8}H_{18}$ | $Si_3BC_{7.5}N_{4.9}$ | 80% |
| **H-N6** | $Si_3BC_{10}N_{6.2}H_{18}$ | $Si_3B_{1.1}C_{7.2}N_6$ | 75% |

[a] Formula refered to $Si_3$
[b] Ceramic yields after pyrolysis at 1400°C ($N_2$ atmosphere, 3 h).

Investigations with solid state NMR, IR, and Raman spectroscopy confirmed the proposed molecular structures and pointed to the fact that during dehydrocoupling, the cyan amide unit of the starting compound isomerized to carbodiimide motifs. $^{13}$C NMR signals at unexpected low fields (~165 ppm) and IR bands at 1610 cm$^{-1}$ and 3380 cm$^{-1}$ assigned C=N-H end groups.

Thermogravimetric analysis in combination with elemental analysis proved that the Si:N ratio adjusted in the polymers was maintained in the ceramic materials (Table 18.1). The mass loss during pyrolysis of all precursors was in the range of 15 to 20%, except that of nitrogen-rich H-N6 (25%). The more or less identical ceramic yields and the similar progression of the polymer-to-ceramic conversion allowed for correlation of the nitrogen content in Si-B-C-N ceramics with their thermal stability. The influence of the nitrogen content on the onset of thermal degradation was proven by high-temperature TGA (Figure 18.8) and XRD investigations of annealed ceramic powders (not shown here).[143]

From the results of the high-temperature TGA investigations shown in Figure 18.8, it is evident that the nitrogen content of the ceramic materials strongly determines their thermal stability. Whereas ceramics derived from H-N1 to H-N3.5 do not decompose at temperatures below 1900°C, there is prompt degradation observed at around 1600°C in the case of nitrogen-rich ceramics derived from H-N5 and H-N6, which is accompanied with a weight loss of 24% and 28%, respectively, at 2000°C. Obviously H-N4 ceramics, which possess 24.6 wt% of nitrogen, represent a borderline case. Even though the onset of thermal decomposition is similar to that of H-N5 and H-N6, degradation is retarded and the weight loss of approximately 7% at 2000°C is lower than thermodynamically calculated. It was therefore stated that 25 wt% of nitrogen represents a critical value with respect to obtaining thermally stable Si-B-C-N ceramics.

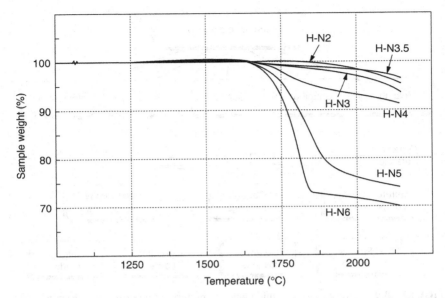

**FIGURE 18.8** High-temperature TGA of ceramics derived from H-N2 to H-N6. Heating rate: T < 1400°C, 10°C/min; T > 1400°C, 2°C/min; Ar atmosphere.[142]

## VII. PERSPECTIVE

The above examples show that preparative organometallic chemistry allows for the production of a wide variety of silicon-based molecular precursors for high-temperature ceramics. The desired physical chemical properties and appropriate thermolysis chemistry can be realized by an intelligent precursor design. Nevertheless, there is still a need for further development; for example, investigations into the synthesis of precursors that release phase-pure ceramics or composites with tunable composition and properties. The focus will also be on designing preceramic polymers, which release functional materials. In this field, very little investigation has been performed so far.

The main advantage of the concept of producing ceramics from molecular precursors over sintering or melting is the possibility of realizing shapes that are impossible or difficult to achieve using the classical procedures. Possible approaches are shown in Figure 18.9.

Several publications have appeared that deal with the production of high-modulus refractory nonoxide ceramic fibers. In this regard, the pioneering work of Yajima and co-workers should be mentioned.[22,144] As pointed out in Figure 18.10, SiC fibers were obtained by melt-spinning of poly(methylmethylenesilylene) which was synthesized according to Scheme 18.5. To render the green fibers infusible, they were cured and finally pyrolyzed.

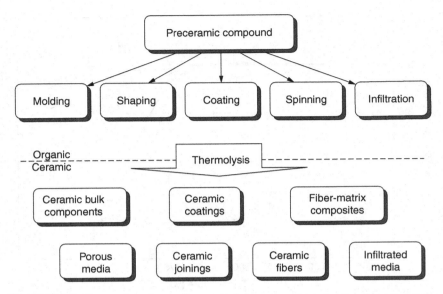

**FIGURE 18.9** Preparation of ceramic materials by processing of molecular precursors.

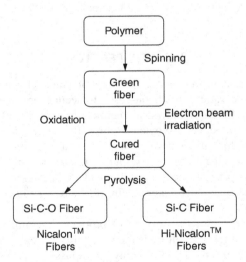

**FIGURE 18.10** Flow scheme for the preparation of Nicalon and Hi Nicalon fibers.[22,144]

Curing was initially performed by surface oxidation of the fibers in humid oxygen atmosphere at approximately 150°C. This processing step caused partial oxidation/hydrolysis of Si-H units and formation of Si-O-Si linkages. Oxygen incorporated by this process was not released during thermolysis and remained in the fibers. Heating such fibers caused the formation of a glassy silica phase, which degraded high-temperature mechanical properties. Fibers obtained by this procedure are commercially available under the trademark Nicalon™.[145] Alternatively, curing was carried out using electron beam (2 MeV) or γ-ray irradiation ($^{60}$Co). Oxygen contamination of the fiber surfaces could be avoided and high-temperature mechanical properties improved. Nevertheless, researchers today are very much focused on the preparation of multinary silicon-based ceramic fibers, that is, in the systems Si-C-N[146–149] and Si-B-C-N.[123,150,151] Such materials crystallize at significantly higher temperatures and possess improved oxidation resistance compared with binary carbon/silicon carbide fibers, which may, depending on the oxygen partial pressure, degrade by either passive or active oxidation at comparably low temperatures.[152]

Thermolysis of preceramic polymers is also a unique method for the preparation of ceramic coatings. In this regard, the dip-coating and pyrolysis process[153] has to be mentioned. Compared with chemical (CVD)[154] or physical vapor deposition (PVD),[155] it is a rather simple, inexpensive, and thus economic procedure. A substrate is introduced into a polymer solution or slurry, pulled out, dried, and pyrolyzed. Film thickness is determined by a number of parameters, such as the viscosity and concentration of the polymer solution/slurry, pull-out velocity, and ceramic yield. Usually filler materials, either inert or reactive, are added to inhibit crack formation during thermolysis. Even though promising first results were obtained in the protection of carbon fiber-reinforced carbon and silicon carbide composites, there is still the need for further development of such precursor-derived ceramic coatings; for example, for the protection of lightweight structures that possess a potential for aerospace applications.

Last but not least, CMCs[156] are of great interest. Such materials have been shown to exhibit extraordinary thermomechanical properties combined with low density. Besides developing refractory high modulus ceramic fibers (e.g., Si-C, Si-C-N, and Si-B-C-N fibers; see above), the focus is on the development of suitable matrix materials. A common procedure for the formation of silicon carbide-based CMCs is liquid silicon infiltration.[157] Molten silicon may react either with carbonized carbon fibers or with extra graphite which is introduced into the fiber prepregs prior to the infiltration of liquid silicon. Even though this procedure is rather cheap, it has several disadvantages: molten silicon readily wets carbon and reacts to silicon carbide, often at the expense of the carbon fibers. To avoid such degradation, it is necessary to protect the fibers, for example, by depositing boron nitride protective coatings, which significantly increases costs. Precursor-derived CMCs, which can be obtained by polymer impregnation and pyrolysis (PIP), do not suffer from these drawbacks.[134,158] However, to allow for efficient polymer infiltration, adequate viscosity/fluidity and controllable cross-linking and pyrolysis chemistry are required. Both aspects are commonly contradictory: low viscosity/high fluidity requires pendent organic groups and low cross-linking density of the polymeric precursor, whereas high ceramic yields require high cross-linking density and the absence of pendent side groups.

Recent investigations have therefore focused on fabrication of two-component matrix precursors, such as in Scheme 18.31, that possess low viscosity and, analogous to epoxy resins, can be easily thermally cross-linked to duroplastic high-ceramic yield polymer matrices. By using inert fillers such as silicon carbide, shrinkage of the matrix during thermolysis can be minimized and the fabrication of comparably cheap CMCs be realized.[159]

## ACKNOWLEDGMENT

The author greatly acknowledges financial support from the Max Planck Society and the German Research Foundation (DFG). Much of the work described in this chapter was performed at the Powder Metallurgical Laboratory (PML) of the Max Planck Institute for Metal Research in Stuttgart, Germany. Thanks to all my co-workers for their contributions.

## REFERENCES

1. J. Bill, F. Aldinger, *Adv. Mater.* 7, 775 (1995).
2. F. S. Kipping, J. E. Sands, *J. Chem. Soc.* 119, 830 (1921).
3. C. A. Burkhardt, *J. Am. Chem. Soc.* 71, 963 (1949).
4. Wurtz-Fittig reactions are C-C coupling reactions of alkyl halides with alkali metals. This synthetic procedure is often applied for Si-Si coupling of silicon halides. See: (a) A. Wurtz, *Ann. Chim. Phys.* 44, 275 (1855). (b) B. Tollens, R. Fittig, *Ann.* 131, 303 (1864).
5. (a) P. A. Bianconi, F. C. Schilling, T. W. Weidman, *Macromolecules* 22, 1697 (1989). (b) P. A. Bianconi, T. W. Weidman, *J. Am. Cem. Soc.* 110, 2342 (1988).
6. J. Maxka, J. Chrusciel, M. Sasaki, K. Matyjaszewski, *Macromol. Symp.* 77, 79 (1994).
7. M. Ishikawa, M. Kumada, H. Sakurai, *J. Organomet. Chem.* 23, 63 (1970).
8. H. Gilman, C. L. Smith, *J. Organomet. Chem.* 8, 245 (1967).
9. R. West, *J. Organomet. Chem.* 300, 327 (1986) and literature cited therein.
10. (a) C. L. Schilling, T. C. Williams, *Am. Chem. Soc., Polym. Prepr.* 1, 25 (1986). (b) C. L. Schilling, B. *Polym. J.* 18, 355 (1986). (c) C. L. Schilling, J. P. Wesson, T. C. Williams, *Am. Ceram. Soc. Bull.* 62, (1983) 912. (d) C. L. Schilling, *Brit. Polymer Journal* 18, 355 (1986).
11. (a) B. Boury, L. Carpenter, R. J. P. Corriu, *Angew. Chem., Int. Ed. Engl.* 29, 785 (1990). (b) B. Boury, R. J. P. Corriu, W. E. Douglas, *Chem. Mater.* 3, 487 (1991).
12. L. S. Chang, J. Y. Corey, *Organometallics* 8, 1885 (1989).
13. J. F. Harrod, *Coord. Chem. Rev.* 206 – 207, 493 (2000) and literature cited therein.
14. H.-G. Woo, J. F. Walzer, T. Don Tilley, *J. Am. Chem. Soc.* 114, 7047 (1992).
15. F. Gauvin, J. F. Harrod, H. G. Woo, *Adv. Organomet. Chem.* 42, 363 (1998).
16. E. Hengge in *Organosilicon Chemistry*, N. Auner, J. Weiss (Eds.) VCH Weinheim, p. 280 (1994).
17. E. Hengge, M. Weinberger, *J. Organomet. Chem.* 443, 167 (1993).
18. Y. Kimata, H. Suzuki, S. Satoh, A. Kuriyama, *Organometallics* 14, 2506 (1995).
19. U. Herzog, R. Richter, W. Brendler, G. Roewer, *J. Organomet. Chem.* 507, 221 (1996).
20. (a) R. F. Trandell, G. Urry, *J. Inorg. Nucl. Chem.* 40, 1305 (1978). (b) G. Urry, *J. Inorg. Nucl. Chem.* 26, 409 (1964).

21. (a) M. Kumada, K. Tamao, *Adv. Organomet. Chem.* 6, 19 (1968). (b) H. Watanabe, M. Kobayashi, Y. Koike, S. Nagashima, Y. Nagai, *J. Organomet. Chem.* 128, 173 (1977).

22. S. Yajima, J. Hayashi, M. Omori, *Chem. Lett.*, 931 (1975).

23. B. van Aefferden, W. Habel, P. Sartori, *Chemiker Ztg.* 114, 367 (1990).

24. Q. Liu, H.-J. Wu, R. Lewis, G. E. Maciel, L. V. Interrante, *Chem. Mater.* 11, 2038 (1999).

25. (a) A. Tazi Hemida, M. Birot, J. P. Pillot, Dunogues, R. Pailller, *J. Mater. Sci.* 32, 3475 (1997). (b) A. Tazi Hemida, M. Birot, J. P. Pillot, Dunogues, R. Pailller, R. Naslain, *J. Mater. Sci.* 32, 3485 (1997).

26. (a) H.-J. Wu, L. V. Interrante, *Chem. Mater.* 1, (1989) 564. (b) C. K. Whitmarsh, L. V. Interrante, *Organometallics* 10, 1336 (1991).

27. (a) J. L. Speier, US Patent 2,626,272 (1953). (b) J. L. Speier, J. A. Webster, G. H. Barnes, *J. Am. Chem. Soc.* 79, 974 (1957).

28. A. J. Chalk, J. F. Harrod, *J. Am. Chem. Soc.* 87, 16 (1965).

29. R. J. P. Corriu, D. Leclerq, P. H. Mutin, J.-M. Planeix A. Vioux, *Organometallics* 12, 454 (1993).

30. Y. Pang, S. Ijadi-Maghsoodi, T. J. Barton, *Macromolecules* 26, 5671 (1993).

31. U. Lay, Ph.D. Thesis, Universität Heidelberg (1991).

32. (a) R. J. P. Corriu, C. Guérin, B. J. L. Henner, T. Kuhlmann, A. Jean, *Chem. Mater.* 2, 351 (1990). (b) J. L. Bréfort, R. J. P. Corriu, P. Gerbier, C. Guérin, B. J. L. Henner, A. Jean, T. Kuhlmann, *Organometallics* 11, 2500 (1992). (c) R. J. P. Corriu, N. Devylder, C. Guérin, B. J. L. Henner, A. Jean, *Organometallics* 13, 3194 (1994).

33. R. J. P. Corriu, C. Guérin, B. J. L. Henner, A. Jean, H. Mutin, *J. Organomet. Chem.* 396, C35 (1990).

34. R. J. P. Corriu, P. Gerbier, C. Guérin, B. J. L. Henner, R. Fourcade, *J. Organomet. Chem.* 449, 111 (1993).

35. R. J. P. Corriu, P. Gerbier, C. Guérin, B. J. L. Henner, H. Mutin, *Organometallics* 11, 2507 (1992).

36. D. Seyferth in *Inorganic and Organometallic Polymers* (Eds.: M. Zeldin, K. J. Wynne, H. R. Allcock) ACS Symposium Series 360, American Chemical Society, Washington DC, p. 21 (1988).

37. D. H. Ballard, H. Gilman, *J. Organomet. Chem.* 15, 321 (1968).

38. A. B. Holmes, C. L. D. Jennings-White, A. H. Schultess, B. Akinde, D. R. M. Walton, *J. Chem. Soc., Chem. Commun.*, 840 (1979).

39. S. Ijadi-Maghsoodi, T. J. Barton, *Macromolecules* 23, 4485 (1990).

40. R. Bortolin, S. S. D. Brown, B. Parbhoo, *Macromolecules* 23, 2465 (1990).

41. E. Hengge, A. Baumegger, *J. Organomet. Chem.* 369, C39 (1989).

42. M. Ishikawa, Y. Hasegawa, A. Kunai, T. Yamanaka, *J. Organomet. Chem.* 381, C57 (1990).

43. T. Iwahara, S. Hayase, R. West, *Macromolecules* 23, 1298 (1990).

44. (a) M. Ishikawa, Y. Hasegawa, T. Hatano, A. Kunai, *Organometallics* 8, 2741 (1989). (b) M. Ishikawa, T. Hatano, Y. Hasegawa, T. Horoi, A. Kunai, A. Miyai, T. Ishida, T. Tsukihara, T. Yamanaka, T. Koike, J. Shioya, *Organometallics* 11, 1604 (1992).

45. R. J. P. Corriu, P. Gerbier, C. Guérin, B. Henner, *Chem. Mater.* 12, 805 (2000).

46. A. Kunai, E. Toyoda, I. Nagamoto, T. Horio, M. Ishikawa, *Organometallics* 15, 75 (1996).

47. U. Wannagat, *Adv. Inorg. Chem. Radiochem.* 6, 225 (1964).
48. Ammonolysis describes a reaction involving ammonia.
49. H. Schutzenberger, C. R. Colson, *R. Acad. Sci.* 92, 1508 (1885).
50. (a) O. Glemser, K. Beltz, P. Naumann, Z. *Anorg. Allg. Chem.* 291, 4 (1957). (b) O. Glemser, P. Naumann, Z. *Anorg. Allg. Chem.* 298, 134 (1959).
51. D. Seyferth, G. H. Wiseman, C. Prud'homme, *J. Am. Ceram. Soc.* 66, C-13 (1983).
52. A. Stock, K. Somieski, *Ber. Dtsch. Chem. Ges.* 54, 740 (1921).
53. T. Isoda, H. Kaya, H Nishii, O. Funayama, T. Suzuki, Y. Tashiro, *J. Inorg. Organomet. Polym.* 2, 151 (1992).
54. C. R. Blanchard, S. T. Schwab, *J. Am. Ceram. Soc.* 77, 1729 (1994).
55. J. Haug, P. Lamparter, M. Weinmann, F. Aldinger, *Chem. Mater.* 16, 72 (2004).
56. D. M. Narsavage, L. V. Interrante, P. S. Marchetti, G. E. Maciel, *Chem. Mater.* 3, 721 (1991).
57. Nitrogen-rich in this context means N:Si > 4:3.
58. VT 50 is a commercial polyvinylsilazane produced by Hoechst AG, Germany.
59. NCP200 is a commercial polyperhydridosilazane produced by Nichimen Corporation, Japan.
60. J. Peng, Ph.D. Thesis, Universität Stuttgart (2002).
61. H. J. Seifert, Habilitation Thesis, Universität Stuttgart (2002).
62. (a) W. Fink, *Chem. Ber.* 96, 1071 (1963). (b) W. Fink, *Helv. Chim. Acta* 47, 498 (1964). (c) W. Fink, *Angew. Chem.* 78, 803 (1966).
63. K. A. Andrianov, B. A. Ismailov, A. M. Kononov, G. V. Kotrelev, *J. Organomet. Chem.* 3, 129 (1965).
64. C. R. Krüger, E. C. Rochow, *J. Polym. Sci.* A2, 3179 (1964).
65. D. L. Bailey, L. H. Sommer, F. C. Whitmore, *J. Am. Chem. Soc.* 70, 435 (1948).
66. U. Wannagat, F. Brandmair, *Z. Anorg. Allg. Chem.* 280, 223 (1955).
67. J. P. Mooser, H. Nöth, W. Tinhof, *Z. Naturforsch.* 29b, 166 (1974).
68. J. Wilden, A. Wank, M. Asmann, J. V. Heberlein, M. I. Boulos, F. Gitzhofer, *Appl. Organomet. Chem.* 15, 841 (2001).
69. (a) D. Seyferth, G. H. Wiseman, *J. Am. Ceram. Soc.* 67, C-132 (1984). (b) D. Seyferth in *Transformation of Organometallics into Common and Exotic Materials: Design and Activation*, NATO ASI Ser. E: Appl. Sci.-No. Vol. 141, R. M. Laine (Ed.), Kluwer Publ. Dordrecht, p. 133 (1988). (c) D. Seyferth, G. H. Wiseman, *Ultra Structure Processing of Ceramics, Glasses and Composites*, Wiley, New York, p. 265 (1984). (d) D. Seyferth, G. H. Wiseman, *Science of Ceramic Chemical Processing*, Wiley, New York, p. 354 (1986).
70. M. Weinmann, S. Nast, F. Berger, K. Müller, F. Aldinger, *Appl. Organomet. Chem.* 15, 867 (2001).
71. (a) A. Lavedrine, D. Bahloul, P. Goursat, N. S. Choong Kwet Yive, R. J. P. Corriu, D. Leclerq, H. Mutin, A. Vioux, *J. Europ. Ceram. Soc.* 8, 221 (1991). (b) N. S. Choong Kwet Yive, R. J. P. Corriu, D. Leclerq, H. Mutin, A. Vioux, *Chem. Mater.* 4, 141 (1992). (c) N. S. Choong Kwet Yive, R. J. P. Corriu, D. Leclerq, H. Mutin, A. Vioux, *Chem. Mater.* 4, 1263 (1992). (d) O. Delverdier, M. Monthioux, A. Oberlin, A. Lavedrine, D. Bahloul, P. Goursat, *High Temp. Processes* 1, 139 (1992). (e) D. Bahloul, M. Pereira, P. Goursat, N. S. Choong Kwet Yive, R. J. P. Corriu, *J. Am. Ceram. Soc.* 76, 1156 (1993).
72. N. S. Choong Kwet Yive, R. J. Corriu, D. Leclerq, P. H. Mutin, A. Vioux, *New. J. Chem.* 15, 85 (1991).

73. J. Bill, H. Seitz, G. Thurn, H. Dürr, J. Canel, B. Z. Janos, A. Jalowiecki, A. Sauter, S. Schempp. H. P. Lamparter, J. Mayer, F. Aldinger, *Phys. Stat. Sol.* 166A, 269 (1998).

74. Poly(silsesquiazane)s are polymeric silazanes in which three silicon-bonded substituents are nitrogen atoms and in which N:Si = 1.5.

75. K. A. Andrianov, G. V. Kotrelev, *J. Organomet. Chem.* 7, 217 (1967).

76. M. G. Voronkov, V. I. Lavent'yev, *Top. Curr. Chem.* 102, 199 (1982).

77. V. W. Day, W. G. Klemperer, V. V. Mainz, D. M. Millar, *J. Am. Chem. Soc.* 107, 8262 (1985).

78. G. Calzaferri, D. Herren, R. Imhof, *Helv. Chim. Acta* 74, 1278 (1991).

79. B. Räke, H. W. Roesky, I. Usón, P. Müller, *Angew. Chem.* 110, 1508 (1998).

80. E. Kroke Y.-L. Li, C. Konetschny, E. Lecomte, C. Fasel, R. Riedel, *Mat. Sci. Eng.* R26, 97 (2000).

81. M. Weinmann, A. Zern, F. Aldinger, *Adv. Mater.* 13, 1704 (2001).

82. M. Hörz, H. Kummer, A. Zern, F. Berger, K. Müller, F. Aldinger, M. Weinmann, *J. Europ. Ceram. Soc.*, 25, 99 (2005).

83. M. Weinmann, Habilitation Thesis, Universität Stuttgart (2003).

84. M. Weinmann, M. Hörz, A. Müller, Chemiedozententagung 2004, Dormund (Germany).

85. M. Hörz, Ph.D. Thesis, Universität Stuttgart (in progress).

86. Y. D. Blum, R. M. Laine, *Organometallics* 5, 2081 (1986).

87. R. M. Laine, Y. D. Blum, A. Chow, R. D. Hamlin, K. B. Schwartz, D. J. Rowcliffe, *Polym. Prep.* 28, 393 (1987).

88. a) K. A. Youngdahl, R. M. Laine, R. A. Kennish, T. R. Cronin, G. G. Balavoine, *Mat. Res. Soc. Symp. Proc.* 121, 489 (1988). b) Y. D. Blum, K. B. Schwartz, E. J. Crawford, R. D. Hamlin, *Mat. Res. Soc. Symp. Proc.* 121, 565 (1988).

89. Y. Blum, personal information.

90. H. Q. Liu, J. F. Harrod, *Organometallics* 11, 822 (1992).

91. W. Fink, *Angew. Chem.* 78, 803 (1966).

92. D. Seyferth, J. M. Schwark, R. M. Steward, *Organometallics* 8, 1980 (1989).

93. E. Duguet, M. Schappacher, A. Soum, *Macromolecules* 25, 4835 (1992).

94. S. Bruzaud, A. Soum, *Macromol. Chem. Phys.* 197, 2379 (1996).

95. S. Bruzaud, A.-F. Mingotaud, A. Soum, *Macromol. Chem. Phys.* 198, 1873 (1996).

96. C. Cazalis, A.-F. Mingotaud, A. Soum, *Macromol. Chem. Phys.* 198, 3441 (1996).

97. M. Bouquey, C. Brochon, S. Bruzaud, A.-F. Mingotaud, M. Schappacher, A. Soum, *J. Organomet. Chem.* 521, 21 (1996).

98. H. Ishikawa, M. Kumada, H. Sakurai, *J. Organomet. Chem.* 23, 63 (1970).

99. M. Rivière-Baudet, P. Rivière, J. Satgé, G. Lacrampe, *Neth. Chem. Soc.* 98, 42 (1979).

100. (a) E. A. Ebsworth, M. J. Mays, *J. Chem. Soc.*, 4879 (1961). (b) E. A. Ebsworth, M. J. Mays, *Angew. Chem.* 74, 117 (1962).

101. (a) J. Pump, U. Wannagat, *Angew. Chem.* 74, 117 (1962). (b) J. Pump, U. Wannagat, *Ann. Chem.* 652, 21 (1962). (c) J. Pump, E. G. Rochow, U. Wannagat, *Monatsh. Chem.* 94, 588 (1963).

102. L. Birkofer, A. Ritter, P. Richter, *Tetrahedron Lett.* 5, 195 (1962).

103. J. Pump, E. G. Rochow, *Z. Anorg. Allg. Chem.* 330, 101 (1964).

104. J. F. Klebe, J. G. Murray, US Patent 3,352,799 (1968).

105. (a) A. O. Gabriel, R. Riedel, *Angew. Chem. Int. Ed. Engl.* 36, 384 (1997). (b) A. O. Gabriel, R. Riedel, S. Storck, W. F. Maier, *Appl. Organomet. Chem.* 11, 833 (1997).

106. (a) A. Kienzle, J. Bill, F. Aldinger, R. Riedel, *Nanostruct. Mater.* 6, 349 (1995). (b) C. Balan, K. W. Völger, E. Kroke, R. Riedel, *Macromolecules* 33, 3404 (2000).

107. A. Kienzle, A. Obermeyer, R. Riedel, F. Aldinger, A. Simon, *Chem. Ber.* 126, 2569 (1993).

108. J. Schuhmacher, Ph.D. Thesis, Universität Stuttgart (2000).

109. R. Riedel, A. Greiner, G. Miehe, W. Dreßler, H. Fueß, J. Bill, F. Aldinger, *Angew. Chem. Int. Ed. Engl.* 36, 603 (1997).

110. L. Jäger, H, Köhler, *Sulfur Reports* 12, 159 (1992).

111. (a) M. Takamizawa, T. Kobayashi, A. Hayashida, Y. Takeda, US Pat. 4,550,151 (1985). (b) M. Takamizawa, T. Kobayashi, A. Hayashida, Y. Takeda, US Pat. 4,604,367 (1986). (c) M. Takamizawa, T. Kobayashi, A. Hayashida, Y. Takeda, N. Joetsu, Deutsches Patent DE 344 430 6 (1986).

112. H. Nöth, Z. *Naturforsch.* B16, (1961) 618.

113. (a) D. Seyferth, H. Plenio, *J. Am. Ceram. Soc.* 73, 2131 (1990). (b) D. Seyferth, H. Plenio, W. S. Rees Jr., K. Büchner, in *Frontiers of Organosilicon Chemistry*, A. R. Bassindale, P. P. Gaspar (Eds.) The Royal Society of Chemistry, Cambridge, UK, p. 15 (1991).

114. (a) K. Su, E. E. Remsen, G. A. Zank, L. G. Sneddon, *Polym. Prep.* 34, 334 (1993). (b) K. Su, E. E. Remsen, G. A. Zank, L. G. Sneddon, *Chem. Mater.* 5, 547 (1993).

115. (a) T. Wideman, K. Su, E. E. Remsen, G. A. Zank, L. G. Sneddon, *Chem. Mater.* 7, 2203 (1995). (b) T. Wideman, K. Su, E. E. Remsen, G. A. Zank, L. G. Sneddon, *Mat. Res. Soc. Symp. Proc.* 410, 185 (1996).

116. T. Wideman, E. Cortez, E. E. Remsen, P. J. Carroll, L. G. Sneddon, *Chem. Mater.* 9, 2218 (1997).

117. D. Srivastava, E. N. Duesler, R. T. Paine, *Eur. J. Inorg. Chem.* 855 (1998).

118. J. Haberecht, A. Krummland, F. Breher. B. Gebhardt, H. Rüegger, R. Nesper, H. Grützmacher, *Dalton Trans* 11, 2126 (2003).

119. J. Haberecht, F. Krumeich, H. Grützmacher, R. Nesper, *Chem. Mater.* 16, 418 (2004).

120. R. Nesper, J. Haberecht, H. Grützmacher, Pat. CH-0149/03 (2003).

121. M. Jansen, T. Jäschke, Z. *Anorg. Allg. Chem.* 625, 1957 (1999).

122. (a) M. Jansen, H.-P. Baldus, Ger. Offen. DE 41 07 108 A1 (1992). (b) H.-P. Baldus, O. Wagner, M. Jansen, *Mat. Res. Soc. Symp. Proc.* 271, 821 (1992). (c) H.-P. Baldus, M. Jansen, O. Wagner, *Key Eng. Mat.* 89-91, 75 (1994). (d) H.-P. Baldus, M. Jansen, *Angew. Chem. Int. Ed. Engl.* 36, 328 (1997).

123. P. Baldus, M. Jansen, D. Sporn, *Science* 285, 699 (1999).

124. (a) R. Riedel, A. Kienzle, G. Petzow, M. Brück, T. Vaahs, Ger. Offen. DE 43 20 783 A1 (1994). (b) R. Riedel, A. Kienzle, G. Petzow, M. Brück, T. Vaahs, Ger. Offen. DE 43 20 786 A1 (1994).

125. (a) R. Riedel, A. Kienzle, W. Dressler, L. Ruwisch, J. Bill, F. Aldinger, *Nature* 382, 796 (1996). (b) R. Riedel, J. Bill, A. Kienzle, *Appl. Organomet. Chem.* 10, 241 (1996).

126. A. Kienzle, Ph.D. Thesis, Universität Stuttgart (1994).

127. P. R. Jones, J. K. Myers, *J. Organomet. Chem.* 34, C9 (1972).

128. M. Weinmann, T. W. Kamphowe, P. Fischer, F. Aldinger, *J. Organomet. Chem.* 592, 115 (1999).

129. (a) L. M. Ruwisch, P. Dürichen, R. Riedel, *Polyhedron* 19, 323 (2000). (b) L. M. Ruwisch, Ph.D. Thesis, Technische Universität Darmstadt (1998). (c) R. Riedel, L. M. Ruwisch, in *McGraw-Hill Yearbook of Science & Technology 1999*, McGraw Hill, New York, p. 70 (1998).

130. (a) F. Aldinger, M. Weinmann, J. Bill, *Pure Appl. Chem.* 70, 439 (1998). (b) M. Weinmann, in *Precursor-Derived Ceramics*, J. Bill, F. Wakai, F. Aldinger (Eds.) Wiley-VCH, Weinheim, p. 83 (1999). (c) M. Weinmann, F. Aldinger, In *New Properties from Atomic Level Processing — Proc. Symposium on International Joint Project Ceramics Superplasticity*, Tokyo, p. 16 (1999). (d) M. Weinmann, H. J. Seifert, F. Aldinger, In *Contemporary Boron Chemistry*, Eds.: M. G. Davidson, A. K. Hughes, T. B. Marder, K. Wade, The Royal Society of Chemistry, Cambridge, p. 88 (2000).

131. M. Weinmann, J. Schuhmacher, H. Kummer, S. Prinz, J. Peng, H. J. Seifert, M. Christ, K. Müller, J. Bill, F. Aldinger, *Chem. Mater.* 12, 623 (2000).

132. M. Weinmann, S. Nast, F. Berger, K. Müller, F. Aldinger, *Appl. Organomet. Chem.* 15, 867 (2001).

133. M. Weinmann, J. Bill, F. Aldinger, Ger. Offen. DE 197 41 459 A 1 (1999).

134. (a) M. Weinmann, J. Bill, F. Aldinger, Ger. Offen. DE 197 41 459 A 1 (1999). (b) M. Weinmann, T. W. Kamphowe, J. Schuhmacher, K. Müller, F. Aldinger, *Chem. Mater.* 12, 2112 (2000). (c) M. Weinmann, T. W. Kamphowe, S.-H. Lee, F. Aldinger, in *Verbundwerkstoffe und Werkstoffverbunde*, B. Wielage, G. Leonhardt (Eds.) Wiley-VCH, Weinheim p. 268 (2001).

135. See for example: (a) B. Marciniec, J. Gulinski, W. Urbaniak, Z. W. Kornetka, *Comprehensive Handbook on Hydrosilylation Chemistry*, Pergamon Press, Oxford (1992). (b) M. A. Brook, *Silicon in Organic, Organometallic and Polymer Chemistry*, John Wiley & Sons, New York, p. 401 (2000). (c) S. Pawlenko, *Organosilicon Chemistry*, Walter de Gruyter, Berlin – New York, p. 35 (1986). (d) I. Ojima, Z. Li, J. Zhu, in *The Chemistry of Organic Silicon Compounds*, Vol. 2, Z. Rappoport, Y. Apeloig (Eds.), J. Wiley & Sons, London (1998).

136. B. Wrackmeyer, O. L. Tok, Y. N. Bubnov, *Angew. Chem.* 111, 214 (1999).

137. S.-H. Lee, Ph.D. Thesis, Universität Stuttgart (2004).

138. E. R. Generazio, *Advanced Ceramic Matrix Composites: Design Approaches, Testing & Life Prediction Methods*, Technomic Pub Co., Lancaster, PA (1995).

139. (a) J. J. Ebelmen, *C. R. Acad. Sci.* 19, 398 (1844). (b) J. J. Ebelmen, *Ann.* 57, 331 (1846).

140. See for example: (a) C. F. Brinker, G. W. Scherer, *Sol-Gel Science, The Physics and Chemistry of Sol-Gel Processing*, Academic Press, San Diego, CA (1990). (b) L. C. Klein (Ed.), *Sol-Gel Technology for Thin Films, Fibers, Preforms, Electronics and Speciality Shapes*, Noyes Publications, Park Ridge, NJ (1988). (c) C. K. Narula, *Ceramic Precursor Technology and its Applications*, Marcel Dekker, New York (1995). (d) U. Schubert, N. Hüsing, Synthesis of inorganic materials, in *Sol-Gel Processing of Silicate Materials*, Wiley-VCH, Weinheim, p. 200 (2000).

141. (a) M. Weinmann, R. Haug, J. Bill, F. Aldinger, J. Schuhmacher, K. Müller, *J. Organomet. Chem.* 541, 345 (1997). (b) M. Weinmann, R. Haug, J. Bill, M. De Guire, F. Aldinger, *Appl. Organomet. Chem.* 12, 725 (1998). (c) J. Schuhmacher, K. Müller, M. Weinmann, J. Bill, F. Aldinger in Proc. Werkstoffwoche 1998, Band VII, Keramik/Simulation, J. Heinrich, G. Ziegler. W. Hermel, H. Riedel (Eds.) Wiley-VCH, Weinheim p. 321 (1999).

142. (a) M. Weinmann, A. Zern, M. Hörz, F. Berger, K. Müller, F. Aldinger, *J. Met. Nano. Mater.* 386-388, 335 (2002). (b) M. Weinmann, M. Hörz, F. Berger, A. Müller, K. Müller, F. Aldinger, *J. Organomet. Chem.* 659, 29 (2002).

143. M. Weinmann et al., to be published.

144. (a) S. Yajima, J. Hayashi, M. Omori, *Chem. Lett.*, 1209 (1975). (b) S. Yajima, *Ceram. Bull.* 62, 893 (1983). (c) S. Yajima, J. Hayashi, M. Omori, K. Okamura, *Nature* 261, 683 (1976). (d) S. Yajima, J. Hayashi, M. Omori, K. Okamura, *J. Am. Ceram. Soc.* 59, 324 (1976).

145. Nicalon™ and Hi-Nicalon™ are trademarks of Nippon Carbon Co., Tokyo, Japan.

146. Y. Nakaido, Y. Otani, N. Kozakai, S.Otani, *Chem. Lett.*, 705 (1987).

147. (a) G. E. LeGrow, T. F. Lim, J. Lipowitz, R. S. Reaoch, *Am. Ceram. Soc. Bull.* 66, 363 (1987). (b) J. Lipowitz, *Am. Ceram. Soc. Bull.* 70, 1888 (1991).

148. G. Motz, J. Hacker, G. Ziegler, *Ceramic Engineering & Science Proc.* 21, 307 (2000).

149. O. Delverdier, M. Monthioux, D. Mocaer, R. Pailler, *J. Europ. Ceram. Soc.* 14, 313 (1994).

150. H.-P. Baldus, G. Pasing, D. Sporn, A. Thierauf, *Ceram. Trans.* 58, 75 (1995).

151. (a) S. Bernard, M. Weinmann, D. Cornu, P. Miele, F. Aldinger, *J. Europ. Ceram. Soc.*, 25, 251 (2005). (b) S. Bernard, M. Weinmann, P. Gerstel, P. Miele, F. Aldinger, *J. Mater. Chem.* 15, 289 (2005).

152. J. Livage, C. Sanchez, F. Babonneau in *Chemistry of Advanced Materials*, L. V. Interrante, M. J. Hampden- Smith (Eds.), Wiley-VCH, New York. ISBN 0-471-18590-6 (1998).

153. (a) J. Bill, D. Heimann, *J. Europ. Ceram. Soc.* 16, 1115 (1996). (b) D. Heimann, Ph.D. Thesis, Universität Stuttgart (Germany) 1997.

154. (a) M. L. Hitchman, K. F. Jensen (Eds.), Chemical Vapor Deposition: Principles and Applications, Academic Press (1993). (b) W.S. Rees, CVD of Nonmetals, Wiley-VCH Verlag (1998). (c) R. U. Claessen, Understanding Key Molecular Properties in a CVD Process Feasibility Study, BoD GmbH, Norderstedt, Germany (2002).

155. J. E. Mahan, Physical Vapor Deposition of Thin Films, Wiley Interscience (2000).

156. (a) W. Krenkel, R. Naslain, H. Schneider (Eds.), *High Temperature Ceramic Matrix Composites,* Wiley-VCH, Weinheim (2001). (b) T. G. Gutowski (Ed.), *Advanced Composites Manufacturing,* John Wiley & Sons. (1997). (c) B. Wielage, G. Leonhardt (Eds.), *Verbundwerkstoffe und Werkstoffverbunde,* Wiley-VCH, Weinheim (2001).

157. W. Krenkel, Ph.D. Thesis, Universität Stuttgart (2000).

158. T. Kamphowe, Ph.D. Thesis, Universität Stuttgart (2000).

159. S.-H. Lee, M. Weinmann, F. Aldinger, to be published.

# 19 Polymer Pyrolysis

*Masaki Narisawa*

## CONTENTS

## I. INTRODUCTION

The development of processing ceramics from polymer precursors has attracted great attention. In particular, inorganic polymers containing silicon are important for SiC-based ceramic synthesis. SiC ceramics have the advantage of high-temperature stability in an oxidation atmosphere. SiC is not readily sintered, and thus is difficult to obtain in either fiber or film form by traditional inorganic processes.

SiC fiber was produced from polycarbosilane (PCS) by Yajima et al.[1-4] in 1975, which is the earliest case of organosilicon polymer utilization for an industrial structural material. In the Yajima process, PCS was mainly synthesized from polydimethylsilane (PDS) by a thermal conversion process using an autoclave or an open reflux system. This is the commonly available PCS.[5-7] Its melt spinability, solubility in various organic solvents, and stability for storing at room temperature are critically important for industrial uses.

The chemical composition of PCS is reported to be $SiC_{1.77}H_{3.70}O_{0.035}$.[7] The molecular structure of PCS is, however, difficult to represent precisely. $^{29}Si$ nuclear magnetic resonance (NMR) studies indicate that the PCS structure is represented by three simple units. These are silicon bonded to four carbon atoms ($SiC_4$), silicon bonded to three carbon atoms ($SiC_3H$), and silicon bonded to $x$ carbon atoms and 4-$x$ silicon atoms ($SiC_xSi_{4x}$, $x = 1, 2,$ or 3). In a typical PCS synthesized without catalysts, $SiC_4$ and $SiC_3H$ units are dominant, with the

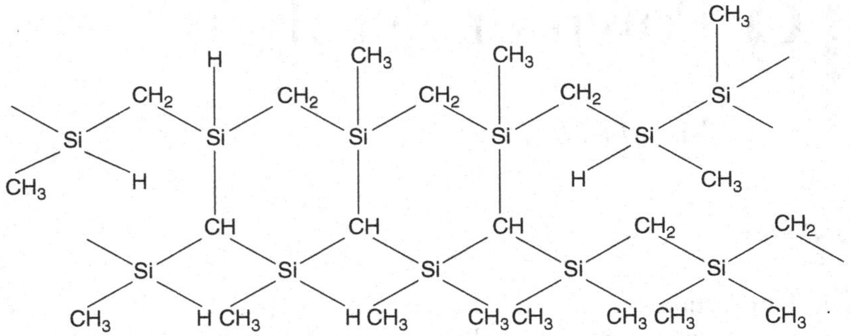

**FIGURE 19.1** Chemical structure of PCS.

$SiC_xSi_{4x}$ unit being minor.[7,8] Therefore PCS is considered to possess a two-dimensional ladder or three-dimensional cage Si-C structure, as partially shown in Figure 19.1.

Yajima's innovation directly contributed to Si-C-O fiber (Nicalon; NCK, Toyama, Japan) and Si-Ti-C-O fiber (Tyranno; Ube Industries, Tokyo, Japan) production in the early years.[9,10] The Tyranno fiber production process showed the possibility of introducing various alkoxides in the starting PCS to modify the resulting SiC grain boundaries.

Takeda et al.[11] and Okamura and Seguchi[12] investigated the radiation-curing process on spun PCS fibers. The oxygen content in the pyrolyzed fibers is successfully diminished by this curing method. The oxygen in conventional Si-C-O fibers is evolved in the form of CO and SiO at 1573 K to 1773 K (Figure 19.2). Such gas evolution is generally accompanied by surface pore and coarse SiC crystallite growth on the fiber surface, which should reduce the fiber strength. Thus the tensile strength of conventional Si-C-O fibers rapidly decreases at high temperatures. The radiation curing process, however, contributes Si-C fiber production (Hi-Nicalon by NCK) with high heat resistance, which contains a considerable amount of excess carbon. Near-stoichiometric SiC fiber was developed from radiation-cured PCS fibers using a hydrogen-rich atmosphere for fiber pyrolysis.[13] Thus tailored near-stoichiometric SiC fiber with high modulus is also commercially available (Hi-Nicalon-S; NCK).

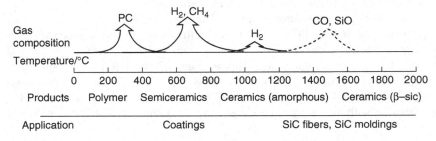

**FIGURE 19.2** Gas evolution from oxidation-cured PCS during thermal decomposition.

On the other hand, Ishikawa et al.[14] produced highly heat resistant SiC fibers by introducing a small amount of aluminum in a starting PCS precursor. After the elimination of oxygen in the form of CO and SiO by heat treatment, SiC grain boundaries can keep sufficient contact to give sintered SiC fiber. This fiber maintained strength even with heat treatment at 2273 K (Tyranno SA; Ube Industries).

These advanced SiC-based fibers have been widely investigated and used for reinforcement in ceramic matrix composites because of their high tensile strength, high heat resistance, and thin diameters, appropriate for being shaped into fabrics.

Besides the continuous fibers, application of metallorganic polymers to heat-resistant coatings, dense ceramic moldings, porous bodies, and SiC matrix sources in advanced ceramics via polymer infiltration pyrolysis (PIP) have been developed. Novel precursor polymers have been synthesized and investigated for ceramics in addition to PCS (Table 19.1).[15–19] For SiC ceramics, various Si-C backbone polymers have been synthesized.[20–25] Their polymer nature (e.g., viscosity, stability, cross-linking mechanism, and ceramic yield) are, however, fairly different from PCS. On the other hand, polysilazane, perhydropolysilazane, polyborazine, aluminum nitride polymers, and their copolymers have been investigated

## TABLE 19.1
## Ceramics from Polymer Precursors

| Decade | Polymer | Ceramics |
|---|---|---|
| 1970 | Polysilazanes | Si-C-N |
| | Polycarbosilane | Si-C-O (SiC like microstructure) |
| | | Si-N-O, $Si_3N_4$ [Decade: 1980] |
| | | Si-C [Decade: 1990] |
| | | SiC (stoichiometric) [Decade: 1990] |
| | Polysilanes | Si-C |
| | Polyborosiloxane | Si-B-C-O |
| 1980 | Polytitanocarbosilane | Si-Ti-C-O |
| | Polymetallocarbosilanes | Si-M-C-O |
| | Polycarbosilanes (linear, branched, or with substituted side groups) | Si-C, Si-C-O, SiC (stoichiometric) |
| | Polysilazanes | Si-N-C (containing $Si_3N_4$ phase) |
| | Perhydropolysilazane | $Si_3N_4$ (stoichiometric) |
| | B-N-C polymers | BN-$B_4C$ |
| 1990 | Polyborazine | BN |
| | Aluminum nitride polymer | AlN |
| | Hybrid co-polymers | $Si_3N_4$/AlN, $Si_3N_4$/BN, etc. |
| | Polyborosilazanes | Si-C-B-N amorphous and crystalline |
| | Polycarboxilane modified with Al | SiC-(Al) |

as ceramic precursors to yield SiC-$Si_3N_4$, BN-$B_4C$, $Si_3N_4$, BN, AlN, Si-C-B-N, and AlN-BN ceramics after pyrolysis.[26-35] Riedel et al.[36] and Chang et al.[37] synthesized a polyborosilazane precursor for Si-C-B-N materials ($Si_{3.0}B_{1.0}C_{4.3}N_{2.0}$). The obtained Si-C-B-N material does not show any mass loss or chemical composition change at 1273 K to 2273 K.

As mentioned above, the development of organometallic chemistry has contributed to molecular structure arrangement of ceramic precursors. In order to control chemical composition, ceramic yield, and thermosetting properties, various monomer units can be built in a single polymeric precursor as desired. In industry today, however, various polymer-base materials have been produced from rather limited kinds of fundamental low-price polymers. Blending of the starting polymers is one of the key technologies. From the investigation of various organometallic precursors, it became apparent that the polymer nature of the synthesized precursors is also diverse. Precursor manufacturing (molding, impregnation, curing, pyrolysis, etc.) must be tuned to the starting polymer. As compared with novel precursor synthesis, investigation of polymer blend techniques in the precursor method is at a quite early stage of development.[38-41] It is probable, however, that the polymer blend technique will be used in a wide range of future applications of the precursor method (Figure 19.3). The high cost and complicated steps now required for the control of the precursor composition, chemical nature, and nanomicrostructure are expected to be drastically reduced.

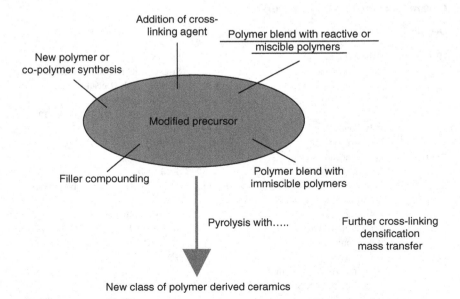

**FIGURE 19.3** Precursor modification processes and the polymer blend technique.

## II. POLYMETHYLSILANE AND PCS: PROMISING PRECURSORS FOR SIC-BASED MATERIALS VIA THE POLYMER BLEND TECHNIQUE

In the following sections, the ceramization process of polymethylsilane (PMS) and polyvinylsilane (PVS) is described from the viewpoint of the thermal history of the starting polymers and spectroscopic analysis. After these descriptions, simple applications of these polymers for SiC ceramic fibers via the polymer blend technique are discussed. The major modification of PCS with PMS or PVS is carried out not in an atomic scale, but sufficiently homogeneous in a polymer chain scale.

### A. CERAMIZATION PROCESS OF PMS

Polymethylsilane has been studied for a long time as a promising precursor to SiC because of its ideal stoichiometry in the chemical composition.[42-44] PMS, however, possesses a chemically active nature (rapid gas evolution during heat treatment, thermosetting property at relatively low temperature, inflammable nature in an oxidation atmosphere, etc.) as compared with PCS. This active nature of PMS is probably due to the large number of Si-H groups on PMS not surrounded by larger chemical groups and facile Si-Si chain cleavage with heat or ultraviolet (UV) radiation. Thus chain cleavage often causes the low ceramic yield for PMS. Some effective cross-linking agents for PMS have been reported on the basis of the reactivity of Si-H groups.[43-46]

Recently detailed investigations were performed on the reflux treatment process and pyrolysis process of PMS-based precursors without any cross-linking agents using a vertical furnace with a reflux condenser.[47,48] By adjusting the reflux conditions, various kinds of condensed resins with various ceramic yields were obtained. The starting PMS was synthesized by a Wurtz condensation reaction. The PMS obtained was a pale yellow, viscous liquid ($M_w = 1780$, $M_w/M_n = 2.0$). The molecular structure was characterized to be $[-(CH_3SiH)_{0.60}-(CH_3Si)_{0.40}=]_n$ by $^1H$ NMR. In order to investigate the reflux effect on the resulting ceramic yield, two different preheat treatments were performed on the as-synthesized PMS:

1. About 1.0 g of the as-synthesized PMS was placed in a quartz crucible suspended in a furnace and heat treated at 523 to 723 K (heating rate 5 K/min, holding time 2 h). The argon flow rate was 1 L/min. All the volatile compounds that formed from the PMS during the heat treatment were flowed out under this condition.

2. As-synthesized PMS was placed in a Pyrex tube (diameter 20 mm) to be refluxed. The Pyrex tube was then placed inside a larger Pyrex cell (diameter 40 mm) equipped with a reflux condenser and a vacuum pump. After sufficient evacuation, argon gas was introduced in the cell and flowed out at a rate of 300 ml/min. The cell was heated at 423 to 723 K (heating rate 5 K/min, holding time 2 h).

**TABLE 19.2**
**PMS Resin Recovery and Ceramic Yield of the Cross-Linked PMS Resin at 1273 K**

| Pre-Heat Treatment Temperature (K) | Reflux | | Open Ar Gas Flow | |
|---|---|---|---|---|
| | Recovery (%) | Ceramic Yield at 1273 K (%) | Recovery (%) | Ceramic Yield at 1273 K (%) |
| 423 | 96 | 40 | — | — |
| 473 | 92 | 62 | — | — |
| 523 | 87 | 77 | 85 | 60 |
| 573 | 81 | 90 | 68 | 76 |
| 623 | 78 | 96 | 48 | 88 |
| 673 | 75 | 97 | 47 | 91 |
| 723 | 74 | 98 | 43 | 96 |

After these preheat treatment processes, the two treated (cross-linked) PMS resins were pyrolyzed at 1273 K. Resin recovery after preheat treatment and ceramic yield of these cross-linked resins at 1273 K are summarized in Table 19.2. Use of the reflux system is effective to increase resin recovery and ceramic yield. Figure 19.4 shows the overall ceramic yield of the starting PMS with different thermal histories. The dotted line indicates the intrinsic ceramic yield from PMS by direct pyrolysis up to 1273 K (about 30%). The filled circle indicates the ceramic yield at 1273 K with two-step pyrolysis, in which reflux treatment is the first step. Overall ceramic yield begins to increase at 423 to 523 K, and is saturated (approximately 75%) at 623 K. Even when a preheat treatment step on PMS is a simple heat treatment in an open argon gas flow (open circle), the overall

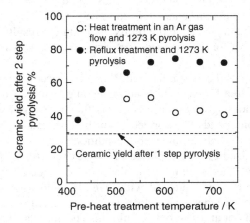

**FIGURE 19.4** Ceramic yield of PMS after two-step pyrolysis.[48]

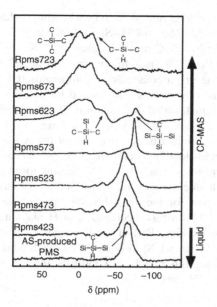

**FIGURE 19.5** ²⁹Si NMR of reflux-treated PMS.[47]

ceramic yield increases at 523 to 573 K (approximately 50%). Beyond this temperature range, the effect of simple heat treatment (as a first step) on the ceramic yield is considerably reduced. These results suggest that the effective cross-linking between PMS macromolecules occurs at temperatures of 523 to 573 K. The reflux heat treatment process possibly plays a role in trapping methylsilane oligomers to be cross-linked again with the PMS resin at the cell bottom.

Figure 19.5 shows ²⁹Si NMR spectra of condensed PMS resins after reflux treatment. At reflux temperatures up to 573 K, the intensity of the signal assigned to $SiC(H)Si_2$ (−67 ppm) decreases, while the intensity of the signal assigned to $SiCSi_3$ (−75 ppm) increases. At 573 K, the sharp signal of $SiCSi_3$ becomes dominant. Above 573 K, the signals assigned to $SiC_2(H)Si$ (−38 ppm), $SiC_3Si$ or $SiC_3H$ (−15 ppm), and $SiC_4$ (0 ppm) appear. At 723 K, the intensity of these signals shows a marked increase, while the signals assigned to $SiCSi_3$ disappear.

¹³C NMR spectra of the same PMS resins supplied additional information on the molecular structure after reflux treatment. Up to 573 K, the signal assigned to $C(H_3)Si$ (methyl groups: −10 ppm) was dominant, showing a continuous downfield shift with the reflux temperature rising. Above 573 K, the signal assigned to $C(H_3)Si$ almost disappeared, while the broad signal assigned to $C(H_2)Si_2$ (−15 to −20 ppm) became dominant. At 723 K, the signal was broadened and became symmetrical in signal shape. The downfield shift (13 ppm in peak position) was also observed. These changes probably correspond to $CSi_4$ formation.

From these observed spectra, it is apparent that the Si-Si cross-linking at 523 to 573 K accompanied by Si-H consumption plays a main role in ceramic yield improvement. Such Si-Si cross-linking is converted to Si-CH₂-Si bridges at 623 K with the insertion of methyl groups to Si-Si bonds, which is known as the Kumada rearrangement process.[5,49] The Si-CH$_2$-Si bridging thus formed is moderately converted to inorganic-like Si-C networks above 673 K.

It is curious that the Si-Si cross-linking process at 523 to 573 K is accompanied by an evolution of gaseous silicon compounds (CH$_3$-SiH$_3$, (CH$_3$)$_2$SiH$_2$, etc.). The methylsilane oligomer is also formed in the same temperature range, which can be seen as small liquid droplets (mist) in the reflux cell. These silicon compounds must be formed by Si-Si bond cleavage in the PMS main chains. Such conflict suggests a redistribution reaction for the cross-linking mechanism, which intrinsically yields low molecular weight oligomers as in the cross-linking process.[45] The redistribution reaction should be composed of various chemical reactions including chain cleavages, rearrangements, and substitutions. We show the probable mechanism for the cross-linking process based on silyl radical behavior in Scheme 19.1. In this mechanism, Si-H plays a role in silyl radical trapping to form cross-linking. Part of the hydrogen radicals formed after silyl radical trapping probably attack Si-Si bonds in the PMS main chains to form new silyl radicals, while the other hydrogen radicals form pairs to yield gaseous compounds, H$_2$, SiH$_3$-CH$_3$, and SiH$_4$.

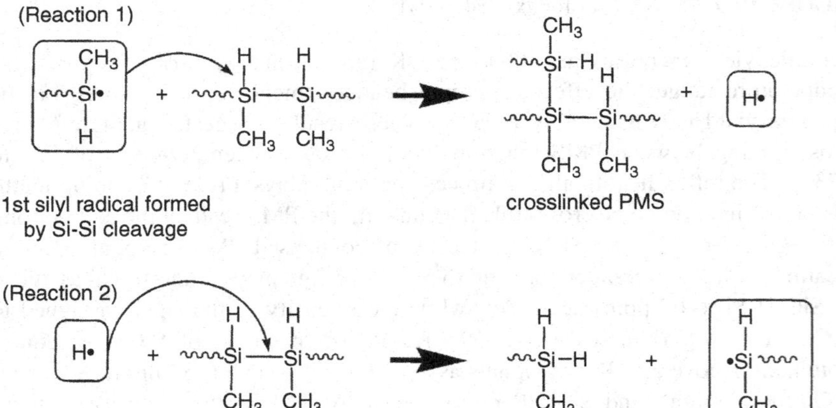

**SCHEME 19.1** Redistribution reaction via silyl radicals (chain reaction).[48]

## B. UTILIZATION OF PMS FOR CERAMIC PRECURSOR

The cross-linking nature of PMS, indicated in NMR and thermogravimetric (TG) analysis, is potentially useful to modify PCS base ceramic precursors. In particular, utilization for ceramic fiber production has attracted considerable interest. The conventional organic fiber production process is composed of the various polymer blend techniques for controlling melt spinability, fiber morphology,

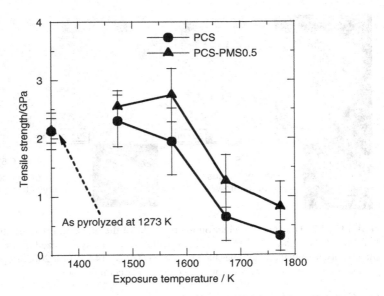

**FIGURE 19.6** Tensile strength of the Si-C-O fibers after high-temperature exposure in an argon gas flow.

resulting functionality, etc. We prepared melt spinable precursor for SiC fibers by merely blending PMS to PCS at 0.5 mass%.[50] This PCS-0.5 PMS polymer blend possesses almost the same spinnability as PCS at 573 K. Oxidation curing is also available. The ceramic yield after pyrolysis is substantially increased, while the oxygen content in the resulting fiber is not increased by the PMS addition. Detailed investigation of the heat treatment condition on ceramic yield reveals that the PMS in the system traps the volatile species (silane or carbosilane oligomers) at approximately 573 K to increase the ceramic yield.

The fiber degradation derived from the PCS-0.5 PMS precursor after high-temperature heat treatment is also reduced. Figure 19.6 shows the tensile strength of the PCS-0.5 PMS fibers after high-temperature exposures. The PCS-0.5 PMS fiber shows the highest strength after 1573 K exposure, while the PCS fiber shows the highest strength after 1473 K exposure.

The mass loss with CO and SiO evolutions is also reduced in the PCS-0.5 PMS fiber. Probably the high degree of cross-linking at relatively low temperature contributes to the dense Si-C-O amorphous formation even after high-temperature pyrolysis. Mass transfer phenomena (evolution of $H_2$, evolutions of SiO and CO, segregation of carbon domains, etc.) corresponding to fiber degradation are considered to be reduced. Such a reduced degradation process exactly influences the SiC nucleation and growth process in the Si-C-O fibers. Figure 19.7 shows transmission electron micrographs of the fibers after heat treatment at 1773 K. Heterogeneous SiC nucleation and growth is observed in PCS fibers, while homogeneous crystallite growth is observed in PCS-0.5 PMS fibers.

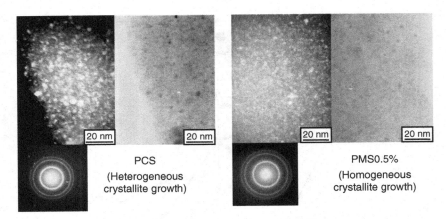

**FIGURE 19.7** Transmission electron micrographs of the obtained fiber after treatment at 1773 K. (From Narisawa, M. et al., SiC ceramic fibers synthesized from polycarbosilane–polymethylsilane polymer blends, *Ceram. Trans.*, 144A, 173, 2002. Reprinted with permission of The American Ceramic Society, www.ceramics.org.)

## C. Ceramization Process of PVS

Polyvinylsilane was synthesized from vinylsilane ($CH_2$=CH-$SiH_3$) by radical polymerization using 2,2-azobis(isobutylbenzene) (AIBN) as a catalyst.[51,52] This is a type of organosilicon polymer containing -$CH_2$-CH($SiH_3$)- and -$SiH_2$-$CH_2$-$CH_2$- units in a backbone. In previous studies by Boury et al.[24] and Corriu et al.,[53] -$SiH_2$-$CH_2$-$CH_2$- backbone polymers were suggested to be promising as SiC ceramic precursors. They polymerized vinyldichlorosilane ($CH_2$=CH-$SiHCl_2$) to obtain [-$SiCl_2$-$CH_2$-$CH_2$-]$_n$, which can be reduced to [-$SiH_2$-$CH_2$-$CH_2$-]$_n$ (polysilyethylene) by using $LiAlH_4$. As compared with the reduction process, direct polymerization of vinylsilane gives some varieties on the chain structure. In a usual case, the obtained polymer possesses 50% of -$CH_2$-CH($SiH_3$)- units besides the -$SiH_2$-$CH_2$-$CH_2$- units. The obtained chain structure is controllable, depending on the kind of catalyst and the synthesis temperature.[51]

The gaseous nature of the starting monomer, vinylsilane, requiring an autoclave for polymerization, causes some difficulty in targeted polymer production on a small scale. Such a synthesis procedure, however, is analogous to polyethylene or polypropylene production. Therefore PVS directly derived from vinylsilane is potentially promising for large-scale production and widespread use in industry.

For typical applications in ceramic materials, PVS has been investigated as a matrix for SiC/SiC composite.[54] The interface properties in the synthesized composites were improved, possibly due to the liquid nature of PVS during impregnation at room temperature. High stability in environments containing oxygen and moisture is another attractive property of PVS for industrial uses.

As in the case of the PMS pyrolysis process, two kinds of preheat treatment were performed on the prepared PVS to investigate the effect on the resulting ceramic yield:[55] (1) simple heat treatment at 500 to 700 K, and (2) reflux treatment

**TABLE 19.3**
**PVS Resin Recovery and Ceramic Yield of the Cross-Linked PVS Resin at 1273 K[a]**

| Pre-Heat Treatment Temperature (K) | Reflux | | Open Ar Gas Flow | |
|---|---|---|---|---|
| | Recovery (%) | Ceramic Yield at 1273 K (%) | Recovery (%) | Ceramic Yield at 1273 K (%) |
| 500 | 100 | 39 | 90 | 41 |
| 550 | 78 | 71 | 81 | 45 |
| 600 | 74 | 80 | 68 | 53 |
| 650 | 65 | 87 | 47 | 78 |
| 700 | 58 | 91 | 40 | 84 |

[a] Reference 55.

at 500 to 700 K was performed on PVS. The conditions were the same as those for PMS preheat treatment.

After the preheat treatment, the obtained cross-linked PVS resins were pyrolyzed at 1273 K. Resin recovery after preheat treatment and the ceramic yields of these cross-linked resins at 1273 K are summarized in Table 19.3. After simple heat treatment, the ceramic yield of the cross-linked PVS resin increased as compared with the starting PVS (ceramic yield 36 mass%). The increasing rate in the ceramic yield is, however, counterbalanced by low resin recovery.

Figure 19.8 shows the overall ceramic yield of PVS. Simple heat treatment in an argon gas flow does not increase the resulting ceramic yield. Even in the case of reflux conditions, the reflux treatment is not effective to increase the

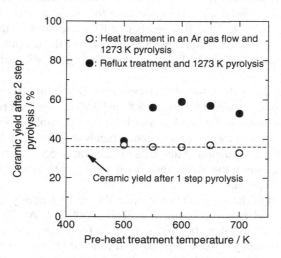

**FIGURE 19.8** Ceramic yield of PVS after two-step pyrolysis.

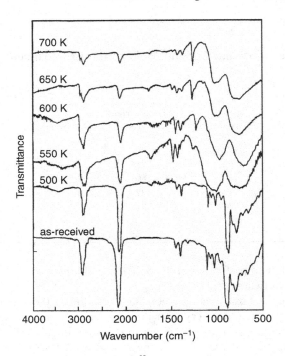

**FIGURE 19.9** IR spectra of refluxed PVS.[55]

overall ceramic yield up to 500 K. At 550 K, resin recovery and ceramic yield of the PVS resin begin to increase. The increase in the ceramic yield is substantial, while the increase in polymer recovery is minor for overall ceramic yield improvement, as shown in Table 19.3. It suggests that PVS itself cannot trap vinylsilane oligomers, even in reflux conditions. The increase in the overall ceramic yield mainly originates from the high degree of cross-linking in PVS resin during reflux treatment.

Figure 19.9 shows the infrared (IR) spectra of PVS resins after heat treatment. A major change was observed between 500 and 550 K. The PVS resin preheated at 500 K shows small spikes at 1130 and 1020 cm$^{-1}$ possibly assigned to the Si-CH$_2$-CH$_2$-Si bond, and the strong absorption bands at 945-940 and 870-830 cm$^{-1}$ assigned to SiH$_2$ deformation. After reflux treatment at 550 K, these absorption bands disappear, while the broad absorption band at 1100 cm$^{-1}$ assigned to the Si-CH$_2$-Si bond appears.

Investigation of the gas evolution profiles during PVS pyrolysis revealed that silane, ethane, and ethylene begin to evolve at 600 K.[55] At 700 K, methane, ethylsilane, and diethylsilane (a typical molecule for SiC$_4$H$_{12}$ composition) begin to evolve as dominant gaseous compounds. Beyond 700 K, the evolution of silicon containing gaseous compounds has stopped, while the evolution of methane and hydrogen continues. Hydrogen evolution possesses a broad tail even at 1400 K, which would correspond to C-H bond scission in the inorganic Si-C amorphous. These gas evolution profiles are qualitatively consistent with the results of TG-mass spectrometry analysis on polysilyethylene (consisting of 100% -$SiH_2$-$CH_2$-$CH_2$- units).[53] The $SiH_4$ evolution is, however, remarkable because of -$SiH_3$ groups on the PVS side chain.

In the case of polysilyethylene, synthesized by Corriu et al.,[25] it has been mentioned that the silylene intermediate formation with hydrogen evolution from -$SiH_2$- plays a role in Si-Si cross-linking. Such Si-Si bonds are converted to Si-$CH_2$-Si bridges by the Kumada rearrangement process at approximately 623 K. PVS, however, contains a large number of -$SiH_3$ groups. Decomposition of -$SiH_3$ groups yields $SiH_3$ radicals, which are easily converted to gaseous $SiH_4$, not to be trapped in the reflux system. Thus the system is mainly filled with remaining carbon radicals during the PVS condensation reaction. Si-H bonds are probably forced to trap the carbon radicals in such an environment, although it is not favored in terms of bond rearrangement energies as compared with silyl radical or silylene intermediate trapping.[56]

The 550 to 600 K temperature range of the PVS ceramic, where yield increases, is considerably higher than the required reflux temperature for PMS. This temperature range is consistent with the Si-$CH_2$-Si bridge formation accompanied by silane, ethane, and ethylene evolution.

In any case, various chemical reactions with different activation energies proceed competitively at such high temperatures. Perhaps the direct trapping of -Si-$CH_2$ radicals by Si-H groups is favored to explain Si-$CH_2$-Si formation in the case of PVS, while the Kumada rearrangement process, requiring many Si-Si bonds for methyl group insertion, is a minor reaction.

## D. UTILIZATION OF PVS FOR CERAMIC PRECURSOR

Study of the heat treatment process on PVS reveals that PVS remains stable in an inert atmosphere as compared with PMS, in spite of a large amount of -$SiH_2$-. The absence of a radical source at relatively low temperature may contribute to the high stability of PVS. Thus a PCS-PVS polymer blend can retain its thermoplastic nature up to the melting state. Thermosetting appears at high temperature, 600 to 700 K.[55] Figure 19.10 shows the melt viscosity of the PCS-PVS polymer blend.[57] The relationship between the viscosity and the temperature can be formulated by the classic Andrade equation: ($\eta = A \exp(U/RT)$, where $\eta$ is the viscosity, $A$ is a constant, $R$ is the gas constant, and $T$ is the absolute temperature. Plots between $\ln(\eta)$ and $1/T$ show a linear relationship in the viscosity range of $1 - 10^6$ Pa s. The apparent activation energy ($U$), thus estimated from the linearity, is lowered by the PVS addition, as shown in Figure 19.11. The low activation energy of the PCS-PVS polymer blend indicates that the viscosity of the melt increases more gradually during the cooling process from the spinneret to the rotation drum. Therefore the PCS-PVS polymer blend can be drawn into thin fibers without breaking.

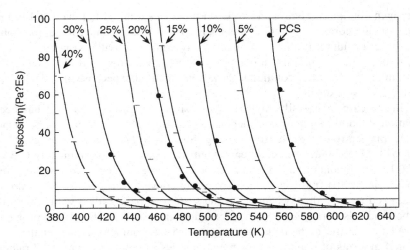

**FIGURE 19.10** Relationship between melt viscosity and temperature of PCS-PVS blend precursors.[57]

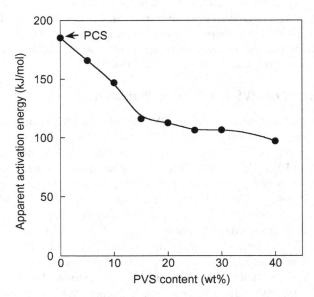

**FIGURE 19.11** Apparent activation energy of PCS-PVS melt viscosity.[57]

In real applications, thin precursor fibers using a PCS-20 mass% PVS polymer blend have been synthesized.[57,58] A 20 mass% is sufficiently high to lower the activation energy. A 25 mass% is not available for spinning at room temperature, about 300 K, because of the tacking between monofilament. The obtained precursor fiber can be converted to SiC fibers after pyrolysis with thermal oxidation curing or radiation curing. SiC fibers with a diameter of 6.0 μm have been obtained, which is far thinner than the fiber derived from PCS (approximately 9.0 μm). Such thin fiber derived from PCS-PVS precursor can be shaped into fabrics easily and has promising applications for reinforcement in ceramic matrix composites.

## III. DISCUSSION

The major difference between PMS and PVS in the reflux process possibly originates in the chemical structure of the fragments (high molecular weight radicals) formed during reflux treatment. In PMS, silyl radicals formed by main chain cleavage probably play a main role in Si-Si cross-linking. The role of silylene intermediates formed at terminal $SiH_2$ groups is considerable. On the other hand, PVS contains a large number of $SiH_2$ units and $SiH_3$ groups. The required high reflux temperature for improvement in the ceramic yield, however, suggests that carbon radicals play a role in the cross-linking process. The high stability of Si-C or C-C main chains and the chemical environments of $SiH_2$ units in PVS must be taken into account.

Use of the polymer blend technique for PMS and PVS with a large amount of PCS provide us with a polymer precursor method with a wide variety of conventional applications. PMS plays a role as the cross-linking agent for PCS. During the melt spinning and the pyrolysis process, the PMS structure is considered to be built in the dominant PCS structures (Figure 19.12). When the PMS content is increased, a PCS-PMS polymer blend suitable for PIP is expected. By adjusting the hybridized chain structure in PCS-PMS resins, thermosetting or viscoelastic properties of the resins can be controlled. On the other hand, PVS is merely miscible with PCS. There is no strong chemical reaction during melt spinning. Therefore the physical properties of PCS-PVS polymer blends are easily predictable, at least in the low-temperature thermoplastic region.

## ACKNOWLEDGMENT

I am grateful to Dr. Kiyohito Okamura, an honorary professor at Osaka Prefecture University, for his advice and inspiration in the area of the polymer precursor method for ceramic synthesis. I am also grateful for kind permissions from Blackwell Publishing (Figure 19.4, Scheme 19.1), American Chemical Society (Figure 19.5), American Ceramic Society (Figure 19.7), and Springer Science and Business Media (Table 19.3, Figures 19.9 through 19.11).

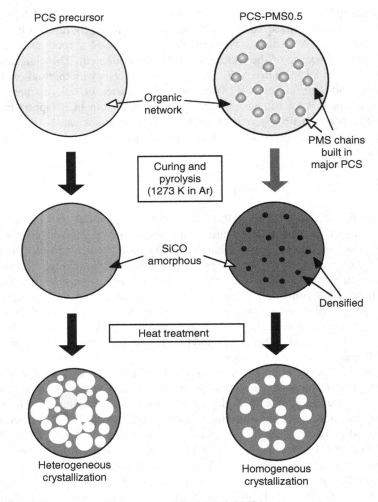

**FIGURE 19.12** Ceramization process of PCS-0.5% PMS precursor.

# REFERENCES

1.  Yajima, S., Hayashi, J., and Omori, M., *Chem. Lett.*, 931, 1975.
2.  Yajima, S., Okamura, K., Hayashi, J., and Omori, M., *J. Am. Ceram. Soc.*, 59, 324, 1976.
3.  Yajima, S., Hayashi, J., Omori, M., and Okamura M., *Nature*, 261, 686, 1976.
4.  Yajima, S., Okamura, K., Matsuzawa, T., Hasegawa, Y., and Shishido, T., *Nature*, 279, 706, 1979.
5.  Yajima, S., Hasegawa, Y., Hayashi, J., and Iimura, M., *J. Mater. Sci.*, 13, 2569, 1978.
6.  Hasegawa, Y., Iimura, M., and Yajima, S., *J. Mater. Sci.*, 15, 720, 1980.

7. Hasegawa, Y. and Okamura, K., *J. Mater. Sci.*, 18, 3633, 1983.
8. Okamura, K., *Composites*, 18, 107, 1987.
9. Yajima, S., Iwai, T., Yamamura, T., Okamura, K., and Hasegawa, Y., *J. Mater. Sci.*, 16, 1349, 1981.
10 Yamamura, T., Ishikawa, T., Shibuya, M., Hisayuki, T., and Okamura, K., *J. Mater. Sci.*, 23, 2589, 1988.
11. Takeda, M., Imai, Y., Ichikawa, H., Ishikawa, T., Seguchi, T., and Okamura, K., *Ceram. Eng. Sci. Proc.*, 12, 1007, 1991.
12. Okamura, K. and Seguchi, T., *J. Inorg. Organomet. Polym.*, 2, 171, 1992.
13. Takeda, M., Saeki, A., Sakamoto, J., Imai, Y., and Ichikawa, H., *J. Am. Ceram. Soc.*, 83, 1063, 2000.
14. Ishikawa, T., Kohtoku, Y., Kumagawa, K., Yamamura, T., and Nagasawa, T., *Nature*, 391, 773, 1998.
15. Cooke, T.F., *J. Am. Ceram. Soc.*, 74, 2959, 1991.
16. Laine, R.M. and Babonneau, F., *Chem. Mater.*, 5, 260, 1993.
17. Birot, M., Pillot, J.-P., and Dunogues, J., *Chem. Rev.*, 95, 1443, 1995.
18. Bill, J. and Aldinger, F., *Adv. Mater.*, 7, 775, 1995.
19. Baldus, H.-P., and Martin, J., *Angew. Chem. Int. Ed. Engl.*, 36, 328, 1997.
20. Schilling, C.L., Wesson, J.P., and Williams, T.C., *Am. Ceram. Soc. Bull.*, 62, 912, 1983.
21. Bacque, E., Pillot, J.-P., Birot, M., and Dunogues, J., *Macromolecules*, 21, 30, 1988.
22. Wu, H.J. and Interrante, L.V., *Chem. Mater.*, 1, 564, 1989.
23. Interrante, L.V., Rushkin, I., and Shen, Q., *Appl. Organometal. Chem.*, 12, 695, 1998.
24. Boury, B., Corriu, R.J.P., Leclercq, D., Mutin, H., Planeix, J.-M., and Vioux, A., *Organometallics*, 10, 1457, 1991.
25. Corriu, R.J.P., Gerbier, P., Guerin, C., Herrner, B.J.L., Jean, A., and Mutin, P.H., *Organometallics*, 11, 2507, 1991.
26. Penn, B.G., Ledbetter, F.E., III, Clemons, J.M., and Daniels, J.G., *J. Appl. Polym. Sci.*, 27, 3751, 1982.
27. Seyferth, D. and Wiseman, G.H., *J. Am. Ceram. Soc.*, 67, C-132, 1984.
28. Wada, H., Ito, S., Kuroda, K., and Kato, C., *Chem. Lett.*, 691, 1985.
29. Arai, M., Sakurada, S., Isoda, T., and Tomizawa, T., *Polym. Preprints*, 28, 407, 1987.
30. Funayama, O., Tashiro, Y., Kamo, A., Okumura, M., and Isoda, T., *J. Mater. Sci.*, 29, 4883, 1994.
31. Fazen, P.J., Beck, J.S., Lynch, A.T., Remsen, E.E., and Sneddon, L.G., *Chem. Mater.*, 2, 96, 1990.
32. Hashimoto, N., Sawada, Y., Bando, T., Yoden, H., and Deki, S., *J. Am. Ceram. Soc.*, 74, 1282, 1991.
33. Mocaer, D., Pailler, R., Naslain, R., Richard, C., Pillot, J.-P., Dunogues, J., Gerardin, C., and Taulelle, F., *J. Mater. Sci.*, 28, 2615, 1993.
34. Funayama, O., Kato, T., Tashiro, Y., and Isoda, T., *J. Am. Ceram. Soc.*, 76, 717, 1993.
35. Xiao, T.D., Gonsalves, K.E., and Strutt, P.R., *J. Am. Ceram. Soc.*, 76, 987, 1993.
36. Riedel, R., Lienzle, A., Dressler, W., Ruwisch, L., Bill, J., and Aldinger, F., *Nature*, 382, 796, 1996.
37. Chang, Z.-C., Aldinger, F., and Riedel, R., *J. Am. Ceram. Soc.*, 84, 2179, 2001.

38. Toreki, W., Creed, N.M., Sacks, M.D., and Batich, C.D., Saleem, M., Choi, G. J., and Morrone, A.A., *Composites Sci. Technol.*, 51, 145, 1994.
39. Kho, J.-G., Min, D.-S., and Kim, D.-P., *J. Mater. Sci. Lett.*, 19, 303, 2000.
40. Interrante, L.V., *Pure Appl. Chem.*, 74, 2247, 2002.
41. Moraes, K., Vosburg, J., Wark, D., Interrante L.V., Puerta A.R., Sneddon L.G., and Narisawa M., *Chem. Mater.*, 16, 125, 2004.
42. Zhang, Z., Babonneau, F., Laine, R.M., Mu., Y., Harrod, J.F., and Rahn, J.A., *J. Am. Ceram. Soc.*, 74, 670, 1991.
43. Seyferth, D., Wood, T.G., Tracy, H.J., and Robisoe, J.L., *J. Am. Ceram. Soc.*, 75, 1300, 1991.
44. Kobayashi, T., Sakakura, T., Hayashi, T., Yumura, M., and Tanaka, M., *Chem. Lett.*, 1157, 1992.
45. Boury, B., Bryson, N., and Soula, G., *Chem. Mater.*, 10, 297, 1998.
46. Kho, J.-G., Min, D.-S., and Kim, D.-P., *J. Mater. Sci. Lett.*, 19, 303, 2000.
47. Iseki, T., Narisawa, M., Katase, Y., Okamura, K., Oka, K., and Dohmaru, T., *Chem. Mater.*, 13, 4163, 2001.
48. Narisawa, M., Iseki, T., Katase, Y., Okamura, K., Oka, K., and Dohmaru, T., *J. Am. Ceram. Soc.*, 86, 227, 2003.
49. Shiina, K. and Kumada, K., *J. Org. Chem.*, 23, 139, 1958.
50. Narisawa, M., Nishioka, M., Okamura, K., Oka, K., and Dohmaru, T., *Ceram. Trans.*, 144A, 173, 2002.
51. Itoh, M., Iwata, K., Kobayashi, M., Takeuchi, R., and Kabeya, T., *Macromolecules*, 31, 5609, 1998.
52. Seyferth, D., Tasi, M., and Woo, H.-G., *Chem. Mater.*, 7, 236, 1995.
53. Corriu, R.J.P., Leclercq, D., Mutin, P.H., Planeix, J.M., and Vioux, A., *Organometallics*, 12, 454, 1991.
54. Kotani, M., Inoue, T., Kohyama, A., Okamura, K., and Katoh, Y., *Composites Sci. Technol.*, 62, 2179, 2002.
55. Idesaki, A., Miwa, Y., Katase, Y., Narisawa, M., Okamura, K., and Itoh, M., *J. Mater. Sci.*, 38, 2591, 2003.
56. Tachibana, A., Kurosaki, Y., Yamaguchi, K., and Yamabe, T., *J. Phys. Chem.*, 95, 6849, 1991.
57. Idesaki, A., Narisawa, M., Okamura, K., Sugimoto, M., Tanaka, S., Morita, Y., Seguchi, T., and Itoh, M., *J. Mater. Sci.*, 36, 5565, 2001.
58. Narisawa, M., Kitano, S., Okamura, K., and Itoh, M., *J. Am. Ceram. Soc.*, 78, 3405, 1995.

# Section V

## Processing of Specialty Ceramics

# 20 Chemical Vapor Deposition of Ceramics

*Guozhong Cao and Ying Wang*

## CONTENTS

## I. INTRODUCTION

Chemical vapor deposition (CVD) is a process of volatile compounds reacting in gas phase or on a growth surface, producing a nonvolatile solid film that deposits atomistically on a substrate.[1] This mature method can deposit single-layer, multilayer, composite, nanostructured, and functionally graded coating materials of high purity and good conformal coverage in single-crystal, polycrystalline, and amorphous forms at a relatively low processing temperature and in general under vacuum. It has wide applications in the area of microelectronics, optoelectronics, energy conversion devices, and ceramic materials.[2-6] This chapter offers an overview on fundamental aspects of CVD, including heterogeneous nucleation, reaction kinetics, vacuum science, transport phenomena, and CVD methods. Variant CVD methods based on different system designs, types of precursor used, and reaction mechanism are briefly discussed. More detailed discussion is devoted to aerosol-assisted CVD, electrochemical vapor deposition

(EVD), chemical vapor infiltration (CVI), CVD of diamond films, and atomic layer deposition (ALD).

## II. NUCLEATION OR INITIAL DEPOSITION

Growth of thin films, as in all phase transformation, involves the processes of nucleation and subsequent growth. Nucleation in film formation is a heterogeneous process. The size and the shape of the initial nuclei are solely dependent on the change in Gibb's free energy, due to supersaturation, and the combined effect of surface and interface energies governed by Young's equation. No other interaction between the film or nuclei and the substrate was taken into consideration. In practice, the interaction between film and substrate plays a very important role in determining the initial nucleation and film growth. Many experimental observations have revealed that there are three basic nucleation modes: island or Volmer-Weber growth, layer or Frank-van der Merwe growth, and island-layer or Stranski-Krastonov growth.

Figure 20.1 illustrates these three basic modes of initial nucleation in film growth. Island growth occurs when the growth species are more strongly bonded to each other than to the substrate. Many systems of metals on insulator substrates, alkali halides, graphite, and mica substrates display this type of nucleation during the initial film deposition. Subsequent growth results in the coalescence of islands to form a continuous film. Layer growth is the opposite of island growth, where

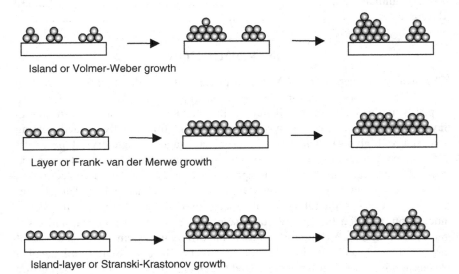

Island or Volmer-Weber growth

Layer or Frank- van der Merwe growth

Island-layer or Stranski-Krastonov growth

**FIGURE 20.1** Schematic illustrating three basic modes of initial nucleation in film growth. Island growth occurs when the growth species are more strongly bonded to each other than to the substrate.

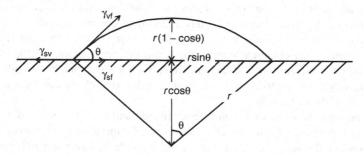

**FIGURE 20.2** Schematic illustrating heterogeneous nucleation process with all related surface energy in equilibrium.

growth species are equally more strongly bound to the substrate than to each other. First, a complete monolayer is formed, then the deposition of a second layer occurs. The most important examples of layer growth are the epitaxial growth of single crystal films. Island-layer growth is a combination of layer growth and island growth. Such a growth mode typically involves stress, which is developed during the formation of nuclei or films.

For the initial formation of a solid phase on a substrate surface from vapor precursors through heterogeneous nucleation, as is schematically illustrated in Figure 20.2, the critical nucleus size, $r^*$, and the corresponding energy barrier, $\Delta G^*$, are given by the following equations:

$$r^* = (2\pi\gamma_{vf}/\Delta G_v)\{(\sin^2\theta\cos\theta + 2\cos\theta - 2)/(2 - 3\cos\theta + \cos^3\theta)\} \quad (20.1)$$

$$\Delta G^* = \{16\pi\gamma_{vf}/3(\Delta G_v)^2\}\{(2 - 3\cos\theta + \cos^3\theta)/4\}, \quad (20.2)$$

where $\gamma_{vf}$ is the surface energy of the solid-vapor surface, $\Delta G_v$ is the change of bulk Gibb's free energy, and $\theta$ is the contact angle defined by Young's equation. For island growth, the contact angle must be larger than zero, that is, $\theta > 0$. According to Young's equation, we have

$$\gamma_{sv} < \gamma_{fs} + \gamma_{vf}. \quad (20.3)$$

If the deposit does not wet the substrate at all or $\theta = 180°$, the nucleation is a homogeneous nucleation. For layer growth, the deposit wets the substrate completely and the contact angle equals zero; the corresponding Young's equation becomes

$$\gamma_{sv} = \gamma_{fs} + \gamma_{vf}. \quad (20.4)$$

The most important layer growth is the deposition of single-crystal films through either homoepitaxy, in which the deposited film has the same crystal structure and chemical composition as that of the substrate, or heteroepitaxy, in which the deposited film has a close matching crystal structure to that of the substrate. Homoepitaxy is a simple extension of the substrate, and thus there is virtually no interface between the substrate and the depositing film and no nucleation process. Although the deposit has a chemical composition different from that of the substrate, the growth species prefers to bind to the substrate. Because of the difference in chemical composition, the lattice constants of the deposit most likely differ from those of the substrate. Such a difference commonly leads to the development of stress in the deposit; stress is one of the common reasons for the island-layer growth.

Island-layer growth is a bit more complicated and involves *in situ* developed stress. Initially the deposition proceeds following the mode of layer growth. When the deposit is elastically strained due to, for example, lattice mismatch between the deposit and the substrate, strain energy is developed. As each layer of deposit is added, more stress is developed and so is the strain energy. Such strain energy is proportional to the volume of the deposit, assuming there is no plastic relaxation. Therefore the change in Gibb's free energy should include the strain energy and Equation 20.2 is modified accordingly:

$$\Delta G^* = \{16\pi\gamma_{vf}/3(\Delta G_v + \omega)^2\}\{(2 - 3\cos\theta + \cos^3\theta)/4\}, \qquad (20.5)$$

where $\omega$ is the strain energy per unit volume generated by the stress in the deposit. Because the sign of $\Delta G_v$ is negative, and the sign of $\omega$ is positive, the overall energy barrier to nucleation increases. When the stress exceeds a critical point and can't be released, the strain energy per unit area of deposit is large with respect to $\gamma_{vf}$, permitting nuclei to form above the initial layered deposit. In this case, the surface energy of the substrate exceeds the combination of both the surface energy of the deposit and the interfacial energy between the substrate and the deposit:

$$\gamma_{sv} > \gamma_{fs} + \gamma_{vf}. \qquad (20.6)$$

It should be noted that there are other situations where the overall Gibb's free energy may change. For example, initial deposition or nucleation on substrates with cleavage steps and screw dislocations would result in a stress release and thus an increased change of the overall Gibb's free energy. As a result, the energy barrier for the initial nucleation is reduced and the critical size of nuclei becomes small. Substrate charge and impurities affect the $\Delta G^*$ through the change in surface, electrostatic, and chemical energies in a similar manner.

It should be noted that the aforementioned nucleation models and mechanisms are applicable to the formation of single-crystal, polycrystalline, and amorphous deposits, and of inorganic, organic, and hybrid deposits. Whether the deposit is single crystalline, polycrystalline, or amorphous depends on the growth conditions and the substrate. Deposition temperature and the impinging rate of growth species are the two most important factors and are briefly summarized below:

1. Growth of single-crystal films is most difficult and requires (1) a single-crystal substrate with a close lattice match, (2) a clean substrate surface so as to avoid possible secondary nucleation, (3) high growth temperature so as to ensure sufficient mobility of the growth species, and (4) low

impinging rate of growth species so as to ensure sufficient time for surface diffusion and incorporation of growth species into the crystal structure and structural relaxation before the arrival of the next growth species.

2. Deposition of amorphous films typically occurs (1) when a low growth temperature is applied and there is insufficient surface mobility of growth species, and/or (2) when the influx of growth species onto the growth surface is very high and growth species do not have enough time to find the growth sites with the lowest energy.

The conditions for the growth of polycrystalline crystalline films fall between the conditions of single-crystal growth and amorphous film deposition. In general, the deposition temperature is moderate, ensuring reasonable surface mobility of the growth species and the impinging flux of growth species is moderately high.

Figure 20.3, as an example, shows the growth conditions for single crystalline, polycrystalline, and amorphous films of silicon by CVD.[7] The above discussion is applicable to single-element films; however, the growth process is complex in the presence of impurities and additives and in the case of multicomponents.

Epitaxy is a very special process and refers to the formation or growth of a single crystal on top of a single-crystal substrate or seed. Epitaxial growth can

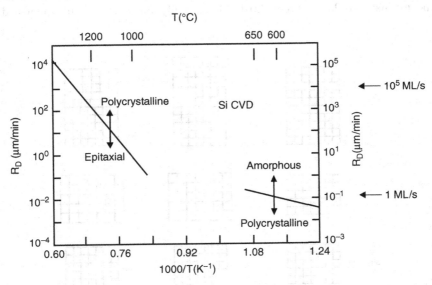

**FIGURE 20.3** The growth conditions for the single-crystalline, polycrystalline, and amorphous films of silicon by CVD. (From Bloem, J., in *Proceedings of the Seventh Conference on CVD*, T.O. Sedgwick and H. Lydtin, Eds., *Electrochem. Soc. Proc.*, 79-3, 41, 1979.)

be further divided into homoepitaxy and heteroepitaxy. Homoepitaxy is when the growth film and the substrate are the same material. Homoepitaxial growth is typically used to grow better quality film or introduce dopants into the grown film. Heteroepitaxy refers to the case where films and substrates are different materials. One obvious difference between homoepitaxial films and heteroepitaxial films is the lattice match between the films and substrates. There is no lattice mismatch between the films and substrates by homoepitaxial growth. On the contrary, there will be a lattice mismatch between films and substrates in heteroepitaxial growth. The lattice mismatch is also called misfit, and is given by

$$f = (a_s - a_f)/a_f, \tag{20.7}$$

where $a_s$ is the unstrained lattice constant of the substrate and $a_f$ is the unstrained lattice constant of the film. If $f > 0$, the film is strained in tension, whereas if $f < 0$, the film is strained in compression. Strain energy, $E_s$, develops in strained films:

$$E_s = 2\mu_f\{(1 + v)/(1 - v)\}\varepsilon^2 hA, \tag{20.8}$$

where $\mu_f$ is the shear modulus of the film, $v$ is the Poisson's ratio ($<1/2$ for most materials), $\varepsilon$ is the plane or lateral strain, $h$ is the thickness, and $A$ is the surface area. It should be noted that strain energy increases with thickness. The strain energy can be either accommodated by straining both film and substrate when the mismatch is relatively small, or relaxed by formation of dislocations when the mismatch is large. Figure 20.4 schematically illustrates a lattice-matched

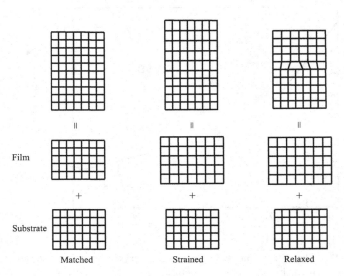

**FIGURE 20.4** Schematic illustrating the lattice-matched homoepitaxial film and substrate, and strained and relaxed heteroepitaxial structures.

homoepitaxial film and substrate, and strained and relaxed heteroepitaxial structures. Both homoepitaxial and heteroepitaxial growth of films has been a well-established technique and found wide applications, particularly in the electronics industry.

## III. FILM GROWTH

Crystal growth can be generally considered as a heterogeneous reaction, and typical crystal growth proceeds in the following sequence, as illustrated in Figure 20.5:

1. Diffusion of growth species from the bulk (such as the vapor or liquid phase) to the growing surface, which, in general, is considered to proceed rapidly enough and thus is not a rate limiting process.
2. Adsorption and desorption of growth species onto and from the growing surface. This process can be rate limiting if the supersaturation or concentration of the growth species is low.
3. Surface diffusion of adsorbed growth species. During surface diffusion, an adsorbed species may either be incorporated into a growth site, which contributes to crystal growth, or escape from the surface.
4. Surface growth by irreversibly incorporating the adsorbed growth species into the crystal structure. When a sufficient supersaturation or a high concentration of growth species is present, this step will be the rate-limiting process and determines the growth rate.
5. If by-product chemicals are generated on the surface during growth, by-products will desorb from the growth surface, so that growth species can adsorb onto the surface and the growth process can continue.
6. By-product chemicals diffuse away from the surface so as to vacate the growth sites for continued growth.

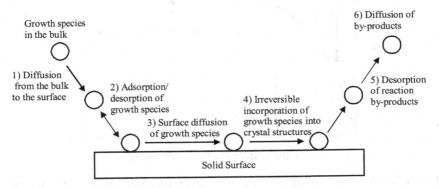

**FIGURE 20.5** Schematic illustrating the six steps in crystal growth, which can be generally considered as a heterogeneous reaction: typical crystal growth proceeds in this fashion.

For most crystal growth, the rate-limiting step is either adsorption-desorption of growth species on the growth surface (step 2) or surface growth (step 4). When step 2 is rate limiting, the growth rate is determined by the condensation rate, $J$ (atoms/cm$^2$·sec), which depends on the number of growth species adsorbed onto the growth surface, which is directly proportional to the vapor pressure or concentration, $P$, of the growth species in the vapor as given by

$$J = \alpha \sigma P_0 (2\pi mkT)^{-\frac{1}{2}}, \tag{20.9}$$

where $\alpha$ is the accommodation coefficient, $\sigma = (P - P_0)/P_0$ is the supersaturation of the growth species in the vapor in which $P_0$ is the equilibrium vapor pressure of the crystal at temperature $T$, $m$ is the atomic weight of the growth species, and $k$ is the Boltzmann constant. $\alpha$ is the fraction of impinging growth species that becomes accommodated on the growing surface, and is a surface-specific property. A surface with a high accommodation coefficient will have a high growth rate as compared with low $\alpha$ surfaces. A significant difference in accommodation coefficients in different facets will result in anisotropic growth. When the concentration of the growth species is very low, the adsorption is more likely a rate-limiting step. For a given system, the growth rate increases linearly with the increase in the concentration of growth species. Further increases in the concentration of growth species result in a change from an adsorption-limited to surface growth-limited process. When the surface growth becomes a limiting step, the growth rate becomes independent of the concentration of growth species, as schematically shown in Figure 20.6. A high concentration or vapor pressure of

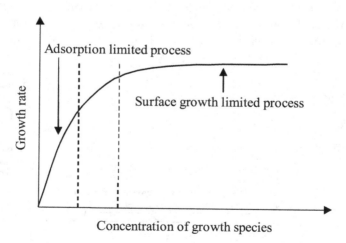

**FIGURE 20.6** Relation between growth rate and reactant concentration. At low concentration, growth is diffusion limited and thus increases linearly with increasing reactant concentration. At high concentration, surface reaction is the limiting step and thus the growth rate becomes independent of reactant concentration.

growth species in the vapor phase increases the probability of defect formation, such as impurity inclusion and stack faults. Further, a high concentration may result in a secondary nucleation on the growth surface or even homogeneous nucleation, which would effectively terminate epitaxial or single-crystal growth.

An impinging growth species onto the growth surface can be described in terms of the residence time and/or diffusion distance before escaping back to the vapor phase. The residence time, $\tau_s$, for a growth species on the surface is described by

$$\tau_s = \frac{1}{\nu} \exp\left(\frac{E_{des}}{kT}\right), \qquad (20.10)$$

where $\nu$ is the vibrational frequency of the adatom, that is, the adsorbed growth species, on the surface (typically $10^{12}$ sec$^{-1}$), and $E_{des}$ is the desorption energy required for the growth species escaping back to the vapor. While residing on the growth surface, a growth species will diffuse along the surface with the surface diffusion coefficient, $D_s$, given by

$$D_s = \frac{1}{2} a_0 \nu \exp\left(\frac{-E_s}{kT}\right), \qquad (20.11)$$

where $E_s$ is the activation energy for surface diffusion and $a_0$ is the size of the growth species. So the mean diffusion distance, $X$, for a growth species from the site of incidence is

$$X = \sqrt{2 D_s \tau_s} = a_0 \exp\left(\frac{E_{des} - E_s}{kT}\right). \qquad (20.12)$$

It is clear that in a crystal surface, if the mean diffusion distance is far longer than the distance between two growth sites, such as with kinks or ledges, all adsorbed growth species will be incorporated into the crystal structure and the accommodation coefficient will be unity. If the mean diffusion distance is far shorter than the distance between growth sites, all adatoms will escape back to the vapor and the accommodation coefficient will be zero. The accommodation coefficient is dependent on the desorption energy, the activation energy of surface diffusion, and the density of growth sites.

When step 2 proceeds sufficiently rapidly, surface growth, that is, step 4, becomes a rate-limiting process. In a given crystal, different facets have different atomic density and atoms on different facets have a different number of unsatisfied bonds (also referred to as broken or dangling bonds), leading to a different surface energy. Such a difference in surface energy or the number of broken chemical bonds leads to different growth mechanisms and varied growth rates. According to the periodic bond chain (PBC) theory developed by Hartman and Perdok,[8] all

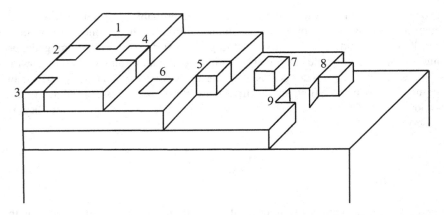

**FIGURE 20.7** Schematic illustrating the step growth mechanism, considering a {100} surface of a simple cubic crystal as an example and each atom as a cube with a coordination number of six (six chemical bonds) in bulk crystal.

crystal facets can be categorized into three groups based on the number of broken periodic bond chains on a given facets: flat surface, stepped surface, and kinked surface. The number of broken periodic bond chains can be understood as the number of broken bonds per atom on a given facet. Let's first review the growth mechanisms on a flat surface.

For a flat surface, the classic step-growth theory was developed by Kossel, Stranski, and Volmer, which is also called the KSV theory.[9] They recognized that the crystal surface, on the atomic scale, is not smooth, flat, or continuous, and such discontinuities are responsible for crystal growth. To illustrate the step-growth mechanism, we consider a {100} surface of a simple cubic crystal as an example and each atom as a cube with a coordination number of six (six chemical bonds), as schematically illustrated in Figure 20.7. When an atom adsorbs onto the surface, it diffuses randomly on the surface. When it diffuses to an energetically favorable site, it will be irreversibly incorporated into the crystal structure, resulting in growth of the surface. However, it may escape from the surface back to the vapor. On a flat surface, an adsorbed atom may find different sites with different energy levels. An atom adsorbed on a terrace will form one chemical bond between the atom and the surface; such an atom is called an adatom, which is in a thermodynamically unfavorable state. If an adatom diffuses to a ledge site, it will form two chemical bonds and become stable. If an atom is incorporated to a ledge-kink site, three chemical bonds will be formed. An atom incorporated into a kink site will form four chemical bonds. Ledge, ledge-kink, and kink sites are all considered as growth sites; incorporation of atoms into these sites is irreversible and results in growth of the surface. The growth of a flat surface is due to advancement of the steps (or ledges). For given crystal facets and a given growth condition, the growth rate will be dependent on the step density. A misorientation results in an increased density of steps and consequently leads to

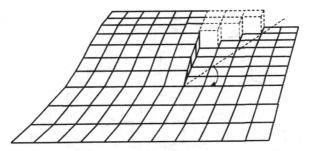

**FIGURE 20.8** Crystal growth proceeds in a spiral fashion, known as BCF theory, in which screw dislocation serves as a continuous source to generate growth sites so that the stepped growth will continue.

a high growth rate. An increased step density favors the irreversible incorporation of adatoms by reducing the surface diffusion distance between the impinging site and the growth site, before adatoms escape back to the vapor phase.

The obvious limitation of this growth mechanism is the regeneration of growth sites, when all available steps are consumed. Burton et al.[10] proposed that screw dislocation serves as a continuous source to generate growth sites so that the stepped growth would continue (as shown in Figure 20.8). The crystal growth proceeds in a spiral growth pattern, and this crystal growth mechanism is now known as BCF theory. The presence of screw dislocation not only ensures the continuing advancement of the growth surface, but also enhances the growth rate. The growth rate of a given crystal facet under a given experimental condition will increase with an increased density of screw dislocations parallel to the growth direction. It is also known that different facets can have significantly different abilities to accommodate dislocations. The presence of dislocations on a certain facet can result in anisotropic growth, leading to the formation of nanowires or nanorods.

The PBC theory offers a different perspective in understanding the different growth rates and behaviors in different facets. Let us take a simple cubic crystal as an example to illustrate the PBC theory, as shown in Figure 20.9. According to the PBC theory, {100} faces are flat surfaces (denoted as F-faces) with one PBC running through one such surface, {110} are stepped surfaces (S-faces) that have two PBCs, and {111} are kinked surfaces (K-faces) that have three PBCs. For the {110} surfaces, each surface site is a step or ledge site, and thus any impinging atom is incorporated wherever it adsorbs. For the {111} facets, each surface site is a kink site and will irreversibly incorporate any incoming atom adsorbed onto the surface. For both the {110} and {111} surfaces, the above growth is referred to as a random addition mechanism and no adsorbed atoms will escape back to the vapor phase. It is obvious that both the {110} and {111} faces have a faster growth rate than that of the {100} surface in a simple cubic crystal. In general, S-faces and K-faces have a higher growth rate than F-faces. For both S- and K-faces, the growth process is always adsorption limited, since

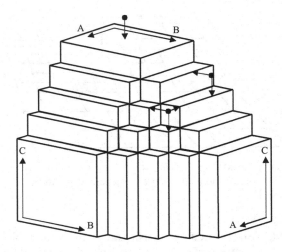

**FIGURE 20.9** Schematic illustrating the PBC theory. In a simple cubic crystal, {100} faces are flat surfaces (denoted as F-faces) with one PBC running through one such surface, {110} are stepped surfaces (S-faces) that have two PBCs, and {111} are kinked surfaces (K-faces) that have three PBCs. (From Hartman, P., and Perdok, W.G., *Acta Crystl.*, 8, 49, 1955.)

the accommodation coefficients on these two type surfaces are unity, that all impinging atoms are captured and incorporated into the growth surface. For F faces, the accommodation coefficient varies between zero (no growth at all) and unity (adsorption limited), depending on the availability of kink and ledge sites.

## IV. VACUUM SCIENCE

Most film deposition and processing are carried out in a vacuum. In addition, almost all the characterization of films is performed in a vacuum. Although there is much rich literature on vacuums, it seems that a brief discussion of relevant subjects is necessary. Specifically, some of the most commonly encountered concepts in thin film deposition and characterization, such as mean free path and flow regimes and their dependence on pressure and temperature, will be introduced. Readers who want to learn more fundamentals and technique details of vacuums are referred to References 11 through 13.

In a gas phase, gas molecules are constantly in motion and colliding among themselves as well as with the container walls. The pressure of a gas is the result of momentum transfer from the gas molecules to the walls, and is the most widely quoted system variable in vacuum technology. The mean distance traveled by molecules between successive collisions is called the mean free path, and it is an important property of the gas that depends on pressure, given by

$$\lambda_{mfp} = 5 \times 10^3/P, \tag{20.13}$$

where $\lambda_{mfp}$ is the mean free path in centimeters and $P$ is the pressure in torr. When the pressure is below $10^3$ torr, the gas molecules in typical film deposition and characterization systems virtually collide only with the walls of the vacuum chamber, that is, no collisions among the gas molecules.

The gas impingement flux in film deposition is a measure of the frequency with which gas molecules impinge on or collide with a surface, and is the most important parameter. This is because, for film deposition, only molecules imping-ing onto the growth surface are able to contribute to the growth process. The number of gas molecules that strike a surface per unit time and area is defined as the gas impingement flux, $\Phi$:

$$\Phi = 3.513 \times 10^{22} P/(MT)^{1/2}, \tag{20.14}$$

where $P$ is the pressure in torr, $M$ is the molecular weight, and $T$ is temperature. It should be noted that gas flow is different from the restless motion and collision of gas molecules. Gas flow is defined as a net directed movement of gas in a system, and occurs when there is a pressure drop. Depending on the geometry of the system involved, as well as the pressure, temperature, and type of gas in question, gas flow can be divided into three regimes: molecular flow, intermediate flow, and viscous flow. Free molecular flow occurs at low gas densities or high vacuum, when the mean free path between intermolecular collisions is larger than the dimensions of the system and the molecules collide only with the walls of the system. At high pressure, intermolecular collisions predominate since the mean free path is reduced and the gas flow is referred to as in the viscous flow regime. Between free molecular flow and viscous flow, there is a transition regime: intermediate flow. The above gas flow can be defined by the magnitude of the Knudsen number, $K_n$, given by

$$K_n = D/\lambda_{mfp}, \tag{20.15}$$

where $D$ is the characteristic dimension of the system, for example, the diameter of a pipe, and $\lambda_{mfp}$ is the gas mean free path. Figure 20.10 shows the gas flow regimes in a tube as a function of system dimensions and pressure; whereas the range of Knudsen numbers corresponding to gas flow regimes are summarized below:

| Gas Flow Regimes | Knudsen Number | $D \cdot P$[a] |
|---|---|---|
| Molecular flow | $K_n < 1$ | $D \cdot P < 5 \times 10^3$ cm.torr |
| Intermediate flow | $1 < K_n < 110$ | $5 \times 10^3 < D.P. < 5 \times 10^1$ cm.torr |
| Viscous flow | $K_n > 110$ | $D \cdot P > 5 \times 10^1$ cm.torr |

[a] $D$ is the characteristic dimension of the system and $P$ is the pressure.

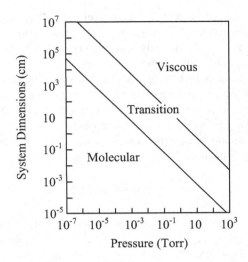

**FIGURE 20.10** Gas flow regimes in a tube as a function of system dimensions, pressure, and the range of Knudsen numbers corresponding to gas flow regimes (summarized in Table 20.1). (From Ohring, M., *The Materials Science of Thin Films*, Academic Press, San Diego, 1992.)

Viscous flow is a bit complex and can be further divided into laminar flow, turbulent flow, and transition flow. At a low gas flow velocity, the flow is laminar where layered, parallel flow lines may be visualized, no perpendicular velocity is present, and mixing of the gas is by diffusion only. In this flow, the velocity is zero at the gas-wall interface and gradually increases as one moves away from the interface, reaching a maximum at the center when gas is flowing inside a pipe. Viscous flow behavior can be defined by the so-called Reynolds number, *Re*, given below taking gas flow inside a pipe:

$$Re = D \cdot v \cdot \rho / \eta \qquad (20.16)$$

where $D$ is the diameter of the pipe, $v$, is the velocity, $\rho$, is the density, and $\eta$, is the viscosity of the gas. Laminar flow corresponds to a small $Re < 2100$. At a high gas velocity, the flow is turbulent flow, where the gas is constantly intermixing, where $Re > 4000$. At $2100 < Re < 4000$, a transition from laminar to turbulent flow occurs and is referred to as transition flow. There is always laminar flow near to the solid surface in both turbulent and transition flows, since the viscous friction forces a deceleration of the gas at the surface.

Diffusion is one of the mass transfer mechanisms in gases, which also occurs in liquids and solids. Diffusion is the movement of atoms or molecules from regions of higher to lower concentration, thus increasing the entropy of the system. Another mechanism is convection, a bulk gas flow process. Convection arises from the response to gravitational, centrifugal, electric, and magnetic forces. Convection can play an important role in high-pressure film deposition.

For example, a hotter and less dense gas above a hot substrate will rise, whereas a cooler and denser gas will sink. Such a situation is often encountered in cold-wall CVD reactors.

## V. TYPICAL CHEMICAL REACTIONS

The CVD process involves various chemical reactions of gaseous compounds in an activated environment (with heat, light, or plasma), followed by the deposition of a stable solid product. Because of the versatile nature of CVD, the chemistry is very rich and various types of chemical reactions are involved. Gas phase (homogeneous) reactions and surface (heterogeneous) reactions are intricately mixed. Gas phase reactions become progressively important with increasing temperature and partial pressure of the reactants. An extreme high concentration of reactants will make gas phase reactions predominant, leading to homogeneous nucleation. For deposition of good-quality films, homogeneous nucleation should be avoided. The wide variety of chemical reactions can be grouped into the following types: pyrolysis, reduction, oxidation, compound formation, disproportionation, and reversible transfer, depending on the precursors used and the deposition conditions applied. Examples of the above chemical reactions are given below:[1]

(A) Pyrolysis or thermal decomposition:

$$SiH_4 \text{ (g)} === Si \text{ (s)} + 2H_2 \text{ (g) at } 650°C \qquad (20.17)$$

$$Ni(CO)_4 \text{ (g)} === Ni \text{ (s)} + 4CO \text{ (g) at } 180°C \qquad (20.18)$$

(B) Reduction:

$$SiCl_4 \text{ (g)} + 2H_2 \text{ (g)} === Si \text{ (s)} + 4HCl \text{ (g) at } 1200°C \qquad (20.19)$$

$$WF_6 \text{ (g)} + 3H_2 \text{ (g)} === W \text{ (s)} + 6HF \text{ (g) at } 300°C \qquad (20.20)$$

(C) Oxidation:

$$SiH_4 \text{ (g)} + O_2 \text{ (g)} === SiO_2 \text{ (s)} + 2H_2 \text{ (g) at } 450°C \qquad (20.21)$$

$$4PH_3 \text{ (g)} + 5O_2 \text{ (g)} === 2P_2O_5 \text{ (s)} + 6H_2 \text{ (g) at } 450°C \qquad (20.22)$$

(D) Compound formation:

$$SiCl_4 \text{ (g)} + CH_4 \text{ (g)} === SiC \text{ (s)} + 4HCl \text{ (g) at } 1400°C \qquad (20.23)$$

$$TiCl_4 \text{ (g)} + CH_4 \text{ (g)} === TiC \text{ (s)} + 4HCl \text{ (g) at } 1000°C \qquad (20.24)$$

(E) Disproportionation:

$$2GeI_2 \text{ (g)} === Ge \text{ (s)} + GeI_4 \text{ (g) at } 300°C \qquad (20.25)$$

(F) Reversible transfer:

$$As_4 \text{ (g)} + As_2 \text{ (g)} + 6GaCl \text{ (g)} + 3H_2 \text{ (g)} === 6GaAs \text{ (s)} + 6HCl \text{ (g) at } 750°C \qquad (20.26)$$

The versatile chemical nature of the CVD process is further demonstrated by the fact that for deposition of a given film, many different reactants or precursors can be used and different chemical reactions may apply. For example, silica film is attainable through any of the following chemical reactions using various reactants:[14–17]

$$SiH_4 \text{ (g)} + O_2 \text{ (g)} === SiO_2 \text{ (s)} + 2H_2 \text{ (g)} \qquad (20.27)$$

$$SiH_4 \text{ (g)} + 2N_2O \text{ (g)} === SiO_2 \text{ (s)} + 2H_2 \text{ (g)} + 2N_2 \text{ (g)} \qquad (20.28)$$

$$SiH_2Cl_2 \text{ (g)} + 2N_2O \text{ (g)} === SiO_2 \text{ (s)} + 2HCl \text{ (g)} + 2N_2 \text{ (g)} \qquad (20.29)$$

$$Si_2Cl_6 \text{ (g)} + 2N_2O \text{ (g)} === SiO_2 \text{ (s)} + 3Cl_2 \text{ (g)} + 2N_2 \text{ (g)} \qquad (20.30)$$

$$Si(OC_2H_5)_4 \text{ (g)} === SiO_2 \text{ (s)} + 4C_2H_4 \text{ (g)} + 2H_2O \text{ (g)} \qquad (20.31)$$

From the same precursors and reactants, different films can be deposited when the ratio of reactants and the deposition conditions are varied. For example, both silica and silicon nitride films can be deposited from a mixture of $Si_2Cl_6$ and $N_2O$, and Figure 20.11 shows the deposition rates of silica and silicon nitride as a function of the ratio of reactants and deposition conditions.[16]

# VI. REACTION KINETICS

Although CVD is a nonequilibrium process controlled by chemical kinetics and transport phenomena, equilibrium analysis is still useful in understanding the CVD process. The chemical reaction and phase equilibrium determine the feasibility of a particular process and the final state attainable. In a given system, multistep complex reactions are often involved. The fundamental reaction pathways and kinetics have been investigated for only a few well-characterized industrially important systems. We will take the reduction of chlorosilane by hydrogen as an example to illustrate the complexity of the reaction pathways and kinetics involved in such a seemingly simple system and deposition process. In this Si-Cl-H system, there exist at least eight gaseous species: $SiCl_4$, $SiCl_3H$, $SiCl_2H_2$,

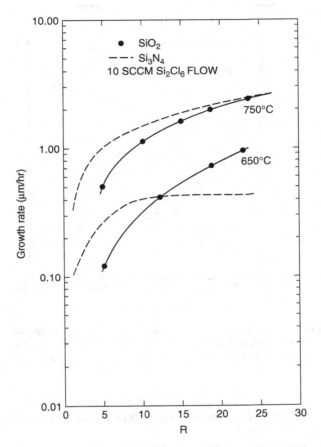

**FIGURE 20.11** Deposition rates of silica and silicon nitride as a function of the ratio of reactants and deposition conditions. (From Taylor, R.C., and Scott, B.A., *J. Electrochem. Soc.*, 136, 2382, 1989.)

$SiClH_3$, $SiH_4$, $SiCl_2$, HCl, and $H_2$. These eight gaseous species are in equilibrium under the deposition conditions governed by six equations of chemical equilibrium. With the available thermodynamic data, the composition of the gas phase as a function of reactor temperature for a molar ratio of Cl/H = 0.01 and a total pressure of 1 atm was calculated and is presented in Figure 20.12.[18]

## VII. TRANSPORT PHENOMENA

Transport phenomena play a critical role in CVD by governing access of film precursors to the substrate and by influencing the degree of desirable and unwanted gas phase reactions taking place before deposition. The complex reactor geometries and large thermal gradient characteristics of CVD chambers lead to

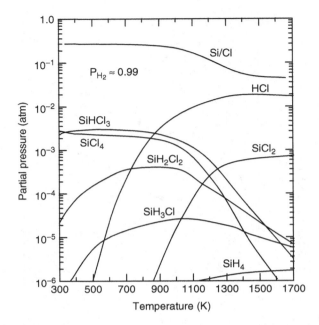

**FIGURE 20.12** Composition of gas phase as a function of reactor temperature for a molar ratio of Cl/H = 0.01 and a total pressure of 1 atm, calculated using the available thermodynamic data. (From Sirtl, E., Hunt, L.P., and Sawyer, D.H., *J. Electrochem. Soc.*, 121, 919, 1974.)

a wide variety of flow structures that affect film thickness, compositional uniformity, and impurity levels.[4]

For CVD reactors operating at a low pressure, where the mean free path of gas molecules is 10 times larger than the characteristic length of the reactor, there is no collision between gas molecules and thus the transport of gas is in the free molecular flow regime. For most CVD systems, the characteristic pressure is 0.01 atm and above, and the mean free paths are far larger than the characteristic system dimension. In addition, the gas velocities are low in most CVD reactors, typically tens of centimeters per second, the Reynolds number is typically less than 100, and the flows are laminar. As a result, a stagnant boundary layer of thickness $\delta$ adjacent to the growth surface is developed during the deposition. In this boundary layer, the composition of growth species decreases from the bulk concentration, $P_i$, to the surface concentration above the growing film, $P_{io}$, and the growth species diffuses through the boundary layer prior to depositing onto the growth surface. When the perfect gas laws are applied, since the gas composition in the typical CVD systems is reasonably dilute, the diffusion flux of gas or growth species through the boundary layer is given by

$$J_i = D(P_i - P_{io}) / \delta \, RT, \tag{20.32}$$

where $D$ is the diffusivity and is dependent on pressure and temperature:

$$D = D_0(P_0/P)(T/T_0)^n, \qquad (20.33)$$

where $n$ is experimentally found to be approximately 1.8. The quantity $D_0$ is the value of $D$ measured at standard temperature $T_0$ (273 K) and pressure $P_0$ (1 atm), and depends on the gas combination in question. As Equation 20.33 indicates, the gas diffusivity varies inversely with pressure, and thus the diffusion flux of gas through the boundary layer can be enhanced simply by reducing the pressure in the reactor.

When the growth rate is high and the pressure in the reactor chamber is high, diffusion of growth species through the boundary layer can become a rate-limiting process. For deposition of large-area films, depletion of growth species or reactants above the growth surface can result in nonuniform deposition of films. To overcome such nonuniformity in deposited films, various reactor designs have been developed to improve the gas mass transport through the boundary layer. Examples include using low pressure and new reactor chamber designs and substrate susceptors.

## VIII. CVD METHODS

A variety of CVD methods and CVD reactors have been developed, depending on the types of precursors used, the deposition conditions applied, and the forms of energy introduced to the system to activate the chemical reactions desired for the deposition of solid films on substrates. For example, when metalorganic compounds are used as precursors, the process is generally referred to as MOCVD (metalorganic CVD), and when plasma is used to promote chemical reactions, this is called plasma-enhanced CVD (PECVD). There are many other modified CVD methods, such as LPCVD (low-pressure CVD), laser-enhanced or assisted CVD, and aerosol-assisted CVD (AACVD).

The CVD reactors are generally divided into hot-wall and cold-wall types. Figure 20.13 depicts a few common setups for CVD reactors. Hot-wall CVD reactors are usually tubular in form, and heating is accomplished by surrounding the reactor with resistance elements.[19] In typical cold-wall CVD reactors, substrates are directly heated inductively with graphite susceptors, while chamber walls are air- or water-cooled.[20] LPCVD differs from conventional CVD in the low gas pressure of ~ 0.5 to 1 torr that is typically used; low pressure enhances the mass flux of gaseous reactants and products through the boundary layer between the laminar gas stream and substrates. In PECVD processing, plasma is sustained within chambers where simultaneous CVD reactions occur. Typically the plasma is excited either by a radiofrequency (RF) field, with frequencies ranging from 100 kHz to 40 MHz at gas pressures between 50 mtorr and 5 torr, or by microwaves with frequencies of 2.45 GHz. Often microwave energy is coupled to the natural resonant frequency of the plasma electrons in the presence of a static magnetic field, and such plasma is referred to as electron cyclotron

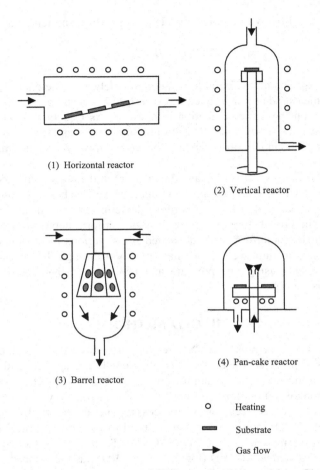

(1) Horizontal reactor

(2) Vertical reactor

(3) Barrel reactor

(4) Pan-cake reactor

○     Heating

▬     Substrate

➤     Gas flow

**FIGURE 20.13** A few common setups of CVD reactors. (Adapted from Ohring, M., *The Materials Science of Thin Films*, Academic Press, San Diego, 1992.)

resonance (ECR) plasma.[21] The introduction of plasma results in much enhanced deposition rates, thus permitting the growth of films at relatively low substrate temperatures. Figure 20.14 compares the growth rate of polycrystalline silicon films deposited with and without plasma enhancement.[22] MOCVD, also known as organometallic vapor phase epitaxy (OMVPE), differs from other CVD processes in the chemical nature of the precursor gases; metalorganic compounds are employed.[23,24] Lasers have also been employed to enhance or assist in the chemical reactions or deposition, and two mechanisms are involved: pyrolytic and photolytic processes.[25,26] In the pyrolytic process, the laser heats the substrate to decompose gases above it and enhance the rates of chemical reactions there, whereas in the photolytic process, laser photons are used to directly dissociate the precursor molecules in the gas phase. AACVD is used for systems in which

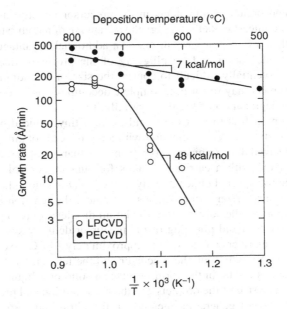

**FIGURE 20.14** Growth rate of polycrystalline silicon films deposited with and without plasma enhancement. (From Hajjar, J.J., Reif, R., and Adler, D., *J. Electron. Mater.*, 15, 279, 1986.)

no gaseous precursors are available and the vapor pressures of the liquid and solid precursors are too low.[27-29]

## IX. AEROSOL-ASSISTED CVD

Despite the many advantages it offers, CVD suffers from the limited availability of suitable precursors for high-Z elements. The vapor pressure of most high-Z element precursors is too low (e.g., below 1 mtorr at room temperature) to deliver a sufficient amount of material to the deposition chamber. Consequently it's difficult to achieve a uniform large-area film due to depletion of the precursors. Many efforts have been made to overcome this obstacle. For example, a CVD process with a liquid or solid as transport precursors has been developed, and such CVD processes are generally referred to as liquid-injection, aerosol-assisted, or "mist" CVD.[30,31]

Aerosol-assisted CVD is different from other CVD techniques in its technical details. In the AACVD process, the precursors are delivered in the form of either a liquid or solid by a carrier gas, while in conventional CVD processes, precursors are in the form of a gas. Solid or liquid precursors are first dissolved in a solvent to form a liquid precursor. Such a liquid precursor can be directly injected by a high-pressure carrier gas into a CVD reaction chamber. Once the liquid is injected into the deposition chamber, the solvent in the tiny liquid droplets evaporates immediately and organic components in the precursor undergo decomposition. The direct liquid injection method offers the advantage of a high rate of precursor delivery and thus a high film growth rate. An additional advantage is that there is no liquid condensation and no plugging of the precursor delivery system. However, it is potentially difficult to control the size of the liquid droplets and large liquid droplets may result in incomplete decomposition of organic components and thus degradation of the film crystallinity.

Another approach is to mist the liquid to form tiny liquid droplets through either ultrasonication or decentrification with a high-speed rotating disc or using a commercially available atomizer. Submicron (less than 1 μm in diameter) liquid droplets are mixed with a carrier gas (thus forming an aerosol or mist). The aerosol may be transported either directly to the CVD reaction chamber, or first to the so-called evaporizer before the aerosol enters the reaction chamber. The evaporizer evaporates the solvent from the liquid droplets so as to prevent liquid condensation and to avoid plugging of the precursor delivery system. The temperature in the evaporizer can range from approximately 150°C to several hundred degrees Celsius. Depending on the precursor and the temperature in the evaporizer, the precursor may be in the form of either a solid or a liquid when exiting the evaporizer. The size of the droplets and their distribution and production rates depend on the aerosol generation method. In the ultrasonic aerosol generation method, the diameter of the droplets is related to the ultrasonic frequency. AACVD has demonstrated its great potential in the growth of ferroelectric thin film as well as other multicomponent films. With aerosol-assisted precursor delivery, ferroelectric films with good crystallinity and the desired stoichiometric composition have been reported.

Aerosol-assisted CVD introduces rapid evaporation of the precursor and short delivery time of vapor precursor to the reaction zone. The small diffusion distance between the reactant and intermediates leads to higher deposition rates at relatively low temperatures. Single precursors are more inclined to be used in AACVD; therefore, due to good molecular mixing of precursors, the stoichiometry in the synthesis of multicomponent materials can be well controlled. In addition, AACVD can be preformed in an open atmosphere to produce thin or thick oxide films, hence its cost is low compared to sophisticated vacuum systems. CVD methods have also been modified and developed to deposit solid phase from gaseous precursors on highly porous substrates or inside porous media. The two most used deposition methods are known as electrochemical vapor deposition (EVD) and chemical vapor infiltration (CVI).

## X. ELECTROCHEMICAL VAPOR DEPOSITION AND CHEMICAL VAPOR INFILTRATION

Electrochemical vapor deposition has been explored for making gas-tight, dense, solid electrolyte films on porous substrates,[32,33] and the most studied system has been the yttria-stabilized zirconia films on porous alumina substrates for solid oxide fuel cell applications and dense membranes.[32-35] In the EVD process for growing solid oxide electrolyte films, a porous substrate separates the metal precursor(s) and oxygen source. Typically chlorides are used as metal precursors, whereas water vapor, oxygen, or air, or a mixture of these is used as the source of oxygen. Initially the two reactants interdiffuse in the substrate pores and react with each other only when they meet to deposit the corresponding solid oxides. When the deposition conditions are appropriately controlled, the solid deposition can be located at the entrance of the pores at the side facing the metal precursors and plug the pores. The location of the solid deposit depends mainly on the diffusion rate of the reactants inside the pores as well as the concentrations of the reactants inside the deposition chamber. Under typical deposition conditions, reactant molecules diffusing inside pores are in the Knudsen diffusion region, in which the diffusion rate is inversely proportional to the square root of the molecular weight. Oxygen precursors diffuse much faster than metal precursors, and consequently the deposition normally occurs near the entrance of the pores facing the metal precursor chamber. If the deposit solid is an insulator, deposition by the CVD process stops when the pores are plugged by the deposit, since no further direct reaction between the two reactants occurs. However, for solid electrolytes, particularly ionic-electronic mixed conductors, the deposition proceeds further by means of EVD, and the film may grow on the surface exposed to the metal precursor vapor. In this process, the oxygen or water is reduced at the oxygen-film interface, and the oxygen ions transfer in the film, as the oxygen vacancies diffuse in the opposite direction and react with the metal precursors at the film-metal precursor interface to continuously form metal oxide.

Chemical vapor infiltration involves the deposition of solid products inside a porous medium, and the primary focus of CVI is on the filling of voids in porous graphite and fibrous mats to make carbon-carbon composites[36,37] and has been applied to other materials, such as depositing $ZrO_2$ into porous bodies of $MoSi_2$.[38] Various CVI techniques have been developed for infiltrating porous substrates, with the main goals being to shorten the deposition time and achieve homogeneous deposition. These techniques include isothermal and isobaric infiltration, thermal gradient infiltration,[36] pressure gradient infiltration,[36] forced flow infiltration,[39] pulsed infiltration,[40] and plasma enhanced infiltration.[39]

Various hydrocarbons have been used as precursors for CVI and typical deposition temperatures range from 850 to 1100°C and deposition times range from 10 to 70 h, and is rather long compared to other vapor deposition methods. The long deposition time is due to the relatively low chemical reactivity and gas diffusion into porous media. Furthermore, the gas diffusion will get progressively smaller as more solid is deposited inside the porous substrates. To enhance the gas diffusion, various techniques have been introduced, including forced flow, thermal, and pressure gradient. Plasma has been used to enhance the reactivity, however, preferential deposition near surfaces results in inhomogeneous filling. Complete filling is difficult and takes a very long time, since the gas diffusion becomes very slow in small pores.

## XI. DIAMOND FILMS BY CVD

Diamond is a thermodynamically metastable phase at room temperature,[41] so synthetic diamonds are made at high temperatures under high pressures with the aid of transition metal catalysts such as nickel, iron, and cobalt.[42,43] The growth of diamond films under low pressure (1 atm) and low temperatures (~800°C) is not a thermodynamic equilibrium process and differs from other CVD processes. The formation of diamond from gas phase at low pressure was initially reported in late 1960s.[44,45] The typical CVD process of diamond films is illustrated schematically in Figure 20.15.[46] A gaseous mixture of hydrocarbon (typically methane) and hydrogen is fed into an activation zone of the deposition chamber, where activation energy is introduced to the mixture and causes the dissociation of both hydrocarbon and hydrogen molecules to form hydrocarbon free radicals and atomic hydrogen. Many different activation schemes have been found to be effective in depositing diamond films and include hot filament, RF and microwave

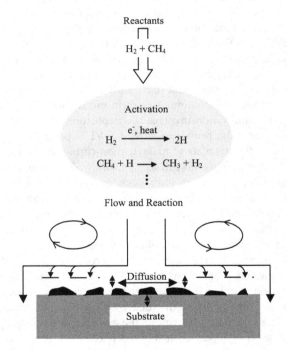

**FIGURE 20.15** Schematic showing the principal elements in the complex diamond CVD process: flow or reactants into the reactor, activation of the reactants by the thermal and plasma processes, reaction and transport of the species to the growing surface, and surface chemical processes depositing diamond and other forms of carbon. (From Pehrsson, P.E., Celii, F.G., and Butler, J.E., in *Chemical Mechanisms of Diamond CVD*, Davis, R.F., Ed., Noyes Publications, New Jersey, 1993.)

plasma, and flames. Upon arrival on the growth surface, a generic set of surface reactions occurs:[46]

$$C_DH + H\cdot \rightarrow C_D\cdot + H_2 \qquad (20.34)$$

$$C_D\cdot + \cdot CH_3 \rightarrow C_D\text{-}CH_3 \qquad (20.35)$$

$$C_D\cdot + C_xH_y \rightarrow C_D\text{-}C_xH_y. \qquad (20.36)$$

Reaction 34 activates a surface site by removal of a surface hydrogen atom linked to a carbon atom on the diamond surface. An activated surface site readily combines with either a hydrocarbon radical (reaction 20.35) or an unsaturated hydrocarbon molecule (e.g., $C_2H_2$; reaction 20.36). A high concentration of atomic hydrogen has proved to be a key factor in the successful growth of diamond films, and atomic hydrogen is believed to constantly remove graphite deposits on the diamond growth surface, so as to ensure continued deposition of diamond.[45] Oxygen species have also proved to be important in the deposition of diamond films by atmospheric combustion flames using oxygen and acetylene.[47,48] Other hydrocarbon fuels including ethylene, propylene, and methyl acetylene can all be used as precursors for the growth of diamond films.[49–52]

## XII. ATOMIC LAYER DEPOSITION

Atomic layer deposition (ALD) is a unique thin-film growth method and differs significantly from other thin-film deposition methods. The most distinctive feature of ALD is its self-limiting growth; each time only one atomic or molecular layer can grow. Therefore ALD offers the best possibility of controlling the film thickness and surface smoothness in the nanometer or subnanometer range. Excellent reviews on ALD have been published by Ritala and Leskelä.[53,54] In the literature, ALD is also called atomic layer epitaxy (ALE), atomic layer growth (ALG), atomic layer CVD (ALCVD), and molecular layer epitaxy (MLE). In comparison with other thin-film deposition techniques, ALD is a relatively new method and was first employed to grow ZnS film.[55] More publications appeared in the literature in the early 1980s.[56–58] ALD can be considered as a special modification of CVD, or a combination of vapor phase self-assembly and surface reaction. In a typical ALD process, the surface is first activated by chemical reaction. When precursor molecules are introduced into the deposition chamber, they react with the active surface species and form chemical bonds with the substrate. Since the precursor molecules do not react with each other, no more than one molecular layer can be deposited at this stage. Next, the monolayer of precursor molecules that chemically bonded to the substrate is activated again through surface reaction. Either the same or different precursor molecules are subsequently introduced to the deposition chamber and react with the activated monolayer previously deposited. As the steps repeat, more molecular or atomic layers are deposited one layer at a time.

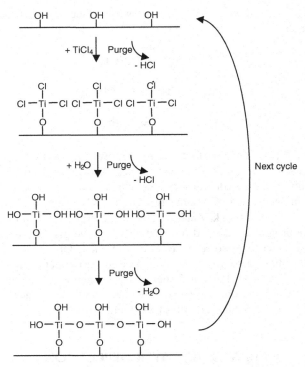

**FIGURE 20.16** Schematic illustrating the principal reactions and processing steps for the formation of titania film by ALD.

Figure 20.16 schematically illustrates the process of titania film growth by ALD. The substrate is hydroxylated first, prior to the introduction of titanium precursor, titanium tetrachloride. Titanium tetrachloride reacts with the surface hydroxyl groups through a surface condensation reaction:

$$TiCl_4 + HO\text{-}Me \rightarrow Cl_3Ti\text{-}O\text{-}Me + HCl, \qquad (20.37)$$

where Me represents metal or metal oxide substrates. The reaction stops when all the surface hydroxyl groups react with titanium tetrachloride. Then the gaseous by-product, HCl, and excess precursor molecules are purged, and water vapor is introduced to the system. Trichloride titanium clusters chemically bonded onto the substrate surface undergo a hydrolysis reaction:

$$Cl_3Ti\text{-}O\text{-}Me + H_2O \rightarrow (HO)_3Ti\text{-}O\text{-}Me + HCl. \qquad (20.38)$$

Neighboring hydrolyzed titanium precursors subsequently condensate to form a Ti-O-Ti linkage:

$$(HO)_3Ti\text{-}O\text{-}Me + (HO)_3Ti\text{-}O\text{-}Me \rightarrow Me\text{-}O\text{-}Ti(OH)_2\text{-}O\text{-}Ti\,(HO)_2\text{-}O\text{-}Me + H_2O \qquad (20.39)$$

The by-product HCl and excess $H_2O$ are then removed from the reaction chamber. One layer of $TiO_2$ has been grown by the completion of one cycle of chemical reactions. The surface hydroxyl groups are ready to react with titanium precursor molecules again in the next cycle. By repeating the above steps, two or more $TiO_2$ layers can be deposited in a very precisely controlled way.

The growth of ZnS film is another often used classical example for illustrating the principles of the ALD process. $ZnCl_2$ and $H_2S$ are used as precursors. First, $ZnCl_2$ is chemisorbed on the substrate, then $H_2S$ is introduced to react with $ZnCl_2$ to deposit a monolayer of ZnS on the substrate and HCl is released as a by-product. A wide spectrum of precursor materials and chemical reactions has been studied for the deposition of thin films by ALD. Thin films of various materials including various oxides, nitrides, fluorides, elements, and II-VI, II-VI, and III-V compounds in epitaxial, polycrystalline, and amorphous form deposited by ALD are summarized in Table 20.1.[53,54]

The choice of proper precursors is the key issue in a successful design of an ALD process. Table 20.2 summarizes the requirements for ALD precursors.[53,54] A variety of precursors have been used in ALD. For example, elemental zinc and sulfur were used in the first ALD experiments for the growth of ZnS.[55] Metal chlorides were studied soon after the first demonstrations of ALD.[59] Metallorganic compounds including both organometallic compounds and metal alkoxides are widely used. For nonmetals, mostly simple hydrides have been used: $H_2O$, $H_2O_2$, $H_2S$, $H_2Se$, $H_2Te$, $NH_3$, $N_2H_4$, $PH_3$, $AsH_3$, $SbH_3$, and HF.

## TABLE 20.1
### Thin-Film Materials Deposited by ALD[53,54]

| | |
|---|---|
| II-VI compounds | ZnS, ZnSe, ZnTe, $ZnS_{1-x}Se_x$, CaS, SrS, BaS, $SrS_{1-x}Se_x$, CdS, CdTe, MnTe, HgTe, $Hg_{1-x}Cd_x$Te, $Cd_{1-x}Mn_x$Te |
| II-VI based phosphors | ZnS:M (M = Mn, Tb, Tm), CaS:M (M = Eu, Ce, Tb, Pb), SrS:M (M = Ce, Tb, Pb, Mn, Cu) |
| III-V compounds | GaAs, AlAs, AlP, InP, GaP, InAs, $Al_xGa_{1-x}As$, $Ga_xIn_{1-x}As$, $Ga_xIn_{1-x}P$ |
| Nitrides | AlN, GaN, InN, $SiN_x$, TiN, TaN, $Ta_3N_5$, NbN, MoN, $W_2N$, Ti-Si-N |
| Oxides | $Al_2O_3$, $TiO_2$, $ZrO_2$, $HfO_2$, $Ta_2O_5$, $Nb_2O_5$, $Y_2O_3$, MgO, $CeO_2$, $SiO_2$, $La_2O_3$, $SrTiO_3$, $BaTiO_3$, $Bi_xTi_yO_z$, $In_2O_3$, $In_2O_3$:Sn, $In_2O_3$:F, $In_2O_3$:Zr, $SnO_2$, $SnO_2$:Sb, ZnO, ZnO:Al, $Ga_2O_3$, NiO, $CoO_x$, $YBa_2Cu_3O_{7-x}$, $LaCoO_3$, $LaNiO_3$ |
| Fluorides | $CaF_2$, $SrF_2$, $ZnF_2$ |
| Elements | Si, Ge, Cu, Mo, Ta, W |
| Others | $La_2S_3$, PbS, $In_2S_3$, $CuGaS_2$, SiC |

**TABLE 20.2**
**Requirements for ALD Precursors**[53]

| Requirement | Comments |
| --- | --- |
| Volatility | For efficient transportation, a rough limit of 0.1 Torr at the applicable maximum source temperature |
| | Preferably liquids or gases |
| No self-decomposition | Would destroy the self-limiting film growth mechanism |
| Aggressive and complete reactions | Ensure fast completion of the surface reactions and thereby short cycle times |
| | Lead to high film purity |
| | No problems of gas phase reactions |
| No etching of the film or substrate material | No competing reaction pathways |
| | Would prevent the film growth |
| No dissolution to the film | Would destroy the self-limiting film growth mechanism |
| Unreactive by-product | To avoid corrosion |
| | By-product readsorption may decrease the growth rate |
| Sufficient purity | To meet the requirements specific to each process |
| Inexpensive | |
| Easy to synthesize and handle | |
| Nontoxic and environmentally friendly | |

In comparison to other vapor phase deposition methods, ALD offers advantages, including precise control of film thickness and conformal coverage. Precise control of film thickness is due to the nature of the self-limiting process, and the thickness of a film can be set digitally by counting the number of reaction cycles. Conformal coverage is due to the fact that the film deposition is immune to variations caused by nonuniform distribution of vapor or temperature in the reaction zone. Figure 20.17 shows the x-ray diffraction spectrum and the cross-sectional scanning electron micrograph image of a 160 nm TiN film on patterned silicon wafer.[60] However, it should be noted that excellent conformal coverage can only be achieved when the precursor doses and pulse time are sufficient for reaching the saturated state at all surfaces at each step and no extensive precursor decomposition takes place.

Atomic layer deposition is an established technique for the production of large-area electroluminescent displays,[61] and is a likely future method for the production of the very thin films needed in microelectronics.[62] However, many other potential applications of ALD are discouraged by its low deposition rate, typically less than 0.2 nm (less than half a monolayer) per cycle. For silica deposition, completing a cycle of reactions typically requires more than 1

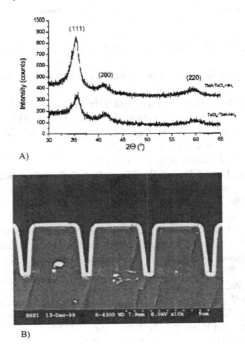

**FIGURE 20.17** (A) X-ray diffraction spectra and (B) cross-sectional scanning electron micrograph image of 160 nm Ta(Al)N(C) film on a patterned silicon wafer. (From Allén, P., Juppo, M., Ritala, M., Sajavaara, T., Keinonen, J., and Leskelä, M., *J. Electrochem. Soc.*, 148, G566, 2001.)

minute.[63,64] Some recent efforts have been directed toward the development of a rapid ALD deposition method. For example, highly conformal layers of amorphous silicon dioxide and aluminum oxide nanolaminates were deposited at rates of 12 nm, or more than 32 monolayers per cycle, and the method has been referred to as "alternating layer deposition."[65] The exact mechanism for such a multilayer deposition in each cycle is unknown, but it is obviously different from the self-limiting growth discussed above. The precursor employed in this experiment, tris(tert-butoxy)silanol, can react and thus the growth is not self-limiting.

## XIII. SUMMARY

This chapter reviewed the fundamentals of CVD processing and discussed general CVD methods. CVD is a common technique for the formation of thin films of all kinds of materials and has been studied extensively. However, the EVD, CVI, and ALD methods are more specifically focused on the deposition of ceramic films. CVD of diamond films is a unique example of deposition of metastable phase at low temperature under vacuum.

## REFERENCES

1. Ohring, M., *The Materials Science of Thin Films*, Academic Press, San Diego, 1992.
2. Vossen, J.L. and Kern, W., Eds., *Thin Film Processes II*, Academic Press, San Diego, 1991.
3. Nalwa, H.S., Ed., *Handbook of Thin Film Materials*, vol. 1, *Deposition and Processing of Thin Films*, Academic Press, San Diego, 2002.
4. Jensen, K.F. and Kern, W., in *Thin Film Processes II*, J.L. Vossen and W. Kern, Eds., Academic Press, San Diego, 1991.
5. Choy, K.L., *Prog. Mater. Sci.*, 48, 57, 2003.
6. Ser, P., Kalck, P., and Feurer, R., *Chem. Rev.*, 102, 3085, 2002.
7. Bloem, J., in *Proceedings of the Seventh Conference on CVD*, T.O. Sedgwick and H. Lydtin, Eds., *Electrochem. Soc. Proc.*, 79-3, 41, 1979.
8. Hartman, P. and Perdok, W.G., *Acta Crystl.*, 8, 49, 1955.
9. Vere, A.W., *Crystal Growth: Principles and Progress*, Plenum, New York, 1987.
10. Burton, W., Cabrera, N., and Frank, F.C., *Philos. Trans. R. Soc.*, 243, 299, 1951.
11. Roth, A., *Vacuum Technology*, North-Holland, Amsterdam, 1976.
12. Dushman, S., *Scientific Foundations of Vacuum Techniques*, John Wiley & Sons, New York, 1962.
13. Glang, R., in *Handbook of Thin Film Technology*, L.I. Maissel and R. Glang, Eds., McGraw-Hill, New York, 1970.
14. Goldsmith, N. and Kern, W., *RCA Rev.*, 28, 153, 1967.
15. Rosler, R., *Solid State Technol.*, 20, 63, 1977.
16. Taylor, R.C. and Scott, B.A., *J. Electrochem. Soc.*, 136, 2382, 1989.
17. Jordon, E.L., *J. Electrochem. Soc.*, 108, 478, 1961.
18. Sirtl, E., Hunt, L.P., and Sawyer, D.H., *J. Electrochem. Soc.*, 121, 919, 1974.
19. Adams, A.C., in *VLSI Technology*, 2nd ed., S.M. Sze, Ed., McGraw-Hill, New York, 1988.
20. Sze, S.M., *Semiconductor Devices: Physics and Technology*, John Wiley & Sons, New York, 1985.
21. Matuso, S., *Handbook of Thin Film Deposition Processes and Techniques*, Noyes, Park Ridge, NJ, 1982.
22. Hajjar, J.J., Reif, R., and Adler, D., *J. Electron. Mater.*, 15, 279, 1986.
23. Dupuis, R.D., *Science*, 226, 623, 1984.
24. Stringfellow, G.B., *Organo Vapor-Phase Epitaxy: Theory and Practice*, Academic Press, New York, 1989.
25. Osgood, R.M. and Gilgen, H.H., *Annu. Rev. Mater. Sci.*, 15, 549, 1985.
26. Abber, R.L., in *Handbook of Thin-Film Deposition Processes and Techniques*, K.K. Schuegraf, Ed., Noyes, Park Ridge, NJ, 1988.
27. McMillan, L.D., de Araujo, C.A., Cuchlaro, J.D., Scott, M.C., and Scott, J.F. *Integr. Ferroelectr.*, 2, 351, 1992.
28. Xia, C.F., Ward, T.L., and Atanasova, P., *J. Mater. Res.*, 13, 173, 1998.
29. Van Buskirk, P.C., Roeder, J.F., and Bilodeau, S., *Integr. Ferroelectr.*, 10, 9, 1995.
30. Isobe, C., Ami, T., Hironaka, K., Watanabe, K., Sugiyama, M., Nagel, N., Katori, K., Ikeda, Y., Gutleben, C.D., Tanaka, M., Yamoto, H., and Yagi, H., *Integr. Ferroelectr.*, 14, 95, 1997.
31. Solayappan, N., Derbenwick, G.F., McMillan, L.D., Paz de Araujo, C.A., and Hayashi, S., *Integr. Ferroelectr.*, 14, 237, 1997.

32. Isenberg, A.O., in *Electrode Materials and Processes for Energy Conversion and Storage*, J.D.E. McIntyre, S. Srinivasan, and F.G. Will, Eds., *Electrochem. Soc. Proc.*, 77-6, 572, 1977.

33. Carolan, M.F. and Michaels, J.M., *Solid State Ionics*, 25, 207, 1987.

34. Lin, Y.S., de Haart, L.G.J., de Vries, K.J., and Burggraaf, A.J., *J. Electrochem. Soc.*, 137, 3960, 1990.

35. Cao, G.Z., Brinkman, H.W., Meijerink, J., de Vries, K.J., and Burggraaf, A.J., *J. Am. Ceram. Soc.*, 76, 2201, 1993.

36. Kotlensky, W.V., *Chem. Phys. Carbon*, 9, 173, 1973.

37. Delhaes, P., in *Proceedings of Fourteenth Conference on Chemical Vapor Deposition, Electrochem. Soc. Proc.*, 97-25, 486, 1997.

38. Yoshikawa, N. and Evans, J.W., *J. Am. Ceram. Soc.*, 85, 1477, 2002.

39. Vaidyaraman, S., Lackey, W.J., Freeman, G.B., Agrawal, P.K., and Langman, M.D., *J. Mater. Res.*, 10, 1469, 1995.

40. Dupel, P., Bourrat, X., and Pailler, R., *Carbon*, 33, 1193, 1995.

41. Berman, R., in *Physical Properties of Diamond*, R. Berman, Ed., Clarendon Press, Oxford, 1965.

42. Bovenkerk, H.P., Bundy, F.P., Hall, H.T., Strong, H.M., and Wentorf, R.H., *Nature*, 184, 1094, 1959.

43. Wilks, J. and Wilks, E., *Properties and Applications of Diamonds*, Butterworth-Heinemann, Oxford, 1991.

44. Derjaguin, B.V. and Fedoseev, D.V., *Sci. Am.*, 233, 102, 1975.

45. Angus, J.C., Will, H.A., and Stanko, W.S., *J. Appl. Phys.*, 39, 2915, 1968.

46. Pehrsson, P.E., Celii, F.G., and Butler, J.E., in *Chemical Mechanisms of Diamond CVD*, Davis, R.F., Ed., Noyes Publications, New Jersey, 1993.

47. Hanssen, L.M., Carrington, W.A., Butler, J.E., and Snail, K.A., *Mater. Lett.*, 7, 289, 1988.

48. Rosner, D.E., *Annu. Rev. Mater. Sci.*, 2, 573, 1972.

49. Schermer, J.J., de Theije, F.K., and Elst, W.A.L.M., *J. Crystl. Growth*, 243, 302, 2002.

50. Harris, S.J., Shin, H.S., and Goodwin, D.G., *Appl. Phys. Lett.*, 66, 891, 1995.

51. Yarina, K.L., Dandy, D.S., Jensen, E., and Butler, J.E., *Diamond Relat. Mater.*, 7, 1491, 1998.

52. Gruen, D.M., *Annu. Rev. Mater. Sci.*, 29, 211, 1999.

53. Ritala, M. and Leskelä, M., in *Handbook of Thin Film Materials*, vol. 1, *Deposition and Processing of Thin Films*, H.S. Nalwa, Ed., Academic Press, San Diego, 2002.

54. Ritala, M. and Leskelä, M., *Nanotechnology*, 10, 19, 1999.

55. Suntola, T. and Antson, J., U.S. Patent no. 4,058,430, 1977.

56. Ahonen, M. and Pessa, M., *Thin Solid Films*, 65, 301, 1980.

57. Pessa, M., Mäkelä, R., and Suntola, T., *Appl. Phys. Lett.*, 38, 131, 1981.

58. Suntola, T. and Hyvärinen, J., *Annu. Rev. Mater. Sci.*, 15, 177, 1985.

59. Suntola, T., Antson, J., Pakkala, A., and Lindfors, S., *SID 80 Dig.*, 11, 108, 1980.

60. Allén, P., Juppo, M., Ritala, M., Sajavaara, T., Keinonen, J., and Leskelä, M., *J. Electrochem. Soc.*, 148, G566, 2001.

61. Suntola, T. and Simpson, M., Eds., *Atomic Layer Epitaxy*, Blackie, London, 1990.

62. Kingon, A.I., Maria, J.P., and Streiffer, S.K., *Nature*, 406, 1032, 2000.

63. Ferguson, J.D., Weimer, A.W., and George, S.M., *Appl. Surf. Sci.*, 162, 280, 2000.

64. Morishita, S., Gasser, W., Usami, K., and Matsumura, M., *J. Non-Crystl. Solids*, 187, 66, 1995.

65. Hausmann, D., Becker, J., Wang, S., and Gordon, R.G., *Science*, 298, 402, 2002.

# 21 Ceramic Photonic Crystals: Materials, Synthesis, and Applications

*Jeffrey DiMaio and John Ballato*

## CONTENTS

## I. INTRODUCTION

Modern communication is based on light. Regardless of whether it is electromagnetic waves of gigahertz frequency used for wireless telephony or near-infrared radiation used in terrestrial optical fiber communications, the higher bandwidth associated with a light-based carrier compared to an electrical carrier means that the "photonics age" will continue for the foreseeable future. Exemplifying this point, consider the technological progress made in light-confining glass fibers and ancillary components, which has resulted in a million-fold increase in information capacity over the past 30 years alone (a growth of approximately 50% per year). By way of comparison, it took the previous 120 years to achieve the same bandwidth increase using purely electronic means.[1] However, given the progressively more stringent requirements on reducing device size, weight, power consumption, and cost, academic and industrial researchers are always looking for more efficient approaches for the creation, routing, processing, and detection of light.

Spatial periodicities in an optical material's dielectric function promote coherent interference through multiple scattering when the wavelength of incident light is comparable in size to the scale of the periodicity. Allowed and forbidden directions for light of certain energies to propagate can result and may provide added control over the properties of the light interacting with such a *photonic crystal*. For a photonic crystal to exhibit a true or complete band gap requires that a band of energies be forbidden to propagate irrespective of direction or polarization in reciprocal space.[2] Although the use of band theory, including Brillouin zones, Bloch functions, and the reciprocal lattice, to describe light in a periodic structure is analogous to that for electrons in an atomic lattice, there are some important differences:[3–5]

The dispersion relationship for photons is linear, whereas for electrons it is parabolic.

Due to electron-electron interactions, the band theory for electrons is approximate, whereas for photons it is exact, since photon-photon interactions are considered negligible.

For electrons (being fermions of spin 1/2) a scalar wave approximation is often used, whereas for photons (being bosons of spin 1), the full vector field needs to be taken into consideration (see *Optics Express*, vol. 8, no. 3, 2001, for a focus issue on photonic band gap calculations).

These structured dielectrics can effectively control the propagation characteristics of light and therefore yield a myriad of useful optical effects ranging from enhanced or inhibited light generation,[3] omnidirectional reflection,[6] and ultra-low-threshold lasing,[7] to name just a few. Due to both academic as well as commercial interest, research and publications relating to photonic crystals have grown remarkably since the first works of Yablonovitch[3] and John.[4]

The purpose of this chapter is to provide an overview of ceramic materials used for photonic crystals, their synthesis, and macroscopic structures and architectures. Particularly close attention is given to the fabrication of silica colloidal crystals, since these forms are the most commonly studied. Initial efforts into devices are discussed, as are newer ceramic photonic crystal structures, including an overview of work in photonic crystal optical fibers. For completeness, non-oxide and organic photonic crystals also are included briefly.

## II. CERAMIC COLLOIDAL CRYSTALS

Scientists have spent a great deal of effort in recent years learning to synthetically emulate nature's ability to assemble particles into ordered arrays so as to more efficiently control the characteristics of light via reflection or transmission. From these efforts, significant scholarship has been realized on forming such colloidal crystals.

### A. SYNTHESIS OF COLLOIDAL PHOTONIC CRYSTALS

It is important to have a strong understanding of the effects of processing on the optical properties of a material or structure. In this case, we look to understand how structure affects properties and how processing affects structure; one then can decide the best approach for preparing a photonic crystal with a desired performance.

Colloidal particles, typically with sizes ranging from a few tens of nanometers to a few micrometers, are the preferred precursor to an artificial opal. Particles in suspension are influenced by a number of forces, including Brownian motion, gravity, viscous drag, and particle-particle and particle-liquid interactions. The reader is referred to Reference 8 for seminal discussions on colloidal and surface chemistry. Most inorganic photonic crystals are formed by steric packing of spherical particles that result from a destabilization of the aforementioned colloid. The diameter of the sphere dictates the lattice constant for the crystal. For photonic crystals in the visible and near-infrared range, our interests lie in the production of ceramic particles with diameters ranging from 300 to 900 nm. For a high-quality crystal to form, it is necessary that the spheres be monodispersed, as a crystal is constructed from the regular repetition of identical building blocks.[9] As with most materials that are described as crystals or crystalline, order/disorder inherently plays a strong role in the resulting properties. While a photonic crystal's repeating units are orders of magnitude larger than atomic and molecular crystal dimensions, the same types of defects are present. The control of defects is of paramount importance, as it will be an enabling technology in the fabrication of photonic crystal filters and waveguides. For this reason, the discussion of the processing of photonic crystals will be organized around how different processing techniques introduce disorder. It is important to have control over the type of structure formed, whether a random hexagonal close-packed (RHCP) structure or a diamond structure. The type of crystal and the shape of the individual repeat units, for example, spheres, ellipsoids, etc., will have a great effect in determining whether a band gap is a pseudo-band gap or a complete band gap. Having materials with a large enough dielectric mismatch is also necessary in the search for a complete photonic band gap. This is discussed in greater detail below.

The simplest photonic crystal is nature's opal. The artificial opal is composed of monodispersed spheres of a dielectric, usually silica. Considerable work has been done using latex or polystyrene spheres,[10] but we largely will restrict ourselves here to ceramics. In producing high-quality photonic crystals, care must be taken in each of the three main steps: particle synthesis, sedimentation, and sintering.

## B.  PARTICLE SYNTHESIS

There is a wealth of knowledge on the synthesis of colloidal particles, so only an overview of processes will be given. The reader is referred to the many reviews on the subject.[8,11,12] The most well-established inorganic colloid system is that of silica,[13] but many other hydrous and anhydrous metal oxides and nonoxides can be formed in a range of sizes and morphologies. There are two general techniques for forming colloidal particles. The first and simplest is a chemical reaction in an aerosol. In this process a high-purity vapor is condensed and the resultant liquid droplets are reacted with a vapor, usually water, to yield a solid. This procedure produces spherical particles of high purity. The particles may then be dispersed in water as long as the pH is away from the isoelectric point. While this technique is good for producing spherical metal oxides, it is limited in the number of variables that can be controlled.

Homogeneous precipitation is a second technique for the preparation of colloidal particles with narrow size distributions. When working with homogeneous precipitation it is important to remember that many "recipes" for colloidal particles exist, but an underlying physical explanation cannot be found that would allow for an a priori determination of what the resultant product will be. With this in mind, it is important for the researcher to carefully control the variables in a process in order to scale-up production for manufacturing. Generally the variables of interest are the pH, concentration of the reactants, temperature, concentration of salts, method of mixing, etc.[11] In order to achieve monodispersed spheres, it is important that there is a short period of nucleation, which produces the nascent particles for the reaction, followed by a growth phase. In this way, all the particles will grow at the same rate to the same size. If nucleation occurs during the growth regime, there will be a broader distribution of particle diameters, which will tend to decrease the degree of ordering of the particles during sedimentation/crystallization.

There are several ways in which homogeneous precipitation may occur: forced hydrolysis, controlled release of hydroxide ions, and decomposition of organometallic compounds. In forced hydrolysis, aqueous solutions of metal salts are heated from 80 to 100°C for a period to deprotonate the metal salts, lowering the pH of the solution, and yielding solid hydrous oxides.[11] Similarly, the controlled release

**TABLE 21.1**
## Commercial Sources of Monodispersed Colloidal Spheres[a,b]

| Company | Contact Information | Size Range | Comments |
|---------|--------------------|------------|----------|
| Bangs Laboratories[c] | www.bangslabs.com | 0.020–5.0 µm (polystyrene) 0.3–5.0 µm (silica) | Polystyrene and silica spheres |
| Duke Scientific[c] | www.dukesci.com | 0.020–1.0 µm (polystyrene) 0.5–1.6 µm (silica) | Polystyrene and silica spheres |
| Dyno Particles AS | www.pss.aus.net | 0.5–20 µm | Polystyrene spheres |
| Interfacial Dynamics[c] | www.idclatex.com | 0.020–10.0 µm | Polystyrene spheres |
| Nissan Chemicals | www.snowtex.com | 0.003–0.100 µm | Colloidal silica and antimony pentoxide. |
| Polyscience[c] | www.polysciences.com | 0.05–90 µm (polystyrene) 0.05–0.45 µm (silica) | Polystyrene, silica, and glass spheres |
| Seradyn | www.seradyn.com | 0.05–5.0 µm | Polystyrene spheres |

[a] For a more complete list see Xia, Y., Gates, B., Yin, Y., et al., Monodispersed colloidal spheres: old materials with new applications, *Adv. Mater.*, 12, 693, 2000.
[b] Information contained here is in accordance with knowledge of these companies at the time of the referred to publication (2000).
[c] Custom synthesis available from these companies.

of hydroxide ions can be used to initiate precipitation from metal salt solutions. In this case, urea or formamide is often used. As the organic compound decomposes, hydroxide ions are slowly released, raising the pH of the solution, and precipitating metal nonoxides, which may be calcined to form metal oxides. As with forced hydrolysis, the presence of anions can have a dramatic effect on the end product. The decomposition of organometallic compounds usually begins by hydrolyzing an alkoxide by the addition of water. The properties of the precipitates can be controlled by the alcohol:water ratio, pH, and the method of mixing.[11] One of the most well-known processes for creating monosized silica particles is the Stöber-Fink process.[14] In this procedure, tetraethoxysilane (TEOS) is hydrolyzed in ethanol at a high pH. The resulting particles are monodispersed silica spheres ranging from 300 nm to 2 µm. Since synthetic approaches are often time consuming, there are presently a number of commercial vendors for purchasing silica colloids (see Table 21.1).

## C. Ordering of Particles

Once the building blocks for the photonic crystal have been made (or purchased), they must be placed in an ordered arrangement. Different crystal structures will yield varying degrees of success in obtaining a photonic band structure. A case in point is, since we are dealing with monodispersed silica colloidal crystals, a complete gap cannot be obtained.[2] Later we will discuss inverse opals, where this may be possible. Further, in monodispersed systems, only simple metal-like structures can be obtained: simple cubic, FCC, and BCC (though FCC is the thermodynamically preferred structure). The simplest method for ordering colloidal spheres is controlled sedimentation. The advantage of sedimentation is thick, high-quality, self-assembled photonic crystals. The disadvantage is the slow processing time, which potentially can be on the order of months.

When creating photonic crystals for colloidal particles, the first step is to create a suspension of particles for sedimentation. A suspension of particles can have liquid and crystalline characteristics depending on their volume fraction, $\varphi$. (Most of the research leading to the discussion of this section was conducted with polymer spheres, but the phenomenon extends to all colloids that exhibit hard sphere behavior.) It has been demonstrated that for $\varphi < 0.494$, the suspension is below its "freezing transition," $\varphi_f$, and the particles in suspension behave as a fluid. When $\varphi > 0.545 = \varphi_m$, the "melting concentration," the suspension will become fully crystalline over time as secondary attractive forces overwhelm Brownian motion and assemble the particles into their lowest energy solid state.[15] In the range $\varphi_f < \varphi < \varphi_m$, there is an equilibrium formed between the crystalline phase and the fluid phase. In competition with crystallization of the suspension is sedimentation, which is due to gravity. It has been shown that by slowly rotating a suspension, crystallization will occur even if the time-averaged gravitational force is zero.[16] This is especially suited to binary component systems where there would be a phase separation due to particle size or density.[17,18]

For the following discussion, a suspension of silica spheres in water is assumed. While forming fully crystallized samples in suspension requires $\varphi < \varphi_m$, a more dilute suspension is usually used in order to obtain high-quality crystals upon sedimentation; $\varphi$ can vary from 0.002 to 0.30, depending upon applications. As is shown in Figure 21.1, four distinct regions can be observed as sedimentation occurs. At the top is clear fluid, below which is a layer of suspension with a uniform dispersion. Below the suspension, a layer with increasing $\varphi$ with depth is found (called the "fan"). The bottom is the sediment, which should have a crystal structure. The sediment is necessarily iridescent when $\varphi R^3 < 5\pi\ \mu m^3$.[16]

Sedimentation is a result of the gravitational force dominating other forces on the particles in suspension, such as Brownian motion. The particles will fall out of suspension depending on their size and density. As the particles pass from the fan into the sediment, they must self-assemble into an ordered structure. This is accomplished in a manner that is consistent with Edward-Wilkinson (EW) behavior, which is characterized by a particle falling randomly onto a crystal surface and diffusing laterally to find its lowest energy state.[19] If the velocity of the particles is high, diffusion of the particles to an order state will not occur, as crystallization will be slower than sedimentation. The Peclet number, $Pe = m_b gR/kT$, gives the balance between diffusion and sedimentation, where $m_b$ is the buoyant mass, $R$ is the radius of the particle, $g$ is the acceleration due to gravity,

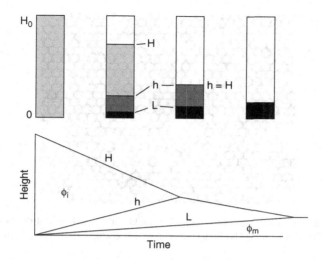

**FIGURE 21.1** The four distinct regions during sedimentation: the clear fluid is $x > H$, uniform dispersion from $h < x < H$, fan from $L < x < h$, and sediment for $x < L$.[15] (From Davis, K. et al., *Science*, 245, 507, 1989. With permission.)

$k$ is the Boltzmann constant, and $T$ is the absolute temperature.[20] Restated, the Peclet number describes Brownian forces relative to gravitational forces. If sedimentation is too fast, due to large particles or an artificially large gravitational force via centrifuge, then EW behavior will not be followed. Instead, a ballistic model would be used where a particle remains where it falls. This yields a noncrystalline structure instead of a crystalline one. A suspension with a low Peclet number will crystallize much easier than those with high Peclet number;[21] therefore greater care must be taken when trying to crystallize particles with a diameter greater than about 550 nm.[22]

When creating a colloidal crystal photonic structure, different crystal symmetries will yield different band gaps. For this reason, one would like to have control over the type of structure that is formed when processing a photonic crystal. There are three possible crystal structures that can occur upon sedimentation of monodispersed particles: FCC, HCP, or a combination of the two. Both FCC and HCP have a $\varphi = 0.74$ and 12 nearest neighbors. The only difference in the arrangement is the FCC structure stacks as ABCABC... and the HCP phase stacks as ABABAB.... The first layer deposited is the A site with a close-packed triangular arrangement in two dimensions, as shown in Figure 21.2. The second layer is the B site. The next layer can go in an A or C site. If the crystal structure is FCC, then the (111) plane is perpendicular to the growth direction; while the (1000) plane is perpendicular to the growth direction for HCP. This is because these planes have the highest packing density and are most likely to be nucleated at the interface on forming from solution. A stacking fault occurs when a layer that should occupy a C site occupies an A site, or vice versa. If it is equally probable that a layer will occupy an A site as a C site, the structure is RHCP.

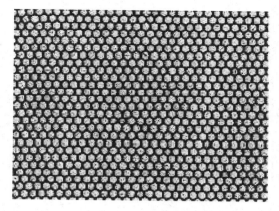

**FIGURE 21.2** The close-packed trigonal face can be either the (111) of the FCC or (100) of the HCP.[25] (From Míguez, H. et al., *Langmuir*, 13, 6009, 1997. With permission.)

This should be the case for the A site and C site because they have the same energy level with respect to gravity, so RHCP would be favored in colloidal sedimentation. In order to quantitatively define the amount of the FCC phase versus the HCP phase, an overall stacking parameter, $\alpha$, is used. This stacking parameter is defined such that $\alpha = 0$ for the HCP, $\alpha = 1$ for the FCC phase, and a RHCP structure would have an $\alpha = 0.5$. This has been used to describe colloidal crystal structure and stacking fault density.[23]

There have been many reports on the crystal structures that occur in a colloidal crystal. Near $\varphi_m$, the RHCP structure can be found, but deviations from $\varphi_m$ lead to FCC crystallization.[24–27] Figure 21.3 shows the planes of a truncated octahedral

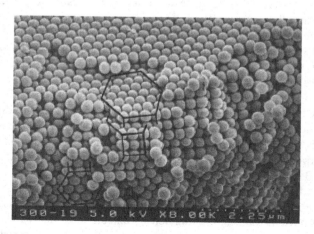

**FIGURE 21.3** Micrograph showing the boundaries of the cleaved facets of a truncated octahedron for FCC crystal.[26] (From Cheng, B. et al., *Opt. Commun.*, 170, 41, 1999. With permission.)

typical of the FCC phase. Crystallization experiments done in microgravity have shown complete RHCP, leading researchers to conclude that gravity has some effect in producing the FCC phase.[28] Mathematical modeling of a colloidal crystal by Woodcock[29] found that the FCC phase is thermodynamically favored over the HCP due to entropic considerations. The difference in energy is approximately $10^4 \cdot kT$ per particle, so that the FCC is only slightly favored.[30,31] The initial development of the RHCP phase typically had been attributed to kinetics, but simulations by Auer and Frenkel[32] suggest the RHCP phase is actually thermodynamically stable. As the crystal grows, the FCC phase becomes more stable and over a period of time the metastable RHCP phase will transform into the FCC phase. This result supports their earlier findings that in crystallites smaller than $3 \cdot 10^4$ particles, an FCC phase may not be formed. Larger crystallites, smaller than $10^6$ particles, will transform over time to the FCC phase, but will have a significant concentration of stacking faults.[33] For larger crystals, a reorientation from RHCP to FCC occurs, but the time scale is on the order of months.

There is much interest in understanding how to make high-quality FCC colloidal crystals for photonic applications. As such, it is convenient to have a way of characterizing a suspension quantitatively so that comparisons can be made to the final structure. Kegel and Dhont[23] use a "gravitational length" scale, $h$, which is defined as $h = kT/m_b g = R/Pe$. Using this term, samples formed in microgravity, Earth's gravity, and under centrifugal force can be compared. It is found that samples with a larger gravitational length (i.e., lower gravitational stress) form an FCC phase at a rate much slower than samples with smaller gravitational lengths. This supports the aforementioned finding that the structures formed in microgravity were RHCP. The time required to transition to the thermodynamically stable FCC phase was not days, but months, and could not be reached on a typical space mission. Another parameter used to characterize a suspension was developed by van Blaaderen et al.[34] Taking into account the maximum rate of crystallization, the rate of sedimentation, and the initial $\varphi$ yields a factor of the form $kT/(\varphi_o \Delta \rho g R^4) = kT/\varphi_o m_b R = (\varphi_o Pe)^1$.[22] An increase in $(\varphi_o Pe)^{-1}$ corresponds to a greater likelihood of achieving an FCC crystal structure.

It has been shown that high-quality bulk FCC colloidal crystals can be formed over a period extending from days to months. By increasing the "gravitational" force, the processing time to form a crystal from a suspension can be significantly decreased. Gravitational effects can be accounted for by using parameters such as $(\varphi_o Pe)^{-1}$ and the gravitational length, however, the question arises as to the effect of a high gravitational force on the quality of the colloidal photonic crystal and the resultant band structure. One would expect that the crystal growth mode goes from that of the EW model to a ballistic model. As discussed previously, the ballistic model tends to yield a glassy structure. A comparison of such structures is shown in Figure 21.4; the centrifugally cast sample is disordered and the transmission spectra is broad (Figure 21.4; left). The sample that was allowed to sediment undisturbed is highly ordered and has a considerably narrower band gap (Figure 21.4; right).[35] While disorder increases with increasing "gravitational" force, Shelekhina et al.[36] have reported that further increasing the centrifugal force

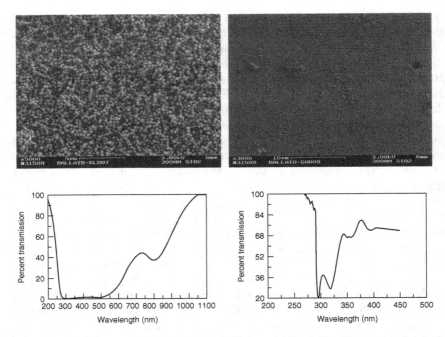

**FIGURE 21.4** Micrograph of disordered colloidal crystal (top left) with associated optical transmission spectrum (bottom left) and that for an ordered colloidal crystal (micrograph: top right; spectrum: bottom right).[35] (From Ballato, J. et al., *Appl. Phys. Lett.*, 75, 1497, 1999. With permission.)

improves the quality of packing. Their reports show essentially no band gap when processed at an acceleration of 4500 m/s², the formation of a band gap at 7000 m/s², and a maximum band gap at 8750 m/s². There are techniques that will be discussed later that can be used to give a quantitative description of the effect of the degree of order/disorder on the optical properties of a colloidal crystal.[37]

Another way to increase the rate of sedimentation is by using electrophoretic deposition. An advantage of electrophoretic deposition is that the sedimentation rate for large and small particles can be controlled so that both are deposited in the same amount of time and the same quality crystal can be formed. Electrophoretic deposition takes advantage of the surface charge density of a particle away from its zero point of charge (ZPC). By changing the pH of the solution further from or closer to the ZPC, the surface charge density can be controlled. By applying an electric field to the suspension, the field can oppose the gravitational force and dominate Brownian motion and slow the sedimentation of particles so that crystallization can occur. By reversing the electric field, the velocity of the particles can be increased.[23] This process is ideal for particles with diameters smaller than 300 nm or larger than 550 nm.

Another degree of control over crystallization lies in the surface of the substrate on which the suspended particles are crystallized. Crystallization on a smooth wall is considered to be spontaneous heterogeneous nucleation, sometimes referred to as "prefreezing." Courtemanche and van Swol[38] reported crystallization rates five times faster for a wall structured with an FCC (111) face versus a smooth wall for a colloidal suspension. However, much of the experimentation, including that by Courtemanche and van Swol, focuses on colloidal polymers that crystallize in suspension, not sedimentation. While resultant structures are similar, the growth kinetics are not. Suspensions crystallize via a layer-by-layer method, while sedimentation features continuous growth. Heni and Löwen[39,40] found that walls with FCC (111) and HCP (110) planes crystallized into the bulk, but the FCC (100) and (110) planes do not completely wet the surface and the crystalline structure does not continue into the bulk crystal. Conversely, van Blaaderen et al.[34] found that using an FCC (100) template in sedimentation can result in an FCC structure with the crystal growth direction along <100>. Using slow sedimentation and $\varphi \ll 0.49$, layer-by-layer growth was achieved ($Pe \approx 1$). Unlike the (111) face, which stacks as ABC... or ABA..., the (100) face has only one possible position for the next layer. As a result, it was found that a pure FCC phase was formed that was single crystalline across the whole template. If however there is any mismatch in the lattice spacing, $a$, after at least 10 layers, the structure has reverted to an RHCP structure. The exception to this is if the particles have a diameter $\sqrt{2}a$; the square structure of the (100) face will be maintained but rotated 45°.[34] Using a template for sedimentation decreases not only the stacking faults near the wall, but yields a 20% increase in the FCC content. Further, the orientation for large crystals remains the same as that of the template, while for smaller crystallites there is no specific orientation.[22]

Once sedimentation and crystallization have occurred and the sample has been dried to remove any remnant solvent, an artificial opal can be sintered to increase its mechanical strength.[41] Heating the compact to 100°C will initiate neck formation between the particles. Necking occurs more completely between 700 and 950°C. At temperatures of less than 700°C, there is shrinkage of the particles and a change in the refractive index that causes a blue shift in the spectral placement of the band gap, $\lambda_c$, as the lattice spacing decreases. Above 950°C, $\lambda_c$ rapidly shifts to lower wavelengths as the structure's density approaches the monolithic value and the attenuation dip diminishes.[43] Kuai et al.[43] reported an increase in crystal quality as a crystal is sintered, attributing this to the transition from a loose compact of spheres to that of a tight arrangement, which will decrease defects. Figure 21.5 shows the change in optical spectra of a silica colloidal crystal as it sinters.[42]

## D. INVERSE OPALS

Although many (perhaps most) applications do not require the existence of a complete band gap, significant efforts to develop such structures have been undertaken.[44-49] For a given symmetry, the characteristics of an induced band gap depend upon the refractive index contrast and filling factor of the two phases (e.g., silica spheres and air interstices). Inverse opals are negative replicates of the sterically packed structures that have been described above. In this case, one generally has

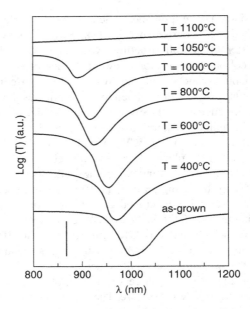

**FIGURE 21.5** Transmission of a photonic crystal with incident light normal to the growth surface (111 face) after annealing. The samples were heated with a ramp rate of 1°C/min and held at the annealing temperature for 3 hours.[42] (From Míguez, H., et al., *Adv. Mater.*, 10, 480, 1998. With permission.)

formed a conventional synthetic opal and infiltrated the interstices with a new material that, once solidified, forms the periodic structure after the initial opal is etched away (e.g., polymer or some other material where the air interstices once were and air now where the initial silica colloidal spheres were). Inverse opals of high index materials can exhibit a full photonic band gap, which is more difficult to realize in a conventional opal.[44] Figure 21.6 shows a carbon inverse opal where a carbon-yielding organic precursor was infiltrated into a silica opal, pyrolyzed, and the silica removed via subsequent hydrofluoric acid etching.[50]

## E. BINARY AND ANISOTROPIC PHOTONIC CRYSTALS

Since monosized particles lead to the highest degree of order in colloidal crystal photonic band gap structures, the resultant "crystallography" necessarily resembles that of a simple single-component metal (e.g., FCC, BCC, or HCP structures). As with atomic crystals, interesting optical and electronic properties arising from band theory result when the structure is binary (e.g., MX, $M_2$, $M_2X$, $M_2X_3$, Ö structures, where M is a particle of one size—analogous to a cation—and X is a particle of a difference size—analogous to an anion). The structures now take on different point group symmetries than do conventional (single material) colloidal crystals and, as such, their properties can take on greater degrees of anisotropy. This opens the door to a wider range of band gap characteristics generated by the multiple periodicity

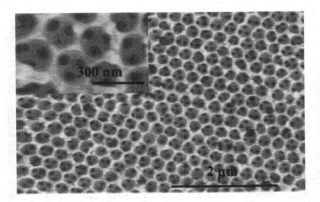

**FIGURE 21.6** Micrograph of an inverse carbon opal (magnification 25K; inset: magnification 150K).[50]

scales. To date, this remains a nascent area of photonic crystal resear[17,18,51,52] but shows great promise for value-added future applications.

## F. Effect of Order/Disorder of PBG

Numerous studies have developed approaches to quantifying the level of order or disorder in photonic crystals and the resultant effect on the band gap characteristics (i.e., defining the structure/optical property relationships). Qualitatively it is obvious that the higher the degree of order, the more well defined the band gap will be for a given symmetry and number of scattering centers (e.g., particles). Many approaches developed to quantify the effects of order/disorder rely on an order parameter defined using the photon localization length, being itself quantified in terms of the system's optical transmission over the various configurations.[53] Others have made use of the scale-invariance of Maxwell's equations to draw analogies between photonic crystals and x-ray diffraction and use optical scattering to decipher crystallographic structure based on pair correlation functions; shown in Figure 21.7.[54,55] Great strides have been made in defining the structure-property relationships in photonic crystals,[20,56] although, as the complexity of structures continues to increase, this topical area will only increase in importance.

## G. Other-Than-Silica Ceramic Colloidal Photonic Crystals

The thermal, mechanical, and chemical robustness of $SiO_2$ coupled with well-established synthetic approaches to preparing monodispersed colloids (i.e., Stöber-Fink) has led to silica being the dominant ceramic material used in photonic crystal research and application. However, an increasing amount of photonic crystal research is being performed using nonceramic materials, including metallodielectrics, semiconductors, and polymers. Although this chapter focuses on

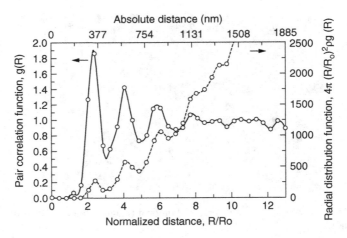

**FIGURE 21.7** Pair correlation and radial distribution function (RDF) for Figure A micro-structure. Results analogous to RDF for an atomic glass.[55]

ceramic materials, the following section is offered as a brief reference from which the reader can explore further.

Interestingly, some of the first experimental validations of photonic band gap structures were performed using an alumina ($Al_2O_3$) block that was tediously drilled at specific angles so that the resultant alumina/air structure possessed a specified symmetry with periodicities promoting a band gap in the microwave spectral region (unit cell length ~13 mm and a band gap centered at 15 GHz).[2] At these frequencies, alumina has a microwave refractive index of approximately 3. This experimental validation using alumina spurred follow-on research into other inorganic materials for radar applications, particularly GaAs (microwave refractive index ~3.6), since photonic crystal substrates were shown to yield considerably higher antenna power radiating into the air than into the substrate (a problem for conventional antennae).[57]

## 1. Titania

Among oxide ceramics, titania ($TiO_2$) possesses one of the largest refractive indices (between 2.4 and 2.6 depending on orientation and crystal structure). For this reason, the synthesis of titania particles,[58] inverse opals,[59–62] and polymer composites[63,64] have all received attention. The preparation of monodispersed titania colloids tends to be more difficult than it is for silica. For these reasons, despite the considerably higher refractive index, there is considerably less work in titania than silica. For completeness, it should be noted that other higher-index materials also have been made into photonic crystals, including lithium niobate ($LiNbO_3$),[65] lanthanum-doped lead zirconate titanate (PLZT),[66] and chalcogenide glass.[67]

## 2. Carbon

Inorganic carbon has also received some attention as an interesting material for photonic crystal structures and applications. Although not often considered a high-quality optical material, there are advantages to carbon over more canonical ceramics like silica. Of particular note is carbon's chemical inertness in certain acidic environments that would negate the use of oxides, a higher (complex) refractive index, thermoelectric and electrostrictive properties of interest,[68] as well as the potential for a plasmon resonance in the infrared region (i.e., potentially leading to a plasmon-defined photonic band gap), which is beyond the frequency range typically permitted by metallic or semiconducting systems. Dense and porous[69] carbon opals and inverse opals have been prepared using CVD[70] and polymer precursor[50] routes as well as through replication approaches.[71]

## 3. Polymers

Much of the early work in the synthesis and optical properties of organic colloidal crystals, as well as their subsequent stabilization in polymers, was performed by Prof. S. Asher (University of Pittsburgh)[10,72,73] prior to photonic crystals being en vogue. From these seminal works, a great amount of photonic crystal research into new organic materials, assembly techniques, and applications has resulted and include efforts utilizing block copolymers,[74,75] thermally and electrically tunable optical properties,[76–78] and amplified spontaneous emission in dye-doped[79,80] or conducting polymers[81–83] and liquid crystals,[84,85] mechanochromic band gap tuning in all-polymeric composites,[86–89] nonlinear optical polymers,[90] as well as holographic[91] and two-photon[92] forming and electropolymerization approaches, to name just a few.

## 4. Metallodielectrics and Other Multiphasic Composites Using Multiple Materials

Considerable interest also has been directed at the use of multicomponent composites where, in theory, the most useful properties from each phase can be realized in the whole. This includes metallodielectric structures where a metallic phase[93] imparts, for example, a high index or more exotic effect (e.g., plasmon resonance) and a low-loss or property-tunable dielectric phase.[94–99] The dielectric phase can be ceramic or polymeric and also has included ferroelectric polymers,[100] embedded nanoparticles,[101] and organic/inorganic hybrids.[102,103]

## III. PHOTONIC CRYSTAL OPTICAL FIBERS

In terms of ceramic photonic crystals, as noted, opaline silica structures have dominated research activities and the publication and patent literature. In more recent years, significant efforts have developed in the fabrication, modeling, and application of optical fibers in which some degree of dielectric periodicity is engineered into the structure. Typically this periodicity is found along the fiber's cross section and so the transverse components of the waveguide mode are most strongly affected.

Although not initially conceived for use as novel optical fibers, work on nanochannel glasses provided insight into the structure-property relationships resulting from drawn glass arrays possessing a periodicity along the fiber axis.[104,105] Around the same time, what can be considered the first work specifically on optical fiber structures was being performed. Initially efforts focused on a solid core surrounded by a cladding that contained a periodic ordering of air channels parallel to the fiber axis (later called microstructured fibers).[106,107] In this case, light guidance is still via total internal reflection, since the (solid) core has a higher refractive index than the (periodically "holey") cladding, even though they are made from the same native material (i.e., silica). This is because the presence of the holes that run the length of the fiber reduces the effective refractive index such that the cladding has a lower index than the core. Figure 21.8 shows some different microstructured photonic crystal fibers.[108] These structures, especially in achieving endlessly single-mode propagation,[109] opened the door to great creativity among fiber designers and optical physicists, leading to a remarkable growth in photonic crystal fiber publications. It was not long after these initial successes that researchers fabricated and discussed the extraordinary properties of hollow core (also called "air-filled") photonic crystal fibers.[110–112] In these "holey fibers," guidance arises from interference of the optical waves with an engineered periodicity in the cladding parallel to the fiber axis such that the light is confined to the core (i.e., Bragg-like diffraction of propagating modes in the [lower index, hollow] core by the periodic [higher index, glass] cladding as opposed to total internal reflection, which requires the core to be of a higher index than the cladding and that the critical angle be exceeded).

With respect to conventional fiber, the potential benefits of a hollow-core photonic crystal fiber are lower transmission losses (due to both reduced absorption and Raleigh scattering) and lower nonlinearities, since the core is air. In only a few years of research, interest in this area has led to remarkably low-loss fibers (13 dB/km at 1500 nm; 100 m length).[113] With continued advances toward lower-loss photonic crystal fibers, efforts also have begun to address field-level issues, including microbend losses[114,115] and fiber splicing.[116]

## A. Unique Properties of Photonic Crystal Fibers

The added control over the characteristics of light that is imposed by the periodicity in a photonic crystal fiber has been found to permit other highly useful effects in addition to the above-noted potential for subsilica attenuations. Of particular note are the advantages relating to dispersion management and optical nonlinearity.

**FIGURE 21.8** Scanning electron micrographs of several microstructured optical fibers including (a) high delta, (b) photonic crystal, (c) "grapefruit," and (d) air-clad structures.[108]

## 1.  Dispersion

As is well known for conventional optical fibers, the group velocity dispersion is governed principally by the dispersion of bulk silica, which has its zero near 1.3 μm.[117] Modifications to this can be made by tailoring the core/clad index profile, which has led to considerable progress in dispersion compensation.[118,119] Generally though, there are higher-order dispersion characteristics in these modified conventional fibers that limit their degree of spectra "flatness" (for dispersion flattened fiber). The periodic holes present in photonic crystal fibers offer an added degree of design flexibility with respect to conventional dispersion-modified fibers.[120–123] The result has been the development of ultraflat dispersion over large spectral ranges,[124,125] large anomalous dispersion while maintaining low loss and single-mode propagation,[126,127] as well as a suite of related characteristics.[128,129]

## 2.  Nonlinearities

Modifications to the electromagnetic field distribution and dispersion characteristics for light propagating down a photonic crystal fiber also can lead to unique nonlinear optical properties, including broadband supercontinuum generation[130] and parametric processes[131] such as second harmonic generation.[132] Supercontinuum generation arises from a combination of enhanced nonlinearity in a small core photonic crystal fiber and the fiber's ability to exhibit near-zero chromatic dispersion over an expanded spectral range. In this case, ultrashort pulses with high peak intensities are maintained over long interaction lengths such that ultrabroadband light is generated nonlinearly (e.g., 550 THz bandwidth; from violet into the infrared).[130,133] Initial applications under consideration include optical coherence tomography[134,135] and metrology.[136,137]

## 3.  Other Devices

For completeness, it should be noted that silica-based photonic crystal fibers have also received attention for lasers and amplifiers (predominantly $Yb^{3+}$-doped systems, although $Nd^{3+}$ and $Er^{3+}$ have been studied),[138–144] as well as polarization-maintaining structures.[145,146]

## B.  Nonsilica Photonic Crystal Fibers

Much like conventional optical fibers, photonic crystal fibers are most often made from silicate glasses. However, there has been some work to date on nonsilica fibers, including soft glasses (Schott's SF6 glass, index of 1.801 at 590 nm),[147,148] polymers (polymethyl methacrylate [PMMA]),[149,150] and chalcogenides (Ga-La-S).[151] Although this is still a nascent area of research, the reduced fabrication temperatures associated with polymers and soft glasses, the high indices of chalcogenides, and the large nonlinearities of chalcogenides and chromophore-doped polymers, with respect to silica, mark these types of specialty materials

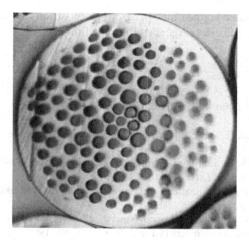

**FIGURE 21.9** Cross-section of an extruded bicomponent polymer photonic crystal fiber.[152]

as strong candidates for continued attention. As an example of unconventional materials and processing, Figure 21.9 shows the cross-section of a bicomponent polymer photonic crystal fiber prepared by coextrusion.[152]

## IV. CONCLUSIONS

In a period of about 15 years, the field of photonic crystals has grown into an international effort integrating materials scientists, chemists, physicists, electrical engineers, and even artists. Commercial applications of photonic crystals, although still several years from realization in most cases, are clearly envisioned and will produce significant progress in a diverse range of technologies including the fields of optical, radio, and cellular telecommunications and biomedicine, to name just a few. This chapter has aimed to provide an overview of ceramic photonic crystals focusing mostly on the synthesis of silica-based colloidal crystalline structures (synthetic opals), but including, for completeness, other materials, structure/property considerations, and photonic crystal optical fibers.

Lastly, given the immense interest in photonic crystals, several websites have been established to chronicle publications, patents, monoliths, special reports, computer programs, research programs, and the like. Although we cannot say for certain how long these sites will remain active, we do note them for consideration: photonic and acoustic band gaps, http://www.pbglink.com/; photonic crystal fibers, http://www.ece.utexas.edu/~yqjiang/PCFpaper.htm.

## ACKNOWLEDGMENTS

The authors wish to thank Sheryl Gonzales for editorial assistance and Michelle DiMaio and Heidi Ballato for their ceaseless support. The authors are also grateful

to the National Science Foundation (J.D.), 3M Corporation (J.B.), and the Defense
Advanced Research Projects Agency (J.B.) for funding.

## REFERENCES

1. Keck, D., A future full of light, *IEEE J. Select. Top. Quant. Electron.*, 6, 1254, 2000.
2. Yablonovitch, E., and Gmitter, T., Photonic band structure: the face-centered cubic case, *Phys. Rev. Lett.*, 63, 1950, 1989.
3. Yablonovitch, E., Inhibited spontaneous emission in solid-state physics and electronics, *Phys. Rev. Lett.*, 58, 2059, 1987.
4. John, S., Strong localization of photons in certain disordered dielectric superlattices, *Phys. Rev. Lett.*, 58, 2486, 1987.
5. Joannopoulos, J., Meade, R., and Winn, J., *Photonic Crystals: Molding the Flow of Light*, Princeton University Press, Princeton, NJ, 1995.
6. Fink, Y., Winn, J., Fan, S. et al., A dielectric omnidirectional reflector, *Science*, 282, 1679, 1998.
7. Hirayama, H., Hamano, T., and Aoyagi, Y., Novel surface emitting laser diode using photonic band-gap crystal cavity, *Appl. Phys. Lett.*, 69, 791, 1996.
8. Russell, W., Saville, D., and Schowalter, W., *Colloidal Dispersions*, Cambridge University Press, New York, 1989.
9. Kittel, C., *Introduction to Solid State Physics*, 2nd ed., John Wiley & Sons, New York, 1956.
10. Asher, S., Holtz, J., Weissman, J. et al., Mesoscopically periodic photonic crystal materials for linear and nonlinear optics and chemical sensing, *MRS Bull.*, 23, 44, 1998.
11. Matijevic, E., Preparation and properties of uniform size colloids, *Chem. Mater.*, 5, 412, 1993.
12. Sacks, M., and Tseng, T., Preparation of $SiO_2$ glass from model powder compacts: I, formation and characterization of powders, suspensions, and green compacts, *J. Am. Ceram. Soc.*, 67, 526, 1984.
13. Iler, R., *The Chemistry of Silica*, John Wiley & Sons, New York, 1979.
14. Stöber, W., Fink, A., and Bohn, E., Controlled growth of monodispersed silica spheres in the micron sized range, *J. Colloid Interface Sci.*, 26, 62, 1968.
15. Davis, K., Russel, W., and Glantschnig, W., Disorder-to-order transitions in settling suspensions of colloidal silica: x-ray measurements, *Science*, 245, 507, 1989.
16. Bartlett, P., Pusey, P., and Ottewill, R., Colloidal crystallization under time-averaged zero gravity, *Langmuir*, 7, 213, 1991.
17. Bartlett, P., and Ottewill, R., Freezing of binary mixtures of colloidal hard spheres, *J. Chem. Phys.*, 93, 1299, 1990.
18. Underwood, S., van Megen, W., and Pusey, P., Observations of colloidal crystals with the cesium chloride structure, *Phys. A*, 221, 438, 1995.
19. Salvarezza, R., Vàzquez, L., Míguez, H. et al., Edward-Wilkinson behavior of crystal surfaces grown by sedimentation of $SiO_2$ nanospheres, *Phys. Rev. Lett.*, 77, 4572, 1996.
20. Vlasov, Y., Astratov, V., Baryshev, A. et al., Manifestation of intrinsic defects in optical properties of self-organized opal photonic crystals, *Phys. Rev. E*, 61, 5784, 2000.

21. Hoogenboom, J., Derks, D., Vergeer, P. et al., Stacking faults in colloidal crystals grown by sedimentation, *J. Chem. Phys.*, 117, 11320, 2002.
22. Holgado, M., García-Santamaría, F., Blanco, A. et al., Electrophoretic deposition to control artificial opal growth, *Langmuir*, 15, 4701, 1999.
23. Kegel, W., and Dhont, J., "Aging" of the structure of crystals of hard colloidal spheres, *J. Chem. Phys.*, 112, 3431, 2000.
24. Pusey, P., van Megen, W., Bartlett, P. et al., Structure of crystals of hard colloidal spheres, *Phys. Rev. Lett.*, 63, 2753, 1989.
25. Míguez, H., Meseguer, F., López, C. et al., Evidence of FCC crystallization of $SiO_2$ nanospheres, *Langmuir*, 13, 6009, 1997.
26. Cheng, B., Ni, P., Jin, C. et al., More direct evidence of the FCC arrangement for artificial opal, *Opt. Commun.*, 170, 41, 1999.
27. Míguez, H., López, C., Meseguer, F. et al., Photonic crystal properties of packed submicrometer $SiO_2$ spheres, *Appl. Phys. Lett.*, 71, 1148, 1997.
28. Zhu, J., Li, M., Rogers, R. et al., Crystallization of hard-sphere colloids in microgravity, *Nature*, 387, 883, 1997.
29. Woodcock, L., Entropy difference between the face-centred cubic and hexagonal close-packed crystal structures, *Nature*, 385, 141, 1997.
30. Bolhuis, P., Frenkel, D., Mau, S. et al., Entropy difference between crystal phases, *Nature*, 388, 235, 1997.
31. Bruce, A., Wilding, N., and Ackland, G., Free energy of crystalline solids: a lattice-switch Monte Carlo method, *Phys. Rev. Lett.*, 79, 3002, 1997.
32. Auer, S., and Frenkel, D., Prediction of absolute crystal-nucleation rate in hard-sphere colloids, *Nature*, 409, 1020, 2001.
33. Pronk, S., and Frenkel, D., Can stacking faults in hard-sphere crystals anneal out spontaneously? *J. Chem. Phys.*, 110, 4589, 1999.
34. van Blaaderen, A., Ruel, R., and Wiltzius, P., Template-directed colloidal crystallization, *Nature*, 385, 321, 1997.
35. Ballato, J., DiMaio, J., and James, A., Photonic band engineering through tailored microstructural order, *Appl. Phys. Lett.*, 75, 1497, 1999.
36. Shelekhina, V., Prokhorav, O., Vityaz, P. et al., Towards 3D photonic crystals, *Synthet. Met.*, 124, 137, 2001.
37. Ballato, J., DiMaio, J., Taylor, T. et al., Structure determination in colloidal crystal photonic bandgap structures, *J. Am. Ceram. Soc.*, 85, 1366, 2002.
38. Courtemanche, D., and van Swol, F., Wetting state of a crystal-fluid system of hard spheres, *Phys. Rev. Lett.*, 69, 2078, 1992.
39. Heni, M., and Löwen, H., Precrystallization of fluids induced by patterned substrates, *J. Phys. Condensed Matter*, 13, 4675, 2001.
40. Heni, M., and Löwen, H., Surface freezing on patterned substrates, *Phys. Rev. Lett.*, 85, 3668, 2000.
41. Sachs, M., and Tseng, T., Preparation of $SiO_2$ glass from model powder compacts: II, sintering, *J. Am. Ceram. Soc.*, 67, 532, 1984.
42. Míguez, H., Meseguer, F., López, C. et al., Control of the photonic crystal properties of the FCC-packed submicrometer $SiO_2$ spheres by sintering, *Adv. Mater.*, 10, 480, 1998.
43. Kuai, S., Zhang, Y., Truong, V. et al., Improvement of optical properties of silica colloidal crystals by sintering, *Appl. Phys. A*, 74, 89, 2002.
44. John, S., and Busch, K., Photonic bandgap formation and tunability in certain self-organizing systems, *J. Lightwave Technol.*, 17, 1931, 1999.

45. Johnson, N., McComb, D., Richel, A. et al., Synthesis and optical properties of opal and inverse opal photonic crystals, *Synthet. Met.*, 116, 469, 2001.

46. Ni, P., Cheng, B., and Zhang, D., Inverse opal with an ultraviolet photonic gap, *Appl. Phys. Lett.*, 80, 1879, 2002.

47. Schroden, R., Al-Daous, M., Blanford, C. et al., Optical properties of inverse opal photonic crystals, *Chem. Mater.*, 14, 3305, 2002.

48. Meseguer, F., Blanco, A., Miguez, H. et al., Synthesis of inverse opals, *Colloid Surf. A*, 202, 281, 2002.

49. Zeng, F., Sun, Z., Wang, C. et al., Fabrication of inverse opal via ordered highly charged colloidal spheres, *Langmuir*, 18, 9116, 2002.

50. Perpall, M., Perera, K., DiMaio, J. et al., A novel network polymer for templated carbon photonic crystal structures, *Langmuir*, 19, 7153, 2003.

51. Velikov, K., van Dillen, T., Polman, A. et al., Photonic crystals of shape-anisotropic colloidal particles, *Appl. Phys. Lett.*, 81, 838, 2002.

52. Velikov, K., Christova, C., Dullens, R. et al., Layer-by-layer growth of binary colloidal crystals, *Science*, 296, 106, 2002.

53. Sigalas, M., Soukoulis, C., Chan, C. et al., Localization of electromagnetic waves in two-dimensional disordered systems, *Phys. Rev. B*, 53, article no. 8340, 1996.

54. Ballato, J., Tailorable visible photonic bandgaps through microstructural order and coupled material effects in $SiO_2$ colloidal crystals, *J. Opt. Soc. Am. B*, 17, 219, 2000.

55. Ballato, J., DiMaio, J., Gulliver, E. et al., A simple *a priori* determination of optical transmission gaps in photonic crystals of weak symmetry, *Opt. Mater.*, 20, 51, 2002.

56. Astratov, V., Adawi, A., Fricker, S. et al., Interplay of order and disorder in the optical properties of opal photonic crystals, *Phys. Rev. B*, 66, article no. 165215, 2002.

57. Brown, E., Parker, C., and Yablonovitch, E., Radiation properties of a planar antenna on a photonic-crystal substrate, *J. Opt. Soc. Am. B*, 10, 404, 1993.

58. Jiang, X., Herricks, T., and Xia, Y., Monodispersed spherical colloids of titania: synthesis, characterization, and crystallization, *Adv. Mater.*, 15, 1205, 2003.

59. Wijnhoven, J., and Vos, W., Preparation of photonic crystals made of air spheres in titania, *Science*, 281, 802, 1998.

60. Wijnhoven, J., Bechger, L., and Vos, W., Fabrication and characterization of large macroporous photonic crystals in titania, *Chem. Mater.*, 13, 4486, 2001.

61. Richel, A., Johnson, N., and McComb, D., Observation of Bragg reflection in photonic crystals synthesized from air spheres in a titania matrix, *Appl. Phys. Lett.*, 76, 1816, 2000; erratum: *Appl. Phys. Lett.*, 77, 1062, 2000.

62. Dong, W., Bongard, H., and Marlow, F., New type of inverse opals: titania with skeleton structure, *Chem Mater.*, 15, 568, 2003.

63. Subramania, G., Constant, K., Biswas, R. et al., Visible frequency thin film photonic crystals from colloidal systems of nanocrystalline titania and polystyrene microspheres, *J. Am. Ceram. Soc.*, 85, 1383, 2002.

64. Kirihara, S., Miyamoto, Y., and Kajiyama, K., Fabrication of ceramic-polymer photonic crystals by stereolithography and their microwave properties, *J. Am. Ceram. Soc.*, 85, 1369, 2002.

65. Wang, D., and Caruso, F., Lithium niobate inverse opals prepared by templating colloidal crystals of polyelectrolyte-coated spheres, *Adv. Mater.*, 15, 205, 2003.

66. Li, B., Zhou, J., Li, Q. et al., Synthesis of (Pb,La)(Zr,Ti)O$_3$ inverse opal photonic crystals, *J. Am. Ceram. Soc.*, 85, 867, 2003.

67. Astratov, V., Adawi, A., Skolnick, M. et al., Opal photonic crystals infiltrated with chalcogenide glasses, *Appl. Phys. Lett.*, 78, 4094, 2001.

68. Zakhidov, A., Baughman, R., Iqbal, Z. et al., Carbon structures with three-dimensional periodicity at optical wavelengths, *Science*, 282, 897, 1998.

69. Kajii, H., Take, H., and Yoshino, K., Novel properties of periodic porous nanostructured carbon materials, *Synthet. Met.* 121, 1315, 2001.

70. Zakhidov, A., Khayrullin, I., Baughman, R. et al., CVD synthesis of carbon-based metallic photonic crystals, *Nanostruct. Mater.*, 12, 1089, 1999.

71. Kajii, H., Kawagishi, Y., Take, H. et al., Optical and electrical properties of opal carbon replica and effect of pyrolysis, *J. Appl. Phys.*, 88, 758, 2000.

72. Reese, C., and Asher, S., Emulsifier-free emulsion polymerization produces highly charged, monodisperse particles for near infrared photonic crystals, *J. Colloid. Interf. Sci.*, 248, 41, 2002.

73. Asher, S., Peteu, S., Reese, C. et al., Polymerized crystalline colloidal array chemical-sensing materials for detection of lead in body fluids, *Anal. Bioanal. Chem.*, 373, 632, 2002.

74. Urbas, A., Fink, Y., and Thomas, E., One-dimensionally periodic dielectric reflectors from self-assembled block copolymer-homopolymer blends, *Macromolecules*, 32, 4748, 1999.

75. Urbas, A., Sharp, R., Fink, Y. et al., Tunable block copolymer/homopolymer photonic crystals, *Adv. Mater.*, 12, 812, 2000.

76. Yoshino, K., Satoh, S., Shimoda, Y. et al., Tunable optical properties of conducting polymers infiltrated in synthetic opal as photonic crystal, *Synthet. Met.*, 121, 1459, 2001.

77. Satoh, S., Kajii, H., Kawagishi, Y. et al., Temperature and voltage dependent optical properties of conducting polymer in synthetic opal as photonic crystal, *Synthet. Met.*, 121, 1503, 2001.

78. Satoh, S., Kajii, H., Kawagishi, Y. et al., Tunable optical stop band utilizing thermochromism of synthetic opal infiltrated with conducting polymer, *Jpn. J. Appl. Phys.*, 238, L1475, 1999.

79. Yoshino, K., Tatsuhara, S., Kawagishi, Y. et al., Amplified spontaneous emission and lasing in conducting polymers and fluorescent dyes in opals as photonic crystals, *Appl. Phys. Lett.*, 74, 2590, 1999.

80. Romanov, S., Maka, T., Torres, C. et al., Emission properties of dye-polymer-opal photonic crystals, *J. Lightwave Technol.*, 17, 2121, 1999.

81. Nagata, T., Matsui, T., Ozaki, M. et al., Novel optical properties of conducting polymer-photochromic polymer systems, *Synthet. Met.*, 119, 607, 2001.

82. Deutsch, M., Vlasov, Y., and Norris, D., Conjugated-polymer photonic crystals, *Adv. Mater.*, 12, 1176, 2000.

83. Cassagneau, T., and Caruso, F., Semiconducting polymer inverse opals prepared by electropolymerization, *Adv. Mater.*, 14, 34, 2002.

84. Busch, K., and John, S., Liquid-crystal photonic-band-gap materials: the tunable electromagnetic vacuum, *Phys. Rev. Lett.*, 83, 967, 1999.

85. Ozaki, M., Shimoda, Y., Kasano, M. et al., Electric field tuning of the stop band in a liquid-crystal-infiltrated polymer inverse opal, *Adv. Mater.*, 14, 514, 2002.

86. Foulger, S., Kotha, S., Sweryda-Krawiec, B. et al., Robust polymer colloidal crystal photonic bandgap structures, *Opt. Lett.*, 25, 1300, 2000.

87. Foulger, S., Lattam, A., Jiang, P. et al., Optical and mechanical properties of poly(ethylene glycol) methacrylate hydrogel encapsulated crystalline colloidal arrays, *Langmuir*, 17, 6023, 2001.

88. Foulger, S., Jiang, P., Ying, Y. et al., Photonic bandgap composites, *Adv. Mater.*, 13, 1898, 2001.

89. Foulger, S., Lattam, A., Jiang, P. et al., Photonic bandgap composites exhibiting tunable color characteristics, *Adv. Mater.*, 15, 685, 2003.

90. Inoue, S., Kajikawa, K., and Aoyagi, Y., Dry-etching method for fabricating photonic-crystal waveguides in nonlinear-optical polymers, *Appl. Phys. Lett.*, 82, 2966, 2003.

91. Tondiglia, V., Natarajan, L., Sutherland, R. et al., Holographic formation of electro-optical polymer-liquid crystal photonic crystals, *Adv. Mater.*, 14, 187, 2002.

92. Sun, H., Tanaka, T., Takada, K. et al., Two-photon photopolymerization and diagnosis of three-dimensional microstructures containing fluorescent dyes, *Appl. Phys. Lett.*, 79, 1411, 2001.

93. Zhou, J., Zhou, Y., Ng, S. et al., Three-dimensional photonic band gap structure of a polymer-metal composite, *Appl. Phys. Lett.*, 76, 3337, 2000.

94. Sievenpiper, D., Yablonovitch, E., Winn, J. et al., 3D metallo-dielectric photonic crystals with strong capacitive coupling between metallic islands, *Phys. Rev. Lett.*, 80, 2829, 1998.

95. Sibilia, C., Scalora, M., Centini, M. et al., Electromagnetic properties of periodic and quasi-periodic one-dimensional, metallo-dielectric photonic band gap structures, *J. Opt. A Pure Appl. Opt.*, 1, 490, 1999.

96. Larciprete, M., Sibilia, C., Paoloni, S. et al., Accessing the optical limiting properties of metallo-dielectric photonic band gap structures, *J. Appl. Phys.*, 93, 5013, 2003.

97. Scalora, M., Bloemer, M., Pethel, A. et al., Transparent, metallo-dielectric, one-dimensional, photonic band-gap structures, *J. Appl. Phys.*, 83, 2377, 1998.

98. Chan, C., Zhang, W., Wang, Z. et al., Photonic band gaps from metallo-dielectric spheres, *Physica B*, 279, 150, 2000.

99. Moroz, A., Metallo-dielectric diamond and zinc-blende photonic crystals, *Phys. Rev. B*, 66, article no. 115109, 2002.

100. Xu, T., Cheng, Z., Zhang, Q. et al., Fabrication and characterization of three-dimensional periodic ferroelectric polymer-silica opal composites and inverse opals, *J. Appl. Phys.*, 88, 405, 2000.

101. Zhou, J., Zhou, Y., Buddhudu, S. et al., Photoluminescence of ZnS : Mn embedded in three-dimensional photonic crystals of submicron polymer spheres, *Appl. Phys. Lett.*, 76, 3513, 2000.

102. Segawa, H., Yoshida, K., Kondo, T. et al., Fabrication of photonic crystal structures by femtosecond laser-induced photopolymerization of organic-inorganic film, *J. Sol-Gel Sci. Technol.*, 26, 1023, 2003.

103. Serbin, J., Egbert, A., Ostendorf, A. et al., Femtosecond laser-induced two-photon polymerization of inorganic-organic hybrid materials for applications in photonics, *Opt. Lett.*, 28, 301, 2003.

104. Lin, H., Tonucci, R., and Campillo, A., Observation of two-dimensional photonic band behavior in the visible, *Appl. Phys. Lett.*, 68, 2927, 1996.

105. Rosenberg, A., Tonucci, R., and Bolden, E., Photonic band-structure effects in the visible and near ultraviolet observed in solid state dielectric arrays, *Appl. Phys. Lett.*, 69, 2639, 1996.

106. Knight, J., Birks, T., and Atkin, D., All-silica single-mode optical fiber with photonic crystal cladding, *Opt. Lett.*, 21, 1547, 1996.

107. Barkou, S., Broeng, J., and Bjarklev, A., Silica-air photonic crystal fiber design that permits waveguiding by a true photonic bandgap effect, *Opt. Lett.*, 24, 46, 1999.

108. Eggleton, B., Kerbage, C., Westbrook, P. et al., Microstructured optical fiber devices, *Opt. Express*, 9, 698, 2001.

109. Birks, T., Knight, J., and Russell, P., Endlessly single-mode photonic crystal fiber, *Opt. Lett.*, 22, 961, 1997.

110. Knight, J., Broeng, J., Birks, T. et al., Photonic band gap guidance in optical fibers, *Science*, 282, 1476, 1998.

111. Cregan, R., Mangan, B., Knight, J., Birks, T., Russell, P., Roberts, P., and Allan, D., Single-mode photonic band gap guidance of light in air, *Science*, 285, 1537, 1999.

112. Broeng, J., Photonic crystal fibers: a new class of optical waveguides, *Opt. Fiber Technol.*, 5, 305, 1999.

113. Smith, C., Venkataraman, N., Gallagher, M. et al., Low-loss hollow-core silica/air photonic crystal fibre, *Nature (Lond)*, 424, 657, 2003.

114. Sorensen, T. Broeng, J., Bjarklev, A. et al., Macro-bending loss properties of photonic crystal fibre, *Electron. Lett.*, 37, 287, 2001.

115. Sorensen, T., Broeng, J., Bjarklev, A. et al., Spectral macro-bending loss considerations for photonic crystal fibres, *IEEE P-Optoelectron.*, 149, 206, 2002.

116. Lizier, T., and Town, G., Splice losses in holey optical fibers, *IEEE Photon. Technol. Lett.*, 13, 794, 2001.

117. Payne, D., and Gambling, W., Zero material dispersion in optical fibres, *Electron. Lett.*, 11, 176, 1975.

118. Ainslie, B., and Day, C., A review of single-mode fibers with modified dispersion characteristics, *J. Lightwave Technol.*, LT-3, 958, 1985.

119. Grüner-Nielsen, L., Knudsen, S., Edvold, B. et al., Dispersion compensating fibers, *Opt. Fiber Technol.*, 6, 164, 2000.

120. Mogilevtsev, D., Birks, T., and Russell, P., Group-velocity dispersion in photonic crystal fibers, *Opt. Lett.*, 23, 1662, 1998.

121. Gander, M., McBride, R., Jones, J. et al., Experimental measurement of group velocity dispersion in photonic crystal fibre, *Electron. Lett.*, 35, 63, 1999.

122. Ferrando, A., Silvestre, E., Miret, J. et al., Designing a photonic crystal fibre with flattened chromatic dispersion, *Electron. Lett.*, 35, 325, 1999.

123. Birks, T., Mogilevtsev, D., Knight, J. et al., Dispersion compensation using single-material fibers, *IEEE Photon. Technol. Lett.*, 11, 674, 1999.

124. Reeves, W., Knight, J., Russell, P. et al., Demonstration of ultra-flattened dispersion in photonic crystal fibers, *Opt. Express*, 10, 609, 2002.

125. Ferrando, A., Silvestre, E., Miret, J. et al., Nearly zero ultraflattened dispersion in photonic crystal fibers, *Opt. Lett.*, 25, 790, 2000.

126. Hasegawa, T., Hole-assisted lightguide fiber for large anomalous dispersion and low optical loss, *Opt. Express*, 9, 681, 2001.

127. Knight, J., Arriaga, J., Birks, T. et al., Anomalous dispersion in photonic crystal fiber, *IEEE Photon. Technol. Lett.*, 12, 807, 2000.

128. Ferrando, A., Silvestre, E., Andres, P. et al., Designing the properties of dispersion-flattened photonic crystal fibers, *Opt. Express*, 9, 687, 2001.

129. Kuhlmey, B., Renversez, G., and Maystre, D., Chromatic dispersion and losses of microstructured optical fibers, *Appl. Opt.*, 42, 634, 2003.

130. Ranka, J., Windeler, R., and Stentz, A., Visible continuum generation in air-silica microstructure optical fiber with anomalous dispersion at 800 nm, *Opt. Lett.*, 25, 25, 2000.

131. Sharping, J., Fiorentino, M., Kumar, P. et al., Optical parametric oscillator based on four-wave mixing in microstructure fiber, *Opt. Lett.*, 27, 1675, 2002.

132. Monro, T., Pruneri, V., Broderick, N. et al., Broad-Band Second-Harmonic Generation in Holey Optical Fibers, *IEEE Photon. Technol. Lett.*, 13, 981, 2001.

133. Wadsworth, W., Ortigosa-Blanch, A., Knight, J. et al., Supercontinuum generation in photonic crystal fibers and optical fiber tapers: a novel light source, *J. Opt. Soc. Am. B*, 19, 2148, 2002.

134. Hartl, I., Li, X., Chudoba, C. et al., Ultrahigh-resolution optical coherence tomography using continuum generation in an air-silica microstructure optical fiber, *Opt. Lett.*, 26, 608, 2001.

135. Wang, Y., Zhao, Y., Nelson, J. et al., Ultrahigh-resolution optical coherence tomography by broadband continuum generation from a photonic crystal fiber, *Opt. Lett.*, 28, 182, 2003.

136. Jones, D., Diddams, S., Ranka, J. et al., Carrier-envelope phase control of femtosecond mode-locked lasers and direct optical frequency synthesis, *Science*, 288, 635, 2000.

137. Fedotov, A., Zheltikov, A., Ivanov, A. et al., Supercontinuum-generating holey fibers as new broadband sources for spectroscopic applications, *Laser Phys.*, 10, 723, 2000.

138. Wadsworth, W., Percival, R., Bouwmans, G. et al., High power air-clad photonic crystal fibre laser, *Opt. Express*, 11, 48, 2003.

139. Limpert, J., Schreiber, T., Nolte, S. et al., High-power air-clad large-mode photonic crystal fiber laser, *Opt. Express*, 11, 818, 2003.

140. Wadsworth, W., Knight, J., Reeves, W. et al., $Yb^{3+}$-doped photonic crystal fibre laser, *Electron. Lett.*, 36, 1452, 2000.

141. Furusawa, K., Malinowski, A., Price, J. et al., Cladding pumped ytterbium-doped fiber laser with holey inner and outer cladding, *Opt. Express*, 9, 714, 2001.

142. Price, J., Furusawa, K., Monro, T. et al., Tunable, femtosecond pulse source operating in the range 1.06-1.33 mu m based on an $Yb^{3+}$-doped holey fiber amplifier, *J. Opt. Soc. Am. B*, 19, 1286, 2002.

143. Glas, P., and Fischer, D., Cladding pumped large-mode-area Nd-doped holey fiber laser, *Opt. Express*, 10, 286, 2002.

144. Cregan, R., Knight, J., Russell, P. et al., Distribution of spontaneous emission from an $Er^{3+}$-doped photonic crystal fiber, *J. Lightwave Technol.*, 17, 2138, 1999.

145. Suzuki, K., Kubota, H., Kawanishi, S. et al., High-speed bi-directional polarisation division multiplexed optical transmission in ultra low-loss (1.3 dB/km) polarisation-maintaining photonic crystal fibre, *Electron. Lett.*, 37, 1399, 2001.

146. Suzuki, K., Kubota, H., Kawanishi, S. et al., Optical properties of a low-loss polarization-maintaining photonic crystal fiber, *Opt. Express*, 9, 676, 2001.

147. Kumar, V., George, A., Reeves, W. et al., Extruded soft glass photonic crystal fiber for ultrabroad supercontinuum generation, *Opt. Express*, 10, 1520, 2002.

148. Kiang, K., Frampton, K., Monro, T. et al., Extruded single mode non-silica glass holey optical fibres, *Electron. Lett.*, 38, 546, 2002.

149. Van Eijkelenborg, M., Large, M., and Argyros, A., Microstructured polymer optical fibre, *Opt. Express*, 9, 319, 2001.
150. Argyros, A., Bassett, I., van Eijkelenborg, M. et al., Ring structures in microstructured polymer optical fibres, *Opt. Express*, 9, 813, 2001.
151. Monro, T., West, Y., Hewak, D. et al., Chalcogenide holey fibres, *Electron. Lett.*, 36, 1998, 2000.
152. Brown, P., private communication, 2003.

# 22 Tailoring Dielectric Properties of Perovskite Ceramics at Microwave Frequencies

*Eung Soo Kim, Ki Hyun Yoon, and Burtrand I. Lee*

## CONTENTS

## I. INTRODUCTION

With the progress in microwave telecommunication technology, dielectric materials have come to play an important role in the miniaturization and compactness of microwave passive components. The dielectric materials available for microwave devices are required to have predictable properties with respect to a high dielectric constant ($K$), high quality factor ($Qf$), and small temperature coefficient of resonant frequency (TCF). Numerous microwave dielectric materials have been prepared and investigated for their microwave dielectric properties and for satisfying these requirements. In particular, complex perovskite compounds A(B,B′)O$_3$

have been extensively studied because of their superior dielectric properties at microwave frequencies.

The dielectric properties of materials are strongly dependent on the composition of the materials, the chemical nature of the constituent ions, the distance between cations and anions, and the structural characteristics originating from the bonding type. Therefore it is necessary that the intrinsic properties of materials should be controlled and designed. In preparing the materials, the dielectric properties are affected by the processing conditions as well as extrinsic factors such as pores, grain boundaries, and secondary phases, which are inevitable in polycrystalline ceramics. Knowledge of the fundamental relationship between the structural characteristics and the dielectric properties is also necessary to find new microwave dielectric materials effectively. In this chapter, the control and design of dielectric properties of complex perovskite compounds, typically lead-based and calcium-based perovskite compounds, will be discussed based on the relationships between the structural characteristics and the dielectric properties at microwave frequencies, along with the processing variable effects on dielectric properties.

## II. PEROVSKITE CRYSTAL STRUCTURE

### A. CRYSTAL STRUCTURE STABILITY

The number of possible compounds of perovskite compounds, $ABO_3$, is greatly expanded when multiple ions are substituted for one or more of the original ions. This substitution occurs on the cation sites and leads to a large number of compounds known as complex perovskite, $AA'BB'O_6$. The substitution ions can occupy the original cation site of the simple structure in a random or an ordered type.

The ideal perovskite structure has cubic ($Pm\bar{3}m$) symmetry, which is composed of a three-dimensional framework of corner-sharing $BO_6$ octahedra. As shown in Figure 22.1, the A-site cation is surrounded by 12 oxygen ions in the dodecahedral environment. The B-site cations are coordinated by six oxygen ions, and the oxygen ions are coordinated by four A-site cations and two B-site cations. However, there are several structural deviations from the ideal cubic structure, both for simple and ordered perovskites.

For most $ABO_3$ perovskite compounds, the octahedral distortion and tilting result from the significant changes of the first oxygen coordination to the A-site cation. However, the first oxygen coordination to the B-site cation is left virtually unchanged. As Goldschmidt[1] pointed out, it is reasonable to assume that the octahedral distortion and tilting depend on the coordination of the oxygen to the A-site cation. With the assumption of the lattice as a close-packed array of hard spheres, the edge length and the face diagonal length of the unit cell are equal to twice the A-O bond distance and twice the B-O bond distance, respectively. From this geometric relationship, the Goldschmidt tolerance factor,[1] $t$, can be obtained by Equation 22.1:

$$t = (R_A + R_O)/\sqrt{2}\,(R_B + R_O). \qquad (22.1)$$

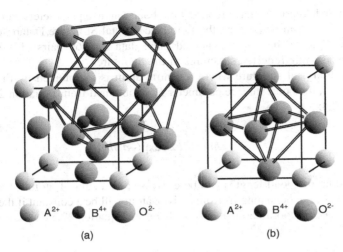

**FIGURE 22.1** The $ABO_3$ perovskite structure: (a) $AO_{12}$ cuboctahedra, (b) $BO_6$ octahedra.

Generally the perovskite structure will be formed if the value of $t$ is close to 1, and the crystallographic symmetry will be lower if $t$ is far from 1. The perovskite compounds are found to have a tolerance factor range of $0.78 < t < 1.05$.[2] For complex perovskites, where more than one ion occupies the A- and/or B-sites, the average radius of the ions on each site is used for $R_A$ and $R_B$, respectively.

## B. BOND VALENCE

Although the atoms are incorporated in the same structure, their bond length and strength are changed due to the interactions with their surrounding atoms. These interactions lead to crystal structural modifications and determine the physical properties of materials. Relationships between atomic interactions and bond length can be evaluated quantitatively by the bond valence, which is the actual valence of the atoms affected by the surrounding atoms in the crystal structure. Hence the bond length is a unique function of bond valence.

If an atom $i$ is surrounded by several other atoms $j$, the valence of atom $i$ to atom $j$, $Vi$, is defined as the sum of all of the valences from the given atom $i$, and is expressed by Equation 22.2 and Equation 22.3:[3]

$$Vi = \Sigma_j v_{ij} \qquad (22.2)$$

$$v_{ij} = \exp[(R_{ij} - d_{ij})/b'], \qquad (22.3)$$

where $R_{ij}$ is the bond valence parameter, $d_{ij}$ is the length of a bond between atom $i$ and $j$, and $b'$ is commonly taken to be a universal constant equal to 0.37 Å.[4] The parameter $R_{ij}$ should be known to calculate the bond valances of a given atom in the crystal structure.

Brown and Altermatt[3] have reported the bond valence parameters from bond lengths for 750 atom pairs using the Inorganic Crystal Structure Database. Also Brese and O'Keefe[4] have supplemented these data for 969 pairs of atoms by critically examining reported structures in the literature.

For each crystal structure where a central atom is surrounded by the same kind of atoms, the bond valence parameter $R_{ij}$ is obtained from Equation 22.3 by Equation 22.4:

$$R_{ij} = b\ln\left[V_i / \Sigma_j \exp(-d_{ij}/b)\right] \tag{22.4}$$

Assuming that the bond lengths can be expressed as a sum of atomic radii, the difference between bond lengths from a given atom will be a constant if the given atom has a specific coordination number. Therefore Equation 22.5 can be obtained by rearranging Equation 22.4:

$$d_{ij} = R_{ij} - b\ln(v_{ij}) \tag{22.5}$$

## III. MICROWAVE DIELECTRIC PROPERTIES

### A. DIELECTRIC PROPERTIES AT MICROWAVE FREQUENCY

The application of an alternating electric field causes polarization of dielectrics in the low-frequency regions of the field. As the frequency increases, the polarization does not follow the changes in the electric field. The dielectric constant of the dielectric materials decreases with the frequency increase by the space charge, dipole, ionic, and/or electronic polarization mechanisms. The dielectric loss is the maximum at the dispersed frequencies ($f_d$), as shown in Figure 22.2.

The dielectric properties at the microwave frequencies are mainly determined by the ionic polarization, which can be explained by the lattice vibration model.[5] According to this model, two types of ions with masses $m_1$ and $m_2$ with an electric charge of $\pm Ze$ are one-dimensionally coupled, and the complex dielectric constants $\varepsilon'(\omega)$ is expressed by Equation 22.6 through Equation 22.8.

$$\varepsilon = \varepsilon' - j\varepsilon'' \tag{22.6}$$

$$\varepsilon'(\omega) - \varepsilon(\infty) = \frac{D(Ze)^2}{m} \frac{\omega_r^2 - \omega^2}{\left(w_r^2 - \omega^2\right)^2 + r^2\omega^2} \tag{22.7}$$

$$\varepsilon''(\omega) = \frac{D(Ze)^2}{m} \frac{\omega_r}{\left(\omega_r^2 - \omega^2\right)^2 + r^2\omega^2} \cdot \tag{22.8}$$

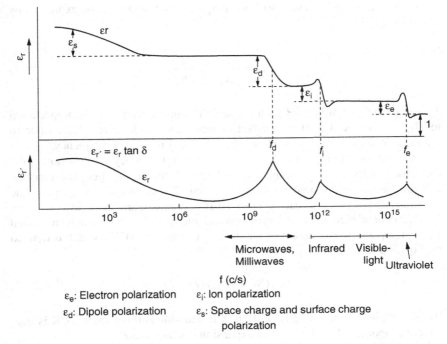

**FIGURE 22.2** Frequency dispersion behavior of dielectric materials with various polarization mechanisms.

At the microwave frequencies ($10^9 \sim 10^{10}$ Hz), the optical mode of the transverse wave is higher than the frequencies ($\omega_r \gg \omega$) and Equation 22.7 can be reduced to Equation 22.9:

$$\varepsilon'(\omega) - 1 = \frac{D(Ze)^2}{m\omega_r^2} = \varepsilon'(0) - 1 \tag{22.9}$$

$$\tan\delta = \frac{\varepsilon''(\omega)}{\varepsilon'(\omega)} = \frac{r}{\omega_r^2}\omega \tag{22.10}$$

From Equation 22.9 and Equation 22.10, the dielectric constant of materials is independent of frequency and $\tan\delta$ is proportional to the frequency. Also, the product of the quality factor $Q$ ($\approx 1/\tan\delta$) and frequency $\omega$ is an inherent property of materials, and it can be considered as a reference for evaluation of the materials.

For a dielectric material in dielectric resonator (DR) applications, the wavelength of the standing wave $\lambda_d$ equals the diameter $D$ of the DR, and the velocity $V_d = c/\varepsilon_r^{1/2}$, the wavelength in dielectrics, $\lambda_d$ is conversely proportional to the $\varepsilon_r^{1/2}$, $(\lambda_d = \lambda_o/\varepsilon_r^{1/2})$. The resonance frequency of the DR can be expressed by Equation 22.11:[6]

$$f_0 = \frac{C}{\lambda_d \varepsilon_r^{1/2}} = \frac{C}{D\varepsilon_r^{1/2}}. \qquad (22.11)$$

Therefore the size of the DR with the same resonance frequency is proportional to $1/\sqrt{\varepsilon_r}$, and the dielectric constant of materials should have a high value to miniaturize the microwave components for a fixed resonance frequency.

As confirmed in Equation 22.11, the resonance frequency is a function of the dielectric constant and the dimensions of DR. However, the dielectric constant and dimensions of the materials can be changed with temperature, and so the resonance frequency also can be changed if the temperature changes.

The dependence of the dielectric constant on the temperature can be indicated by the temperature coefficient of the dielectric constant (TCK), which is defined by Equation 22.12:

$$TCK = (1/\varepsilon_r)\cdot(\delta\varepsilon/\delta T). \qquad (22.12)$$

By rearranging Equation 22.12 with the Clausius-Mosotti equation, TCK is negatively proportional to the thermal expansion coefficient, $\alpha_L$:

$$TCK = \frac{\varepsilon_r}{3}\left(\frac{1}{\alpha}\frac{\delta\alpha}{\delta T} - 3\alpha_L\right) = -\alpha_L\varepsilon_r. \qquad (22.13)$$

Finally, the TCF can be obtained by differentiating Equation 22.11 with respect to temperature. From Equation 22.14, the temperature dependence of resonance frequency largely depends on the TCK, TCF, and $\alpha_L$:

$$TCF = \frac{1}{f_0}\frac{\delta f_0}{\delta T} = \frac{-1}{D}\frac{\delta D}{\delta T} - \frac{1}{2}\frac{1}{\varepsilon}\frac{\delta\varepsilon}{\delta T} = -\alpha_L - \frac{1}{2}TCK. \qquad (22.14)$$

Therefore the dielectric constant is changed with temperature and the resonant frequency will change with temperature, and the microwave devices cannot respond at a specific frequency if the dielectric materials in microwave frequencies show a large TCK and thermal expansion coefficient $\alpha_L$ due to the thermal expansion of dielectric materials and the temperature dependence of polarizability. In general, the $\alpha_L$ of dielectric ceramics, which is well known as the slope of the Cockbain equation,[7] is about 10 ppm/°C. Therefore control of TCF can be achieved by adequate manipulations of TCK. It is an important requirement for practical applications to control the stable TCF, nearly zero, which is available to temperature-stable microwave devices.

When a dielectric resonator is coupled with microwave circuits, the dielectric material responds to the frequency. The frequency selectivity of the microwave device depends on the loss quality of the materials. The selectivity ($Q_o$) of the dielectric materials is defined as the ratio of $f_o$ to $\Delta f$, and the $Q_o$ approximates the reciprocals of the loss factor ($\tan\delta$). The loss in DR ($1/Q_o$) is the sum of the loss of dielectric materials ($1/Q_u$), surface conduction ($1/Q_c$), and radiation loss ($1/Q_r$):

$$1/Q_o = 1/Q_u + 1/Q_c + 1/Q_r. \tag{22.15}$$

The quality factors $Q_u$, $Q_c$, and $Q_r$ are very small and could be neglected. Therefore the most important thing for the frequency selectivity is the control of dielectric loss of materials.

Based on the above considerations, a large dielectric constant for small-size devices, a small TCF for a stable resonant frequency, and a low loss of dielectrics for stable selectivity of frequency are three key requirements for microwave dielectric materials.

## B. DIELECTRIC POLARIZABILITY AND ADDITIVE RULE

As discussed in the previous section, the dielectric properties of materials at microwave frequencies are strongly dependent on the ionic polarization. Theoretical dielectric constants of materials can be obtained from the dielectric polarizabilities of composing ions through the understanding of crystal structure. Let us consider the basic relationships between the dielectric polarizabilities and dielectric constants and how the control of dielectric properties and the search for new materials can be achieved by the additive rule.

From the Clausius-Mosotti equation (Equation 22.16), the relationship between the dielectric polarizabilities and the measured dielectric constant can be expressed by Equation 22.17.

$$\frac{\varepsilon_r - 1}{\varepsilon_r + 2} = \frac{N\alpha}{3\varepsilon_0} \tag{22.16}$$

$$\alpha_{mea.} = \frac{V_m\left(\varepsilon_r^{mea.} - 1\right)}{b\left(\varepsilon_r^{mea.} + 2\right)} \tag{22.17}$$

where $\alpha_{mea.}$, $V_m$, and $\varepsilon_r^{mea.}$ represent the dielectric polarizabilities, molar volume ($\text{Å}^3$), and the measured dielectric constant, respectively, and $b$ equals 4/3.

Shannon and Subramanian[8] applied this relationship to calculate the dielectric polarizabilities of materials practically. Real polarizabilities of materials can be obtained from the additivity rule of the polarizabilities, which are a sum of the polarizabilities of composing ions as given in Equation 22.18:

$$\alpha_D(ABO_3) = \alpha_D(AO) + \alpha_D(BO_2) = \alpha(A^{2+}) + \alpha(B^{4+}) + 3\alpha(O^{2-}) \tag{22.18}$$

According to this additivity rule of polarizabilities, the dielectric polarizabilities of the complex compositions can be obtained from the polarizabilities of the composing molecules, and the dielectric polarizabilities of molecules can be obtained from the sum of polarizabilities of composing ions. Heydweiier[9] applied the additivity rule of the polarizabilities first to calculate the polarizabilities of anhydrous salts and water molecules, in which the dielectric polarizability of water molecules $\alpha_D(H_2O)$ was 3.2 ~3.5Å$^3$ similar to the value of 3.25 Å$^3$ obtained for ice by Wilson et al.[10] Also, Jonker and Van Santen[11] analyzed the polarizabilities of $MTiO_3$ (M = Mg, Ca, Sr, Ba) and Lasaga and Cygan[12] derived the polarizabilities of 24 silicate minerals by this additivity rule.

Finally, Shannon[13] obtained 61 sets of ionic polarizabilities for 129 oxides and 25 fluorides using the Clausius-Mosotti equation and least square refinements, and suggested the periodic table of ionic polarizabilities. Therefore the dielectric constant of materials with compositional changes can be successfully predicted by Equation 22.17 and Equation 22.18. From another arrangement of Equation 22.16, the theoretical dielectric constant can be obtained from the total ionic polarizabilities in Equation 22.19. From Equation 22.17 through Equation 22.19, the theoretical values of dielectric constant and polarizabilities can be obtained as well as the measured values:

$$\varepsilon_r^{theo.} = \frac{\left(3V_m + 8\pi\alpha_D^T\right)}{\left(3V_m - 4\pi\alpha_D^T\right)}, \tag{22.19}$$

where $\varepsilon_r^{theo}$, $V_m$, and $\alpha_D^T$ denote the theoretical dielectric constant, $V_{unit\ cell}/Z$, and total dielectric polarizabilities, respectively.

## C. Intrinsic and Extrinsic Dielectric Losses

Dielectric losses at microwave frequencies are originated from three major loss mechanisms: anharmonic lattice forces interacting between the crystals phonons in perfect (ideal) crystals; periodic defects in real homogeneous crystals; point defects such as dopant atoms, vacancies, etc., and grain boundaries, pores, inclusions, and secondary phases in real inhomogeneous ceramics. The losses in perfect crystals and real homogeneous crystals can be assigned to the intrinsic dielectric loss, while the losses in real inhomogeneous ceramics are to the extrinsic dielectric losses.

In microwave dielectric ceramics, the extrinsic dielectric loss dominates. However, the intrinsic losses should be studied to control and predict the dielectric properties at microwave frequencies.

Extrinsic dielectric loss in the microwave frequency range is related to microstructure, secondary phases, and processing conditions. Intrinsic loss, however, represents the minimum loss related to the lattice anharmonicity that can be expected for a particular material composition and crystal structure. It plays a fundamental role in the infrared (IR) region. Far-IR reflectivity spectra measurements of dielectrics[7–9] are useful to understand the origin of the dielectric properties at microwave frequencies. Far-IR reflectivity spectra can be analyzed by a combined method of the Kramers-Kronig relation[10] and the classical oscillator model. Dielectric losses at microwave frequencies can be estimated by the dispersion parameters obtained by the classical oscillator model.

According to the reports,[14,15] there is no hopping of point defects such as oxygen vacancies in the IR region. Therefore the intrinsic dielectric loss of materials can be obtained from the dielectric loss in the IR region, in which the effects of extrinsic loss on the dielectric loss could be neglected. Far-IR spectra measurements of dielectric materials or the solid solution systems have been shown to be a useful method to gain an understanding of the intrinsic dielectric loss mechanism at microwave frequencies.[15–17]

## IV. LEAD-BASED MICROWAVE DIELECTRIC MATERIALS

Several kinds of dielectric materials have been widely investigated to improve their properties and to meet the requisites of high dielectric constant ($K$), low dielectric loss ($Qf$), and low TCF. Based on these requisites, complex perovskite compound, $A(B,B')O_3$, was extensively studied from the viewpoint of the compositional and structural dependence on their microwave dielectric properties. Among them, much attention has been paid to lead-based ceramics with complex perovskite structures because of their superior dielectric properties required for microwave devices.

Most of the search for improved microwave materials has been mainly empirical, and it is necessary that the intrinsic properties of materials be known to control and design the dielectric properties of materials. Recently there have been reports on the intrinsic properties of materials by the IR reflectivity spectra[18] and calculation of theoretical polarizability.[19] However, there are discrepancies between the intrinsic properties obtained from these methods and the measured properties due to grain, grain boundary, and pores. Therefore the effects of porosity on the dielectric properties should be considered to evaluate the intrinsic dielectric properties and to predict the dielectric properties of ceramics with pores.

Let's discuss modification of the theoretical polarizability and prediction of the dielectric loss of $(Pb_{0.5}Ca_{0.5})(Fe_{0.5}Ta_{0.5})O_3$ ceramics with porosity as an example. The powder x-ray diffraction (XRD) patterns of the $(Pb_{0.5}Ca_{0.5})(Fe_{0.5}Ta_{0.5})O_3$ specimens[20] indicated that a single perovskite phase of cubic structure was obtained for the specimens sintered at 1100 to 1250°C. As the lattice parameter was not changed remarkably with the sintering temperatures, the average lattice parameter and molar volume were 3.946 Å and 61.46 Å³, respectively. The porosity of the specimens drastically decreased with sintering temperature, and showed a minimum value of 3.8% for the specimens sintered at temperatures above 1200°C. The dielectric constant ($K$) and $Qf$ value increased with sintering

**TABLE 22.1**
**Porosity, Dielectric Constant, and Qf of $(Pb^{0.5}Ca^{0.5})(Fe^{0.5}Ta^{0.5})O^3$ Specimens Sintered at Different Temperatures**

| Sintering Temperature (°C) | Porosity (%) | Dielectric Constant (K) | Qf (GHz) |
|---|---|---|---|
| 1100 | 15.2 | 67.9 | 5681 |
| 1150 | 6.8 | 79.6 | 6153 |
| 1200 | 3.8 | 84.0 | 6652 |
| 1250 | 3.8 | 83.9 | 6685 |

temperature, and a $K$ of 82 and $Qf$ of 6700 GHz were obtained for the specimens sintered at 1200°C for 30 min, as confirmed in Table 22.1.[20]

Bosman and Havinga[21] reported that the theoretical dielectric constant could be obtained from the measured dielectric constant of porous samples by the experimental equation of the dielectric constant and porosity, as in Equation 22.20:

$$K_W = K_{mea.}(1 + 1.5\ P) \tag{22.20}$$

where $K_w$, $K_{mea.}$, and $P$ are the theoretical dielectric constant, measured dielectric constant, and porosity, respectively. Also, the dielectric constant of polycrystalline structures can be evaluated by the Maxwell's equations in Equation 22.21 through Equation 22.24, assuming the mixture of dielectrics and spherical pores with 3-0 connectivity:

$$K_m = K_2\left(1 + \frac{3V_f(K_1 - K_2)}{K_1 + 2K_2 - V_f(K_1 - K_2)}\right), \tag{22.21}$$

where $K_m$, $K_2$, and $K_1$ are the dielectric constants of the mixture sample, the matrix dielectrics, and the pores, respectively, and $V_f$ is the volume fraction of the dispersed phase.

Since $K_2 \gg K_1$ (= 1), Equation 22.22 is obtained from the rearrangement of Equation 22.21.

$$K_m = K_2\left(\frac{2 + V_f - 3V_f}{2 + V_f}\right). \tag{22.22}$$

$$K_m = K_2\left(\frac{4 - 2V_f - 4V_f + 2V_f^2}{4 - V_f^2}\right), \tag{22.23}$$

Neglecting $V_f^2$, which corresponds to the square of the porosity,

$$K_m \approx K_2 \frac{(2 - 3V_f)}{2}$$

where $K_m$ and $K_2$ correspond to $K_{mea.}$ of the measured dielectric constant and $K_M$ of theoretical dielectric constant, respectively.

$$K_M = K_{mea.}\left(\frac{2}{2 - 3P}\right), \tag{22.24}$$

where $K_M$, $K_{mea}$, and $P$ are the theoretical dielectric constant, the measured dielectric constant, and porosity, respectively.

For the specimens sintered at 1200°C, the observed dielectric polarizability calculated from Equation 22.17 showed 14.090 Å, while the theoretical dielectric polarizability calculated from Equation 22.18 showed 14.41 Å, and the relative deviation $((\alpha_{mea.} - \alpha_{theo.})/(\alpha_{mea.} \times 100))$ was $-2.91\%$. However, the theoretical dielectric polarizabilities modified by Equation 22.20 and Equation 22.24 are 14.187 Å (deviation $-1.57\%$) and 14.189 Å (deviation $-1.56\%$), respectively, as shown in Figure 22.3. These results agree with the reports of Shannon,[13] within 0.5 to 1.5%

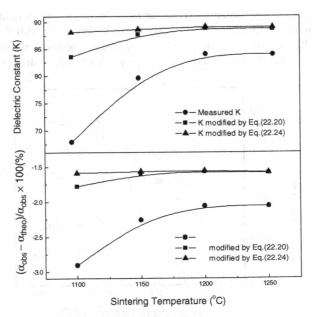

**FIGURE 22.3** Dielectric constant deviation from theoretical ionic polarizability of $(Pb_{0.5}Ca_{0.5})(Fe_{0.5}Ta_{0.5})O_3$ specimen sintered at different temperatures: -•- measured $K$ ($K_{mea.}$), -■- modified by Wiener's Equation ($K^W$), and -▲- by Maxwell's Equation ($K^M$)

deviations for the ionic polarizabilities obtained from the additivity rule. Also, the relative deviations modified by Maxwell's Equation, Equation 22.24, do not change with porosity, while those modified by Wiener's Equation, Equation 22.20, are only valid for the specimens with a porosity lower than 4%. Therefore the dielectric constant of the specimens with a single phase can be predicted by Equation 22.24 from the ionic polarizabilities of composing elements and porosity.

Far-IR reflectivity spectra of the $(Pb_{0.5}Ca_{0.5})(Fe_{0.5}Ta_{0.5})O_3$ specimens sintered at 1250°C for 30 min were taken to calculate the intrinsic dielectric loss at microwave frequencies. The spectra of the specimens were fitted by 10 resonant modes. The calculated reflectivity spectra are well fitted with the measured ones, as shown in Figure 22.4 and Table 22.2. The dispersion parameters of the specimens in Table 22.2 were determined by the Kramers-Kronig analysis and the classical oscillator model. The calculated $Qf$ values were higher than the measured ones by Hakki and Coleman's method, which is due to extrinsic effects such as grain size and porosity. Assuming the mixture of dielectrics and spherical pore with 3-0 connectivity, the measured loss quality also depends on the porosity as well as the intrinsic loss of materials, and Equation 22.24 may be modified for the loss quality, as in Equation 22.25:

$$Q \cdot f_{pred.} = Q \cdot f_{theo.}^{IR} \left( \frac{2 - 3P}{2} \right). \tag{22.25}$$

With the intrinsic loss obtained from far-IR reflectivity and porosity, the loss quality predicted by Equation 22.25 is shown in Figure 22.5, with the measured ones by Hakki and Coleman's method for comparison. The predicted loss qualities are consistent with the measured ones. In general, $Qf$ values do not follow the

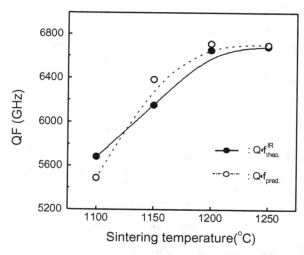

**FIGURE 22.4** $Qf$ of $(Pb_{0.5}Ca_{0.5})(Fe_{0.5}Ta_{0.5})O_3$ specimens sintered at different temperatures.

**TABLE 22.2**
**Dispersion Parameters, Measured and Calculated Dielectric Properties of $(Pb_{0.5}Ca_{0.5})(Fe_{0.5}Ta_{0.5})O_3$ Ceramics Obtained from the Best Fit to the Reflectivity Data**

| | Dispersion Parameters | | | |
|---|---|---|---|---|
| j | $\omega_j(cm^{-1})$ | $\gamma_j(cm^{-1})$ | $\Delta\varepsilon_j'(cm^{-1})$ | $\tan\delta_j(\times10^{-4})$ |
| 1 | 70 | 34 | 26.0 | 4.0540 |
| 2 | 93 | 61 | 8.0 | 0.7793 |
| 3 | 125 | 74 | 17.3 | 2.4199 |
| 4 | 162 | 66 | 5.7 | 0.3678 |
| 5 | 210 | 45 | 19.0 | 0.3924 |
| 6 | 252 | 70 | 1.0 | 0.0223 |
| 7 | 293 | 36 | 1.0 | 0.0057 |
| 8 | 577 | 48 | 1.2 | 0.0040 |
| 9 | 610 | 47 | 0.05 | 0.0006 |
| 10 | 727 | 140 | 0.08 | 0.0006 |

| Dielectric Properties | Measured | Calculated |
|---|---|---|
| K | 83 | 85 |
| Qf | 6700 | 7111 |

*Note:* $\varepsilon_\infty = 5.16$.

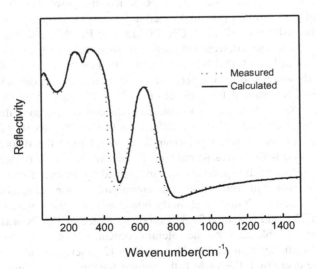

**FIGURE 22.5** Infrared reflectivity spectra of $(Pb_{0.5}Ca_{0.5})(Fe_{0.5}Ta_{0.5})O_3$ specimen sintered at 1250°C for 30 min.

dielectric mixing rule for the specimens with two or more phases, which is due to the dependence of the $Qf$ value on the microstructure resulted from the reactivity and sinterability of each phase. Based on the dielectric mixing rule with 3-0 connectivity, however, the $Qf$ values could be predicted by Equation 22.25 for single-phase specimens with porosity.

Assuming the mixture of dielectrics and spherical pore with 3-0 connectivity, the dielectric constant ($K$) and loss quality ($Q$) of $(Pb_{0.5}Ca_{0.5})(Fe_{0.5}Ta_{0.5})O_3$ with different porosity were evaluated by the dielectric mixing rule. For the specimens with porosity, the ionic polarizabilities modified by Maxwell's Equation were closer to the theoretical values than those modified by Wiener's Equation. The predicted loss quality obtained from intrinsic ones and Maxwell's Equation agree with the observed ones.

As mentioned in a previous section, the dielectric properties are largely affected by the structural characteristics of solid solution. Since the bond valence is a function of bond strength and bond length,[4] the structural characteristics largely depend on bond valence. Therefore the dielectric properties could effectively be estimated by bond valence. Let's examine the effects of A-site and B-site bond valence on the microwave dielectric properties of lead-based complex perovskite compounds.

$(Pb_{1-x}Ca_x)(Ca_{0.33}Nb_{0.67})O_3$ (PCCN) ($0.6 \leq x \leq 0.8$), $(Pb_{1-x}Ca_x)(Mg_{0.33}Ta_{0.67})O_3$ (PCMT) ($0.45 \leq x \leq 0.65$), and $(Pb_{0.45}Ca_{0.55})[Fe_{0.5}(Nb_{1-x}Ta_x)_{0.5}]O_3$ (PCFNT) ($0.0 \leq x \leq 1.0$) ceramics were prepared by the conventional mixed oxide method using the respective high-purity oxides (>99.9%) via the columbite route. Raw materials were ball milled and calcined at 900°C for 3 h followed by sintering at 1150 to 1400°C for 3 h. To inhibit the loss of PbO and decomposition during high-temperature sintering, the specimens were buried in a powder of the same composition and placed into a platinum crucible.

From the XRD patterns of PCCN, PCMT, and PCFNT, a single phase with perovskite structure was detected in each compositional range. Relative densities of PCCN, PCMT, and PCFNT are greater than 94% after each sintering. Table 22.3 showed the observed ($\alpha_{obs.}$) and theoretical ($\alpha_{theo.}$) dielectric polarizability of PCCN obtained from Equation 22.17 and Equation 22.18. Both $\alpha_{obs.}$ and $\alpha_{theo.}$ of PCCN decreased with the calcium content because the dielectric polarizability of $Ca^{2+}$ (3.16 Å) is smaller than that of $Pb^{2+}$ (6.58 Å). However, the deviation of the observed polarizabilities ($\alpha_{obs.}$) from the theoretical value ($\alpha_{theo.}$) decreased with increasing calcium content in PCCN. This means that the dielectric constant of the specimen was influenced by another factor.

With the change in composition, the atomic interactions of materials should be changed, which result in changes in the bond valence of the material. According to Equation 22.2 and Equation 22.3, the bond valence of PCCN was calculated and is shown in Table 22.4. As the calcium content, whose atomic size (1.34 Å, C.N. = 12) is smaller than lead (1.34 Å, C.N. = 12), increased, the bond valence of the A-site decreased. For PCMT, the similar tendency of the dielectric polarizabilities and A-site bond valence with calcium content was confirmed. The bond

**TABLE 22.3**
**Comparison of Observed and Theoretical Polarizabilities**
**of $(Pb_xCa_{1-x})(Ca_{0.33}Nb_{0.67})O_3$ Specimens**

| Composition | Theoretical | | Observed | | | $\Delta, \%$ $(\alpha_{obs.}-\alpha_{theo.})/$ $\alpha_{obs.}\times100$ |
|---|---|---|---|---|---|---|
| | $\alpha_{theo.}$ | K | $V_{unit\ cell}$ | Z | $\alpha_{obs.}$ | |
| $(Pb_{0.40}Ca_{0.60})$ $(Ca_{0.33}Nb_{0.67})O_3$ | 14.2670 | 55.75 | 64.349 | 1 | 14.5642 | 2.04 |
| $(Pb_{0.35}Ca_{0.65})$ $(Ca_{0.33}Nb_{0.67})O_3$ | 14.0929 | 51.45 | 64.329 | 1 | 14.4954 | 2.78 |
| $(Pb_{0.30}Ca_{0.70})$ $(Ca_{0.33}Nb_{0.67})O_3$ | 13.9187 | 47.05 | 64.371 | 1 | 14.4275 | 3.53 |
| $(Pb_{0.25}Ca_{0.75})$ $(Ca_{0.33}Nb_{0.67})O_3$ | 13.7477 | 41.97 | 64.657 | 1 | 14.3826 | 4.41 |
| $(Pb_{0.20}Ca_{0.80})$ $(Ca_{0.33}Nb_{0.67})O_3$ | 13.5767 | 39.58 | 64.771 | 1 | 14.3473 | 5.37 |

**TABLE 22.4**
**Bond Valence $(V_{A-O})$ of $(Pb_xCa_{1-x})(Ca_{0.33}Nb_{0.67})O_3$ Specimens**

| Composition | $R_{(A-O)}$ | $d_{(A-O)}$ | B | $v_{A-O}$ | $V_{A-O}$ |
|---|---|---|---|---|---|
| $(Pb_{0.40}Ca_{0.60})$ $(Ca_{0.33}Nb_{0.67})O_3$ | 2.025 | 2.788 | 0.37 | 0.1124 | 1.349 |
| $(Pb_{0.35}Ca_{0.65})$ $(Ca_{0.33}Nb_{0.67})O_3$ | 2.018 | 2.833 | 0.37 | 0.1104 | 1.325 |
| $(Pb_{0.30}Ca_{0.70})$ $(Ca_{0.33}Nb_{0.67})O_3$ | 2.011 | 2.834 | 0.37 | 0.1080 | 1.296 |
| $(Pb_{0.25}Ca_{0.75})$ $(Ca_{0.33}Nb_{0.67})O_3$ | 2.003 | 2.838 | 0.37 | 0.1048 | 1.257 |
| $(Pb_{0.20}Ca_{0.80})$ $(Ca_{0.33}Nb_{0.67})O_3$ | 1.996 | 2.834 | 0.37 | 0.1039 | 1.247 |

valence of the A-site decreased and the dielectric constant of the specimens increased with calcium content.

Figure 22.6 shows the dependence of relative deviations $((\alpha_{mea.} - \alpha_{theo.})/(\alpha_{mea.} \times 100))$ on the A-site bond valence of PCCN and PCMT. As the A-site bond valence decreases, A-site $Pb^{2+}$ and $Ca^{2+}$ ions behaved like the rattling cation, which resulted in higher dielectric constant, and in turn the differences between $\alpha_{obs.}$ and $\alpha_{theo.}$ were decreased.

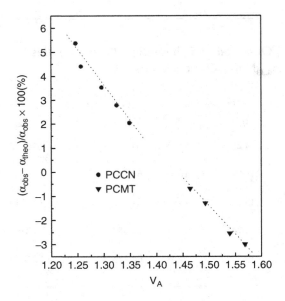

**FIGURE 22.6** Dielectric polarizability deviation of PCCN, PCMT specimens with A-site bond valence.

**TABLE 22.5**
**Bond Valance ($V_{B-O}$) of ($Pb_{0.45}Ca_{0.55}$)[$Fe_{0.5}(Nb_xTa_{1-x})_{0.5}$]$O_3$ Specimens**

| Composition | $R_{(B-O)}$ | $d_{(B-O)}$ | b | $v_{B-O}$ | $V_{B-O}$ |
|---|---|---|---|---|---|
| ($Pb_{0.45}Ca_{0.55}$)($Fe_{0.5}Nb_{0.5}$)$O_3$ | 1.8350 | 1.9710 | 0.37 | 0.6924 | 4.1545 |
| ($Pb_{0.45}Ca_{0.55}$)[$Fe_{0.5}(Nb_{0.8}Ta_{0.2})_{0.5}$]$O_3$ | 1.8359 | 1.9686 | 0.37 | 0.6985 | 4.1912 |
| ($Pb_{0.45}Ca_{0.55}$)[$Fe_{0.5}(Nb_{0.6}Ta_{0.4})_{0.5}$]$O_3$ | 1.8368 | 1.9683 | 0.37 | 0.7009 | 4.2053 |
| ($Pb_{0.45}Ca_{0.55}$)[$Fe_{0.5}(Nb_{0.4}Ta_{0.6})_{0.5}$]$O_3$ | 1.8386 | 1.9693 | 0.37 | 0.7020 | 4.2144 |
| ($Pb_{0.45}Ca_{0.55}$)($Fe_{0.5}Ta_{0.5}$)$O_3$ | 1.8395 | 1.9700 | 0.37 | 0.7028 | 4.2167 |

It has been reported[22] that even if TCF is related to the tolerance factor ($t$) in the complex perovskite, differences among compositions with the same value of $t$ can be observed. These differences can be explained by the bond valences of the A- and B-sites in the $ABO_3$ perovskite structure. The bond valence and TCF of PCFNT as well as PCCN and PCMT were investigated to evaluate these relations because the ionic radii of $Nb^{5+}$ and $Ta^{5+}$ are the same value of 0.64 Å at C.N. = 6. Table 22.5 shows the B-site bond valence of PCFNT obtained from

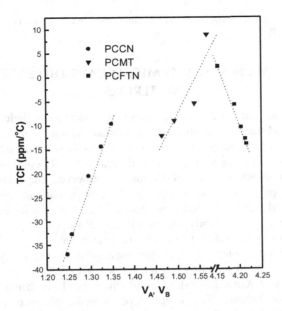

**FIGURE 22.7** Temperature coefficient of resonant frequency (TCF) of PCCN, PCMT, PCFTN specimens with A-site and B-site bond valence.

Equation 22.2 and Equation 22.3. The bond valence of the B-site increased with tantalum content, even though the tolerance factor is the same value of 0.9712 for all the compositions of PCFNT.

It is found that the TCF behavior of lead-based complex perovskites are related to the bond valence of the A- and B-site with the relative composition of A- and B-site ions, as shown in Figure 22.7. As the bond valence of the A-site increases, the TCFs of PCCN and PCMT increase, while the TCF of PCFNT decreases with B-site bond valence increases.

The tolerance factor, $t$, of the perovskite structure has been reported to be closely related to the tilting of oxygen octahedra.[22] However, it is not enough to explain the change in TCF by the tolerance factor because the effective size of the ion in the center of the oxygen octahedra changes with tilting, as reported by Shannon.[23] This suggests that even if the tilting is introduced because the ionic radii of the A-site are too small to occupy fully the available volume, $d$ (A–O) is closely parallel to the effective ionic radii of the A-site. Therefore the TCF of complex perovskite compounds is associated with the bond valences of the A- and B-sites, as well as the tolerance factor.

Therefore the dielectric constants and TCFs of PCCN and PCMT with compositional variation of the A-site cations are closely related to the bond valence of the A-site as well as ionic polarizability. With an increase in calcium content, the bond valence of the A-site decreases and A-site ions rattle easily, which results in an increase in the dielectric constant of specimens. Also, the TCF of PCCN and PCMT increases with an increase in the bond valence of the A-site.

For PCFNT with the same $t$, the TCF decreases with an increase in the bond valence of the B-site.

## V. CALCIUM-BASED MICROWAVE DIELECTRIC MATERIALS

Extensive studies have been carried out on the microwave dielectric materials with high $K$ and thermal stability for the miniaturization of microwave passive component. In particular, $CaTiO_3$-based materials have attracted considerable interest due to their high $K$. These titanates can be easily combined with other perovskite compounds to form solid solutions. However, they have a large TCF for practical applications. Various attempts have been made to control the TCFs of $CaTiO_3$-based materials. However, most of them are mainly empirical, such as the addition of the materials with negative TCF values.

It is well known that $ABO_3$ perovskite compounds have oxygen octahedra composed of octahedral cations with six oxygen ions, and distortion of oxygen octahedra arises if the ionic radii of the A-site are too small to fully occupy the available volume.[22] TCFs of $ABO_3$ perovskites result from the TCK, being related to the structural change. Although TCK largely depends on the tolerance factor and tilting of the oxygen octahedral of $ABO_3$ perovskite, the change of TCK with the tolerance factor could not be fully explained by tilting of the oxygen octahedral due to changes in the effective ionic size in the center of the oxygen octahedra with tilting.[13]

It could be expected that the microwave dielectric properties of materials with oxygen octahedra are related to the oxygen octahedra. In particular, the $K$ and the TCF of $ABO_3$ perovskite compounds strongly depend on the structural characteristics of $BO_6$ octahedra, such as the bond strength and bond angle between the octahedral-site cation and oxygen, respectively.[22,24] Moreover, bond strength and bond angle are closely related with the octahedral-site bond valence because the bond valence is a function of bond length and bond strength between cation and anion.[4] Therefore the octahedral-site bond valence could be affected by the substitution of A-site cation due to the change of bond length between the B-site ion and the oxygen resulting from the change in unit-cell volume.

In this section, the effect of bond valence on the microwave dielectric properties of $CaTiO_3$-based ceramics such as $Ca_{1x}Sm_{2x/3}TiO_3$, $CaTiO_3$-$Li_{1/2}Sm_{1/2}TiO_3$, and $Ca_xSm_{2x/3}TiO_3$-$Li_{1/2}Sm_{1/2}TiO_3$ are discussed. Reagent grade $CaCO_3$, $Sm_2O_3$, $Li_2CO_3$, and $TiO_2$ powders were used as starting materials. They were weighed according to the compositions of $Ca_{1x}Sm_{2x/3}TiO_3$ ($0 \leq x \leq 0.6$, CST), $(1-y)$ $CaTiO_3$-$yLi_{1/2}Sm_{1/2}TiO_3$ ($0 \leq y \leq 0.7$, CT-LST), and $(1-y)$ $Ca_xSm_{2x/3}TiO_3$-$yLi_{1/2}Sm_{1/2}TiO_3$ ($x = 0.6$, $0.2 \leq y \leq 1.0$, CST-LST), respectively. The mixtures were milled with $ZrO_2$ balls for 24 h in ethanol followed by drying. The mixed powders of CST were calcined at 1250°C for 3 h, and CT-LST and CST-LST were calcined at 1100°C for 2 h. After pressing at 1450 kg/cm² isostatically, the pressed specimens were sintered at 1450°C/3 h for CST and sintered at 1300°C/3 h for CT-LST and CST-LST, respectively.

From the XRD patterns of CST, CT-LST, and CST-LST, a single phase with perovskite structure was detected in each composition range. With an increase of $Sm^{3+}$ and/or $(Li_{1/2}Sm_{1/2})^{2+}$ content, unit-cell volume decreased due to the smaller ionic radii of $Sm^{3+}$ (0.958 Å) and $(Li_{1/2}Sm_{1/2})^{2+}$ (0.9995 Å) than that of $Ca^{2+}$ (1.12 Å) at the same coordination number in the A-site cation of the $ABO_3$ perovskite structure.[23] The relative densities of all the specimens were greater than 95%. Therefore $K$ was not significantly affected by the porosity, nor by the secondary phase.[25]

Table 22.6 summarizes the observed dielectric polarizabilities ($\alpha_{obs.}$) determined from the measured $K$'s using the Clausius-Mosotti equation, and the theoretical dielectric polarizabilities ($\alpha_{theo.}$) calculated by Equation 22.17 and Equation 22.18, respectively. With an increase of $Sm^{3+}$ and/or $(Li_{1/2}Sm_{1/2})$ content, the $K$'s decreased, as well as the deviations of $\alpha_{obs.}$ from $\alpha_{theo.}$. These results are due to the decrease in the rattling effect dependent on the bond strength between cation and anion.[13] Because the bond valence is a function of bond strength,[4] the decrease of the rattling effect could be explained by the bond valence. The octahedral site bond valence was calculated from the bond valence parameter of the B-site cation and distance between the B-site cation and oxygen ($d_{B-O}$). For the case of orthorhombic perovskite, $d_{B-O}$ was defined as half the cube root on

## TABLE 22.6
## Comparison of Observed and Theoretical Polarizabilities of the CaTiO₃-Based Ceramics

| Composition | Theoretical $\alpha_{theo.}$ | Observed | | | | $\Delta$, % $(\alpha_{obs.} - \alpha_{theo.})/ \alpha_{obs.} \times 100$ |
|---|---|---|---|---|---|---|
| | | K | $V_{unit\ cell}$ | Z | $\alpha_{obs.}$ | |
| $CaTiO_3$ | 12.1200 | 170 | 223.8 | 4 | 13.1229 | 7.64 |
| $Ca_{0.8}Sm_{0.13}TiO_3$ | 12.1200 | 119.3 | 223.7 | 4 | 13.0211 | 6.92 |
| $Ca_{0.6}Sm_{0.27}TiO_3$ | 12.1200 | 101 | 223.5 | 4 | 12.9500 | 6.41 |
| $Ca_{0.4}Sm_{0.4}TiO_3$ | 12.1200 | 94.5 | 223.2 | 4 | 12.9057 | 6.09 |
| $Ca_{0.7}(Li_{1/2}Sm_{1/2})_{0.3}$ $TiO_3$ | 12.0630 | 150 | 223.1 | 4 | 13.5000 | 7.56 |
| $Ca_{0.5}(Li_{1/2}Sm_{1/2})_{0.5}$ $TiO_3$ | 12.0250 | 130.5 | 222.6 | 4 | 12.9835 | 7.38 |
| $Ca_{0.3}(Li_{1/2}Sm_{1/2})_{0.7}$ $TiO_3$ | 11.9870 | 114 | 222.1 | 4 | 12.9123 | 7.17 |
| $(Ca_{0.4}Sm_{0.4})_{0.8}(Li_{1/2}$ $Sm_{1/2})_{0.2}TiO_3$ | 12.0820 | 94.9 | 222.8 | 4 | 12.8846 | 6.23 |
| $(Ca_{0.4}Sm_{0.4})_{0.6}(Li_{1/2}$ $Sm_{1/2})_{0.4}TiO_3$ | 12.0440 | 90.8 | 222.3 | 4 | 12.8371 | 6.18 |
| $(Ca_{0.4}Sm_{0.4})_{0.2}(Li_{1/2}$ $Sm_{1/2})_{0.8}TiO_3$ | 11.9680 | 71.7 | 221.8 | 4 | 12.6977 | 5.75 |
| $Li_{1/2}Sm_{1/2}TiO_3$ | 11.9300 | 56 | 221.4 | 4 | 12.5309 | 4.80 |

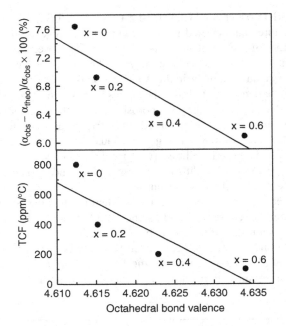

**FIGURE 22.8** TCF and ionic polarizability deviation of CST specimens with octahedral bond valence.

the cell volume per formula. As shown in Table 22.6, the octahedral bond valence increased with an increase in $Sm^{3+}$ and/or $(Li_{1/2}Sm_{1/2})^{2+}$ content. These results are due to the decrease in unit-cell volume from the substitution of $Sm^{3+}$ and/or $(Li_{1/2}Sm_{1/2})^{2+}$ for $Ca^{2+}$ in the A-site, because the effective average ionic radii of $Sm^{3+}$ (0.958 Å) and $(Li_{1/2}Sm_{1/2})^{2+}$ (0.9995 Å) are smaller than $Ca^{2+}$ (1.12 Å) at the same coordination number.[23] Figure 22.8, Figure 22.9, and Figure 22.10 show that the deviations of $\alpha_{obs.}$ from $\alpha_{theo.}$ decreased with an increase of octahedral bond valence in $CaTiO_3$-based materials. Therefore the $K$'s depend not only on the ionic polarizabilities, but also on the rattling effect, which can be evaluated by the octahedral bond valence of $CaTiO_3$-based materials.

As the TCF is related to the TCK, as shown in Equation 22.14, the magnitude of $\alpha_L$ is generally constant in the ceramics, and TCK can be divided into three terms of A, B, and C, as shown in Equation 22.26:[21]

$$TCK = \frac{(\varepsilon-1)(\varepsilon+2)}{\varepsilon} \times \left[ \frac{1}{\alpha_m}\left(\frac{\partial \alpha_m}{\partial T}\right)_v + \frac{1}{\alpha_m}\left(\frac{\partial \alpha_m}{\partial V}\right)_T \left(\frac{\partial V}{\partial T}\right)_P - \frac{1}{V}\left(\frac{\partial V}{\partial T}\right)_P \right]$$

$$= \frac{(\varepsilon-1)(\varepsilon+2)}{\varepsilon}(A+B+C)$$

(22.26)

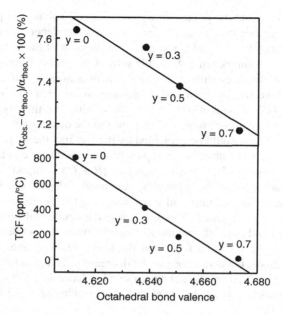

**FIGURE 22.9** TCF and ionic polarizability deviation of CT-LST specimens with octahedral bond valence.

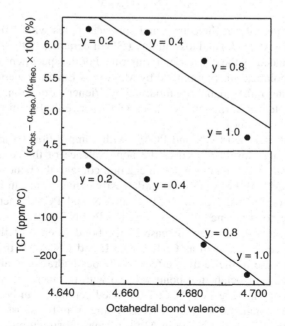

**FIGURE 22.10** TCF and ionic polarizability deviation of CST-LST specimens with octahedral bond valence.

where $\alpha_m$ and $V$ denote the polarizability and unit cell volume, respectively. The term $A$, commonly negative values, represents the direct dependence of the polarizability on temperature. $B$ and $C$ represent the increase of the polarizability and the decrease of the number of polarizable ions in the unit-cell, respectively. The unit cell volume increases with an increase in temperature. $B$ and $C$ are normally the large, but have similar values with opposing signs. Hence $B + C$ is a small positive value. TCK increases with an increase in tilting on the oxygen octahedra in the perovskite structure,[26] which corresponds to the decrease in TCF, following Equation 22.14. These could be explained by the fact that the increase in thermal energy is supposed to be absorbed completely in recovering the octahedral tilting rather than in restoring the $A$ term. Therefore $(B + C)$ is supposed to be larger than $A$. With an increase in octahedral site bond valence, the bond strength between octahedra site cations and oxygen and/or the degree of tilting on the oxygen octahedra is increased. This is because the octahedral bond valence is a function of the bond strength and the distance between the octahedra cation and oxygen. Finally, the restoring force and the tilting recovery increased with the increase in the B-site bond valence and the decrease in TCF. Therefore TCF could be effectively evaluated by the octahedra bond valence in perovskite structures with the dependence on B-site bond valence, as shown in Figure 22.8, Figure 22.9, and Figure 22.10.

## VI. SUMMARY

Assuming the mixture of dielectrics and spherical pores with 3-0 connectivity, the dielectric constant $(K)$ and loss quality $(Qf)$ of ceramics with different porosity can be evaluated by the dielectric mixing rule. For the specimens with porosity, the dielectric polarizabilities modified by Maxwell's Equation were closer to the theoretical values rather than those modified by Wiener's Equation. The predicted loss quality obtained from intrinsic ones and Maxwell's Equation was agreed with the observed values.

The $K$ and TCF of PCCN and PCMT with compositional variations of the A-site cations are closely related to the bond valence of the A-site as well as ionic polarizability. With an increase in calcium content, the bond valence of the A-site decreased and the A-site ions rattle easily, which result in an increase in the dielectric constant. Also, the TCF of PCCN and PCMT increase with an increase in the bond valence of the A-site. For PCFNT with the same tolerance factor, TCF decreases with an increase in the bond valence of the B-site. For $CaTiO_3$-based ceramics, such as CST, CT-LST, and CST-LST, the $K$ decreases due to the decrease in the rattling effect on the octahedral site cations resulting from the increase of octahedral bond valence with the increase in $Sm^{3+}$ and/or $(Li_{1/2}Sm_{1/2})^{2+}$ content. The TCF of $CaTiO_3$-based ceramics can be controlled by a change in the octahedral bond valence resulting from the substitution of $Sm^{3+}$ and/or $(Li_{1/2}Sm_{1/2})^{2+}$ for the A-site in $ABO_3$ perovskite structures.

## ACKNOWLEDGMENT

This work was partially supported by Kyonggi University.

## REFERENCES

1. Goldschmidt, V.M., *Naturwissenschaften*, 14, 477, 1926.
2. Randall, C.A., Bhalla, A.S., Shrout, T.R., and Cross, L.E., *J. Mat. Res.*, 5, 829, 1990.
3. Brown, I.D., and Altermatt, D., *Acta. Crystl.*, B41, 244, 1985.
4. Brese, N.B., and O'Keefe, M., *Acta. Crystl.*, B47, 192, 1991.
5. Cochran, W., *The Dynamics of Atoms in Crystals*, Edward Arnold, London, 1973.
6. Moulson, A.J., and Herbert, J.M., *Electroceramics: Materials, Properties, Applications*, Chapman & Hall, London, 1990.
7. Cockbain, A.G., and Harrop, P.J., *Br. J. Appl. Phys. J. Phys. D*, 1, 1109, 1968.
8. Shannon, R.D., and Subramanian, M.A., *Phys. Chem. Miner.*, 16, 747, 1989.
9. Heydweiier, A., *Z. Phys.*, 3, 308, 1920.
10. Wilson, G.I., Chan, R.K., Davidson, D.W., and Whalley, E., *J. Chem. Phys.*, 43, 2384, 1965.
11. Jonker, G.H., and Van Santen, J.H., *Chem. Week.*, 43, 672, 1947.
12. Lasaga, A.C., and Cygan, R.T., *Am. Miner.*, 67, 328, 1982.
13. Shannon, R.D., *J. Appl. Phys.*, 73, 348, 1993.
14. Petzelt, J., and Setter, N., *Ferroelectrics*, 150, 89, 1993.
15. Petzelt, J., Pacesova, S., Fousek, J. et al., *Ferroelectrics*, 93, 77, 1989.
16. Wakino, K., Murata, M., and Tamura, H., *J. Am. Ceram. Soc.,* 69, 34, 1986.
17. Tochi, K., and Takeuchi, N., *J. Mater. Sci. Lett.*, 7, 1080, 1988.
18. Kim, W.S., Yoon, K.H., and Kim, E.S., *J. Am. Ceram. Soc.*, 83, 2327, 2000.
19. Spitzer, W.G., Miller, R.C., Kleinman, D.A., and Howarth, L.E., *Phys. Rev.*, 126, 1710, 1962.
20. Kim, E.S., Park, H.S., and Yoon, K.H., *Mat. Chem. Phys.*, 79, 213, 2003.
21. Bosman, A.J., and Havinga, E.E., *Phys. Rev.*, 129, 1593, 1963.
22. Reaney, I.M., Colla, E.L., and Setter, N., *Jpn. J. Appl. Phys.*, 33, 3984, 1994.
23. Shannon, R.D., *Acta Crystl.*, A32, 751, 1976.
24. Kim, E.S., Choi, W., Yoon, K.H., and Kim, Y.T., *Ferroelectrics*, 257, 169, 2001.
25. Iddles, D.M., Bell, A.J., and Moulson, A.J., *J. Mater. Sci.*, 27, 6303, 1992.
26. Colla, E.L., Reaney, I.M., and Setter, N., *J. Appl. Phys.*, 74, 3414, 1993.

# 23 Synthesis and Processing of High-Temperature Superconductors

## Toshiya Doi

## CONTENTS

## I. HISTORICAL BACKGROUND

In 1986 Bednorz and Muller[1] opened the new stage of superconductivity. (La,Ba)$_2$CuO$_4$, a new type of superconducting material, now known as high-temperature superconductors (HTSCs), drastically improved the superconducting transition or transition temperature ($T_c$). Following this discovery, YBa$_2$Cu$_3$O$_7$ was discovered by a group of researchers at the University of Houston in 1987.[2] Since the $T_c$ of YBa$_2$Cu$_3$O$_7$ (92 K) exceeded the boiling point of liquid nitrogen, the superconducting products were expected to be operable using liquid nitrogen, which is cheap and easy to handle. Moreover, the next year new superconducting materials were discovered in the Bi(Pb)-Sr-Ca-Cu-O system[3] with a $T_c$ of 110 K and in the Tl-Ba-Ca-Cu-O system[4] with a $T_c$ of 125 K. In 1993 HgBa$_2$Ca$_2$Cu$_3$O$_9$, with a $T_c$ of 134 K was discovered,[5] and at present, $T_c$ has reached 164 K (under high pressure).[6]

At present, the fierce competition has cooled down. However, research and development are steadily advancing, and superconducting magnets, superconducting quantum interfering devices (SQUIDs), superconducting filters, etc., are being developed. We can foresee these products becoming commercially available in the near future.

## II. PHYSICAL PROPERTIES AND CRYSTAL STRUCTURES

The high-temperature superconducting materials are a group having many types with the structural features of ionic crystals. The layered structures of the HTSCs are shown in Figure 23.1. All HTSCs contain the $CuO_2$ plane in their crystal structure. The layers between the $CuO_2$ planes are called the charge reservoir layers. Features of the crystal structure of HTSCs include the following:

The crystal structure is layered and the $CuO_2$ layer and charge reservoir layer are laminated periodically.

The parent material is the antiferromagnetic insulator. By doping electrons or holes to the $CuO_2$ plane from the charge reservoir layer, the $CuO_2$ plane becomes metallic and the superconductivity appears.

At least one $CuO_2$ plane where the superconducting current flows must be included in the unit cell.

The carrier concentration, when superconductivity appears, is 0.15 to 0.2 in the $CuO_2$ plane per the number of copper ions, and then the effect of the antiferromagnetism is strongly exhibited.

In addition, it is possible to consider the crystal structures as being constructed by the stacking of layers consisting of metal ions and oxygen ions in the direction of the c-axis. That is, all HTSCs are composed of only four kinds of layers, shown in Figure 23.2. In the MO layer, oxygen ions and metal ions (copper, bismuth, thallium, mercury, etc.) exist in which the ionic radius is almost equal to copper. The AO layer consists of oxygen ions and lanthanum, strontium, barium, etc., in which the ionic radius is relatively large and in which ionicity is strong. In CuO layer, of course, only copper and oxygen ions exist. In the B layer consists of metal ions in which the ionic radius is relatively small and ionicity is strong, such as yttrium, rare earth elements, or calcium. In Figure 23.3, the crystal structure models[7-11] of $YBa_2Cu_3O_7$, $TlBa_2Ca_2Cu_3O_9$, $Bi_2Sr_2CaCu_2O_8$, $Bi_2Sr_2Ca_2Cu_3O_{10}$, $Tl_2Ba_2CaCu_2O_8$, and $Tl_2Ba_2Ca_2Cu_3O_{10}$, which have a $T_c$ greater than 77 K, are shown. The crystal structure, space group, and lattice constants of representative HTSCs are summarized in Table 23.1.

High-temperature superconductors have unique features such as a very high upper critical field, very high anisotropy, and an extremely short coherence length, $\xi$, in addition to the $T_c$ being very high. The physical properties of representative HTSCs are shown in Table 23.2. For a comparison, those of $(Ba_{0.6}K_{0.4})BiO_3$, which does not belong to the HTSCs group, but is one of the oxide superconducting materials, is also shown. The coherence length of HTSCs is very short, and this produces many problems for practical applications of HTSCs. That is, it is not possible for the superconductive electrons to pass through the barrier of the grain boundary of HTSCs crystals behaving as weak junctions, since the coherence length is very short. Figure 23.4 shows the relationship between the grain boundary angle which two grains of $YBa_2Cu_3O_7$ produce and the $J_c$ (current density flowing without electrical resistance) across the grain boundary of $YBa_2Cu_3O_7$.[12] It has been proved

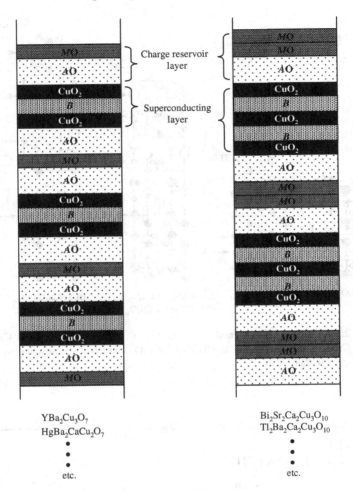

**FIGURE 23.1** The layered structures of typical high-temperature superconductors.

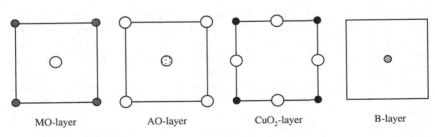

**FIGURE 23.2** Four kinds of layers comprising the crystal structures of various high-temperature superconductors.

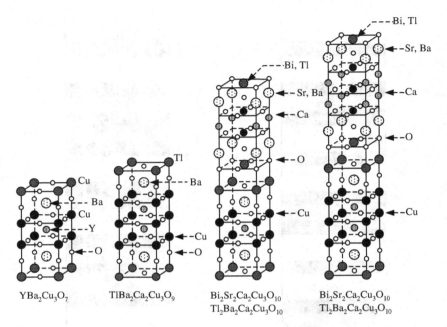

**FIGURE 23.3** Crystal structure models of $YBa_2Cu_3O_7$, $TlBa_2Ca_2Cu_3O_9$, $Bi_2Sr_2CaCu_2O_8$, $Bi_2Sr_2Ca_2Cu_3O_{10}$, $Tl_2Ba_2CaCu_2O_8$, and $Tl_2Ba_2Ca_2Cu_3O_{10}$, which are representative HTSCs with a $T_c$ greater than 77 K.

## TABLE 23.1
## Crystal Data of Various HTSC

| Material | Crystal Structure | a | b (nm) | c | Space Group |
|---|---|---|---|---|---|
| $La_2CuO_4$ | Orthorhombic | 0.5336 | 0.5421 | 1.316 | Fmmm |
| $Nd_2CuO_4$ | Tetragonal | 0.3942 | — | 1.212 | |
| $YBa_2Cu_3O_7$ | Orthorhombic | 0.38177 | 0.38836 | 1.16827 | Pmmm |
| $TlSr_2CaCu_2O_7$ | Tetragonal | 0.37859 | — | 1.2104 | P4/mmm |
| $TlBa_2CaCu_2O_7$ | Tetragonal | 0.3847 | — | 1.2771 | P4/mmm |
| $TlSr_2Ca_2Cu_3O_9$ | Tetragonal | 0.38093 | — | 1.5273 | P4/mmm |
| $TlBa_2Ca_2Cu_3O_9$ | Tetragonal | 0.3853 | — | 1.5913 | P4/mmm |
| $HgBa_2CuO_5$ | Tetragonal | 0.38791 | — | 0.95159 | P4/mmm |
| $HgBa_2CaCu_2O_7$ | Tetragonal | 0.3860 | — | 1.2693 | P4/mmm |
| $HgBa_2Ca_2Cu_3O_9$ | Tetragonal | 0.3854 | — | 1.5855 | P4/mmm |
| $Bi_2Sr_2CaCu_2O_8$ | Tetragonal | 0.541 | — | 3.079 | A2aa |
| $(Bi, Pb)_2Sr_2Ca_2Cu_3O_{10}$ | Tetragonal | 0.541 | — | 3.705 | Bbmb |

**TABLE 23.2**
**Physical Properties of HTSCs**

| Material Name | $(La_{0.25}Sr_{0.75})_2CuO_4$ | $YBa_2Cu_3O_7$ | $Bi_2Sr_2CaCu_2O_8$ | $Tl_2Ba_2CaCu_2O_8$ | $(Ba_{0.6}K_{0.4})BiO_3$ |
|---|---|---|---|---|---|
| Critical temperature $T_c$ [K] | 38 | 92 | 85 | 106 | 30 |
| Upper critical field $B_{c2}$ (at 0 K) [T] | 125 (//ab) | 674 (//ab) | 533 (//ab) | 1400 (//ab) | 22.7 |
| | 24 (//c) | 122 (//c) | 22 (//c) | 42 (//c) | |
| Lower critical field $B_{c1}$ (at 0 k) [T] | 0.007 (//ab) | 0.025 (//ab) | 0.085 (//c) | — | 0.011 |
| | 0.03 (//c) | 0.085 (//c) | | | |
| Carrier concentration n [cm$^{-3}$] | $6 \times 10^{21}$ | $1.5 \times 10^{22}$ | $3 \times 10^{21}$ | $4 \times 10^{21}$ | $2.1 \times 10^{22}$ |
| Coherence length $\xi_{GL}$ (at 0 K) [nm] | 3.2 (//ab) | 1.15 (//ab) | 3.8 (//ab) | 2.8 (//ab) | 3.68 |
| | 0.27 (//c) | 0.15 (//c) | 0.16 (//c) | 0.08 (//c) | |
| Penetration depth $\lambda$ (at 0 K) [nm] | 250 | 142 (//ab) | 300 (//ab) | 221 (//ab) | 220 |
| | | >700 (//c) | $1.0 \times 10^5$ (//c) | | |

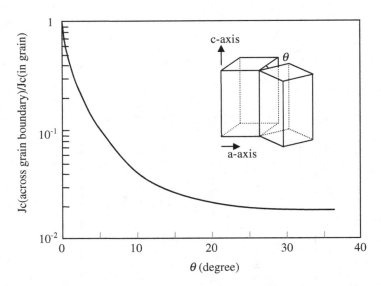

**FIGURE 23.4** Relationship between grain boundary angle that two grains of $YBa_2Cu_3O_7$ produce and the $J_c$ across the grain boundary. (From Dimos, D., Chaudhari, P., Manhart, J., and Legoues, F.K., *Phys. Rev. Lett.*, 61, 219, 1988.)

that the grain boundary $J_c$ is one order of magnitude lower than the $J_c$ in the grains. Therefore the crystals of HTSCs must be biaxially aligned in order to improve the $J_c$ to a sufficient level for practical use. Moreover, HTSCs are greatly affected by the thermal fluctuations from the very large anisotropy and extremely short coherence length.[13–18] This makes the behavior of the flux line lattice in HTSCs complicated. The temperature changes of irreversible fields (practical upper critical field) of single-crystal samples of representative HTSCs are shown in Figure 23.5. The irreversible fields of $Bi_2Sr_2CaCu_2O_8$, in which anisotropy is much stronger than other HTSCs, become very low at temperatures greater than 20 K. It is necessary to pay careful attention to these unique characteristics of HTSC materials when developing practical superconducting applications.

## III. YBa$_2$Cu$_3$O$_7$ SUPERCONDUCTOR

$YBa_2Cu_3O_7$ has a relatively high $T_c$ of 92 K and is easy to synthesize. Therefore research on $YBa_2Cu_3O_7$ is the most widely carried out for the development of practical applications at 77 K (the boiling point of liquid nitrogen). Usual sintered samples are synthesized by the standard high-temperature solid phase reaction method. A general procedure is as follows:

1. The powder of CuO, $BaCO_3$, and $Y_2O_3$ are weighed and mixed with the stoichiometric composition of Y:Ba:Cu = 1:2:3.
2. The mixed powders are put into an alumina crucible and fired at 850 to 900°C for 10 to 20 h in air (prefiring step).

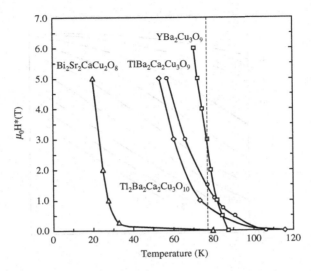

**FIGURE 23.5** The temperature changes of irreversible fields of single crystal samples of $YBa_2Cu_3O_7$, $TlBa_2Ca_2Cu_3O_9$, $Bi_2Sr_2CaCu_2O_8$, and $Tl_2Ba_2Ca_2Cu_3O_{10}$.

3. The prefired powder is ground and pressed into pellets that are fired at 900 to 930°C in $O_2$ gas for 5 to 10 h.
4. The fired samples are then cooled slowly to room temperature in an oxygen atmosphere.

The prefiring is carried out to decompose $BaCO_3$. The $YBa_2Cu_3O_7$ must be cooled slowly from the final firing temperature down to room temperature in an oxygen atmosphere to absorb oxygen. One of the most important parameters in synthesizing $YBa_2Cu_3O_7$ is the oxygen content. Figure 23.6 shows the relationship between the oxygen deficiency $\delta$ and the oxygen partial pressure at various temperatures.[19] It can be seen that the oxygen deficiency $\delta$ changes greatly from 0 to 1.0 depending on the heat treatment condition. The lattice constants are also changed by oxygen deficiency $\delta$. The $T_c$ drastically decreases with increases in the oxygen deficiency $\delta$, as shown in Figure 23.7.[20] The $T_c$ remains at over 90 K with $0 < \delta < 0.2$, and decreases rapidly with an increase in $\delta$ from 0.2 to 0.6. $YBa_2Cu_3O_7$ does not show superconductivity even at 4.2 K, and becomes an antiferromagnetic insulator at $\delta > 0.6$. As described above, the preparation conditions must be chosen carefully in order for the oxygen deficiency $\delta$ to approach zero as close as possible.

Next we describe the phase diagram of $YBa_2Cu_3O_7$. Figure 23.8 shows the pseudoternary phase diagram of the $YO_{1.5}$-BaO-CuO compound system at 900°C and an oxygen partial pressure of 0.21 atm.[21] There are four kinds of quaternary compounds—$YBa_2Cu_3O_7$ (123), $Y_2BaCuO_5$ (211), $Y_2Ba_8Cu_6O_{18}$ (143), $YBa_6Cu_3O_{11}$ (163)—and five kinds of ternary compounds—$Y_2BaO_4$ (210), $Y_3Ba_3O_9$ (340), $Ba_2CuO_3$ (021), $BaCuO_2$ (011), and $Y_2Cu_2O_5$ (101). The degree of freedom in Gibbs' phase rule $F = C - P + 1$ becomes $5 - P$, since $C = 4$. Since

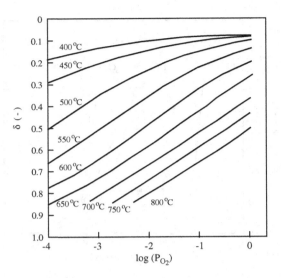

**FIGURE 23.6** Relationship between the oxygen deficiency and oxygen partial pressure at various temperatures in YBa$_2$Cu$_3$O$_7$. (Kishio, K., Shimoyama, J., Hasegawa, T., Kitazawa, K., and Fueki, K., *Jpn. J. Appl. Phys.*, 26, L1228, 1987.)

$F = 2$ at the equilibrium state with three phases shown in Figure 23.8, the three-phase equilibrium region changes both by changing the temperature and oxygen partial pressure. Figure 23.9 shows the vertical cross-sectional diagram including 123 and 211 compounds.[21] The YBa$_2$Cu$_3$O$_7$ phase is produced by the following ternary peritectic reaction at 1010°C:

$$L + 211 + 143 => 123. \tag{23.1}$$

The materials derived from YBa$_2$Cu$_3$O$_7$ by replacing yttrium with other rare earth elements (lutetium, ytterbium, thulium, erbium, holmium, dysprosium, gadolinium, europium, samarium, neodymium, lanthanum) are also superconductors, with $T_c$'s of 88 to 96 K. The crystal structures of RBa$_2$Cu$_3$O$_7$ are almost the same as those of YBa$_2$Cu$_3$O$_7$. The lattice constant is slightly different for the different ionic radii of the rare earth elements, and yet their chemical and physical properties are almost the same as those of YBa$_2$Cu$_3$O$_7$.

The main commercial products using zero resistance, which is the most important feature of superconductivity, are magnets, power cables, generators, and motors. As the first step of this development, superconducting wire has to be formed. Until 1990, research and development for HTSC wire was carried out using the powder-in-tube method, which is a relatively simple production technique. However, recently a few researchers have moved away from powder-in tube wires because of their very poor $J_c$. In order to achieve a high $J_c$, the YBa$_2$Cu$_3$O$_7$ crystals should be aligned biaxially. Therefore special metal tapes with biaxial crystal orientation of an oxide material or a metal have been used as substrates since 1992. On these special metal tapes, YBa$_2$Cu$_3$O$_7$ thin films have been epitaxially grown using the

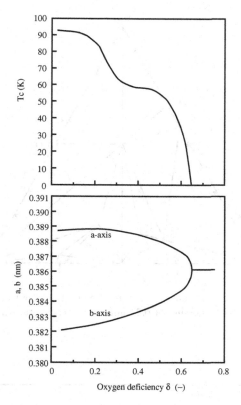

**FIGURE 23.7** $T_c$ and lattice constant changes with the oxygen deficiency δ. (From Jorgensen, J.D., et al., *Phys. Rev.*, B36, 3608, 1987.)

template effect. The structures of various $YBa_2Cu_3O_7$ wires (coated conductor) are shown in Figure 23.10.[22–27] These techniques can be classified into two types. One is the method of using a non-crystal-aligned metal tape, on which biaxially orientated oxide buffer layers are epitaxially grown by the flow of ions or plasma particles. One of the representative techniques is called ion beam assisted deposition (IBAD).[22] This technique enables alignment of the yttrium-stabilized zirconia (YSZ) biaxially by argon ion beam irradiation, while the YSZ is deposited on non-crystal-oriented metal tapes. By the template effect, $YBa_2Cu_3O_7$ thin films tend to have a biaxial crystal orientation, resulting in high $J_c$ values (at 77 K and self-field) which exceed 1 MA/cm².

The other technique is to use a rolling and recrystallization texture. Nickel is normally used in this method, since the heat resistance of nickel is good and the {100}<001> texture is easily obtained. A nickel ingot is rolled with a reduction ratio of more than 90%, then the rolled nickel tape is annealed in an inert atmosphere at a temperature near 1000°C for recrystallization.[25] After recrystallization, nickel crystals align biaxially, and {100}<001> texture was achieved. A $CeO_2$ buffer layer is deposited heteroepitaxially on the textured nickel in order to prevent a chemical

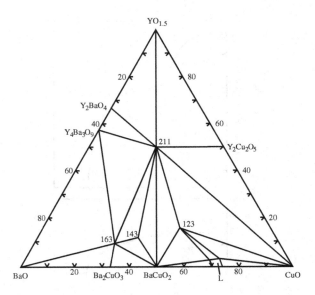

**FIGURE 23.8** Pseudoternary phase diagram of $YO_{1.5}$-BaO-CuO compound system at 900°C and oxygen partial pressure of 0.21 atm. (From Osamura, K., et al., *Z. Metllkd.*, 84, 408, 1991.)

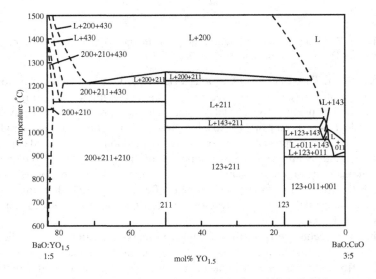

**FIGURE 23.9** Vertical cross-sectional diagram including 123 and 211 compounds in pseudoternary phase diagram. (From Osamura, K., et al., *Z. Metllkd.*, 84, 408, 1991.)

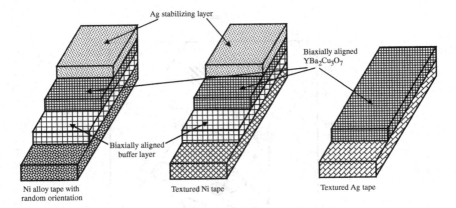

**FIGURE 23.10** Structures of various $YBa_2Cu_3O_7$ wires (coated conductor).

reaction between nickel and $YBa_2Cu_3O_7$. Finally, the $YBa_2Cu_3O_7$ layer is epitaxially grown on the biaxially oriented $CeO_2$ buffer layer using various film deposition techniques such as pulse laser deposition (PLD), chemical vapor deposition (CVD), or metalorganic deposition (MOD) methods. A high $J_c$ value over 1 MA/cm$^2$ (at 77 K and self-field) was also obtained by this technique. It was also possible to produce a grain-aligned $YBa_2Cu_3O_7$ layer that is directly deposited on the metal tape of textured silver tape, as shown in Figure 23.10.

Another commercial use for superconductors is in electronic devices. Using the macroquantum phenomena and zero electric resistance of superconductors, ultrasensitive sensors, ultrasensitive wireless communication devices, ultra-high-speed processors, etc., are expected to be practical. High-quality superconducting thin films are necessary for these applications. To achieve high performance, $YBa_2Cu_3O_7$ thin films must be single crystals. In general, oxide single crystals, such as $SrTiO_3$ (100), LaAlO3 (100), MgO (100), etc., with lattice constants matching that of $YBa_2Cu_3O_7$ are selected as substrates. PLD is the most popular technique for fabricating $YBa_2Cu_3O_7$ thin films.[28] This is because the chemical composition of the film prepared by PLD is the same as that of the target, which is a sintered $YBa_2Cu_3O_7$ ceramic. Thus control of the chemical composition of the film is very easy in the case of PLD. Another advantage is that in PLD the film can be deposited under high oxygen pressure. This is advantageous to oxidize the $YBa_2Cu_3O_7$ film sufficiently during the deposition. High-quality $YBa_2Cu_3O_7$ thin films are also prepared by sputtering,[29] vacuum deposition, molecular beam epitaxy (MBE),[30] and metalorganic chemical vapor deposition (MOCVD).[31] For exact control of the film composition, precise control of both the substrate temperature and the oxygen partial pressure during deposition is required. Figure 23.11 shows the relationship between the substrate temperature and the oxygen partial pressure for the preparation of $YBa_2Cu_3O_7$ thin films.[32]

Products closest to commercial production are SQUIDs[33,34] and a microwave passive device.[35] A SQUID is a superconducting device containing one or two

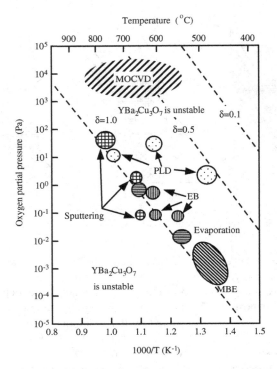

**FIGURE 23.11** Relationship between the substrate temperature and the oxygen partial pressure for the preparation of $YBa_2Cu_3O_7$ thin film reported by various researchers. (From Hammond, R.H., and Bormann, R., *Phys. C*, 162, 703, 1989.)

Josephson junctions in the superconducting loop. It is an extremely sensitive magnetic sensor that can detect a magnetic field of about $10^{-18}$ Wb. At present, HTSCs SQUIDs operable in liquid nitrogen using $YBa_2Cu_3O_7$ have been developed and are used in magnetoencephalographs and magnetocardiographs.[35,36] The performance of microwave passive devices, such as band-pass filters and antennas utilized for communication equipment, are expected to be improved greatly by using HTSCs materials because the surface resistance of HTSCs is two to three orders of magnitude lower than that of copper at the microwave frequency band. Presently a field test of band-pass filters using $YBa_2Cu_3O_7$ for the base station for cellular phone systems is ongoing.[37]

## IV. $Bi_2Sr_2Ca_2Cu_3O_{10}$ AND $Bi_2Sr_2CaCu_2O_8$ SUPERCONDUCTORS

It is more difficult to synthesize $Bi_2Sr_2Ca_2Cu_3O_{10}$ as a single phase, thus $(Bi_{1-x}Pb_x)_2Sr_2Ca_2Cu_3O_{10}$ (here $x = 0.1$–$0.2$) composition is generally used. Usual sintered samples are synthesized by a standard solid state reaction method. A general procedure is as follows:

1. The powders of $Bi_2O_3$, PbO, SrO, CaO, and CuO are weighed and mixed in the proportion of Bi:Pb:Sr:Ca:Cu = (1.6–1.8):(0.4–0.2):2:2:3.
2. The mixed powder is put into an alumina crucible and fired at 700 to 800°C for 10 to 20 h in air (prefiring step).
3. The prefired powder is ground and pressed into pellets. Then the pellets are fired at 836°C in a 20% $O_2$ atmosphere for 50 h.
4. The fired pellets are again ground and the ground powder is pressed into pellets, then the pellets are fired at 836°C in a 20% $O_2$ atmosphere for 50 h. This step is repeated several times to obtain the sufficient volume fraction of $(Bi_{1-x}Pb_x)_2Sr_2Ca_2Cu_3O_{10}$.

The temperature range where single-phase $(Bi_{1-x}Pb_x)_2Sr_2Ca_2Cu_3O_{10}$ is obtained is very narrow and the furnace must be controlled within a few degrees. The $T_c$ of $(Bi_{1-x}Pb_x)_2Sr_2Ca_2Cu_3O_{10}$ is 110 K higher than that of $YBa_2Cu_3O_7$. $(Bi_{1-x}Pb_x)_2Sr_2Ca_2Cu_3O_{10}$ superconducting wires are mostly fabricated by the powder-in-tube method. A general manufacturing procedure is shown in Figure 23.12. To improve the volume fraction of $(Bi_{1-x}Pb_x)_2Sr_2Ca_2Cu_3O_{10}$ phase and crystal orientation (in one direction), the rolling and firing steps are repeated. The present performance of $J_c$ in a short-length sample in the laboratory reaches 80 kA/cm$^2$ at 77 K and without magnetic fields, and 30 kA/cm$^2$ in long lengths of wire more than 1 km long.[38] At present, many types of products, such as superconducting magnetic energy storage (SMES) systems, transmission cable, transformers, and generators, are being tested for practical use. In the meantime, it is very difficult to obtain good-quality $(Bi_{1-x}Pb_x)_2Sr_2Ca_2Cu_3O_{10}$ thin films, thus little research into device applications has been reported.

The $T_c$ of $Bi_2Sr_2CaCu_2O_8$ is 85 K, which is not so attractive compared with $(Bi_{1-x}Pb_x)_2Sr_2Ca_2Cu_3O_{10}$. However, it is easy to synthesize and the irreversible field at temperatures less than 20 K is very high, therefore the research and development focusing on ultra-high-field magnets is extensive. $Bi_2Sr_2CaCu_2O_8$ superconducting wire is fabricated by the powder-in-tube method, as is $(Bi_{1-x}Pb_x)_2Sr_2Ca_2Cu_3O_{10}$ wire. Repeating the rolling and firing steps is not necessary for the case of $Bi_2Sr_2CaCu_2O_8$ superconducting wire. Figure 23.13 shows the typical heat treatment pattern of the last heat treatment step. The wire is heated above the temperature at which the liquid phase forms and then is cooled slowly to room temperature. During this slow cooling step, the crystals of $Bi_2Sr_2CaCu_2O_8$ grow and become large plate-like crystals. These plate-like crystals grow parallel along the internal face of the silver sheath, and $Bi_2Sr_2CaCu_2O_8$ crystals are aligned uniaxially, as a result, a high $J_c$ is obtained in $Bi_2Sr_2CaCu_2O_8$ superconducting wire. The $J_c$ at 4.2 K and 23 T reaches to 180 kA/cm$^2$. At present, wire with lengths longer than 1 km are being produced.[39] Little research into device applications has been reported because of the difficulty in obtaining good-quality $Bi_2Sr_2CaCu_2O_8$ thin films and lower $T_c$ of 85 K than $YBa_2Cu_3O_7$.

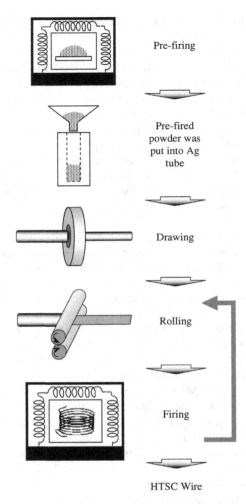

**FIGURE 23.12** Typical manufacturing procedure for the powder-in-tube method.

The only weak point of $(Bi_{1-x}Pb_x)_2Sr_2Ca_2Cu_3O_{10}$ and $Bi_2Sr_2CaCu_2O_8$ is the low-flux pinning force at temperature greater than 20 K, when they are used as superconducting wires. Although wires more than 1 km in length are already being produced, the application products are limited. That is, the applications are limited to products that do not use magnetic fields when they are cooled by liquid nitrogen, since these wires cannot generate strong magnetic fields. Even then, $(Bi_{1-x}Pb_x)_2Sr_2Ca_2Cu_3O_{10}$ and $Bi_2Sr_2CaCu_2O_8$ have to be cooled to temperatures less than 20 K in high magnetic fields. Presently extensive research is being conducted to raise the pinning force.

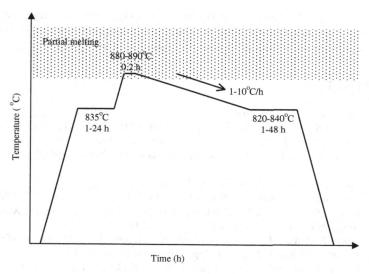

**FIGURE 23.13** A typical pattern of the final heat treatment step for $Bi_2Sr_2CaCu_2O_8$ powder-in-tube wire.

## V. $Tl_2Ba_2Ca_2Cu_3O_{10}$ AND $TlBa_2Ca_2Cu_3O_9$-RELATED SUPERCONDUCTORS

$Tl_2Ba_2Ca_2Cu_3O_{10}$ has a $T_c$ of 125 K, which is higher than that of $YBa_2Cu_3O_7$ and $(Bi_{1-x}Pb_x)_2Sr_2Ca_2Cu_3O_{10}$. $Tl_mBa_2Ca_{n-1}Cu_nO_{2n+m+2}$ are also superconductors ($m = 1$ or 2, $n = 1$, 2, 3, 4, or 5). The crystal is very flexible, especially in the case of $m = 1$, and the thallium site can be partially substituted by lead, bismuth, mercury, copper, etc., and the barium site can be partially or fully substituted by strontium. The highest $T_c$ in the combinations of any elements is given by $n = 3$ or 4. For example, the $T_c$ of $TlBa_2Ca_2Cu_3O_9$ and $Tl_{0.5}Pb_{0.5}Sr_2Ca_2Cu_3O_9$ is 110 K and 120 K, respectively.

Although the $T_c$ of $Tl_mBa_2Ca_{n-1}Cu_nO_{2n+m+2}$ is very high, the research for practical applications is not so active. This is because of not only the high vapor pressure of thallium oxide when firing the sample to synthesize $Tl_mBa_2Ca_{n-1}Cu_nO_{2n+m+2}$, but also because of the toxicity of thallium.

## VI. $HgBa_2Ca_2Cu_3O_9$ SUPERCONDUCTORS

At present, the highest $T_c$ is from $HgBa_2Ca_2Cu_3O_9$ superconductors. The chemical compositions of a series of mercury-containing HTSCs are expressed as $HgBa_2Ca_{n-1}Cu_nO_{2n+3}$. The crystal structures are very similar to those of $TlBa_2Ca_{n-1}Cu_nO_{2n+3}$. $T_c$ under the normal pressure is 135 K, which is above the boiling point of LNG, and the $T_c$ at 32 GPa is surprisingly high, 164 K. $HgBa_2Ca_2Cu_3O_9$ is the most attractive material for practical use because of its high $T_c$, however, the research on $HgBa_2Ca_{n-1}Cu_nO_{2n+3}$ remains very basic. This is because mercury is very toxic for humans and because of the high vapor pressure of mercury oxide, as in case of $Tl_mBa_2Ca_{n-1}Cu_nO_{2n+m+2}$.

## REFERENCES

1. Bednorz, J.G., and Muller, K.A., *Z. Phys.*, B64(1986)189.
2. Wu, M.K., Ashburn, J.R., Torng, C.J., Hor, P.H., Meng, R.L., Gao, L., Huang, Z.J., Wang, Y.Q., and Chu, C.W., *Phys. Rev. Lett.*, 58, 908, 1987.
3. Maeda, H., Tanaka, Y., Fujimoto, M., and Asano, T., *Jpn. J. Appl. Phys.*, 27, L209, 1988.
4. Sheng, Z.Z., and Hermann, A.M., *Nature*, 332, 138, 1988.
5. Schilling, A., Cantoni, M., Guo, J.D., and Ott, H.R., *Nature*, 363, 56, 1993.
6. Chu, C.W., Rameriz, D., Chen, F., Mao, H.K., Eggert, J., Gao, L., Xiong, Q., Meng, R.L., and Xue, Y.Y., *Phys. C*, 235, 1493, 1994.
7. Izumi, F., Asano, H., Ishigaki, T., Takayama-Muromachi, E., Uchida, Y., Watanabe, N., and Nishikawa, T., *Jpn. J. Appl. Phys.*, 26, L649, 1987.
8. Martin, C., Michel, C., Maigman, A., Hervieu, M., and Raveau, B., *C. R. Acad. Sci. Paris*, 307, 27, 1988.
9. Sunshine, S.A., Siegrit, T., Schneemeyer, L.F., Murphy, D.W., Cava, R.J., Batlogg, B., van Dover, R.B., Fleming, R.M., Glarum, S.H., Nakahara, S., Farrow, R., Krajewski, J.J., Zahurak, S. M., Waszczak, J.V., Marshall, J.H., Marsh, P., Rupp, Jr., L.W., and Peck, W.F., *Phys. Rev.*, B38, 893, 1988.
10. Subramanian, M.A., Calabrese, J.A., Trardi, C.C., Gopalakrishnan, J., Askew, T.R., Flippen, R.B., Morrissey, K.J., Chowdhry, U., and Sleight, A.W., *Nature*, 332, 420, 1998.
11. Trardi, C.C., Subramanian, M.A., Calabrese, J.C., Gopalakrishnan, J., Morrissey, K.J., Askew, T.R., Flippen, R.B., Chowdhry, U., and Sleight, A.W., *Science*, 240, 631, 1988.
12. Dimos, D., Chaudhari, P., Manhart, J., and Legoues, F.K., *Phys. Rev. Lett.*, 61, 219, 1988.
13. Nelson, D.R., *Phys. Rev. Lett.*, 60, 1973, 1988.
14. Worthington, T.K., Gallapher, W.J., and Dinger, T.R., *Phys. Rev. Lett.*, 59, 1160, 1987.
15. Yesurun, Y., and Malozemoff, A.P., *Phys. Rev. Lett.*, 60, 2202, 1988.
16. Fisher, M.P.A., *Phys. Rev. Lett.*, 62, 1415, 1989.
17. Kes, P.H., Aarts, J., Vinokur, V.M., and van der Beek, C.J., *Phys. Rev. Lett.*, 64, 1063, 1990.
18. Clem, J.R., *Phys. Rev.*, B43, 7837, 1991.
19. Kishio, K., Shimoyama, J., Hasegawa, T., Kitazawa, K., and Fueki, K., *Jpn. J. Appl. Phys.*, 26, L1228, 1987.
20. Jorgensen, J.D., et al., *Phys. Rev.*, B36, 3608, 1987.
21. Osamura, K., et al., *Z. Metllkd.*, 84, 408, 1991.
22. Iijima, Y., Tanabe, N., Kohno, O., and Ikeno, Y., *Appl. Phys. Lett.*, 60, 769, 1992.
23. Hasegawa, K., Fujino, K., Mukai, H., Konishi, M., Hayashi, K., Sato, K., Honjo, S., Sato, Y., Ishii, H., and Iwata, Y., *Appl. Supercond.*, 4, 487, 1998.
24. Fukutomi, M., Aoki, S., Komori, K., Chatterjee, R., and Maeda, H., *Phys. C*, 219, 333, 1994.

25. Goyal, A., Norton, D.P., Budai, J.D., Paranthaman, M., Specht, E.D., Kroeger, D.M., Christen, D.K., He, Q., Saffian, B., List, F.A., Lee, D.F., Martin, P.M., Klabunde, C.E., Hartfield, E., and Sikka, V.K., *Appl. Phys. Lett.*, 69, 1795, 1996.
26. Matsumoto, K., Kim, S.B., Wen, J.G., Hirabayashi, I., Watanabe, T., Uno, N., and Ikeda, M., *IEEE Trans. Appl. Supercond.*, 9, 1539, 1999.
27. Doi, T.J., Yuasa, T., Ozawa, T., and Higashiyama, K., in *Advances in Superconductivity VII*, K. Yamafuji and T. Morishita, Eds., Springer-Verlag, Tokyo, 1995, p. 817.
28. Wu, X.D., Muenchausen, R.E., Foltyn, S., Estler, R.C., Dye, R.C., Flamme, C., Noger, N.S., Garcia, A.R., Martin, J., and Tesmer, J., *Appl. Phys. Lett.*, 56, 1481, 1990.
29. Tanaka, S., and Itozaki, H., *Jpn. J. Appl. Phys.*, 27, L622, 1988.
30. Terashima, T., Iijima, K., Yamamoto, K., Bando, Y., and Mazaki, H., *Jpn. J. Appl. Phys.*, 27, L91, 1988.
31. Ushida, T., Higashiyama, K., Hirabayashi, I., and Tanaka, S., *Jpn. J. Appl. Phys.*, 30, L35, 1991.
32. Hammond, R.H., and Bormann, R., *Phys. C*, 162, 703, 1989.
33. Dilorio, M.S., Yoshizuki, S., Yang, K.-Y., Zhang, J., Maung, M., and Poter, B., in *Advances in Superconductivity V*, Y. Bando and H. Yamauchi, Eds., Springer-Verlag, Tokyo, 1993.
34. Tanaka, S., Itozaki, H., and Nagaishi, T., *Jpn. J. Appl. Phys.*, 32, L662, 1993.
35. Zhang, D., Liang, G.-C., Shin, C.-F., Johnson, M.R., and Withers, R.S., *IEEE Trans. MTT*, 43, 3030, 1995.
36. Hitachi Ltd., http://www.hqrd.hitachi.co.jp/global/newsrelease.html, January 10, 2003.
37. Superconductor Technologies Inc., http://www.suptech.com.
38. Han, Z., Bodin, P., Wang, W.G., Bentzon, M.D., Skov-Hansen, P., Goul, J., and Vase, P., *IEEE Trans. Appl. Supercond.*, 9, 2537, 1999.
39. Okada, M., Tanaka, K., Fukushima, K., Sato, J., Kitaguchi, H., Kumakura, H., Kiyoshi, T., Inoue, K., and Togano, K., *Jpn. J. Appl. Phys.*, 35, L623, 1996.

# 24 Synthesis of Bone-Like Hydroxyapatite/Collagen Self-Organized Nanocomposites

*Masanori Kikuchi*

## CONTENTS

## I. INTRODUCTION

Hydroxyapatite (HAp) and collagen are the main inorganic and organic components of the hard tissues, bones, and teeth of vertebrates, and those in bone show an oriented structure in which HAp nanocrystals 20 to 40 nm in size are aligned along collagen fibrils.[1] Therefore HAp and related calcium phosphate ceramics are widely used in the medical and dental fields as bone filler and coating materials for metal prosthesis due to their high osteoconductive and direct bone-bonding properties. These ceramics are too hard and brittle to endure living stresses and need to be substituted by natural bone, therefore their applications are restricted.[2]

In the last decade, many researchers have prepared HAp/polymer composites to solve the problems of hardness and brittleness. There are two strategies to solve the problem: One is a preparation of composites having bone-like mechanical properties by combining with nonbiodegradable, high-strength, biotolerant polymers. One of the most successful composite from this strategy is HAPEX®, developed by Wang et al.[3] Another is a preparation of composites to be substituted natural bone before serious problems occur by combining with biodegradable polymers such as polylactide (PLLA) and collagen. PLLA and its related polymers are biotolerant and have good mechanical and biodegradable properties; therefore some composites with HAp or related calcium phosphates have recently become commercially available.[4,5] However, these kinds of composites are not degraded by cellular reactions, but purely chemical reactions. Cooperative degradation (hopefully osteoclastic resorption) followed by osteogenesis is expected for the materials prepared from collagen and hydroxyapaite.

Many researchers have reported hydroxyapatite/collagen (HAp/Col) composites by conjugating HAp and collagen or deposited HAp on collagen fibers/sponges.[6–8] Although the main organic and inorganic components of the bone and HAp/Col composites are almost the same, and cells, cytokines, and physical conditions are also provided *in vivo*, the composites demonstrated no obvious osteoclastic resorption or cooperative new bone formation. They even showed highly biocompatible and bone bonding properties; further, some of them indicated osteoclastic resorption *in vitro*. The structural difference between the bone and the composites should be the reason that the composites do not to show osteoclastic resorption and cooperative bone formation. In this chapter, the synthesis of bone-like HAp/Col nanocomposites is described from the viewpoint of the self-organization process.[9–11]

## II. REPRODUCTION OF BONE NANOSTRUCTURE THROUGH THE SELF-ORGANIZATION PROCESS: HOW CAN WE MIMIC THE NANOSTRUCTURE OF BONE?

Osteogenesis is the only guidance to achieve this goal. Osteogenesis starts from secretion of collagen molecules followed by their fibrillogenesis and then HAp nanocrystals start to deposit on hole zones of the collagen fibrils by releasing calcium and phosphate ions (or HAp nanocrystals themselves from matrix vesicles). However, the process/chemical conditions of formation of the oriented structure *in vivo* have not been investigated in detail. Although our knowledge about the process is very limited, the following hypotheses have evolved:

1. The formation process occurs *in vivo*, that is, the conditions are not that different from physiological conditions.
2. The orientation of HAp and collagen should not be achieved by any cellular functions because HAp nanocrystals and collagen molecules

are too small to manipulate with cells' pseudopodium, but by a self-organizing mechanism between them via their interfacial interactions, because many kinds of biotic substances, including cell membrane, are formed by self-organization and self-assembly processes.

In particular, the second hypothesis, that the nanostructure is constructed by the materials themselves, is important because it means the bone nanostructure can be reproduced by controlling chemical and physical conditions. Further, HAp does not deposit on most collagen fibrils in life forms in a healthy condition. Even the body fluid is supersaturated to HAp, and collagen has residual carboxyl groups that are reported as nucleation centers of HAp in physiological conditions. That is, the formation of bone structure is assumed to depend on the concentration of calcium and phosphate ions in surrounding body fluid. Therefore HAp/collagen (HAp/Col) composites were synthesized from $Ca(OH)_2$, $H_3PO_4$ (reagent grade, Wako Pure Chemicals Inc., Osaka, Japan) and atelocollagen (biomaterial grade, Nitta Gelatin Inc., Osaka, Japan). The $Ca(OH)_2$ used was prepared by hydration of CaO which was obtained by firing of $CaCO_3$ (alkaline analysis grade, Wako Pure Chemicals Inc.) to prevent the reaction from trace alkaline-earth elements, especially magnesium. The atelocollagen was extracted from porcine dermis and treated with pepsin to remove the main antigenic telopeptide regions and was dispersed in dilute $H_3PO_4$ solution instead of ordinary HCl. To simplify the conditions, physiological saline, which is generally important for fibrillogenesis of collagen, was not used for the synthesis, because fibrillogenesis is equivalent to a salting out of colloidal substance and the saline should be compensated with other ions. The synthesis was carried out in an apparatus as shown in Figure 24.1 by a simultaneous

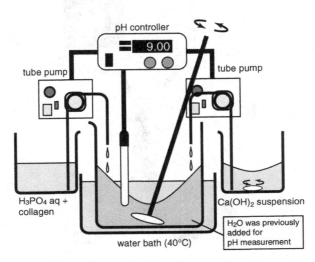

**FIGURE 24.1** Schematic drawing of a simultaneous titration synthesis system.

titration method. The starting solutions, 2 dm³ of 99.6 mM Ca(OH)₂ suspension and 2 dm³ of 59.7 mM H₃PO₄ solution with 5 g of collagen, were stored in reservoirs and titrated into a central reaction vessel through tube pumps controlled by a pH controller to maintain a pH in the reaction vessel of a preset pH, 7, 8, or 9. The amounts of the starting materials were determined from the ideal reaction of HAp formation to be an HAp/Col mass ratio of 80/20. One cubic decimeter of distilled deionized water was previously added to the vessel for measuring pH from the same time to start synthesis. The reaction temperature was maintained by a water bath at 25°C, 30°C, 35°C, and 40°C. The composite fibrils obtained were scooped with corodion membrane on the copper grid to observe the nanostructure with a transmission electron microscope (TEM).

From TEM observation, fibrous structures composed of HAp nanocrystals 20 nm to 40 nm in length were observed in several conditions. The typical fibril is shown in Figure 24.2, and an inlet is a selected area electron diffraction (SAD) pattern of the fibril. The crescent-shaped diffraction of 002 of HAp and circled diffraction just outside the crescent is observed. Although the circle diffraction is a mixture of the three strongest diffractions of HAp—211, 112, and 300—and cannot be separated to respective diffractions because of the resolution of the SAD, the crescent shape of the 002 diffraction means the $c$-axis of HAp nanocrystals is aligned in a longitudinal direction to the fibril with some deviations. This fibrous structure is very similar to the nanostructure of bone and is never seen in wet syntheses of HAp without any templates; therefore the structure is prepared by collagen molecules. Further, the fibrils grow up to 20 μm in length; even the collagen organic template used in this study is monomeric, which has

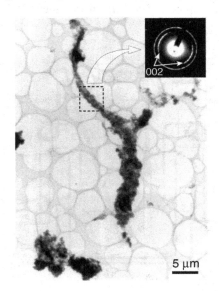

**FIGURE 24.2** Transmission electron micrograph of a typical HAp/Col fiber.

**TABLE 24.1**
**Degree of Self-Organization Estimated from TEM and SAD**

| Temperature (°C) | pH | | |
|---|---|---|---|
| | 7 | 8 | 9 |
| 25 | Not tested | Poor | Fair |
| 30 | Not tested | Poor | Fair |
| 35 | Poor | Fair | Good |
| 40 | Fair | Excellent | Excellent |

a length of 300 nm. This means the presence of HAp (and/or Ca and $PO_4$ ions) helps fibrillogenesis of collagen under these conditions. The degree of self-organization estimated by the orientation of HAp nanocrystals and the length of the fibrils is summarized in Table 24.1.

The self-organization of HAp and collagen is reproducible at 40°C, pH 8–9, which is weakly alkaline in comparison to body fluid, but agreed with activation of alkaline phosphatase in osteoblasts just before osteogenesis.

## III. CHARACTERIZATION OF HAp/COl COMPOSITES

The composites obtained contain large amounts of water; thus they were filtered and compacted by dehydration with 20 MPa uniaxial pressing, followed by 200 MPa cold isostatic pressing (CIP), because heating of the composite is restricted by the chemical properties of collagen, which easily decomposes to gelatin by heating in a wet condition. The compacts prepared are white and hard, as shown in Figure 24.3. Water and collagen amounts in the compacts were determined by

**FIGURE 24.3** Photograph of a composite compact.

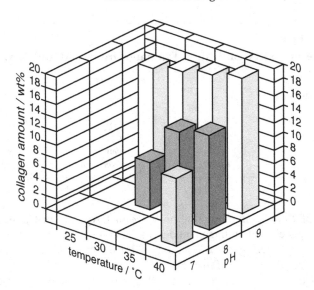

**FIGURE 24.4** Relations between Col/(HAp + Col) mass ratio of HAp/Col composites and synthesis conditions.

thermal analysis with a water carbon determinator (RC-412, LECO, St. Joseph, MI). The water amounts after CIP were in the range of 10 to 20 mass% and the compacts from composites synthesized at pH 9 contained less water than the others. The collagen mass ratios in the compacts, Col/(HAp+Col), are shown in Figure 24.4.

The collagen amounts decreased with decreasing pH and temperature. These results well agreed with the degree of self-organization; therefore the reasons why ideal collagen amounts and good self-organization occurred should be explained as follows:

1. Differential thermal analysis and thermogravimetry demonstrate collagen releases its hydrated water at about 40°C.
2. The body temperature of a pig is 40°C.
3. A weakly alkaline pH of 8 to 9 is the isoelectric point of collagen according to the literature as measured by the PAGI method. This means the electric repulsion between collagen molecules is at a minimum.
4. This pH helps stable formation of HAp under the simultaneous titration condition used in this synthesis.

To investigate the influence of self-organization on the mechanical properties of the composites, the compacts were cut into test pieces $5 \times 3 \times 20$ mm$^3$ in size with a diamond saw for a three-point bending test. The three-point bending tests were performed at a crosshead speed of 500 µm/min and a span of 15 mm with

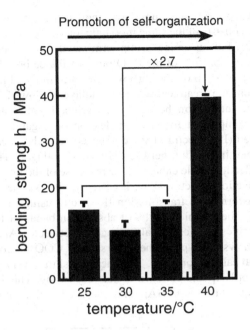

**FIGURE 24.5** Three-point bending strength of composite compacts synthesized at various temperatures and pH 9.

a universal testing machine (AGS-1kN, Shimadzu, Kyoto, Japan). Young's moduli were calculated from stress-strain curves. The three-point bending strengths of the composite compacts synthesized at pH 9 are shown in Figure 24.5. All the composite compacts showed higher mechanical strength than pure HAp compacts (data not shown) prepared the same as the HAp/Col composite, and they are already reinforced by collagen molecules of 300 nm-long fibrils. Moreover, the composite compacts synthesized at 40°C demonstrated 2.7 times greater strength than other compacts. These compacts contained almost the same amounts of collagen and water; therefore the difference between a 40°C compact and others derived from the self-organization. It is concluded that the self-organized oriented structure effects the high strength of our bone.

## IV. CHEMICAL INTERACTION BETWEEN HAP AND COLLAGEN: THE DRIVING FORCE OF SELF-ORGANIZATION

The chemical interactions between HAp and collagen are very important to consider in the mechanism of self-organization. In particular, functional groups on the collagen molecules affect the nucleation/adsorption of HAp. An investigation of HAp nucleation on functional group terminated monolayers using the Langmuir-Blodgett method has been reported by Sato et al.[12] The results show that a carboxyl group-terminated monolayer induces HAp formation in simulated body fluid, which consists of the same inorganic ions as body fluid and has similar ion concentrations, but an amino group-terminated monolayer does not. They also reported energy splitting of the C–O bond of the carboxyl group, observed after the formation of HAp on the carboxyl group-terminated monolayer, is also induced by calcium ion treatment of the monolayer followed by phosphate ion treatment, that is, interaction between the carboxyl group and calcium with phosphate ions is important for HAp nucleation on organic substances.[13] The reflectance infrared (IR) spectra of the compacts were also measured for observation of interfacial interaction between HAp and collagen. The reflectance IR at 75° or more reflection angle enhanced the response of the interfacial interaction by a multiple reflection effect, making it easier to observe the interaction.[14] The spectra were transformed to transmission IR spectra using the Kramers-Kronig equation to clarify the chemical shifts of absorption bands of functional groups on collagen molecules by a chemical interaction between HAp and collagen.

Figure 24.6 shows the typical chemical shift of COO on collagen molecules by interaction with calcium on HAp. The chemical shift is very small, that is, the interaction between HAp and collagen is not very strong. This is the reason why collagen has no potential to form HAp on its surface in SBF, but this is enough for

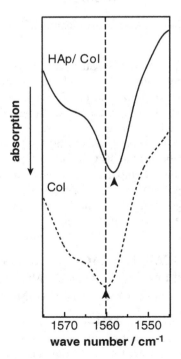

**FIGURE 24.6** Fourier transformed infrared spectra of HAp/Col composite and pure collagen.

HAp nucleation under present conditions, that is, no presence of HAp growth inhibitor, sodium ions. Concerning the surface structure and coordination number of calcium in HAp, the direction of the carboxyl groups on collagen molecules for effective interaction with HAp is tightly restricted.[15] Even if one carboxyl group can freely rotate on the collagen molecules after interacting with an HAp nanocrystal, once the next carboxyl group interacts with the same HAp nanocrystal, its direction is fixed and the direction of the HAp nanocrystal is restricted to the same longitudinal direction as the collagen molecule/fibrils.

## V. CROSS-LINKAGE OF COLLAGEN: CONTROL OF RESORPTION BY OSTEOCLASTIC CELLS

The results of animal tests using rats, rabbits, and dogs have demonstrated high biocompatibility and high osteoconductivity. Moreover, the most impressive bone tissue reaction of the HAp/Col composites is incorporation of the composites into the bone-remodeling process, that is, they are resorbed by osteoclastic cells.*

To control the resorption rate, two possible ways are proposed. One is to change the HAp crystal size and the other is to introduce cross-linkage to collagen. Crystal growth of HAp is out of the question because it generally requires high temperatures and it is difficult to control the solubility.

There are three possible methods for introducing cross-linkage:

1. Cross-linking agent was added to $Ca(OH)_2$ suspension and cross-linkage was introduced at the same time as the self-organization.
2. Cross-linking agent was added into self-organized composite fibrils.
3. The composite compacts were soaked in cross-linking agent solution for 16 h.

Method 1 is the most homogeneous and method 3 is the most heterogeneous. Preliminary tests using glutaraldehyde (GA) as a cross-linking agent were carried out to determine the best method. TEM observation of the composites from method 1 demonstrated randomly oriented structure, though the composites from method 2 indicated randomly oriented, large, membrane-like particles composed of the self-organized fibrils. These results suggest that random collagen cross-linkage by GA overrides the slow and delicate self-organization process through high reactivity of GA. Cross-linked compacts by method 3 demonstrated a gradient of cross-linkage from the surface to the center. Therefore method 2 is the best way to introduce cross-linkage.

The HAp/Col composites were synthesized and the precipitates obtained were aged at 40°C for 3 h and cross-linked with GA (first grade, Wako Pure Chemicals

---

* Osteoclastic cells are the same cells as "osteoclasts," but they are artificial cells that resorb bone. Therefore the cells that resorb the HAp/Col composites are not osteoclasts if one uses the histologic terminology correctly.

Inc.) for 10 min under vigorous stirring. The concentrations of GA per 1 g of collagen were varied between 0.0191 and 13.5 mmol. The concentration of GA, 0.191 mmol/$g_{Col}$, corresponds to an ideal value by which all ε-amino groups on collagen molecules were cross-linked. The precipitates were filtered, washed three times with pure $H_2O$ to remove excess GA, and compacted under a uniaxial pressure of 20 MPa for 24 h. The GA cross-linked composites changed color from pale yellow to brown with increasing GA content due to polymerization of GA. This is the special characteristic of GA and makes possible cross-linking of discontiguous ε-amino groups on collagen.

## VI. CHARACTERIZATION OF CROSS-LINKED COMPOSITES

The amount of cross-linkage was estimated from the analysis of -amino groups by the modified sulfo-succinimidyl-4-O-(4,4'-dimethoxytrityl)-butyrate (S-SDTB, Pierce, Rockford, IL) method according to the manufacturer's instructions. Freeze-dried HAp precipitates, atelocollagen powder, and non-cross-linked HAp/Col composite particles were used as controls. The three-point bending strength, the amounts of collagen and water in the composite compacts, and the swelling ratio were measured at 7 and 21 days after soaking in Dulbecco's phosphate buffered saline (PBS, Dainippon Pharmaceutical Co., Ltd., Japan) at pH 7.4 and a temperature of 37°C.

Cross-linkage ratios calculated from S-SDTB measurement support the results of colorization, that is, the excess GA molecules were self-polymerized and used not only for intrafibril (the self-organized units) cross-linkage, but also for interfibril cross-linkage. Subsequently this excess cross-linkage had the side effect that it stored interfibril water.

The three-point bending strength of the composite increased with the GA content and showed a maximal value at 1.35 mmol/$g_{Col}$, as shown in Figure 24.7. This value is in good agreement with the value of S-SDTB measurement, that is, the GA amount is sufficient for complete cross-linkage of collagen molecules in the self-organized fibrils via self-polymerized GA. At 0.675 mmol/$g_{Col}$, GA was sufficient to react with all ε-amino groups on the collagen; however, it was not sufficient for cross-linkage between the collagen molecules. The HAp/Col weight ratio in the GA cross-linked composite was almost constant, at about 80/20. The water content normalized by the collagen amount in the compact increased only at high GA concentrations. These results indicate that the short-range cross-linkage that occurred in the intrafibril restricts water intrusion into the fibrils. Therefore less water in the compacts at low GA concentrations is caused by consumption of GA molecules mainly in the intrafibril, and even if they are used for interfibril cross-linkage, they only form short chains due to the insufficiency of GA. However, the interfibril cross-linkage with long GA chains increases with increasing GA concentrations, and it holds large amounts of water even after pressure dehydration.

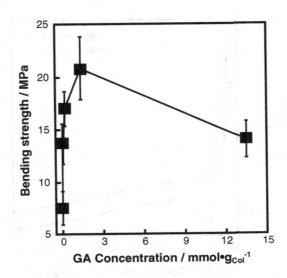

**FIGURE 24.7** Relationships between three-point bending strength of the composite compacts and GA concentrations.

Figure 24.8 shows the collagen-normalized swelling ratio of the GA cross-linked composite at 1 and 21 days after soaking in PBS. The collagen-normalized swelling ratio decreased with GA concentration and increased with time. The decrease in the swelling ratio originated from both intrafibril and interfibril cross-linkages. Although excess GA is not good for increasing mechanical strength, the interfibril cross-linkage is useful for controlling the resorption/degradation rate of the HAp/Col composites. In fact, the animal tests using rats and beagles demonstrated that the resorption/degradation rate of the HAp/Col composite was reduced with increasing GA content, observed both the naked eye and light microscopic observations. The GA cross-linked composite with less resorbability showed direct bone bonding ability the same as the sintered HAp, with neither toxic nor inflammatory reactions. The cross-linked composites were still mainly resorbed by osteoclastic cells, as demonstrated in the non-cross-linked HAp/Col composites.

## VII. FORMATION OF HAp/Col LONG FIBERS

The length of HAp/Col nanocomposite fibrils in bone is reported as approximately 400 μm, while those of *in vitro* self-organized HAp/Col fibrils are a maximum of 20 μm. Generally collagen fibrillogenesis in physiological saline is explained by an analogy to crystal growth, even though they do not demonstrate long fibers grown in diluted collagen solution. As mentioned above, the self-organization of HAp and collagen compensates the ion concentrations of the saline solution. This allows growing the fibrils at much thinner solution that is believed not to occur

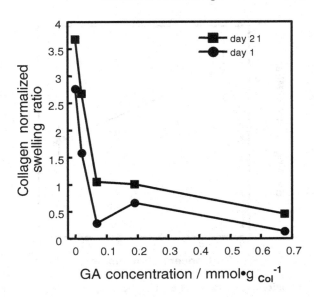

**FIGURE 24.8** Collagen normalized swelling ratio of composite compacts and GA concentrations.

in collagen fibril formation. Therefore the self-organization system can be used for fibril growth of the HAp/Col nanocomposites. According to this hypothesis, the HAp/Col nanocomposites were synthesized with changes in titration rates ranging from 8 to 50 cm$^3$/min. The lengths and aspect ratios of the fibrils obtained were measured with Rapid Vue (Beckman-Coulter, Fullerton, CA) in the range of 10 to 2500 μm.

The three-point bending strengths of the composites compacted only by uniaxial pressing at 20 MPa were measured. The fibrils were grown with decreasing titration rates, and the length obtained for titration rates of 10 to 20 cm$^3$/min was almost the same, 57 ± 2 μm. The strongest compact in three-point bending strength was four times greater than the others, and was prepared from the fibrils synthesized at a titration rate of 15 cm$^3$/min. Considering these results, the titration rate for the synthesis with changes in starting material concentrations was fixed at 15 cm$^3$/min. The starting material concentrations and amounts are shown in Table 24.2. The composites fibrils grew with decreasing starting material concentrations and the composites demonstrated thick thread-like fibers at the starting calcium concentration of 100 mM, as shown in Figure 24.9. On the contrary, maintaining of the physiological saline condition at the synthesis of the composite fibrils inhibited the fibril growth, even with the same starting material concentrations. Moreover, calcium concentration measured with a calcium ion meter indicated oscillation of the concentration become wider at 200 mM of starting calcium than at lower concentrations.

**TABLE 24.2**
**Synthesis Conditions for HAp/Col Self-Organized Long Fibers**

| Ca(OH)₂ suspension | Conc. [mM] | 50 | 100 | 200 | 300 | 400 |
|---|---|---|---|---|---|---|
| | Amount [cm³] | 1600 | 800 | 400 | 166.7 | 200 |
| H₃PO₄ solution | Conc. [mM] | 15 | 30 | 60 | 90 | 120 |
| | Amount [cm³] | 3200 | 1600 | 800 | 333.3 | 400 |
| Collagenlagen | Amount [g] | 2 | | | | |

*Note:* These values were determined by obtaining 10 g of the HAp/Col composite by ideal reaction.

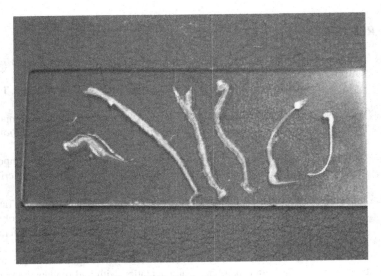

**FIGURE 24.9** Thread-like fiber of HAp/Col composite.

Thus, the mechanism of the fiber growth is assumed as follows. At low concentrations of starting materials, HAp nucleation only occurs on the carboxyl groups on the collagen molecules because of too low concentration for homogeneous nucleation of HAp and collagen fibrillogenesis needs the presence of the HAp nanocrystals as a binder because low ionic concentrations do not allow usual fibrillogenesis. Therefore, fiber formation of the composite behaves similar to crystal growth. In the physiological saline condition, fibrillogenesis is promoted even without HAp nanocrystals, and fiber formation does not behave as crystal growth. At high concentrations of starting materials, the ion concentration is enough to form HAp crystals and collagen fibrils, and the precipitates are conjugates of self-organized fibrils, HAp crystal aggregates, and collagen fibrils. As a result, the composite fibrils cannot grow well.

## VIII.  SUMMARY

In this chapter, the synthesis and properties of HAp/Col composites having bone-like nanostructure and chemical composition are introduced. This material is the first material that was incorporated into the bone-remodeling metabolism and this property could be induced by nanostructure. Further, the HAp/Col long fiber, which has never been observed in natural bone, is synthesized by controlling the concentration in the reaction vessel. These kinds of soft solution biomimetic processes are achieved under very simple conditions with simple apparatuses; however, the conditions are very critical for the HAp/Col fibers to grow well. This kind of process gives us better materials without the consumption of large amounts of energy.

## REFERENCES

1.  Neuman, W.F., and Neuman, M.W., *The Chemical Dynamics of Bone Mineral*, University of Chicago Press, Chicago, 1958.
2.  Aoki, H., *Medical Applications of Hydroxyapatite*, Ishiyaku Euro America, Tokyo, 1994.
3.  Wang, M., Joseph, R., and Bonfield, W., Hydroxyapatite-polyethylene composites for bone substitution: effects of ceramic particle size and morphology, *Biomaterials*, 19, 2357, 1998.
4.  Sikinami, Y., and Okuno, M., Bioresorbable devices made of forged composites of hydroxyapatite particles and poly-L-lactide (PLLA): part I, basic characteristics, *Biomaterials*, 20, 859, 1999.
5.  Kikuchi, M., Koyama, Y., Takakuda, K., Miyairi, H., Shirahama, N., and Tanaka, J., In vitro change in mechanical strength of β-tricalcium phosphate/copolymer-ized poly-L-lactide composites and their application for guided bone regeneration, *J. Biomed. Mater. Res.*, 62, 265, 2002.
6.  Mehlisch, D.R., Leider, A.S., and Roberts, W.E., Histologic evaluation of the bone/graft interface after mandibular augmentation with hydroxylapatite/purified fibrillar collagen composite implants. *Oral Surg. Oral Med. Oral Pathol.*, 70, 685, 1990.
7.  TenHuisen, K.S., Martin, R.I., Klimkiewicz, M., and Brown, P.W., Formation and properties of a synthetic bone composite: hydroxyapatite-collagen, *J. Biomed. Mater. Res.*, 29, 803, 1995.
8.  Miyamoto, Y., Ishikawa, K., Takechi, M., Toh, T., Yuasa, T., Nagayama, M., and Suzuki, K., Basic properties of calcium phosphate cement containing atelocollagen in its liquid or powder phases, *Biomaterials*, 19, 707, 1998.
9.  Kikuchi, M., Itoh, S., Ichinose, S., Shinomiya, K., and Tanaka, J., Self-organiza-tion mechanism in a bone-like hydroxyapatite/collagen nanocomposite synthe-sized in vitro and its biological reaction in vivo, *Biomaterials*, 22, 1705, 2001.
10. Kikuchi, M., Matsumoto, M., Yamada, T., Koyama, Y., Takakuda, K., and Tanaka, J., Glutaraldehyde cross-linked hydroxyapatite/collagen self-organized nanocom-posites, *Biomaterials*, 25, 63, 2004.

11. Kikuchi, M., Itoh, S., Matsumoto, H.N., Koyama, Y., Takakuda, K., Shinomiya, K., and Tanaka, J., Fibrillogenesis of hydroxyapatite/collagen self-organized nanocomposites, *Key Eng. Mater.*, 567, 240–242, 2003.

12. Sato, K., Kumagai, Y., and Tanaka, J., Apatite formation on organic monolayers in simulated body environment. *J. Biomed. Mater. Res.*, 50, 16, 2000.

13. Sato, K., Kogure, T., Kumagai, Y., and Tanaka, J., Crystal orientation of hydroxyapatite induced by ordered carboxyl groups, *J. Collegen Interf. Sci.*, 240, 133, 2001.

14. Kikuchi, M. and Tanaka, J., Chemical interaction in β-tricalcium phosphate/copolymerized poly-L-lactide composites, *J. Ceram. Soc. Jpn.*, 108, 642, 2000.

15. Ikoma, T. and Tanaka, J., Electron status in the interface between organic functional group and hydroxyapatite [in Japanese], *J. DV-Xα Res. Assoc.*, 13, 120, 2000.

# 25 Ceramic Membrane Processing: New Approaches in Design and Applications

*André Ayral, Anne Julbe, and Christian Guizard*

## CONTENTS

Due to their remarkable intrinsic properties, ceramic membranes have enjoyed rapid development over the two last decades. In order to update the previous overview on ceramic membrane processing,[1] this chapter is devoted to the new approaches in their design and applications. This description is preceded by background information for better understanding of current research in the ceramic membrane area.

## I. BACKGROUND

A membrane can be defined as a thin and selective barrier that enables the transport or the retention of compounds between two media. In the case of ceramic membranes, the usual driving force for transport is a pressure gradient between the feed and strip compartments (transmembrane pressure). The treated phases can be liquid or gas. For porous membranes, the pore size mainly manages the cutoff of the membrane. However, for retention of the smallest entities by the smallest pores, the transport mechanisms are more complex than simple sieving.[2] Specific physical and chemical interactions (electrostatic repulsion, physisorption, capillary condensation, etc.) become preponderant and determine the membrane selectivity. Table 25.1 summarizes the characteristics of the main processes in which ceramic membranes are involved.

As already stated in the introduction, the interest in ceramic membranes is first related to the intrinsic characteristics of the materials: mechanical strength, allowing large pressure gradients without significant strain; chemical resistance, which permits applications in corrosive aqueous media or in organic solvents; refractoriness for use at high temperatures. Other specific properties are their ability to be counterpressure cleaned and sterilized, and the fact that they are also completely resistant to bacterial attacks. Moreover, the conventional ceramic processing can easily produce macroporous supports and layers, whereas mesoporous or microporous layers can be achieved by the sol-gel route. The pore size classification is that recommended by the International Union of Pure and Applied Chemistry (IUPAC) (Table 25.2).

The overall performance of membranes is related to two main characteristics: their permeability and their permselectivity (separation ability). For porous membranes, the selectivity and the membrane cutoff depend on the pore size and the pore size distribution of the separative layer. The membrane permeability and the membrane thickness fix the viscous flux for a given transmembrane pressure. The viscous flux of a liquid, $J$, across a porous medium is given by Darcy's law:[3]

$$J = -\frac{D}{\eta} \nabla P_L ,$$

where $J$ is the flux (volume per unit of porous material area and per unit of time; m/s), $\nabla P_L$ is the pressure gradient in the liquid (Pa/s), $\eta$ is the liquid viscosity (Pa/s), and $D$ is the permeability of the porous material (m$^2$).

**TABLE 25.1**

**Characteristics of the Main Processes in which Ceramic Membranes Can Be Used**

| Process | Nature of Feed/Strip | Pore Size | Origin of Selectivity | Pressure Gradient | Elemental Operation |
|---|---|---|---|---|---|
| Microfiltration | | 0.1–10 μm | Sieving effect | 1–3 bars | Clarification, debacterization, separation |
| Ultrafiltration | Liquid/liquid | 1 nm–0.1 μm | Sieving effect | 3–10 bars | Clarification, purification, concentration |
| Nanofiltration | | <2 nm | Sieving + specific interactions with the membrane | 10–40 bars | Purification, water softening, separation, concentration |
| Pervaporation | Liquid/gas | <2 nm | Sieving + additional specific interactions | 1 bar | Separation |
| Gas filtration | | 100 μm–0.01 μm | Sieving effect | 0.1–5 bars | Separation, dusting |
| Gas separation | Gas/gas | 50 nm–<2 nm | Sieving + additional specific interactions | 0.1–50 bars | Separation, extraction, purification |
| Gas separation | | Dense | Ionic conduction of $O^{2-}$ by oxides | $\Delta p(O_2)$ | Air separation, transport of $O_2$ |

**TABLE 25.2**
**IUPAC Classification of the Pores as a Function of Their Size**

| Micropores <2 nm | | Mesopores | Macropores |
|---|---|---|---|
| | | 2–50 nm | >50 nm |
| Ultramicropores | Supermicropores | | |
| <0.7 nm | >0.7 nm | | |

This relation, which is analogous to Poiseuille's relation, gave rise to various models taking into account the irregularity of the porous medium (tortuosity, noncircular sections, etc.). Carman-Kozeny's model is a simple and usually precise model which leads to the following expression of $D$:[3]

$$D = \frac{\varepsilon^2}{5[(1-\varepsilon)S\rho_s]^2},$$

where $\varepsilon$ is the porosity, $S$ is the specific surface area, and $\rho_s$ is the skeleton density.

In the case of the smallest pores (mesopores and micropores), the developed area is very large and the permeability is very low. Thus the thickness of the separative layer must be thin enough to reach attractive fluxes with experimentally acceptable transmembrane pressure. On the other hand, the mechanical strength of the membrane must be large enough to withstand the applied pressure. These considerations led to the concept of asymmetric structure based on macroporous support and successive layers with decreasing thickness and pore size[4] (Table 25.3; Figure 25.1).

In addition to the considerations on permeability and permselectivity, another important parameter is the ratio of filtering surface to membrane volume, because

**TABLE 25.3**
**Characteristics of the Intermediate and Separative Top Layers**

| Process | Average Number of Layers | Thickness of the Separative Layer | Pore Size in the Separative Layer | Nature of the Porosity |
|---|---|---|---|---|
| Microfiltration | 1–3 | Few tens of μm | 5–0.1 μm | Macroporosity |
| Ultrafiltration | 3–4 | A few μm | 5 nm | Mesoporosity |
| Nanofiltration/gas separation | 4–5 | <1 μm | 1 nm | Microporosity |

From Siskens, C.A.M., in *Fundamentals of Inorganic Membrane Science and Technology*, Burgraff, A.J. and Cot, L., Eds., Elsevier, Amsterdam, 1996.

**FIGURE 25.1** Scanning electron microscope (SEM) image of the cross-section of a commercial ultrafiltration alumina membrane (Pall Exekia). The average pore size of the support, of the two intermediate layers, and of the separative top layer are 10 μm, 0.8 μm, 0.2 μm, and 5 nm, respectively.

it defines the final size of the membrane units. The stiffness of the ceramic materials hinders the preparation of compact spiral modules accessible with flexible organic membranes. The ceramic membranes are usually tubular membranes, well adapted for tangential filtration, which limits fouling phenomena. Multichannel membranes, honeycomb structures, and recently, ceramic hollow fibers have been used to improve the compactness of filtration units using ceramic membranes.

The main drawbacks to the large-scale use of ceramic membranes are their higher price and also the deficiency of commercial products. The first problem is compensated for by higher performances in terms of durability compared to organic membranes, which lead to lower operating costs. Different types of microfiltration and ultrafiltration ceramic membranes have been commercially available for the last 15 years or so, whereas the first nanofiltration ceramic membranes have been commercially available for only a few years. This explains the greater use of ceramic membranes for microfiltration and ultrafiltration of food liquids and industrial effluents.[5] Besides the current industrial applications, several potential applications exist whose emergence depends on the advancement of academic research and on the adoption of new and more restrictive environmental rules. Among these emerging applications, many are associated with the use of microporous membranes for the separation and/or filtration of gases and vapors requiring thermal and chemical stability, and/or coupled functionalities like catalytic activity. This purpose can be illustrated by the treatment of automotive exhaust with simultaneous retention of soots, reduction of $NO_x$ compounds, and oxidation of volatile organic compounds (VOCs).[6]

The latest developments in ceramic membranes are closely related to recent advances in materials science;[7] in particular, in the development of nanomaterials by innovative sol-gel or hydrothermal routes. In correlation with chemical engineering and transport modeling considerations, several complementary strategies have been adopted in terms of materials engineering. The first one is the selection of the most suitable solid phase to manage the fluid-membrane interactions. Layers exhibiting specific physical or chemical properties can be advantageously prepared. Multifunctional membranes coupling separation with another functionality such as catalysis, photocatalysis, or adsorption can also be designed. A second aspect deals with tailoring the nanoporous texture (porosity, pore size and pore size distribution, connectivity and tortuosity of the pore network). The third point is concerned with the design of the membrane shape to increase the surface:volume ratio to promote antifouling and hydrodynamics properties. Various examples will be now detailed to illustrate these new approaches in the design of ceramic membranes with properties tailored to the requirements of the final applications.

## II. CHOICE OF CERAMIC MATERIALS

### A. NANOFILTERS FOR THE TREATMENT OF AQUEOUS SOLUTIONS

Nanofiltration is a membrane process involving microporous membranes, with a cutoff lower than 1000 Da. This process is used for the selective retention of small neutral molecules or of small dissociated ions found in solvents with a high dielectric constant, like water.

The preparation of the required microporous ceramic layers is possible by the sol-gel route from stable colloidal dispersions with individual nanoparticles of less than 10 nm. Different types of ceramic nanofilters have been prepared from such aqueous or organic sols of the following oxides: $\gamma$-alumina,[8] zirconia,[9,10] hafnia,[11] and titania.[12]

Electron microscope images of a zirconia nanofilter prepared from an organosol are shown in Figure 25.2.[9] This sol was obtained using a chelating agent, acetylacetone, which acts in blocking the functional group in the condensation of the zirconia precursor, zirconium isopropoxide, and sterically stabilizes the nanoparticles in the organic medium. After deposition by slip casting on an asymmetric ceramic substrate and sintering at 500°C, a zirconia layer is obtained that exhibits a tetragonal crystalline structure with very fine nanocrystallites (Figure 25.2b). In this top layer, the mean pore size is about 1.4 nm, with a porosity of 18%.

Due to the amphoteric behavior of the used oxides, the separative properties of these ceramic nanofilters for ionic solutes in aqueous solutions will depend on both sieving and electrical effects. Complex electrokinetic phenomena occur during the forced flow of the ionic solutions through the confined volume of the micropores because the thickness of the double layer formed on the charged pore surface and the pore size have the same order of magnitude. Figure 25.3 illustrates

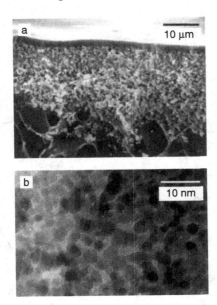

**FIGURE 25.2** Electron microscope images of a zirconia nanofilter. (a) SEM cross-section view of the multilayer asymmetric structure with the thin zirconia top layer. (b) Transmission electron microscope (TEM) surface view of the sintered packing of 5 nm to 10 nm zirconia grains.

the correlation that is experimentally observed between the electrophoretic mobility of hafnia particles (proportional to zeta potential, that is, the electrical potential at the surface of the particles) and the rejection rate of the corresponding hafnia nanofilter. Each oxide exhibits an intrinsic value of zero point of charge (ZPC) and isoelectric point (IEP) values associated with specific conditions in terms of the nature of the aqueous solution in contact.[13,14] The use of mixed oxides like silica-alumina[15] is an additional way to tune the electrical surface properties of the ceramic oxides. These data clearly highlight the importance of the choice of the nanofilter for a given application in the treatment of aqueous solutions.

## B. NANOFILTERS FOR THE TREATMENT OF ORGANIC SOLUTIONS

Application of membrane processes to nonaqueous liquids appears to be very promising. Ceramic membranes offer the utmost advantage, exhibiting a very good stability with practically all organic solvents and over a wide temperature range. Recent results obtained with nanofiltration ceramic membranes[16,17] show that permeation of organic solvents does not simply obey conventional Darcy's law. Fluxes obtained with different organic solvents (ethanol, hexane, heptane, toluene) using three sol-gel-derived mixed oxide membranes with different compositions based on $Al_2O_3$, $ZrO_2$, $TiO_2$, and $SiO_2$, are shown in Figure 25.4. The data are presented as histograms comparing solvent fluxes for each membrane

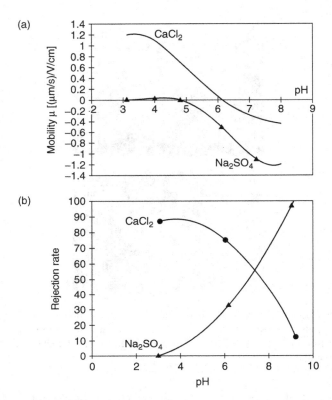

**FIGURE 25.3** (a) HfO$_2$ electrophoretic mobility vs. pH in $10^{-3}$ M Na$_2$SO$_4$ and CaCl$_2$ solutions; (b) Na$_2$SO$_4$ and CaCl$_2$ rejections versus pH using a hafnia nanofilter. (From Blanc, P., Larbot, A., Palmeri, J., Lopez, M., and Cot, L., *J. Membr. Sci.*, 149, 151, 1998.)

(Figure 25.4a) and membrane fluxes for each solvent (Figure 25.4b). For the sake of understanding and comparison of the membranes, fluxes are normalized with ethanol taking an arbitrary value of 100 at a transmembrane pressure of 10 bars for each membrane. This is justified by the fact that pore sizes, porous volume, and membrane thickness are not exactly identical for the three membranes. A number of parameters (capillary pressure, disjoining pressure, acid-base properties of ceramic oxide materials) have been considered to account for deviations from Darcy's law.[17] The specific permeability observed for each solvent/membrane pair (Figure 25.4a) can be explained by the relation existing between the acid-base surface properties of ceramic oxides and the solid surface energies,[18] from the more basic (alumina-zirconia) to the more acidic (silica-titania). The utilization of an adequate acid-base scale for the evaluation of the specific interactions between solid surfaces and liquids should be of interest to classify ceramic oxide membranes, in view of their applications with nonaqueous systems. It can be anticipated from Figure 25.4b that the silica-titania membrane is well adapted to paraffinic solvents, whereas the silica-zirconia membrane is more suitable for

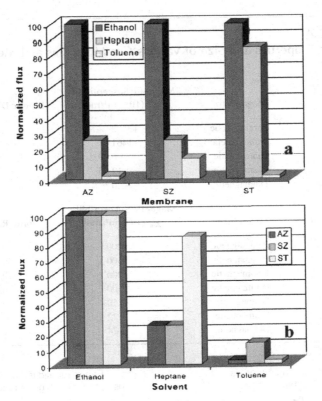

**FIGURE 25.4** Comparison of normalized fluxes for the different investigated solvents and nanofiltration membranes: $AZ = 3Al_2O_3\text{-}2ZrO_2$; $SZ = SiO_2\text{-}ZrO_2$; $ST = SiO_2\text{-}TiO_2$.

aromatic solvents. The alumina-zirconia membrane would be preferred for polar solvents.

## C. Porous Ceramic Membranes Exhibiting Specific Magnetic Properties for Separation

As a new concept of inorganic membranes, the potential interest of porous magnetic films as separative barriers based on magnetic selectivity toward para- and diamagnetic species has been recently pursued. The idea is to promote magnetic interactions within the pores of the membrane in order to separate gas molecules or metallic cations with very close size, but different magnetic behavior (Table 25.4).

A superconductor becomes diamagnetic below its critical temperature (greater than the liquid nitrogen temperature for the copper oxide-based superconductors). So if a porous membrane is prepared with superconducting material, it should preferentially repel the paramagnetic species below its critical temperature (Figure 25.5a). The separation mechanism can be explained from the mirror

**TABLE 25.4**

**Magnetic Properties and Size of Various Gas Molecules and Metal Cations[19–21]**

| Molecule | Magnetism | Magnetic Susceptibility, $\chi_m$ ($10^{-6}cm^3/mol$) | Kinetic Diameter (Å) |
|---|---|---|---|
| $O_2$ | Paramagnetic | +3402 | 3.46 |
| NO | Paramagnetic | +1461 | ~3.8 |
| Ar | Diamagnetic | −6.99 | 3.35 |
| $N_2$ | Diamagnetic | −12.05 | 3.64 |
| $CO_2$ | Diamagnetic | −20 | ~4.0 |

| Metal Cation | Magnetism | Magnetic Susceptibility, $\chi_m$ ($10^{-6}cm^3/mol$) | Ionic Radius (Å) |
|---|---|---|---|
| $Fe^{3+}$ | Paramagnetic | 15,000 | 0.64 |
| $Co^{2+}$ | Paramagnetic | 9500 | 0.78 |
| $Ni^{2+}$ | Paramagnetic | 4200 | 0.69 |
| $Cu^{2+}$ | Paramagnetic | 1500 | 0.69 |
| $Zn^{2+}$ | Diamagnetic | −10 | 0.74 |
| $Cd^{2+}$ | Diamagnetic | −22 | 0.97 |
| $Ag^+$ | Diamagnetic | −24 | 1.26 |

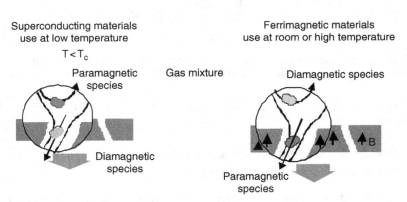

**FIGURE 25.5** Schematic representation of the magnetic separation of diamagnetic and paramagnetic species using (a) superconducting membranes and (b) ferri or ferromagnetic membranes.

principle of superconductors, suggested by Reich and Cabasso.[22] This principle is based on the magnetic interaction between a diamagnetic superconductor and a molecular magnet. Superconducting asymmetric porous membranes of YBCO and BSCCO were prepared by the sol-gel route and electrophoretic deposition of superconducting colloids, respectively (Figures 25.6 and 25.7). In order to avoid problems of reactions between superconducting films and support materials,

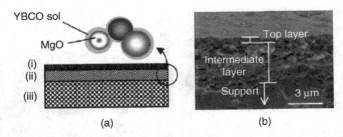

**FIGURE 25.6** (a) Schematic representation of the YBCO membrane: (i) top layer of sol-gel derived YBCO; (ii) YBCO-MgO composite intermediate layer; (iii) YBCO macroporous support. (b) SEM image of the YBCO membrane (cross-section).

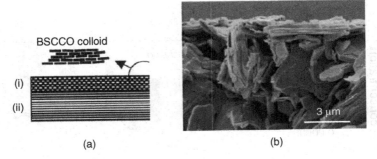

**FIGURE 25.7** (a) Schematic representation of the BSCCO membrane: (i) top layer of BSCCO colloids; (ii) BSCCO macroporous support. (b) SEM image of the BSCCO membrane (cross-section).

superconducting YBCO and BSCCO themselves were used as porous supports. Gas permeation and separation experiments were performed at 77 K with the superconducting membranes using an $O_2/N_2$ gas mixture. The prepared membranes did not show a significant selectivity effect (selectivity very close to one). High superconductivity and microporosity, which are required on the basis of theoretical considerations, cannot be simultaneously obtained. It was concluded that, from oxide superconducting membranes, it will be difficult to reach the high selectivity values required for valuable technological applications from oxide superconductors.

Ferrimagnetic and ferromagnetic membranes which can both be magnetized, appear as an attractive alternative to superconducting membranes for the separation of para-/diamagnetic species at room or high temperature (Figure 25.5b). [26,27] Ferrimagnetic membranes exhibiting small pore sizes (meso-/microporosity) required to enhance the magnetic short-distance interactions can be easily produced by the sol-gel route from colloidal ferrofluids. Thus $\gamma$-$Fe_2O_3$ and $Fe_2CoO_4$ membranes were prepared as supported porous thin layers or composite membranes (Figure 25.8). Such composite membranes obtained by impregnation of

**FIGURE 25.8** SEM images of the ferromagnetic composite membranes prepared by impregnation of porous substrates: (a) silica-coated $\gamma$-$Fe_2O_3$; (b) $Fe_2CoO_4$.

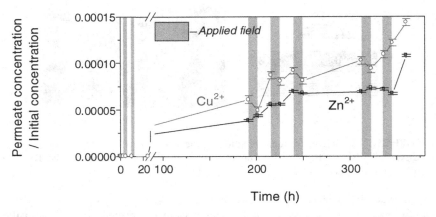

**FIGURE 25.9** Permeate concentration ratio as a function of time across the $\gamma$-$Fe_2O_3$ ferrimagnetic composite membrane.

porous substrates are interesting for applications in gas separation, since they minimize the risk of the presence of defects or cracks without disadvantageous limitation of flux. The $\gamma$-$Fe_2O_3$ and $Fe_2CoO_4$ composite membranes were tested for air separation applications. The experiments revealed a very low selectivity effect resulting from the application of an external magnetic field, but the measured selectivity values $\alpha_{N_2O_2}$ are always close to 1. Preliminary experiments on the separation of metal cations in an aqueous solution showed a slight magnetic effect on selectivity between para- and diamagnetic species (Figure 25.9).[26] The optimization of the porosity of the tested membranes remains possible (decrease the pore size to increase the short-range magnetic interactions), and additional studies are required to definitively determine the potential uses of such membranes.

## D. Dense Membranes for Oxygen Transport

Dense metallic membranes, in particular those based on palladium alloys, have been extensively studied for the selective transport of $H_2$.[28,29] In the case of $O_2$, the use of silver-based metallic membranes has also been investigated.[30] However, dense oxide ceramic membranes are the most attractive in terms of durability and reliability. An exhaustive review on this topic is presented in Bouwmeester and Burggraaf.[31] Transport of $O_2$ (oxidation number 0) occurs via the conduction of the ions $O^{2-}$ (oxidation number $-2$) through the oxide network (ceramic ion conducting membranes [CICMs]). A sufficient ion conductivity requires high working temperatures, usually $\geq 800°C$. In the case of an oxide exhibiting both ion and electron conductivity (mixed ions-electrons conducting [MIEC]), the electron countercurrent enables the membrane electroneutrality. Thus the driving force associated with oxygen transmembrane gradient pressure is sufficient to induce oxygen transport across the membrane (Figure 25.10a). If the electron conduction is not large enough, an external electrical circuit is required with electrodes on both sides (Figure 25.10b). Another strategy is to design a composite membrane adding an electron conductor as a second phase (Figure 25.10c). This last approach is not easy in terms of practical application. In the case of the external circuit, a voltage can be applied by an electric generator to favor $O_2$ transport (oxygen pump).

Oxides exhibiting only high ion conductivity are mainly fluorite-related structures based on zirconia or ceria. Zirconia-based electrolytes are currently used in solid oxide fuel cells (SOFCs). The MIEC oxides are more attractive for separative membrane applications, and these oxides mainly belong to the following types: fluorite-related oxides doped to improve their electron conduction,[31,32]

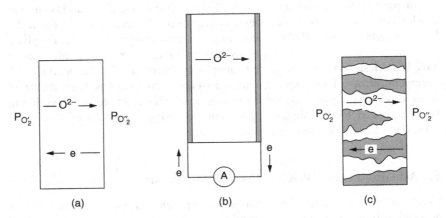

**FIGURE 25.10** Different membrane concepts incorporating an oxygen ion conductor: (a) mixed conducting oxide, (b) solid electrolyte cell (oxygen pump), and (c) dual-phase membrane. (From Bouwmeester, H.J.M., and Burggraaf, A.J., in *Fundamentals of Inorganic Membrane Science and Technology*, A.J. Burgraff and L. Cot, Eds., Membrane Science and Technology Series 4, Elsevier, Amsterdam, 1996, Chapter 10.)

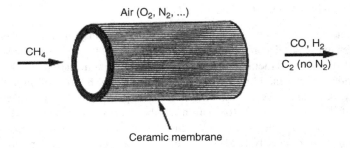

Air (O$_2$, N$_2$, ...)

CH$_4$

CO, H$_2$
C$_2$ (no N$_2$)

Ceramic membrane

**FIGURE 25.11** Schematic representation of the application of an oxygen-conductive membrane for the direct conversion of methane to syngas. (From Balachandran, U., Dusek, J.T., Sweeney, S.M., Poeppel, R.B., Mieville, R.L., Maiya, P.S., Kleefisch, M., Pei, S., Kobylinski, T.P., Udovich C.A., and Bose, A.C., *Bull. Am. Ceram. Soc.*, 74, 71, 1995.)

pervoskite-related oxides based on lanthanides with the generic formula $Ln_{1-x}A_xCo_{1-y}B_yO_3$ (Ln = La, Pr, Nd, Sm, Gd; A = Sr, Ba, Ca, Na; B = Fe, Cr, Ni, Cu, Mn),[32,33] brownmillerite-related oxides like $Ca_2Fe_2O_5$, $Ba_2In_2O_5$, and $Sr_2Fe_2O_5$,[34-36] or BiMeVOx.[37] The most studied oxides for membrane applications are pervovskite-related oxides, in particular the system La-Sr-Fe-Co-O (LSFC). Figure 25.11 schematically shows such a dense membrane applied for the direct conversion of methane in syngas. High conversion efficiencies can be reached (>99%), however, fracture of certain LSFC tubes is observed which is a consequence of an oxygen gradient that introduced a volumetric lattice difference between the inner and outer sides of the membrane.[32]

In parallel to the concept of fully dense membranes, applications of ion-conducting mesoporous membranes has been recently investigated.[38] From the literature data dealing with the influence on oxygen transport of porous nanophase ion-conducting ceramics, mesoporous and nanophase ceria-based membranes have been designed, containing palladium or platinum metallic nanoparticles. A synergetic effect of the nanoparticles on oxygen transport has been shown in relation with the triple phase boundary concept.[38] This is a potential direction to be investigated for membrane applications requiring higher fluxes and lower selectivity than dense ion-conductive membranes.

## E. MULTIFUNCTIONAL POROUS CERAMIC MEMBRANES

Coupling two operations like membrane separation and a catalytic reaction or adsorption in a given process of synthesis, purification, or decontamination of effluents is intrinsically interesting from a general technical-economical point of view. Ceramic membranes are ideal solid-fluid contactors, which can be efficiently used to couple separation and heterogeneous catalysis for membrane reactor applications.[39]

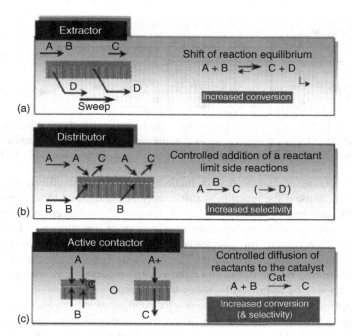

**FIGURE 25.12** The three main membrane functions in a membrane reactor. (From Julbe, A., Farrusseng, D., and Guizard, C., *J. Membr. Sci.*, 181, 3, 2001.)

## 1. Ceramic Membranes for Catalytic Membrane Reactors

The concept of combining membranes and reactors is being explored in various configurations, which can be classified into three groups, related to the role of the membrane in the process.[40] As shown in Figure 25.12, the membrane can act as: (a) an extractor, where the removal of the product(s) increases the reaction conversion by shifting the reaction equilibrium; (b) a distributor, where the controlled addition of reactant(s) limits side reactions; and (c) an active contactor, where the controlled diffusion of reactants to the catalyst can lead to an engineered catalytic reaction zone. In the first two cases, the membrane is usually catalytically inert and is coupled with a conventional fixed bed of catalyst placed on one of the membrane sides.

In the active contactor mode, the membrane acts as a diffusion barrier and does not need to be permselective, but catalytically active. The concept can be used with a forced-flow mode or with an opposing reactant mode. The forced-flow contactor mode, largely investigated for enzyme-catalyzed reactions,[41] has also been applied to the total oxidation of volatile organic compounds.[42,43] The selectivity of alkene hydrogenation triphasic reactions can also be greatly improved when both the alkene and $H_2$ are forced to pass through a microporous catalytic membrane, because back mixing of the initial products is prevented.[44,45]

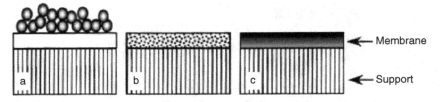

**FIGURE 25.13** The main membrane/catalyst combinations: (a) bed of catalyst on an inert membrane; (b) catalyst dispersed in an inert membrane; and (c) inherently catalytic membrane. (From Julbe, A., Farrusseng, D., and Guizard, C., *J. Membr. Sci.*, 181, 3, 2001.)

The opposing reactant contactor mode applies to both equilibrium and irreversible reactions, if the reaction is sufficiently fast compared to transport resistance (diffusion rate of reactants in the membrane). This concept has been demonstrated experimentally for reactions requiring strict stoichiometric feeds, such as the Claus reaction, or for kinetically fast, strongly exothermic heterogeneous reactions, such as partial oxidations.[46–49] Triphasic (gas/liquid/solid) reactions, which are limited by the diffusion of the volatile reactant (e.g., olefin hydrogenation), can also be improved by using this concept.

The different types of membrane reactor configurations can also be classified according to the relative placement of the two most important elements of this technology: the membrane and the catalyst. Three main configurations can be considered (Figure 25.13): the catalyst is physically separated from the membrane; the catalyst is dispersed in the membrane; or the membrane is inherently catalytic. The first configuration is often called the inert membrane reactor (IMR), in contrast to the two other ones, which are catalytic membrane reactors (CMRs).[50]

When the catalyst is immobilized within the pores of an inert membrane (Figure 25.13b), the catalytic and separation functions are engineered in a very compact fashion. In classical reactors, the reaction conversion is often limited by the diffusion of reactants into the pores of the catalyst or catalyst carrier pellets. If the catalyst is inside the pores of the membrane, the combination of the open pore path and transmembrane pressure provides easier access for the reactants to the catalyst. Two contactor configurations—forced-flow mode or opposing reactant mode—can be used with these catalytic membranes, which do not necessarily need to be permselective. It is estimated that a membrane catalyst could be 10 times more active than in the form of pellets,[51] provided that the membrane thickness and porous texture, as well as the quantity and location of the catalyst in the membrane, are adapted to the kinetics of the reaction.[46] For biphasic applications (gas/catalyst), the porous texture of the membrane must favor gas-wall (catalyst) interactions to ensure a maximum contact of the reactant with the catalyst surface. In the case of catalytic consecutive-parallel reaction systems, such as the selective oxidation of hydrocarbons, the gas-gas molecular interactions must be limited because they are nonselective and lead to a total oxidation of reactants and products. For these reasons, small-pore mesoporous or microporous membranes, in which the dominant gas transport is Knudsen or micropore activated diffusion, are typically favored for contactor applications in biphasic reactions. Larger pore sizes (10 to 25 nm) are preferred for triphasic contactor applications (e.g., hydrogenation of liquid alkenes) with an opposing reactant mode,[52] although microporous membranes can be attractive for increasing the selectivity of such reactions with a forced-flow mode.[44,45] Catalysts such as palladium and platinum have been impregnated or dispersed in various types of meso- and microporous membrane matrices (e.g., $\gamma$-$Al_2O_3$, $SiO_2$, $TiO_2$, carbon, glass) and largely used for low-temperature hydrogenation or for dehydrogenation reactions up to 675°C.[41] The combustion of volatile organic compounds has also been investigated with a Pt-$\gamma$-$Al_2O_3$ membrane (flow-through contactor mode).[42] Other types of catalysts have also been impregnated or dispersed in inert membrane supports, for example, $Fe_2O_3$ in $SiO_2$,[53] and $V_2O_5$,[54–56] $MoS_2$,[57] $Os$,[58] $Cr_2O_3$,[58] $SmO_2$,[59] or $NiO$[60] in $Al_2O_3$.

Other membranes are called inherently catalytic membranes (Figure 25.13c). In this highly challenging case, the membrane material serves as both the separator and catalyst, and controls the two most important functions of the reactor. As in the previous case, such porous catalytic membranes are used as active contactors to improve the access of the reactants to the catalyst. A number of meso- and microporous inorganic membrane materials have been investigated for their intrinsic catalytic properties, such as alumina,[51] titania,[54] zeolites with acid sites,[61,62] V-ZSM-5 zeolite,[63] rhenium oxide,[64] $LaOCl$,[65] $RuO_2$-$TiO_2$ and $RuO_2$-$SiO_2$,[66] VMgO,[67,68] and La-based perovskites.[43] Depending upon the type of material and catalytic reaction, the efficiency of such membrane contactors is not always obvious. The membrane does not need to be permselective, but needs to be highly active for the intended reaction, contain a sufficient quantity of active sites, have a sufficiently low overall permeability, and operate in the diffusion-controlled regime.[46] In most cases, new synthesis methods have to be developed for preparing these catalytically active membranes, particularly when the optimum catalyst composition is complex. The catalytic membrane composition, activity, and porous texture have to be optimized for each considered reaction and remain stable upon use. This challenge explains the limited number of examples given in the literature for the development of inherently catalytic membranes.

## 2. Membranes Exhibiting Photocatalytic Properties

Some semiconducting single or mixed oxides such as $TiO_2$, $ZnO$, and $CaIn_2O_4$ are known to exhibit photoactivity under ultraviolet (UV)-visible irradiation.[69–71] Titania is currently the most used oxide due to its attractive photoactivity properties under UV irradiation, particularly in its anatase form. These properties gave impetus to various technological applications such as photovoltaic cells, self-cleaning layers, and sensors.[69–70] Several papers have been published in the past concerning the use of membranes in photocatalytic reactors. In most cases, the membranes do not present photoactivity, they are used to separate reactants and to retain the titania particles dispersed inside the reactor loop.

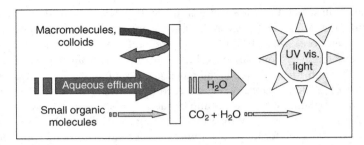

**FIGURE 25.14** Schematic representation of the coupled separation and photodegradation of VOCs in the treatment of wastewater using a photoactive low-ultrafiltration membrane.

Only a few recent studies have dealt with the use of photoactive membranes prepared by immobilization of titania particles in polymeric membranes[72] or deposition of porous titania coatings.[73-77] They were concerned with photooxidation applications like antifouling or elimination of small organic molecules that cannot be stopped by conventional membrane treatments, but which are very harmful to the environment, like VOCs.

Ceramic membranes are the most often used asymmetric membranes. When the separative layer, which is usually in contact with the feed, is also photoactive, irradiation must be applied on this top layer. A second configuration can also be considered. It consists of a conventional asymmetric membrane without photoactive separative layers but with a photoactive coating deposited on the surface of the grains of the support. In this case, the irradiation is applied on the opposite side of the membrane, in contact with the permeate. Such a configuration could be used for instance in the final treatment of wastewater with a low-ultrafiltration membrane which provides retention of colloids and macromolecules, whereas small unretained molecules like VOCs would be photo-oxidized on the other side of the membrane (Figure 25.14).

Preliminary experiments on such a strategy of coupled separation and photodegradation were carried out on a simplified experimental device schematically represented in Figure 25.15a.[76] It consists of two tanks separated by a 1.8 μm pore size alumina microfiltration symmetric membrane with grains coated with an anatase layer. The feed tank contained methylene blue (MB) and the reception tank was initially filled with pure water. Under continuous UV irradiation, MB is completely destroyed as it arrives in the reception tank (Figure 25.15b). The quantity of destroyed MB per surface area unit is equal to ~$1.0 \times 10^7$ mol.s$^{-1}$.m$^2$.

## 3. Ceramic Membranes with Adsorptive Properties

Ceramic membranes with adsorptive properties is a new field of application for multifunctional membranes. Single or mixed reactive oxides can be used for removing toxic gas from gas mixtures by selective chemisorption.[78-82] For instance, sulfur compounds like hydrogen sulfide, $H_2S$, or $SO_2$ can be chemisorbed as sulfide or

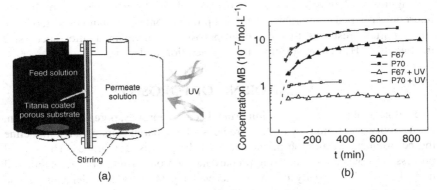

**FIGURE 25.15** Preliminary experiments on coupling separation and photodegradation: (a) experimental dialysis device; (b) methylene blue concentration versus time in the reception tank. P70 and F67 are two types of anatase coatings. (From Bosc, F., Ayral, A., and Guizard, C., in *Proceedings of the ICIM8*, J. Lin, Ed., Cincinnati, OH, 2004.)

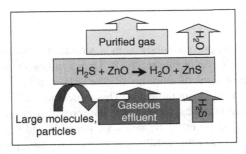

**FIGURE 25.16** Schematic representation of the coupled separation and chemisorption of $H_2S$ in the treatment of gaseous effluent using a microporous membrane containing ZnO nanoparticles.

sulfate, respectively. Figure 25.16 illustrates such a concept of adsorptive membrane with the treatment of gaseous effluent using a microporous ZnO-based membrane for $H_2S$ from a gas effluent by selective and reversible chemisorption as ZnS.[82]

In addition to the design of the solid-gas contactor device, the yield of a desulfurization process directly depends on the physicochemical properties of the used adsorbent: crystallite size of the active phase, specific surface area, and porous texture. In the case of a microporous membrane, if the gas flow is forced across the microporosity, it can be expected that the retention will be highly efficient. In return, the low amount of adsorbent restricts the potential applications to the elimination of traces in high-purity gas or to the design of integrated filters for miniaturized devices like micro fuel cells.

Figure 25.17 illustrates some experimental results about a sol-gel-derived microporous silica membrane and related powder containing nanodispersed ZnO. An important result is that the microporosity is maintained after successive treatments of $H_2S$ chemisorption and regeneration under air.

## III. TAILORING OF POROSITY

The tailoring of porosity is very important because porosity, pore size distribution, connectivity, and tortuosity of the pore network are parameters which both define the permselectivity and the permeability of the porous membranes. The sol-gel process is a convenient method to prepare mesoporous or microporous supported membranes. The porosity of sol-gel-derived layers drastically depends on the various synthesis parameters (Table 25.5).[83,84]

With conventional sol-gel routes, the pore size distribution is usually broad and the tortuosity of the pore network is important with the presence of constrictions. Thus ordered interconnected pore networks with constant pore size are strongly attractive. Hierarchical porosity and adaptive porosity are also fascinating approaches to increase or manage the permeability of ceramic membranes.

### A. ZEOLITE MEMBRANES

Zeolites are ultramicroporous solids with a structural porosity.[85] Examples of zeolite structures are given in Figure 25.18. Their synthesis is usually carried out by hydrothermal treatment of alumino-silicate gels in a basic aqueous medium in the presence of a templating molecule. This molecule is often a tetrasubstituted quaternary ammonium such as, for instance, tetrapropyl ammonium. Substitutions of aluminum or silicon by other elements are possible and pure silica structures can also be obtained. Zeolite exhibits a set of attractive properties including regular pore sizes with molecular dimensions (enabling shape- or size-selective catalysis or separation), a high thermal stability, acid or basic properties, hydrophilic or organophilic properties, possibility of ion exchange, dealumination-realumination, isomorphous substitution, and insertion of catalytically active guests (transition-metal ions, complexes or chelates, basic alkali metal or metal oxide clusters, and enzymes).

The specific properties of zeolites, coupled with the separation properties of membranes, open the field to many areas of research for the future. This explains why the preparation and application of zeolite membranes is the subject of intensive research.[87–89] By combining their adsorption and molecular sieving properties, zeolite membranes have been used for the separation of mixtures containing nonadsorbing molecules, different organic compounds, permanent gas-vapor mixtures, or water-organic mixtures.

MFI zeolite membranes (silicalite-1, ZSM-5), on either flat or tubular porous supports, have been the most investigated for gas separation, catalytic reactors, and pervaporation applications. The structural porosity of MFI zeolite consists of channels of about 5.5 Å, in diameter, the silica-rich compositions induce

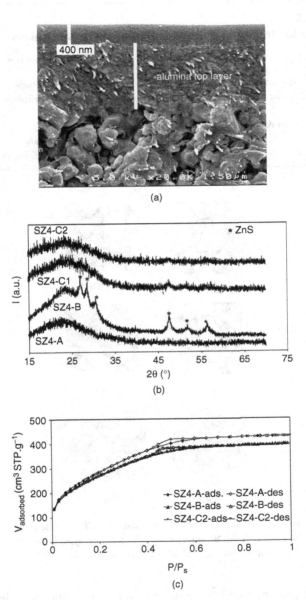

**FIGURE 25.17** (a) SEM cross-section image of a microporous silica membrane with nanodispersed ZnO (designated as SZ4). (b) Evolution of the x-ray pattern for an SZ4 powder. (c) Evolution of the $N_2$ adsorption-desorption isotherm for an SZ4 powder. A: initial state; B: after contact with $H_2S$; C1 and C2: after thermal regeneration. (From Goswamee, R., Bosc, F., Cot, D., El Mansouri, A., Lopez, M., Morato, F., and Ayral, A., *J. Sol-Gel Sci. Technol.*, 29, 97, 2004.)

**TABLE 25.5**
**Tools to Tailor the Initial Porosity of Sol-Gel-Derived Layers**

| Stage | Parameters |
|---|---|
| Choice of the elemental bricks | Size, shape, size distribution |
| Preparation of the sol before deposition | Conditions of polymerization/destabilization |
|  | Sol aging |
| Reinforcement after deposition and before drying | Thermal aging |
|  | Chemical posttreatments |
| Drying method | Freeze-drying, hypercritical drying |

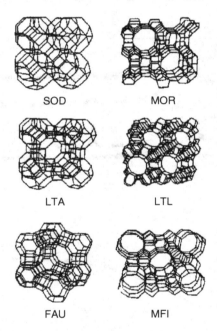

**FIGURE 25.18** Main zeolite structures. SOD: sodalite; LTA: Linde type A zeolite; FAU: faujasite, zeolites X or Y; MOR: mordenite; LTL: Linde type L zeolite; MFI: ZSM-5. (From Ozin, G.A., Kuperman, A., and Stein, A., *Angew. Chem. Int. Ed. Engl.*, 28, 359, 1989.)

hydrophobic properties and also a stability at high temperature (at least 600°C). A number of other supported zeolite membranes have also been prepared in the system: LTA (NaA), FAU (NaY, NaX), MOR (Mordenite), FER (ZSM-35, fer-rierite), KFI (P) and ANA (Analcime), and Sodalite (SOD, clathrasil). The hydro-thermal synthesis method has also been used for preparing zeolite-related materials such as AlPO-5 and SAPO-34.

Starting from the synthesis method developed in Giroir-Fendler et al.,[90] it has been possible to prepare "composite MFI:$\alpha$Al$_2$O$_3$ zeolite membranes" in which most of the zeolite material was grown *in situ* by hydrothermal treatment inside specific pores of a commercial $\alpha$-Al$_2$O$_3$ support. The method utilized hydrothermal treatment at 175 to 190°C of a sol (SiO$_2$: TPAOH:H$_2$O) in contact with the alumina support. The synthesis method has been extended to the synthesis of composite MOR (mordenite) zeolite membranes.[91] Compared with thin films supported on a porous support, these infiltrated composite membranes are highly attractive for a number of reasons, including good thermochemical resistance, low sensitivity to the presence of defects, sufficient flow resistance (barrier effect), and good reproducibility. Furthermore, in the case of the catalytic applications of these membranes (used as contactors), the quantity of active sites is higher than with thin films.

The insertion of catalytically active guests, such as transition metal ions, is an example of the potentialities of zeolite membranes for applications in catalytic membrane reactors.[92] The well-known catalytic properties of supported vanadium oxides for selective oxidations have recently prompted a number of studies on the possibility of inserting vanadium in the framework of crystalline microporous silica and aluminosilicate powders.[93,94]

The performance of both MFI and V-MFI membranes have been compared for the oxidative dehydrogenation of propane (ODHP) in a stainless steel membrane reactor designed to operate with two separated inlet flows.[95] Both MFI and V-MFI membranes were found to be active for ODHP at 550°C, but behaved differently according to the O$_2$ partial pressure, leading to slightly better propane yield for the V-MFI. Acid sites related to some aluminum insertion during the hydrothermal synthesis are likely to be responsible for the MFI membrane activity. The vanadyl sites present in the V-MFI contribute to the improved performance of this membrane (higher O$_2$ and C$_3$H$_8$ conversions with a selectivity in the range of 40 to 50%).

In spite of the progress made in the field of supported zeolite membranes during the last decade, a number of points still need to be explored or studied further, including

Improve synthesis control and reproducibility (large scale).

Control membrane thickness and location (at the surface or in the support pores).

Control membrane quality, detection of microdefects, and influence on membrane performance.

Control zeolite crystal orientation.

Extension of membrane synthesis to zeolite structures with smaller pores (e.g., 3 Å).

Modification of existing membrane performance (ion exchange, plugging defects, insertion of catalytically active guests).

Clarify transport mechanisms (multicomponent systems at low and high temperature).

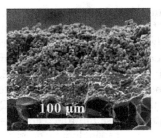

**FIGURE 25.19** SEM observation of a sodalite/$\alpha$Al$_2$O$_3$ membrane obtained by secondary growth of a seeded membrane (cross-section). (From Julbe, A., Motuzas, J., Cazevielle, F., Volle, G., and Guizard, C., *Separ. Purif. Technol.*, 32, 139, 2003.)

The main problem remains to reproducibly obtain a continuous filtering barrier with a high permselectivity. It is essential to minimize the presence of defects and transport through the grain boundaries. In order to overcome these problems, a promising strategy has recently been considered: the preseeding/secondary growth method (Figure 25.19).[96]

## B.  MEMBRANES WITH AN ORDERED MESOPOROSITY

Extension of the molecular sieves to the mesoporosity range is possible using lyotropic liquid crystal mesophases (Figure 25.20) as removable templates. These mesophases result from the self-assembly of surfactants or amphiphilic molecules and can be thermally or chemically eliminated after the formation of the inorganic network. This approach enables the preparation of materials exhibiting an ordered

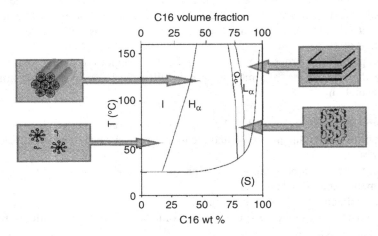

**FIGURE 25.20** Water-hexadecyltrimethylammonium bromide. I: micellar solution; H$\alpha$: two-dimensional hexagonal mesophase; Q$\alpha$: bicontinuous cubic mesophase; L$\alpha$: lamellar mesophase. (From Wärnheim, T., Jönsson, A., and Sjöberg, M., *Progr. Colloid Polym. Sci.*, 82, 271, 1990.)

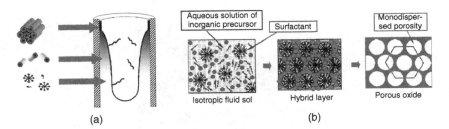

**FIGURE 25.21** Schematic representation of the formation of membranes with an ordered mesoporosity resulting from self-assembly of amphiphilic molecules. (a) Deposition by slip-casting in a tubular substrate and solvent evaporation. (b) Various stages of the synthesis process.

mesoporosity, with pores usually ranging from 2 to more than 10 nm. The pioneering work of Kresge et al.[98] and Beck et al.[99] detailed the use of the templating effect to produce ordered mesoporous aluminosilicates. Different materials have been prepared by phase separation and precipitation under hydrothermal conditions. Since these first articles, many investigations have been carried out on this new class of materials, in particular for the preparation of sol-gel-derived silica layers exhibiting hexagonal, cubic, or lamellar structures using cationic surfactant of alkyltrimethylammonium halide type.[100–108] At the same time, this synthesis method was extended to the use of nonionic surfactants[109,110] and block copolymers.[111–113] The preparation of other mesoporous oxides was also demonstrated.[114,115]

The formation of film is based on "evaporation-induced self-assembly" of the surfactant molecules.[116] The synthesis process is schematically shown in Figure 25.21. The synthesis rules to prepare continuous layers without extraporosity were initially investigated in the case of silica.[108,117] The two main parameters are (1) the size of the inorganic clusters or nanoparticles, which must be small enough to enable the self-assembly process, and (2) the surfactant volume fraction in the dried layer, which must be in agreement with the aimed mesostructure. This approach was successfully extended to the preparation of thin layers and membranes of oxides like $Al_2O_3$ and $TiO_2$ (Figure 25.22 and Figure 25.23).[117–121]

The resulting mesoporous layers don't usually exhibit extraporosity at a larger scale, but as smaller pores in the oxide walls (Figure 25.24).[108,121] Due to limitations associated with intrinsic mesostructure characteristics, anisotropy resulting from preferential orientations, or boundaries between ordered domains, the templated mesoporosity is usually not directly interconnected.[122–124] In such situations, it does directly define the selectivity of the membrane, which depends on the pore size of the oxide walls.[75–77] However, the resulting hierarchical porosity (templated mesopores and smaller pores of the oxide walls) favors a decrease in layer permeability. The templated mesoporosity can also be used to functionalize the membrane.

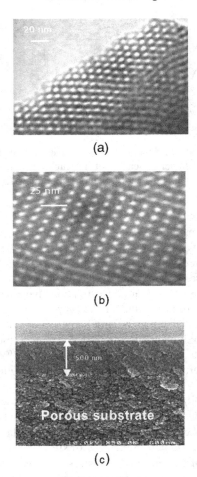

**FIGURE 25.22** TEM images of TiO$_2$ anatase with ordered mesoporosity: (a) two-dimensional hexagonal; (b) cubic; (c) SEM image of the cross-section of a two-dimensional hexagonal mesoporous layer deposited on a porous substrate. (From Bosc, F., Ayral, A., Albouy, P.A., and Guizard, C., *Chem. Mater.*, 15, 2463, 2003; Bosc, F., Ayral, A., Albouy, P.A., Datas, L., and Guizard, C., *Chem. Mater.*, 16, 2208, 2004.)

## C. POTENTIALITIES OF THE HIERARCHICAL POROUS STRUCTURES

As previously explained, it can be advantageous to generate extraporosity at a larger scale in the separative layer. The main condition that has to be respected is that the additional porosity must not be directly interconnected in order to preserve the cutoff fixed by the porosity of the continuous phase. Templating by polystyrene latex was used to produce individual macropores inside the silica layer (Figure 25.25).[83,84,125] This route can be applied to prepare membranes of other oxides[126,127] with various possible strategies in terms of the synthesis process (Figure 25.26). In addition, the presence of dispersed micron-size or submicron-size

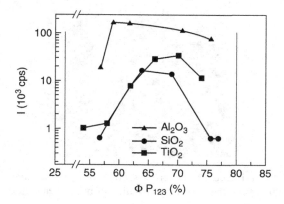

**FIGURE 25.23** Evolution of the intensity of the main diffraction peak associated with ordered mesoporosity as a function of the volume fraction of triblock copolymer (P123). The limits of the hexagonal phase at 30°C in the water-P123 binary diagram are reported as vertical lines. (From Bosc, F., Ayral, A., Albouy, P.A., Datas, L., and Guizard, C., *Chem. Mater.*, 16, 2208, 2004.)

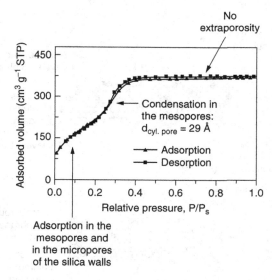

**FIGURE 25.24** Nitrogen adsorption-desorption isotherm of two-dimensional hexagonal silica layer.

particles inside the starting suspensions modify their rheology and decrease their ability to infiltrate the porous substrates. It can be used to reduce the number of intermediate layers of the asymmetric ceramic membranes. It must be noted that dispersion of dense and unremovable particles such as oxide powders inside conventional sols enabled us to deposit homogeneous thick layers (from a few μm to a few tens of μm) on macroporous substrates.[25,26,128]

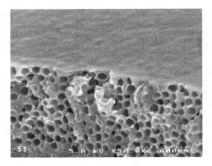

**FIGURE 25.25** SEM image of the cross-section of a silica layer with spherical macropores resulting from the thermal degradation of polystyrene latex.

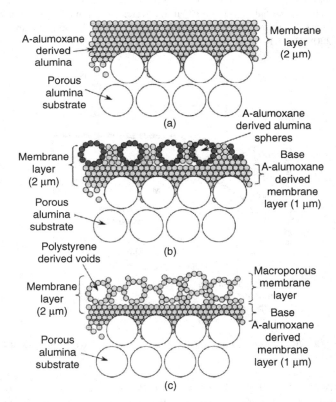

**FIGURE 25.26** Schematic representation of (a) a simple asymmetric filter and alternative designs of a hierarchical asymmetric filter using (b) preformed alumina spheres and (c) polystyrene-derived voids. (From DeFriend, K.A., and Barron, A.R., *J. Membr. Sci.*, 212, 29, 2003.)

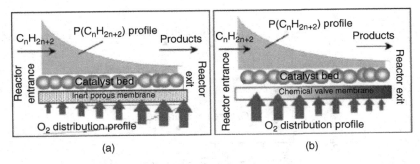

**FIGURE 25.27** Schematic representation of the $O_2$ profile generated in a tubular reactor for the partial oxidation of alkanes: (a) by a classical inert uniform porous membrane, and (b) by the suggested "chemical valve" membrane. (From Julbe, A., Farrusseng, D., and Guizard, C., *J. Membr. Sci.*, 181, 3, 2001.)

## D. The Concept of a Chemical Valve

In certain applications it is important to be able to manage the porosity not only across the membrane, but also along the membrane. This situation is schematically illustrated in Figure 25.27 with the case of a tubular porous membrane reactor used as an $O_2$ distributor for the partial oxidation of an alkane. A porous membrane with a uniform thickness and porous structure is schematically shown in Figure 25.27a. An innovative method has been proposed to better regulate the $O_2$ flux in the reactor: the concept is based on the use of a chemical valve membrane whose permeability can be controlled by the oxidation-reduction (redox) properties of the gas phase.[39,129] $V_2O_5$ was found to be an attractive key constituent for this membrane because of its reversible redox behavior ($V_2O_5/V_2O_3$) and related textural variations potentially able to regulate the membrane permeance behavior, as schematically shown in Figure 25.27b.

The results obtained with single gases confirmed that the membrane behaves as a chemical valve: its permeance is higher when it is reduced and lower when it is oxidized. The ratio between the permeances of pure n-$C_4H_{10}$ and pure $O_2$ is about 70 at 500°C (Figure 25.28).

## IV. DESIGN OF THE SHAPE

### A. Improvement of Hydrodynamics

Numerical simulations and experiments with turbulence promoters or pulsating flux confirm that promoting turbulence close to the membrane surface can partially overcome damageable phenomena like fouling and concentration polarizations in tangential filtration processes. An original approach consisted of the design and preparation of tubular ceramic membranes with helical relief stamps (Figure 25.29). Permeation measurements have demonstrated the interest of such geometries compared to conventional smooth membranes.[130,131] In addition to

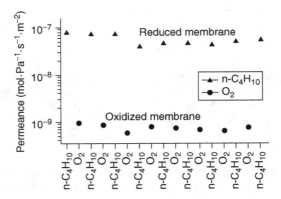

**FIGURE 25.28** Evolution of $O_2$ and $n$-$C_4H_{10}$ single gas permeances through the "chemical valve" membrane at 500°C during $n$-$C_4H_{10}/O_2$ cycling experiments. (From Julbe, A., Farrusseng, D., and Guizard, C., *J. Membr. Sci.*, 181, 3, 2001.)

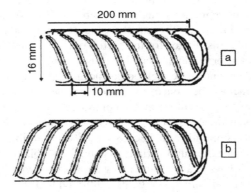

**FIGURE 25.29** Examples of membrane configurations. Longitudinal cross sections: (a) simple helix, (b) helix direction inversion. (From Broussous, L., Schmitz, P., Boisson, H., Prouzet, E., and Larbot, A., *Chem. Eng. Sci.*, 55, 5049, 2000.)

improvements in the hydrodynamics of the feed flow, it must be emphasized that such geometry also gives rise to high surface:volume ratios.

## B. IMPROVEMENT OF THE SURFACE:VOLUME RATIO: CERAMIC HOLLOW FIBERS

Due to their high stiffness and brittleness, it is not possible to extend to ceramic membranes geometries applicable with organic membranes like spirals, which give rise to high surface:volume ratios. The ceramic membranes used for tangential filtration are usually multichannel tubes or, in some applications, honeycomb monoliths. A very attractive type of membrane is the ceramic hollow fiber with an external diameter of less than 1 mm and ceramic walls with a thickness of a

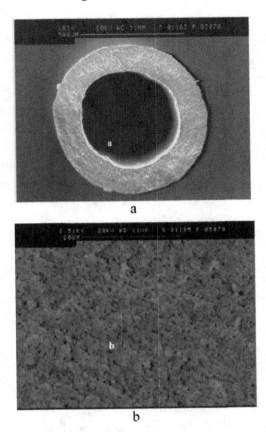

**FIGURE 25.30** Images for -Al₂O₃ hollow fibers obtained by ceramic paste extrusion: (a) cross-section morphology, (b) surface morphology. (From Pan, X.L., Stroh, N., Brunner, H., Xiong, G.X., and Sheng, S.S., *Separation & Purification Technology*, 32, 265, 2003.)

few hundred μm. An increasing number of papers have been published during the last decade on the preparation and applications of such ceramic hollow fibers.[132–140] The preparation of these hollow fibers has been carried out using different methods such as extrusion of ceramic pastes, vapor deposition on degradable wires, pyrolysis of polymeric fibers, and more recently using a technique derived from the spinning of organic hollow fibers: the phase inversion method. Alumina hollow fibers with pore sizes from a few microns to a few tens of nanometers were thus prepared from alumina powders (Figure 25.30 and Figure 25.31). As a function of the synthesis method, the overall microstructures of the fibers differ significantly, as illustrated in Figure 25.30 and Figure 25.31.

Such ceramic hollow fibers can be assembled in fully ceramic modules of a thousand fibers exhibiting surface:volume ratios of about 2 m² L-1.[135] Different types of mesoporous, microporous, or dense separative layers have been deposited usually on the outer surface of macroporous hollow fibers using γ-alumina,[133,137,138]

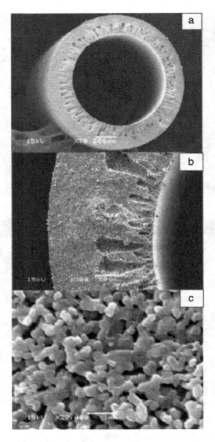

**FIGURE 25.31** SEM images of sintered hollow fiber prepared by phase inversion spinning from 0.3 μm $Al_2O_3$ particles: (a) cross-section; (b) membrane wall; (c) membrane surface. (From Liu, S., and Li, K., *J. Membr. Sci.*, 218, 269, 2003.)

titania,[139] zeolite,[140] or palladium[138] layers. Mixed proton- and electron-conducting hollow fibers based on $SrCe_{0.95}Yb_{0.05}O_{2.975}$ were also prepared by spinning and sintering at high temperatures (1300 to 1600°C).[136]

Other types of multifunctional miniaturized devices using ceramic hollow fibers can and should be developed in the near future, as suggested by Figure 25.32, which illustrates the case of coupling separation and photocatalysis.

## V. CONCLUSION

Multifunctional and adaptive membranes, and miniaturized and integrated separative devices are current trends in membrane science research and development. Increasing societal requests in terms of environmental protection, health, energy savings, etc., are long-term driving forces for such activities.

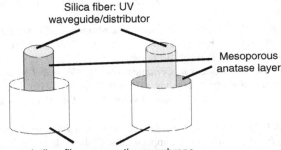

**FIGURE 25.32** Two possible configurations for coupling separation and photocatalysis in the case of a compact device based on the use of hollow fibers for separation and optical fibers for light distribution.

Taking advantage of the advances in materials science, it is possible to prepare ceramic-based membranes with improved properties with regard to the specifications of the intended final applications. The choice of materials, the tailoring of porosity, and the shape of the membrane are the main variables at the disposal of designers of these membranes.

## REFERENCES

1. Guizard, C., Julbe, A., Larbot, A., and Cot, L., *Chemical Processing of Ceramics*, B.I. Lee and E.J.A. Pope, Eds., Marcel Dekker, New York, 1994.
2. Burgraff, A.J., and Cot, L., Eds., *Fundamentals of Inorganic Membrane Science and Technology*, Membrane Science and Technology Series 4, Elsevier, Amsterdam, 1996.
3. Brinker, C.J., and Scherer, G.W., *Sol-Gel Science, the Physics and Chemistry of Sol-Gel Processing*, Academic Press, Boston, 1990.
4. Bonekamp, B.C., in *Fundamentals of Inorganic Membrane Science and Technology*, A.J. Burgraff and L. Cot, Eds., Membrane Science and Technology Series 4, Elsevier, Amsterdam, 1996.
5. Siskens, C.A.M., in *Fundamentals of Inorganic Membrane Science and Technology*, A.J. Burgraff and L. Cot, Eds., Membrane Science and Technology Series 4, Elsevier, Amsterdam, 1996.
6. Bishop, B.A., Higgins, R.J., Abrams, R.F., and Goldsmith, R.L., in *Proceedings of the Third International Conference on Inorganic Membranes*, Y.H. Ma, Ed., Worcester Polytechnic Institute, Worcester, MA, 1994, pp. 355–364.
7. Cot, L., Ayral, A., Durand, J., Guizard, C., Hovnanian, N., Julbe A., and Larbot, A., *Solid State Sci.*, 3, 313, 2000.
8. Larbot, A., Alami-Younssi, S., Persin, M., Sarrazin, J., and Cot, L., *J. Membr. Sci.*, 97, 167, 1994.
9. Vacassy, R., Guizard, C., Thoraval, V., and Cot, L., *J. Membr. Sci.*, 132, 109, 1998.
10. Benfer, S., Popp, U., Richter, H., Siewert, C., and Tomandl, G., *Sep. Purif. Technol.*, 22, 231, 2001.

11. Blanc, P., Larbot, A., Palmeri, J., Lopez, M., and Cot, L., *J. Membr. Sci.*, 149, 151, 1998.
12. Van Gestel, T.V., Vandecasteele, C., Buekenhoudt, A., Dotremont, C., Luyten, J., Leysen, R., van der Bruggen, B., and Maes, G., *J. Membr. Sci.*, 207, 73, 2003.
13. Parks, G.A., *Chem. Rev.*, 65, 177, 1965.
14. James, R.O., in *Advances in Ceramics*, vol. 21, *Ceramic Powder Science*, American Ceramic Society, Colombus, OH, 1987, pp. 349–410.
15. Schwarz, J.A., Driscoll, C.T., and Bhanot, A.K., *J. Colloid Interface Sci.*, 97, 55, 1984.
16. Guizard, C.G., Ayral, A., and Julbe, A., in *Proceedings of the 3rd Nanofiltration and Applications Workshop*, Lappeenranta, Finland, June 26–28, 2001.
17. Guizard, C., Ayral, A., and Julbe, A., *Desalination*, 147, 275, 2002.
18. Clint, J.H., *Curr. Opin. Colloid Interface Sci.*, 6, 28, 2001.
19. Lide, D.R., Ed., *Handbook of Chemistry and Physics*, 78th ed., CRC Press, Boca Raton, FL, 1997.
20. Fujiwara, M., Kodoi, D., Duan, W., and Tanimoto, Y., *J. Phys. Chem. B*, 105, 3343, 2001.
21. Tonneau, J., *Tables de Chimie*, 2nd ed., De Boeck Université, Bruxelles, 2000.
22. Reich, S., and Cabasso, I., *Nature*, 338, 330, 1989.
23. Yoon, J.B., Jang, E.S., Ayral, A., Cot, L., and Choy, J.H., *Bull. Korean Chem. Soc.*, 22, 1149, 2001.
24. Yoon, J.B., Jang, E.S., Ayral, A., Cot, L., and Choy, J.H., *Bull. Korean Chem. Soc.*, 22, 1111, 2001.
25. Gwak, J., Ayral, A., Rouessac, V., Cot, L., Grenier, J.C., Jang, E.S., and Choy, J.H., *Mater. Chem. Phys.*, 84, 348, 2004.
26. Gwak, J., Synthesis and characterization of porous inorganic membranes exhibiting specific magnetic properties, Ph.D. dissertation, University Montpellier II, Montpellier, France, 2003.
27. Gwak, J., Ayral, A., Rouessac, V., Cot, L., Grenier, J.C., and Choy, J.H., *Microporous Mesoporous Mater.*, 63, 177, 2003.
28. Gryaznov, V.M., *Platinum Met. Rev.*, 30, 68, 1986.
29. Gryaznov, V.M., *Z. Phys. Chem. NF*, 147, 123, 1986.
30. Gryaznov, V.M., et al., *Russ. J. Phys. Chem.*, 47, 1517, 1973.
31. Bouwmeester, H.J.M., and Burggraaf, A.J., in *Fundamentals of Inorganic Membrane Science and Technology*, A.J. Burgraff and L. Cot, Eds., Membrane Science and Technology Series 4, Elsevier, Amsterdam, 1996, Chapter 10.
32. Balachandran, U., Dusek, J.T., Sweeney, S.M., Poeppel, R.B., Mieville, R.L., Maiya, P.S., Kleefisch, M., Pei, S., Kobylinski, T.P., Udovich, C.A., and Bose, A.C., *Bull. Am. Ceram. Soc.*, 74, 71, 1995.
33. Teraoka, Y., Nobunaga, T., Okamoto, K., Miura, N., and Yamazoe, N., *Solid State Ionics*, 48, 207, 1991.
34. Zhang, G.B., and Smith, D.M., *Solid State Ionics*, 82, 162, 1995.
35. Kendall, K.R., Navas, C., Thomas, J.K., and zur Loye, H., *Solid State Ionics*, 82, 215, 1995.
36. Wissman, S., and Becker, K.D., *Solid State Ionics*, 85, 279, 1996.
37. Boivin, J.C., Pirovano, C., Nowogrocki, G., Mairesse, G., Labrune, P., and Lagrange, G., *Solid State Ionics*, 113, 639, 1998.
38. Guizard, C., Levy, C., Dalmazio, L., and Julbe, A., *Mater. Res. Soc. Symp. Proc.*, 752, 131, 2003.

39. Julbe, A., Farrusseng, D., and Guizard, C., *J. Membr. Sci.*, 181, 3, 2001.
40. Dalmon, J.A., in *Handbook of Heterogeneous Catalysis*, G. Ertl, H. Knözinger, and J. Weitkamp, Eds., VCH Publishers, Weinheim, Germany, 1997, Chapter 9.3.
41. Hsieh, H.P., in *Inorganic Membranes for Separation and Reaction*, Membrane Science and Technology Series 3, Elsevier, Amsterdam, 1996.
42. Pina, M.P., Menendez, M., and Santamaria, J., *Appl. Catal. B Environ.*, 11, L19, 1996.
43. Irusta, S., Pina, M.P., Menendez, M., and Santamaria, J., *Catal. Lett.*, 54, 69, 1998.
44. Maier, W.F., Lange, C., Tilgner, I., and Tesche, B., in *Proceedings of the 3rd International Conference on Catalysis in Membrane Reactors*, P.E. Hojlund, Nielsen Haldor, and A.S. Topsoe, Eds., The Danish Society of Chemical Engineers, Copenhagen, 1998, p. 22.
45. Lange, C., Storck, S., Tesche, B., and Maier, W.F., *J. Catal.*, 175, 280, 1998.
46. Harold, M.P., Lee, C., Burggraaf, A.J., Keizer, K., Zaspalis, V.T., and De Lange, R.S.A., *MRS Bull.*, 9, 44, 1994.
47. Sloot, H.J., Versteeg, G.F., and van Swaaij, W.P.M., *Chem. Eng. Sci.*, 45, 2415, 1990.
48. Veldsink, J.W., van Damme, R.M.J., Versteeg, G.F., and Van Swaaij, W.P.M., *Chem. Eng. Sci.*, 47, 2939, 1992.
49. Veldsink, J.W., Versteeg, G.F., and van Swaaij, W.P.M., *Ind. Eng. Chem. Res.*, 34, 2833, 1995.
50. Zaman, J., and Chakma, A., *J. Membr. Sci.*, 92, 1, 1994.
51. Zaspalis, V.T., Van Praag, W., Keizer, K., Van Ommen, J.G., Ross, J.R.H., and Burggraaf, A.J., *Appl. Catal.*, 74, 205, 1991.
52. Peureux, J., Torres, M., Mozzanega, H., Giroir-Fendler, A., and Dalmon, J.A., *Catal. Today*, 25, 409, 1995.
53. Terry, P.A., Anderson, M., and Tejedor, I., *J. Porous Mater.*, 6, 267, 1999.
54. Zaspalis, V.T., Van Praag, W., Keizer, K., Van Ommen, J.G., Ross, J.R.H., and Burggraaf, A.J., *Appl. Catal.*, 74, 249, 1991.
55. Capanelli, G., Carosini, E., Monticelli, O., Cavani, F., and Trifiro, F., *Catal. Lett.*, 39, 241, 1996.
56. Alfonso, M.J., Julbe, A., Farrusseng, D., Menendez, M., and Santamaria, J., *Chem. Eng. Sci.*, 54, 1265, 1999.
57. Abe, F., European Patent application no. 228,885, 1987.
58. Furneaux, R.C., Davidson, A.P., and Ball, M.D., European Patent application no. 244, 1987.
59. Binkered, C.R., Ma, Y.H., Moser, W.R., and Dixon, A.G., in *Proceedings of the ICIM4*, D.E. Fain, Ed., Gatlinburg, TN, 1996, p. 441.
60. Zhao, H.B., Draelants, D.H., and Baron, G.V., *Catal. Today*, 56, 229, 2000.
61. Suzuki, H., U.S. Patent no. 4,699,892, 1987.
62. Haag, W.O., and Tsikoyiannis, J.G., U.S. Patent no. 5,019,263, 1991.
63. Julbe, A., Farrusseng, D., Guizard, C., Jalibert, J.C., and Mirodatos, C., *Catal. Today*, 56, 199, 2000.
64. Seok, D.R., and Hwang, S.T., *Stud. Surf. Sci. Catal.*, 54, 248, 1990.
65. Borges, H., Giroir-Fendler, A., Mirodatos, C., Chanaud, P., and Julbe, A., *Catal. Today*, 25, 377, 1995.
66. Pârvulescu, V., Pârvulescu, V., Popescu, G., Julbe, A., Guizard, C., and Cot, L., *Catal. Today*, 25, 385, 1995.

67. Julbe, A., Albaric, L., Hovnanian, N., Guizard, C., Jalibert, J.C., Pantazidis A., and Mirodatos, C., in *Proceedings of the ICIM5*, S. Nakao, Ed., University of Tokyo, Nagoya, Japan, 1998.

68. Ramos, R., Alfonso, M.J., Menendez, M., and Santamaria, J., in *Proceedings of the ICIM5*, S. Nakao, Ed., Nagoya, Japan, 1998.

69. Kaneko, M., and Okura, I., *Photocatalysis*, University of Tokyo, Kodansha, Tokyo, 2002.

70. Fujishima, A., Hashimoto, K., and Watanabe, T., in *TiO$_2$ Photocatalysis, Fundamentals and Applications*, BKC Inc., Tokyo, 2001.

71. Tang, J., Zou, Z., Yin, J., and Ye, J., *Chem. Phys. Lett.*, 383, 175, 2003.

72. Molinari, R., Palmisano, L., Drioli, E., and Schiavello, M., *J. Membr. Sci.*, 206, 399, 2002.

73. Tsuru, T., Toyosada, T., Yoshioka, T., and Asaeda, M., *J. Chem. Eng. Jpn.*, 34, 844, 2001.

74. Tsuru, T., Kan-no, T., Yoshioka, T., and Asaeda, M., *Catal. Today*, 82, 41, 2003.

75. Bosc, F., Synthesis and characterization of anatase-based mesostructured thin layers and membranes with photocatalytic properties, Ph.D. dissertation, University Montpellier II, Montpellier, France, 2004.

76. Bosc, F., Ayral, A., and Guizard, C., in *Proceedings of the ICIM8*, J. Lin, Ed., Adams Press, Cincinnati, OH, 2004.

77. Bosc, F., Ayral, A., and Guizard, C., *J. Membr. Sci.*, submitted.

78. Gibson, J.B., III, and Harrison, D.P., *Ind. Eng. Chem. Process Des. Dev.*, 19, 231, 1980.

79. Carnell, P.J.H., in *Catalyst Handbook*, 2nd ed., M.V. Twigg, Ed., Wolfe Publishing Ltd., London, 1989, p. 191.

80. Baird, T., Campbell, K.C., Holliman, P.J., Hoyle, R., Stirling, D., and Williams, B.P., *J. Chem. Soc. Faraday Trans.*, 91, 3219, 1995.

81. Goswamee, R.L., Ayral, A., Bhattacharyya, K.G., and Dutta, D.K., *Mater. Lett.*, 46, 105, 2000.

82. Goswamee, R., Bosc, F., Cot, D., El Mansouri, A., Lopez, M., Morato, F., and Ayral, A., *J. Sol-Gel Sci. Technol.*, 29, 97, 2004.

83. Ayral, A., Julbe, A., Guizard, C., and Cot, L., *J. Korean Chem. Soc.*, 41, 566, 1997.

84. Klotz, M., Ayral, A., Guizard, C., and Cot, L., *Bull. Korean Chem. Soc.*, 20, 879, 1999.

85. Occelli, M.L. and Robson, H.E., Eds., *Synthesis of Microporous Materials*, vol. 1, *Molecular Sieves*, Van Nostrand Reinhold, New York, 1992.

86. Ozin, G.A., Kuperman, A., and Stein, A., *Angew. Chem. Int. Ed. Engl.*, 28, 359, 1989.

87. Nakao, S., Ed., *Proceedings of the International Workshop on Zeolitic Membranes and Films*, Japan Association of Zeolite, Gifu, Japan, 1998.

88. Coronas, J. and Santamaria, J., *Sep. Purif. Meth.*, 28, 127, 1998.

89. Tavolaro, A. and Drioli, E., *Adv. Mater.*, 11, 975, 1999.

90. Giroir-Fendler, A., Julbe, A., Ramsay, J.D.F., and Dalmon, J.A., French Patent (CNRS) no. 94-05562-PCT, Int. Appl. FR 94-0429.

91. Tavolaro, A., Basile, A., Julbe, A., and Drioli, E., in *Proceedings of the Ravello Conference on New Frontiers for Catalytic Membrane Reactors and Other Membrane Systems*, Ravello, Italy, 1999, p. 274.

92. Julbe, A. and Dalmon, J.A., in *Proceedings of the International Workshop on Zeolitic Membranes and Films*, S. Nakao, Ed., Japan Association of Zeolite, Gifu, Japan, 1998.

93. Bellusi, G. and Rigutto, M.S., in *Advanced Zeolite Science and Application, Studies in Surface Science and Catalysis*, vol. 85, J.C. Jansen, M. Stöcker, H.G. Karge, and J. Weitkamp, Eds., Elsevier, Amsterdam, 1994.

94. Julbe, A., Farrusseng, D., Volle, G., Sanchez, J., and Guizard, C., in *Proceedings of the ICIM 98*, S. Nakao, Ed., University of Tokyo, Nagoya, Japan, 1998.

95. Julbe, A., Farrusseng, D., Guizard, C., Jalibert, J.C., and Mirodatos, C., *Catal. Today*, 56, 199, 2000.

96. Julbe, A., Motuzas, J., Cazevielle, F., Volle, G., and Guizard, C., *Separ. Purif. Technol.*, 32, 139, 2003.

97. Wärnheim, T., Jönsson, A., and Sjöberg, M., *Progr. Colloid Polym. Sci.*, 82, 271, 1990.

98. Kresge, C.T., Leonowicz, M.E., Roth, W.J., Vartuli, J.C., and Beck, J.S., *Nature*, 359, 710, 1992.

99. Beck, J.S., Vartuli, J.C., Roth, W.J., Leonowicz, M.E., Kresge, C.T., Schmitt, K.D., Chu, C.T.-W., Olson, D.H., Sheppard, E.W., McCullen, S.B., Higgins, J.B., and Schlenker, J.L., *J. Am. Chem. Soc.*, 114, 10,834, 1992.

100. Ogawa, M., *J. Am. Chem. Soc.*, 116, 7941, 1994.

101. Dabadie, T., Ayral, A., Guizard, C., Cot, L., Robert, J.C., and Poncelet, O., *Mater. Res. Soc. Symp. Proc.*, 346, 849, 1994.

102. Dabadie, T., Ayral, A., Guizard, C., Cot, L., Robert, J.C., and Poncelet, O., in *Proceedings of the Third International Conference on Inorganic Membranes*, Y.H. Ma, Ed., Worcester Polytechnic Institute, Worcester, MA, 1994.

103. Ayral, A., Balzer, C., Dabadie, T., Guizard, C., and Julbe, A., *Catal. Today*, 25, 219, 1995.

104. Ogawa, M., *Chem. Commun.*, 10, 1149, 1996.

105. Dabadie, T., Ayral, A., Guizard, C., Cot, L., and Lacan, P., *J. Mater. Chem.*, 6, 1789, 1996.

106. Bruinsma, P.J., Hess, N.J., Bontha, J.R., Liu, J., and Baskaran, S., *Mater. Res. Soc. Symp. Proc.*, 443, 105, 1997.

107. Lu, Y., Ganguli, R., Drewien, C.A., Anderson, M.T., Brinker, C.J., Gong, W., Guo, Y., Soyez, H., Dunn, B., Huang, M.H., and Zinks, J.I., *Nature*, 389, 364, 1997.

108. Klotz, M., Ayral, A., Guizard, C., and Cot, L., *J. Mater. Chem.*, 10, 663, 2000.

109. Tanev, P.T., and Pinnavaia, T.J., *Science*, 267, 865, 1995.

110. Bagshaw, S.A., Prouzet, E., and Pinnavaia, T.J., *Science*, 269, 1242, 1995.

111. Templin, M., Franck, A., Du Chesne, A., H. Leist, Y. Zhang, R. Ulrich, V. Schädler and U. Wiesner, *Science*, 278, 1795, 1997.

112. Zhao, D., Yang, P., Melosh, N., Feng, J., Chmelka, B.F., and Stucky, G.D., *Adv. Mater.*, 10, 1380, 1998.

113. Zhao, D., Huo, Q., Feng, J., Chmelka, B.F., and Stucky, G.D., *J. Am. Chem. Soc.*, 120, 6024, 1998.

114. Sayari, A. and Liu, P., *Microporous Mater.*, 12, 149, 1997.

115. Yang, P., Zhao, D., Margolese, D.I., Chmelka, B.F., and Stucky, G.D., *Chem. Mater.*, 11, 281, 1999.

116. Brinker, C.J., Lu, Y., Sellinger, A., and Fan, A., *Adv. Mater.*, 11, 579, 1999.

117. Klotz, M., Idrissi Kandri, N., Ayral, A., and Guizard, C., *Mater. Res. Soc. Symp. Proc.*, 628, CC7.4.1, 2000.

118. Idrissi Kandri, N., Ayral, A., Klotz, M., Albouy, P.A., El Mansouri, A., van der Lee, A., and Guizard, C., *Mater. Lett.*, 50, 57, 2001.

119. Ayral, A., and Guizard, C., *Mater. Trans.*, 42, 1641, 2001.

120. Bosc, F., Ayral, A., Albouy, P.A., and Guizard, C., *Chem. Mater.*, 15, 2463, 2003.
121. Bosc, F., Ayral, A., Albouy, P.A., Datas, L., and Guizard, C., *Chem. Mater.*, 16, 2208, 2004.
122. Klotz, M., Albouy, P.A., Ayral, A., Menager, C., Grosso, D., van der Lee, A., Cabuil, V., Babonneau, F., and Guizard, C., *Chem. Mater.*, 12, 1721, 2000.
123. Klotz, M., Ayral, A., Guizard, C., and Cot, L., *Separ. Purif. Technol.*, 25, 71, 2001.
124. Klotz, M., Besson, S., Ricolleau, C., Bosc, F., and Ayral, A., *Mater. Res. Soc. Symp. Proc.*, 752, 123, 2003.
125. Ayral, A., Guizard, C., and Cot, L., *J. Mater. Sci. Lett.*, 13, 1538, 1994.
126. DeFriend, K.A., and Barron, A.R., *J. Membr. Sci.*, 212, 29, 2003.
127. Guliants, V.V., Carreon, M.A., and Lin, Y.S., *J. Membr. Sci.*, 235, 53, 2004.
128. Pintault, B., and Ayral, A., *J. Eur. Ceram. Soc.*, in preparation.
129. Farruseng, D., Julbe, A., and Guizard, C., *J. Am. Chem. Soc.*, 112, 12592, 2000.
130. Broussous, L., Schmitz, P., Boisson, H., Prouzet, E., and Larbot, A., *Chem. Eng. Sci.*, 55, 5049, 2000.
131. Broussous, L., Schmitz, P., Prouzet, E., Becque, L., and Larbot, A., *Separ. Purif. Technol.*, 25, 333, 2001.
132. Okubo, T., Haruta, K., Kusakabe, K., Morooka, S., Anzai, H., and Akiyama, S., *Ind. Eng. Chem. Res.*, 30, 614, 1991.
133. Smid, J., Avci, C.G., Gilnay, V., Terpstra, R.A., and Van Eijk, J.P.G.M., *J. Membr. Sci.*, 112, 85, 1996.
134. Brinkman, H.W., van Eijik, J.P.G.M., Meinema, H.A., and Terpstra, A., *Bull. Am. Ceram. Soc.*, 78, 51, 1999.
135. Tudyka, S., in *Advanced Membrane Technology Conference*, Barga, Italy, October 14–19, 2001.
136. Liu, S., Tan, X., Li, K., and Hughes, R., *J. Membr. Sci.*, 193, 249, 2001.
137. Pan, X.L., Stroh, N., Brunner, H., Xiong, G.X., and Sheng, S.S., *J. Membr. Sci.*, 226, 111, 2003.
138. Pan, X.L., Stroh, N., Brunner, H., Xiong, G.X., and Sheng, S.S., *Separation & Purification Technology*, 32, 265, 2003.
139. Liu, S., and Li, K., *J. Membr. Sci.*, 218, 269, 2003.
140. Xu, X., Yang, W., Liu, J., Lin, L., Stroh, N., and Brunner, H., *J. Membr. Sci.*, 229, 81, 2004.

# 26 Ceramic Materials for Lithium-Ion Battery Applications

*Jeffrey P. Maranchi, Oleg I. Velikokhatnyi,
Moni K. Datta, Il-Seok Kim, and
Prashant N. Kumta*

## CONTENTS

Reduction in the size of electronic components (due to improved integrated circuit technology and fabrication processes) has led to the miniaturization of electronic devices and related peripherals. The consumer market has readily embraced the miniaturization of products and manufacturers have continuously come up with new products and marketing concepts. All of the ubiquitous miniaturized devices are in need of efficient, lightweight, and rechargeable power sources. One power source that meets all of these requirements is the bulk lithium-ion battery. Continuous improvements in the areas of battery performance, weight and size reduction, and cost reduction are desired. Incremental improvements can be made through improved battery package design and materials selection. However, substantial breakthrough improvements will only be achieved through better design and engineering of the essential material components of the lithium-ion battery, namely, the cathode, electrolyte, and anode. The present review will outline and describe the salient features and state of the art with regard to the synthesis, processing, and characterization of ceramic electrode and electrolyte materials used in lithium-ion batteries. Accordingly, the review is divided into three sections: cathode materials, electrolytes, and anode materials.

## I. CATHODE MATERIALS

At present there are three intercalation materials that are used commercially as positive electrode materials for lithium-ion rechargeable batteries: $LiCoO_2$, $LiNiO_2$, and $LiMn_2O_4$. Of these, $LiCoO_2$ and $LiNiO_2$ have the layered $\alpha$-$NaFeO_2$ rock salt-type crystal structure exhibiting the $R\bar{3}m$ space group,[1-3] where lithium and transition metal cations reside on the octahedral (3a) and (3b) sites, respectively, and oxygen anions occupy (6c) sites. $LiMn_2O_4$ adopts the spinel structure ($MgAl_2O_4$ type) exhibiting the $Fd3m$ space group[4] and is typically referred to as the framework structure because it has three-dimensional interstitial spaces for deintercalation and intercalation of lithium ions. All of the materials exhibit characteristic intercalation potentials of about 4 V with respect to lithium anode and are reasonably stable when exposed to air and moisture.[5-10] Typical crystal structures of cathode materials are shown in Figure 26.1.

## A.  $LiCoO_2$

The first investigation of $Li_xCoO_2$ was carried out by Mizushima et al. in 1980,[11] where the material was suggested as a possible positive electrode for lithium-ion rechargeable batteries. In 1991 Sony Corporation commercialized the first lithium-ion battery in which lithium cobalt oxide was used as the positive electrode and graphite (carbon) as the negative electrode.[12] Since then, $LiCoO_2$ has been the most widely used cathode material in commercial lithium-ion batteries and retains its industrial importance as a cathode material.

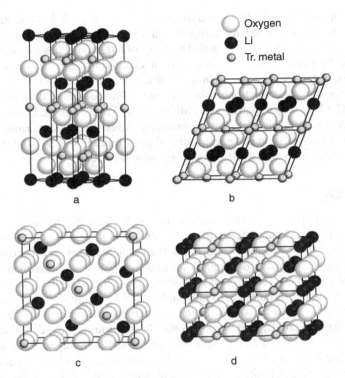

**FIGURE 26.1** Typical representative cathode crystal structures: (a) rhombohedral $R\bar{3}m$; (b) monoclinic $C2/m$; (c) spinel $Fd3m$; (d) orthorhombic $Pmmn$.

The essential advantage of $LiCoO_2$ is the relative ease and simplicity of preparation. $LiCoO_2$ can be prepared conveniently using both solid state and chemical approaches (see, for example, Mizushima et al.,[11] Reimers and Dahn,[13] and Kumta et al.[14]). Essentially any precursor of lithium and cobalt can be mixed and heat-treated to generate the oxide. For example, Mizushima et al.[11] prepared the oxide by heating a pelletized mixture of lithium carbonate and cobalt carbonate in air at 900°C for 20 h followed by two further heat treatments. The ease of fabrication allows the oxide to be the most popular among the possible cathode materials.

Numerous studies in the $Li_xCoO_2$ system have been conducted thus far.[9,13,15,16] The $Li_xCoO_2$ system demonstrates excellent rechargeability at room temperature within the concentration interval $1 > x > 0.5$. This limit restricts the specific capacity of the material to the range of 137 to 140 mAh/g (although the theoretical capacity of $LiCoO_2$ is 273 mAh/g). There are different reasons for the capacity to be restricted to only $x = 0.5$. This is because of an undesirable phase transformation to the monoclinic phase known to occur near $x = 0.5$ in a small concentration interval of $x$, depending on the temperature, after which the matrix restores the original $R\bar{3}m$ symmetry. In particular, at room temperature the monoclinic phase exists approximately in the range $0.48 < x < 0.52$.[13] Furthermore, with the chemical instability of $Li_xCoO_2$ at deep charge levels of $x < 0.5$, the oxide begins to lose oxygen ions from its lattice because of a chemical reaction with the electrolyte.[17,18] It should also be noted that $Li_xCoO_2$ demonstrates excellent cyclability on the order of tens of thousands of charge-discharge cycles without a noticeable reduction in capacity in the lithium range of $1 > x > 0.5$.

Despite the good electrochemical characteristics and simplicity of preparation, $Li_xCoO_2$ is very expensive and highly toxic. These drawbacks have stimulated wide research activity targeted toward the development of alternate cathode materials for lithium-ion batteries. In this regard, lithium nickel oxide ($Li_xNiO_2$) could be more attractive due to its lower cost and toxicity.

## B. LiNiO$_2$

As mentioned above, $LiNiO_2$ is isostructural to $LiCoO_2$. Synthesis of the oxide is, however, more complicated than that of $LiCoO_2$. The Li-Ni-O system is characterized by a deviation from the normal stoichiometry and is represented by $Li_{1-y}Ni_{1+y}O_2$, consisting of the presence of additional nickels ions on the lithium sites, and vice versa. This property makes the synthesis of the stoichiometric oxide with all the lithium sites completely filled by lithium very difficult and small deviations from the stoichiometry are very much apparent. For $y > 0$, the structure has a partially distorted cationic distribution at the lithium sites and lithium/nickel mixing of cations appears within the lithium layers, leading to the gradual deterioration of the electrochemical performance of the material (low capacity and high polarization). Therefore obtaining material with a very accurate composition (near stoichiometry) is quite a difficult technical task. Traditionally this material, as well as $LiCoO_2$ and other compounds such as $LiCo_xNi_{1-x}O_2$, have been synthesized using solid state processes employing different precursors and atmospheres. Various chemical and solid state methods of synthesizing $LiNiO_2$ have been described in the literature.[19–21]

Most solid-state and chemical approaches result in antisite defects with misposition of nickel and lithium sites, causing significant lowering of capacity due to the nickel on lithium sites resulting in a hindrance to lithium diffusion. Novel particulate sol-gel approaches developed by Chang et al.[22,23] have demonstrated the possibility of creating a $LiNiO_2$ gel that can be converted to stoichiometric oxide following a 2-step heat treatment in air and oxygen. The resultant oxide has less than 1% antisite defects, exhibiting a capacity of more than 200 mAh/g in the first charge. Figure 26.2a shows the capacity versus cycle plot for the first 30 cycles. A high capacity of 205 mAh/g is obtained in the first cycle when the material is charged and discharged using a current density of 500 $\mu$A/cm$^2$. However, it should be noted that although the capacity has been increased, the capacity fade is still high (0.71% per cycle). The reason for this fade arises from the decomposition of the material when charged to high voltages. A plot of the first cycle voltage versus lithium content is shown in Figure 26.2b. Chang et al.[24,25] have also demonstrated that the stoichiometry has a far more dominating effect on the oxide. Furthermore, Chang et al.[26–28] have demonstrated that doping the

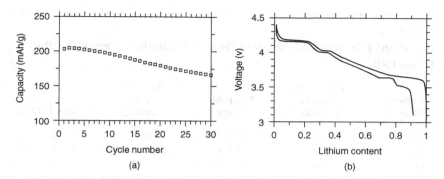

**FIGURE 26.2** (a) The capacity versus cycle number showing the first 30 cycles for $LiNiO_2$ synthesized using the particulate sol-gel process, and (b) the first cycle voltage vs. lithium content for the low defect concentration of $LiNiO_2$.

oxide with a small amount of divalent cations such as $Mg^{2+}$ has a significant influence on the voltage durability and cyclability of the oxide.

$Li_xNiO_2$ has a higher rechargeable capacity than $Li_xCoO_2$, since the amount of lithium that can be removed or intercalated during cycling is around 0.55 (instead of 0.5 for $LiCoO_2$), allowing the specific capacity to be more than 150 mAh/g with appropriate cyclability. Depending upon the external conditions of the charging process (such as the charge end voltage, which must be less than 4.2 V), it is possible to obtain rechargeability with $x < 0.3$ and reversible capacity of up to 200 mAh/g.[21] However, maintenance of the charge end voltage at the appropriate level is very difficult for lithium battery applications, and attaining a capacity of 200 mAh/g for the material is almost impractical.

Improvement in the chemical stability (resistance to loss of oxygen at deep charge) of $LiNiO_2$ oxide occurs after a substitution of nickel by a small amount of cobalt. The oxide $LiNi_{0.85}Co_{0.15}O_2$ exhibits a rechargeable capacity near 180 mAh/g with removal/intercalation of 0.65 lithium ion per formula unit.[17,18,29] Despite the fact that the material demonstrates attractive electrochemical properties, it contains the expensive and toxic element cobalt, which cannot be considered an ideal cathode material for rechargeable batteries, especially for electric vehicles, since the cost of such a battery would be very high, rendering it unattractive.

## C. $LiMn_2O_4$

The third most popular cathode material for lithium-ion batteries is $LiMn_2O_4$, with a framework spinel crystal structure (*Fd3m* symmetry) where the oxygen ions form a cubic close-packed array and occupy the (32e) sites. Lithium ions reside on the tetrahedral (8a) sites and manganese ions on the octahedral (16d) sites. The main advantages of $LiMn_2O_4$ with respect to $LiCoO_2$ and $LiNiO_2$ are that the material is nontoxic, comes from an abundant materials source,[4,30] and provides a high oxidation potential of 3 V to 4 V with respect to pure lithium. This material can be synthesized, for example, by conventional solid state reaction of mixed $Li_2CO_3$ and $MnCO_3$ in ambient atmospheres at elevated temperature

**TABLE 26.1**
**Important Properties of the Most Popular Cathode Materials for the Li-Ion Cells**

| Cathode Material | Theoretical Capacity (mAh/g) | Reversible Range | Practical Capacity (mAh/g) | Voltage (V) Versus Lithium |
|---|---|---|---|---|
| $LiCoO_2$ | 274 | 0–0.5 | 130–140 | 3.9 |
| $LiNiO_2$ | 275 | 0–0.7 | 150–190 | 3.7 |
| $LiMn_2O_4$ | 148 | 0–1.0 | 120–148 | 4.0 |

near 750°C.[31] Aqueous and nonaqueous wet chemical approaches have also been used for synthesizing the spinel compound.[32–34]

In principle, $Li_xMn_2O_4$ enables the insertion/extraction of lithium ions in the range of $0 < x < 2$, with an intercalation potential near 4 V within the $0 < x < 1$ interval of lithium concentration, and near 3 V within $1 < x < 2$.[30,35] In this interval, lithium ions move from tetrahedral (8a) sites to octahedral (16c) sites to yield the composition $Li_2Mn_2O_4$.[30] Obviously the insertion of lithium ions effectively decreases the average valency of manganese ions from 3.5 for $LiMn_2O_4$ to 3.0 for $Li_2Mn_2O_4$. This leads to a pronounced cooperative Jahn-Teller effect in which the cubic spinel crystal becomes distorted tetragonally with a c/a $\approx 1.16$ and the volume of the unit cell increases by 6.5%. For intermediate values of $x$ between 1 and 2 the material consists of two different phases—cubic in bulk and tetragonal at the surface. Due to the high c/a ratio, the integrity of the material is lost and its electrochemical properties are drastically reduced.[36] As a result, the spinel $Li_xMn_2O_4$ can be used as a cathode material only in the range of $0 < x < 1$, with an intercalation potential near 4 V and a rechargeable capacity not exceeding 148 mAh/g being practically restricted to only 120 to 125 mAh/g. Even with this limited capacity, significant capacity fade is observed at elevated temperatures in the range of 50 to 70°C. Table 26.1 summarizes the most important properties of the above mentioned popular cathode materials.

## D. LiMnO₂

### 1. Orthorhombic

$LiMnO_2$ is very attractive as a cathode material because it combines a high theoretical capacity (285 mAh/g) with the nontoxicity and low cost of manganese. However, this oxide is not an isostructural analog of $LiCoO_2$ and $LiNiO_2$. The thermodynamic equilibrium state of the oxide is the orthorhombic structure with a space group, *Pmmn*.[37] This structure is characterized as a modified rock salt type with a distorted cubic close-packed oxygen array. Lithium and manganese ions form alternate crimped zig-zag layers with oxygen ions between them.

Depending on the method used to synthesize orthorhombic $LiMnO_2$, the material demonstrates different electrochemical properties. For example, the material obtained from $Li_2O$ and $Mn_2O_3$ at 750°C under argon atmosphere shows unimpressive electrochemical activity. But the product obtained at 300 to 450°C by reacting γ-MnOOH with LiOH delivers a specific capacity of 190 mAh/g between 2 and 4.5 V vs. metallic lithium as described by Ohzuku et al.[38] The initial electrochemical extraction of lithium from orthorhombic $LiMnO_2$ prepared at low temperature can reach a charge capacity of more than 200 mAh/g.[38,39] However, all of the materials are unstable to lithium-ion deintercalation and demonstrate a limited cyclability due to irreversible phase transformation to a spinel-type structure.[40]

## 2. Monoclinic

This is another crystallographic form of $LiMnO_2$. It has a monoclinic layered structure with $C2/m$ group symmetry. This symmetry results from a rhombohedral $R\overline{3}m$ symmetry distorted by a Jahn-Teller cooperative atomic displacement induced by $Mn^{3+}$ active ions. This structure is not thermodynamically stable and can only be synthesized by ion exchange of sodium by lithium from the monoclinic form of α-$NaMnO_2$ via a metastable chimie douce reaction.[41,42] Although this layered structure provides easy diffusion of lithium during intercalation and most of the lithium can be removed from the host crystal during charge (up to 0.95 lithium ions per formula unit, which corresponds to 270 mAh/g), the reinsertion of lithium during discharge, however, is not a reversible process. As more than half of the lithium is extracted from $LiMnO_2$, there is displacement of the manganese ions into the lithium layers, which leads to the formation of regions exhibiting the spinel structure ($LiMn_2O_4$).[43,44] This irreversible transformation makes the oxide unstable and hence questionable for use as a cathode material in rechargeable lithium-ion batteries.

## 3. Layered O2 Structure

The new layered $Li_{2/3}(Li_xMn_{1-x})O_2$ ($x$ = 1/6, 1/18) with O2-type crystal structure has been synthesized and reported by Paulsen et al.[45] This structure was obtained by the ion exchange of $Na^+$ with $Li^+$ in the layered sodium manganese bronzes having a P2 crystal structure. The O2-type structure completely differs from that attributed to monoclinic $C2/m$ or spinel $Fd3m$ symmetry. As a result, during cycling these materials do not convert to spinel and demonstrate a reversible charge capacity near 150 mAh/g. However, fabrication of the materials involves a complicated, multistep and expensive process that also makes these oxides inconvenient for use as cathode materials.

Therefore there is a need to identify dopants that can stabilize the layered structure of $LiMnO_2$, while preventing its transformation to the spinel phase during insertion and removal of lithium. Considerable experimental and theoretical studies have been conducted to examine the structural stability of doped $LiMnO_2$, and consequently several dopants have been explored.

## E. LiMnO₂ WITH DOPANTS

Several attempts have been made to stabilize the layered $LiMnO_2$ phase by substitution of manganese ions with small amounts of chromium[46] and aluminum[47] by conventional solid state reaction and the introduction of cobalt[48] and nickel[49] on the manganese sites via the chimie douce reaction.[48,49] Although these substituted layered compounds resulted in improvements in capacity retention in comparison to pure $LiMnO_2$, they have been found to be thermodynamically unstable with respect to their transformation to the less desired spinel structure.

Lately, numerous studies of manganese oxides doped with substantial amounts of nickel have been carried out.[50-55] The $LiMn_{0.5}Ni_{0.5}O_2$ material was suggested by Ohzuku and Makimura[50] as a mixture of $LiMnO_2$ and $LiNiO_2$ with an oxide ratio of 1:1 synthesized via a metal double-hydroxide route. This new material adopted the layered $R\bar{3}m$ crystal structure and exhibited very good cyclability without any signs of transformation into spinel during charge-discharge cycling, and exhibited a specific capacity of about 150 mAh/g after 30 cycles within a 2.5 to 4.5 V interval.[50,51] Also, the $Li_xMn_{0.5}Ni_{0.5}O_2$ compound was prepared by Johnson et al.,[52] demonstrating possible cyclability within the range $0 < x < 2$ and showing that a rechargeable capacity >500 mAh/g could be obtained in a large voltage interval between 1.0 and 4.6 V. The layered crystal structure of the $Li_2Mn_{0.5}Ni_{0.5}O_2$ oxide is hexagonally close-packed with a $P3m1$ space group, which contains up to 2 lithium ions per formula unit, providing an unusually large specific capacity.

Another layered material, $Li_{2/3}Mn_{2/3}Ni_{1/3}O_2$, with unconventional O2-type structure was synthesized by the solid state reaction followed by the ion exchange procedure.[53,54] Similar to the above mentioned layered O2-type $Li_{2/3}(Li_xMn_{1-x})O_2$, this material does not undergo any transformation into the undesirable spinel during charge-discharge cycling and demonstrates a large specific capacity near 180 mAh/g.

Thus lithiated manganese oxides with a substantial amount of nickel and other elements could play a crucial role in stabilizing the layered crystal structure, thereby improving the electrochemical characteristics.

## F. LITHIUM-VANADIUM OXIDES

The layered structure of $LiVO_2$ is the same as that of $LiCoO_2$ and $LiNiO_2$ and allows lithium ions to be removed or intercalated between oxygen layers. However, when more than 0.3 lithium ions per formula unit are removed from the host structure, the vanadium ions migrate from their own layers into the lithium planes, filling empty sites created due to lithium deinsertion.[56] Such a structural instability destroys the two-dimensional oxygen framework, resulting in poor electrochemical characteristics. The same situation is observed with $LiV_2O_4$ spinel, where the vanadium ions also migrate from their standard positions during lithium insertion-deintercalation.[57] Other vanadium oxides such as $VO_2$,[58] $V_2O_5$,[59] and $V_6O_{13}$[60] provide high capacities, but are not attractive materials for rechargeable lithium-ion cells with carbon anodes. In this regard, the materials $Li_xV_3O_8$ and $Li_xNaV_3O_8$ demonstrate a specific capacity of more than 300 mAh/g,[61,62] and due to chemical lithiation, these materials can contain more than three lithium ions per formula unit, allowing them to be used as cathodes in the rechargeable cells containing carbon as the negative electrode.

## G. Iron-Based Oxides

From the viewpoint of minimal cost, low toxicity, and natural abundance, the compound $LiFeO_2$ with a rock salt layered structure (isostructural with $LiCoO_2$ and $LiNiO_2$) could be an attractive cathode material. This material can be obtained by ion exchange reaction from $\alpha$-$NaFeO_2$,[63] however, previous investigations of this oxide showed unimpressive battery performance[64] because of the metastable nature of the layered structure of $LiFeO_2$. In this structure the iron ions move from their octahedral sites to the lithium layers, thereby worsening the electrochemical characteristics. These considerations motivated Goodenough's group to investigate a series of compounds with framework structures built with large polyanions having inexpensive and less toxic transition metals. The compounds $M_2(XO_4)_3$ (M = Ti, Fe, Nb, V; X = S, P, As, Mo, W) have the NASICON[65] (sodium superionic conductor) three-dimensional framework structure built from corner-sharing $MO_6$ octahedra and $XO_4$ tetrahedral anions, which allows for the high mobility of lithium ions even with smaller transition metal atoms for M.[66–74] The polyanion oxides $M_2(SO_4)_3$,[69] $Li_xM_2(PO_4)_3$,[69] and $Li_3Fe_2(XO_4)_3$[71] are recognized as good materials that are able to accept/donate lithium ions between 2.5 and 4 V vs. metallic lithium anode yielding a specific capacity of about 100 to 110 mAh/g.

The open NASICON structure allows easy diffusion of lithium ions, however, a separation of the $FeO_6$ octahedra by polyanions reduces the electronic conductivity, which is associated with the Fe-O-X-O-Fe bonds, leading to low rate capability. There is also another alternative olivine-type structure in which the $FeO_6$ octahedra share common corners. Different materials exhibiting the ordered olivine structure, for example, $LiMPO_4$ compounds (M = Fe, Mn, Co, Ni)[68] have been synthesized. The electrochemical activity of the materials was investigated by Padhi et al.[68] and only $LiFePO_4$ was identified as a promising material for use in the lithium ion cells with 3.5 V vs. lithium and specific capacity near 100 to 110 mAh/g. For all of the other materials, the electrochemical deintercalation of lithium was unsuccessful.

Since the pioneering work of Padhi et al.,[68] numerous studies dedicated to improving the electrochemical characteristics of the phospho-olivines have been undertaken.[75–80] Among the different phospho-olivine materials studied for use in lithium-ion rechargeable cells, the most promising phosphate is $LiFePO_4$, and recently Chung et al.[81] have shown that it is possible to enhance the electronic conductivity ($\sigma_e$) by doping the oxide. They[81] have demonstrated a $10^8$-fold increment in $\sigma_e$ by doping $LiFePO_4$ with different metal ions such as $Mg^{2+}$, $Al^{3+}$, $Ti^{4+}$, $Zr^{4+}$, $Nb^{5+}$, and $W^{6+}$. Thus iron-phospho-olivine is a serious candidate cathode material for use in the next generation of lithium-ion cells.

Wide experimental investigations of the cathode materials for lithium-ion batteries have stimulated intensive theoretical studies using various *ab-initio* methods, most of which are based on the Density Functional Theory.[82–93] Researchers have used different *ab-initio* approaches dedicated to conducting so-called "numerical experiments" in these materials. The intercalation voltage and structural stability of cathode materials have been the chief characteristics of interest in these theoretical studies.[94–102]

## II. DOPED LITHIUM TITANIUM PHOSPHATE-BASED CERAMIC ELECTROLYTES FOR LITHIUM ION BATTERIES

Solid electrolytes for lithium-ion batteries are expected to offer several advantages over traditional, nonaqueous liquid electrolytes. A solid electrolyte would give a longer shelf life, along with an enhancement in specific energy density. A solid electrolyte may also eliminate the need for a distinct separator material, such as the polypropylene or polyethylene microporous separators commonly used in contemporary liquid electrolyte-based batteries. Solid electrolytes are also desirable over liquid electrolytes in certain specialty applications where bulk lithium-ion batteries as well as thin-film lithium-ion batteries are needed for primary and backup power supplies for systems, devices, and individual integrated circuit chips.

While research on a diverse group of solid state ion conductors has been conducted for many decades, lithium-ion-based solid electrolytes have only recently been studied. For example, $Li_{3-x}-H_xN$, a crystalline material with the highest room temperature lithium-ion conductivity (~$1 \times 10^{-3}$ S·cm$^{-1}$) was reported in 1983 by Lapp et al.[103] For detailed review articles on the wide range of solid state lithium-ion conductors, the reader is directed to two books, *Solid State Electrochemistry*[104] and *Solid State Batteries: Materials Design and Optimization*.[105] Several of the most commonly reported solid state lithium-ion conductors are amorphous LiPON,[106,107] $Li_4SiO_4$ derivatives,[108] $LiI/Al_2O_3$ composites,[109,110] and $Li_2S-P_2S_5$ glasses.[111] To the best of our knowledge, no review article has been published solely focused on LISICON (NASICON derivatives) lithium titanium phosphate-based lithium-ion conductors. Therefore the focus of this article is on the synthesis and lithium-ion conducting properties of lithium titanium phosphate-doped compounds.

Following a period of intense research on NASICON ($Na_3Zr_2Si_2PO_{12}$) three-dimensional sodium-ion conductors, research interests drifted toward identifying similar LISICON-type three-dimensional lithium-ion conductors. A direct analogue compound ($Li_3Zr_2Si_2PO_{12}$) was a stable phase, but exhibited poor ionic conductivity.[112] Researchers also analyzed the ionic conductivity of various

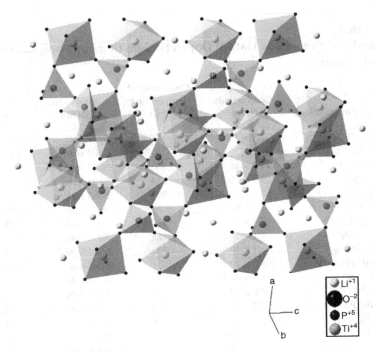

**FIGURE 26.3** Representation of the $R\bar{3}c$ NASICON-type $LiTi_2(PO_4)_3$ crystal structure.

$LiTi_2(PO_4)_3$ and doped $LiTi_2(PO_4)_3$ compounds. A representation of the $R\bar{3}c$ NASICON-type $LiTi_2(PO_4)_3$ crystal structure is shown in Figure 26.3. While the undoped $LiTi_2(PO_4)_3$ material exhibited low ionic conductivity, some of the doped materials displayed very high lithium-ion conductivities at room temperature that were comparable to those of $Li_3N$. Table 26.2 shows the ionic conductivities of various doped lithium titanium phosphate electrolyte materials. Since 1982, research on doped lithium titanium phosphates has been conducted by many groups around the world. Thus doped lithium titanium phosphates are a good example of a class of ceramic materials that have been widely studied for their outstanding lithium-ion conductivity. The remainder of this article will focus on reviewing the electrochemical properties of doped lithium titanium phosphates synthesized by solid state, sol-gel, and hydrothermal methods. A small section at the end of the article is devoted to more recent research on bulk glass and glass-ceramics with similar constituent materials as the crystalline doped lithium titanium phosphates.

## A. SOLID STATE SYNTHESIS AND PROPERTIES

Materials in the $Li_{1+x}Ti_{2-x}M_xP_3O_{12}$ (M = In, Ga, Mg, and Cr) system were synthesized by a solid state reaction using $Li_2CO_3$ or $Li_3PO_4$, $TiO_2$, $NH_4H_2PO_4$, and either $In_2O_3$,[113] $Ga_2O_3$,[114] MgO,[114] or $Cr_2O_3$.[115] The powders were weighed,

**TABLE 26.2**
**Lithium-Ion Conductivity Data of Doped Lithium Titanium Phosphate-Type Ceramic Electrolytes**

| Material (Composition) | Conductivity (S/cm) | Measurement Temperature (°C) | Synthesis Method | Ref. |
|---|---|---|---|---|
| $Li_{1.4}Ti_{1.6}In_{0.4}P_3O_{12}$ | $1.9 \times 10^{-4}$ | 23 | Solid state | 110 |
| $Li_{1.4}Ti_{1.6}In_{0.4}P_3O_{12}$ | $5.5 \times 10^{-2}$ | 300 | Solid state | 110 |
| $Li_{1.6}Ti_{1.4}Ga_{0.6}P_3O_{12}$ | $8.4 \times 10^{-3}$ | 300 | Solid state | 110 |
| $Li_{1.3}Ti_{1.7}Mg_{0.3}P_3O_{12}$ | $5.3 \times 10^{-3}$ | 300 | Solid state | 110 |
| $Li_{1.8}Ti_{1.2}Cr_{0.8}P_3O_{12}$ | $4.8 \times 10^{-2}$ | 300 | Solid state | 110 |
| $LiZr_{0.2}Ti_{1.8}(PO_4)_3$ | $8.5 \times 10^{-8}$ | 23 | Solid state | 113 |
| $LiZr_{0.2}Ti_{1.8}(PO_4)_3$ | $7.4 \times 10^{-3}$ | 300 | Solid state | 113 |
| $LiSc_{0.3}Ti_{1.7}(PO_4)_3$ | $4.1 \times 10^{-4}$ | 23 | Solid state | 113 |
| $LiSc_{0.3}Ti_{1.7}(PO_4)_3$ | $6.3 \times 10^{-2}$ | 300 | Solid state | 113 |
| $Li_{1.2}In_{0.2}Hf_{1.8}(PO_4)_3$ | $4.0 \times 10^{-6}$ | 23 | Solid state | 113 |
| $Li_{1.2}In_{0.2}Hf_{1.8}(PO_4)_3$ | $8.0 \times 10^{-2}$ | 300 | Solid state | 113 |
| $Li_{1.35}Ti_{1.65}In_{0.35}P_3O_{12}$ | $2.0 \times 10^{-2}$ | 300 | Solid state | 114, 115 |
| $Li_{2.8}Ti_{0.2}In_{1.8}P_3O_{12}$ | $8.0 \times 10^{-3}$ | 300 | Solid state | 114, 115 |
| $Li_{1.3}Al_{0.3}Ti_{1.7}(PO_4)_3$ | $7 \times 10^{-4}$ | 23 | Solid state | 116, 117 |
| $Li_{1.3}Sc_{0.3}Ti_{1.7}(PO_4)_3$ | $7 \times 10^{-4}$ | 23 | Solid state | 116, 117 |
| $LiTi_2(PO_4)_3\text{-}0.2Li_3BO_3$ | $3 \times 10^{-4}$ | 23 | Solid state | 118 |
| $LiSnZr(PO_4)_3$ | $7.3 \times 10^{-3}$ | 327 | Solid state | 123 |
| $Li_{1.2}In_{0.2}Sn_{0.9}Zr_{0.9}(PO_4)_3$ | $3.2 \times 10^{-2}$ | 327 | Solid state | 123 |
| $Li_{1.4}Ti_2(PO_4)_3$ | $2.7 \times 10^{-3}$ | 300 | Solid state | 127 |
| $LiTaAl(PO_4)_3$ | $1.0 \times 10^{-2}$ | 350 | Solid state | 136 |
| $Li_{1.4}Y_{0.4}Ti_{1.6}(PO_4)_3$ | $1.8 \times 10^{-3}$ | 23 | Solid state | 139 |
| $Li_{1.3}Al_{0.3}Ti_{1.7}(PO_4)_3$ | $6.0 \times 10^{-5}$ | 23 | Sol-gel | 141 |
| $Li_{1.3}Al_{0.3}Ti_{1.7}(PO_4)_3$ | $4.0 \times 10^{-3}$ | 300 | Sol-gel | 142 |
| $14Li_2O\text{-}9Al_2O_3\text{-}38TiO_2\text{-}39P_2O_5$ | $1.3 \times 10^{-3}$ | 23 | Melt quenching | 146 |
| $2[Li_{1.4}Ti_2Si_{0.4}P_{2.6}O_{12}]\text{-}AlPO_4$ | $1.5 \times 10^{-3}$ | 23 | Melt quenching | 147 |
| $14Li_2O\text{-}9Al_2O_3\text{-}38GeO_2\text{-}39P_2O_5$ | $1 \times 10^{-4}$ | 23 | Melt quenching | 148 |
| $14Li_2O\text{-}9Ga_2O_3\text{-}38TiO_2\text{-}39P_2O_5$ | $9 \times 10^{-4}$ | 23 | Melt quenching | 149 |

mixed in alcohol, and heat-treated at 500°C in air for 4 h. The synthesized powders were isostatically pressed into disks and heat-treated at 900 to 1200°C for 24 h. In order to determine bulk electrical resistivities, the sintered pellets were coated with gold electrodes. Bulk and grain boundary conductivities were determined by the complex impedance method. In the indium system, $Li_{1.4}Ti_{1.6}In_{0.4}P_3O_{12}$ was found to have the highest $Li^+$ ion conductivities of $5.5 \times 10^{-2}$ S/cm at 300°C and $1.9 \times 10^{-4}$ S/cm at room temperature. The authors also determined a breakdown voltage for $Li_{1.4}Ti_{1.6}In_{0.4}P_3O_{12}$ to be 1.8 to 2.2 V. Substitution of $Ga^{3+}$ and $Mg^{2+}$ led to maximum conductivities at $x = 0.6$ and $x = 0.3$, with values equal to $8.4 \times 10^{-3}$ and $5.3 \times 10^{-3}$ S/cm at 300°C, respectively. $Cr^{3+}$ substitution at $x = 0.8$ composition yielded a conductivity of $4.8 \times 10^{-2}$ S/cm at 300°C, comparable to that of the In-doped system. Li and Lin[113] and Lin et al.[114,115] found that the ionic radius and polarizability of the substituent ion had strong effects on the resultant conductivity. They speculated that the combination of an ionic radius close to that of $Ti^{4+}$ and high polarizability may yield the highest conductivity.

Subramanian et al.[116] studied the effect of doping the parent $LiTi_2(PO_4)_3$ phase with zirconium, hafnium, indium, and scandium. Stoichiometric mixtures of the corresponding oxides were mixed with $Li_3PO_4$ and transferred to a platinum tube for annealing. The tube was heated at 800°C for 6 h and 1000 to 1100°C in air for 20 h. The resultant powders were pressed into pellets and annealed at 1150 to 1200°C for 24 h in a closed platinum crucible. The pellets were polished using 400 grit emery paper and the faces were coated with conductive platinum paint. After heating to 600°C to remove all of the solvent (not specified), the samples were tested using the complex impedance method. $LiZr_{0.2}Ti_{1.8}(PO_4)_3$ exhibited the highest lithium-ion conductivity for the zirconium-titanium system, with values of $7.4 \times 10^{-3}$ S/cm at 300°C and $8.5 \times 10^{-8}$ S/cm at room temperature. $Sc^{3+}$ doping for $Ti^{4+}$ showed maximum conductivities at $x = 0.3$ of $6.3 \times 10^{-2}$ S/cm at 300°C and $4.1 \times 10^{-4}$ S/cm at room temperature. The authors also observed fairly high conductivity values for the system $Li_{1+x}In_xHf_{2-x}(PO_4)_3$ at $x = 0.2$ with values of $8.0 \times 10^{-2}$ at 300°C and $4 \times 10^{-6}$ S/cm at room temperature.

Hamdoune et al.[117,118] studied the crystal structures and conductivities of materials in the $Li_{1+x}Ti_{2-x}In_xP_3O_{12}$ system for the composition range of $x = 0$ to $x = 2$. Stoichiometric mixtures of $Li_2CO_3$, $TiO_2$, $In_2O_3$, and $NH_4H_2PO_4$ were finely ground and heated in air progressively to 1000°C in platinum crucibles for 2 to 5 h. Repeated heating and grinding steps were used to improve the crystallinity and obtain the desired single phase. In the low-dopant regime from $x = 0$ to $x = 0.4$, the material possesses the rhombohedral $R\bar{3}c$ crystal structure. Compounds in the range from $x = 0.4$ to $x = 1.0$ are orthorhombic, Pbca. Increasing the indium concentration to values between $x = 1.0$ and $x = 2.0$ results in the formation of a monoclinic $P2_1/n$ phase. Conductivity data at 300°C indicates two conductivity maxima, one at $x = 0.35$ and one at $x = 1.8$, yielding values of $2 \times 10^{-2}$ S/cm and $8 \times 10^{-3}$ S/cm, respectively. The authors did not establish a clear connection between the observed conductivity maxima and structural results.

The most experimental work to date on doped lithium titanium phosphates has been carried out by Aono et al.[119-125] Aono et al. have examined the effects of numerous dopants, including aluminum, scandium, yttrium, lanthanum, chromium, iron, indium, lutetium, germanium, tin, hafnium, and zirconium. Aono et al. have also studied improving the sinterability of lithium titanium phosphate-based compounds by utilizing lithium salts as binding agents. Stoichiometric mixtures of $Li_2CO_3$, $TiO_2$, $(NH_4)_2HPO_4$, and corresponding metal oxide were ground and heated in air at 900°C for 2 h in a platinum crucible. The resulting material was reground using wet ball milling for 6 h. The same annealing step was repeated, followed by another ball milling step for 12 h, which resulted in a particle size of less than 1 μm. A 3% polyvinyl alcohol (PVA) solution was used as a binding agent to produce pellets of the material. The pellets were sintered in air at 900 to 1250°C for 2 h. Finally, the pellets were polished with 800 grit emery paper and coated with gold on both sides by vacuum evaporation. The complex impedance method was used to evaluate the ionic conductivities of the materials.

Aono et al. found that dopant atoms (M = Al, Sc, Y, and La) all similarly increased the conductivity when compared with that of undoped $LiTi_2(PO_4)_3$, the maxima in the conductivities appearing at compositions of $x = 0.3$ in the respective compounds, $Li_{1+x}M_xTi_{2-x}(PO_4)_3$.[119] Aono et al. also found that the conductivities of the aluminum- and scandium-doped systems at $x = 0.3$ rivaled that of $Li_3N$ at room temperature, with values of ~$7 \times 10^{-4}$ S/cm.[120] In the same work, Aono et al. also found that adding excessive lithium to the system in the form of $Li_2O$ or $Li_4P_2O_7$ served to increase the lithium-ion conductivity by enhancing the densification of the sintered pellets as well as increasing the lithium content at the grain boundaries. In another study, Aono et al.[121] observed that $Li_3PO_4$, $Li_3BO_3$, $Li_2SO_4$, $LiCl$, and $LiNO_3$ could all be used as binders to improve the densification and conductivity of the parent $LiTi_2(PO_4)_3$ phase. The sample that provided the maximum room temperature conductivity of $3 \times 10^{-4}$ S/cm was found to belong to the $LiTi_2(PO_4)_3$-$0.2Li_3BO_3$ system. In a related study focused only on binder additions of $Li_3PO_4$ or $Li_3BO_3$, Aono et al.[124] reported similar conductivity improvements, while also including scanning electron micrographs which highlight the effect of binder addition on the surface microstructure of the sintered pellets. In the cases of aluminum, scandium, and yttrium doping, Aono et al.[122] found that the total conductivity enhancement was due solely to a decrease in the activation energy for grain boundary conduction. The partial substitution of a large variety of $M^{3+}$ atoms (M = Al, Cr, Ga, Fe, Sc, In, Lu, Y, or La) for $Ti^{4+}$ in the formula $Li_{1+x}M_xTi_{2-x}(PO_4)_3$, along with careful pellet density studies, allowed Aono et al.[123] to show that a lower porosity leads to higher ionic conductivity.

Finally, Aono et al.[125] showed that the addition of $Li_2O$ as a binder also enhanced the conductivity and sinterability, but that doping with germanium, tin, hafnium, or zirconium actually decreased the conductivity and increased the bulk grain component activation energy. The large amount of research on lithium titanium phosphate-based lithium-ion conductors by Aono et al. has highlighted the effects of doping many diverse atoms for titanium, while also illustrating the importance of pellet density, sample drying, binder phases, and lithium concentration at the grain boundaries.

The solid state synthesis process has also been used to study many other variations of doped lithium titanium phosphate solid ionic conductors.[126–143] The ionic conductivities and compositions of the most promising lithium-ion ceramic electrolytes are shown in Table 26.2.

## B. SOL-GEL AND WET CHEMISTRY SYNTHESIS AND PROPERTIES

Cretin and Fabry[144] used a sol-gel approach to synthesize aluminum-doped $LiTi_2(PO_4)_3$ and $LiGe_2(PO_4)_3$. The reactants used were $Ti(OC_4H_9)_4$, $Ge(OC_2H_5)_4$, $Al(OC_2H_5)_3$, $LiOOCCH_3 \cdot 2H_2O$, and $NH_4H_2PO_4$. The alkoxide components were dissolved in ethanol at 70°C, while the lithium and phosphorus reactants were dissolved in water at 70°C. The two solutions were added together and water was added in the appropriate molar ratio to instigate hydrolysis. The colloidal precipitates were collected and freeze dried. The resultant powder was calcined at ~500°C, pressed into pellets, and sintered at temperatures ranging from 850 to 1050°C (in air). The room temperature ionic conductivity of the sol-prepared $Li_{1.3}Al_{0.3}Ti_{1.7}(PO_4)_3$ was ~6 × $10^{-5}$ S/cm, slightly less than the conductivity reported by Aono et al.[120] for the same composition prepared by conventional processing.

Rather than use a conventional phosphorus reactant, Schmutz et al.[145] relied on the polymeric reaction product of $P_2O_5$ + ethanol. Schmutz et al. also used $LiOOCCH_3 \cdot 2H_2O$, $Ti(OC_2H_5)_4$, and aluminum di-sec butoxide acetoacetic ester chelate to synthesize a precursor sol. The researchers used spin-coating techniques with the precursor sol to produce $Li_{1.3}Al_{0.3}Ti_{1.7}(PO_4)_3$ thin films on glass substrates at 650°C. The 300°C ionic conductivity was determined to be 4 × $10^{-3}$ S/cm.

Casciola et al.[146] also used a solution chemistry technique to produce $LiTi_xZr_{2-x}(PO_4)_3$ compounds. First, amorphous precipitates of nominal composition, $LiTi_xZr_{2-x}(PO_4)_3 \cdot nH_2O$, were formed by adding an aqueous solution of $ZrOCl_2$ and $TiCl_4$ to phosphoric acid. The precipitates were dried and placed in a vacuum oven at 150°C to remove HCl fumes. Titration with 1 M LiOH was used to obtain the lithium salts. The samples were crystallized at 900°C in air for various times. For all of the compositions studied, the 300°C conductivity was within the range of 4 × $10^{-3}$ to 3 × $10^{-4}$ S/cm and no discernable dependence of the conductivity on composition was observed.

$LiTi_2(PO_4)_3$ was also synthesized hydrothermally by Yong and Wenqin.[147,148] $TiO_2$, 85% $H_3PO_4$, and LiOH were used as the reactants in a Teflon-lined autoclave. It was sealed and heated in an oven at 250°C for 5 to 7 days. The product was confirmed to be pure $LiTi_2(PO_4)_3$ by x-ray diffraction (XRD), Raman spectroscopy, and nuclear magnetic resonance (NMR) studies. The resultant particles were well-faceted cuboids of ~50 μm in size. No data are available on the ionic conductivity of the low-temperature, hydrothermally produced $LiTi_2(PO_4)_3$ materials.

## C. GLASS AND GLASS-CERAMIC SYNTHESIS AND PROPERTIES

The lithium-ion conductivities of glasses and glass-ceramics in the $Li_2O-Al_2O_3$-$TiO_2-P_2O_5$ system were investigated by Fu.[149] The starting materials used in the study were $Li_2CO_3$, $Al(OH)_3$, $TiO_2$, and $NH_4H_2PO_4$. Thirty gram batches corresponding to $14Li_2O-9Al_2O_3-38TiO_2-39P_2O_5$ (molar ratio) were loaded into platinum crucibles and melted in an electric furnace in air. Melt quenching from 1450 to 1550°C was used to produce the parent glass. The glass transition temperature ($T_g$) and crystallization temperature ($T_c$) were determined using differential thermal analysis (DTA). The glass-ceramics were formed by annealing at temperatures above $T_c$. The major crystalline phase was identified as $Li_{1+x}Al_xTi_{2-x}(PO_4)_3$. Surprisingly, the maximum room temperature conductivity (glass-ceramic sample) observed was $1.3 \times 10^{-3}$ S/cm, one of the highest conductivities for a solid electrolyte. Fu[150] extended the research by conducting similar experiments on $SiO_2$-containing glass-ceramics. Fu found the highest room temperature conductivity to be $1.5 \times 10^{-3}$ S/cm at around $x = 0.4$ in the glass-ceramic system $2[Li_{1+x}Ti_2Si_xP_{3-x}O_{12}]$-$AlPO_4$. The substitution of germanium for titanium in the $Li_2O$-$Al_2O_3$-$TiO_2$-$P_2O_5$ system led to a slightly reduced room temperature conductivity of just over $10^{-4}$ S/cm.[151] The use of $Ga_2O_3$ rather than $Al_2O_3$ also decreased the maximum conductivity slightly, to $9 \times 10^{-4}$ S/cm at room temperature.[152] The high conductivity in Fu's samples was not experimentally verified in the similar work carried out by Abrahams and Hadzifejzovic,[153] who also investigated the $Li_2O$-$Al_2O_3$-$TiO_2$-$P_2O_5$ glass and glass-ceramic systems. Abrahams and Hadzifejzovic saw a maximum room temperature conductivity of $3.98 \times 10^{-6}$ S/cm in their crystallized cast-glass pellet sample. Fu[154] extended the research even further by investigating the incorporation of several other $M_2O_3$-type constituents where M = Y, Dy, Gd, and La. Fu found that the ionic conductivity of the parent glass did not vary much with increasing ionic radius of $M^{3+}$. However, the ionic conductivities of the glass-ceramics decreased substantially as the ionic radius of the $M^{3+}$ atom increased.

## III. ANODE MATERIALS FOR LITHIUM-ION BATTERIES

### A. Introduction

In early ambient-temperature rechargeable lithium cells, the negative electrode (anode) was metallic lithium. Lithium, the lightest of all metals, has the lowest standard redox potential ($-3.04$ V vs. a standard hydrogen electrode) and provides the highest specific capacity (3860 mAh/g). Rechargeable batteries using lithium metal anodes combined with a positive electrode material (cathode), such as a transition metal oxide or chalcogenide (e.g., $TiS_2$, $MnO_2$, $NbSe_3$, or $MoS_2$), are capable of providing both high voltage and excellent capacity, resulting in an extraordinarily high energy density.[155–158] However, the periodic dissolution and deposition of metallic lithium during cycling of the negative electrode are the main roots of serious problems, such as low charge-discharge cyclability and poor safety, that prevented the commercialization of these systems.[155,157–159] The most difficult issue is the "dendritic deposition" of lithium during the repetitive charge-discharge process. After several tens of cycles, dendritic growth occurs in the direction toward the cathode, perforating the separator and reaching the cathode, leading to an internal short circuit inside of the battery accompanied by a sudden increase in temperature.[155,157–159] Due to the low melting point of metallic lithium (~453 K), this local overheat can trigger a violent reaction called "venting with flame." The occurrence in several practical cases gave evidence of these processes and forced lithium battery manufacturers to stop the production of lithium anode secondary cells such as the $Li/MoS_2$ technology by MOLI Canada and $Li/MnO_2$ by Tadiran Israel.

Many researchers have performed studies to prevent this dendritic deposition in an effort to improve the cyclability and reliability. These attempts include using a less reactive liquid electrolyte and polymer electrolyte, coating lithium with a lithium-ion-conducting membrane, using scavengers in the electrolyte that dissolve the dendritic lithium filaments, and using mechanical pressure to suppress dendritic lithium growth.[155,160–164] However, all of the above attempts brought only partial improvements in the cycle life and safety of the lithium electrode.

Due to the inherent instability of lithium metal, especially during charging, research shifted to a nonmetallic lithium battery employing a lithium alloy-based electrode.[165–168] In lithium alloy-based cells, only the lithium ion ($Li^+$) moves between the two electrodes upon charge and discharge, and therefore these are known as "lithium-ion" cells or batteries. Although slightly lower in energy density than batteries using lithium metal, lithium-ion batteries overcome the problems related to dendrite formation and safety. In contrast to metallic lithium, the cyclability of a lithium alloy-based electrode depends mainly on the structural integrity of the host material during alloying and dealloying of the lithium ion. For most of the lithium alloy-based electrodes, the volume difference between the lithium alloy and the corresponding lithium free matrix is significant and can range from 100% for LiAl to as high as 325% for $Li_{4.4}Si$ (see Figure 26.4).[168] The severe crystallographic and accompanying volume changes that occur during charge and discharge of the cells leads to mechanical failure of the intermetallic structure, resulting in crumbling of the electrodes. This effect seriously compromises the cycling efficiency and cycle life of lithium-ion cells due to the formation of insulating layers on the surface of the electrochemically pulverized electrode particles, leading to a loss of electronic particle-to-particle contact and the consequential increase of electrical resistivity.

## B. Carbon-Based Anode Materials

In contrast to most of the lithium alloy-based electrodes, carbonaceous materials such as graphite experience small volume changes ($\sim$10%) during alloying and dealloying with lithium ions, and due to their structural stability, they show excellent cyclability.[168–172] Lithium ions can be reversibly alloyed with a majority of carbonaceous materials, and the resulting lithiated carbons show a large negative potential close to that of metallic lithium that is very useful for high-voltage batteries.[169,170] In 1991, Sony Corporation commercialized the first lithium-ion battery by substituting the lithium metal electrode with carbonaceous materials. In this battery, the thermal decomposition product of polyfurfuryl alcohol resin (PFA) was used as the carbon anode. Since then, many kinds of carbonaceous materials, from crystalline to strongly disordered carbon, have been tested as anodes in lithium-ion batteries.[168–174]

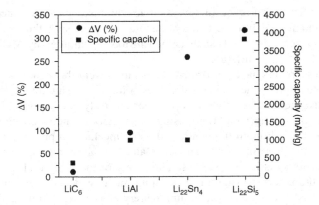

**FIGURE 26.4** Volume expansion and theoretical capacity of lithium-based compounds resulting from the alloying of different elements with lithium.

## 1. Crystal Structure of Carbonaceous Materials

Carbons that are capable of reversible lithium accommodation can roughly be classified as graphitic and nongraphitic (disordered).[168–174] The graphitic structure consists of carbon atoms arranged in hexagonal rings that are stacked in an orderly fashion (see Figure 26.5). Only weak van der Waals forces exist between these layer planes. The usual stacking sequence of the carbon layers is ABAB for hexagonal graphite (2H). The stacking sequence ABCABC is found less frequently (i.e., usually a few percent) and is called rhombohedral graphite (3R).[169–172] Natural graphite is typically comprised of these two crystal structures, with the rhombohedral ratio being less than 3 to 4%. Graphite has two kinds of

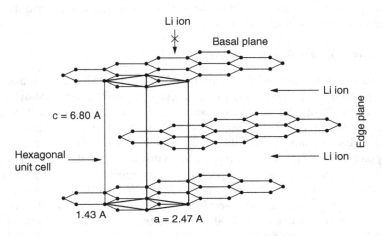

**FIGURE 26.5** Crystal structure of hexagonal (2H) graphite showing the stacking sequence and unit cell of graphite.

distinct surfaces, normal and parallel to its $c$-axis, which are called the basal and the edge plane, respectively.[169-172] During an electrochemical reaction, lithium ion is incorporated from the electrolyte only into the edge plane to form lithium-graphite-based intermetallics (see Figure 26.5). The lithium-graphite-based intermetallics are layered compounds in which the lithium ions are accommodated between the graphite interlayers, and therefore the alloying behavior is known as an "intercalation reaction." Consequently the reaction product is known as a "lithium-graphite intercalation compound" (Li-GIC).

Nongraphitic carbonaceous materials consist of carbon atoms that are mainly arranged in a planar hexagonal network with no significant ordering in the $c$-axis. Nongraphitic carbons are mostly prepared by pyrolysis of organic polymers or hydrocarbon precursors.[169-174] Heat treatment of most nongraphitic (disordered) carbons at temperatures from ~1773 to ~3273 K allows one to distinguish between the two different carbon types, graphitizable carbons and nongraphitizable carbons. The structural models proposed for graphitizable and nongraphitizable carbons are shown in Figure 26.6. Graphitizable carbons (see Figure 26.6a) develop the graphite structure during the heating process, as cross-linking between the carbon layers is weak and therefore the layers are mobile enough to form graphite-like crystallites. In contrast, nongraphitizable carbons (Figure 26.6b) show no true development of the graphite structure, even at high temperatures (~3273 K), since the carbon layers are immobilized by cross-linking. Since nongraphitizable carbons are mechanically harder than graphitizable ones, it is common to divide the nongraphitic carbons into "soft" and "hard" carbons. Examples of soft carbon are petroleum coke and carbon black, while examples of hard carbons are glassy carbon and activated carbon.

## 2. Electrochemical Response of Various Carbons

### a. Graphitic Carbon

At present, graphite is most widely used as the negative electrode in commercial rechargeable lithium-ion batteries because of its high specific capacity, low electrode potential, small irreversible loss, high columbic efficiency, good cyclability, and high level of safety.[169-172] In addition, these graphitic carbon electrodes provide a flatter discharge voltage curve and offer a sharp knee bend at the end of discharge. A fully lithiated graphite electrode, $LiC_6$, provides a theoretical capacity of 372 mAh/g, which translates to a volumetric capacity of 818 mAh/L based on a density of 2.2 g/ml. The charge and discharge characteristics of synthetic graphite, measured at a $C$ rate of $C/32$ in a three-electrode hockey puck cell, is shown in Figure 26.7. As shown in Figure 26.7, the charge consumed in the first cycle (~460 mAh/g) significantly exceeds the theoretical specific capacity for the first stage. The subsequent discharge recovers only about 70% of this charge. The capacity loss in the first cycle that cannot be recovered is frequently called the irreversible capacity loss, which is observed in the first charge/discharge cycle. However, in the second and subsequent cycles, graphite shows good reversibility where the charge consumption and charge recovery is close to 100%.

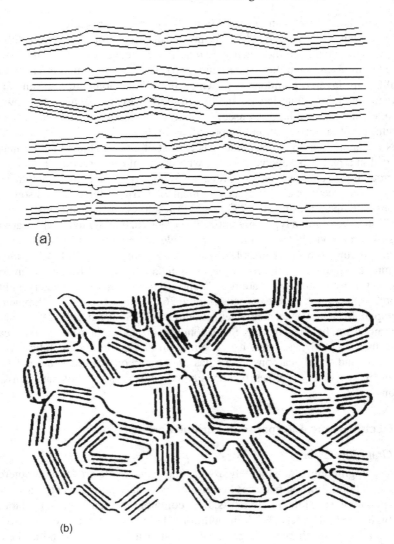

**FIGURE 26.6** Schematic structural models of (a) graphitizable and (b) nongraphitizable carbons.

    The reversible lithium intercalation is called the reversible specific capacity, which is typically in the range of 300 to 370 mAh/g for different kinds of graphite.[169–172] The excess charge consumed in the first cycle generally arises due to the formation of a solid electrolyte interface (SEI) on the graphite surface via reductive decomposition of electrolyte solution in the initial state of charging.[169–171,175–177] Slow-scan cyclic voltammograms obtained on natural graphite electrode, shown in Figure 26.8, clearly indicate that all the intercalation/deintercalation

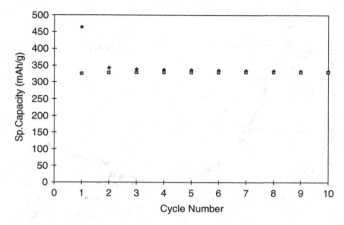

**FIGURE 26.7** Variation of specific capacity with cycle number of synthetic graphite cycled at a *C*-rate of *C*/32 measured on a three-electrode hockey-puck cell.

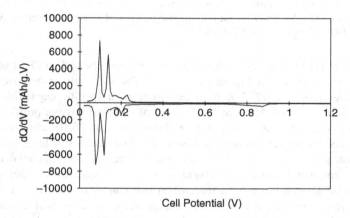

**FIGURE 26.8** Slow-scan cyclic voltammograms obtained on synthetic graphite indicate that all the intercalation/deintercalation processes proceed within a potential range of 0.25 V vs. Li/Li⁺.

processes proceed within a potential range of 0.25 V versus Li/Li⁺, one that is significant for a high-voltage battery. In addition, during the first charge, a significant potential at ~0.85 V positive to lithium has been noticed, which is mainly related to the solvent decomposition and formation of solid electrolyte interface.[170,175–177]

### b. Nongraphitic Carbon

As mentioned earlier, the nongraphitic carbons can be classified into two categories, soft and hard carbons. Different levels of graphitization can be obtained from the nongraphitic carbon materials, depending essentially on the heat treatment

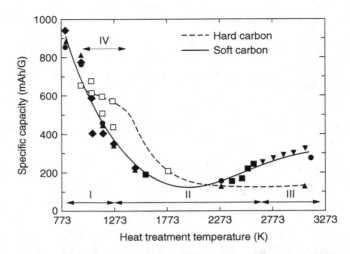

**FIGURE 26.9** The variation of the reversible specific capacity with heat treatment temperature of soft and hard carbons. (Reprinted with permission from J.R. Dahn, T. Zheng, Y. Liu, and J.S. Xue, Mechanisms for lithium insertion in carbonaceous materials, *Science*, 270, 590, 1995. Copyright 1995 AAAS.)

temperature (HTT). The variation of the specific capacity with HTT of soft and hard carbons is shown in Figure 26.9.[170,173,174] The soft carbons heat-treated at temperatures less than 1273 K have extremely high specific capacities in the range of 500 to 1000 mAh/g (region I). Typical charge/discharge profiles of the soft carbons corresponding to region I are shown in Figure 26.10, which is mainly characterized by a large irreversible loss in the first cycle with large hysteresis in their potential profiles.[169,170,173,174] The large hysteresis is a significant detriment for use in commercial lithium-ion batteries because it requires a higher voltage for the removal of lithium from the carbon that causes a reduction in the cell potential (e.g., the lithium battery will charge at say 4 V, but will discharge at only 3 V). The presence of the large hysteresis also leads to a loss of stored electrical energy, which will be dissipated as heat during charge/discharge cycles. In addition, this type of soft carbon shows poor cyclability, and the capacity decreases to half of the initial capacity within several cycles.[170–173] Therefore, in order for soft carbons to be useful for practical applications (shown in region I), the serious drawbacks, such as low density (1.5 g/cm³), large irreversible loss in the first cycle, large hysteresis, and poor cyclability all need to be solved.

With increasing HTT above 1273 K, the specific capacity decreases, yielding a minimum at about 1973 K and then increases again (region II). The soft carbons in this region contain many imperfections such as unorganized carbon and turbostratic disorder, where the graphene hexagonal layers are rotated, but still remain parallel with respect to each other.[169–174] The graphene sheets with turbostratic disorder will most likely be pinned, making them unable to accommodate greater lithium amounts between the graphene sheets. Therefore the reversible capacity

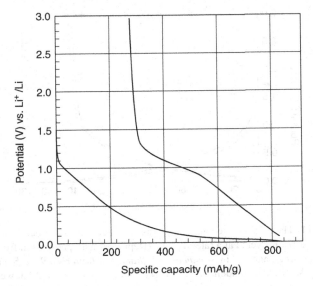

**FIGURE 26.10** Typical charge/discharge profiles of the MCMB 700 soft carbon. (Reprinted with permission from Z. Ogumi and M. Inaba, Carbon anodes, in *Advances in Lithium-Ion Batteries*, W.A. Schalkwijk and B. Scrosati, Eds., Kluwer Academic, New York, 2002. Copyright 2002 Kluwer Academic Publishers.)

in this region mainly arises due to the presence of unorganized carbon. With increasing temperature, the unorganized fraction decreases and almost disappears after thermal treatment at ~1973 K, but the turbostratic part is gradually removed only by heat treatment above 1973 K. Therefore the reversible capacity shows the minimum at ~1973 K. Figure 26.11 shows the first and second charge discharge curves of mesocarbon microbeads heat-treated at 2073 K cycled at a rate of C/36. The potential profile differs considerably from that of graphite, as the reversible intercalation of lithium ion begins at around 1.2 V vs. Li/Li$^+$ without any potential plateaus. These monotonic profiles indicate that lithium ions are intercalated randomly between graphene sheets without the formation of any stage structure. The turbostratic disordered carbons have found considerable interest as lithium host materials because cross-linking of the graphene layers mechanically suppresses the formation of solvated graphite intercalation compound Li(Solv)$_y$C$_n$, which can reduce the irreversible capacity loss in the first cycle and improve the cyclability.[169,170]

With increasing temperature above 2273 K, the probability of turbostratic disorder decreases and the fraction of the stacked layer with graphitic order increases; hence the reversible capacity also increases.[169–173] Figure 26.12 shows the variation of reversible capacity of carbons in region III with the probability of turbostratic disorder (P),[170,173] emphasizing the fact that the reversible capacity decreases with increasing turbostratic disorder with a relation of 372(1 − P). Thermal treatments of soft carbons at temperatures greater than 2673 K form

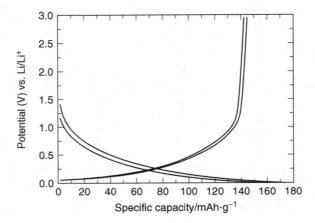

**FIGURE 26.11** The first- and second-charge discharge curves of mesocarbon microbeads heat-treated at 2073 K cycled at a rate of $C/36$. (Reprinted with permission from Z. Ogumi and M. Inaba, Carbon anodes, in *Advances in Lithium-Ion Batteries*, W.A. Schalkwijk and B. Scrosati, Eds., Kluwer Academic, New York, 2002. Copyright 2002 Kluwer Academic Publishers.)

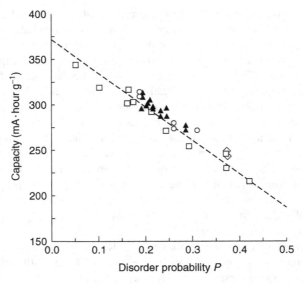

**FIGURE 26.12** Variation of reversible capacity of region III carbons with the probability of turbostratic disorder ($P$). (Reprinted with permission from J.R. Dahn, T. Zheng, Y. Liu, and J.S. Xue, Mechanisms for lithium insertion in carbonaceous materials, *Science*, 270, 590, 1995. Copyright 1995 AAAS.)

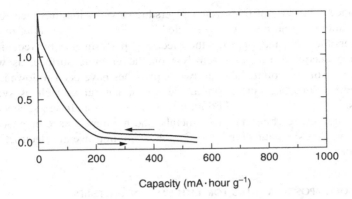

**FIGURE 26.13** The second-charge discharge curves of novalac resin heat-treated at 1273 K. (Reprinted with permission from J.R. Dahn, T. Zheng, Y. Liu, and J.S. Xue, Mechanisms for lithium insertion in carbonaceous materials, *Science*, 270, 590 1995. Copyright 1995 AAAS.)

highly graphitized carbons without significant turbostratic disorder, having specific capacities in the range 300 to 370 mAh/g (region III in Figure 26.9). The electrochemical response of graphitized carbon belongs to region III and is quite similar to natural graphite, which was discussed in Section III.B.2.a.

Hard carbons thermally treated at temperatures greater than 2273 K generally have specific capacities lower than those of soft carbons; however, they have recently attracted considerable attention from many researchers because some types of hard carbons heat-treated at around 1273 K exhibit high specific capacities in the range 500 to 700 mAh/g (region IV) with little hysteresis.[169–174,178] For example, carbonaceous materials derived from an epoxy novolac resin at 1273 K exhibit a reversible capacity of 570 mAh/g with minimal hysteresis in the cycling profiles (Figure 26.13).[173,178] These kinds of hard carbons are mainly prepared from furan resin, poly(furfuryl alcohol) resin, and phenolic resins.[173,174,178] It seems that the high capacity is brought about due to absorption of lithium ion on both sides of single-layer sheets, which is called the house of cards model. However, the high specific charge of these types of hard carbons are obtained at a very low and large potential plateau of about 0.05 V, which is very close to the lithium deposition potential. The region IV carbons are thus very promising candidates as high-capacity anodes for use in the next generation of lithium-ion battery systems. Their drawbacks are high hygroscopicity, low density (1.5 to 1.8 g/cm³), degradation of the capacity at high current densities, and the risk of lithium metal deposition during charging.

## C. MODIFICATION OF CARBONACEOUS MATERIALS

The irreversible capacity loss in the first cycle of carbon-based anode materials through consumption of a considerable amount of lithium is detrimental to the specific energy of the whole cell and has to be minimized for optimum cell performance. The issue of first-cycle irreversible capacity loss has received considerable attention and been widely studied.[169–173,175–177,179–194] In order to reduce the irreversible loss and improve the electrode performance by reducing the irreversible decomposition of electrolyte on the graphite surface, as well as reducing the solvent cointercalation, two approaches have been followed: structural modifications and surface modifications of natural as well as synthetic graphites.[172,176,179–181,185–188] In addition, a wide variety of additives such as polymers, nongraphitic carbon materials, metals, and nonmetals have been added to natural as well as synthetic graphite[172,179–184,189–194] to improve the electrochemical performance.

## D. NANOCOMPOSITE ANODES FOR LITHIUM-ION BATTERIES

Lithium-alloy anodes were first introduced in the late 1970s by Matsushita-Panasonic,[195] who designed the cell using Bi-Pb-Sn-Cd alloy as the anode, and similar types of materials were also reported in the literature.[196,197] These systems could attain reasonable cyclability only when the cycling was limited to very "shallow" reactions, utilizing less than 10% of its electrochemical capability (capacity) in order to prevent any mechanical failure during the alloying and dealloying process. Therefore the resultant capacities are very low with respect to their theoretical capacities. Recently Fuji Photo Film Celltec Co. reported tin-based amorphous composite oxide (TCO), which is claimed to have a specific capacity of 800 mAh/g with good reversibility.[198] The TCO is made from $SnO$, $B_2O_3$, $Sn_2P_2O_7$, $Al_2O_3$, and other oxide precursors.

According to their results, Sn(II) compounds only react with lithium, whereas other oxides including boron, phosphorus, and aluminum are electrochemically inactive. Two important aspects become evident from the discovery of scientists at Fuji Photo. The first aspect is that this system shows reasonably good cyclability in spite of the large volume expansion of tin during the charge/discharge process. The reason for good reversibility can be attributed to the fine particle size and the employment of electrochemically inactive components, as mentioned above. Nano-sized active materials are less prone to subdivision because of the smaller absolute volumetric changes and reduced effective mechanical stresses within the particles.[199,200] In addition, ultrafine particles also have a relatively low density of atoms per grain, which leads to relatively smaller volume changes during cycling. The atoms are segregated to the grain boundaries in nanosized materials and hence the grain boundary dynamics play an important role with regard to stabilizing the material. Furthermore, the mechanical stresses caused by volume expansion of tin during cycling are reduced or relieved by the presence of electrochemically inert materials, which play a very important role in maintaining the mechanical integrity of the electrode system. Another aspect of selecting a nanosized inactive matrix is that the matrix could undergo plastic deformation due to the stress resulting from volume expansion of the active element. It is known that some ceramic materials such as $TiO_2$ or $ZrO_2$ exhibit superplastic deformation due to their nanoparticle size.[201] A similar response can thus be expected from the nanosized systems selected.

The specific systems selected for this study are described later. The major problem of TCO is the first irreversible capacity loss, resulting from the reaction of lithium with SnO and the formation of $Li_2O$. Nevertheless, the development of TCO was the first demonstration of the potential for creating the so-called active-inactive nanocomposite anode materials by the electrochemical insertion of lithium into the tin oxide-based amorphous glass.

Following the discovery and reports of TCO, the concept of "active-inactive" nanocomposites has been studied extensively. The main concept behind active-inactive nanocomposites is that the cycling performance of materials reactive to lithium (active) can be improved significantly when materials that are inert or nonreactive to lithium (inactive) act as the host or matrix. The inactive component plays a very important role as a "buffer" or a "matrix" that endures the large volumetric stresses related to the active species, thereby alleviating the mechanical stress arising from volume expansion/contraction during electrochemical cycling, as mentioned earlier. The inactive component can also act as a skeleton for the resulting microstructure, which provides the sites for the reaction at the interface between the active and inactive components. In other words, the microstructure of the nanocomposite should be able to accommodate any structural change during cycling in order to retain good cyclability. The ratio between active and inactive components is also important in obtaining the desired cyclability because the aggregation of active particles is inversely proportional to the amount of the matrix components used.[202] However, with an increase in the amount of the inactive component, there is a significant reduction in the utilizable capacity. The matrix component is also required to allow the transport of lithium ions and provide electron conduction.

Mao et al.[203–206] reported studies on Sn-Fe-C-based nanocomposites consisting of active $Sn_2Fe$ and almost inactive $SnFe_3C$. The real reactant in this system, however, is tin, while iron is the inactive element. They claim that the anode has a capacity of ~200 mAh/g with good reversibility. A similar concept of creating the active-inactive composite upon lithium insertion has been explored in the $Cu_6Sn_5$, InSb, and MnSb systems by Kepler et al.[207], Vaughey et al.,[208] and Fransson et al.[209] Similarly, other systems such as SnSb- have also been examined.[210–212]

As was seen earlier, most examples of nanocomposite anode materials are based on the use of alloys or compounds (e.g., $Sn_2Fe$, InSb, SnO, SnSb, etc.) which form an inactive component during lithiation. Therefore the selection of inactive components is relatively limited because of the availability of such alloys. From this point of view, direct synthesis of active/inactive composite, which does not undergo any alteration of each component, is advantageous. In order to generate nanocomposite anodes without inducing any reaction between the active and inactive components, selected inactive matrixes need to be chemically inert against active elements such as tin and silicon. In addition to the chemical inertness, electrochemical inactivity is also required to achieve structural integrity, which essentially eliminates oxides and most metals barring transition metals.

Kim et al.[213–218] have studied the use of TiN, TiB$_2$ SiC, and carbon as inactive matrix components in order to develop silicon- or tin-based novel anode systems. The inactive matrix materials are chosen based on their intrinsic material properties, such as good electronic and/or ionic conductivity, good mechanical strength, and light weight (see Table 26.3).[219,220,221] Examples of nanocomposite anode and their electrochemical characteristics are presented in the following section.

## 1. Silicon-Based Nanocomposites Synthesized Using High-Energy Mechanical Milling

Si/TiN nanocomposites comprising Si:TiN = 1:2 have been generated using the high-energy mechanical milling (HEMM) approach. In order to analyze the phases present after milling, XRD was conducted on the as-milled powders obtained after milling for various time periods (see Figure 26.14). All the peaks in the patterns correspond to TiN and the broad nature of the peaks are clearly indicative of the nanocrystalline nature of the nitride. The nonobservance of any silicon-related peak in XRD indicates that silicon exists in an x-ray amorphous form well dispersed inside the powder even after milling for only 6 h. This suggests that the HEMM process provides enough energy to generate the nanocomposite powder of silicon and TiN. Based on the XRD patterns, it can be convincingly argued that the composites are composed of nanosized TiN containing a uniform dispersion of silicon independent of the composition. The very small size of the silicon crystals prevents the identification of a distinct phase boundary between silicon and TiN. As a result, it is possible that there is a continuous change in volume rather than the abrupt discrete transitions normally observed. Thus its influence on the neighboring inactive phase is minimized. This is in fact an important requirement for achieving good cyclability.

The specific gravimetric and equivalent volumetric capacities of the electrode prepared with these powders are shown in Figure 26.15. The gravimetric capacity was converted to the equivalent volumetric form using the calculated density of the Si/TiN composite. The overall capacity appears to decrease as the milling time is increased, indicating a reduction in the amount of the active silicon phase. The exact reason for the decrease in capacity is still not clear, and further detailed characterization studies are needed. At present, the reasons can be speculated. One possible reason could be that the silicon nanoparticles are embedded or enclosed by TiN during milling, thereby preventing their reaction with lithium. The composite obtained after milling for 6 h shows a fade in capacity, while the samples milled for a longer time exhibit moderate to good capacity retention. Li et al.[222] have investigated silicon/carbon black composite for potential use as an anode. The material exhibits an initial capacity as high as ≈3000 mAh/g, however, it shows poor retention characteristics. This could be attributed to the poor binding between the active and inactive components.

The composite obtained after milling for 12 h exhibits an initial discharge capacity of 422 mAh/g, but a good stable average capacity of 300 mAh/g when

**TABLE 26.3**

**Electronic and Mechanical Properties of Active and Inactive Components Used for This Study**

| | Electronic Resistivity ($\mu\Omega$cm) | Vickers Hardness (kg/mm$^2$) | Tensile Strength ($10^3$psi) | Young's Modulus (GPa) | Ultimate Tensile Stress (MPa) | Flexural Strength (MPa) | Fracture Toughness (MPa/m$^{1/2}$) |
|---|---|---|---|---|---|---|---|
| Si | $3.3 \times 10^6$ | 820 | 0.4 | 12.4 | 300 | n.a. | 0.75. |
| Sn | 12.6 | 255 | 2.1. | 49.9 | 14.5 | n.a. | n.a. |
| TiN | 23 | 1900 | 9.0 | 248 | n.a | n.a. | 8.7 |
| TiB$_2$ | 15 | 3370 | 18.4 | 372–551 | 131 | 240 | 6.7 |
| β-SiC | 107–200 | 2400–2500 | 5–20 | 262–468 | 550 | 131 | 3.4–4.4. |
| Vitreous cCarbon®$^{2\ a}$ | 5500 | 225 | n.a. | 28 | n.a. | 100 | n.a. |

*Note:* n.a.: not available.

$^a$ Commercial amorphous carbon product by Le Carbone Lorraine, Paris, France.

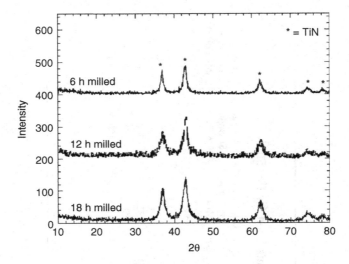

**FIGURE 26.14** X-ray diffraction patterns of Si:TiN = 1:2 composites milled for 6 h, 12 h, and 18 h, respectively.

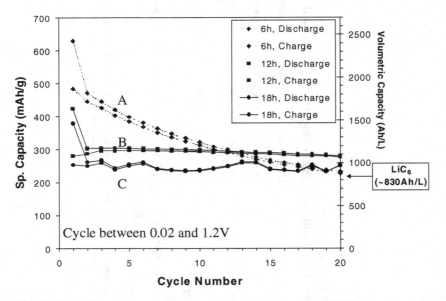

**FIGURE 26.15** Capacity as a function of cycle number for Si/TiN nanocomposites obtained after milling for 6 h, 12 h, and 18 h each.

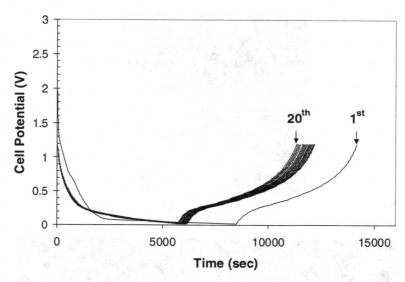

**FIGURE 26.16** Cell potential profile for the first 20 cycles of the 12 h milled Si/TiN composite with Si:TiN = 1:2 molar ratio obtained after milling for 12 h.

tested at a C-rate of ~C/10 (see Figure 26.15). The overall capacity is still lower than the theoretical capacity of 776 mAh/g, calculated assuming complete reaction of silicon with 4.4 lithium atoms. Although these data correspond to the reaction of only 1.7 lithium atoms per single atom of silicon, this does not necessarily suggest the formation of $Li_xSi$ ($x = 1.7$) due to a large ($\approx 44\%$) fraction of inactive silicon, as mentioned above. The composite exhibits a lower gravimetric capacity in comparison to conventional carbon, however, it exhibits an almost 30% higher volumetric capacity, reflecting its promising nature.

Figure 26.16 shows the variation in the cell potential with time for all the 20 cycles for the composite containing 33.3 mol% silicon obtained after milling for 12 h. The plot indicates that this anode composition exhibits a smooth plateau in the low-voltage range without exhibiting significant fade in capacity. The difference on the time axis between the first and subsequent cycles suggests the first irreversible capacity loss (~30%), possibly caused by the formation of a lithium-containing passivation layer and/or possible oxidation of the surface of the composite.

Figure 26.17a shows the high-resolution transmission electron microscope (HR-TEM) image of the composite containing 33 mol% silicon obtained after milling for 12 h. The dark areas represent the TiN nanocrystallites, while the bright regions correspond to amorphous silicon. Lattice fringes from the TiN particles can be observed conforming to the nanocrystalline (< 20 nm) nature of TiN. Figure 26.17b,c show the high-resolution elemental maps corresponding to silicon and titanium, respectively, obtained using electron energy-loss spectroscopy (EELS). The bright regions in the map indicate the high intensity reflecting

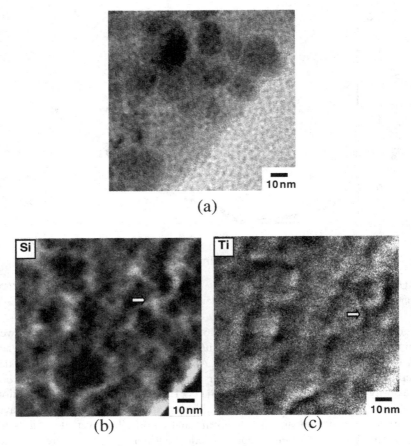

**FIGURE 26.17** (a) HR-TEM micrographs of the Si/TiN nanocomposite containing 33 mol% silicon obtained after milling for 12 h, (b) elemental map of silicon, and (c) elemental map of titanium analyzed by electron energy-loss spectroscopy (EELS). (The arrows represent the same position.)

the amount of each component. The elemental maps therefore show that TiN is distributed homogeneously in the powder and silicon appears to surround the TiN particles. This suggests the influence of mechanical milling on the two materials exhibiting different hardnesses. Silicon exhibits a lower hardness ($\cong$820 kg/mm$^2$) in comparison to TiN ($\cong$1900 kg/mm$^2$) and thus undergoes significant pulverization. The TiN particles thus appear to be uniformly coated with finely milled amorphous silicon forming the nanocomposite. Generally the samples need to be sufficiently thin to obtain good electron energy loss spectra.[223] This involves elaborate specimen preparation techniques normally used for bulk samples, which are not applicable for fine powders. Hence, due to the difficulties associated with sample preparation, the as-milled powder was used directly in this case. As a result, overlapping of the elemental maps can be expected and the boundaries are also possibly diffused. Nevertheless, the elemental maps do reveal two different regions separated distinctly to represent the distribution of the silicon and TiN phases. These results are therefore indicative of the stability of the Si/TiN nano-composite arising from the nanoscale distribution of the two phases achieved by mechanical milling.

Si/TiB$_2$ and Si/SiC nanocomposite systems, which are similarly generated using HEMM, have also been tested for their electrochemical responses as potential anode materials. These systems consist of a very homogeneous distribution of nanoparticles of silicon in the nanocomposites exhibiting a stable capacity of 350 to ~400 mAh/g when tested at a *C*-rate of ~*C*/25.[214,215] These nanocomposite systems are promising; however, there are still problems related to the loss in capacity after prolonged mechanical milling. There is thus a need for further improvement in the demonstrated concept of the active/inactive nanocomposite electrode materials for lithium-ion batteries using viable synthetic approaches.

## IV. SUMMARY

A review of the different primary material components used in lithium-ion battery systems is presented. The salient material characteristics and methods used to generate the different cathode, electrolyte, and anode materials are outlined. Lithium-ion batteries represent one of the primary rechargeable battery sources in existence today. Considerable advances have been made in both the selection and synthesis of the various material systems. The technology has certainly witnessed meteoric advances in all the various categories discussed in this review over the past decade. The continuing need for these novel energy storage systems with the increasing demand for miniaturized high-power devices is obvious. However, there are still several challenges to be overcome, as outlined in the review, and future research will certainly dictate the nature of these advances for meeting the ever-increasing demand.

## V. ACKNOWLEDGMENTS

The authors would like to acknowledge the support of NSF (CTS-0000563), NASA (NAG3–2640), and ONR (grant N00014-00-1-0516) for this work.

## REFERENCES

1. Johnston, W.D., Heikes, R.R., and Sestrich, D., The preparation, crystallography, and magnetic properties of the Li$_x$CoO$_2$ system, *J. Phys. Chem. Solids*, 7, 1, 1958.
2. Goodenough, J.B., Wickham, D.G., and Croft, W.J., Some ferrimagnetic properties of the system Li$_x$Ni$_{1-x}$O, *J. Appl. Phys.*, 29, 382, 1958.
3. Besenhard, J.O., Ed., *Handbook of Battery Materials*, Wiley-VCH, Weinheim, Germany, 1999.

4. Thackeray, M.M., Manganese oxides for lithium batteries, *Prog. Solid. State Chem.*, 25, 1, 1997.

5. Ozawa, K., Lithium-ion rechargeable batteries with $LiCoO_2$ and carbon electrodes: the $LiCoO_2$/C system, *Solid State Ionics*, 69, 212, 1994.

6. Ohzuku, T., Kitagawa, M., and Hirai, T., Electrochemistry of manganese dioxide in lithium nonaqueous cell (III) x-ray diffractional study on the reduction of spinel-related manganese dioxide, *J. Electrochem. Soc.*, 137, 769, 1990.

7. Guyomard, D., and Tarascon, J.M., The carbon/$Li_{1+x}Mn_2O_4$ system, *Solid State Ionics*, 69, 222, 1994.

8. Ebner, W., Fouchard, D., and Xie, L., The $LiNiO_2$/carbon lithium-ion battery, *Solid State Ionics*, 69, 238, 1994.

9. Ohzuku, T., and Ueda, A., Why transition metal (di)oxides are the most attractive materials for batteries, *Solid State Ionics*, 69, 201, 1994.

10. Ohzuku, T., Ueda, A., and Kouguchi, M., Synthesis and characterization of $LiAl_{1/4}Ni_{3/4}O_2$ (R3m) for lithium-ion (shuttlecock) batteries, *J. Electrochem. Soc.*, 142, 4033, 1995.

11. Mizushima, K., Jones, P.C., Wiseman, P.J., and Goodenough, J.B., $Li_xCoO_2$ (0<x<1): a new cathode material for batteries of high energy density, *Mater. Res. Bull.*, 15, 783, 1980.

12. Nagaura, T., and Tozawa, K., Lithium ion rechargeable battery, *Prog. Batt. Solar Cells*, 9 209, 1990.

13. Reimers, J.N., and Dahn, J.R., Electrochemical and in situ x-ray diffraction studies of lithium intercalation in $Li_xCoO_2$, *J. Electrochem. Soc.*, 139, 2091, 1992.

14. Kumta, P.N., Gallet, D., Waghray, A., Blomgren, G.E., and Setter, M.P., Synthesis of $LiCoO_2$ powders for lithium-ion batteries from precursors derived by rotary evaporation, *J. Power Sources*, 72, 91, 1998.

15. Ohzuku, T., Ueda, A., Nagayama, N., Iwakoshi, Y., and Komori, H., Comparative study of $LiCoO_2$, $LiNi_{1/2}Co_{1/2}O_2$ and $LiNiO_2$ for 4 volt secondary lithium cells, *Electrochim. Acta*, 38, 1159, 1993.

16. Amatucci, G.G., Tarascon, J.M., and Klein, L.C., $CoO_2$, the end member of the $Li_xCoO_2$ solid solution, *J. Electrochem. Soc.*, 143, 1114, 1996.

17. Chebiam, R.V., Prado, F., and Manthiram, A., Soft chemistry synthesis and characterization of layered $Li_{1-x}Ni_{1-y}Co_yO_{2\delta}$ (0 Σ x Σ 1 and 0 Σ y Σ 1), *Chem. Mater.*, 13, 2951, 2001.

18. Chebiam, R.V., Kannan, A.M., Prado, F., and Manthiram, A., Comparison of the chemical stability of the high energy density cathodes of lithium-ion batteries, *Electrochem. Commun.*, 3, 624, 2001.

19. Dahn, J.R., von Sacken, U., and Michal, C.A., Structure and electrochemistry of $Li_{1+y}NiO_2$ and a new $Li_2NiO_2$ phase with the $Ni(OH)_2$ structure, *Solid State Ionics*, 44, 87, 1990.

20. Rougier, A., Gravereau, P., and Delmas, C., Optimization of the composition of the $Li_{1-z}Ni_{1+z}O_2$ electrode materials: structural, magnetic, and electrochemical studies, *J. Electrochem. Soc.*, 143, 1168, 1996.

21. Ohzuku, T., Ueda, A., and Nagayama, N., Electrochemistry and structural chemistry of $LiNiO_2$ (R3m) for 4 volt secondary lithium cells, *J. Electrochem. Soc.*, 140, 1862, 1993.

22. Chang, C.-C., Kim, J.Y., and Kumta, P.N., Influence of crystallite size on the electrochemical properties of chemically synthesized stoichiometric $LiNiO_2$, *J. Electrochem. Soc.*, 149, A1114, 2002.

23. Chang, C.-C., Kim, J.Y., and Kumta, P.N., Implications of reaction mechanism and kinetics on the synthesis of stoichiometric $LiNiO_2$, *J. Electrochem. Soc.*, 149, A331, 2002.

24. Chang, C.-C., Scarr, N., and Kumta, P.N., Synthesis and electrochemical characterization of $LiMO_2$ (M = Ni, $Ni_{0.75}Co_{0.25}$) for rechargeable lithium-ion batteries, *Solid State Ionics*, 112, 329, 1998.

25. Chang, C.-C., and Kumta, P.N., Particulate sol-gel synthesis and electrochemical characterization of $LiMO_2$ (M = Ni, $Ni_{0.75}Co_{0.25}$) powders, *J. Power Sources*, 75, 44, 1998.

26. Chang, C.-C., Kim, J.Y., and Kumta, P.N., Divalent cation incorporated $Li_{(1+x)}MMg_xO_{2(1+x)}$ (M = $Ni_{0.75}Co_{0.25}$): viable cathode materials for rechargeable lithium-ion batteries, *J. Power Sources*, 89, 56, 2000.

27. Chang, C.-C., Kim, J.Y., and Kumta, P.N., Synthesis and electrochemical characterization of divalent cation-incorporated lithium nickel oxide, *J. Electrochem. Soc.*, 147, 1722, 2000.

28. Chang, C.-C., Particulate sol-gel (PSG) synthesis, structural and electrochemical characterization of $LiMO_2$ (M = Ni, $Ni_{0.75}Co_{0.25}$), Ph.D. dissertation, Carnegie Mellon University, Pittsburgh, PA, 1999.

29. Li, W., and Curie, J., Morphology effects on the electrochemical performance of $LiNi_{1-x}Co_xO_2$, *J. Electrochem. Soc.*, 144, 2773, 1997.

30. Thackeray, M.M., David, W.I.F., Bruce, P.G., and Goodenough, J.B., Lithium insertion into manganese spinels, *Mater. Res. Bull.*, 18, 461, 1983.

31. Thackeray, M.M., Manganese oxides for lithium batteries, *Prog. Solid State Chem.*, 25, 1, 1997.

32. Buhrmester, T., and Martin, M., X-ray absorption investigation on the ternary system lithium manganese oxide, *Solid State Ionics*, 135, 267, 2000.

33. Hon, Y.M., Fung, K.Z., Lin, S.P., and Hon, M.H., Effects of metal ion sources on synthesis and electrochemical performance of spinel $LiMn_2O_4$ using tartaric acid gel process, *J. Solid State Chem.*, 163, 231, 2002.

34. Zhang, Y.C., Wang, H., Xu, H.Y. et al., Low-temperature hydrothermal synthesis of spinel-type lithium manganese oxide nanocrystallites, *Solid State Ionics*, 158, 113, 2003.

35. Thackeray, M.M., Johnson, P.J., Adendorff, L.A., and Goodenough, J.B., Electrochemical extraction of lithium from $LiMn_2O_4$, *Mater. Res. Bull.*, 19, 179, 1984.

36. Thackeray, M.M., Structural considerations of layered and spinel lithiated oxides for lithium ion batteries, *J. Electrochem. Soc.*, 142, 2558, 1995.

37. Hoppe, R., Brachtel, G., and Jansen, M., Zeitschrift fur anorganische und allgemeine Chemie, *Z. Anorg. Allg. Chem.*, 417, 1, 1975.

38. Ohzuku, T., Ueda, A., and Hirai, T., Lithium manganese oxide ($LiMnO_2$) as cathode for secondary lithium cell, *Chem. Express*, 7, 193, 1992.

39. Koetschau, I., Richard, M.N., Dahn, J.R., Soupart, J.B., and Rousche, J.C., Orthorhombic $LiMnO_2$ as a high capacity cathode for Li-ion cells, *J. Electrochem. Soc.*, 142, 2906, 1995.

40. Gummow, R.J., Liles, D.C., and Thackeray, M.M., Lithium extraction from orthorhombic lithium manganese oxide and the phase transformation to spinel, *Mater. Res. Bull.*, 28, 1249, 1993.

41. Capitaine, F., Gravereau, P., and Delmas, G., A new variety of $LiMnO_2$ with a layered structure, *Solid State Ionics*, 89, 197, 1996.

42. Armstrong, A.R. and Bruce, P.G., Synthesis of layered LiMnO$_2$ as an electrode for rechargeable lithium batteries, *Nature*, 381, 499, 1996.

43. Bruce, P.G., Armstrong, A.R., and Gitzendanner, R.L., New intercalation compounds for lithium batteries: layered LiMnO$_2$, *J. Mater. Chem.*, 9, 193, 1999.

44. Armstrong, A.R., Robertson, A.D., and Bruce, P.G., Structural transformation on cycling layered Li(Mn$_{1-y}$Co$_y$)O$_2$ cathode materials, *Electrochim. Acta*, 45, 285, 1999.

45. Paulsen, J.M., Thomas, C.L., and Dahn, J.R., Layered Li-Mn-oxide with the O$_2$ structure: a cathode material for Li-ion cells which does not convert to spinel, *J. Electrochem. Soc.*, 146, 3560, 1999.

46. Davidson, I.J., McMillan, R.S., Slerg, H., Luan, B., Kargina, I., Murray, J.J., and Swainson, I.P., Electrochemistry and structure of Li$_{2-x}$Cr$_y$Mn$_{2-y}$O$_4$ phases, *J. Power Sources*, 81, 406, 1999.

47. Chiang, Y.-M., Sadoway, D.R., Jang, Y.-J., Huang, B., and Wang, H., High capacity, temperature-stable lithium aluminum manganese oxide cathodes for rechargeable batteries, *Electrochem. Solid State Lett.*, 2, 107, 1999.

48. Robertson, A.D., Armstrong, A.R., Fowkes, A.J., and Bruce, P.G., Li$_x$(Mn$_{1-y}$Co$_y$)O$_2$ intercalation compounds as electrodes for lithium batteries: influence of ion exchange on structure and performance, *J. Mater. Chem.*, 11, 113, 2001.

49. Quine, T.E., Duncan, M.J., Armstrong, A.R., Robertson, A.D., and Bruce, P.G., Layered Li$_x$Mn$_{1-y}$Ni$_y$O$_2$ intercalation electrodes, *J. Mater. Chem.*, 10, 2838, 2000.

50. Ohzuku, T., and Makimura, Y., Layered lithium insertion material of LiNi$_{1/2}$Mn$_{1/2}$O$_2$: a possible alternative to LiCoO$_2$ for advanced lithium-ion batteries, *Chem. Lett.*, 8, 744, 2001.

51. Lu, Z., MacNeil, D.D., and Dahn, J.R., Layered cathode materials Li[Ni$_x$Li$_{(1/3-2x/3)}$Mn$_{(2/3-x/3)}$]O$_2$ for lithium-ion batteries, *Electrochem. Solid State Lett.*, 4, A191, 2001.

52. Johnson, C.S., Kim, J., Kropf, A.J., Kahaian, A.J., Vaughey, J.T., Fransson, L.M.L., and Edstrom, K., Structural characterization of layered Li$_x$Ni$_{0.5}$Mn$_{0.5}$O$_2$ $0 \le x \le 2$ oxide electrodes for Li batteries, *Chem. Matter.*, 15, 2313, 2003.

53. Paulsen, J.M., Thomas, C.L., and Dahn, J.R., O2 structure Li$_{2/3}$[Ni$_{1/3}$Mn$_{2/3}$]O$_2$: a new layered cathode material for rechargeable lithium batteries. I. Electrochemical properties, *J. Electrochem. Soc.*, 147, 861, 2000.

54. Paulsen, J.M., and Dahn, J.R., O2-type Li$_{2/3}$[Ni$_{1/3}$Mn$_{2/3}$]O$_2$: a new layered cathode material for rechargeable lithium batteries II. Structure, composition, and properties, *J. Electrochem. Soc.*, 147, 2478, 2000.

55. Kang, S.-H., Kim, J., Stoll, M.E., Abraham, D., Sun, Y.K., and Amine, K., Layered Li(Ni$_{0.5-x}$Mn$_{0.5-x}$M$_{2x}$)O$_2$ (M = Co, Al, Ti; $x = 0$, 0.025) cathode materials for Li-ion rechargeable batteries, *J. Power Sources*, 112, 41, 2002.

56. de Picotto, L.A., Thackeray, M.M., David, W.I.F., Bruce, P.G., and Goodenough, J.B., Structural characterization of delithiated LiVO$_2$, *Mater. Res. Bull.*, 19, 1497, 1984.

57. de Picotto, L.A., and Thackeray, M.M., Insertion/extraction reactions of lithium with LiV$_2$O$_4$, *Mater. Res. Bull.*, 20, 1409, 1985.

58. Tsang, C., and Manthiram, A., Synthesis of nanocrystalline VO$_2$ and its electrochemical behavior in lithium batteries, *J. Electrochem. Soc.*, 144, 520, 1997.

59. Tipton, A.L., Passerini, S., Owens, B.B., and Smyrl, W.H., Performance of lithium/V$_2$O$_5$ xerogel coin cells, *J. Electrochem. Soc.*, 143, 3473, 1996.

60. Lampe-Onnerud, C., Thomas, J.O., Hardgrave, M., and Yde-Andersen, S., The performance of single-phase $V_6O_{13}$ in the lithium/polymer electrolyte battery, *J. Electrochem. Soc.*, 142, 3648, 1995.

61. Pistoia, G., Panero, S., Tocci, M., Moshtev, R.V., and Manev, V., Solid solutions $Li_{1+x}V_3O_8$ as cathodes for high rate secondary Li batteries, *Solid State Ionics*, 13, 311, 1984.

62. Spahr, M.E., Novak, P., Scheifele, W., Haas, O., and Nesper, R., Characterization of layered lithium nickel manganese oxides synthesized by a novel oxidative coprecipitation method and their electrochemical performance as lithium insertion electrode materials, *J. Electrochem. Soc.*, 145, 1113, 1998.

63. Nalbandyan, V.B., and Shukaev, I.L., New modification of lithium ferrite and the morphotropic series $AFeO_2$, *Russ. J. Inorg. Chem.*, 32, 808, 1987.

64. Takeda, Y., Nakahara, K., Nishijama, M., Imanashi, N., Yamamoto, O., Takano, M., and Kanno, R., Sodium deintercalation from sodium iron oxide, *Mater. Res. Bull.*, 29, 659, 1994.

65. Hong, H.Y.P., Crystal structures and crystal chemistry in the system $Na_{1+x}Zr_2Si_x P_{3-x}O_{12}$, *Mater. Res. Bull.*, 11, 173, 1976.

66. Manthiram, A. and Goodenough, J.B., Lithium insertion into $Fe_2(MO_4)_3$ frameworks: comparison of M = tungsten with M = molybdenum, *J. Solid State Chem.*, 71, 349, 1987.

67. Manthiram, A. and Goodenough, J.B., Lithium insertion into $Fe_2(SO_4)_3$ frameworks, *J. Power Sources*, 26, 403, 1989.

68. Padhi, A.K., Nanjundaswamy, K.S., and Goodenough, J.B., Phospho-olivines as positive-electrode materials for rechargeable lithium batteries, *J. Electrochem. Soc.*, 144, 1188, 1997.

69. Nanjundaswamy, K.S., Padhi, A.K., Goodenough, J.B., Okada, S., Ohtsuka, H., Arai, H., and Yamaki, J. Synthesis, redox potential evaluation and electrochemical characteristics of NASICON-related-3D framework compounds, *Solid State Ionics*, 92, 1, 1996.

70. Padhi, A.K., Nanjundaswamy, K.S., Masquelier, C., Okada, S., and Goodenough, J.B., Effect of structure on the $Fe^{3+}/Fe^{2+}$ redox couple in iron phosphates, *J. Electrochem. Soc.*, 144, 1609, 1997.

71. Masquelier, C., Padhi, A.K., Nanjundaswamy, K.S., and Goodenough, J.B., New cathode materials for rechargeable lithium batteries: the 3-D framework structures $Li_3Fe_2(XO_4)_3(X=P, As)$, *J. Solid State Chem.*, 135, 228, 1998.

72. Padhi, A.K., Manivannan, V., and Goodenough, J.B., Tuning the position of the redox couples in materials with NASICON structure by anionic substitution, *J. Electrochem. Soc.*, 145, 1518, 1998.

73. Gaubicher, J., Angenault, J., Chabre, Y., Le Mercier, T., and Quarton, M., Lithium electrochemical intercalation/deintercalation in rhombohedral $V_2(SO_4)_3$, *Mol. Crystl. Liq. Crystl.*, 311, 453, 1998.

74. Padhi, A.K., Nanjundaswamy, K.S., Masquelier, C., and Goodenough, J.B., Mapping of transition-metal redox energies in phosphates with NASICON structure by lithium intercalation, *J. Electrochem. Soc.*, 144, 2581, 1997.

75. Andersson, A.S., Thomas, J.O., Kalska, B., and Haggstrom, L., Thermal stability of $LiFePO_4$-based cathodes, *Electrochem. Solid State Lett.*, 3, 66, 2000.

76. Amine, K., Yasuda, H., and Yamachi, M., Olivine $LiCoPO_4$ as 4.8 V electrode material for lithium batteries, *Electrochem. Solid State Lett.*, 3, 178, 2000.

77. Li, G., Azuma, H., and Tohda, M., LiMnPO$_4$ as the cathode for lithium batteries, *Electrochem. Solid State Lett.*, 5, A135, 2002.

78. Yamada, A., Chung, S.C., and Hinokuma, K., Optimized LiFePO$_4$ for lithium battery cathodes, *J. Electrochem. Soc.*, 148, A224, 2001.

79. Yamada, A., Kudo, Y., and Liu, K-Y., Reaction mechanism of the olivine-type Li$_x$(Mn$_{0.6}$Fe$_{0.4}$)PO$_4$ ($0 \le x \le 1$), *J. Electrochem. Soc.*, 148, A747, 2001.

80. Yamada, A., and Chung, S.-C., Crystal chemistry of the olivine-type Li(Mn$_y$Fe$_{1-y}$)PO$_4$ and (Mn$_y$Fe$_{1-y}$)PO$_4$ as possible 4 V cathode materials for lithium batteries, *J. Electrochem. Soc.*, 148, A960, 2001.

81. Chung, S.-Y., Bloking, J.T., and Chiang, Y.-M., Electronically conductive phospho-olivines as lithium storage electrodes, *Nat. Mater.*, 1, 123, 2002.

82. Hohenberg, P., and Kohn, W., Inhomogeneous electron gas, *Phys. Rev.*, 136, B864, 1964.

83. Kohn, W., and Sham, L.J., Self-consistent equations including exchange and correlation effects, *Phys. Rev.*, 140, A1133, 1965.

84. Andersen, O.K., Linear methods in band theory, *Phys. Rev. B*, 12, 3060, 1975.

85. Methfessel, M., Elastic constants and phonon frequencies of Si calculated by a fast full-potential linear-muffin-tin-orbital method, *Phys. Rev. B*, 38, 1537, 1988.

86. Weinert, M., Wimmer, E., and Freeman, A.J., Total-energy all-electron density functional method for bulk solids and surfaces, *Phys. Rev. B*, 26, 4571, 1982.

87. Phillips, J.C., Energy-band interpolation scheme based on a pseudopotential, *Phys. Rev.*, 112, 685, 1958.

88. Cohen, M.L. and Heine, V., The fitting of pseudopotentials to experimental data and their subsequent application, *Solid State Phys.*, 24, 37, 1970.

89. Kresse, G. and Hafner, J., *Ab initio* molecular-dynamics simulation of the liquid-metal-amorphous-semiconductor transition in germanium, *Phys. Rev. B*, 49, 14251, 1994.

90. Kresse, G., and Furthmuller, J., Efficiency of ab-initio total energy calculations for metals and semiconductors using a plane-wave basis set, *Comput. Mater. Sci.*, 6, 15, 1996.

91. Blaha, P., Schwarz, K., Sorantin, P., and Trickey, S.B., Full-potential, linearized augmented plane wave programs for crystalline systems, *Comput. Phys. Commun.*, 59, 399, 1990.

92. Savrasov, S.Y., and Savrasov, D.Y., Full-potential linear-muffin-tin-orbital method for calculating total energies and forces, *Phys. Rev. B*, 46, 12181, 1992.

93. Savrasov, S.Y., Linear-response theory and lattice dynamics: a muffin-tin-orbital approach, *Phys. Rev. B*, 54, 16470, 1996.

94. Ceder, G., Van der Ven, A., and Aydinol, M.K., Lithium-intercalation oxides for rechargeable batteries, *J. Metals*, 50(9), 35, 1998.

95. Mishra, S.K., and Ceder, G., Structural stability of lithium manganese oxides, *Phys. Rev. B*, 59, 6120, 1999.

96. Ceder, G., and Mishra, S.K., The stability of orthorhombic and monoclinic-layered LiMnO$_2$, *Electrochem. Solid State Lett.*, 2, 550, 1999.

97. Reed, J., and Ceder, G., Charge, potential, and phase stability of layered Li(Ni$_{0.5}$Mn$_{0.5}$)O$_2$, *Electrochem. Solid State Lett.*, 5, A145, 2002.

98. Arroyo y de Dompablo, M.E., and Ceder, G., First-principles calculations on Li$_x$NiO$_2$: phase stability and monoclinic distortion, *J. Power Sources*, 119, 654, 2003.

99. Landa, A.I., Chang, C.-C., Kumta, P.N., Vitos, L., and Abrikosov, I.A., Phase stability of $Li(Mn_{1-x}Co_x)O_2$ oxides: an ab initio study, *Solid State Ionics*, 149, 209, 2002.

100. Velikokhatnyi, O.I., Chang, C.-C., and Kumta, P.N., Phase stability and electronic structure of $NaMnO_2$, *J. Electrochem. Soc.*, 150, A1262, 2003.

101. Morgan, D., Ceder, G., Saidi, M.Y., Barker, J., Swoyer, J., Huang, H., and Adamson, G., Experimental and computational study of the structure and electrochemical properties of monoclinic $Li_xM_2(PO_4)_3$ compounds, *J. Power Sources*, 119, 755, 2003.

102. Wu, E.J., Tepesch, P.D., and Ceder, G., Size and charge effect on the structural stability of $LiMO_2$ (M = transition metal) compounds, *Philos. Mag. B*, 77, 1039, 1998.

103. Lapp, T., Skaarup, S., and Hooper, A., Ionic conductivity of pure and doped $Li_3N$, *Solid State Ionics*, 11, 97, 1983.

104. Bruce, P., *Solid State Electrochemistry*, Cambridge University Press, Cambridge, 1997.

105. Julien, C., and Nazri, G.-A., *Solid State Batteries: Materials Design and Optimization*, Kluwer Academic, Boston, 1994.

106. Dudney, N., and Neudecker, B., Solid state thin-film lithium battery systems, *Curr. Opin. Solid State Mater. Sci.*, 4, 479, 1999.

107. Bates, J., Dudney, N., Neudecker, B., Ueda, A., and Evans, C., Thin-film lithium and lithium-ion batteries, *Solid State Ionics*, 135, 33, 2000.

108. Irvine, J. and West, A., Crystalline lithium ion conductors, in *High Conductivity Solid Ionic Conductors*, T. Takahashi, Ed., World Scientific, Singapore, 1989.

109. Liang, C., Conduction characteristics of the lithium iodide-aluminum oxide solid electrolytes, *J. Electrochem. Soc.*, 120, 1289, 1973.

110. Shahi, K., Wagner, J., and Owens, B., Solid electrolyte lithium cells, in *Lithium Batteries*, J. Gabano, Ed., Academic Press, London, 1983, pp. 407–448.

111. Souquet, J., and Kone, A., Glasses as electrolytes and electrode materials in advanced batteries, in *Materials for Solid State Batteries*, B. Chowdari and S. Radhakrishna, Eds., World Scientific, Singapore, 1986, pp. 241–258.

112. Hong, H., Crystal structures and crystal chemistry in the system $Na_{1+x}Zr_2 Si_xP_{3-x} O_{12}$, *Mater. Res. Bull.*, 11, 173, 1976.

113. Li, S., and Lin, Z., Phase relationship and ionic conductivity of $Li_{1+x}Ti_{2-x}In_xP_3O_{12}$ system, *Solid State Ionics*, 9, 835, 1983.

114. Lin, Z., Yu, H., Li, S., and Tian, S., Phase relationship and electrical conductivity of $Li_{1+x}Ti_{2-x}Ga_xP_3O_{12}$ and $Li_{1+x}Ti_{2-x}Mg_xP_3O_{12}$, *Solid State Ionics*, 18, 549, 1986.

115. Lin, Z., Yu, H., Li, S., and Tian, S., Lithium ion conductors based on $LiTi_2P_3O_{12}$ compound, *Solid State Ionics*, 31, 91, 1988.

116. Subramanian, M., Subramanian, R., and Clearfield, A., Lithium ion conductors in the system $AB(IV)_2(PO_4)_3$ (B = Ti, Zr and Hf), *Solid State Ionics*, 18, 562, 1986.

117. Hamdoune, S., Qui, D., and Schouler, E., Ionic conductivity and crystal structure of $Li_{1+x}Ti_{2-x}In_xP_3O_{12}$, *Solid State Ionics*, 18, 587, 1986.

118. Hamdoune, S., Gondrand, M., and Qui, D., Synthese et caracterisation cristallographique d'un systeme conducteur ioniqe $Li_{1+x}Ti_{2-x}In_x(PO_4)_3$, *Mater. Res. Bull.*, 21, 237, 1986.

119. Aono, H., Sugimoto, E., Sadaoka, Y., Imanaka, N., and Adachi, G., Ionic conductivity of the lithium titanium phosphate $Li_{1+x}M_xTi_{2-x}(PO_4)_3$, M = Al, Sc, Y, and La, *J. Electrochem. Soc.*, 136, 591, 1989.

120. Aono, H., Sugimoto, E., Sadaoka, Y., Imanaka, N., and Adachi, G., Ionic conductivity of solid electrolytes based on lithium titanium phosphate, *J. Electrochem. Soc.*, 137, 1023, 1990.

121. Aono, H., Sugimoto, E., Sadaoka, Y., Imanaka, N., and Adachi, G., Ionic conductivity of $LiTi_2(PO_4)_3$ mixed with lithium salts, *Chem. Lett.*, 3, 331–334, 1990.

122. Aono, H., Sugimoto, E., Imanaka, N., Sadaoka, Y., and Adachi, G., Electrical properties of sintered lithium titanium phosphate ceramics $(Li_{1+x}M_xTi_{2-x}(PO_4)_3,$ $M^{3+} = Al^{3+}, Sc^{3+}, Y^{3+})$, *Chem. Lett.*, 10, 1825–1828, 1990.

123. Aono, H., Sugimoto, E., Sadaoka, Y., Imanaka, N., and Adachi, G., Ionic conductivity and sinterability of lithium titanium phosphate system, *Solid State Ionics*, 40, 38, 1990.

124. Aono, H., Sugimoto, E., Sadaoka, Y., Imanaka, N., and Adachi, G., Electrical property and sinterability of $LiTi_2(PO_4)_3$ mixed with lithium salt $(Li_3PO_4$ or $Li_3BO_3)$, *Solid State Ionics*, 47, 257, 1991.

125. Aono, H., Sugimoto, E., Sadaoka, Y., Imanaka, N., and Adachi, G., The electrical properties of ceramic electrolytes for $LiM_xTi_{2-x}(PO_4)_3 + yLi_2O$, M = Ge, Sn, Hf, and Zr systems, *J. Electrochem. Soc.*, 140, 1827, 1993.

126. Winand, J., Rulmont, A., and Tarte, P., Nouvelles solution solides $L^I(M^{IV})_{2-x}(N^{IV})_x(PO_4)_3$ (L = Li, Na M, N = Ge, Sn, Ti, Zr, Hf) Synthese et etude par diffraction x et conductivite ionique, *J. Solid State Chem.*, 93, 341, 1991.

127. Saito, Y., Ado, K., Asai, T., Kageyama, H., and Nakamura, O., Grain boundary ionic conductivity in nominal $Li_{1+x}M_xTi_{2-x}(PO_4)_3$ (M = $Sc^{3+}$ or $Y^{3+}$), *J. Mater. Sci. Lett.*, 11, 888, 1992.

128. Wang, S. and Hwu, S., $Li_{3-x}Ti_2(PO_4)_3$ $(0 \leq x \leq 1)$: a new mixed valent titanium (III/IV) phosphate with a NASICON-type structure, *J. Solid State Chem.*, 90, 377, 1991.

129. Wang, S. and Hwu, S., A new series of mixed-valence titanium (III/IV) phosphates, $Li_{1+x}Ti_2(PO_4)_3$ $(0 \leq x \leq 2)$, with NASICON-related structures, *Chem. Mater.*, 4, 589, 1992.

130. Wang, S., and Hwu, S., Ionic conductivity of $Li_{1+x}Ti_2(PO_4)_3$ $(0.2 \leq x \leq 1.72)$ with NASICON-related structures, *Chem. Mater.*, 5, 23, 1993.

131. Kuwano, J., Sato, N., Kato, M., and Takano, K., Ionic conductivity of $LiM_2(PO_4)_3$ (M = Ti, Zr, Hf) and related compositions, *Solid State Ionics*, 70, 332, 1994.

132. Zhao, W., Chen, L., Xue, R., Min, J., and Cui, W., Ionic conductivity and luminescence of $Eu^{3+}$-doped $LiTi_2(PO_4)_3$, *Solid State Ionics*, 70, 144, 1994.

133. Paris, M., Martinez-Juarez, A., Rojo, J., and Sanz, J., Lithium mobility in the NASICON-type compound $LiTi_2(PO_4)_3$ by nuclear magnetic resonance and impedance spectroscopies, *J. Phys. Condens. Mater.*, 8, 5355, 1996.

134. Paris, M. and Sanz, J., Structural changes in the compounds $LiM^{IV}_2(PO_4)_3$ $(M^{IV}$ = Ge, Ti, Sn, and Hf) as followed by $^{31}P$ and $^7Li$ NMR, *Phys. Rev. B*, 55, 14270, 1997.

135. Kobayashi, Y., Tabuchi, M., and Nakamura, O., Ionic conductivity enhancement in $LiTi_2(PO_4)_3$-based composite electrolyte by the addition of lithium nitrate, *J. Power Sources*, 68, 407, 1997.

136. Martinez-Juarez, A., Pecharroman, C., Iglesias, J., and Rojo, J., Relationship between activation energy and bottleneck size for $Li^+$ conduction in NASICON materials of composition $LiMM'(PO_4)_3$; M, M' = Ge, Ti, Sn, Hf, *J. Phys. Chem. B*, 102, 372, 1998.

137. Wong, S., Newman, P., Best, A., Nairn, K., MacFarlane, D., and Forsyth, M., Towards elucidating microstructural changes in Li-ion conductors $Li_{1+y}Ti_{2-y}Al_y(PO_4)_3$ and

$Li_{1+y}Ti_{2-y}Al_y(PO_4)_{3-x}(MO_4)_x$ (M = V and Nb): x-ray and $^{27}Al$ and $^{31}P$ NMR studies, *J. Mater. Chem.*, 8, 2199, 1998.

138. Best, A., Newman, P., MacFarlane, D., Nairn, K., Wong, S., and Forsyth, M., Characterization and impedance spectroscopy of substituted $Li_{1.3}Al_{0.3}Ti_{1.7}(PO_4)_{3-x}(ZO_4)_x$ (Z = V, Nb) ceramics, *Solid State Ionics*, 126, 191, 1999.

139. Thangadurai, V., Shukla, A., and Gopalakrishnan, J., New lithium-ion conductors based on the NASICON structure, *J. Mater. Chem.*, 9, 739, 1999.

140. Nuspl, G., Takeuchi, T., Weib, A., Kageyama, H., Yoshizawa, K., and Yamabe, T., Lithium ion migration pathways in $LiTi_2(PO_4)_3$ and related materials, *J. Appl. Phys.*, 86, 5484, 1999.

141. Forsyth, M., Wong, S., Nairn, K., Best, A., Newman, P., and MacFarlane, D., NMR studies of modified NASICON-like, lithium conducting solid electrolytes, *Solid State Ionics*, 124, 213, 1999.

142. Sobiestianskas, R., Dindune, A., Kanepe, Z. et al., Electrical properties of $Li_{1+x}Y_yTi_{2-y}(PO_4)_3$ (where x,y = 0.3; 0.4) ceramics at high frequencies, *Mater. Sci. Eng. B*, 76, 184, 2000.

143. Takada, K., Tansho, M., Yanase, I., Inada, T., Kajiyama, A., Kouguchi, M., Kondo, S., and Watanabe, M., Lithium ion conduction in $LiTi_2(PO_4)_3$, *Solid State Ionics*, 139, 241, 2001.

144. Cretin, M., and Fabry, P., Comparative study of lithium ion conductors in the system $Li_{1+x}Al_xA_{2-x}^{IV}(PO_4)_3$ with $A^{IV}$ = Ti or Ge and $0 \le x \le 0.7$ for use as $Li^+$ sensitive membranes, *J. Eur. Ceram. Soc.*, 19, 2931, 1999.

145. Schmutz, C., Basset, E., and Barboux, P., Couches minces de phosphates de titane par voie sol-gel, *J. Phys. III France*, 3, 757, 1993.

146. Casciola, M., Costantino, U., Andersen, I., and Andersen, E., Preparation, structural characterization and conductivity of $LiTi_xZr_{2-x}(PO_4)_3$, *Solid State Ionics*, 37, 281, 1990.

147. Yong, Y., and Wenqin, P., Hydrothermal synthesis and characterization of $LiTi_2(PO4)_3$, *J. Mater. Sci. Lett.*, 9, 1392, 1990.

148. Yong, Y., and Wenqin, P., Hydrothermal synthesis of $MTi_2(PO_4)_3$ (M = Li, Na, K), *Mater. Res. Bull.*, 25, 841, 1990.

149. Fu, J., Superionic conductivity of glass-ceramics in the system $Li_2O-Al_2O_3-TiO_2-P_2O_5$, *Solid State Ionics*, 96, 195, 1997.

150. Fu, J., Fast $Li^+$ ion conduction in $Li_2O-Al_2O_3-TiO_2-SiO_2-P_2O_5$ glass-ceramics, *J. Am. Ceram. Soc.*, 80, 1901, 1997.

151. Fu, J., Fast $Li^+$ ion conducting glass-ceramics in the system $Li_2O-Al_2O_3-GeO_2-P_2O_5$, *Solid State Ionics*, 104, 191, 1997.

152. Fu, J., Fast $Li^+$ ion conduction in $Li_2O-(Al_2O_3-Ga_2O_3)-TiO_2-P_2O_5$ glass-ceramics, *J. Mater. Sci.*, 33, 1549, 1998.

153. Abrahams, I. and Hadzifejzovic, E., Lithium ion conductivity and thermal behaviour of glasses and crystallised glasses in the system $Li_2O-Al_2O_3-TiO_2-P_2O_5$, *Solid State Ionics*, 134, 249, 2000.

154. Fu, J., Effects of $M^{3+}$ ions on the conductivity of glasses and glass-ceramics in the system $Li_2O-M_2O_3-GeO_2-P_2O_5$ (M = Al, Ga, Y, Dy, Gd, and La), *J. Am. Ceram. Soc.*, 83, 1004, 2000.

155. Yamaki, J., and Tobishima, S., Rechargeable lithium anodes, in *Handbook of Battery Materials*, J.O. Besenhard, Ed., Wiley-VCH, Weinheim, Germany, 1999, p. 339–357.

156. Brandt, K., Historical development of secondary lithium batteries, *Solid State Ionics*, 69, 173, 1994.

157. Abraham, K.M., Directions in secondary lithium battery research and development, *Electrochim. Acta.*, 38, 1233, 1993.
158. Levy, S.C. and Bro, P., *Battery Hazards and Accident Prevention*, Plenum Press, New York, 1994, p. 257–272.
159. Aurbach, D., Zinigrad, E., Teller, H., and Dan, P., Factors which limit the cycle life of rechargeable lithium (metal) batteries, *J. Electrochem. Soc.*, 147, 1274, 2000.
160. Takehara, Z., Future prospects of lithium metal anode, *J. Power Sources*, 68, 82, 1997.
161. Yamaki, J., Design of the lithium anode and electrolytes in lithium secondary batteries with a long cycle life, in *Lithium Ion Batteries: Fundamentals and Performance*, M. Wakihara and Y. Yamamoto, Eds., Kodansha Ltd., Tokyo, 1998, p. 67–97.
162. Matsuda, Y., Ishikawa, M., Yoshitake, S., and Moria, M., Characterization of lithium-organic electrolyte interface containing inorganic and organic additives by in situ techniques, *J. Power Sources*, 54, 301, 1995.
163. Aurbach, D., Zinigard, E., Teller, H., Cohen, Y., Salitra, G., Yamin, H., Dan P., and Elster, E., Attempts to improve the behavior of Li electrodes in rechargeable lithium batteries, *J. Electrochem. Soc.*, 149, A1267, 2002.
164. Borghini, M.C., Mastragostino, M., and Zanelli, A., Investigation on lithium/polymer electrolyte interface for high performance lithium rechargeable batteries, *J. Power Sources*, 68, 52, 1997.
165. Huggins, R.A., Lithium alloy anodes, in *Handbook of Battery Materials*, J.O. Besenhard, Ed., Wiley-VCH, Weinheim, Germany, 1999, pp. 359–381.
166. Wachtler, M., Winter, M., and Besenhard, J.O., Anodic materials for rechargeable Li-batteries, *J. Power Sources*, 105, 151, 2002.
167. Thackeray, M.M., Vaughey, J.T., and Fransson, L.M.L., Recent developments in anode materials for lithium batteries, *J. Metals*, 54, 20, 2002.
168. Tirado, J.L., Inorganic materials for the negative electrode of lithium-ion batteries: state-of-the-art and future prospects, *Mater. Sci. Eng. R*, 40, 103, 2003.
169. Winter, M., Besenhard, J.O., Spahr, M.E., and Novak, P., Insertion electrode materials for rechargeable lithium batteries, *Adv. Mater.*, 10, 725, 1998.
170. Ogumi, Z. and Inaba, M., Carbon anodes, in *Advances in Lithium-Ion Batteries*, W.A. Schalkwijk and B. Scrosati, Eds., Kluwer Academic, New York, 2002, pp. 79–101.
171. Yazami, R., Carbon-lithium negative electrode for lithium ion batteries: main characteristics and features, in *Materials for Lithium-Ion Batteries*, C. Julien and Z. Stoynov, Eds., Kluwer Academic, Amsterdam, 2000, pp. 105–159.
172. Imanishi, N., Takeda, Y., and Yamamoto, O., Development of the carbon anode in lithium ion batteries, in *Lithium Ion Batteries: Fundamentals and Performance*, M. Wakihara and Y. Yamamoto, Eds., Kodansha Ltd., Tokyo, 1998, pp. 98–126.
173. Dahn, J.R., Zheng, T., Liu, Y., and Xue, J.S., Mechanisms for lithium insertion in carbonaceous materials, *Science*, 270, 590, 1995.
174. Buiel, E., and Dahn, J.R., Li-insertion in hard carbon anode materials for Li-ion batteries, *Electrochim. Acta*, 45, 121, 1999.
175. Peled, E., Golodnitsky, D., and Penciner, J., The anode/electrolyte interface, in *Handbook of Battery Materials*, J.O. Besenhard, Ed., Wiley-VCH, Weinheim, Germany, 1999, pp. 419–456.

176. Aurbach, D., The role of surface films on electrodes in Li-ion batteries, in *Advances in Lithium-Ion Batteries*, W.A. van Schalkwijk and B. Scrosati, Eds., Kluwer Academic, New York, 2002, pp. 7–77.

177. Besenhard, J.O., Winter, M., Yang, J., and Biberacher, W., Filming mechanism of lithium-carbon anodes in organic and inorganic electrolytes, *J. Power Sources*, 54, 228, 1995.

178. Liu, Y., Xue, J.S., Zheng, T., and Dahn, J.R., Mechanism of lithium insertion in hard carbons prepared by pyrolysis of epoxy resins, *Carbon*, 34, 193, 1996.

179. Noel, M. and Suryanarayana, V., Role of carbon host lattices in Li-ion intercalation/de-intercalation, *J. Power Sources*, 111, 193, 2002.

180. Wu, Y.P., Rahm, E., and Holze, R., Carbon anode materials for lithium ion batteries, *J. Power Sources*, 114, 228, 2003.

181. Mabuchi, A., Tokumitsu, K., Fujimoto, H., and Kasuh, T., Charge-discharge characteristics of the mesocarbon microbeads heat-treated at different temperatures, *J. Electrochem. Soc.*, 142, 1041, 1995.

182. Pan, Q., Guo, K., Wang, L., and Fang, S., Ionic conductive copolymer encapsulated graphite as an anode material for lithium ion batteries, *Solid State Ionics*, 149, 193, 2000.

183. Yoshio, M., Wang, H., Fukuda, K., Hara, Y., and Adachi, Y., Effect on carbon coating on electrochemical performance of treated natural graphite as lithium-ion battery anode materials, *J. Electrochem. Soc.*, 147, 1245, 2000.

184. Yu, P., Ritter, J.A., White, R.E., and Popov, B.N., Ni-composite microencapsulated graphite as the negative electrode in lithium ion batteries: I. Initial irreversible capacity study, *J. Electrochem. Soc.*, 147, 1280, 2000.

185. Yamaguchi, K., Suzuki, J., Saito, M., Sekine, K., and Takamura, T., Stable charge/discharge of Li at a graphitized carbon fiber electrode in a pure PC electrolyte and the initial charging loss, *J. Power Sources*, 97, 159, 2001.

186. Xing, W., and Dahn, J.R., Study of irreversible capacities for Li insertion in hard and graphitic carbons, *J. Electrochem. Soc.*, 144, 1195, 1997.

187. Yuqin, C., Hong, L., Lie, W., and Tianhong, L., Irreversible capacity loss of graphite electrode in lithium ion batteries, *J. Power Sources*, 68, 187, 1997.

188. Chung, G.C., Jun, S.H., Lee, K.Y., and Kim, M.H., Effect of surface structure on the irreversible capacity of various graphitic carbon electrodes, *J. Electrochem. Soc.*, 146, 1664, 1999.

189. Veeraraghavan, B., Paul, J., Haran, B., and Popov, B., Study of polypyrrole graphite composite as anode material for secondary lithium-ion batteries, *J. Power Sources*, 109, 377, 2002.

190. Guo, K., Pan, Q., and Fang, S., Poly(acrylonitrile) encapsulated graphite as anode materials for lithium ion batteries, *J. Power Sources*, 111, 350, 2002.

191. Gaberscek, M., Bele, M., Drofenik, J., Dominko, R., and Pejovnik, S., Improved carbon anode for Li ion batteries: pretreatment of carbon particles in a polyelectrolyte solution, *Electrochem. Solid State Lett.*, 3, 171, 2000.

192. Yoon, S., Kim, H., and Oh, S.M., Surface modification of graphite by coke coating for reduction of initial irreversible capacity in lithium secondary batteries, *J. Power Sources*, 94, 68, 2001.

193. Lee, H., Baek, J., Jang, S., Lee, S., Hong, S., Lee K., and Kim, M., Characteristics of carbon coated graphite prepared from a mixture of graphite and polyvinylchloride as anode materials for lithium ion batteries, *J. Power Sources*, 101, 206, 2001.

194. Nishimura, K., Honbo, H., Takeuchi, S., Horiba, T., Oda, M., Koseki, M., Muranaka, Y., Kozono Y., and Miyadera, H., Design and performance of 10Wh rechargeable lithium batteries, *J. Power Sources*, 68, 436, 1997.

195. Winter, M. and Besenhard, J.O., Electrochemical lithiation of tin and tin-based intermetallics and composites, *Electrochim. Acta*, 45, 31, 1999.

196. Sanyo Electric, U.S. Patent no. 4,820,599, 1989.

197. Nohma, T., Yoshimura, S., Nishio, K., Yamamoto, Y., Fukuoka, S., and Hara, M., Development of coin-type lithium secondary batteries containing manganese dioxide/Li-Al, *J. Power Sources*, 58, 205, 1996.

198. Idota, Y., Kubota, T., Matsufuji, A., Maekawa, Y., and Miyasaka, T., Tin-based amorphous oxide: a high-capacity lithium-ion-storage material, *Science*, 276, 1395, 1997.

199. Yang, J., SnSb$_x$-based composite electrodes for lithium ion cells, *Solid State Ionics*, 135, 175, 2000.

200. Beaulieu, L.Y., Larcher, D., Dunlap, R.A., and Dahn, J.R., Reaction of Li with grain-boundary atoms in nanostructured compounds, *J. Electrochem. Soc.*, 147, 3206, 2000.

201. Mayo, M.J., High and low temperature superplasticity in nanocrystalline materials, *Nanostruct. Mater.*, 9, 717, 1997.

202. Courtney, I.A., McKinnon, W.R., and Dahn, J.R., On the aggregation of tin in SnO composite glasses caused by the reversible reaction with lithium, *J. Electrochem. Soc.*, 146, 59, 1999.

203. Mao, O., Turner, R.L., Courtney, I.A., Fredericksen, B.D., Buckett, M.I., Krause, L.J., and Dahn, J.R., Active/inactive nanocomposites as anodes for Li-ion batteries, *Electrochem. Solid State Lett.*, 2, 3, 1999.

204. Mao, O., and Dahn, J.R., Mechanically alloyed Sn-Fe(-C) powders as anode materials for Li-ion batteries: I. the Sn$_2$Fe-C system, *J. Electrochem. Soc.*, 146, 405, 1999.

205. Mao, O., and Dahn, J.R., Mechanically alloyed Sn-Fe(-C) powders as anode materials for Li-ion batteries: II. The Sn-Fe system, *J. Electrochem. Soc.*, 146, 414, 1999.

206. Mao, O., and Dahn, J.R., Mechanically alloyed Sn-Fe(-C) powders as anode materials for Li-ion batteries: III. Sn$_2$Fe:SnFe$_3$C active/inactive composites, *J. Electrochem. Soc.*, 146, 423, 1999.

207. Kepler, K.D., Vaughey, J.T., and Thackeray, M.M., Li$_x$Cu$_6$Sn$_5$ (0 < $x$ < 13): an intermetallic insertion electrode for rechargeable lithium batteries, *Electrochem. Solid State Lett.*, 2, 307, 1999.

208. Vaughey, J.T., O'Hara, J., and Thackeray, M.M., Intermetallic insertion electrodes with a zinc blende-type structure for Li batteries: a study of Li$_x$InSb (0 ≤ $x$ ≤ 3), *Electrochem. Solid State Lett.*, 3, 13, 2000.

209. Fransson, L.M.L., Vaughey, J.T., Edstrom, K., and Thackeray, M.M., Structural transformations in intermetallic electrodes for lithium batteries: an in situ x-ray diffraction study of lithiated MnSb and Mn$_2$Sb, *J. Electrochem. Soc.*, 150, A86, 2003.

210. Yang, J., Wachtler, M., Winters, M., and Besenhard, J.O., Sub-microcrystalline Sn and Sn-SnSb powders as lithium storage materials for lithium-ion batteries, *Electrochem. Solid State Lett.*, 2, 161, 1999.

211. Ehrlich, G.M., Durand, C., Chen, X., Hugener, T.A., Spiess, F., and Suib, S.L., Metallic negative electrode materials for rechargeable nonaqueous batteries, *J. Electrochem. Soc.*, 147, 886, 2000.

212. Kim, H., Park, B., Sohn, H., and Kang, T., Electrochemical characteristics of Mg-Ni alloys as anode materials for secondary Li batteries, *J. Power Sources*, 90, 59, 2000.

213. Kim, I.S., Blomgren, G.E., and Kumta, P.N., Si/TiN nanocomposites, novel anode materials for Li-ion batteries, *Electrochem. Solid State Lett.*, 3, 493, 2000.

214. Kim, I.S., Blomgren, G.E., and Kumta, P.N., Nanostructured Si/TiB$_2$ nanocomposites anode materials for Li-ion batteries, *Electrochem. Solid State Lett.*, 6, A157, 2003.

215. Kim, I.S., Blomgren, G.E., and Kumta, P.N., Si-SiC nanocomposite anodes synthesized using high-energy mechanical milling, *J. Power Sources,* 130, 275, 2004.

216. Kim, I.S., Blomgren, G.E., and Kumta, P.N., Sn/C composite anodes for Li-ion batteries, *Electrochem. Solid State Lett.*, 7, A44, 2004.

217. Kim, I.S., Blomgren, G.E., and Kumta, P.N., Si/TiN nanocomposite anodes by high-energy mechanical milling, *Ceram. Trans.*, 249, 35, 2002.

218. Kim, I.S., Blomgren, G.E., and Kumta, P.N., New nanostructured silicon and titanium nitride composite anodes for Li-ion batteries, *Ceram. Trans.*, 249, 127, 2002.

219. Shackelford, J.F., and Alexander, W., *Materials Science and Engineering Handbook*, 3rd ed., CRC Press, Boca Raton, FL, 2001.

220. Lynch, C.T., *CRC Handbook of Materials Science*, vol. I, CRC Press, Cleveland, OH, 1974.

221. Kim, I.S., Synthesis, structure and properties of electrochemically active nanocomposites, Ph.D. dissertation, Carnegie Mellon University, 2003, p. 54.

222. Li, H., Huang, X., Chen, L., Wu, Z., and Liang, Y., A high capacity nano-Si composite anode material for lithium rechargeable batteries, *Electrochem. Solid State Lett.*, 2, 547, 1999.

223. Williams, D.B., and Carter, C.B., *Transmission Electron Microscopy*, Plenum Press, New York, 1996.

# 27 Chemical Solution Deposition of Ferroelectric Thin Films

*Robert Schwartz, Theodor Schneller, Rainer Waser, and Harold Dobberstein*

## CONTENTS

# I. OVERVIEW

## A. HISTORICAL PERSPECTIVE

Sol-gel processing of materials has been investigated for more than 100 years,[1] but the use of related chemical solution deposition (CSD) techniques for the preparation of ferroelectric thin films is significantly more recent. The process[2–5] involves the synthesis of a multicomponent solution that may be applied as a coating onto a substrate, with the formation of the desired crystalline phase being induced by heat treatment. A variety of solution synthesis methods have been developed, the vast majority of which are based on metallo-organic precursors. Three general classes of approaches are today extensively utilized, including sol-gel, chelate (molecularly modified precursors), and metallo-organic decomposition (MOD). Noteworthy activities in CSD began about 20 years ago with the work of Fukushima et al.[6] and Budd and co-workers.[7,8] These investigations are significant because they represent some of the first studies that demonstrated it was possible to attain the bulk material properties associated with ferroelectric compositions in thin-film form. These studies, and the work of others on the chemical preparation of materials,[9–11] led to the rapid growth of this field and investigations around the world on ferroelectric thin-film devices.[12,13] Today, commercially available devices include decoupling capacitors,[14,15] smart cards based on ferroelectric random access memory (FeRAM), and microelectromechanical systems (MEMS).

This chapter reviews the general aspects of the CSD method for ferroelectric thin-film preparation, with attention given to precursors, solution chemistry, and process development. An additional focus of the chapter is on the structural evolution of the solution precursor into the crystalline (typically perovskite) state and the impact of precursor chemistry and film fabrication conditions on the transformation process. Lastly, the chapter reviews the advantages and disadvantages of the CSD method and discusses industrial implementation of the technique.

## B. MATERIAL SYSTEMS AND APPLICATIONS

A broad range of electronic ceramic materials have been prepared by CSD, but three material systems have dominated the field of ferroelectric thin films. These include the perovskites $PbZrO_3$-$PbTiO_3$ (lead zirconate titanate; PZT), $BaTiO_3$-$SrTiO_3$ (barium strontium titanate; BST), and the layered perovskite $SrBi_2Ta_2O_9$ (strontium bismuth tantalate; SBT).[16–19] The extensive solid solubility ranges

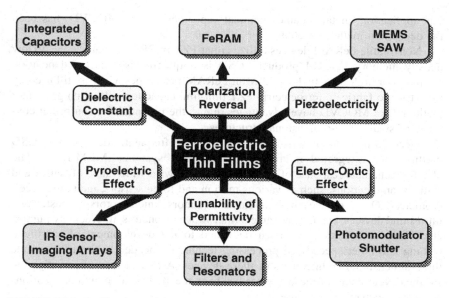

**FIGURE 27.1** Applications of ferroelectric thin films and their enabling properties.

afforded by these materials is one of their key characteristics, as is the ability to add dopant species on either the A- or B-site of the $ABO_3$ perovskite structure, which allows further tailoring and optimization of properties.[20,21] Other material systems of note that have been prepared using CSD include relaxer ferroelectrics, such as $Pb(Mg_{1/3},Nb_{2/3})O_3$ (PMN),[22] high-temperature superconductors, such as $YBa_2Cu_3O_7$,[23] and colossal magnetoresistive (CMR) materials, such as $La_{1-x}(Ca,Sr)_xMnO_3$.[24]

Applications that have received attention, and the material properties that enable them, are shown in Figure 27.1. These applications are reviewed in detail in Waser[25,26] and Ramesh.[27] Decoupling capacitors and filters on semiconductor chips, packages, and polymer substrates (e.g., embedded passives[28]) utilize planar or low aspect ratio oxide films. These films, with thicknesses of 0.1 to 1 μm, are readily prepared by CSD. Because capacitance density is a key consideration, high-permittivity materials are of interest. These needs may be met by morphotropic phase boundary PZT materials, BST, and BTZ ($BaTiO_3$-$BaZrO_3$) solid solutions. Phase shifters (for phase array antennas) and tunable resonator and filter applications are also enabled by these materials because their effective permittivity exhibits a dependence on the direct current (DC) bias voltage, an effect called tunability.[29–32]

On the other hand, there are applications shown in Figure 27.1 that are not effectively met by CSD. For example, integrated capacitors for dynamic random access memory (DRAM) node elements require a much higher capacitance density, extremely small lateral dimensions, and three-dimensional architectures. For this application, metal organic chemical vapor deposition (MOCVD) has been the deposition method of choice.[12,13]

Nonvolatile FeRAM devices utilize either PZT or SBT derivatives.[12,13] In low density memory FeRAM products, CSD is frequently used as the deposition method. For high-density 4- or 32-Mbit FeRAM prototypes, CSD is still used by industry[33] to fabricate ferroelectric PZT thin-film capacitors, although gas phase methods like MOCVD have advantages due to the potential for conformal coverage of small three-dimensional structures.

Piezoelectric and electrostrictive materials are frequently deposited by CSD methods for integrated microelectromechanical systems (MEMS).[34,35] The electromechanical response of these materials allows a wide range of sensor and actuator applications, such as various motion and force sensors, including accelerometers. Other devices include linear actuators, micromotors, acoustic and ultrasound devices, electromechanical microwave resonators, microcavity pumps, etc.[36-38] Great attention has been dedicated to the development of sacrificial etching and other specialized processing techniques needed to form the unique three-dimensional architectures of these devices. A related area of investigation for the development of these devices has been the study of domain contributions to electromechanical response.[39-43]

The pyroelectric response of ferroelectrics may be exploited to detect temperature changes with extremely high sensitivity. The most common devices are uncooled infrared (IR) detectors, which may be used for spectroscopic analysis as well as imaging applications.[44] Pyroelectric thin films based on perovskite-type complex oxides, including $Pb(Sc,Ta)O_3$ have been deposited by CSD for intruder alarms, gas sensors, and IR cameras.[44-46] It is anticipated that these thin-film devices will be substantially less expensive to manufacture than existing bulk polycrystalline devices, which require labor-intensive manufacturing procedures.

## II. FILM FABRICATION VIA CSD: EXPERIMENTAL METHODS

### A. INTRODUCTION

For CSD processing of ferroelectric thin films, a "homogeneous" solution of the necessary cation species that may later be applied to a substrate must be prepared. Overall, the basic process involves the steps of solution preparation, film deposition, pyrolysis for removal of organic constituents, and heat treatment to induce crystallization, as shown in Figure 27.2. This section provides some of the experimental details associated with the CSD process, while the next section discusses the process from a more fundamental perspective.

### 1. Solution Precursors

The requirements of mutual solubility together with pyrolysis behavior that leaves only the cations and oxygen represent a significant challenge regarding the choice

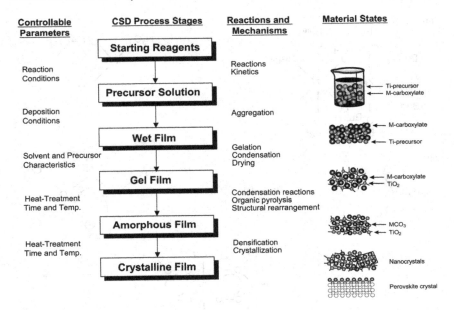

| Controllable Parameters | CSD Process Stages | Reactions and Mechanisms | Material States |
|---|---|---|---|

**FIGURE 27.2** Flow diagram of CSD. Also shown: material states, controllable processing parameters, and reactions and mechanisms. After Schwartz et al.[4]

of precursor compounds. Usually metallo-organic materials are suitable because their solubility in polar and nonpolar solvents can be tuned by modifying the organic part of the molecule, and because the organic moiety pyrolyzes in oxidizing ambient without residue. A variety of metallo-organic precursors have been utilized in CSD processing. The first three classes of compounds—metal alkoxides, carboxylates, and β-diketonates—are discussed below in order of decreasing reactivity. A fourth class of compounds, the mixed-ligand precursors, have reactivity intermediate to the parent alkoxide and carboxylate, or β-diketonate, groups from which they are synthesized. Lastly, specialized precursors with known cation and ligand stoichiometry have been synthesized in-house and are discussed in the section on Structural Evolution. For further information regarding the underlying chemistry, the reader is referred to standard introductory texts,[47] texts dealing specifically with metallo-organic compounds,[48,49] and texts related to sol-gel processing.[50,51]

### a. Alkoxides

The metal salts of alcohols are called alkoxides, $M(OR)_x$, where M is a metal and R is an alkyl group. Typical examples in the synthesis of ferroelectric thin films include titanium isopropoxide [$Ti(OPr^i)_4$; where $Pr^i$ represents isopropyl], tantalum ethoxide [$Ta(OEt)_5$; where Et represents ethyl], and zirconium butoxide butanol [$Zr(OBu^n)_4 \cdot n\text{-}BuOH$; where $Bu^n$ represents $n$-butyl], an alkoxide/alcohol compound. The structure of this compound, shown in Figure 27.3a, illustrates its dimeric nature, that is, the fact that each molecule contains two metal centers.

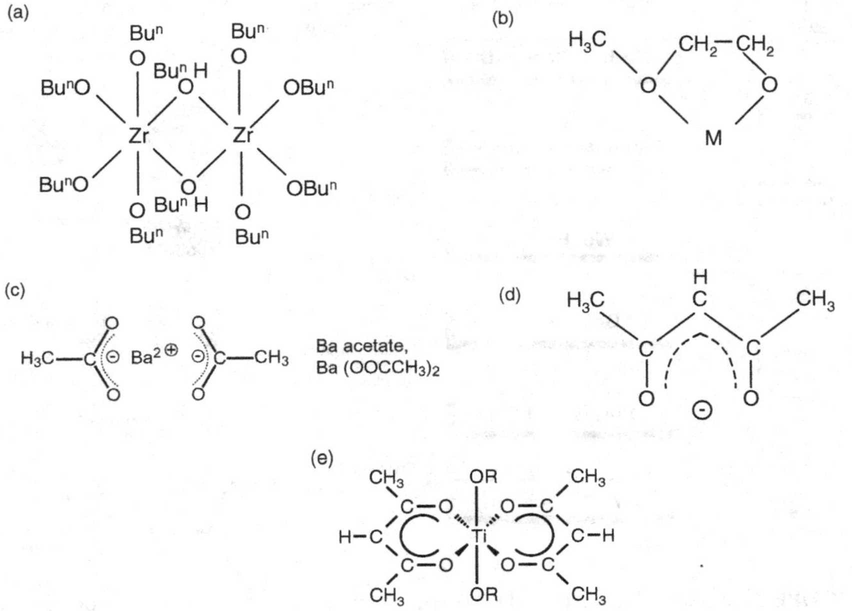

**FIGURE 27.3** Examples of metallo-organic precursors commonly used in the fabrication of ferroelectric perovskite thin films: (a) zirconium butoxide·butanol; (b) schematic of the 2-methoxyethoxy ligand bonded to a metal center; (c) barium acetate; (d) the acac ligand; and (e) the mixed-ligand precursor titanium bis-acac diisopropoxide. After Schwartz et al.[4]

Due to their polar M–O bond, nonstabilized alkoxides are very sensitive toward hydrolysis, leading to the formation of metal hydroxides (M-OH), with the concurrent release of an alcohol molecule. Hydrolysis reactions are common in the use of these compounds and they have been studied in detail for the silica system, with the effects of catalysis type (acid or base), extent of hydrolysis, and resulting oligomeric species being well documented.[50,52,53] The goal in using these precursors for film formation is to control the extent of hydrolysis and subsequent condensation to yield short-chain (a few to a few dozen metal centers) polymeric species, referred to as oligomers.

Alkoxides can be stabilized toward hydrolysis if their organic part R contains further polar groups, such as ether linkages, R-O-R (e.g., 2-methoxyethanol; shown in Figure 27.3b), amine groups, $-NH_2$, keto groups $>C=O$, or further alcohol groups, $>C-OH$ (e.g., 1,3-propanediol). The stabilization results from the fact that the polar groups form additional bonds to the metal center and thus contribute to the complexation of the cation. Steric hindrance may also play a role in accessibility of the polar M–O bond to attack by a water molecule. If more than one polar group of an organic molecule bonds to the same metal cation, the organic species is called a chelating agent (from the Greek word *chele*, "claw"), which further contributes to the stabilization of the alkoxide.

### b. Carboxylates

Metal carboxylates are salts of carboxylic acids, R-COOH, where R is an alkyl group such as methyl, $CH_3$ (acetic acid; salt: acetate), ethyl, $C_2H_5$ (propionic acid; salt: propionate), etc. The polar nature of the salt is indicated by the charges shown in the illustration for barium acetate (typical compound) in Figure 27.3c. Carboxylates can normally be dissolved in their parent carboxylic acids, and for short alkyl chain compounds, some solubility in water and other highly polar solvents is typically observed, due to the polar nature of the salt. Long, nonpolar alkyl chains, such as 2-ethylhexanoate ($R = C_7H_{15}$), lead to carboxylates that are soluble in nonpolar solvents such as xylene. These precursors find wide use in MOD processes. Chemically carboxylates are stable against water and oxygen, and in the processing of ferroelectric thin films, they are often employed as precursors for lower-valent ionic cations such as $Pb^{2+}$, $Sr^{2+}$, and $Ba^{2+}$.

### c. β-diketonates

Organic molecules with two keto groups separated by one methylene group ($>CH_2$) are called β-diketones. These molecules undergo an internal hydrogen exchange reaction (keto-enol tautomerism) that converts one of the keto groups into an alcohol group. Metal alkoxide-based compounds may be formed which are strongly stabilized by the additional keto group due to the chelate effect, and even more so due to the delocalization of the electrons in the six-membered ring (see Figure 27.3d). Typically β-diketonates are monomeric species that can be dissolved in alcohols, ketones, and ethers. Compared to alkoxide and carboxylate precursors, they exhibit higher pyrolysis temperatures and result in the retention of the amorphous nature of the film to higher temperatures, which may be beneficial in improving film density.

### d. Mixed-Ligand Precursors

Mixed-ligand precursors are also frequently employed in CSD processing. For example, titanium tetraisopropoxide, which is too reactive to be directly employed in most CSD routes, may be converted into a more suitable precursor by a reaction with either acetic acid or acetylacetone (Hacac). Such reactions are critical in dictating precursor characteristics and have been studied extensively.[54,55] Using these reactions, crystalline compounds of known stoichiometry and structure have been synthesized that may subsequently be used as "known" precursors for film fabrication.[56,57] Mixed-ligand molecules (carboxylate-alkoxide[58] and diketonate-alkoxide[49]) represent complexes that are not easily hydrolyzed. A typical structure for one of these compounds is shown in Figure 27.3e.

## B. SYNTHETIC APPROACHES

After purchasing commercial metallo-organic reagents or following the in-house synthesis of precursor compounds, one of three families of approaches is usually employed for solution synthesis. These include sol-gel processes, typically based on the use of metal alkoxides; chelate processes, which utilize modified metal alkoxides (typically modification is by acetic acid); and MOD routes that employ long-chain carboxylate compounds.

## 1. Sol-Gel Processes

Although numerous alcohols are employed as solvents in sol-gel processing, for ferroelectric thin-film fabrication, 2-methoxyethanol ($CH_3OCH_2CH_2OH$) has been the most extensively used.[12,13] The starting reagents utilized are typically alkoxide compounds and the key reactions leading to the formation of the metal-oxygen-metal (M-O-M) network that is eventually formed are hydrolysis and condensation:

Hydrolysis

$$M(OR)_x + H_2O \Rightarrow M(OR)_{x-1}(OH) + ROH \qquad (27.1)$$

Condensation (alcohol elimination)

$$2M(OR)_{x-1}(OH) \Rightarrow M_2O(OR)_{2x-3}(OH) + ROH \qquad (27.2)$$

Condensation (water elimination)

$$2M(OR)_{x-1}(OH) \Rightarrow M_2O(OR)_{2x-2} + H_2O \qquad (27.3)$$

Another key reaction associated with the use of this solvent is the alcohol exchange reaction, which results in reduced hydrolysis sensitivity of the starting reagents, such as zirconium $n$-propoxide and titanium $i$-propoxide, frequently used in the production of PZT films. This reaction is represented as

Alcohol exchange

$$M(OR)_x + xR'OH \Rightarrow M(OR')_x + xROH, \qquad (27.4)$$

where OR is a reactive alkoxy group, OR' is the less reactive methoxyethoxy group, and as written, complete exchange has occurred. 2-methoxyethanol represents a chelate-forming, nonbridging bidentate ligand, as shown in Figure 27.3b. Thus 2-methoxyethanol is not simply a solvent in these processes.

In addition to the alcohol exchange reaction, 2-methoxyethanol has also been found to be beneficial in the dissolution of carboxylate precursors, such as lead acetate. In this case, by refluxing the lead acetate precursor in 2-methoxyethanol, one of the acetate groups is replaced, resulting in the formation of the soluble mixed-ligand precursor $Pb(OOCCH_3)(OCH_2CH_2OCH_3) \cdot 0.5H_2O$.[58] Typically carboxylate compounds for lead are employed due to the instability of lead alkoxides and their limited commercial availability.

A typical flow diagram for a 2-methoxyethanol-based process for the preparation of $PbTiO_3$ is shown in Figure 27.4.[7] The general steps involved include

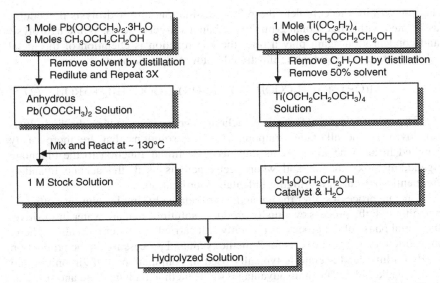

**FIGURE 27.4** Flow diagram for a representative 2-methoxyethanol process, here illustrated for the preparation of PbTiO$_3$.

(i) dissolution of the lead source with the formation of the mixed-ligand precursor noted above; (ii) alcohol exchange through reflux and distillation to form a titanium precursor that is less susceptible to hydrolysis; (iii) combination of the solutions with the possible formation of a mixed-metal compound; and (iv) partial hydrolysis of the solution for the formation of larger oligomeric species. The advantages of the method include excellent control and reproducibility of the process chemistry and low aging rates for nonhydrolyzed solutions. A general disadvantage of the method is the comparatively involved chemical procedures, including several reflux and distillation steps. An additional disadvantage of the method is the toxicity of the most commonly used solvent, 2-methoxyethanol, a known teratogen.

## 2. Chelate Processes

Hybrid, chelate, or molecularly modified precursor routes also utilize alkoxide compounds for the B-site species. Unlike true sol-gel processes, these routes rely on molecular modification of alkoxide compounds through reactions with other reagents, namely chelating ligands, such as acetic acid, acetylacetone, or amine compounds. Despite this difference, chelate processes still share several common attributes with methoxyethanol-based sol-gel processes, most importantly, the formation of oligomeric species during solution synthesis.

The most commonly used chelate processes are based on acetic acid,[55,59-64] and although several steps are typically required for solution synthesis, solution preparation generally requires less than 1 h. This compares to the 1 to 2 days that are required to complete the reflux, distillation, and hydrolysis procedures associated with a traditional 2-methoxyethanol sol-gel process. While hydrolysis and condensation again play a role, the key reaction is chelation of the metal alkoxide precursors, which is illustrated below for acetic acid:

$$M(OR)_n + xCH_3COOH \rightarrow M(OR)_{n-x}(OOCCH_3)_x + xROH. \quad (27.5)$$

The primary reason for using chelating agents is to reduce the hydrolysis sensitivity of the alkoxide compound to form solutions that are more easily handled in air. Chelating agents thus serve a similar function to the bidentate ligand, methoxyethanol, and when acetic acid is used, the acetate ligand is frequently present in a bridging bidentate coordination.

Chelate processes offer the advantages of relatively simple solution synthesis, and although the process eventually produces solutions that are water insensitive, the initial phase of the process is typically still carried out under inert atmosphere, most often, in a glove box. As for 2-methoxyethanol processing, in the production of PZT films, lead acetate is typically used in conjunction with zirconium and titanium alkoxides. The main advantages of the method include rapid and straight-forward solution synthesis. A disadvantage of the method is the relatively high aging rates of the solutions. For further information on chelate processes, the reader is referred to References 59 through 61 for PZT films and References 62 through 64 for BaTiO$_3$-based films.

## 3. MOD Solution Synthesis

Metallo-organic decomposition has historically used long-chain carboxylate or β-diketonate (acac-type) compounds, such as lead 2-ethylhexanoate, zirconium neodecanoate, and titanium dimethoxy dineodecanoate,[65–67] although MOD routes based on short-chain carboxylates for PZT[68] and La$_{1-x}$(Ca,Sr)$_x$MnO$_3$[69] have been reported by Haertling and Hasenkox, respectively. The basic MOD approach consists of simply dissolving the metallo-organic compounds in a common solvent, usually xylene, and combining the individual cation solutions to yield the desired oxide stoichiometry.[65–67] Since the starting compounds are water insensitive, they do not display the oligomerization behavior discussed above, and the precursor species that exist in solution retain a strong resemblance to the starting molecules. This aspect of the process, together with the use of noninteracting solvents, allows for characterization of the solution as a simple mixture of the starting compounds.

The three general families of solution deposition processes possess their own unique attributes, but any can be used effectively to prepare high-quality films, provided adequate attention is given to understanding the chemistry and associated attributes of the method. Generally speaking, sol-gel processes are the most complex to implement, but offer the greatest ability to control film processing behavior. The MOD processes are the simplest, but provide little ability to control film processing behavior from a chemical perspective. And the attributes of chelate methods lie somewhere between these two extremes, but suffer more severe solution aging characteristics due to multiple ligand types and continued esterification reactions.[70]

## C. FILM DEPOSITION

A variety of techniques may be used to deposit the precursor solution onto the substrate, including spin and dip coating, and various spray techniques,[4] though for ferroelectric film fabrication, spin coating[71–73] is the process that has been used almost exclusively. On a laboratory scale, deposition is usually carried out with a photoresist spinner and the substrate is typically either a single crystal oxide or an electroded silicon wafer. The substrate is held in place on the spinner by a vacuum and is typically "flooded" with solution during a static dispense using a syringe with a 0.2 µm filter for elimination of particulates. The wafer is then accelerated rapidly to 1000 rpm to 8000 rpm, with 3000 rpm being the most commonly used speed. The angular velocity and spinning time, together with the solution viscosity, can be used to control the thickness of the wet film.

A variety of defects can arise at this processing stage. Simple defects, such as dust or airborne particulates, may be incorporated in the film, although the density of these defects may be reduced by the use of a clean room or clean tent. Another defect that may arise during deposition is striation formation (i.e., thickness variations) within the layer. Striations typically appear as radial patterns originating at the center of the substrate and are associated with phase segregation during the drying stage of spin casting.[74–76] Other film processing characteristics that are controlled through manipulation of solution chemistry include substrate wetting and film cracking.

Film thickness for most common CSD processes is usually in the range of 0.05 to 0.1 µm per coating, although thinner layers have also been prepared by using more dilute solutions and/or higher spinning rates. For the preparation of thicker coatings, a multilayer approach is often used. In practice, this means the deposition of multiple (usually three or four) layers, with a pyrolysis step after each layer, followed by a crystallization anneal. Films with thicknesses in excess of 10 µm have been prepared by this technique, though some investigators have developed automated dip-coating procedures to achieve the same goal while reducing the manual labor associated with this type of repetitive procedure.[77]

Another method that has been developed for the preparation of thicker layers involves the use of diol solvents, such as 1,3-propanediol. Phillips et al.[78] and Tu et al.[79] have explored this route most thoroughly, and it has also been utilized by manufacturers.[14,15] While the method has not been as intensely investigated as the 2-methoxyethanol process, a number of spectroscopic investigations have been carried out to study the details of the reaction chemistry. Coating thickness per deposited layer is ~0.5 µm, compared to ~0.1 µm for layers prepared by sol-gel methods using more common solvents, or by chelate or MOD processes. The manufacture of 1 µm films is therefore significantly less labor intensive. Lastly, others have prepared suspensions from sol-gel precursor solutions and powders for the deposition of still thicker coatings.[80,81] This "composite" approach has been used to prepare films with acceptable properties with thicknesses between 100 and 200 µm.

## D. FIRING STRATEGIES

Conversion of the as-deposited film into the crystalline state has been carried out by a variety of methods. The most typical approach is a two-step heat treatment process involving separate low-temperature pyrolysis (~300 to 350°C) and high-temperature (~550 to 750°C) crystallization anneals. The times and temperatures utilized depend upon precursor chemistry, film composition, and layer thickness. At the laboratory scale, the pyrolysis step is most often carried out by simply placing the film on a hot plate that has been preset to the desired temperature. Nearly always, pyrolysis conditions are chosen based on the thermal decomposition behavior of powders derived from the same solution chemistry. Thermal gravimetric analysis (TGA) is normally employed for these studies, and while this approach seems less than ideal, it has proved reasonably effective. A few investigators have studied organic pyrolysis in thin films by Fourier transform infrared spectroscopy (FTIR) using reflectance techniques.[82,83] This approach allows for an *in situ* determination of film pyrolysis behavior.

While it might be intuitively felt that the rapid heating rates associated with placing a film on a hot plate set at 300°C would cause cracking of the film, this is in fact rarely observed, at least for the common layer thicknesses that most solution chemistries produce. The suggested origin of this behavior is that the organic constituents (entrapped solvent and nonhydrolyzed organic groups associated with the oligomer backbone) are removed prior to collapse (consolidation) of the film. Ellipsometry has been employed to monitor changes in film thickness during this processing step and shrinkages of 50 to 90% have been observed for $TiO_2$ and $ZrO_2$ thin films.[84]

In the second phase of the standard two-step heat treatment method, the pyrolyzed film, which is amorphous at this stage, is subjected to higher heat treatment temperatures for nucleation and growth into the desired crystalline structure. For PZT, BST, and SBT, the temperatures typically utilized are 550 to 750°C, and most investigators seem to have settled on a heat treatment time of 30 min, though both shorter and longer times have been used. It is virtually certain that nucleation begins during the ramp to the crystallization temperature,[85] with complete conversion to the final crystalline phase occurring during the longer hold period at the maximum temperature.

In addition to the more commonly employed two-step thermal treatment procedure discussed above, a number of investigators have also used a single-step process. This method has generally received more attention for BT- and ST-based materials, while the two-step approach has been more widely applied to lead-based compositions. In the single-step process, the deposited film may be directly inserted into a furnace already at the crystallization temperature, or rapid thermal annealing (RTA) can be used (with or without a separate pyrolysis step).[86,87] One primary consideration in the use of this approach is to densify the film prior to the onset of crystallization.[84] Generally, compared to the slower heating rates inherent in the two-step process, the higher heating rates associated with the single step method delay condensation reactions to higher temperature, resulting in a more compliant film that may be densified more effectively. However, it should also be noted that the complexity of the transformation into the crystalline state is perhaps greater for the single-step process, since many of the processes involved in this conversion (pyrolysis, densification, nucleation, and growth) overlap to a greater extent. Irrespective of the specific effects of RTA on the transformation mechanism, the reduced thermal budget associated with this heat treatment approach can reduce concerns related to degradation of the electrode stack that may arise during the longer thermal cycles associated with standard heat treatment processes.[88]

The goal of the high-temperature heat treatment step is to produce a single-phase microstructure in which transitory phases and porosity have been eliminated. This requires control of the densification, nucleation, and growth behavior of the film, as discussed in Section III. Another issue that must be considered here is compositional variation that might result from the volatility of one of the constituents of the film, such as PbO in PZT[89] or $Bi_2O_3$ in SBT.[90] Frequently the volatility of a component may be compensated through the incorporation of an excess of the parent metallo-organic compound in the precursor solution (e.g., lead acetate for PbO), although other processing strategies have been used effectively to address this issue. Notably, the use of lead overcoat solutions[89] and maintenance of lead oxide activity during firing[91] have received attention.

Heating rates typically range from 10 to 50°C/min when a standard tube or box furnace is used for crystallization, to ~300°C/s when RTA is used.[86,87] Heating rates and the time and temperature of the crystallization anneal are known to play a role in the significance of transitory phases that are formed during heat treatment. These phases may be compositionally related to the final film, as for the formation of fluorite during PZT processing,[91,92] may be decomposition products, such as $BaCO_3$ in BT or BST processing,[93–96] or may involve reaction with the lower electrode, such as the formation of $Pt_3$-Ti or Pb-$Pt_x$ in PZT processing on platinum electrodes.[97,98]

## III. STRUCTURAL EVOLUTION

### A. INTRODUCTION

In this section, the CSD process is explored from a more fundamental perspective, looking at the conversion of the solution species into the crystalline film. This viewpoint is illustrated in Figure 27.2, which shows material characteristics at different processing stages, along with some of the associated properties. Also shown in the figure are the physical and chemical mechanisms leading from one processing stage to the next, together with controllable process parameters. Consideration of CSD processing from this viewpoint is usually referred to as a

**TABLE 27.1**

**Analytical Techniques Used in the Development of CSD Processing Routes**

| Characterization Tools | Properties Studied | Refs. |
|---|---|---|
| **Precursor and Solution** | | |
| Multi-nuclear NMR (H, C, O) – *l, s* | Structure, bonding environment | 50, 55 |
| Elemental analysis | Chemical composition and stoichiometry | |
| Small angle x-ray scattering | Oligomer shape and size | 50, 52 |
| Chromatography | By-products and concentrations | 104 |
| Capillary rheometry | $M_w$ determination and distribution | 104 |
| pH | Solution chemistry/reactions | 70 |
| **Films** | | |
| Ellipsometry | Thickness, index, and consolidation | 84 |
| Profilometry | Thickness and uniformity | |
| SIMS, RBS, atomic emission | Composition vs. depth | 88 |
| EXAFS | Amorphous film structure | 102 |
| Surface acoustic wave devices | Porosity and surface area | 103 |
| Film dissolution behavior | Crosslink density | 85 |
| Pyrolysis behavior | FTIR, TGA | 101, 108 |
| Wafer curvature | Stress magnitude and sign | 133 |

"structural evolution" perspective, and the insight regarding the processes that influence microstructural development has been acquired through the use of a wide variety of analytical methods, which are illustrated in Table 27.1. Techniques that have proven particularly effective are multinuclear nuclear magnetic resonance (NMR), which is frequently utilized to probe the chemical characteristics and the local environment of solution precursor species, and extended x-ray absorption fine structure (EXAFS), which has been used to gain an understanding of the nature of amorphous films.

## B. SOLUTION SPECIES

### 1. Sol-Gel Processing

The coating solution is prepared by mixing or reacting the individual precursors or precursor solutions. Compared to the vast body of work on chemically derived silicate materials, the nature of the species formed during the sol-gel processing of ferroelectric films has received relatively little attention. While it is known that the solution preparation, hydrolysis, and catalysis conditions can impact the nature of the species, detailed studies using analytical techniques such as small angle x-ray scattering (SAXS), which have been widely applied to silica,[50,52] have not been carried out. However, a few reports regarding the influence of solution preparation conditions on the solution species, and the resulting gel characteristics, may be found in the work of Budd et al.,[104,105] Schwartz,[106] and Lakeman et al.[107] Generally speaking, at least for PZT processing, hydrolysis and catalysis conditions display similar effects to those noted for the silica system, but the supporting experimental evidence is not as well developed.

Other notable studies of precursor characteristics include the work of Ramamurthi and Payne,[58] who reported on the nature of the lead precursor in a 2-methoxyethanol processing route and Budd,[104] who investigated the molecular weight of the precursor species in this process, confirming its oligomeric nature. Additional studies of precursor structure are detailed in the report of Coffman and Dey.[108] A particularly important study is that of Arčon et al.,[109] who confirmed the presence of Pb-O-Zr and Pb-O-Ti linkages in sol-gel-derived precursors using EXAFS, while other researchers[64] have suggested that the A-site species may simply be occluded onto the surface of B-site-based oligomers.

A review of the literature reveals that there are many claims regarding the formation of mixed-metal alkoxide species under different reaction conditions, but frequently supporting experimental evidence (spectroscopic or otherwise) is not included. Notable exceptions include the report of Kato et al.[110] on $SrBi_2Nb_2O_9$ and $SrBi_2Ta_2O_9$ precursors and the work of Pelligri et al.[111] on PZT precursors, both of which provide detailed information on precursor structural characteristics.

## 2. Chelate and MOD Processing

In chelate routes, chemical reactions also lead to the formation of oligomeric species. Assink and Schwartz[55] used $^1H$ and $^{13}C$ NMR to determine that the solution species in a typical chelate route exhibited essentially complete replacement of the alkoxy ligands by acetate groups, resulting in the formation of oxo-acetate compounds of unknown stoichiometry. In contrast, for long-chain carboxylate compounds, such as zirconium 2-ethylhexanoate, which have been historically used in MOD routes,[65-67] reactivity is low and the chemical interactions between the different precursor compounds are minimal. Therefore, in MOD processing, solution species resemble those of the starting reagents.

## 3. Mixed-Metal Precursors

To control the extent of intermixing and stoichiometry of the cation species, there have been a number of investigations to synthesize stoichiometric mixed-metal precursors with structures similar to the perovskite crystal structure. The motivation behind these efforts is that stoichiometric precursors with structures similar to the desired crystalline phase should undergo crystallization at lower heat treatment temperatures. Most attempts in this area, however, have resulted in mixed-metal species with a cation stoichiometry different than perovskite.[112,113] The formation of such compounds indicates the importance of thermodynamic sinks in the synthesis of mixed-metal precursors.[57]

The fact that the cation stoichiometry of the precursor species is not identical to the crystalline perovskite phase (and in fact, likely varies from one oligomer to another) has not prohibited the preparation of high-quality ferroelectric thin films. In addition, there is limited evidence that the perovskite formation temperature of films prepared from structurally controlled stoichiometric precursors is significantly lower than for those prepared from single-cation precursors. These observations may be due to the short interdiffusional distances that are operative, even for nonstoichiometric precursors. Or they may be due to the fact that the precursor molecule is destroyed during the creation of the amorphous structure, and therefore perovskite formation temperature may depend more critically on precursor stability, the reaction pathway, or the nature of the intermediate phases rather than on the structure of the original precursor molecule. This said, it should be stressed that further work is required in this area, and the studies of investigators such as Kato et al.[110] suggest such a relationship can lower crystallization temperatures.

## C. FILM CHARACTERISTICS

### 1. The As-Deposited State

The nature of the film immediately after deposition is highly dependent on the characteristics of the solution precursor species, solute concentration, solution viscosity, and spinning speed.[84] For example, for diol-based processes, the film is wet, while for more common solvent systems, the film is better described as a viscoelastic solid that contains organic moieties associated with the gel network and significant amounts of entrapped solvent. The type of film that is formed, that is, the film gelation behavior, is defined by the precursor interactions that take place during the spin-off stage[73] of the deposition process, which is driven by solvent evaporation. For reactive precursor species, such as those that might be present in a sol-gel process, chemical cross-linking reactions occur, resulting in "chemical gel" films. For less reactive oligomers, "physical gel" films are formed that result from the simple aggregation of precursor molecules. It should be noted that even when similar starting compounds are utilized, the choice of modifying ligand can have a significant impact on the characteristics of the gelled film. For example, when acac is used (compared to acetate), due to the greater steric hindrance of this ligand, a physical gel film results rather than a chemical gel film.[84] Lastly, "nongelling" (wet) films are usually obtained from diol solvent systems. Therefore the choice of solvent can also impact film gelation and processing behavior, and the solvent that is selected must be suitable not only for dissolution, stability, and solution aging behavior, but for characteristics such as evaporation rate and surface tension, which can impact wetting behavior.

### 2. The Amorphous State

To obtain the desired oxide phase, solvent remaining entrapped within the pore structure of the film and the organic moieties associated with the gel network must be removed. This is typically achieved via low-temperature heat treatment, which results in a rearrangement of the gel network through a variety of bond reorganizations and structural relaxation processes that eliminate structural free volume, organic constituents, and hydroxyl species within the film.[50] During this process, M–O–C and M–O–H bonds are broken, and as the associated volatile species are removed, the formation of an M–O–M network occurs (or for BST materials, nanocrystalline $TiO_2$ and carbonate phases are formed). The reactions that occur during conversion to the amorphous state depend on the heating rate, time, and temperature, but in general include

Thermolysis (the formation of volatile organic species without oxygen).
Pyrolysis (the formation of volatile organic molecules such as CO, $CO_2$ by combustion).
Dehydration (dehydroxylation; the elimination of OH groups from the network as $H_2O$).
Oxidation (the formation of M-O-$CO_2$ carbonate species).

The processes that occur during this stage of film preparation are understood more in general terms rather than from a detailed perspective. One contributing factor to the limited knowledge of the specific processes involved is the difficulty in characterizing the initial (solution species) and final (amorphous film) states. Despite these challenges, a number of studies have been conducted on the structural changes that are associated with organic removal for PZT, BT, and ST materials.[101,107,114]

For PZT thin films, Lakeman et al.[107] have studied the rearrangements that occur within the film during pyrolysis. As-deposited coatings were found to be amorphous, but possessed short-range order. Following pyrolysis, the development of medium range order was observed and chemical heterogeneity on a nanometer scale was observed.[107] In pyrolyzed films heated to temperatures near 400°C, a nanocrystalline fluorite (or pyrochlore) was typically formed, prior to formation of the perovskite phase, which occurred at slightly higher temperatures.

Coffman and Dey[101,108] published one of the most extensive studies of the processes occurring during conversion into the amorphous state. The techniques employed by these authors included thermal gravimetric and differential thermal analysis (TGA and DTA), dynamic mass spectrometric analysis (DMSA), and FTIR. The results obtained support the general reactions noted above, and in particular, carbon dioxide, acetone, and water were noted as reaction products that were evolved during heat treatment between 100 and 455°C. Generally water was removed at lower temperatures, acetone near 300°C, and $CO_2$ at all temperatures studied, but with greater amounts being formed at 455°C.

Lastly, Sengupta et al.[102] provide some of the most detailed insight into the nature of the amorphous PZT film derived from a 2-methoxyethanol process. In their studies, the authors used EXAFS to probe the local geometry and chemistry of sol-gel-derived amorphous PZT. They found evidence for separate networks of Ti-O-Ti, Zr-O-Zr, and Pb-O-Pb, in contrast to the picture of sol-gel-derived materials being homogeneous on the atomic scale and the work of Arčon et al.[109]

## 3. The Crystalline State

Because of the importance of microstructure on dielectric and ferroelectric properties, the transformation pathway associated with conversion of the amorphous film into the crystalline state has been studied extensively. The basic mechanism involved is one of nucleation and growth, although the formation of intermediate phases that can impact the thermodynamic driving forces associated with the transformation frequently occurs.[91,92,115] Another key aspect of CSD films is that crystallization occurs well below the melting point of the materials.[91] Therefore, compared to standard mixed-oxide processing of bulk materials, the thermodynamic driving forces associated with the transformation are much greater and the kinetics of mass transport are much less.

In analyzing the transformation from the amorphous state to the final crystalline state, standard equations for nucleation and growth in conventionally prepared glasses are usually employed. Expressions for the energy barriers for homogeneous and heterogeneous nucleation are

$$\Delta G^*_{\text{homo}} = \frac{16\pi\gamma^2}{3\left(\Delta G_v\right)^2} \tag{27.6}$$

and

$$\Delta G^*_{\text{hetero}} = \frac{16\pi\gamma^3}{3\left(\Delta G_v\right)^2} \cdot f\left(\theta\right) \tag{27.7}$$

where $\Delta G^*_{\text{homo}}$ is the free energy barrier to homogeneous nucleation, $\Delta G^*_{\text{hetero}}$ is the barrier to heterogeneous nucleation, $\gamma$ is the interfacial energy between the nucleus and the matrix, $\Delta G_v$ is the driving force for crystallization, that is, the free energy difference per unit volume associated with the amorphous film–crystalline film transformation, and $f(\theta)$ is a function related to the contact angle, $\theta$, according to Equation 27.8 for a hemispherical nucleus:

$$f\left(\theta\right) = \frac{2 - 3\cos\theta + \cos^3\theta}{4} \tag{27.8}$$

The driving force for crystallization ($\Delta G_v$) is thus a key thermodynamic variable associated with the transformation process, as is the surface energy. This latter factor has been explored in reasonable depth in other approaches to the problem,[83,116] and in some instances this property is believed to dictate the ability to prepare oriented films by CSD. Other investigators have discussed the impact of electrode reaction layers or decomposition pathways.[97,117–119]

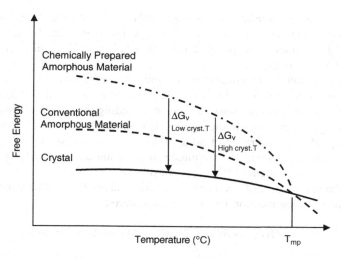

**FIGURE 27.5** Schematic diagram of the free energies of a CSD-derived amorphous film, a conventionally prepare amorphous material, and the crystalline perovskite phase. $\Delta G_v$ is the thermodynamic driving force for crystallization. Also shown is the influence of crystallization temperature on the magnitude of $\Delta G_v$. After Schwartz et al.[4,85] and Roy.[122]

It was noted above that there has been much attention dedicated to the preparation of mixed-metal precursors with a structural similarity to the perovskite state. There has also been keen debate regarding whether sol-gel-derived glasses are different from conventional glasses.[120,121] The general conclusion is that for materials not heated to near the melting point, significant differences can exist. And because of their impact on $\Delta G_v$, as shown in Figure 27.5, these differences can be expected to impact the crystallization behavior and resulting microstructure of the thin film.[50,85,122] Though there have been many observations of solution precursor effects on film microstructure and orientation, the underlying causes of these relationships remain relatively poorly understood. Primarily this is because little is known regarding the associated physical, chemical, and structural characteristics of the amorphous state at the onset of crystallization. Properties of importance may include surface area, residual hydroxyl and organic content, structural free volume (skeletal density), and the atomic organization of the structure.[50] Such information has simply not been acquired for nonsilicate thin-film materials, though there are a few reports regarding these properties in lead titanate powders prepared from different sol chemistries.[119]

To understand why solution precursor characteristics can have such a significant impact on microstructural evolution, it is informative to consider how various features of the amorphous film, which are retained from the solution precursor, might influence the magnitude of $\Delta G_v$. As suggested in Figure 27.5, the free energy of CSD-derived amorphous films is expected to be greater than the analogous conventionally prepared amorphous material. The question that must be answered is how much greater. In the analysis presented here, data reported by Brinker and Scherer[50] for silica are used to address this question.

First, $\Delta G_v$ values for conventionally prepared materials may be estimated from expressions suggested by Turnbull,[123] Hoffmann,[124] and Thompson and Spaeten,[125] though caution must be used in extending these expressions to temperatures significantly below the melting temperature, $T_m$. Using these expressions, $\Delta G_v$ values in the range of 50 to 80 kJ/mol at the crystallization temperatures are estimated. Next, the magnitude of the contribution to $\Delta G_v$ associated with the hydroxyl content of the gel is considered. This is one of three contributions to the excess $\Delta G_v$ of chemically prepared materials. For this contribution, the dehydroxylation reaction for silica is considered:

$$Si(OH)_4 \rightarrow SiO_2 + 2H_2O; \quad \Delta G = -14.9 \text{ kJ/mol.} \quad (27.9)$$

Typical retained hydroxyl levels are estimated at 0.33 to 1.48 per silicon atom, which leads to a contribution to $\Delta G_v$ that is in the range of 1.2 to 5.5 kJ/mol. Surface area measurements of sol-gel-derived silicates[50] and known values of surface energy lead to an estimate of 9 to 30 kJ/mol. And lastly, the reduced skeletal density of sol-gel derived silicates associated with incomplete hydrolysis and condensation leads to a contribution to $\Delta G_v$ of up to 6 kJ/mol. Summing these contributions, a maximum contribution to $\Delta G_v$ of 41.5 kJ/mol may be anticipated for the unique amorphous state features associated with chemically prepared materials. This value represents 50 to 80% of the $\Delta G_v$ expected for a conventionally prepared amorphous material. Therefore, even if the majority of the chemical, physical, and structural differences between the solution and conventionally derived amorphous materials are eliminated during heat treatment prior to crystallization, it may be expected that even small variations in these properties can still lead to kilojoule per mole-level contributions to $\Delta G_v$. These contributions would be of sufficient magnitude to impact the thermodynamics of nucleation and the resulting thin-film microstructure. Therefore the observation of precursor effects on thin-film microstructure should not be surprising.

A final aspect of the transformation process that should also be noted is the impact of the nucleation temperature on $\Delta G_v$, as shown in Figure 27.5. Just as precursor chemistry can impact driving force, so can the temperature at which nucleation occurs. This can affect the active nucleation sites associated with crystallization, as noted previously by Schwartz et al.[85] Frequently in the two-step process, nucleation begins during the ramp up to the annealing temperature; in contrast, when rapid thermal annealing procedures are used, greater control over nucleation and thin-film microstructure may be exerted, since nucleation occurs at the hold temperature. Lastly, as discussed below, the transformation pathway can also exhibit a significant impact on nucleation and growth behavior.

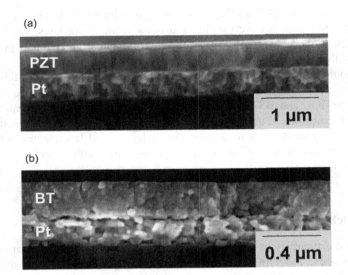

**FIGURE 27.6** Scanning electron microscope photomicrographs of representative microstructures of (a) PZT and (b) BT thin films on platinum prepared by a chelate process.

## 4. Transformation Pathways

Lead zirconate titanate thin films typically undergo a transformation that involves an intermediate fluorite or pyrochlore phase,[91,92,115] while barium titanate and barium strontium titanate thin films crystallize into the perovskite structure via carbonate and titania intermediate phases.[93–96] These pathways contribute to significantly different microstructures for these two families of materials, as illustrated in Figure 27.6. Typically, solution-derived PZT films crystallize with a columnar microstructure that results from the formation of a transient fluorite or pyrochlore phase, followed by nucleation of the perovskite phase at the substrate interface and progression of a growth front associated with this phase toward the surface of the film. The formation of nuclei principally at the substrate interface and the growth front associated with these nuclei lead to the columnar microstructure. The reduction in driving force associated with the formation of the fluorite phase is another key aspect of the process,[91,92,115] and the role of this phase on orientation selection of PZT thin films has recently been investigated.[126]

In contrast to PZT thin films, BT films (Figure 27.6b) are typically noncolumnar in nature. This indicates that nucleation within the bulk of the film has occurred, in addition to nucleation at the interface. From FTIR spectroscopic investigations,[114] it is known that heating of the as-deposited films at an intermediate temperature to remove the organic groups leads to formation of a film that is composed of either alkaline earth carbonate $MCO_3$ (M = Ba, Sr), which may form at temperatures as low as 300 to 350°C and titanium dioxide ($TiO_2$), or nanocrystalline $M_2Ti_2O_5CO_3$. Because the thin film has a homogeneous distribution of the carbonate and oxide phases, the transformation behavior is nearly the same for all parts of the film. Since crystallization starts at numerous sites with identical surroundings and activation energies, a large number of very small grains are produced throughout the film.

It should be noted that these microstructures may be altered significantly by controlling processing conditions. For example, it is possible to prepare columnar $BaTiO_3$, $SrTiO_3$, and BST by reducing layer thickness so that nucleation must occur at the substrate or underlying layer surface.[127] This change in microstructure results in significant improvements in the dielectric and ferroelectric properties of the films, as well as the tunability of the material for filter and phase-shifter applications.

## 5. Modeling of Structural Evolution

Due to the difficulties in acquiring insightful material property data at key processing stages, modeling of microstructural evolution offers an alternative approach to understanding how to control microstructure. While few results have been reported in this area, Dobberstein has completed studies that suggest that the method is effective for predicting both general microstructural differences in PZT and BT films, as well as the impact of processing parameters such as ramp rate, crystallization temperature, and the use of seed layers on microstructure.[116,128,129]

The method is based on classical nucleation and growth equations for amorphous materials and a derived expression for $\Delta G_v$ based on the expressions of Turnbull,[123] Hoffman,[124] and Thompson and Spaeten.[125] Using the calculated $\Delta G_v$ value, published material property data such as modulus and surface energy, and the measured crystallization or glass transition temperature ($T_g$) obtained from differential scanning calorimetry (DSC), analytical expressions for nucleation rate and growth rate can be written. These expressions are then used as the basis for a pixel-by-pixel modeling approach for visualization of the microstructural evolution of the cross-section of a thin film.[116,128,129]

In the modeling studies completed to date, the initial phase is assumed to be a homogeneous amorphous matrix with a cell size of $1000 \times 500$ atoms, and the program that was written includes subroutines for program initialization (input material properties and process conditions), nucleation and growth calculations, and output. Typical results of the approach are shown in Figure 27.7a for PZT and Figure 27.7b for $SrTiO_3$. Considering the use of published bulk material property data, the results are in excellent agreement with observed thin-film microstructures (illustrated in Figure 27.6). (Note that $BaTiO_3$ and $SrTiO_3$ films display nearly identical microstructures; thus comparison of $BaTiO_3$ experimental results to the $SrTiO_3$ simulation is reasonable.) Other results (not included here) confirm the ability of the model to predict process variation effects.[116] It is expected that modeling approaches of this type will have wide utility in process optimization.

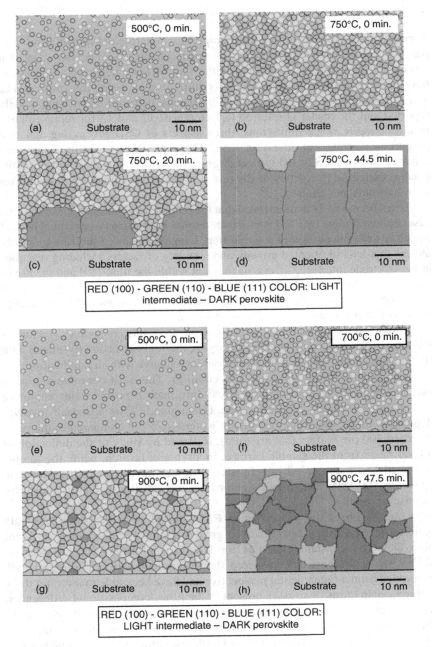

**FIGURE 27.7** Computer simulations of microstructures of (a–d) PZT and (e–h) SrTiO₃ thin-film cross sections illustrating microstructural evolution at various times during the transformation to the perovskite state. Lighter colors associated with intermediate phase; darker colors associated with the perovskite phase.

## 6. Properties

A complete review of the reported properties of ferroelectric thin films prepared by CSD is beyond the scope of this chapter. Suffice it to say that fabrication approaches from each of the three CSD categories noted above have been used to prepare high-quality films in a range of thicknesses. The dielectric response and ferroelectric hysteresis behavior have been widely reported and the reader is referred to References 12 and 13 for representative results. Despite space limitations, three aspects of CSD processing and film properties warrant consideration here. These are (i) the ability to prepare oriented films by CSD; (ii) typical stress levels within the films; and (iii) the general dielectric properties of the thin film materials compared to the corresponding bulk materials.

It has been widely demonstrated that the preparation of oriented, and in some cases epitaxial films by CSD is possible, despite the relatively large thickness of the films that is deposited in a single step. Lange[130] reviewed the various mechanisms that lead to oriented growth, and a variety of factors including reactions at the electrode interface,[97,98] organic content within the film,[117,118] and the use of seed layers[131] to promote homoepitaxy have been discussed. The ability to control film properties (remanent polarization, dielectric constant, etc.) through manipulation of film orientation has also been shown.[132]

Stresses develop within CSD films from a variety of sources, including shrinkage associated with organic pyrolysis, thermal expansion mismatch with the substrate, and phase transformations associated with crystallization (on heating) and the Curie transition (on cooling). The most common approach that has been utilized to acquire stress information is the wafer curvature technique, as detailed by Garino and Harrington.[133] The approach involves deposition of the film onto thin silicon wafers (~75 μm compared to the more standard 300 μm wafers that are used), followed by study of the deflection of a laser beam from the sample surface, in an in situ fashion, while the sample is heated and then cooled in a furnace. Stresses developed in the films can approach 500 MPa, which can significantly impact film dielectric and ferroelectric properties.

Comparison of the properties of CSD thin films to the analogous bulk material properties has also received great attention because of the need for high dielectric constant materials for DRAM applications.[134,135] Basceri et al.[134,135] have thoroughly considered the differences between film and bulk properties from a fundamental perspective and have been able to interpret these differences in terms of stresses present in the films, compositional differences, and the impact of these characteristics on the phenomenological behavior of the material as predicted from a Landau-Ginzburg-Devonshire approach. All observed differences between film and bulk properties were explainable using this approach.

# IV. SUMMARY

## A. Advantages and Disadvantages of Chemical Solution Deposition

The major advantages of the CSD technique are the relatively low capital investment costs and the excellent control of film composition on a molecular level. Other advantages include the opportunity to rapidly develop fabrication routes for the evaluation of new material systems or material systems not previously studied in thin-film form. The high deposition rate, comparative ease of film fabrication over large areas, and compatibility with standard semiconductor fabrication methods represent other advantages of the method.

Perhaps the greatest disadvantage of CSD is the difficulty in depositing ultrathin films (thickness <30 nm). While the possibility has been demonstrated in selected systems, the method is not usually viewed as amenable to that thickness requirement. Another disadvantage of CSD is the fact that three-dimensioanl structures with high aspect ratios cannot be conformally coated. Lastly, while epitaxial films have been prepared by the method,[130,136] generally the films produced are polycrystalline and often not oriented.

A final aspect of the CSD of ferroelectric films is the intensive numbers of variables involved in the process. This can be both an advantage and disadvantage. From the perspective that there are numerous points of process control, the great number of variables allows significant control of film fabrication and properties. However, this also requires a fundamental understanding of a broad range of processing property interrelationships, which is a disadvantage. Because of the large number of parameters associated with the process, statistically designed experimentation has been employed to identify solution synthesis and film fabrication conditions that result in optimal film properties.[137,138]

## B. Industrial Implementation

Chemical solution deposition processes have been used around the world in academic and industrial laboratories for ferroelectric thin-film fabrication. However, CSD has not yet found widespread acceptance in manufacturing environments, aside from a few select instances in which the method has been adopted. Perhaps the greatest challenge to industrial implementation of CSD has been the perception that methods such as MOCVD are more easily controlled and are capable of producing higher-quality materials. While this perception will be difficult to change, it should be noted that the film properties resulting from CSD processes are at least comparable to films produced by other deposition methods. Not only do the dielectric and ferroelectric properties of the films compare favorably, it has been widely demonstrated that highly oriented thin films can be prepared by CSD.[130,131,136,139–144] Hopefully, with increased familiarity with the method, and greater commercial availability of metallo-organic precursors for the most widely investigated material systems, CSD methods will be more widely adopted, as they are suitable for a range of applications that require films in the ~100 nm[145] to 5.0 μm[80,81] thickness range.

## REFERENCES

1. Ebelman, J.J., and Bouquet, M., *Ann. Chim. Phys.*, 17, 54, 1846.
2. Schwartz, R.W., *Chem. Mater.*, 9, 2325, 1997.
3. Sheppard, L.M., *Ceram. Bull.*, 71, 85, 1992.
4. Schwartz, R.W., Schneller, T., and Waser, R., *C. R. Chemie*, 7, 433, 2004.
5. Tuttle, B.A., and Schwartz, R.W., *Mater. Res. Bull.*, 21, 49, 1996.
6. Fukushima, J., Kodaira, K., and Matsushita, T., *J. Mater. Sci.*, 19, 595, 1984.
7. Budd, K.D., Dey, S.K., and Payne, D.A., *Br. Ceram. Proc.*, 36, 107, 1985.
8. Dey, S.K., Budd, K.D., and Payne, D.A., *IEEE Trans. UFFC*, 35, 80, 1988.
9. Blum, J.B., and Gurkovich, S.R., *J. Mater. Sci.*, 20, 4479, 1985.
10. Better ceramics through chemistry (I–VII), *Mater. Res. Soc. Symp. Proc.*, 32, 73, 121, 180, 271, 346, 435, 1984–1996.
11. Mackenzie, J.D., and Ulrich, D.R., Eds., *Ultrastructure Processing of Advanced Ceramics*, John Wiley & Sons, New York, 1988.
12. Ferroelectric thin films (I–X), *Mater. Res. Soc. Symp. Proc.*, 200, 243, 310, 361, 433, 493, 541, 596, 655, 688, 1991–2001.
13. Integrated ferroelectrics, *Proceedings of the International Symposia on Integrated Ferroelectrics*, 1–17, 1990–1997.
14. Liu, D.D.H., and Mevissen, J.P., *Int. Ferro.*, 18, 263, 1997.
15. Liu, D., Kennedy, R.M., Korony, G., Makl, S., Mevissen, J.P., Hock, J.M., and Heistand, R., II, *Proc. SPIE*, 3830, 431, 1999.
16. Araujo, C.A., Cuchiaro, J.D., McMillan, L.D., Scott, M.C., and Scott, J.F., *Nature*, 374, 627, 1995.
17. Jones, R.E., Maniar, P.D., Moazzami, R., Zurcher, P., Witowski, J.Z., Lii, Y.T., Chu, P., and Gillespie, S.J., *Thin Solid Films*, 270, 584, 1995.
18. Jimenez, R., Calzada, M.L., Gonzalez, A., Mendiola, J., Shvartsman, V.V., Kholkin, A.L., and Vilarinho, P.M., *J. Eur. Ceram. Soc.*, 24, 319, 2004.
19. Nagel, N., Mikolajick, T., Kasko, I., Hartner, W., Moert, M., Pinnow, C.-U., Dehm, C., and Mazure, C., *Mater. Res. Soc. Symp. Proc.*, 655, CC1.1.1, 2001.
20. Moulson, A.J., and Herbert, J.M., *Electroceramics: Materials, Properties, and Applications*, John Wiley & Sons, Sussex, England, 2003.
21. Jaffe, B., Cook, W.R., Jr., and Jaffe, H., *Piezoelectric Ceramics*, Academic Press, New York, 1971.
22. Francis, L.F., and Payne, D.A., *Mater. Res. Soc. Symp. Proc.*, 200, 173, 1990.
23. Yao, H.B., Zhao, B., Shi, K., Han, Z.H., Xu, Y.L., Shi, D.L., Wang, S.X., Wang, L.M., Peroz, C., and Villard, C., *Phys. C*, 392(pt 2), 941, 2003.
24. Krishnan, K.M., Ju, H.L., Sohn, H.-C., Nelson, C., and Modak, A.R., *Mater. Res. Soc. Symp. Proc.*, 477, 139, 1997.
25. Waser, R., Ed., *Nanoelectronics and Information Technology: Advanced Electronic Materials and Novel Devices*, Wiley-VCH, Weinheim, Germany, 2003.
26. Waser, R., Guest Ed., Special issue on "Electroceramic thin films—integration technologies and device concepts," *J. Electroceram.*, 3, 103, 1999.

27. Ramesh, R., Ed., *Thin Film Ferroelectric Materials and Devices*, Kluwer Academic, Dordrecht, the Netherlands, 1997.
28. Maria, J.P., Cheek, K., Streiffer, S., Kim, S.H., Dunn, G., and Kingon, A., *J. Am. Ceram. Soc.*, 84, 2436, 2001.
29. Dimos, D., and Mueller, C.H., *Annu. Rev. Mater. Sci.*, 28, 397, 1998.
30. Miranda, R.A., Romanofsky, R.R., Van Keuls, F.W., Mueller, C.H., Treece, R.E., and Rivkin, T.V., *Int. Ferro.*, 17, 231, 1997.
31. Sengupta, L.C., and Sengupta, S., *IEEE Trans. UFFC*, 44, 792, 1997.
32. Jose, K.A., Varadan, V.K., and Varadan, V.V., *Microwave Opt. Technol. Lett.*, 20, 166, 1999.
33. Kim, K., and Lee, S.Y., *Technical Digest — Intl. Electron Dev. Meet.*, 547–550, 2002.
34. Trolier-McKinstry, S., Sabolsky, E., Kwon, S., Duran, C., Yoshimura, T., Park, J.-H., Zhang, Z., and Messing, G.L., in *Piezoelectric Materials and Devices*, Nava Setter, Ed., Ceramic Laboratory, EPFL, Lausanne, Switzerland, 2002.
35. Shin, S.S., Song, S., Lee, Y., Lee, N., Park, J., Park, H., and Lee, J., *Jpn. J. Appl. Phys.*, 42, 6139, 2003.
36. Cui, T., Markus, D., Zurn, S., and Polla, D.L., *Microsyst. Technol.*, 10, 137, 2004.
37. Cao, L., Mantell, S., and Polla, D., *Sens. Act. A Physical*, 94, 117, 2001.
38. Trolier-McKinstry, S., *J. Ceram. Soc. Jpn.*, 109, S76, 2001.
39. Tuttle, B.A., Headley, T., Drewien, C., Michael, J., Voigt, J., and Garino, T., *Ferroelectrics*, 221, 209, 1999.
40. Xu, F., Trolier-McKinstry, S., Ren, W., and Xu, B., *J. Appl. Phys.*, 89, 1336, 2001.
41. Zhang, Q.Q., Zhou, Q.F., and Trolier-McKinstry, S., *Appl. Phys. Lett.*, 80, 3370, 2002.
42. Kalinin, S.V., and Bonnell, D.A., *Phys. Rev. B*, 65, 125408, 2002.
43. Ganpule, C.S., Roytburd, A.L., Nagarajan, V., Stanishevsky, A., Melngailis, J., Williams, E.D., and Ramesh, R., *Mater. Res. Soc. Symp. Proc.*, 655, CC1.5.1, 2001.
44. Watton, R., *Int. Ferro.*, 4, 175, 1994.
45. Whatmore, R.W., *Ferroelectrics*, 118, 241, 1991.
46. Kohli, M., Seifert, A., Willing, B., Brooks, K., and Muralt, P., *Int. Ferro.*, 18, 359, 1997.
47. Morrison, R.T., and Boyd, R.N., *Organic Chemistry*, 3rd ed., Allyn & Bacon, Boston, 1974.
48. Bradley, D.C., Mehrotra, R.C., and Gaur, D.P., *Metal Alkoxides*, Academic Press, New York, 1978.
49. Mehrotra, R.C., Bohra, R., and Gaur, D.P., *Metal β-Diketonates and Allied Derivatives*, Academic Press, New York, 1978.
50. Brinker, C.J., and Scherer, G.W., *Sol-Gel Science: The Physics and Chemistry of Sol-Gel Processing*, Academic Press, San Diego, 1990.
51. Klein, L.C., Ed., *Sol-Gel Technology for Thin Films, Fibers, Preforms, Electronics, Special Shapes*, Noyes Publishers, Park Ridge, NJ, 1988.
52. Keefer, K.D., *Mater. Res. Soc. Symp. Proc.*, 32, 15, 1984.
53. Klemperer, W.D., and Ramamurthi, S.D., *Mater. Res. Soc. Symp. Proc.*, 121, 13, 1988.
54. Schwartz, R.W., Voigt, J.A., Boyle, T.J., Christenson, T.A., and Buchheit, C.D., *Ceram. Eng. Sci. Proc.*, 16, 1045, 1995.
55. Assink, R.A., and Schwartz, R.W., *Chem. Mater.*, 5, 511, 1993.

56. Boyle, T.J., Schwartz, R.W., Doedens, R.J., and Ziller, J.W., *Inorg. Chem.*, 34, 1110, 1995.
57. Boyle, T.J., and Schwartz, R.W., *Comm. Inorg. Chem.*, 16, 243, 1994.
58. Ramamurthi, S.D., and Payne, D.A., *J. Am. Ceram. Soc.*, 8, 2547, 1990.
59. Yi, G., Wu, Z., and Sayer, M., *J. Appl. Phys.*, 64, 2717, 1988.
60. Sayer, M., Lukacs, M., and Olding, T., *Int. Ferro.*, 17, 1, 1997.
61. Schwartz, R.W., Boyle, T.J., Lockwood, S.J., Sinclair, M.B., Dimos, D., and Buchheit, C.D., *Int. Ferro.*, 7, 259, 1995.
62. Hasenkox, U., Hoffmann, S., and Waser, R., *J. Sol-Gel Sci. Technol.*, 12, 67, 1998.
63. Hoffmann, S., and Waser, R., *Int. Ferro.*, 17, 141, 1997.
64. Hennings, D., Rosenstein, G., and Schreinemacher, H., *J. Eur. Ceram. Soc.*, 8, 107, 1991.
65. Mansour, S.A., Liedl, G.L., and Vest, R.W., *J. Am. Ceram. Soc.*, 78, 1617, 1995.
66. Vest, R.W. and Xu, J., *IEEE Trans. UFFC*, 35, 711, 1988.
67. Klee, M., Eusemann, R., Waser, R., Brand, W., and van Hall, H., *J. Appl. Phys.*, 72, 1566, 1992.
68. Haertling, G.H., *Ferroelectrics*, 116, 51, 1991.
69. Hasenkox, U., Mitze, C., and Waser, R., *J. Am. Ceram. Soc.*, 80, 2709, 1997.
70. Boyle, T.J., Dimos, D.B., Schwartz, R.W., Alam, T.M., Sinclair, M.B., and Buchheit, C.D., *J. Mater. Res.*, 12, 1022, 1997.
71. Scriven, L.E., *Mater. Res. Soc. Symp. Proc.*, 121, 717, 1988.
72. Meyerhofer, D., *J. Appl. Phys.*, 49, 3993, 1978.
73. Bornside, D.E., Macosko, C.W., and Scriven, L.E., *J. Imaging Technol.*, 13, 122, 1987.
74. Bailey, J.K., *Mater. Res. Soc. Symp. Proc.*, 271, 219, 1992.
75. Birnie, D.P., *J. Mater. Res.*, 16, 1145, 2001.
76. Haas, D.E., Birnie, D.P., Zecchino, M.J., and Figueroa, J.T., *J. Mater. Sci. Lett.*, 20, 1763, 2001.
77. Haertling, G.H., *Am. Ceram. Soc. Bull.*, 73, 68, 1994.
78. Phillips, N.J., Clazada, M.L., and Milne, S.J., *J. Non-Crystl. Solids*, 147, 285, 1992.
79. Tu, Y.-L., Calzada, M.L., Phillips, N.J., and Milne, S.J., *J. Am. Ceram. Soc.*, 79, 441, 1996.
80. Barrow, D.A., Petroff, T.E., Tandon, R.P., and Sayer, M., *J. Appl. Phys.*, 81, 876, 1997.
81. Sayer, M., Lukacs, M., Holding, T., Pang, G., Zou, L., and Chen, Y., *Mater. Res. Soc. Symp. Proc.*, 541, 599, 1999.
82. Lakeman, C.D.E., Campion, J.-F., Suchicital, C.T.A., and Payne, D.A., *IEEE 7th Int. Symp. Appl. Ferro.*, 681, 1991.
83. Fè, L., Norga, G.J., Wouters, D.J., Maes, H.E., and Maes, G., *J. Mater. Res.*, 16, 2499, 2001.
84. Schwartz, R.W., Boyle, T.J., Voigt, J.A., and Buchheit, C.D., *Ceram. Trans.*, 43, 145, 1994.
85. Schwartz, R.W., Voigt, J.A., Tuttle, B.A., DaSalla, R.S., and Payne, D.A., *J. Mater. Res.*, 12, 444, 1997.
86. Griswold, E.M., Weaver, L., Sayer, M., and Calder, J.D., *J. Mater. Res.*, 10, 3149, 1995.
87. Zai, M.H.M., Akiba, A., Goto, H., Matsumoto, M., and Yeatman, E.M., *Thin Solid Films*, 394, 97, 2001.

88. Schwartz, R.W., Electronic and magnetic ceramics, in *Characterization of Ceramics*, R.E. Loehman, Ed., Butterworth-Heinemann, Boston, 1993, pp. 247–249.

89. Tani, T., and Payne, D.A., *J. Am. Ceram. Soc.*, 77, 1242, 1994.

90. Roeder, J.F., Hendrix, B.C., Hintermaier, F., Desrochers, D.A., Baum, T.H., Bhandari, G., Chappuis, M., Van Buskirk, P.C., Dehm, C., Fritsch, E., Nagel, N., Wendt, H., Cerva, H., Honlein, W., and Mazure, C., *J. Eur. Ceram. Soc.*, 19, 1463, 1999.

91. Lefevre, M.J., Speck, J.S., Schwartz, R.W., Dimos, D., and Lockwood, S.J., *J. Mater. Res.*, 11, 2076, 1996.

92. Polli, A.D., Lange, F.F., and Levi, C.G., *J. Am. Ceram. Soc.*, 83, 873, 2000.

93. Ousi-Benomar, W., Xue, S.S., Lessard, R.A., Singh, A., Wu, Z.L., and Kuo, P.K., *J. Mater. Res.*, 9, 970, 1994.

94. Joshi, V., Dacruz, C.P., Cuchiaro, J.D., Araujo, C.A., and Zuleeg, R., *Int. Ferro.*, 14, 133, 1997.

95. Gust, M.C., Evans, N.D., Momoda, L.A., and Mecartney, M.L., *J. Am. Ceram. Soc.*, 80, 2828, 1997.

96. Mosset, A., Gautier-Luneau, I., Galy, J., Strehlow, P., and Schmidt, H., *J. Non-Crystl. Solids*, 100, 339, 1988.

97. Tani, T., Xu, Z., and Payne, D.A., *Mater. Res. Soc. Symp. Proc.*, 310, 269, 1993.

98. Tu, Y.L., and Milne, S.J., *J. Mater. Res.*, 10, 3222, 1995.

99. Brooks, K.G., Chen, J., Udayakumar, K.R., and Cross, L.E., *J. Appl. Phys.*, 75, 1699, 1994.

100. Schwartz, R.W., Xu, Z., Payne, D.A., DeTemple, T.A., and Bradley, M.A., *Mater. Res. Soc. Symp. Proc.*, 200, 167, 1990.

101. Coffman, P.K., and Dey, S.K., *J. Sol-Gel Sci. Technol.*, 6, 83, 1996.

102. Sengupta, S.S., Ma, L., Adler, D.L., and Payne, D.A., *J. Mater. Res.*, 10, 1345, 1995.

103. Hietala, S.L., Smith, D.M., Hietala, V.M., Frye, G.C., and Martin, S.J., *Langmuir*, 9, 249, 1993.

104. Budd, K.D., Ph.D. dissertation, University of Illinois at Urbana-Champaign, 1986.

105. Budd, K.D., Dey, S.K., and Payne, D.A., *Mater. Res. Soc. Symp. Proc.*, 73, 711, 1986.

106. Schwartz, R.W., Ph.D. dissertation, University of Illinois at Urbana-Champaign, 1989.

107. Lakeman, C.D.E., Xu, Z., and Payne, D.A., *J. Mater. Res.*, 10, 2042, 1995.

108. Coffman, P.K., and Dey, S.K., *J. Sol-Gel Sci. Technol.*, 1, 251, 1994.

109. Arčon, I., Mali, B., Kosec, M., and Kodre, A., *J. Sol-Gel Sci. Technol.*, 13, 861, 1998.

110. Kato, K., Zheng, C., Dey, S.K., and Torii, Y., *Int. Ferro.*, 18(pt 2), 225, 1997.

111. Pelligri, N., Frattini, A., Steren, C.A., Rapp, M.E., Gil, R., Trbojevich, R., Gonzalez, C.J.R., and De Sanctis, O., *Int. Ferro.*, 30, 111, 2000.

112. Chae, H.K., Payne, D.A., Xu, Z., and Ma, L., *Chem. Mater.*, 6, 1589, 1994.

113. Ma, L., and Payne, D.A., *Chem. Mater.*, 6, 875, 1994.

114. Hasenkox, U., Hoffmann, S., and Waser, R., *J. Sol-Gel Sci. Technol.*, 12, 67, 1998.

115. Roberts, M.A., Sankaar, G., Catlow, C.R.A., Thomas, J.M., Greaves, G.N., and Jones, R.H., *Mater. Sci. Forum*, 228(pt 1), 417, 1996.

116. Dobberstein, H., Ph.D. dissertation, Clemson University, Clemson, SC, 2002.

117. Chen, S.-Y., and Chen, I.-W., *J. Am. Ceram. Soc.*, 77, 2332, 1994.

118. Chen, S., and Chen, I., *J. Am. Ceram. Soc.*, 81, 97, 1998.

119. Schwartz, R.W., Lakeman, C.D.E., and Payne, D.A., *Mater. Res. Soc. Symp. Proc.*, 180, 335, 1990.
120. Cooper, A.R., *Mater. Res. Soc. Symp. Proc.*, 73, 421, 1986.
121. Schwartz, R.W., and Payne, D.A., *Mater. Res. Soc. Symp. Proc.*, 121, 199, 1988.
122. Roy, R., *J. Am. Ceram. Soc.*, 52, 344, 1969.
123. Turnbull, D., *J. Appl. Phys.*, 32, 1022, 1950.
124. Hoffmann, J.D., *J. Chem. Phys.*, 29, 1192, 1958.
125. Thompson, C.V., and Spaeten, F., *Acta Metall.*, 27, 1855, 1979.
126. Norga, G.J., Vasiliu, F., Fè, L., Wouters, D.J., and Van der Biest, O., *J. Mater. Res.*, 18, 1232, 2003.
127. Hoffmann, S., Hasenkox, U., Waser, R., Jia, C.L., and Urban, K., *Mater. Res. Soc. Symp. Proc.*, 474, 9, 1997.
128. Dobberstein, H., and Schwartz, R.W., *Proceedings of the 1st Symposium of Adv. Mater. Next Generation*, AIST Chubu, Nagoya, Japan, 2002.
129. Schwartz, R.W., and Dobberstein, H., *Proceedings of the 11th US-Japan Seminar on Dielectric and Piezoelectric Ceramics*, 2003.
130. Lange, F.F., *Science*, 273, 903, 1996.
131. Schwartz, R.W., Clem, P.G., Voigt, J.A., Byhoff, E.R., Van Stry, M., Headley, T.J., and Missert, N.A., *J. Am. Ceram. Soc.*, 82, 2359, 1999.
132. Tuttle, B.A., Garino, T.J., Voigt, J.A., Headley, T.J., Dimos, D., and Eatough, M.O., in *Science and Technology of Electroceramic Thin Films*, Waser, R. and Auciello, O., Eds., Kluwer Academic, 1995, pp. 117–133.
133. Garino, T.J., and Harrington, M., *Mater. Res. Soc. Symp. Proc.*, 243, 341, 1992.
134. Basceri, C., Streiffer, S.K., Kingon, A.I., and Waser, R., *J. Appl. Phys.*, 82, 2497, 1997.
135. Basceri, C., Streiffer, S.K., Kingon, A.I., Bilodeau, S., Carl, R., van Buskirk, P.C., Summerfelt, S.R., McIntyre, P., and Waser, R., *Mater. Res. Soc. Symp. Proc.*, 433, 285, 1996.
136. Seifert, A., Lange, F.F., and Speck, J.S., *J. Mater. Res.*, 10, 680, 1995.
137. Lockwood, S.J., Schwartz, R.W., Tuttle, B.A., and Thomas, V.A., *Mater. Res. Soc. Symp. Proc.*, 310, 275, 1993.
138. Melnick, B.M., Gallegos, R., and Paz de Araujo, C.A., *Proceedings of the 3rd International Symposium Int. Ferro.*, 547, 1991.
139. Langjahr, P.A., Wagner, T., Ruhle, M., and Lange, F.F., *Mater. Res. Soc. Symp. Proc.*, 401, 109, 1996.
140. Suzuki, H., Kondo, Y., Kaneko, S., and Hayashi, T., *Mater. Res. Soc. Symp. Proc.*, 596, 241, 2000.
141. Wu, T.B., Wu, J.M., Wu, M., Shyu, M.J., Chen, M.S., Dorng, J.S., and Yang, C.C., *Mater. Res. Soc. Symp. Proc.*, 433, 169, 1996.
142. Peterson, C.R., Medendorp, N.W., Jr., Slamovich, E.B., and Bowman, K.J., *Mater. Res. Soc. Symp. Proc.*, 433, 297, 1996.
143. Neumayer, D.A., Duncombe, P.R., Laibowitz, R.B., Saenger, K.L., Purtell, R., Ott, J.A., Shaw, T.M., and Grill, A., *Int. Ferro.*, 18, 319, 1997.
144. Norga, G.J., Fè, L., Vasiliu, F., Fompeyrine, J., Loequet, J.P., and Van der Biest, O., *J. Eur. Ceram. Soc.*, 24, 969, 2004.
145. Wouters, D.J., Norga, G.J., and Maes, H.E., *Mater. Res. Soc. Symp. Proc.*, 541, 381, 1999.

# Index

## A

## B

Printed in the United States
by Baker & Taylor Publisher Services